PRINCIPLES OF INDUCTION LOGGING

Methods in Geochemistry and Geophysics

(Volumes 1–28 are out of print)

1. A.S. RITCHIE — CHROMATOGRAPHY IN GEOLOGY
2. R. BOWEN — PALEOTEMPERATURE ANALYSIS
3. D.S. PARASNIS — MINING GEOPHYSICS
4. I. ADLER — X-RAY EMISSION SPECTROGRAPHY IN GEOLOGY
5. THE LORD ENERGLYN AND L. BREADLY — ANALYTICAL GEOCHEMISTRY
6. A.J. EASTON — CHEMICAL ANALYSIS OF SILICATE ROCKS
7. E.E ANGINO AND G.K. BILLINGS — ATOMIC ABSORPTION SPECTROMETRY IN GEOLOGY
8. A. VOLBORTH — ELEMENTAL ANALYSIS IN GEOCHEMISTRY, A: MAJOR ELEMENTS
9. P.K. BHATTACHARYA AND H.P. PATRA— DIRECT CURRENT GEOELECTRIC SOUNDING
10. J.A.S. ADAMS AND P. GASPARINI — GAMMA-RAY SPECTROMETRY OF ROCKS
11. W. ERNST — GEOCHEMICAL FACIES ANALYSIS
12. P.V. SHARMA — GEOPHYSICAL METHODS IN GEOLOGY
13. C.H. CHEN (Editor) — COMPUTER-AIDED SEISMIC ANALYSIS AND DISCRIMINATION

14A. O. KOEFOED — GEOSOUNDING PRINCIPLES, 1. RESISTIVITY SOUNDING MEASUREMENTS

14B. H.P. PATRA AND K. MALLICK — GEOSOUNDING PRINCIPLES, 2. TIME-VARYING GEOELECTRIC SOUNDINGS

15. A.A. KAUFMAN AND G.V. KELLER — THE MAGNETOTELLURIC SOUNDING METHOD
16. A.A. KAUFMAN AND G.V. KELLER — FREQUENCY AND TRANSIENT SOUNDINGS
17. C.H. CHEN (Editor) — SEISMIC SIGNAL ANALYSIS AND DISCRIMINATION
18. J.E. WHITE — UNDERGROUND SOUND APPLICATION OF SEISMIC WAVES
19. M.N. BERDICHEVSKY AND M.S. ZHDANOV — ADVANCED THEORY OF DEEP GEOMAGNETIC SOUNDINGS

20A. A.A. KAUFMAN AND G.V. KELLER — INDUCTIVE MINING PROSPECTING, PART I: THEORY

21. A.W. WYLIE— NUCLEAR ASSAYING OF MINING BOREHOLES - AN INTRODUCTION
22. C.H. CHEN (Editor) — SEISMIC SIGNAL ANALYSIS AND DISCRIMINATION III
23. R.P. PHILP — FOSSIL FUEL BIOMARKERS
24. R.B. JOHNS (Editor) — BIOLOGICAL MARKERS IN THE SEDIMENTARY RECORD
25. J.C. D' ARNAUD GERKENS — FOUNDATION OF EXPLORATION GEOPHYSICS
26. P. TYGEL AND P. HUBRAL — TRANSIENT WAVES IN LAYERED MEDIA
27. A.A. KAUFMAN AND G.V. KELLER — INDUCTION LOGGING
28. J.G.NEGI AND P.D. SARAF — ANISOTROPY IN GEOELECTROMAGNETISM
29. V.P. DIMRI — DECONVOLUTION AND INVERSE THEORY - APPLICATION TO GEOPHYSICAL PROBLEMS
30. K.-M STRACK — EXPLORATION WITH DEEP TRANSIENT ELECTROMAGNETICS
31. M.S. ZHDANOV and G.V. KELLER — THE GEOELECTRICAL METHODS IN GEOPHYSICAL EXPLORATION
32. A.A. KAUFMAN and A.L. LEVSHIN — ACOUSTIC AND ELASTIC WAVE FIELDS IN GEOPHYSICS, I
33. A.A. KAUFMAN and P.A. EATON — THE THEORY OF INDUCTIVE PROSPECTING
34. A.A. KAUFMAN AND P. HOEKSTRA — ELECTROMAGNETIC SOUNDINGS
35. M.S. ZHDANOV AND P.E. WANNAMAKER — THREE-DIMENSIONAL ELECTROMAGNETICS
36. M.S. ZHDANOV — GEOPHYSICAL INVERSE THEORY AND REGULARIZATION PROBLEMS
37. A.A. KAUFMAN, A.L. LEVSHIN AND K.L. LARNER — ACOUSTIC AND ELASTIC WAVE FIELDS IN GEOPHYSICS, II
38. A.A. KAUFMAN and YU. A. DASHEVSKY — PRINCIPLES OF INDUCTION LOGGING

Methods in Geochemistry and Geophysics, 38

PRINCIPLES OF INDUCTION LOGGING

A.A. KAUFMAN
Department of Geophysics
Colorado School of Mines
Golden, CO 80401, U.S.A.

and

YU. A. DASHEVSKY
Institute of Geophysics
Siberian Branch
Russian Academy of Sciences
Novosibirsk, Russia

2003

ELSEVIER
Amsterdam - Boston - Heidelberg - London - New York - Oxford
Paris - San Diego - San Francisco - Singapore - Sydney - Tokyo

ELSEVIER SCIENCE B.V.
Sara Burgerhartstraat 25
P.O. Box 211, 1000 AE Amsterdam, The Netherlands

First edition 2003

Library of Congress Cataloging in Publication Data
A catalog record from the Library of Congress has been applied for.

British Library Cataloguing in Publication Data
A catalogue record from the British Library has been applied for.

ISBN: 0 444 50983 6

♾ The paper used in this publication meets the requirements of ANSI/NISO Z39.48-1992 (Permanence of Paper).
Printed in Hungary.

CONTENTS

ACKNOWLEDGMENTS

In preparation of this volume we were helped by Dr. B. Anderson, Dr. B. Clark, and Dr. S. Davydycheva. We express to all of them our gratitude.

Our special thanks to Dr. V. Druskin. We deeply appreciate his comments and suggestions regarding the part of the book devoted to mathematical modeling of the response of induction logging tools in 3D environment.

It is a pleasure for us to recognize the assistance of Mr. Oleg Dashevsky, who was responsible for the computer make-up of the manuscript.

LIST OF SYMBOLS

a_1	borehole radius
a_2	radius of the invasion zone
$\boldsymbol{A}$	magnetic vector potential, defined by $\boldsymbol{H} = \text{curl}\,\boldsymbol{A}$
$\boldsymbol{B}$	magnetic induction vector, $\boldsymbol{B} = \mu\boldsymbol{H}$
c	ratio of coil moments
$\boldsymbol{D}$	dielectric displacement vector $\boldsymbol{D} = \varepsilon\boldsymbol{E}$
e	charge
$\boldsymbol{E}$	vector electric field, volts/meter
E_n	electric field component normal to surface
$\boldsymbol{E}_0$	primary electric field
$\mathscr{E}$	electromotive force, volts (also EMF)
h	skin depth, $(2/\sigma\mu\omega)^{1/2}$
h	normalized magnetic field strength
$\boldsymbol{H}_0$	primary magnetic field vector, amps/meter
$\boldsymbol{H}_s$	secondary magnetic field vector, amps/meter
H	formation thickness
$I_\nu(), K_\nu()$	modified Bessel functions of the first and second kind of argument ()
In	inphase part of
I	current, amps
$\boldsymbol{j}$	current density vector, amps/meter^2
J_i	component of current density
$J_\nu(), Y_\nu()$	Bessel functions of the first and second kind of argument ()
$k^2 = \mathrm{i}\omega\mu(\sigma + \mathrm{i}\omega\varepsilon)$	square of a complex wave number
M_T	transmitter moment, amps·meter^2
M_R	receiver moment, amps·meter^2
$\boldsymbol{n}$	unit vector normal to a surface
n	number of coil turns
L	inductance, henries
L	probe length
l	linear dimension
$p = L/h$	scaled length
G_1	geometric factor of a borehole
G_2	geometric factor of an invasion zone
G_3	geometric factor of a formation
G_i	geometric factor of i-part of medium
q_r	radial differential response
r_1	coil radius

R	resistance, ohms
$R = (r^2 + z^2)^{1/2}$	radius in spherical coordinates
Q	quadrature part of
S	conductance, mhos
s	ratio of conductivities
t	time, seconds
t_r	ramp time
T	period, seconds
U	scalar potential, volts
δ	volume charge density, coulombs/meter^3
Δ	difference
ρ	resistivity, ohm·meter
σ	electrical conductivity, mhos/meter
σ_a	apparent conductivity
ε	normalized electromotive force
ε	displacement of the probe with respect to borehole axis
ε_0	dielectric permittivity of free space
$\phi()$	probability integral function
Φ, ϕ	magnetic flux or occasionally phase
λ	coefficient of anisotropy
μ	magnetic permeability, henries/meter
μ_0	magnetic permeability of free space, $4\pi \times 10^{-7}$ H/m
$\tau_0 = \rho\varepsilon_0$	time constant of a medium
$\tau = (2\pi\rho t \times 10^7)^{1/2}$	scaled variable used in transient induction logging
ω	angular frequency (radians/second) or occasionally a solid angle

INTRODUCTION

Induction well logging is an established method for surveying the electrical conductivity of rocks surrounding a borehole and proceeded from the early efforts of H. G. Doll (1949, 1952). In its simplest form, an induction well-logging device consists of two coils; one is a transmitter and the other is a receiver. The transmitter coil is energized with an alternating current at frequencies of twenty kilohertz and much higher, while the electromotive force, caused by a change of the magnetic field, is detected at the receiver coil. In almost all cases with some important exceptions, the axes of the coils are coincident with the axis of the borehole. The separation between the transmitter and receiver coils is termed the probe length, and this parameter is commonly used to control the depth of investigation of the logging device away from the borehole axis. The electromotive force, which is detected at the receiver coil, is linearly dependent on the amount of the current provided to the transmitter coil, as well as strengths of currents that are induced in the surrounding medium. The actual distribution of these additional currents depends on the electrical structure of the medium, and in particular, on the conductivity. For this reason, by measuring the electromotive force in the receiver coil one can, in principle, determine the conductivity of the formation opposite which the induction device is located.

In those cases, when the borehole axis is perpendicular to the boundaries between formations, the current flow path in the medium forms a circle, located in a horizontal plane and centered on the borehole axis. Correspondingly, induction logging is very sensitive to thin conductive layers, but it has difficulty in detecting relatively thin and resistive beds.

H. Doll also introduced the differential multi-coil probes, which became very efficient logging tools and defined the path of development and application of induction logging over almost forty years. The use of these differential measurements in induction logging provides a result in which the effect of the borehole fluid, and in many cases also the invasion zone, on measurements is greatly reduced. Such devices are described in detail in this monograph.

H. Doll did not only invent induction probes, but also suggested a very useful though approximate theory for the method, which helped immensely to develop principles of an interpretation and to aid in the design parameters of *focusing* probes. For simplification of the mathematical problem Doll has considered that the induction coils on the logging tool are essentially magnetic dipoles, and for sufficiently low frequencies or a highly resistive medium the skin effect can be neglected. In other words, an interaction between the various induced currents is not strong enough to affect their magnitude appreciably. Respectively, the currents everywhere in the medium are in phase with one another, this phase being ninety degrees shifted with respect to the current in the transmitter coil.

With these approximations the magnitude of the current, induced in the formation at any point, can be calculated by using quite simple formulae. This also allows the definition of a straightforward geometrical factor, which characterizes the relationship between the magnetic fields and the conductivity at an arbitrary point of a medium. According to this approximate theory, the magnetic field, contributed by the induction currents, has only the quadrature (out-of-phase) component, with the in-phase component of the magnetic field being zero.

The concept of the geometric factor for an assembly of elementary rings with centers located on the axis of the borehole plays an essential role in Doll's theoretical approach. By using such geometrical factors Doll was able to calculate the electromotive force, arising in the receiver and caused by various parts of a medium, and to investigate the vertical and radial responses of different induction probes.

The approach, developed by Doll, is so satisfactory that it remains virtually unchanged in developing procedures of interpretation, if the so-called induction parameter is sufficiently small. Of course, this theory is valid when the electric field is tangential to boundaries and, correspondingly, surface charges are absent.

In almost sixty years, since the first development by H. Doll, research on various aspects of induction well-logging has been carried out around the world, and there have been some rather significant advances in theory, interpretation, probe design and equipment. Moreover, completely new modifications of induction logging have been developed and their principles are described in our monograph. As a result of the efforts of scientists and engineers in the United States, former Soviet Union and other countries, induction well-logging has become the most powerful tool for a determination of formation conductivity in uncased wells.

Because much of the development of induction logging was done in proprietary research by logging services and oil companies, the technical articles that appeared in journals do not properly reflect the real volume of research that has been done on the method. For this reason, it is probably impossible to attribute the proper respect to everyone who has contributed in the development of induction well-logging in the western community. Among those who carried through the work started by H. Doll, are J. H. Moran, K. S. Kunetz, W. C. Duesterhoeff, J. L. Dumanoir, M. P. Tixier, M. Martin, A. J. deWitte, and D. A. Lowitz. Later their activity was continued by S. Gianzero, J. Tabanou, B. Anderson, T. Barber, G. Minerbo, B. Clark, S. Chang, V. Druskin, T. Habashy, and many others.

In the USSR, parallel development of theory, interpretation and equipment of induction logging, based on Doll's concepts of the geometric factor and focusing probe, was started at almost the same time. Also, during this research, new modifications of induction logging were introduced and some of them became conventional and are now used over the world.

Theoretical investigations performed by L. Alpin, S. Akselrod, A. Kaufman, Y. Kudravchev, and V. Nikitina allowed us to understand the behavior of the quasistationary electromagnetic fields, caused by the magnetic dipole in a medium with cylindrical as well as horizontal interfaces. These studies helped to design equipment and focusing probes with optimal radial and vertical characteristics (S. Akselrod, M. Plusnin).

Almost from the beginning, the frequency of the transmitter current was chosen much higher than in the west, and it was done in order to improve the vertical responses of probes (Kaufman, 1962). At the same time, it was demonstrated that the quadrature and in-phase components of the secondary magnetic fields deliver a different depth of investigation (Kaufman, 1959). For this reason, it is natural that the conventional equipment of induction logging is able to measure both these components.

As was pointed out, the remarkable simplicity of Doll's theory is related to the fact that interaction of induced currents is neglected. In order to take into account this effect and improve the quality of an interpretation of logging data, a new approach, also an approximate one. was suggested (Kaufman, 1962). This method allows us relatively quickly to evaluate the field, subjected to an influence of the skin effect in a formation, when there are both cylindrical and horizontal interfaces. Much later, this rather complicated problem was solved by V. Dimitriev, L. Tabarovsky, V. Zakharov using the method of integral equations.

At the beginning of the 1960's there was the first attempt to develop induction logging without the use of multi-coil focusing probes. By analogy with the lateral logging soundings with lateral probes, widely used in the former Soviet Union, the induction lateral soundings were suggested (Kaufman, 1962). The patent was applied, several papers were published that described principles of an interpretation of the apparent conductivity curves with two- and three-coil probes of a different length. Additionally, an influence of the in-phase component of the magnetic field on the radial responses was studied, as well as the use of different frequencies for probes of different lengths.

At that time, the logging industry was not ready to accept this approach and rejected it. However, with time, attitude to multi-array systems was completely changed, and during the last twenty years this type of induction logging has been widely used as the conventional method. To a great extent its progress is related to the development of the dielectric logging. At the beginning, borehole measurements of the dielectric constant of formations were performed with a tool that is similar to the capacitor. Then, it was suggested to measure this parameter inductively, that is, using the induction probe (Kaufman, 1963). This approach was successfully developed by D. Daev, who introduced a new approach, namely, measuring the ratio of amplitudes of the magnetic field and the difference of phases with the three-coil probe. It turned out that measurements of these parameters in induction logging also provide excellent radial and vertical responses, if frequencies are properly chosen. For this reason, they are either measured or calculated in the multi-array tool, as well as in the logging while drilling.

At the end of the 1960's serious attention was paid to other modifications to induction logging. One of them is based on the use of transient fields, when measurements are performed in the absence of the primary magnetic field (Kaufman and Sokolov, 1972). The study of the secondary fields, caused by induced currents in a medium with either cylindrical or horizontal interfaces allowed one to describe the radial and vertical responses of the two-coil probe and find the most optimal range of time for measurements.

Finally, theoretical investigations were performed which demonstrate that induction probes with special orientations of coils allow us to evaluate an anisotropy of formations (Kaufman and Kagansky, 1971). This study is also useful for application of induction logging in horizontal wells.

In this monograph we describe the physical principles and theory of almost all possible modifications of induction logging. At the same time, such topics as *inverse problems* are out of the scope of this book.

A need for a fundamental understanding of principles on which induction logging is based does not require any special justification. The first chapter of this monograph is devoted to an explanation of the physical laws of classical electrodynamics and provides this background.

In the last chapter of the book we consider the outstanding results obtained by V. Druskin and L. Knizhnerman in 3D mathematical modeling of the response of induction logging tools in complicated models of a medium.

We would like to note that the theory of induction logging presented in this volume can be applied not only to logging after drilling but to logging while drilling as well.

Chapter 1

BASIC ELECTROMAGNETIC LAWS AND MAXWELL'S EQUATIONS

This chapter describes the principal laws of electromagnetism, which are important in electrical logging methods based on the use of direct and alternating fields. Although these laws are treated in numerous excellent books, examples and models, usually given, are not very appropriate for understanding the behavior of the fields in nonuniform conducting media. The purpose of this chapter is to present the basic laws of electromagnetism from a point of view that will facilitate the application of the theory to problems in electromagnetic logging.

We will first consider the laws of Coulomb, Biot–Savart and Faraday, emphasizing their experimental origin and the areas in which they can be applied. The relationship between these laws and Maxwell's equations will then be described to further explore their physical meaning and especially the precise sources of electric and magnetic fields.

Special attention will be paid to the set of equations which describes the quasistationary or quasistatic fields and provides an accurate model for induction logging, with the exception of dielectric logging where very high frequencies are used.

We will finally consider the formulation of the Helmholtz equations for magnetic and electric vector potentials, which are useful in solving boundary value problems in a conducting medium.

1.1. Coulomb's Law

As a starting point, we will assume that the reader accepts the concept that an electric charge is the source of an electric field. As a consequence, the distribution of electric charges is the main factor in controlling the field. In describing electric fields, we will make use of such functional descriptions of charges as volume, surface and linear densities of charge.

The volume density of charge, δ, is defined by the equation:

$$\delta = \lim_{\mathrm{d}V \to 0} \frac{\mathrm{d}e}{\mathrm{d}V} \tag{1.1}$$

where $\mathrm{d}e$ is the charge in an elementary volume $\mathrm{d}V$. It is clear that as the element of volume $\mathrm{d}V$ decreases, the charge in the elementary volume decreases as well. In the limit, as the ratio of the total charge to the volume remains constant, we obtain a nonzero charge density.

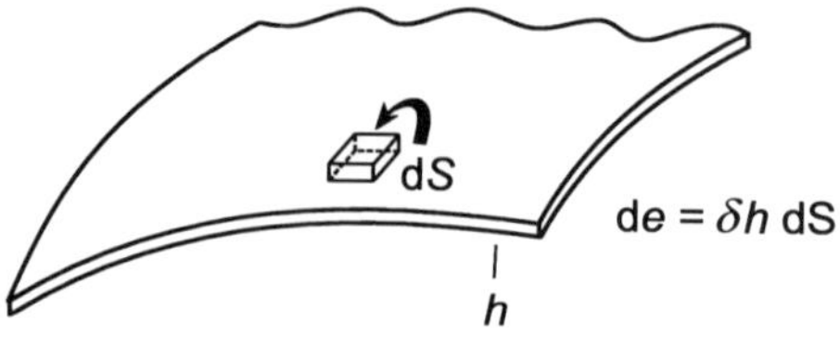

Figure 1.1. Definition of an element of charge within a thin layer.

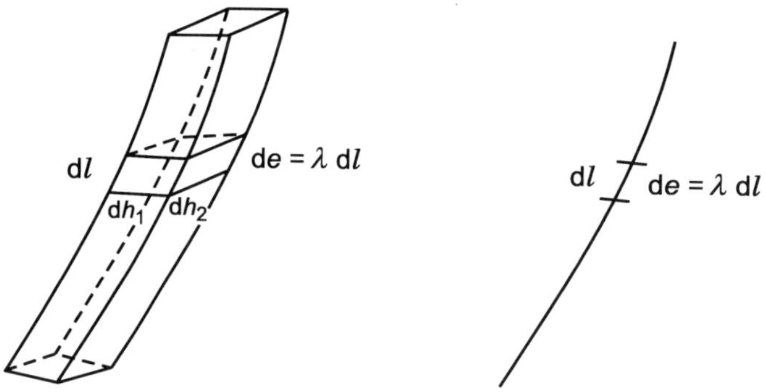

Figure 1.2. Definition of linear charge density.

The volume density of charge is the most general way in which to describe a charge distribution, but in some particular cases, we might also wish to define such functions as a surface or a linear density of charge. Suppose that the volume density δ is invariant in the direction perpendicular to the surface of the thin layer (see Fig. 1.1). The elementary volume charge can then be written as:

$$\mathrm{d}e = \delta h \,\mathrm{d}S$$

where h is the thickness of the thin layer and $\mathrm{d}S$ is an elementary area of its surface. Let the thickness h tend to zero while the charge density δ increases without limit in such a way that the product δh remains constant; we thereby obtain a definition for an elementary surface density of charge:

$$\mathrm{d}e = \Sigma \,\mathrm{d}S \tag{1.2}$$

where Σ is the surface density of charge.

Similarly, when charges are distributed in a rod-like volume of small cross-section, as shown in Fig. 1.2, and we are only concerned with the field at distances which are far greater than its dimensions $\mathrm{d}h_1$ and $\mathrm{d}h_2$, it is often convenient to define a linear elementary change de and a linear density λ as follows:

$$de = \lambda \, dl \tag{1.3}$$

In doing so, we replace the volume within the rod by a line that carries the same amount of charge.

On occasion, it is also convenient to define a point charge e by assuming that the whole charge density under consideration is concentrated within an infinitesimal distance about a single point in the medium.

Elementary volume, surface and linear charges have a common feature in that they are situated within volumes of which at least one characteristic dimension is very small with respect to the distance to the point at which the field is being observed. They differ from each other in unit dimensions. The volume density for an elementary volume charge always remains finite, while for elementary surface and linear charges, the volume density must be assumed to increase without limit within the charged volume. Actually, in accord with eq. 1.2 we have:

$$\delta = \Sigma/h \qquad \text{as } h \to 0$$

Inasmuch as Σ is finite, the volume density of surface charge becomes infinite as the function $1/h$ becomes infinite.

For an elementary linear charge, we have:

$$\delta = \lambda/dh_1 dh_2$$

where dh_1 and dh_2 are the linear dimensions of the cross-section (Fig. 1.2). As dh_1 and dh_2 tend to zero, the volume density of linear charge increases without limit more rapidly than was the case for a surface charge.

The dimensions for charge densities are also different for each type of distribution. The proper units for volume charge density are Coulombs per cubic meter. For surface and linear charge densities, the unit becomes Coulombs per square meter and Coulombs per meter, respectively. These differences in units must be carefully looked after in problems in which these approximations are used. As one might expect, these various degrees of concentration of charge into linear or sheet-like volumes result in different behaviors of the electric field near these charges. A point charge has the distribution characterized by the maximum concentration of charges in a volume, with the volume density of charge going to infinity as $1/h^3$ (here h is taken to be the linear dimension of an elementary volume around the point where the charge is suppose to concentrate).

Now let us discuss the main subject of this section, that is, Coulomb's law. Experimental investigations carried out by Coulomb and other researchers have shown that the force acting between an elementary charge d$e(q)$ situated at point q and another elementary charge d$e(a)$ situated at point a, is described by an extremely simple expression:

$$\boldsymbol{F} = \frac{1}{4\pi\varepsilon_0} \frac{de(q)\, de(a)}{L_{qa}^3} \boldsymbol{L}_{qa} \tag{1.4}$$

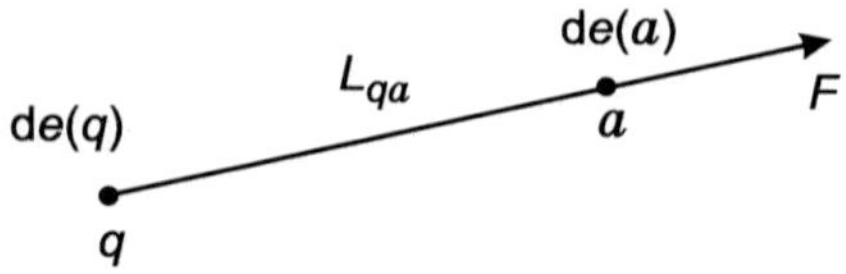

Figure 1.3. Definition of the sign of the force defined by Coulomb's law.

where L_{qa} is the vector:

$$\boldsymbol{L}_{qa} = L_{qa}\boldsymbol{L}_{qa}^0$$

with L_{qa} being the distance between points q and a, while $\boldsymbol{L}_{qa}^0$ is a unit vector directed along the line connecting points q and a, and ε_0 is a constant known as the dielectric permeability of free space. In the practical system of units, this constant is:

$$\varepsilon_0 = \frac{1}{36\pi} \times 10^{-9}\ \mathrm{F/m}$$

also:

$$\frac{1}{4\pi\varepsilon_0} = 9 \times 10^9\ \mathrm{m/F}$$

Equation 1.4 can be rewritten as:

$$\boldsymbol{F} = \frac{1}{4\pi\varepsilon_0}\frac{\mathrm{de}(q)\,\mathrm{de}(a)}{L_{qa}^2}\boldsymbol{L}_{qa}^0 \tag{1.5}$$

The electric force of interaction between two elementary charges is directly proportional to the charge strengths, inversely proportional to the square of the distance between them, and has the same direction as the unit vector $\boldsymbol{L}_{qa}^0$ when charges are of the same sign, or the opposite one when charges are of opposite sign (see Fig. 1.3).

This expression is of course valid only as long as the distance between charges is far greater than the dimensions of the volume wherein the charges are situated. In order to define the electrical force of interaction between charges when one or both are contained in volumes possessing dimensions comparable to the distance between the charges, one must make use of the principle of superposition. According to this principle, each charge exerts a force on every other charge such that the size of the force is independent of the presence of additional charges. Using this principle, an arbitrary volume distribution of charges can be represented as a sum of elementary volumes. For example, the force between an elementary charge at point a, de(a),and a charge distributed in a volume V, as is shown in Fig. 1.4, can be written as:

$$\boldsymbol{F} = \frac{\mathrm{de}(a)}{4\pi\varepsilon_0}\int\limits_V \frac{\delta(q)\,\mathrm{d}V}{L_{qa}^3}\boldsymbol{L}_{qa} \tag{1.6}$$

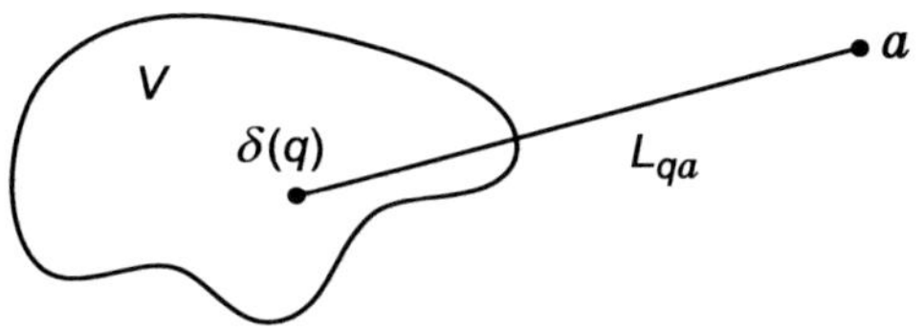

Figure 1.4. Distribution of charge in a volume.

where q indicates the position of any point within the volume V. The total electric force $\boldsymbol{F}(q)$ is the vector sum of all individual forces contributed by the individual elementary charges.

Extending this approach to a more general case in which all types of charges are present (volume, surface, linear and point charges) and again applying the principle of superposition, we obtain the following expression for the electrical force of interaction between an elementary charge $\mathrm{d}e(a)$ and a completely arbitrary distribution of charges:

$$\boldsymbol{F}(a) = \frac{\mathrm{d}e(a)}{4\pi\varepsilon_0}\left[\int\limits_V \frac{\delta(q)\,\mathrm{d}V}{L_{qa}^3}\boldsymbol{L}_{qa} + \int\limits_S \frac{\Sigma(q)\,\mathrm{d}S}{L_{qa}^3}\boldsymbol{L}_{qa} + \int\limits_L \frac{\lambda(q)\,\mathrm{d}l}{L_{qa}^3}\boldsymbol{L}_{qa} + \sum_{i=1}^{N}\frac{e_i(q)}{L_{qa}^3}\boldsymbol{L}_{qa}\right] \tag{1.7}$$

where $\delta\,\mathrm{d}V$, $\Sigma\,\mathrm{d}S$, $\lambda\,\mathrm{d}l$ and e_i are the symbols representing elementary volume, surface, linear and point charges, respectively.

At this point, we will define the strength of the electric field, $\boldsymbol{E}(a)$, as being the ratio of the force of electrical interaction, $\boldsymbol{F}(a)$, to the size of the elementary charge $\mathrm{d}e(a)$ (considered to be a test charge) at point a:

$$\boldsymbol{E}(a) = \frac{\boldsymbol{F}(a)}{\mathrm{d}e(a)} \tag{1.8}$$

For convenience, the strength of the electric field is usually referred to merely by the term *electric field.* It does not have the same dimension as a force, and has, in the practical system of units, the dimension of volts per meter.

The electric field $\boldsymbol{E}$ can be thought of as the force field acting on a test charge de inserted in the region of interest. If the electric field is known, it is a simple matter, using eq. 1.8, to calculate the force of interaction $\boldsymbol{F}$. As follows from eq. 1.7, the expression for the electric field can be written as:

$$\boldsymbol{E}(a) = \frac{1}{4\pi\varepsilon_0}\left[\int\limits_V \frac{\delta(q)\,\mathrm{d}V}{L_{qa}^3}\boldsymbol{L}_{qa} + \int\limits_S \frac{\Sigma(q)\,\mathrm{d}S}{L_{qa}^3}\boldsymbol{L}_{qa} + \int\limits_L \frac{\lambda(q)\,\mathrm{d}l}{L_{qa}^3}\boldsymbol{L}_{qa} + \sum_{i=1}^{N}\frac{e_i(q)}{L_{qa}^3}\boldsymbol{L}_{qa}\right] \tag{1.9}$$

If the distribution of charges is given, the function $\boldsymbol{E}$ only depends on the coordinates at which the test point is located. Because it depends only on position, the function is termed

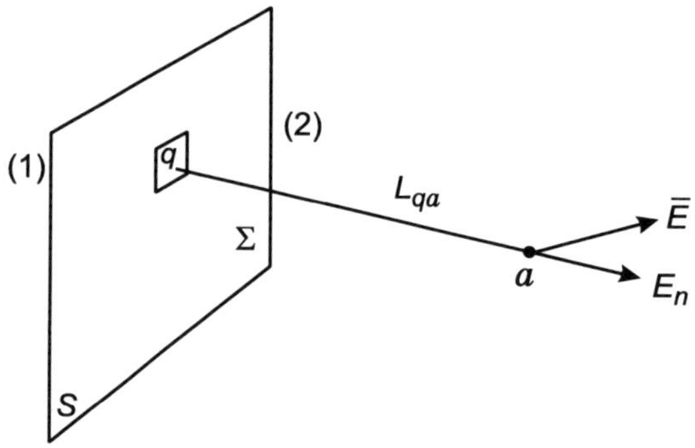

Figure 1.5. A charge on a plane surface.

a *field.* When the electric field does not vary with time, it only depends on distribution of charges in the medium, and its calculation using eq. 1.9 presents no fundamental difficulty. Considering only the portion of the field contributed by charges, a change in the electric field with time indicates that at some place in space there has been a simultaneous change in charge density.

In order to have a complete description of the field behavior, it is necessary to investigate a second source of the electric field, a source which acts when a time-varying magnetic field is present. But before considering this, let us further investigate the nature of the electric field caused by charges only.

First of all, let us consider several examples of fields caused by specific distributions of electric charges.

1.1.1. Example I: Normal Component of the Electric Field Caused by a Planar Charge Distribution

Suppose that there is a surface charge distribution on a plane surface a shown in Fig. 1.5. Let us introduce a vector $\mathrm{d}\boldsymbol{S}$:

$$\mathrm{d}\boldsymbol{S} = \mathrm{d}S\,\boldsymbol{n}$$

where $\boldsymbol{n}$ is the unit vector directed from the back side of the plane (1) toward the front side (2). We need only consider the normal component of the field, that is, the component which is perpendicular to the surface. In accord with Coulomb's law, as expressed in eq. 1.4, every elementary charge $\Sigma(q)\,\mathrm{d}S$ located at point q creates a field described by the equation:

$$\mathrm{d}\boldsymbol{E}(a) = \frac{1}{4\pi\varepsilon_0}\frac{\Sigma(q)\,\mathrm{d}S}{L_{qa}^3}\boldsymbol{L}_{qa} \qquad (1.10)$$

Therefore, the normal component of this field is:

$$\mathrm{d}E_n = \mathrm{d}E\cos(\boldsymbol{L}_{qa},\boldsymbol{n}) = \frac{1}{4\pi\varepsilon_0}\frac{\Sigma(q)\,\mathrm{d}S}{L_{qa}^2}\cos(\boldsymbol{L}_{qa},\boldsymbol{n})$$
$$= \frac{1}{4\pi\varepsilon_0}\frac{\Sigma(q)\,\mathrm{d}S\,L_{qa}}{L_{qa}^3}\cos(\boldsymbol{L}_{qa},\boldsymbol{n}) \tag{1.11}$$

where $(\boldsymbol{L}_{qa},\boldsymbol{n})$ is the angle between directions $\boldsymbol{L}_{qa}$ and $\boldsymbol{n}$. It is clear that the product:

$$\mathrm{d}S\,L_{qa}\cos(\boldsymbol{L}_{qa},\boldsymbol{n})$$

can be written as a scalar product as follows:

$$\mathrm{d}S\,L_{qa}\cos(\boldsymbol{L}_{qa},\boldsymbol{n}) = \mathrm{d}\boldsymbol{S}\cdot\boldsymbol{L}_{qa} = -\mathrm{d}\boldsymbol{S}\cdot\boldsymbol{L}_{aq}$$

because $\boldsymbol{L}_{qa} = -\boldsymbol{L}_{aq}$.

Thus the normal component of the electric field can be written as:

$$\mathrm{d}E_n = -\frac{1}{4\pi\varepsilon_0}\frac{\mathrm{d}\boldsymbol{S}\cdot\boldsymbol{L}_{aq}}{L_{aq}^3}\Sigma(q) \tag{1.12}$$

because $L_{aq} = L_{qa}$.

As can be seen, the quantity $\mathrm{d}\omega_{aq}$ defined as:

$$\mathrm{d}\omega_{aq} = \frac{\mathrm{d}\boldsymbol{S}\cdot\boldsymbol{L}_{aq}}{L_{aq}^3} \tag{1.13}$$

represents the solid angle subtending the element $\mathrm{d}S$ from point a. In a similar manner, the solid angle subtended by the entire surface S as viewed from point a is:

$$\omega_a = \int\limits_S \frac{\mathrm{d}\boldsymbol{S}\cdot\boldsymbol{L}_{aq}}{L_{aq}^3} \tag{1.14}$$

This expression allows us to find the solid angle when the surface S is of arbitrary shape. For example, with an observation point inside a closed surface, the solid angle is 4π. If the observation point is situated outside a closed surface, the solid angle subtended by the surface is zero. This can be derived from the fact that the closed surface could also be represented as two open surfaces, as shown in Fig. 1.6, which are viewed from any external point with the same solid angle by magnitude, but opposite in sign. In so doing, we must remember that the sign for the solid angle is defined by the angle between the direction of the vector $\boldsymbol{L}$ and the vector $\mathrm{d}\boldsymbol{S}$.

Returning again to the calculation of the normal component E_n (Fig. 1.5), we can write it as:

$$E_n = -\frac{1}{4\pi\varepsilon_0}\int\limits_S \Sigma\,\mathrm{d}\omega_{qa} \tag{1.15}$$

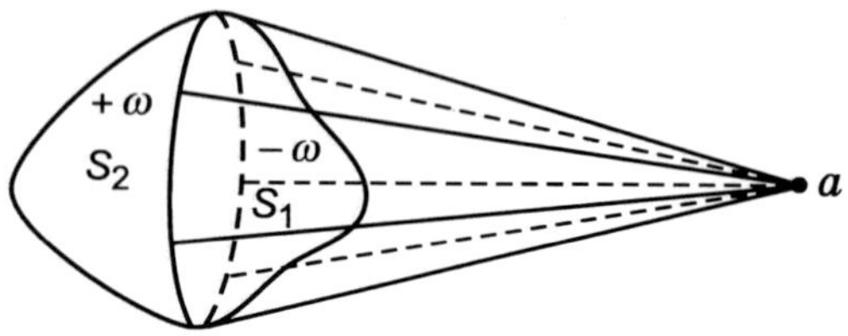

Figure 1.6. Representation of a closed surface by two open surfaces.

In particular, if the charge is distributed uniformly on the surface ($\Sigma = const$) we have:

$$E_n = -\frac{1}{4\pi\varepsilon_0}\omega_a\Sigma \tag{1.16}$$

where ω_a is the solid angle subtended by surface S when viewed from point a. It is obvious (see Fig. 1.7) that the solid angle ω_a is either positive or negative depending on whether the front side or the back side of the surface is viewed.

With increasing distance from the surface S, the solid angle decreases, and correspondingly the normal component of the field becomes smaller. In the opposite case, when the point a is considered to approach the surface, the solid angle increases and in the limit becomes equal to -2π and $+2\pi$ when observation point a is located either on the front side (2) or the back side (1) of the surface, respectively.

Thus we have the following expressions for the normal component of the electric field on either side of the plane surface:

$$\begin{aligned} E_n^{(1)} &= -\Sigma/2\varepsilon_0 \\ E_n^{(2)} &= \Sigma/2\varepsilon_0 \end{aligned} \tag{1.17}$$

These two expressions indicate that the normal component of the electric field is discontinuous across the surface S. Let us examine the normal component in some detail. The normal component of the electric field can be written as the sum of two terms:

$$E_n = E_n^p + E_n^{S-p} \tag{1.18}$$

where E_n^p is the part of the normal component caused by the elementary charge $\Sigma(p)\,\mathrm{d}S$ located in the immediate vicinity of point p, and E_n^{S-p} is the normal component contributed by all the other surface charges.

It is clear that:

$$E_n^{S-p}(a) = -\frac{1}{4\pi\varepsilon_0}\Sigma\int\limits_{S-p}\mathrm{d}\omega = -\frac{1}{4\pi\varepsilon_0}\Sigma\,\omega^{S-p}(a)$$

where $\omega^{S-p}(a)$ is the solid angle subtended by the plane surface S minus the element of surface $\mathrm{d}S(p)$ as viewed from point a. Letting point a approach the element of area

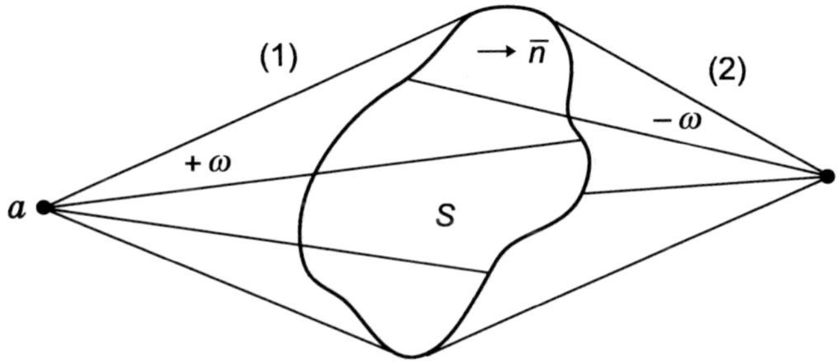

Figure 1.7. Illustrating the fact that the angles subtended by a surface can be either positive or negative depending on the view point.

$dS(p)$, the solid angle subtended by the rest of the surface tends to zero, and the normal component is determined by the charge located on the elementary surface $dS(p)$ only:

$$E^{S-p} \to 0 \quad \text{as } a \to p$$

During the same process, the solid angle subtended by the surface element $dS(p)$, no matter how small its area is, tends to $\pm 2\pi$ when viewed from an infinitesimally small distance from point p:

$$\omega^p \to \pm 2\pi \quad \text{as } a \to p$$

Therefore, the normal component of the field on either side of the surface is determined exclusively by the elementary charge located in the immediate vicinity of the point p:

$$E_n^{(1)}(p) = -\frac{1}{2\varepsilon_0}\Sigma(p)$$
$$E_n^{(2)}(p) = \frac{1}{2\varepsilon_0}\Sigma(p) \tag{1.19}$$

The difference in sign of the field on either side of the surface reflects the fundamental fact that the electric field shows the direction along which an elementary positive charge will move under the force of the field. Therefore, the discontinuity in the normal component as a test point passes through the surface is caused by the elementary charge located near the observation point only. For example if there is a hole in the surface, the component on either side of the surface is E_n^{S-p}, and therefore the field is continuous along a line passing through the hole.

We can generalize these results to the case in which the surface carrying the charge is not planar. Making use of the same approach based on the principle of superposition and the definition of solid angles, we arrive at the following expressions for the normal

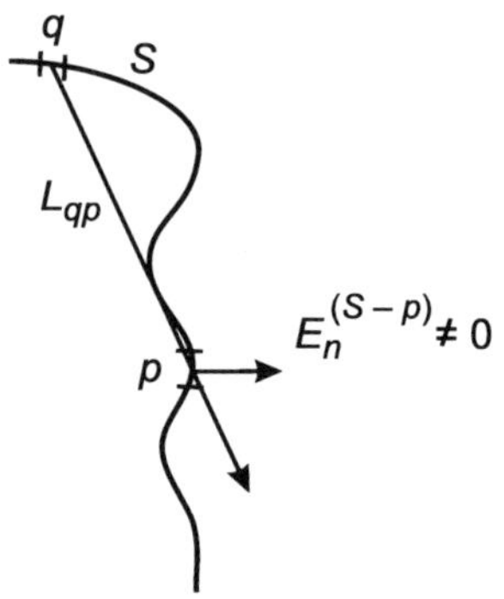

Figure 1.8. Illustrating the fact that the normal component of electric field caused by charges locate on the surface but outside the element $\mathrm{d}\boldsymbol{S}$ is not necessarily zero at a point p.

component on either side of the surface:

$$E_n^{(1)}(p) = -\frac{\Sigma(p)}{2\varepsilon_0} + E_n^{(1)(S-p)}$$
$$E_n^{(2)}(p) = \frac{\Sigma(p)}{2\varepsilon_0} + E_n^{(2)(S-p)} \tag{1.20}$$

In contrast to the previous case, the normal component $E_n^{S-p}(p)$ caused by the charges located on the surface but outside the element $\mathrm{d}\boldsymbol{S}(p)$ is not necessarily zero (see Fig. 1.8). However, we can readily recognize a very important feature of this part of the field. Inasmuch as these changes are located at some distance from point p, their contribution to the field is a continuous function when observation point a passes through element $\mathrm{d}\boldsymbol{S}(p)$, and therefore:

$$E_n^{(1)(S-p)} = E_n^{(2)(S-p)} = E_n^{S-p} \tag{1.21}$$

Correspondingly, eq. 1.20 can be written as:

$$E_n^{(1)}(p) = -\frac{\Sigma(p)}{\varepsilon_0} + E_n^{S-p}$$
$$E_n^{(2)}(p) = \frac{\Sigma(p)}{\varepsilon_0} + E_n^{S-p} \tag{1.22}$$

This means that the discontinuity in the normal component is, as before:

$$E_n^{(2)}(p) - E_n^{(1)}(p) = \frac{\Sigma(p)}{\varepsilon_0} \tag{1.23}$$

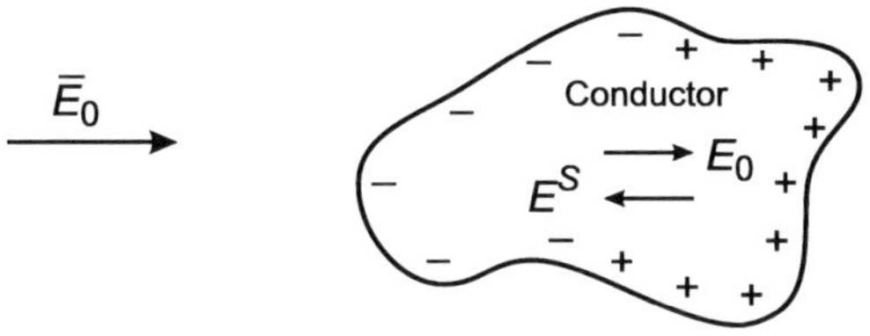

Figure 1.9. Under the action of an applied field, positive and negative charges residing inside a closed conductor move in opposite directions.

and it is caused by charges located within the elementary surface area $\mathrm{d}\boldsymbol{S}(p)$ only.

It should be emphasized that eq. 1.23 is a fundamental equation describing the electromagnetic field behavior and is valid for any rate of change of the field with time. We might say in essence, even though we risk getting ahead of ourselves, that eq. 1.23 is the surface analogy of the third Maxwell equation.

1.1.2. Example II: Effect of a Conductor Situated within an Electric Field

We will now consider a second example which illustrates the electrostatic induction phenomenon. First of all, let us suppose that a conductive body of arbitrary shape is situated within the region of influence of an electric field $\boldsymbol{E}_0$ as shown in Fig. 1.9. Under the action of the field, the positive and negative charges residing inside the conductor move in opposite directions. As consequence of this movement, electric charges accumulate on both sides of the conductor. In so doing, they create a secondary electric field, which is directed in opposite direction to the primary field inside the conductor. The induced surface charges distribute themselves in such a way that the total electric field inside the conductor disappears, that is:

$$\boldsymbol{E}_i = 0 \tag{1.24}$$

where $\boldsymbol{E}_i$ indicates the electric field strength within the conductor. This process is called *electrostatic induction.* At this point it is appropriate to make several comments:

- In our description of this phenomenon, we have given a very approximate explanation of the process in which only the electrostatic field is considered to be present. In fact, the process of accumulation of charges involves other phenomena, as in particular the appearance of a time-varying magnetic field, which plays an important role that will be examined later.

- The phenomenon of electrostatic induction is observed in any conductive body, regardless of its electrical resistivity. For example, the conductive body could be composed of metal, or of an electrolytic solution, of minerals or rocks. It is fundamental, however, that the charges that create the primary field are situated outside the conductor. We will later see that the magnitude of the resistivity plays a role in determining the time which is required for the electric field to vanish inside the conductor, but it does not change the final result of electrostatic induction, that is, the internal electric field goes to zero.

- Considering this effect, we assumed that the conductor has finite dimensions. This condition is not important and electrostatic induction would be observed in an infinite medium as well. For example, suppose that an electric charge e_0 is placed in a nonconducting borehole as is shown in Fig. 1.10. In this case, charges of opposite sign than e_0 appear on the borehole surface and are distributed in such a way that the total electric field of these charges within the borehole is not zero (though it does not contain any information about the distribution of resistivity in the medium). As it concerns induced charges having the same sign as the charge e_0, they are moved to infinity.

- An electric field which does not depend on time within a certain range can be created by various ways. For instance, it can be generated inductively from a current in a close loop, and whose intensity would increase linearly with time. At the same time, electrostatic induction is usually observed when the sources of the primary field are electric charges. Deviations from this rule have very extremal character.

Returning again to the electrostatic induction phenomenon (Fig. 1.9), it should be obvious that the secondary electric field contributed by the surface charge can be defined from the equation:

$$\boldsymbol{E}^S(a) = \frac{1}{4\pi\varepsilon_0} \int\limits_S \frac{\Sigma(q)\,\mathrm{d}S}{L_{qa}^3} \boldsymbol{L}_{qa} \tag{1.25}$$

where $\Sigma(q)$ is the surface density of charge. Correspondingly, condition 1.24 can be rewritten as:

$$\boldsymbol{E}_0(a) + \frac{1}{4\pi\varepsilon_0} \int\limits_S \frac{\Sigma(q)\,\mathrm{d}S}{L_{qa}^3} \boldsymbol{L}_{qa} = 0 \tag{1.26}$$

where $\boldsymbol{E}_0$ is the primary field contributed by external sources.

If, for instance, a single point charge e is situated outside the conductor, electric field at any point inside the conductor is:

$$\boldsymbol{E}_0(a) = \frac{1}{4\pi\varepsilon_0} \frac{e}{L_{0a}^3} \boldsymbol{L}_{0a} \tag{1.27}$$

where a is the point at which $\boldsymbol{E}$ is observed.

It results from this fact that the electric field caused by a given system of charges does not depend on the properties of the medium. If the field changes, this means that new charges develop.

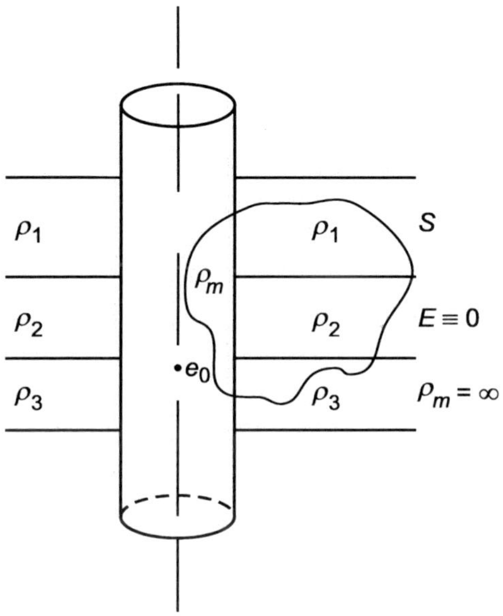

Figure 1.10. Illustration of the electrostatic induction when an electric charge is located within nonconducting borehole (ρ is the resistivity of the medium).

In our case, positive and negative charges arise at the surface of a conductor at the same time, so that the total charge of the neutral conductor remains zero:

$$\Sigma e^S = 0 \tag{1.28}$$

If a conductor has infinite dimensions (Fig. 1.10), charges of one sign appear at infinity and condition 1.28 is still valid.

In conclusion, let us make one more remark. The distribution of charges caused by electrostatic induction is not usually known before-hand, and their determination constitutes one of the classical problems of the theory of electrical fields.

It is appropriate to notice that there are several very well-developed numerical techniques allowing us to solve this problem, such as the method of integral equations, the method of finite differences, and others.

At this point, we can describe some general problems involving the electric fields caused by charges. It is obvious that when the charge distribution is unknown, we cannot make use of Coulomb's law to calculate the field. Unfortunately, in most cases of interest in electric logging, the distribution of charges is unknown, and Coulomb's law is of no help in determining the field. This is why we must consider some general features of the electric field caused by charges.

Proceeding from Coulomb's law, we can obtain fundamental equations for this field. First of all, let us introduce the concept of electric flux N through a given surface S as

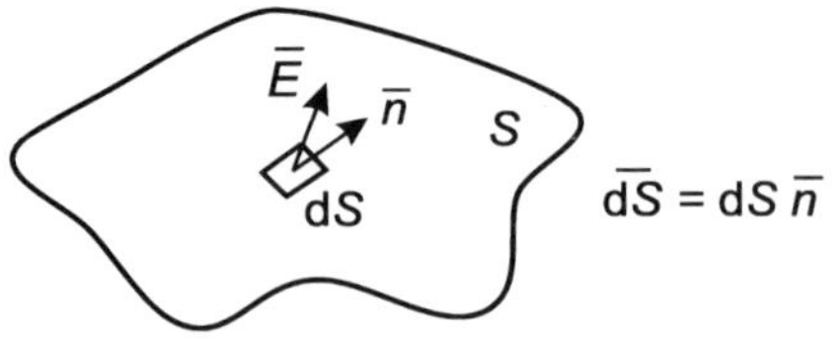

Figure 1.11. Representation of electric flux as being the surface integral of the scalar product of the electric field and the surface.

being the surface integral of the scalar product of the field $\boldsymbol{E}$ to the unit vector $\boldsymbol{n}$ normal to that surface, as illustrated in Fig. 1.11:

$$N = \int_S \boldsymbol{E} \cdot \mathrm{d}\boldsymbol{S} \tag{1.29}$$

where $\boldsymbol{E} \cdot \mathrm{d}\boldsymbol{S} = E\,\mathrm{d}S \cos(\boldsymbol{E}, \mathrm{d}\boldsymbol{S})$.

Suppose that an elementary charge, de, situated at point q, is the sole source of an electric field. In accord with Coulomb's law, the flux of the corresponding electric field through an arbitrary surface S is given by:

$$N = \frac{\mathrm{d}e}{4\pi\varepsilon_0} \int_S \frac{\boldsymbol{L}_{qp} \cdot \mathrm{d}\boldsymbol{S}}{L_{qp}^3} \tag{1.30}$$

where p is an arbitrary point on the surface S. Inasmuch as this integral represents the solid angle ω subtended by the surface S as seen from point q, we can write:

$$\int_S \boldsymbol{E} \cdot \mathrm{d}\boldsymbol{S} = \frac{\mathrm{d}e}{4\pi\varepsilon_0} \omega_S(q)$$

In the particular case in which the surface is closed and the charge de is located inside it, the solid angle $\omega_S(q)$ is 4π and we have:

$$\oint_S \boldsymbol{E} \cdot \mathrm{d}\boldsymbol{S} = \frac{\mathrm{d}e}{\varepsilon_0} \tag{1.31}$$

It should be clear that when the charge is located outside the surface S, the flux of the electric field caused by the charge is zero.

Equation 1.31 has been obtained in the case of an elementary charge. Using the principle of superposition, we can derive the following equation for an arbitrary distribution of charges:

$$\oint \boldsymbol{E} \cdot \mathrm{d}\boldsymbol{S} = \frac{1}{\varepsilon_0} \left[\int_V \delta\,\mathrm{d}V + \int_S \Sigma\,\mathrm{d}S + \int_L \lambda\,\mathrm{d}l + \sum_{i=1}^{M} e_i \right] \tag{1.32}$$

where δ, Σ, and λ are volume, surface and linear charge densities, respectively, and e_i is a point charge, all of which are located inside the surface S. The flux caused by charges outside the surface is zero.

The following comments should be made about eq. 1.32:

- A change in position of the charge within the volume V limited by surface S alters the value of the field $\boldsymbol{E}$ at the surface, but does not affect the value of the flux because it is a function of the total charge within the surface only.
- The surface S can have quite an arbitrary shape and position. In particular, it can intersect portions of the medium characterized by different electrical properties, as shown in Fig. 1.10.

Assuming a distribution of charge described by the volume density δ, we have:

$$\int_S \boldsymbol{E} \cdot \mathrm{d}\boldsymbol{S} = e_V/\varepsilon_0 \tag{1.33}$$

where e_V is the volume charge within the volume V:

$$e_V = \int_V \delta \, \mathrm{d}V$$

Usually, the total charge e_V is the sum of two types of charges, one being *free charges*, which are free to move, and the other being *polarization charges*. The displacement of charges will not be considered here, and therefore we will normally mean that the charge e_V is the free charge.

Equation 1.33, which was developed directly from Coulomb's law, is in fact the third of Maxwell's equations, valid both for constant and time-varying fields. Omitting the subscript V on the charge e, we have:

$$\oint_S \boldsymbol{E} \cdot \mathrm{d}\boldsymbol{S} = e/\varepsilon_0 \tag{1.34}$$

This equation shows the relationship between field values observed on various points of the surface S and can be interpreted from two points of view. If the charge e is known, eq. 1.34 can be considered to be an integral equation in an unknown variable: the normal component of the field. In contrast, when the electric field is known, the use of the flux allows us to determine the sources of the field. If we wish to find the relationship between flux and source within an elementary volume, we can make use of Gauss's theorem:

$$\oint_S \boldsymbol{E} \cdot \mathrm{d}\boldsymbol{S} = \int_V \operatorname{div} \boldsymbol{E} \, \mathrm{d}V \tag{1.35}$$

where S is a closed surface surrounding the volume V. Applying this equation to an elementary volume where the function div $\boldsymbol{E}$ is nearly constant, we have:

$$\oint_S \boldsymbol{E}\cdot \mathrm{d}\boldsymbol{S} \simeq \operatorname{div}\boldsymbol{E}\,\mathrm{d}V$$

or, in the limit:

$$\operatorname{div}\boldsymbol{E} = \frac{1}{\mathrm{d}V}\oint_S \boldsymbol{E}\cdot \mathrm{d}\boldsymbol{S} \tag{1.36}$$

Thus the divergence of the electric field characterizes the flux of the vector $\boldsymbol{E}$ through a surface limiting an elementary volume. In accord with eqs. 1.34–1.36 we have:

$$\operatorname{div}\boldsymbol{E} = \delta/\varepsilon_0 \tag{1.37}$$

The divergence of the electric field along with the flux through an arbitrary closed surface characterizes the distribution of charges. However, eq. 1.37 describes the volume density of charges in the vicinity of any point, that is, it has a differential character, distinct from that of eq. 1.34.

Both eqs. 1.34 and 1.37 are valid for electromagnetic fields regardless of the rate of change of the field with time. Equation 1.37 is the third Maxwell equation in differential form. We must stress that there is a fundamental difference between the two forms presented above for the third Maxwell equation. While the integral form can be applied everywhere, it is necessary to be careful in the use of the differential form. This caution must be exercised because the function div $\boldsymbol{E}$ might not be defined at certain points, lines or surfaces. As a matter of fact, div $\boldsymbol{E}$ is expressed in terms of the first spatial derivatives of the field components. In Cartesian coordinates for example, we have:

$$\operatorname{div}\boldsymbol{E} = \frac{\partial E_x}{\partial x} + \frac{\partial E_y}{\partial y} + \frac{\partial E_z}{\partial z}$$

At points where one of the derivatives is not properly behaved, eq. 1.37 cannot be applied. In other words, it does not permit us to describe the nature of sources at such locations. A very important example from electrical logging where this equation cannot be applied is provided by any model in which electrical charges are distributed at interfaces representing a step-wise change in resistivity. As was shown in the first example of this section, the normal component of the electric field is a discontinuous function of the spatial variable through a surface charge and therefore the normal derivative $\partial E_n/\partial n$ does not exist on the surface. Therefore, in order to characterize sources on such interface, one must use the third Maxwell equation in integral form (eq. 1.34).

Applying it to an elementary cylindrical surface enclosing a small piece interface, as shown in Fig. 1.12, we obtain a well-known relationship:

$$E_n^{(2)} - E_n^{(1)} = \Sigma/\varepsilon_0 \tag{1.38}$$

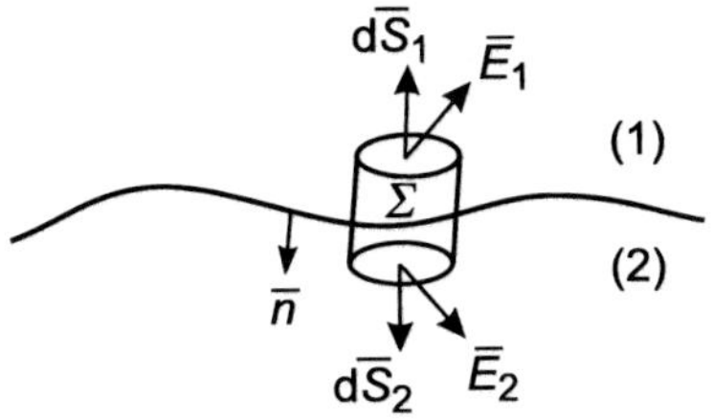

Figure 1.12. Definition of an elementary cylindrical surface that encloses a small piece of an interface between two regions with different resistivity.

where Σ is the surface charge density, $E_n^{(1)}$ and $E_n^{(2)}$ are the normal components of the electric field on either side of the surface, and $\boldsymbol{n}$ is a unit normal vector directed from the reverse side to the front side of the surface. Equation 1.38 can be interpreted as being the surface analogy of eq. 1.37, and is another form the third Maxwell equation as well.

Comparing eq. 1.38 and 1.23, we see that they exactly coincide. This follows directly from the fact that the discontinuity of the normal component E_n is due to the presence of surface charges. In particular, if the surface charge is absent at some point, the normal component of the field is found to be continuous.

By starting with Coulomb's law, we have obtained three useful forms of the third Maxwell third equation:

$$\oint_S \boldsymbol{E} \cdot \mathrm{d}\boldsymbol{S} = e/\varepsilon_0 \qquad \operatorname{div} \boldsymbol{E} = \delta/\varepsilon_0 \qquad E_n^{(2)} - E_n^{(1)} = \Sigma/\varepsilon_0 \tag{1.39}$$

Each of them characterizes the distribution of charges, and one can say that they are the same tool of analysis written in three different ways.

Another highly useful concept that illustrates some of the fundamental characteristics of the electric field can be introduced as follows:

$$\int_a^b \boldsymbol{E} \cdot \mathrm{d}\boldsymbol{l} \tag{1.40}$$

The integral represents the voltage of the electric field between points a and b, measured along some given path L (Fig. 1.13). The scalar product $\boldsymbol{E} \cdot \mathrm{d}\boldsymbol{l}$ can be written as:

$$\boldsymbol{E} \cdot \mathrm{d}\boldsymbol{l} = E\,\mathrm{d}l \cos(\boldsymbol{E}, \mathrm{d}\boldsymbol{l}) = E\,\mathrm{d}l \cos\alpha$$

where α is the angle between the electric field vector and the tangent to the path L at every point. From the physical point of view, the product $\boldsymbol{E} \cdot \mathrm{d}\boldsymbol{l}$ is an element of work performed by the electrical field transporting a unit positive charge along the elementary displacement $\mathrm{d}l$. This product has the dimension of a work per unit charge, and in the

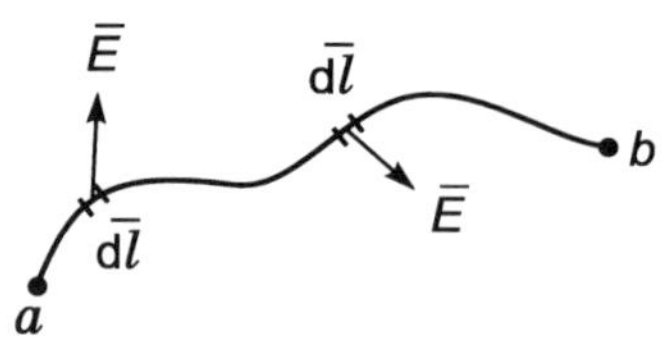

Figure 1.13. An arbitrary path along which the dot product of electric field and direction is integrated.

practical system of units has the dimension of volts. Therefore, the integral in eq. 1.40 represents the work of voltage done in carrying a unit charge between points a and b along the path L. In the general case of any function $\boldsymbol{E}$, this integral depends on the particular path of integration L which is chosen and on the terminal points a and b. Starting from Coulomb's law, it can be demonstrated that the voltage of the electric field caused only by static charges is independent of the path followed and only depends on the terminal points.

Assume that the source for the field is a single elementary charge $\mathrm{d}e$. In accord with Coulomb's law, the electric field is:

$$\boldsymbol{E}(p) = \frac{1}{4\pi\varepsilon_0}\frac{\mathrm{d}e(q)}{L_{qp}^3}\boldsymbol{L}_{qp} \tag{1.41}$$

where $\boldsymbol{L}_{qp}$ is the vector directed from point q to point p. If both terminal points a and b are situated on the same radius vector $\boldsymbol{L}_{qp}$, and the path of integration is along this radius (Fig. 1.14), the voltage between these points is very easily calculated:

$$V = \int_a^b \boldsymbol{E}\cdot \mathrm{d}\boldsymbol{l} = \frac{\mathrm{d}e(q)}{4\pi\varepsilon_0}\int_a^b \frac{\mathrm{d}\boldsymbol{l}\cdot\boldsymbol{L}_{qp}}{L_{qp}^3} = \frac{\mathrm{d}e(q)}{4\pi\varepsilon_0}\int_a^b \frac{\mathrm{d}l}{L_{qp}^2}$$

inasmuch as:

$$\mathrm{d}\boldsymbol{l}\cdot\boldsymbol{L}_{qp} = \mathrm{d}l\, L_{qp}\cos 0 = L_{qp}\,\mathrm{d}l$$

Carrying out the integration as indicated above, we obtain:

$$V = \frac{\mathrm{d}e(q)}{4\pi\varepsilon_0}\left(\frac{1}{L_{qa}} - \frac{1}{L_{qb}}\right) \tag{1.42}$$

Now assume that points a and b are situated on two different radius vectors $\boldsymbol{L}_{qa}$ and $\boldsymbol{L}_{qb}$ as shown in Fig. 1.15. Let us choose the path L_1 as consisting of two parts. The first

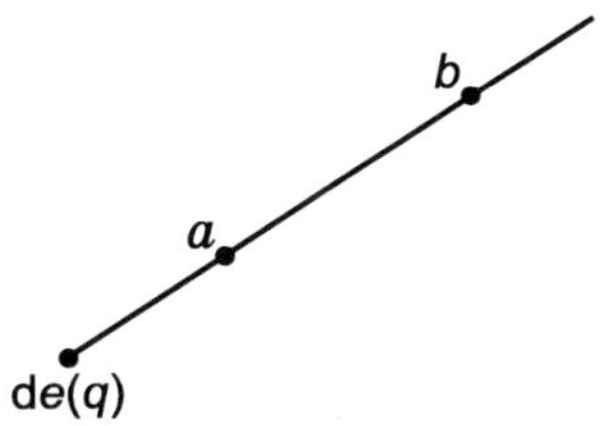

Figure 1.14. Illustration of a case in which the points a and b are on a common radius vector issued from the position of the electric charge.

part is a simple arc ab' and the second one is the segment $b'b$ along the radius vector $\boldsymbol{L}_{qb}$. In this case the voltage can be written as:

$$V = \frac{\mathrm{de}(q)}{4\pi\varepsilon_0}\left[\int\limits_{\text{arc } ab'} \frac{\mathrm{d}\boldsymbol{l}\cdot\boldsymbol{L}_{qp}}{L_{qp}^3} + \int\limits_{b'}^{b} \frac{\mathrm{d}\boldsymbol{l}\cdot\boldsymbol{L}_{qp}}{L_{qp}^3}\right]$$

Since the scalar product $\boldsymbol{L}_{qp}\cdot\mathrm{d}\boldsymbol{l}$ on ab' is zero, the integral along the arc ab' vanishes.

Thus the voltage between points a and b is again equal to:

$$V = \frac{\mathrm{de}(q)}{4\pi\varepsilon_0}\left(\frac{1}{L_{qa}} - \frac{1}{L_{qb}}\right) \tag{1.43}$$

If instead of the path L_1 we consider a more arbitrary path L_2, it is clear that this path can be decomposed as a sum of elements of arcs and of radius vectors as illustrated in Fig. 1.16. All contributions along simple arcs are zero, while the sum of integrals along all the radius vectors is:

$$V = \frac{\mathrm{de}(q)}{4\pi\varepsilon_0}\left(\frac{1}{L_{qa}} - \frac{1}{L_{qb}}\right)$$

that is, it is equal to the voltage along the path L_1.

We have established the second fundamental characteristic of an electric field, namely that the voltage between two points does not depend on the particular path along which integration is carried out, but is determined by the terminal points only. This fact can be expressed formally as follows:

$$\underset{L_1}{\int_a^b} \boldsymbol{E}\cdot\mathrm{d}\boldsymbol{l} = \underset{L_2}{\int_a^b} \boldsymbol{E}\cdot\mathrm{d}\boldsymbol{l} = \cdots\cdot = \underset{L_n}{\int_a^b} \boldsymbol{E}\cdot\mathrm{d}\boldsymbol{l} \tag{1.44}$$

Making use of the principle of superposition, this result can be generalized to a field caused by any distribution of charges. It must be stressed again that this result is valid only for

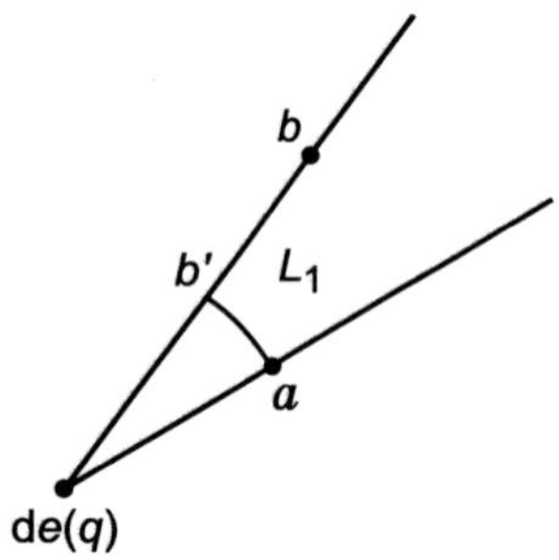

Figure 1.15. Example of a case in which the points a and b are situated on different radius vectors.

Figure 1.16. An arbitrary path that can be represented as the sum of radius vectors and arcs.

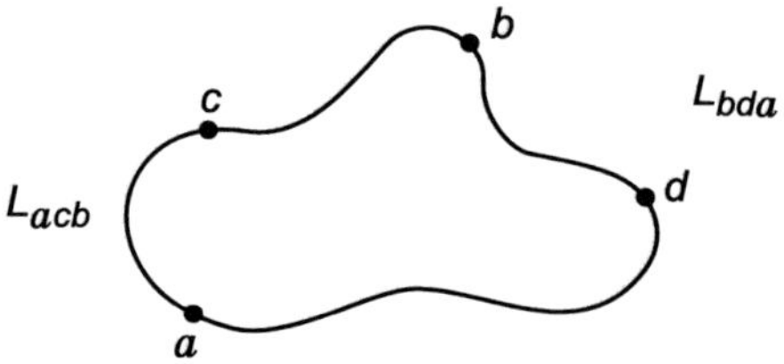

Figure 1.17. A path of integration along a closed contour L which can be broken down into the summations of two open contours.

electric fields caused by constant charges, and that it cannot be applied to time-varying fields.

The independence of the voltage on the path can be written in another form: let us consider a close contour L as shown in Fig. 1.17, which consists of two other contours L_{acb} and L_{adb}. In accord with eq. 1.44, we have:

$$\int\limits_{acb} \boldsymbol{E} \cdot \mathrm{d}\boldsymbol{l} = \int\limits_{adb} \boldsymbol{E} \cdot \mathrm{d}\boldsymbol{l} \tag{1.45}$$

In these integrals, the element $\mathrm{d}\boldsymbol{l}$ is directed from a to b. Changing the direction of integration in the right-hand side of eq. 1.45, we can write:

$$\int\limits_{adb} \boldsymbol{E} \cdot \mathrm{d}\boldsymbol{l} = -\int\limits_{bda} \boldsymbol{E} \cdot \mathrm{d}\boldsymbol{l}, \quad \text{i.e.}$$

$$\int\limits_{acb} \boldsymbol{E} \cdot \mathrm{d}\boldsymbol{l} = -\int\limits_{bda} \boldsymbol{E} \cdot \mathrm{d}\boldsymbol{l}, \text{ or}$$

$$\int\limits_{acb} \boldsymbol{E} \cdot \mathrm{d}\boldsymbol{l} + \int\limits_{bda} \boldsymbol{E} \cdot \mathrm{d}\boldsymbol{l} = 0, \text{ or}$$

$$\oint\limits_{L} \boldsymbol{E} \cdot \mathrm{d}\boldsymbol{l} = 0 \tag{1.46}$$

Thus the voltage along an arbitrary closed path is zero. Sometimes the quantity $\oint_L \boldsymbol{E} \cdot \mathrm{d}\boldsymbol{l}$ is called the *circulation* of the electric field or the *electromotive force* (EMF).

The path L can have an arbitrary shape, and it can intersect media characterized by various physical properties. In particular, it can be completely contained within a conducting medium. Because of the fact that the electromotive force caused by electric charges is zero, a Coulomb force field can cause an electric current by itself. This is the reason why non-Coulomb forces must be considered in order to understand the creation of current flow. Equation 1.46 is the first Maxwell equation for electric fields which do not vary with time, given in its integral form, and relates the values of the field various points in the medium. To obtain eq. 1.46 in differential form, we will make use of Stoke's theorem, according to which for any vector $\boldsymbol{A}$ having first spatial derivatives, the following relationship holds:

$$\oint\limits_{L} \boldsymbol{A} \cdot \mathrm{d}\boldsymbol{l} = \int\limits_{S} \operatorname{curl} \boldsymbol{A} \cdot \mathrm{d}\boldsymbol{S} \tag{1.47}$$

In this expression, the orientations for $\mathrm{d}\boldsymbol{l}$ and $\mathrm{d}\boldsymbol{S}$ are obtained according to right-hand rule convention illustrated in Fig. 1.18.

The function $\boldsymbol{A}$ is a vector expressed in terms of the spatial derivatives of the components of $\boldsymbol{A}$. As an example, in Cartesian coordinates, $\boldsymbol{A}$ expressed as follows:

$$\operatorname{curl} \boldsymbol{A} = \left(\frac{\partial A_z}{\partial y} - \frac{\partial A_y}{\partial z}\right)\boldsymbol{i} + \left(\frac{\partial A_x}{\partial z} - \frac{\partial A_z}{\partial x}\right)\boldsymbol{j} + \left(\frac{\partial A_y}{\partial x} - \frac{\partial A_x}{\partial y}\right)\boldsymbol{k}$$

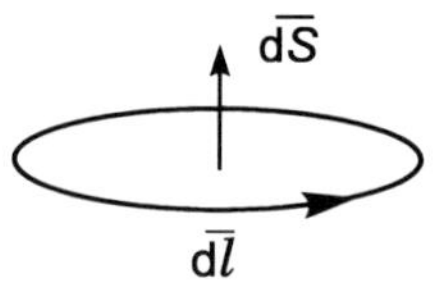

Figure 1.18. Use of the right-hand rule to define the direction of $\mathrm{d}\boldsymbol{l}$ and $\mathrm{d}\boldsymbol{S}$.

Similar relationships can be written in other orthogonal systems of coordinates. We can demonstrate that curl $\boldsymbol{A}$ characterizes the maximum change voltage in the vicinity of a source point, in the direction perpendicular to the field. In accord with eqs. 1.46 and 1.47, we obtain a differential form of the first equation for the electric field:

$$\operatorname{curl}\boldsymbol{E} = 0 \tag{1.48}$$

This equation, as well as eq. 1.46, reflects the fact that the voltage along a closed path must be zero. It is appropriate to emphasize that both of them follow directly from Coulomb's law.

As has been previously mentioned, Stoke's theorem (eq. 1.46) is valid only when the first spatial derivative exist. Thus, this equation cannot be used at points where one of the components is a discontinuous function of position. In order to obtain a differential form of eq. 1.46 valid at such points, we will apply this equation along an elementary path as shown in Fig. 1.19.

Considering that elements $\mathrm{d}\boldsymbol{l}'$ and $\mathrm{d}\boldsymbol{l}''$ are separated by an arbitrarily small distance $\mathrm{d}h$ which tends to zero, we have:

$$\boldsymbol{E}\cdot\mathrm{d}\boldsymbol{l}'' + \boldsymbol{E}\cdot\mathrm{d}\boldsymbol{l}' + 2(\boldsymbol{E}\cdot\mathrm{d}\boldsymbol{h}) = 0$$

and in the limit:

$$E_t^{(2)}\,\mathrm{d}l'' - E_t^{(1)}\,\mathrm{d}l' = 0$$

so that finally:

$$E_t^{(2)} - E_t^{(1)} = 0 \tag{1.49}$$

Thus the tangential component of the field is a continuous function through a charged surface.

We have now derived three forms of the first equation based on Coulomb's law:

$$\oint \boldsymbol{E}\cdot\mathrm{d}\boldsymbol{l} = 0 \qquad \operatorname{curl}\boldsymbol{E} = 0 \qquad E_t^{(2)} - E_t^{(1)} = 0 \tag{1.50}$$

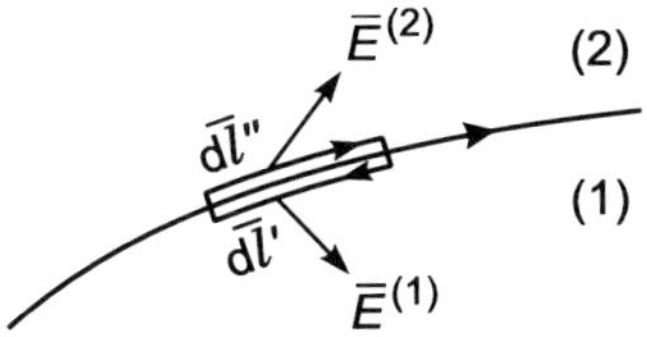

Figure 1.19. Application of Stoke's theorem along an elementary path.

Each of them expresses the same fact, that is, the electromotive force caused by electric charges is zero, or in other words, the voltage between two arbitrary points does not depend on the path of integration.

We must make an important comment about eq. 1.50. The first two relationships are not valid when the field is time-varying, since the second source for an electric field, that is the change of the magnetic field intensity with time, is not taken into account. On the other hand, the surface analog of these equations remains valid for any electromagnetic field. This reflects the fact that this result was obtained assuming that the area surrounded by the path of integration vanishes and, consequently, the flux of the magnetic induction through this area is zero.

Let us note one further thing. Although the equations $\oint \boldsymbol{E} \cdot d\boldsymbol{l} = 0$, $\operatorname{curl} \boldsymbol{E} = 0$ are not valid for time-varying electromagnetic fields, this does not mean that Coulomb's law is inapplicable. In a further analysis, we will have the chance to demonstrate that in many cases time-varying charges create electric fields which are practically described by Coulomb's law.

Returning to the first field equation, let us consider one more important feature of the electric field caused only by charges. In fact, according to eq. 1.48, the field can be written as:

$$\boldsymbol{E} = -\operatorname{grad} U \tag{1.51}$$

inasmuch as:

$$\operatorname{curl} \operatorname{grad} U = 0$$

is an identity relationship.

The scalar function U is called the potential of the electric field. In accord with eq. 1.51, the direction of the field $\boldsymbol{E}$ coincides with the direction of maximum decrease of the potential, and the projection of the field on any direction $\boldsymbol{l}$ can be expressed in terms of the potential as follows:

$$E_l = -\frac{\partial U}{\partial l} \tag{1.52}$$

Equations 1.51 and 1.52 are useful in determining the field when the potential is known.

At this point, we will write an expression for the voltage using the potential. For this purpose, let us write the following equality for the differentiation of the potential:

$$\mathrm{d}U = \frac{\partial U}{\partial l}\,\mathrm{d}l = \operatorname{grad} U \cdot \mathrm{d}\boldsymbol{l} = -\boldsymbol{E} \cdot \mathrm{d}\boldsymbol{l} \tag{1.53}$$

where $\mathrm{d}\boldsymbol{l} = \mathrm{d}l\,\boldsymbol{i}_0$, $\boldsymbol{i}_0$ is a unit vector.

Integrating the last of these terms along any path between two given points a and b, and considering that the voltage does not depend on the path followed, we obtain:

$$\int_a^b \boldsymbol{E} \cdot \mathrm{d}\boldsymbol{l} = -\int_a^b \mathrm{d}U = U(a) - U(b) \tag{1.54}$$

Thus the voltage of the electric field between two points can be expressed as the difference of potential between these points.

We can now use eq. 1.54 to define the potential caused by a distribution of charges. From this equation, we have:

$$U(a) = U(b) + \int_a^b \boldsymbol{E} \cdot \mathrm{d}\boldsymbol{l} \tag{1.55}$$

It is obviously reasonable to assume that at great distances from the source the potential will vanish. Then letting b equal infinity in eq. 1.55 and assuming that the potential at that distance is zero, we have:

$$U(a) = \int_a^\infty \boldsymbol{E} \cdot \mathrm{d}\boldsymbol{l} \tag{1.56}$$

Suppose now that the source of the electric field is a single elementary charge, $\mathrm{d}e$, situated at the point q. Using eqs. 1.41 and 1.56, we obtain:

$$U(a) = \frac{1}{4\pi\varepsilon_0}\frac{\mathrm{d}e}{L_{qa}} \tag{1.57}$$

Making use of the principle of superposition for an arbitrary distribution of volume, surface, linear and point charges, we arrive at the following expression for the potential:

$$U(a) = \frac{1}{4\pi\varepsilon_0}\left[\int_V \frac{\delta\,\mathrm{d}V}{L_{qa}} + \int_S \frac{\Sigma\,\mathrm{d}S}{L_{qa}} + \int_L \frac{\lambda\,\mathrm{d}l}{L_{qa}} + \sum_{i=1}^{N}\frac{e_i}{L_{qa}}\right] \tag{1.58}$$

Comparing this last expression with eq. 1.9, we see that the potential is related with charges in a much simpler way than is the electric field. This simplicity is an important reason for making use of the potential. Having defined U as we have, the electric field is very easily found by applying eq. 1.52.

In conclusion, we will derive equations reflecting the behavior of the potential. Substituting eq. 1.51 into eq. 1.37, we obtain:

$$\text{div grad}\, U = -\delta/\varepsilon_0 \tag{1.59}$$

or

$$\nabla^2 U = \Delta U = -\delta/\varepsilon_0$$

where the operator ∇^2 or Δ is the Laplacian operator. As an example, in Cartesian coordinates, eq. 1.59 becomes:

$$\frac{\partial^2 U}{\partial x^2} + \frac{\partial^2 U}{\partial y^2} + \frac{\partial^2 U}{\partial z^2} = -\frac{\delta}{\varepsilon_0} \tag{1.60}$$

that is, the simple sum of the partial second derivatives of the potential with respect to each of the spatial derivatives is directly proportional to the volume density of charge taken with the opposite sign. Equation 1.59 is most commonly called Poisson's equation for the potential and describes the behavior of the potential at points where the volume density of charge is non-zero. In areas free of charge, it simplifies and becomes Laplace equation for the potential:

$$\frac{\partial^2 U}{\partial x^2} + \frac{\partial^2 U}{\partial y^2} + \frac{\partial^2 U}{\partial z^2} = 0 \tag{1.61}$$

We will derive a general solution for Poisson's equation when the source of the field, and therefore of the potential, is a volume charge only. In accord with eq. 1.58, the potential U caused by such charges is the volume integral:

$$U(a) = \frac{1}{4\pi\varepsilon_0} \int\limits_V \frac{\delta(q)\,\mathrm{d}V}{L_{qa}} \tag{1.62}$$

On the other hand, Poisson's equation describes the potential everywhere, whether a charge is present or not. Therefore, the right-hand side of eq. 1.62 satisfies this equation and is a solution. It is obvious that Poisson's along with Laplace equation describes the potential only when the second spatial derivatives of this function exist, that is, when the first spatial derivatives of the field do exist. Unfortunately, there are many cases when this condition is not met and when consequently eqs. 1.60–1.61 cannot be used. Among these, the most important case is that of a surface distribution of charges. As has been shown before, the tangential component of the electric field is continuous through a charge-carrying surface, while its normal component is discontinuous. Therefore, the derivative in the normal direction does not exist. For this reason, we will define another equation to describe the behavior or the potential near surface charges.

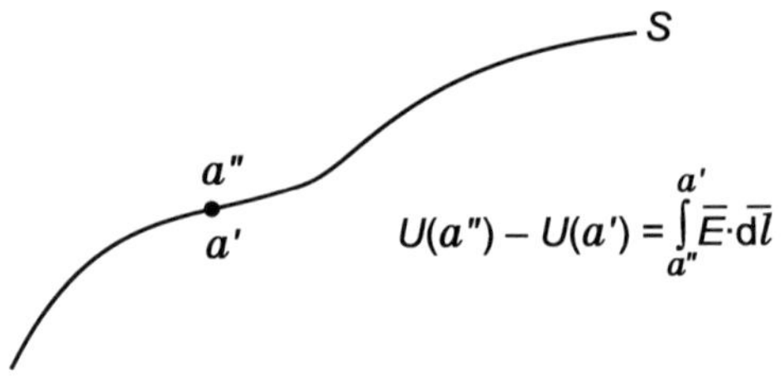

Figure 1.20. Calculations of the difference of potential when passing through a charged surface.

In accord with eq. 1.56, the potential on either side of a surface is (Fig. 1.20):

$$U(a') = \int_{a'}^{\infty} \boldsymbol{E} \cdot \mathrm{d}\boldsymbol{l} \qquad U(a'') = \int_{a''}^{\infty} \boldsymbol{E} \cdot \mathrm{d}\boldsymbol{l}$$

or

$$U(a'') - U(a') = \int_{a''}^{a'} \boldsymbol{E} \cdot \mathrm{d}\boldsymbol{l} \tag{1.63}$$

Inasmuch as the field on both sides is finite and the distance between points a' and a'' is vanishingly small, the difference in potential between the two sides tends to zero. Therefore, the potential of the electric field is a continuous function through any charge-carrying surface:

$$U_1 = U_2 \tag{1.64}$$

This condition can be considered as the surface analogy of Poisson's equation.

We have so far mostly considered electric fields caused by specified charges in free space. We have also investigated the field of charges that accumulate at interfaces between conductors, which along with other source charges create a static field. We will now show that Coulomb's law still manifests itself when there is a current flowing in a conducting medium. In so doing, we will make use of Ohm's law which relates current density to the electric field as follows:

$$\boldsymbol{j} = \sigma \boldsymbol{E} \tag{1.65}$$

where $\boldsymbol{j}$ is current density, which is a vector directed along the current flow and with magnitude equal to the amount of charge passing through a unit area oriented perpendicular to the flow, during a unit time interval. It is clear that the total current I flowing through a surface S is related to the current density as:

$$I = \int_{S} \boldsymbol{j} \cdot \mathrm{d}\boldsymbol{S} \tag{1.66}$$

When the current is measured in amperes, the current density has the units of amperes per square meter.

In Ohm's law, the coefficient of proportionality indicated by the symbol σ, and which is usually determined experimentally, is defined as the conductivity of the medium. The units of conductivity are siemens per meter, or mho per meter. In logging practice, the reciprocal of conductivity, $\rho = 1/\sigma$, is often used and is called resistivity. The units for resistivity are ohm·meters.

In Ohm's law, the electric field need not be a field created by charges only, and it can be written as consisting of two parts:

$$\boldsymbol{E} = \boldsymbol{E}^c + \boldsymbol{E}^{other} \tag{1.67}$$

where $\boldsymbol{E}^c$ is the field contributed by charges only and is governed by Coulomb's law, while $\boldsymbol{E}^{other}$ is the electric field contributed by all other types of sources (electrochemical fields caused by diffusion of ions in rocks, or electric fields induced by time-varying magnetic fields for example). Physical phenomena such as piezoelectricity or thermoelectricity also give rise to electric fields. Assuming that the field does not depend on time and that the observation point is located well away from other sources, we can take the total electric field in Ohm's law as being the Coulomb field:

$$\boldsymbol{E} = \boldsymbol{E}^c \tag{1.68}$$

We will in addition make use of the principle of conservation of the electric charge, reflecting the fact that the flux of direct current density through a closed surface is zero:

$$\oint_S \boldsymbol{j} \cdot \mathrm{d}\boldsymbol{S} = 0 \tag{1.69}$$

This equation is amenable to a direct interpretation. The integral on the left-hand side is the amount of charge passing through a closed surface per unit time, including those charges which enter the volume as well as the ones which leave it. If the total of the two contributions were not zero, we would observe a change in the total charge inside the surface during any interval of time, and consequently the electric field would not remain constant. This is the reason why eq. 1.69 is valid for direct currents.

Applying Gauss's theorem, we obtain the principle of conservation of the charge in differential form:

$$\operatorname{div} \boldsymbol{j} = 0 \tag{1.70}$$

In contrast to the case of eq. 1.69, this equation only applies at points where the first spatial derivatives of current density exist. However, there are places such as interfaces between media with different conductivities where the tangential component of current density is a discontinuous function. According to eq. 1.49, we have (Fig. 1.21):

$$\sigma_i E_{t,i} \neq \sigma_{i+1} E_{t,i+1}$$

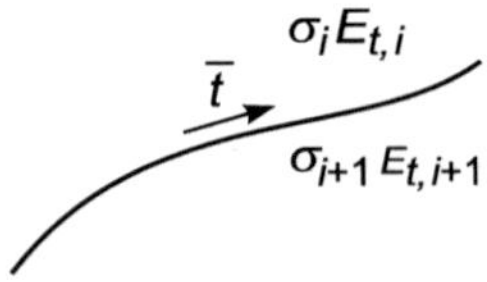

Figure 1.21. Illustration of the fact that the tangential component of the current density is a discontinuous function through an interface between media with different conductivities.

The derivative of the tangential component of the vector $\boldsymbol{j}$ is not defined at such an interface. To obtain a surface analogy of eq. 1.70, we will repeat the algebra carried out in deriving eq. 1.38. In so doing, we obtain:

$$j_n^{(1)} = j_n^{(2)} \tag{1.71}$$

that is, the normal component of current density is a continuous function at interfaces.

We have now obtained three equations describing the conservation of the charge:

$$\oint_S \boldsymbol{j} \cdot d\boldsymbol{S} = 0 \qquad \operatorname{div} \boldsymbol{j} = 0 \qquad j_n^{(1)} = j_n^{(2)} \tag{1.72}$$

It is remarkable that these equations remain valid for time-varying electromagnetic fields so long as the time variation has a quasistationary character.

At this point, we are prepared to demonstrate that in a conducting medium, the current field $\boldsymbol{j}$ is accompanied by the appearance of electric charges, these being the sole source of the electric field at places where $\boldsymbol{E}^{other} = 0$. Let us first of all assume that the conductivity of the medium varies continuously from place to place, and that discontinuous interfaces are absent. In accord with eqs. 1.65 and 1.70 we may write:

$$\operatorname{div} \boldsymbol{j} = \operatorname{div} \sigma \boldsymbol{E} = 0 \tag{1.73}$$

Making use of the rules of derivation of the product of a scalar by a vector we obtain:

$$\operatorname{div} \sigma \boldsymbol{E} = \sigma \operatorname{div} \boldsymbol{E} + \boldsymbol{E} \cdot \operatorname{grad} \sigma = 0$$

and hence

$$\operatorname{div} \boldsymbol{E} = -\frac{\boldsymbol{E} \cdot \operatorname{grad} \sigma}{\sigma} \tag{1.74}$$

Making use of eq. 1.37, we finally have:

$$\delta = -\varepsilon_0 \frac{\boldsymbol{E} \cdot \operatorname{grad} \sigma}{\sigma} \tag{1.75}$$

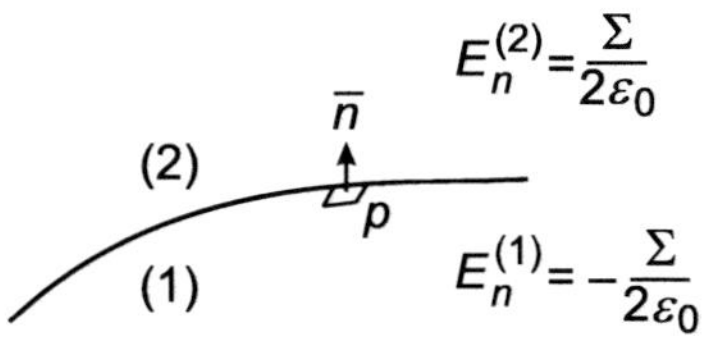

Figure 1.22. Illustration of the behavior of the normal component of the electric field near an interface.

Thus a volume distribution of charge appears in a conducting medium when it is nonuniform and when the electric field is not oriented perpendicularly to the direction of maximum change in conductivity. It is clear that in areas where the medium is uniform, there are no charges and therefore the quantity div $\boldsymbol{E}$ is zero. In accord with eq. 1.75, a motion of charges with zero net-charge density in the medium can be accompanied by the formation of fixed charges. They are in fact the source of a field which in turn governs the behavior of the current density field $\boldsymbol{j}$.

Now we assume that an interface characterized by different conductivities is present. Let us show that surface charges can arise. We will proceed from the third form of eq. 1.72, i.e. the continuity of the normal component of current density:

$$j_n^{(1)} = j_n^{(2)}$$

or

$$\sigma_1 E_n^{(1)} = \sigma_2 E_n^{(2)} \tag{1.76}$$

As was shown earlier, the normal component of the electric field is discontinuous at the interface. This discontinuity is caused by an electric charge with density Σ on the surface, which generates a normal component having opposite sign on either side of the surface. Making use of eq. 1.19 we can write eq. 1.76 as follows:

$$\sigma_1 \left(-\frac{\Sigma}{2\varepsilon_0} + E_n^{S-p} + E_n^0 \right) = \sigma_2 \left(\frac{\Sigma}{2\varepsilon_0} + E_n^{S-p} + E_n^0 \right) \tag{1.77}$$

where $\pm\Sigma/2\varepsilon_0$ is the normal component of the field caused by surface charges situated near point p (Fig. 1.22); E_n^{S-p} is the normal component of the field caused by the rest of the surface charges; E_n^0 is the normal component of the field caused by charges located outside the surface.

It should be noted that components E_n^{S-p} and E_n^0 are continuous at point p. Solving eq. 1.77 we obtain:

$$\Sigma = 2\varepsilon_0 \frac{\sigma_1 - \sigma_2}{\sigma_1 + \sigma_2} E^{av} = 2\varepsilon_0 \frac{\rho_2 - \rho_1}{\rho_2 + \rho_1} E_n^{av} \tag{1.78}$$

where

$$E_n^{av} = E_n^{S-p} + E_n^0 \tag{1.79}$$

is the average value of the normal component of the electric field at point p.

A surface charge thus arises at every point of the interface where the normal component of the field contributed by charges located outside this point is zero. It is appropriate to notice that eq. 1.78 remains valid even when normal component E_n^{av} has a non-Coulomb character.

We might say that the surface and volume charges which develop within conducting medium play a vital role in forming both the electric field and the current field. Without the appearance of these charges, the normal component of the current density could not be a continuous function of the spatial variables at interfaces, and we would observe an accumulation of charges; correspondingly no constant current could occur.

It is evident that volume and surface charges which develop in a conducting medium create an electric field which obeys Coulomb's law. The actual use of Coulomb's law in calculating the electric field in a conducting medium however usually impracticable, inasmuch as the manner of distribution of charges is unknown if the field is unknown.

In conclusion of this section, it is appropriate to notice the following:

The theory of induction logging is mainly based on the assumption that realistic models of a medium and field possess axial symmetry around axis of the borehole. There are however several important exceptions, such as:

- displacement of the transmitter with respect to the axis of the borehole
- presence of caverns
- cases of layers presenting a dip with respect to the axis of the borehole
- the source of the primary field is a transversal magnetic dipole (small coil) located on the borehole axis.

In all these cases, electric charges arise at interfaces. Their field obeys Coulomb's law.

1.2. Biot–Savart Law

In the preceding section it was shown that electric charges create an electric field which behaves in a manner described by Coulomb's law. The next step in our consideration of the behavior of electromagnetic fields will be the analysis of the magnetic fields associated with constant electric currents.

It has been shown experimentally that the magnetic field generated by a direct (constant) current can be described by the equation:

$$\mathrm{d}\boldsymbol{H}(a) = \frac{I}{4\pi}\frac{\mathrm{d}\boldsymbol{l}(p) \times \boldsymbol{L}_{pa}}{L_{pa}^3} \tag{1.80}$$

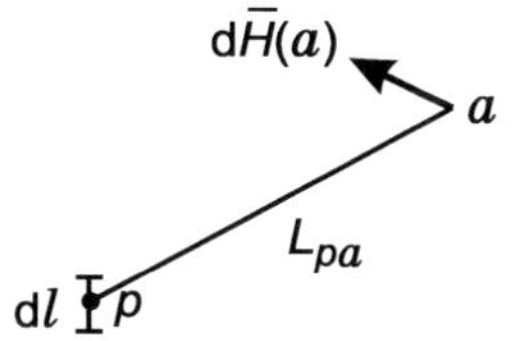

Figure 1.23. Illustration of eq. 1.81.

which is generally known as the Biot–Savart law. In this expression, I is the total current flowing in an element $\mathrm{d}\boldsymbol{l}$ located at point p, as shown in Fig. 1.23, and L_{pa} is the distance from point p to point a. Considering the definition of the vector product, eq. 1.80 can be written as:

$$\mathrm{d}\boldsymbol{H}(a) = \frac{I\,\mathrm{d}l}{4\pi}\frac{\boldsymbol{s}_0}{L_{pa}^2}\sin(\mathrm{d}\boldsymbol{l}, \boldsymbol{L}_{pa}) \tag{1.81}$$

where $\boldsymbol{s}_0$ is a unit vector perpendicular to the plane in which $\mathrm{d}\boldsymbol{l}$ and $\boldsymbol{L}_{pa}$ are located. The magnetic field caused by an elementary linear current is directly proportional to this current and to the sine of the angle between vectors $\mathrm{d}\boldsymbol{l}$ and $\boldsymbol{L}_{pa}$. It can be seen that there is no component of the field along the direction of current flow, $\mathrm{d}\boldsymbol{l}$, because of the presence of the factor $\sin(\mathrm{d}\boldsymbol{l}, \boldsymbol{L}_{pa})$.

By integrating eq. 1.81 along a path L, we obtain an expression for the magnetic field caused by a linear current in a closed loop:

$$\boldsymbol{H}(a) = \frac{I}{4\pi}\oint\limits_L \frac{\mathrm{d}\boldsymbol{l} \times \boldsymbol{L}_{pa}}{L_{pa}^3} \tag{1.82}$$

Let us write the expression for current I as a product:

$$I = \boldsymbol{j} \cdot \mathrm{d}\boldsymbol{S}$$

where $\boldsymbol{j}$ is the current density vector and $\mathrm{d}\boldsymbol{S}$ is the cross-sectional area of an elementary tube. One can then write:

$$I(\mathrm{d}\boldsymbol{l} \times \boldsymbol{L}_{pa}) = (\boldsymbol{j} \cdot \mathrm{d}\boldsymbol{S})(\mathrm{d}\boldsymbol{l} \times \boldsymbol{L}_{pa}) = (\mathrm{d}\boldsymbol{l} \cdot \mathrm{d}\boldsymbol{S})(\boldsymbol{j} \times \boldsymbol{L}_{pa}) = (\boldsymbol{j} \times \boldsymbol{L}_{pa})\,\mathrm{d}V$$

because:

$$\mathrm{d}V = \mathrm{d}\boldsymbol{S} \cdot \mathrm{d}\boldsymbol{l}$$

The magnetic field caused by currents distributed through a volume of conducting medium can therefore be written as:

$$\boldsymbol{H}(a) = \frac{1}{4\pi}\int\limits_V \frac{\boldsymbol{j} \times \boldsymbol{L}_{pa}}{L_{pa}^3}\,\mathrm{d}V \tag{1.83}$$

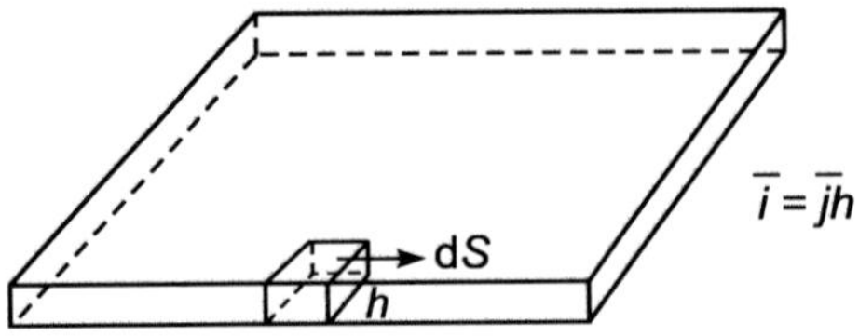

Figure 1.24. Illustration of surface currents flowing inside a relatively thin conducting zone.

In the same way that we considered the distribution of charges on a surface rather than in a volume, let us now assume that we have only surface currents that is, currents flowing inside relatively thin conductive zones, Fig. 1.24. In this case, we replace the product $\boldsymbol{j}\,\mathrm{d}V$ by a surface element of current $\boldsymbol{i}\,\mathrm{d}S$ where $\boldsymbol{i}$ is the surface current density, and we obtain the following equation for the resulting magnetic field:

$$\boldsymbol{H}(a) = \frac{1}{4\pi}\int\limits_S \frac{\boldsymbol{i}\times\boldsymbol{L}_{pa}}{L_{pa}^3}\mathrm{d}S \tag{1.84}$$

Applying the principle of superposition, we obtain an expression for the magnetic field due to the combined effects of linear, volume, and surface currents:

$$\boldsymbol{H}(a) = \frac{1}{4\pi}\left[\int\limits_V \frac{\boldsymbol{j}\times\boldsymbol{L}_{pa}}{L_{pa}^3}\,\mathrm{d}V + \int\limits_S \frac{\boldsymbol{j}\times\boldsymbol{L}_{pa}}{L_{pa}^3}\,\mathrm{d}S + \sum_i I_i\int\limits_{L_i} \frac{\boldsymbol{j}\times\boldsymbol{L}_{pa}}{L_{pa}^3}\right] \tag{1.85}$$

In accord with this expression, we can state that the magnetic field is completely specified by the distribution of currents analogously to the way in which the distribution of electric charges defines the constant electric field.

We should remember that the experimental investigations of the magnetic field behavior were carried out using closed loops of current and therefore eq. 1.80, which deals with currents flowing only between end-points, is actually an assumption which happens to be correct.

We should note again that eq. 1.85 was developed from experiments in which direct currents were used. However, as will be shown later, the equation remains valid for quasistationary fields which are of most interest in induction logging. In the practical system of units, the magnetic field is expressed in amperes per meter.

Comparing eqs. 1.9 and 1.85, we can see that the calculation of the magnetic field will usually be a more complicated procedure than the determination of the electric field caused by charges because of the presence of the vector product in the integrand. In order to simplify such calculation as well as to derive some useful relationships for the magnetic field, we will introduce an auxiliary function for the magnetic field caused by constant currents called vector potential. With this in mind, we will now show that the magnetic field can be represented as being the curl of some vector function. The following identities will be used:

$$\frac{\boldsymbol{L}_{pa}}{L_{pa}^3} = \text{grad}_p \frac{1}{L_{pa}} = -\text{grad}_a \frac{1}{L_{pa}} \tag{1.86}$$

or, in operational notation:

$$\frac{\boldsymbol{L}_{pa}}{L_{pa}^3} = \nabla_p \frac{1}{L_{pa}} = -\nabla_a \frac{1}{L_{pa}} \tag{1.87}$$

Inasmuch as the function L_{pa} can vary as points p and a are changed, one can consider the gradient of this function when either the point a or the point p is fixed. As an example, in Cartesian coordinates, we have:

$$\text{grad}_p \frac{1}{L_{pa}} = \frac{\partial}{\partial x_p}\frac{1}{L_{pa}}\boldsymbol{i} + \frac{\partial}{\partial y_p}\frac{1}{L_{pa}}\boldsymbol{j} + \frac{\partial}{\partial z_p}\frac{1}{L_{pa}}\boldsymbol{k}$$

where $\boldsymbol{i}$, $\boldsymbol{j}$ and $\boldsymbol{k}$ are unit vectors directed along the x, y, and z axes, and:

$$L_{pa} = [(x_a - x_p)^2 + (y_a - y_p)^2 + (z_a - z_p)^2]^{1/2}$$

Carrying out the differentiation, we obtain the result of eq. 1.86. Substituting eq. 1.86 into eq. 1.83, we have:

$$\boldsymbol{H}(a) = \frac{1}{4\pi}\int\limits_V \left(\boldsymbol{j} \times \nabla_p \frac{1}{L_{pa}}\right) \mathrm{d}V = \frac{1}{4\pi}\int\limits_V \left(\nabla_a \frac{1}{L_{pa}} \times \boldsymbol{j}\right) \mathrm{d}V \tag{1.88}$$

because the vector product changes sign when the relative position of the two vectors is changed.

We will now make use of the following identity:

$$\nabla_a \times \frac{\boldsymbol{j}}{L_{pa}} = \nabla_a \frac{1}{L_{pa}} \times \boldsymbol{j} + \frac{\nabla_a \times \boldsymbol{j}}{L_{pa}} \tag{1.89}$$

which can be obtained using the vector identity:

$$\nabla \times (U\boldsymbol{V}) = U\nabla \times \boldsymbol{V} + \nabla U \times \boldsymbol{V}$$

Applying eq. 1.89, we can write eq. 1.88 as:

$$\boldsymbol{H}(a) = \frac{1}{4\pi}\int\limits_V \left(\nabla_a \times \frac{\boldsymbol{j}}{L_{pa}}\right) \mathrm{d}V - \frac{1}{4\pi}\int\limits_V \frac{\nabla_a \times \boldsymbol{j}}{L_{pa}} \mathrm{d}V \tag{1.90}$$

The current density $\boldsymbol{j}$ is a function of the location of the point p and does not depend on the location of the observation point a. The integrand of the second integral is therefore zero and we have:

$$\boldsymbol{H}(a) = \frac{1}{4\pi} \int\limits_V \operatorname{curl} \frac{\boldsymbol{j}(p)}{L_{pa}} \, \mathrm{d}V \tag{1.91}$$

Because the integration and differentiation in eq. 1.91 are carried out with respect to the two mutually independent points p and a, we can interchange the order of the operations and finally obtain:

$$\boldsymbol{H}(a) = \operatorname{curl}_a \left(\frac{1}{4\pi} \int\limits_V \frac{\boldsymbol{j}(p)}{L_{pa}} \, \mathrm{d}V \right) = \operatorname{curl}_a \boldsymbol{A} \tag{1.92}$$

where

$$\boldsymbol{A} = \frac{1}{4\pi} \int\limits_V \frac{\boldsymbol{j}(p)}{L_{pa}} \, \mathrm{d}V \tag{1.93}$$

Thus the magnetic field $\boldsymbol{H}$ caused by constant currents can be expressed in terms of the vector potential $\boldsymbol{A}$ defined in eq. 1.93. The potential $\boldsymbol{A}$ is more simply related to the distribution of currents than is the magnetic field.

We will now derive expressions for the vector potential directly from eq. 1.93 for both surface and linear current flows. Making use of the obvious relationships:

$\boldsymbol{j} \, \mathrm{d}V = \boldsymbol{i} \, \mathrm{d}S$ or $\boldsymbol{j} \, \mathrm{d}V = I \, \mathrm{d}\boldsymbol{l}$

we have:

$$\boldsymbol{A}(a) = \frac{1}{4\pi} \int\limits_S \frac{\boldsymbol{j} \, \mathrm{d}S}{L_{pa}} \text{ and } \boldsymbol{A}(a) = \frac{1}{4\pi} \oint\limits_L \frac{\mathrm{d}\boldsymbol{l}}{L_{pa}} \tag{1.94}$$

Applying the principle of superposition, we obtain the following expression for the vector potential caused by volume, surface and linear currents:

$$\boldsymbol{A}(a) = \frac{1}{4\pi} \left(\int\limits_V \frac{\boldsymbol{j} \, \mathrm{d}V}{L_{pa}} + \int\limits_S \frac{\boldsymbol{j} \, \mathrm{d}S}{L_{pa}} + \sum_i I_i \oint\limits_{L_i} \frac{\mathrm{d}\boldsymbol{l}}{L_{pa}} \right) \tag{1.95}$$

The components of the vector potential can be derived directly from this last expression. In Cartesian coordinates for example, they would be:

$$A_x(a) = \frac{1}{4\pi} \left(\int\limits_V \frac{j_x \, \mathrm{d}V}{L_{pa}} + \int\limits_S \frac{i_x \, \mathrm{d}S}{L_{pa}} + \sum_i I_i \oint\limits_{L_i} \frac{\mathrm{d}x}{L_{pa}} \right)$$

$$A_y(a) = \frac{1}{4\pi}\left(\int\limits_V \frac{j_y\,\mathrm{d}V}{L_{pa}} + \int\limits_S \frac{i_y\,\mathrm{d}S}{L_{pa}} + \sum_i I_i \oint\limits_{L_i} \frac{\mathrm{d}y}{L_{pa}}\right) \tag{1.96}$$

$$A_z(a) = \frac{1}{4\pi}\left(\int\limits_V \frac{j_z\,\mathrm{d}V}{L_{pa}} + \int\limits_S \frac{i_z\,\mathrm{d}S}{L_{pa}} + \sum_i I_i \oint\limits_{L_i} \frac{\mathrm{d}z}{L_{pa}}\right)$$

Similar expressions can be written for the vector potential components in other coordinate systems.

It can be seen from eq. 1.96 that if current flows along a single straight line, the vector potential has but a single component parallel to this line. When currents are situated in a single plane, the vector potential is everywhere parallel to that plane as well.

We will later consider several examples illustrating the behavior of the vector $\boldsymbol{A}$, but at this point we will derive several useful relationships that characterize both the vector potential and the magnetic field.

We will first determine the divergence of the vector potential $\boldsymbol{A}$. In accord with eq. 1.93, we have:

$$\operatorname{div}_a \boldsymbol{A}(a) = \operatorname{div}_a \frac{1}{4\pi}\int\limits_V \frac{\boldsymbol{j}(p)\,\mathrm{d}V}{L_{pa}}$$

Because in this expression the differentiation and integration are performed at independent points in space, we can change the order of operations so that we have:

$$\operatorname{div}_a \boldsymbol{A}(a) = \frac{1}{4\pi}\int\limits_V \operatorname{div}_a \frac{\boldsymbol{j}(p)\,\mathrm{d}V}{L_{pa}} \tag{1.97}$$

The volume over which the integration is carried out includes all the currents that are present and is therefore enclosed by a surface S outside of which there are no currents. The normal component of current density on this surface must in consequence be zero:

$$j_n = 0 \quad \text{on } S \tag{1.98}$$

The integrand in eq. 1.97 can be written as:

$$\nabla_a \frac{\boldsymbol{j}}{L_{pa}} = \frac{\nabla_a \boldsymbol{j}}{L_{pa}} + \boldsymbol{j}\cdot\nabla_a \frac{1}{L_{pa}} = \boldsymbol{j}\cdot\nabla_a \frac{1}{L_{pa}}$$

because $\operatorname{div}_a \boldsymbol{j}(p) = 0$. This last expression can also be written as:

$$\begin{aligned}\boldsymbol{j}\cdot\nabla_a \frac{1}{L_{pa}} &= -\boldsymbol{j}\cdot\nabla_p \frac{1}{L_{pa}} = -\nabla_p\left(\frac{\boldsymbol{j}}{L_{pa}}\right) + \frac{\nabla_p\cdot\boldsymbol{j}}{L_{pa}}\\ &= -\operatorname{div}_p \frac{\boldsymbol{j}}{L_{pa}} + \frac{1}{L_{pa}}\operatorname{div}_p \boldsymbol{j}\end{aligned}$$

According to eq. 1.70, the total charge inside V must be conserved so that we have:

$$\mathrm{div}_p\, \boldsymbol{j} = 0$$

and therefore:

$$\boldsymbol{j} \cdot \nabla_a \frac{1}{L_{pa}} = -\,\mathrm{div}_p \frac{\boldsymbol{j}}{L_{pa}} \tag{1.99}$$

Thus eq. 1.97 can be written as:

$$\mathrm{div}\, \boldsymbol{A} = -\frac{1}{4\pi} \int\limits_V \mathrm{div}_p \frac{\boldsymbol{j}}{L_{pa}} \mathrm{d}V \tag{1.100}$$

Both the integration and differentiation operations carried out on the right-hand side of eq. 1.100 are performed with respect to the same point p, so that one can apply Gauss's theorem, which results in:

$$\mathrm{div}\, \boldsymbol{A} = -\frac{1}{4\pi} \int\limits_V \mathrm{div}_p \frac{\boldsymbol{j}}{L_{pa}}\, \mathrm{d}V = -\frac{1}{4\pi} \oint\limits_S \frac{\boldsymbol{j} \cdot \mathrm{d}\boldsymbol{S}}{L_{pa}} = -\frac{1}{4\pi} \oint\limits_S \frac{j_n\, \mathrm{d}S}{L_{pa}}$$

Considering that the normal component of current density is zero on the surface S surrounding the currents, we obtain:

$$\mathrm{div}\, \boldsymbol{A} = 0 \tag{1.101}$$

that is, the flux lines for the vector potential are closed.

In following chapters when we consider electromagnetic fields, several types of vector potentials will be introduced, and in most cases their divergence will not be zero. In the previous paragraph, it was shown that the potential for the electric field, U, satisfies Poisson's equation:

$$\nabla^2 U = -\delta/\varepsilon_0$$

which has solutions of the form:

$$U = \frac{1}{4\pi\varepsilon_0} \int\limits_V \frac{\delta\, \mathrm{d}V}{L_{pa}} \tag{1.102}$$

As follows from the comparison of eq. 1.96 and 1.102, the components of the vector potential expressed in Cartesian coordinates also satisfy Poisson's equation:

$$\nabla^2 \boldsymbol{A}_x = -j_x \qquad \nabla^2 \boldsymbol{A}_y = -j_y \qquad \nabla \boldsymbol{A}_z = -j_z \tag{1.103}$$

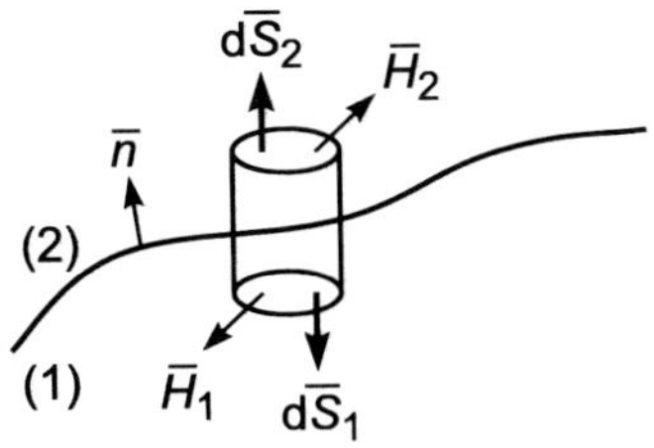

Figure 1.25. Calculation of the magnetic flux through an elementary cylindrical surface.

Multiplying each of these equations by the corresponding unit vectors, $\boldsymbol{i}$, $\boldsymbol{j}$ and $\boldsymbol{k}$, and adding them together, we obtain an equation for the potential $\boldsymbol{A}$:

$$\nabla^2 \boldsymbol{A} = -\boldsymbol{j} \tag{1.104}$$

Equations 1.101 and 1.104 reflect the basic features of the vector potential $\boldsymbol{A}$, and allow one to derive equations for the behavior of the magnetic field. Now, making use of eq. 1.92, we discover that the divergence of the magnetic field is also zero. In fact, we have the following identity:

$$\operatorname{div} \boldsymbol{H} = \operatorname{div} \operatorname{curl} \boldsymbol{A} \tag{1.105}$$

As is well known in vector algebra, the right-hand side of eq. 1.105 is identically zero, therefore:

$$\operatorname{div} \boldsymbol{H} = 0 \tag{1.106}$$

This can be physically interpreted as the indication that magnetic charges do not exist and that magnetic flux lines are closed. Applying Gauss's theorem, we obtain the integral form of this equation:

$$\oint_S \boldsymbol{H} \cdot \mathrm{d}\boldsymbol{S} = 0 \tag{1.107}$$

that is, the total flux of the magnetic field through a closed surface is zero.

Making use of eq. 1.107 in calculating the magnetic flux through an elementary cylindrical surface as shown in Fig. 1.25, we have:

$$H_n^{(2)} = H_n^{(1)} \tag{1.108}$$

As indicated by eq. 1.108, the normal component of the magnetic field is always a continuous function of the spatial variables at an interface between nonmagnetic media.

This behavior is in contrast to that of the normal component of the electric field as specified in eq. 1.76. It is obvious that magnetic charges do not exist.

Thus, we have obtained three forms of the first equation describing the magnetic field caused by constant currents:

$$\operatorname{div} \boldsymbol{H} = 0 \qquad \oint_S \boldsymbol{H} \cdot \mathrm{d}\boldsymbol{S} = 0 \qquad H_n^{(2)} = H_n^{(1)} \tag{1.109}$$

Each of them expresses the same fact, that is, that magnetic charges do not exist. Equations 1.109 have been derived by algebraic manipulation of Biot–Savart law for direct currents, but they actually remain valid for alternating electromagnetic fields, and are in effect the fourth of Maxwell's equations.

At this point, we will derive a second equation for the magnetic field. Making use of the identity:

$$\operatorname{curl}\operatorname{curl} \boldsymbol{M} = \operatorname{grad}\operatorname{div} \boldsymbol{M} - \nabla^2 \boldsymbol{M}$$

from eq. 1.92, we have:

$$\operatorname{curl} \boldsymbol{H} = \operatorname{curl}\operatorname{curl} \boldsymbol{A} = \operatorname{grad}\operatorname{div} \boldsymbol{A} - \nabla^2 \boldsymbol{A}$$

Considering that $\operatorname{div} \boldsymbol{A} = 0$ we obtain:

$$\operatorname{curl} \boldsymbol{H} = -\nabla^2 \boldsymbol{A} = \boldsymbol{j}$$

Thus a second equation for the magnetic field is:

$$\operatorname{curl} \boldsymbol{H} = \boldsymbol{j} \tag{1.110}$$

Physically, this equation expresses the fact that currents are the source of the magnetic field. Making use of Stoke's theorem, we can rewrite this equation in a second form, which is Ampere's law:

$$\oint_L \boldsymbol{H} \cdot \mathrm{d}\boldsymbol{l} = \oint_S \operatorname{curl} \boldsymbol{H} \cdot \mathrm{d}\boldsymbol{S} = \oint_S \boldsymbol{j} \cdot \mathrm{d}\boldsymbol{S} = I$$

or

$$\oint_L \boldsymbol{H} \cdot \mathrm{d}\boldsymbol{l} = I \tag{1.111}$$

where I is the current flowing through the surface S bounded by the path L (see Fig. 1.26). It should be clear that the mutual orientation of the vectors $\mathrm{d}\boldsymbol{l}$ and $\mathrm{d}\boldsymbol{S}$ is not arbitrary, but must be taken in accord with the right-hand rule, the circulation of the magnetic

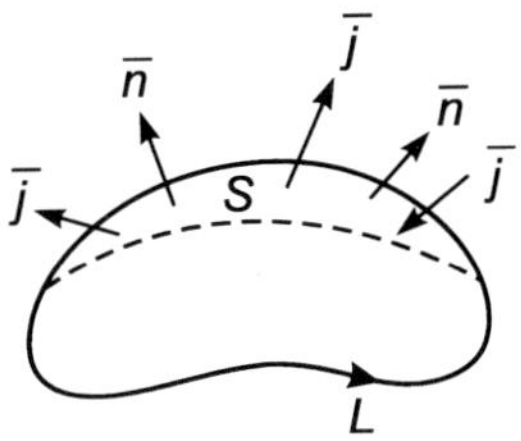

Figure 1.26. Definition of the surface S bounded by the path L used in eq. 1.111.

field is defined by the amount of current piercing through the surface surrounded by the contour L as shown in Fig. 1.26, and it does not depend on currents located outside the perimeter of this area.

It should be obvious that from the fact that the circulation is zero, it does necessarily follow that the magnetic field is also zero at every point along contour L. It is appropriate here to emphasize that the path L can intersect media with different electrical properties. For example, applying eq. 1.111 along a path L enclosed an interface between two media (see Fig. 1.27), we obtain:

$$\oint_L \boldsymbol{H} \cdot \mathrm{d}\boldsymbol{l} = H_t^{(2)}\,\mathrm{d}l - H_t^{(1)}\,\mathrm{d}l + 2j\,\mathrm{d}l\,\mathrm{d}h$$

Letting $\mathrm{d}h$ tend to zero, we have:

$$H_t^{(2)} - H_t^{(1)} = 0 \tag{1.112}$$

We see that the tangential component of the magnetic field is a continuous function of position. At this point, we have again derived three forms of the second equation for the magnetic field caused by direct currents, showing that the circulation of the magnetic field is defined by the current flux through any surface bounded by a path of integration. These forms are:

$$\operatorname{curl}\boldsymbol{H} = \boldsymbol{j} \qquad \oint_L \boldsymbol{H} \cdot \mathrm{d}\boldsymbol{l} = I \qquad H_t^{(2)} - H_t^{(1)} = 0 \tag{1.113}$$

It is interesting to note that the last of these remains valid for any alternating field, and it is usually taken as a boundary condition for the magnetic field. On occasion, it is convenient to assume that there is a surface current density at an interface. Then, repeating the operations carried out above, we find that the tangential component of the magnetic field is discontinuous at such an interface:

$$H_t^{(2)} - H_t^{(1)} = i_1 \tag{1.114}$$

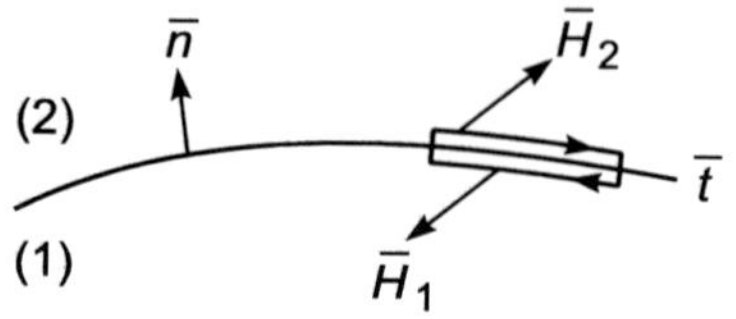

Figure 1.27. Path for the application of the boundary condition in eq. 1.111.

where $\boldsymbol{t}$ and $\boldsymbol{l}$ represent two mutually perpendicular directions tangent to the surface. Referring to eqs. 1.49 and 1.112, we can see that the continuity of the tangential components of the electric and magnetic fields follow directly from Coulomb's and Biot–Savart law, respectively. Although the first two equations of eq. 1.113 were derived from the expression of the magnetic field caused by direct current flows, they remain valid for quasistationary electromagnetic fields such as the ones in use in induction logging, i.e. Biot–Savart and Ampere's laws describe the behavior of quasistationary magnetic fields.

At this point, it may be fruitful to illustrate the use of these equations in terms of several examples.

1.2.1. Example I: The Magnetic Field of a Straight Wire Line

Consider a current I flowing through a vertical line as shown in Fig. 1.28. We will define the magnetic field at an arbitrary point, a, in a cylindrical system of coordinates, r, ϕ, z with the z-axis along the current-carrying line.

Starting with Biot–Savart law, we can say that the magnetic field has axial symmetry and is represented by a single component $\boldsymbol{H}_\phi$. From the principle of superposition, one can say that the total field is the sum of a number of fields contributed by current elements $I\,\mathrm{d}z$. Then we have:

$$H_\phi = \frac{1}{4\pi}\int\limits_{z_1}^{z_2} \frac{\mathrm{d}\boldsymbol{z} \times \boldsymbol{L}_{pa}}{L_{pa}^3} \tag{1.115}$$

where $L_{pa} = (r^2 + z^2)^{1/2}$.

Let β be the angle between the current element $\mathrm{d}\boldsymbol{z}$ and a vector that extends from this element to the observation point. It should be clear that we have:

$$|\mathrm{d}\boldsymbol{z} \times \boldsymbol{L}_{pa}| = \mathrm{d}z\, L_{pa} \sin(\mathrm{d}\boldsymbol{z}, \boldsymbol{L}_{pa}) = \mathrm{d}z\, L_{pa} \sin\beta = \mathrm{d}z L_{pa} \cos\alpha$$

Thus:

$$H_\phi = \frac{I}{4\pi}\int\limits_{z_1}^{z_2} \frac{\mathrm{d}z}{L_{pa}^2} \cos\alpha \tag{1.116}$$

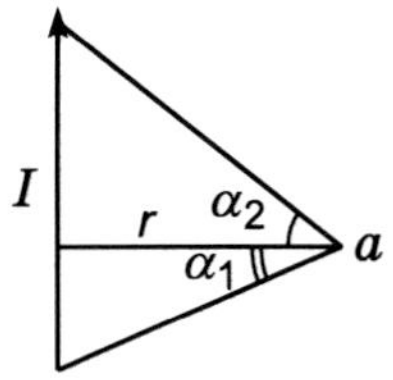

Figure 1.28. Current flowing through a vertical line.

Inasmuch as $z = r \tan\alpha$, we have:

$$\mathrm{d}z = r\sec^2\alpha\,\mathrm{d}\alpha \qquad L_{pa}^2 = r^2(1+\tan^2\alpha) = r^2\sec^2\alpha$$

Substituting these expressions into eq. 1.116, we obtain:

$$H_\phi(a) = \frac{I}{4\pi r}\int_{\alpha_1}^{\alpha_2}\cos\alpha\,\mathrm{d}\alpha = \frac{I}{4\pi r}(\sin\alpha_2 - \sin\alpha_1) \tag{1.117}$$

where α_1 and α_2 are the angles subtended by the radii from point a to the ends of the line. It is readily seen that Ampere's law cannot be applied here because the current flow is not closed.

Let us suppose that the current-carrying line is infinitely long, so that two angles α_1 and α_2 take the values $\pi/2$ and $-\pi/2$, respectively. Then:

$$H_\phi = I/2\pi r \tag{1.118}$$

In this case, one might think that the current is closed at infinity, and Ampere's law can instantly be applied. Considering a closed horizontal circuit and in view of the axial symmetry, we can write:

$$\oint \boldsymbol{H}\cdot\mathrm{d}\boldsymbol{l} = H_\phi\oint\mathrm{d}l = 2\pi r H_\phi = I$$

and hence:

$$H_\phi = I/2\pi r$$

In the case of a long line which is only semi-infinite, i.e. $\alpha_1 = 0$ and $\alpha_2 = \pi/2$ one cannot apply Ampere's law, but using eq. 1.117 we obtain a field which is half that for the case of an infinitely long current-carrying wire, that is:

$$H_\phi = I/4\pi r \tag{1.119}$$

Now suppose that $\alpha_2 = \alpha$ and $\alpha_1 = -\alpha$. Then in accord with eq. 1.117, we have:

$$H_\phi = \frac{I}{2\pi r}\sin\alpha = \frac{I}{2\pi r}\frac{l}{(r^2+l^2)^{1/2}} \tag{1.120}$$

where $2l$ is the length of the current line. If l is significantly larger than the distance r to the wire, the right-hand side of eq. 1.120 can be expanded in a power series of the quantity $(r/l)^2$. Then we obtain:

$$H = \frac{I}{2\pi r}\frac{1}{(1+r^2/l^2)^{1/2}} = \frac{I}{2\pi r}\left[1 - \frac{1}{2}\left(\frac{r}{l}\right)^2 + \frac{3}{8}\left(\frac{r}{l}\right)^4 + \cdots\right]$$

We see that if the length of the current line, $2l$, is four to five times larger than the separation r, the resulting field is practically the same as that from an infinitely long, current-carrying wire.

1.2.2. Example II: The Vector Potential and Magnetic Field of the Current Flowing in a Circular Loop

Assume that the observation point is located on the axis of a loop with radius a as shown in Fig. 1.29, then in accord with equation 1.94, we have:

$$\boldsymbol{A}(p) = \frac{1}{4\pi}\oint_L \frac{\mathrm{d}\boldsymbol{l}(q)}{L_{pq}}$$

Inasmuch as the distance L_{pq} is the same for all points on the loop we have:

$$\boldsymbol{A}(p) = \frac{I}{4\pi L_{pq}}\oint_L \mathrm{d}\boldsymbol{l}$$

By definition, the sum of the elementary vectors $\mathrm{d}\boldsymbol{l}$ along any closed path is zero. The vector potential $\boldsymbol{A}$ is therefore zero on the axis of a current-carrying loop.

We will now calculate the magnetic field on the z-axis. From Fig. 1.29, it can be seen that with a cylindrical system of coordinates, r, ϕ, z, each current element $I\,\mathrm{d}l$ creates two field components $\mathrm{d}H_r$ and $\mathrm{d}H_z$. It is however always possible to find two current elements which contribute the same horizontal component at a point of the z-axis, but of opposite sign. The magnetic field therefore has but a vertical component on the axis. As can be seen from Fig. 1.29, we have:

$$\mathrm{d}H_z = \frac{I}{4\pi}\frac{\mathrm{d}l}{L^2}\frac{a}{L} = \frac{Ia\,\mathrm{d}l}{4\pi L^3}$$

since:

$|\mathrm{d}\boldsymbol{l} \times \boldsymbol{L}| = L\,\mathrm{d}l$ and $L = (a^2+z^2)^{1/2}$

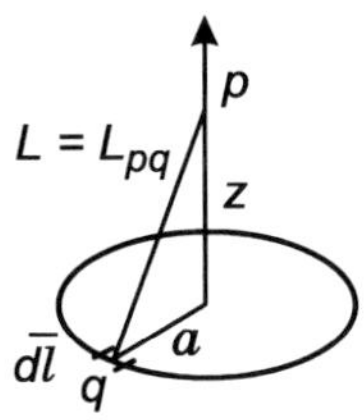

Figure 1.29. An observation point located on the axis of a current-carrying loop with a radius a.

Having integrated along a closed path, we finally obtain:

$$H_z = \frac{Ia\,2\pi a}{4\pi(a^2+z^2)^{3/2}} = \frac{Ia^2}{2(a^2+z^2)^{3/2}} = \frac{M}{2\pi(a^2+z^2)^{3/2}} \tag{1.121}$$

where $M = I\pi a^2 = IS$, with S being the plane area enclosed by the loop.

When the distance z is much greater than the radius of the loop a, we obtain an expression for the magnetic field of a magnetic dipole located on the z-axis. Neglecting a in comparison with z, we have:

$$H_z = \frac{M}{2\pi z^3}$$

We see that a relatively small current loop with a radius a creates the same magnetic field as the magnetic dipole having the moment $M = \pi a^2 I$, oriented along the z-axis.

Let us notice that in most cases, the field created by currents in the coil of an induction probe is equivalent to that of a magnetic dipole.

It can be seen from eq. 1.121 that when the distance z exceeds the radius at least three times, the replacement of the loop by a magnetic dipole locate at its center contributes an error less than 5% for the field. Making use of eq. 1.121, this behavior has been proven on the z-axis only, but it remains in fact valid for any arbitrary position of the observation point provided that distance from the loop is considerably greater than the radius of the loop.

Making use of eq. 1.121, let us explore the influence of the radius of the loop on the magnetic field on the z-axis. This will be useful in understanding the concept of geometrical factor widely used in the theory of induction logging.

As we see from eq. 1.121, for constant currents and for small values of the ratio a/z, the field increases in proportion to a^2 and the current loop behaves as though it were a magnetic dipole. In the case when a/z is much larger than unit, the magnetic field decreases in inverse proportion to a. Therefore, at some critical distance z on the axis, there is an optimum radius for the current loop that provides maximum magnetic field at this point.

So far we have considered the vector potential and the magnetic field along the z-axis only. We will now investigate a more general case by calculating the vector potential at

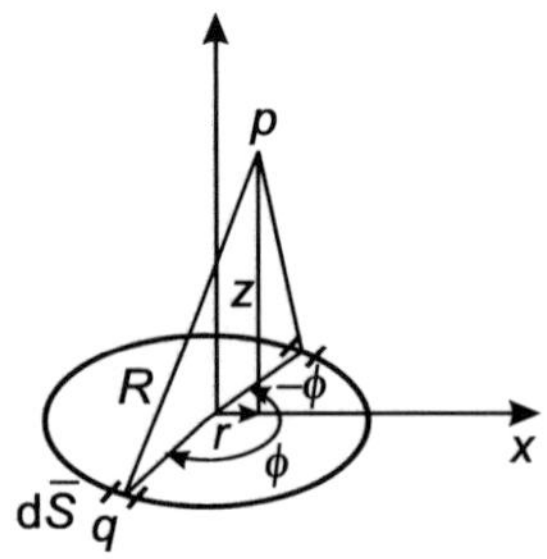

Figure 1.30. An observation point having an arbitrary position with respect to the current-carrying loop.

any point p (see Fig. 1.30). In view of the symmetry, the vector potential $\boldsymbol{A}$, does not depend on the angle ϕ. For simplicity, we can therefore choose the point p in the xz plane where $\phi = 0$. As can be seen from Fig. 1.30, every pair of current-carrying elements $I\,\mathrm{d}\boldsymbol{l}$ equally distant from p and having coordinates ϕ and $-\phi$ creates a vector potential $\mathrm{d}\boldsymbol{A}$ perpendicular to the xz plane. Inasmuch as the whole loop can be presented as the sum of such pairs, we conclude that the vector potential caused by the entire loop has but one component A_ϕ. Therefore, from eq. 1.94 follows that:

$$A_\phi = \frac{I}{4\pi}\oint \frac{\mathrm{d}l_\phi}{R} = \frac{I}{2}\int\limits_0^\pi \frac{a\cos\phi\,\mathrm{d}\phi}{(a^2+r^2+z^2-2ar\cos\phi)^{1/2}} \tag{1.122}$$

where $\mathrm{d}l_\phi$ is the component of $\mathrm{d}\boldsymbol{l}$ along the ϕ direction, and:

$$\mathrm{d}l_\phi = a\cos\phi\,\mathrm{d}\phi \qquad R = (a^2+r^2+z^2-2ar\cos\phi)^{1/2}$$

If the distance from the center of the current-carrying loop to the observation point is considerably greater than the radius of the loop, then:

$$R_0 = (r^2+z^2)^{1/2} \gg a$$

and eq. 1.122 can be simplified into:

$$A_\phi = \frac{Ia}{2\pi}\int\limits_0^\pi \frac{\cos\phi\,\mathrm{d}\phi}{(R_0^2-2ar\cos\phi)^{1/2}} \simeq \frac{Ia}{2\pi R_0}\int\limits_0^\pi \frac{\cos\phi\,\mathrm{d}\phi}{\left(1-\dfrac{2ar}{R_0^2}\cos\phi\right)^{1/2}}$$

$$\simeq \frac{Ia}{2\pi R_0}\int\limits_0^\pi \left(1+\frac{ar}{R_0^2}\cos\phi\right)\cos\phi\,\mathrm{d}\phi$$

$$= \frac{Ia}{2\pi R_0}\int\limits_0^\pi \cos\phi\,\mathrm{d}\phi + \frac{Ia^2r}{2\pi R_0^3}\int\limits_0^\pi \cos^2\phi\,\mathrm{d}\phi$$

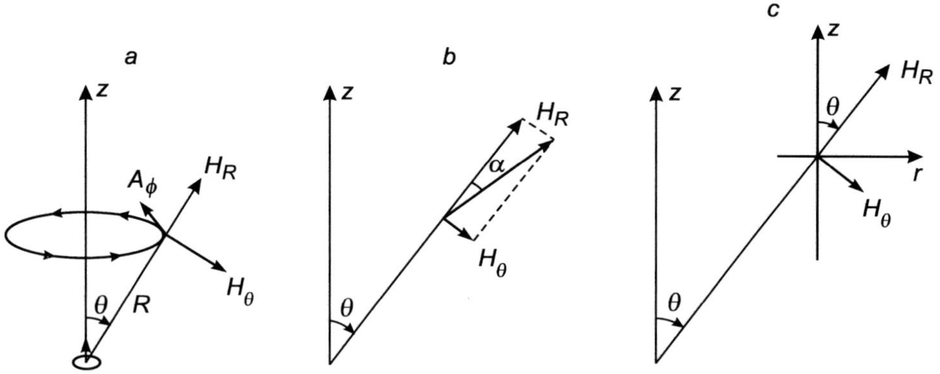

Figure 1.31. Illustration of the behavior of the field of a magnetic dipole.

The first integral in the last expression is zero, so that we obtain:

$$A_\phi = \frac{Ia^2 r}{4R_0^3}$$

or

$$A_\phi = \frac{M}{4\pi R_0^2}\sin\phi \tag{1.123}$$

Thus, at distances considerably greater than the loop radius, the vector potential and the corresponding components of the magnetic field are the same as those for a magnetic dipole located at the center of the loop with its moment directed perpendicularly to the loop.

As was mentioned above, real coils in induction probes can very often be considered as magnetic dipoles. For this reason, let us consider the behavior of the magnetic field of a magnetic dipole in more detail. Suppose that a magnetic dipole is located at the origin of a spherical system of coordinates, as shown in Fig. 1.31a, with its moment oriented along the z-axis. Then, in accord with eq. 1.123, we have:

$$A_\phi = \frac{M}{4\pi R^2}\sin\phi \tag{1.124}$$

where $R = (r^2 + z^2)^{1/2}$. Thus, the vector lines of the function A_ϕ are closed circles located in horizontal planes with centers on z-axis.

Making use of the eq. 1.92 in the spherical system of coordinates R, θ, ϕ:

$$\boldsymbol{H} = \frac{1}{R^2 \sin\theta}\begin{vmatrix} \mathbf{1}_R & R\,\mathbf{1}_\theta & R\sin\theta\,\mathbf{1}_\phi \\ \dfrac{\partial}{\partial R} & \dfrac{\partial}{\partial \theta} & \dfrac{\partial}{\partial \phi} \\ 0 & 0 & R\sin\theta\,A_\phi \end{vmatrix}$$

we have:

$$H_R = \frac{2M}{4\pi R^3}\cos\theta \qquad H_\theta = \frac{M}{4\pi R^3}\sin\theta \qquad H_\phi = 0 \tag{1.125}$$

The magnetic field is therefore located in meridional planes and has two components H_R and H_θ. In particular, at points on the z-axis we have:

$$H_R = H_z = \frac{2M}{4\pi z^3} \qquad H_\theta = 0 \tag{1.126}$$

i.e. the field has the direction of the dipole moment and decreases as $1/z^3$ with increasing z. At the same time, in the equatorial plane ($\theta = \pi/2$), its direction is opposed to that of the moment:

$$H_\theta = \frac{M}{4\pi z^3} \qquad H_R = 0 \tag{1.127}$$

The expression for the field in eq. 1.126 is used to evaluate the primary field magnitude of an induction probe.

Let us now illustrate an interesting feature of the field behavior along an arbitrary radius R (Fig. 1.31b). In accord with eq. 1.135, we have:

$$\frac{H_\theta}{H_R} = \frac{1}{2}\operatorname{tg}\theta \qquad \text{or} \qquad \operatorname{tg}\alpha = \frac{1}{2}\operatorname{tg}\theta \tag{1.128}$$

where θ is the angle between the radius and the magnetic field vector. As follows from eq. 1.128, the orientation of the magnetic field, unlike its magnitude, does not change along radius R and is very simply related to the angle θ. It is also useful to consider the magnetic field in a cylindrical system of coordinates. As is seen from Fig. 1.31c, we have:

$$H_z = H_R\cos\theta - H_\theta\sin\theta \qquad H_r = H_R\sin\theta + H_\theta\cos\theta$$

or

$$H_z = \frac{M}{4\pi R^3}(3\cos^2\theta - 1) \qquad H_r = \frac{3M\sin\theta\,\cos\theta}{4\pi R^3} \tag{1.129}$$

Therefore, the component H_z of the field changes sign along any profile parallel to the z-axis. This takes place when:

$$\cos\theta_0 = \pm\frac{1}{3}\sqrt{3} \tag{1.130}$$

i.e. $\theta_0 = 54.7°$ or $125.3°$.

It is essential that the position of the points where the component of the field H_z is zero are defined by the angle θ_0 only.

This feature of the field is sometimes used in order to control the quality of an induction probe consisting of coils and wires. In the case when the magnetic field is created by one single coil, the points where the component H_z vanishes are easily calculated.

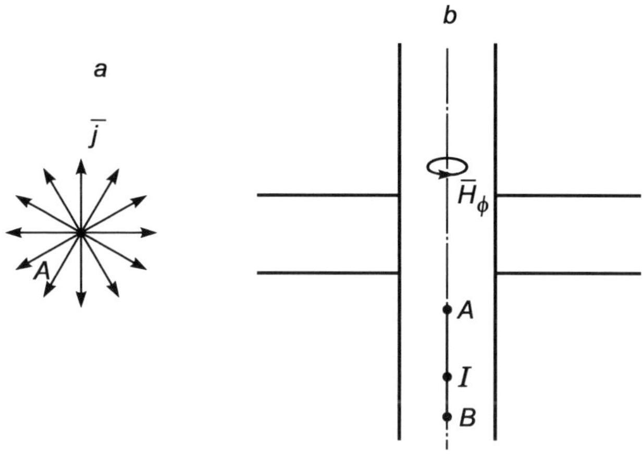

Figure 1.32. Illustration of the behavior of the magnetic field in a conducting medium.

1.2.3. Example III: The Magnetic Field of a Grounded Electrode in a Uniform Conducting Medium

Suppose that a current electrode is placed in a uniform conducting medium so that the distribution of currents possesses the spherical symmetry (Fig. 1.32a). It is then a simple matter to realize that the magnetic field is zero everywhere in the medium. This follows directly from Biot–Savart law and the symmetry of the model. In other words, one can always find two current elements which are located symmetrically with respect to the observation point and of which the magnetic field differ by sign only. Let us notice that Ampere's law does not apply here because the current lines are not closed.

We will now suppose that two current electrodes connected by a wire are located on the axis of a borehole (see Fig. 1.32b). In this case, we have a model characterized by the cylindrical symmetry. Unlike in the previous model, the magnetic field is in general not equal to zero but has one component H_ϕ. However, the field is zero all along the z-axis. In fact, applying Ampere's law as shown in Fig. 1.32b we have:

$$\pi r^2 j_z = 2\pi r H_\phi \text{ or } H_\phi = \frac{r}{2} j_z$$

Correspondingly, with decreasing r the magnitude of the field decreases and in the limit vanishes on the z-axis. For this reason, measuring the magnetic field on the borehole axis, as the source is a grounded line along this axis, can only detect distortions of the cylindrical symmetry, as for example the presence of caverns, nonhorizontal layers, as well as a nonvertical position of the borehole.

In concluding this section it is appropriate to make the following comments:

- According to Biot–Savart law direct currents act as a source of constant magnetic fields.

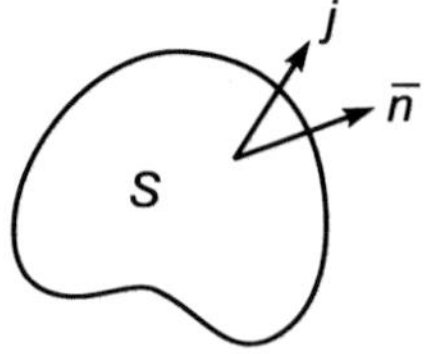

Figure 1.33. Current density at an arbitrary point on an arbitrary closed surface S.

- Starting with Biot–Savart law and making use of the principle of conservation of the charge, we were able to derive two equations describing constant magnetic fields, and each of them can be presented in three forms:

$$\oint_L \boldsymbol{H} \cdot \mathrm{d}\boldsymbol{l} = I \qquad \operatorname{curl} \boldsymbol{H} = \boldsymbol{j} \qquad H_t^{(2)} - H_t^{(1)} = 0 \tag{A}$$

$$\oint_S \boldsymbol{H} \cdot \mathrm{d}\boldsymbol{S} = 0 \qquad \operatorname{div} \boldsymbol{H} = 0 \qquad H_n^{(2)} - H_n^{(1)} = 0 \tag{B}$$

The equations of set (B) reflect the fact that magnetic charges do not exist. This set is also valid for alternating fields.

Equations of set (A) are valid for constant fields, but there will be additional effects to consider for alternating fields. In other words, when time-varying electromagnetic fields are considered, there is another source for the magnetic field in addition to the conduction currents. However, for the so-called quasistationary field, the influence of the second source of the magnetic field (displacement currents) is negligible as this is the case in induction logging, and equations of set (A) can be applied.

1.3. The Postulate of Conservation of Charge and the Distribution of Charges in Conducting Media

This section will show under what conditions electric charges can exist in a conducting medium. In order to investigate this problem, we will make use of the postulate of conservation of the electric charge for time-varying fields:

$$\oint_S \boldsymbol{j} \cdot \mathrm{d}\boldsymbol{S} = -\frac{\partial e}{\partial t} \tag{1.131}$$

where $\boldsymbol{j}$ is the current density vector at any point of an arbitrary surface S as shown in Fig. 1.33, e is the charge distributed within the volume bounded by S, and $\partial e/\partial t$ is the time-derivative of the charge. The scalar product:

$$\boldsymbol{j} \cdot \mathrm{d}\boldsymbol{S} = -\frac{\partial e}{\partial t}$$

represents the amount of charge crossing an element of surface $\mathrm{d}S$ during a unit time period. The integral:

$$\oint\limits_S \boldsymbol{j}\cdot \mathrm{d}\boldsymbol{S} = \oint\limits_S j_n\,\mathrm{d}S$$

defines the flux of electric charges through the surface S during a unit time as well. In general, at some points on the surface S, the vector $\boldsymbol{j}$ is directed outwards while at other points, it is directed inwards. The current density flux given by eq. 1.131 is therefore the algebraic sum of positive and negative fluxes through the surface S. For example, if the flux:

$$\oint\limits_S \boldsymbol{j}\cdot \mathrm{d}\boldsymbol{S}$$

is positive, the physical meaning is that during the time interval, a certain amount of charge leaves the volume V, and the derivative $\partial e/\partial t$ is negative, that is, the total charge e inside the volume decreases. In the opposite case, when the total flux is negative, the derivative $\partial e/\partial t$ is positive and the amount of charge contained inside V increases with time. Moreover, one can imagine a case when the positive and negative fluxes through a closed surface are equal, and the total flux is zero. The derivative $\partial e/\partial t$ vanishes so that the amount of charge inside the volume does not change with time.

We will now write eq. 1.131 in various forms which will be used in further applications. Making use of Gauss's theorem, we obtain:

$$\oint\limits_S \boldsymbol{j}\cdot \mathrm{d}\boldsymbol{S} = \int\limits_V \operatorname{div}\boldsymbol{j}\,\mathrm{d}V$$

At points in the medium where the divergence of the vector $\boldsymbol{j}$ exists, we have:

$$\oint\limits_S \boldsymbol{j}\cdot \mathrm{d}\boldsymbol{S} = \int\limits_V \operatorname{div}\boldsymbol{j}\,\mathrm{d}V = -\frac{\partial e}{\partial t} = -\frac{\partial}{\partial t}\int\limits_V \delta\,\mathrm{d}V = -\int\limits_V \frac{\partial \delta}{\partial t}\,\mathrm{d}V$$

where δ is the charge density. Thus we can write:

$$\int\limits_V \left(\operatorname{div}\boldsymbol{j} + \frac{\partial \delta}{\partial t}\right)\,\mathrm{d}V = 0$$

or

$$\operatorname{div}\boldsymbol{j} = -\frac{\partial \delta}{\partial t} \tag{1.132}$$

This last equation is the differential form of 1.131, and is valid at points where the current density is a continuous function of the spatial variables. It has the same physical meaning

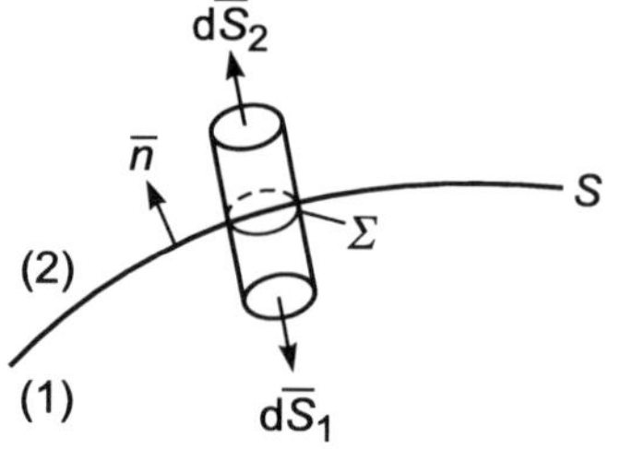

Figure 1.34. Elementary cylindrical surface used to derive eq. 1.133.

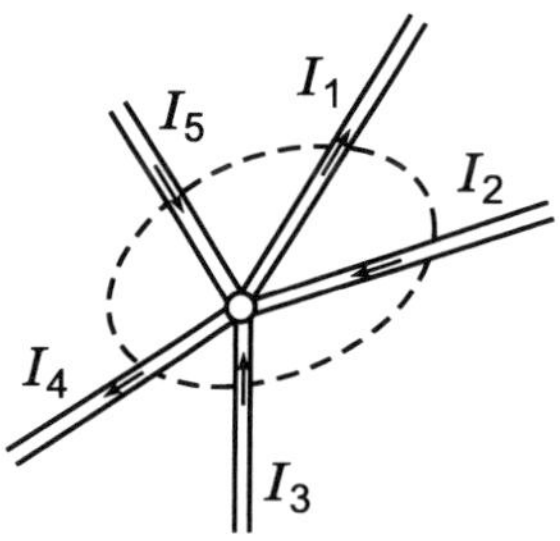

Figure 1.35. A system of linear or quasilinear currents.

as eq. 1.131, but unlike it, it describes the relationship between current flow and charges in the immediate vicinity of a single point.

Assume now that some surface carries a charge with density Σ. Defining the flux of current density through an elementary cylindrical surface as shown in Fig. 1.34 we obtain the surface analog to eq. 1.131:

$$j_n^{(2)} - j_n^{(1)} = \frac{\partial \Sigma}{\partial t} \tag{1.133}$$

Thus, the difference between the normal component of the field on either side of the surface is equal to the time rate of change of the surface charge density, taken with a negative sign. In eq. 1.133, the normal vector $\boldsymbol{n}$ is oriented from side (1) to side (2).

Let us finally assume that we have a system of linear currents as shown in Fig. 1.35. Making use of eq. 1.131, we have:

$$\oint_S \boldsymbol{j} \cdot \mathrm{d}\boldsymbol{S} = \sum_{i=1}^{N} \oint_{S_i} \boldsymbol{j} \cdot \mathrm{d}\boldsymbol{S} = \sum_{i=1}^{N} I_i = -\frac{\partial e}{\partial t}$$

i.e.

$$\sum_{i=1}^{N} I_i = -\frac{\partial e}{\partial t} \qquad (1.134)$$

where I_i is the current in the ith current tube taken with the appropriate sign, and e is the charge at the point where all the tubes intersect. We have now four forms for the postulate of conservation of charge:

$$\oint_S \boldsymbol{j} \cdot \mathrm{d}\boldsymbol{S} = -\frac{\partial e}{\partial t} \qquad \operatorname{div} \boldsymbol{j} = -\frac{\partial \delta}{\partial t}$$
$$j_n^{(2)} - j_n^{(1)} = -\frac{\partial \Sigma}{\partial t} \qquad \sum_{i=1}^{N} I_i = -\frac{\partial e}{\partial t} \qquad (1.135)$$

It should be emphasized that the first equation is the most general, being applicable everywhere. The second one can be used when the current density is a continuous function of space. The third one describes the behavior of the normal component of current density at interfaces that carry a charge, and the fourth expression is appropriate for a system of linear currents. The equations listed in eq. 1.135 are extremely useful in determining under which conditions and with which density, charges arise in a conducting medium.

It is convenient to start our investigation of this problem with a simple case such as a conductive medium in which the electromagnetic field does not depend on time, and therefore all time derivatives are zero. Correspondingly, we will repeat some results obtained in the first section.

Equations 1.135 take the form:

$$\oint_S \boldsymbol{j} \cdot \mathrm{d}\boldsymbol{S} = 0 \qquad \operatorname{div} \boldsymbol{j} = 0$$
$$j_n^{(2)} - j_n^{(1)} = 0 \qquad \sum_{i=1}^{N} I_i = 0 \qquad (1.136)$$

because

$$\frac{\partial e}{\partial t} = \frac{\partial \delta}{\partial t} = \frac{\partial \Sigma}{\partial t} = 0$$

Thus, for a constant field, the flux of current density through a closed surface is always zero, that is, the amount of charge arriving in a volume during a given time period is exactly equal to the amount of charge that leaves that volume in the same period. Let us note that the last equation in set 1.136 is the well-known Kirchoff's law for currents.

In order to determine the distribution of volume charges, we can use equations derived previously:

$$\operatorname{div} \boldsymbol{j} = 0 \qquad \operatorname{div} \boldsymbol{E} = \frac{\delta}{\varepsilon_0} \qquad (1.137)$$

along with Ohm's law:

$$\boldsymbol{j} = \sigma \boldsymbol{E}$$

We will assume here that the electric field $\boldsymbol{E}$ is caused by charges only. In accord with eq. 1.137, we have:

$$\operatorname{div} \boldsymbol{j} = \operatorname{div} \sigma \boldsymbol{E} = \boldsymbol{E} \cdot \operatorname{grad} \sigma + \sigma \operatorname{div} \boldsymbol{E} = 0$$

whence

$$\operatorname{div} \boldsymbol{E} = -\frac{\boldsymbol{E} \cdot \operatorname{grad} \sigma}{\sigma}$$

Comparing this result with the second equation in set 1.137, we have:

$$\delta = -\varepsilon_0 \boldsymbol{E} \cdot \frac{\operatorname{grad} \sigma}{\sigma} = -\varepsilon_0 \frac{\nabla \sigma}{\sigma} \cdot \boldsymbol{E} \tag{1.138}$$

Thus, when a current flows through a conducting medium, electric charges arise at places where the medium is nonuniform, provided that the electric field has a nonzero component along the direction of $\operatorname{grad} \sigma$. The sign of the volume charge depends on the mutual orientation of the electric field and the gradient of conductivity. Electric charges will not appear at points where the medium is uniform in conductivity, and in this case we have:

$$\operatorname{div} \boldsymbol{E} = 0 \tag{1.139}$$

We will now derive expressions for the surface charge. Let us start from the two equations:

$$j_n^{(2)} - j_n^{(1)} = 0 \qquad E_n^{(2)} - E_n^{(1)} = \Sigma / \varepsilon_0 \tag{1.140}$$

where Σ is the surface density of charge. Let us write the first equation in set 1.140 in the form:

$$\sigma_2 E_n^{(2)} - \sigma_1 E_n^{(1)} = \frac{1}{2} \left[(\sigma_1 + \sigma_2)(E_n^{(2)} - E_n^{(1)}) + (\sigma_2 - \sigma_1)(E_n^{(2)} + E_n^{(1)}) \right] = 0$$

Making use of the second equation in set 1.140, we have:

$$(\sigma_2 + \sigma_1) \frac{\Sigma}{2\varepsilon_0} + (\sigma_2 - \sigma_1) E_n^{av} = 0$$

where $E_n^{av} = (E_n^{(1)} + E_n^{(2)})/2$ is the average magnitude of the normal component of the electric field on the surface. Thus, we have the following expression for the surface charge density:

$$\Sigma = -2\varepsilon_0 \frac{\sigma_2 - \sigma_1}{\sigma_2 + \sigma_1} E_n^{av} = -2\varepsilon_0 \, K_{12} E_n^{av} \tag{1.141}$$

where

$$K_{12} = \frac{\sigma_2 - \sigma_1}{\sigma_2 + \sigma_1} = \frac{\rho_1 - \rho_2}{\rho_1 + \rho_2} \tag{1.142}$$

or

$$\Sigma = 2\varepsilon_0 \frac{\rho_2 - \rho_1}{\rho_2 + \rho_1} E_n^{av} \tag{1.143}$$

The quantities ρ_1 and ρ_2 are the resistivities of the two media.

The normal component of the electric field on either side of the interface can be written as:

$$\begin{aligned} E_n^{(1)}(p) &= E_n^0(p) + E_n^{S-p}(p) - \frac{\Sigma(p)}{2\varepsilon_0} \\ E_n^{(2)}(p) &= E_n^0(p) + E_n^{S-p}(p) + \frac{\Sigma(p)}{2\varepsilon_0} \end{aligned} \tag{1.144}$$

where $E_n^0(p) + E_n^{S-p}(p)$ is the normal component of the field at a point p, contributed by all charges except the one at point p. It follows from eq. 1.144 that:

$$E_n^{av} = E_n^0(p) + E_n^{S-p}(p) \tag{1.145}$$

where the normal component is directed from the reverse side (1) to the top side (2) of the interface.

We see that the charge density which arise at the interface is directly proportional to the normal component of the field E_n^{av}, with the coefficient of proportionality represented by the symbol K_{12}. As has been shown, the coefficient K_{12} can vary within the range:

$$-1 \leqslant K_{12} \leqslant +1 \tag{1.146}$$

We should note that due to the presence of the surface electric charge, the normal component of current density is a continuous function of the spatial variable, while the normal component of the electric field is discontinuous. An example of distribution of charges is shown in Fig. 1.36. In this case, charges arise on the electrode surface and at the interface between borehole and formation. There is also a certain amount of charge on the surface of the wire that delivers the current to electrode A.

Let us now consider a general case in which the electromagnetic field varies with time. In determining the charge density, we will make use of eqs. 1.132 and 1.37:

$$\operatorname{div} \boldsymbol{j} = -\frac{\partial \delta}{\partial t} \qquad \operatorname{div} \boldsymbol{E} = \frac{\delta}{\varepsilon_0} \tag{1.147}$$

From these we have:

$$\operatorname{div} \boldsymbol{j} = \operatorname{div} \sigma \boldsymbol{E} = \sigma \operatorname{div} \boldsymbol{E} + \boldsymbol{E} \cdot \operatorname{grad} \sigma = -\frac{\partial \delta}{\partial t}$$

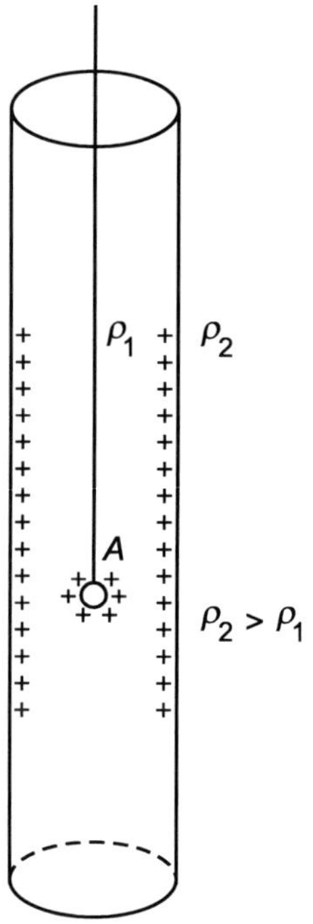

Figure 1.36. Charges arise at the surface of the electrode as well as at the interface between borehole and formation.

or

$$\frac{\sigma}{\varepsilon_0}\delta + \boldsymbol{E} \cdot \operatorname{grad} \sigma = -\frac{\partial \delta}{\partial t}$$

Finally, we obtain the following differential equation for the volume charge density:

$$\frac{\partial \delta}{\partial t} + \frac{1}{\rho \varepsilon_0}\delta = -\boldsymbol{E} \cdot \operatorname{grad} \sigma \tag{1.148}$$

where ρ is resistivity of a medium and ε_0 is the dielectric permittivity of free space. Assume now that the medium is uniform or that the electric field is perpendicular to the gradient of conductivity, in either case, we have:

$$\boldsymbol{E} \cdot \operatorname{grad} \sigma = 0$$

and eq. 1.148 takes a simpler form:

$$\frac{\partial \delta}{\partial t} + \frac{1}{\rho \varepsilon_0}\delta = 0 \tag{1.149}$$

This particular equation has a well known solution:

$$\delta = \delta_0 \, \mathrm{e}^{-t/\rho\varepsilon_0}$$

or

$$\delta = \delta_0 \, \mathrm{e}^{-t/\tau_0} \tag{1.150}$$

where δ_0 is the charge at the initial instant and:

$$\tau_0 = \rho \, \varepsilon_0 \tag{1.151}$$

The quantity τ_0 is a time constant whose value in a conducting medium is usually very small. For example, if $\rho = 100$ ohm·m then $\tau_0 \simeq 10^{-9}$ s.

In accord with eq. 1.150, a charge placed in a conducting medium will disappear very quickly. If we are concerned with charges which exist at times greater than τ_0 only, we can assume that for all practical purposes, such charges will not exist. In addition, it is appropriate to point out that with the kind of excitation used in electrical logging, there is no initial volume charge in the medium, i.e. $\delta_0 = 0$. We can therefore conclude that at points where the medium is uniform or where the condition:

$$\boldsymbol{E} \cdot \nabla\sigma = 0$$

is met, there are no electric charges, and so:

$$\operatorname{div} \boldsymbol{E} = 0 \tag{1.152}$$

A much different situation exists when the medium is nonuniform and:

$$\boldsymbol{E} \cdot \nabla\sigma \neq 0$$

In this case, the right-hand side of eq. 1.148 does not vanish, and we have a first-order nonhomogeneous differential equation of the form:

$$\frac{\mathrm{d}\gamma}{\mathrm{d}t} + \frac{1}{\tau_0}\gamma = f(t) \tag{1.153}$$

where $\gamma = \delta(t)$, $f(t) = -\boldsymbol{E} \cdot \operatorname{grad}\sigma = -\boldsymbol{E} \cdot \nabla\sigma$.

The general solution of eq. 1.153 is known to be of the form:

$$\gamma = \gamma_0 \, \mathrm{e}^{-t/\tau_0} + \mathrm{e}^{-t/\tau_0} \int\limits_0^t \mathrm{e}^{t/\tau_0} f(t) \, \mathrm{d}t \tag{1.154}$$

where γ_0 is the value of the function γ at the instant $t = 0$.

In accord with eq. 1.148, we have:

$$\delta = \delta_0 \, \mathrm{e}^{-t/\tau_0} - \mathrm{e}^{-t/\tau_0} \int_0^t \mathrm{e}^{t/\tau_0} \, (\boldsymbol{E} \cdot \operatorname{grad} \sigma) \, \mathrm{d}t \qquad (1.155)$$

This last equation can also be written as follows:

$$\delta = \delta_0 \, \mathrm{e}^{-t/\tau_0} - \mathrm{e}^{-t/\tau_0} \int_0^t \mathrm{e}^{t/\tau_0} \, E(t) \, \mathrm{d}t \, (\boldsymbol{e}_0 \cdot \nabla \sigma) \qquad (1.156)$$

where $E(t)$ is the magnitude of the electric field and $\boldsymbol{e}_0$ is a unit vector along the direction of the field:

$$\boldsymbol{E} = E \, \boldsymbol{e}_0$$

In general, one can recognize two types of charges which behave quite differently with time. As can be seen from eq. 1.151 or 1.156, the time rate of change of the first kind of charge is independent of the uniformity of the medium, and is only determined by the time constant τ_0. In contrast to the behavior of the first kind of charge, the second type of charge occurs only as a consequence of the existence of an electromagnetic field in a nonuniform medium.

Let us rewrite eq. 1.156 as:

$$\delta = \delta_1 + \delta_2 \qquad (1.157)$$

where:

$$\delta_1 = \delta_0 \, \mathrm{e}^{-t/\tau_0}$$
$$\delta_2 = -\boldsymbol{e}_0 \cdot \nabla \sigma \, \mathrm{e}^{-t/\tau_0} \int_0^t \mathrm{e}^{t/\tau_0} \, E(t) \, \mathrm{d}t \qquad (1.158)$$

Inasmuch as measurements are always performed at times significantly greater than τ_0, and besides δ_0 is frequently zero, we will only consider the second type of charge, δ_2. According to eq. 1.158, a volume charge density will arise in the neighborhood of any point in a nonuniform medium provided that the primary field is not normal to the direction of the gradient of resistivity. Assuming that the condition:

$$t \ll \tau_0 \qquad (1.159)$$

holds, we will expand the expression for δ_2 in a power series of the parameter τ_0. Considering the expression:

$$\int_0^t \mathrm{e}^{t/\tau_0} \, E(t) \, \mathrm{d}t \qquad (1.160)$$

and in integrating it by parts, we obtain:

$$\int_0^t e^{t/\tau_0} E(t)\, dt = \tau_0 \left\{ E(t)\, e^{t/\tau_0} \Big|_0^t - \int_0^t \dot{E}(t)\, e^{t/\tau_0}\, dt \right\}$$

$$= \tau_0 \left\{ E(t)\, e^{t/\tau_0} \Big|_0^t - \tau_0 \left(\dot{E}(t)\, e^{t/\tau_0} \Big|_0^t - \int_0^t \ddot{E}(t)\, e^{t/\tau_0}\, dt \right) \right\}$$

$$= \tau_0 E(t)\, e^{t/\tau_0} - \tau_0 E(0) - \tau_0^2 \dot{E}(t)\, e^{t/\tau_0} + \tau_0^2 \dot{E}(0) + \tau_0^2 \int_0^t \ddot{E}(t)\, e^{t/\tau_0}\, dt$$

Taking into account that the electrical field is absent at $t = 0$ the volume density δ_2 can be written as:

$$\delta_2(t) = -(\boldsymbol{e}_0 \cdot \nabla\sigma) \left\{ \tau_0 E(t) - \tau_0^2 \dot{E}(t) + \tau_0^2\, e^{-t/\tau_0} \int_0^t \ddot{E}(t)\, e^{t/\tau_0}\, dt \right\} \tag{1.161}$$

Continuing this process, it is possible to obtain higher-order terms of the series. Inasmuch as the time constant τ_0 is normally extremely small, and that condition 1.159 usually applies, we can discard all of these terms but the first one and obtain:

$$\delta_2(t) = -(\boldsymbol{e}_0 \cdot \nabla\sigma)\, \tau_0\, E(t) \tag{1.162}$$

In this case, the charge density changes synchronously with the electric field, that is, it is determined by the instantaneous value of the electric field at the same point.

Such a relationship between charge density and electric field at any point of a medium is essential to the definition of the quasistationary behavior which is responsible for many effects in induction logging.

One can conclude that time-varying charges develop in the quasistationary approximation in the same way as they do due to the presence of a constant field. In order to illustrate these results, let us consider two examples.

1.3.1. Example I: Exponential Variation

Assume that an electric field varies exponentially with time as:

$$\boldsymbol{E} = E_0\, e^{-t/\tau}\, \boldsymbol{e}_0 \tag{1.163}$$

where τ is the parameter characterizing the rate of decay of the field with time. Equation 1.158 then becomes:

$$\delta_2(t) = -E_0\, e^{-t/\tau_0} (\boldsymbol{e}_0 \cdot \nabla\sigma) \int_0^t e^{(1/\tau_0 - 1/\tau)t}\, dt$$

After carrying out the indicated integration, one obtains:

$$\delta_2(t) = -\frac{\tau_0\,\mathrm{e}^{-t/\tau_0}}{1-\tau_0/\tau}E_0\left\{\left[1-\mathrm{e}^{-t(1/\tau_0-1/\tau)}\right]\,(\boldsymbol{e}_0\cdot\nabla\sigma)\right\}$$

Further assuming that $\tau_0 \ll \tau$ and $t \gg \tau_0$, i.e. the rate of the field decay is considerably small and only times that widely exceed the time constant τ_0 are considered, we have:

$$\delta_2(t) = -\tau_0\,\mathrm{e}^{-t/\tau}\,(\boldsymbol{e}_0\cdot\nabla\sigma)E_0 \tag{1.164}$$

The volume density of charge δ_2 decays exponentially with time at exactly the same rate as the electric field, regardless of the conductivity of the medium.

1.3.2. Example II: Sinusoidal Variation

Assume now that the primary electromagnetic field varies with time as:

$$\boldsymbol{E} = E_0\sin\omega t\,\boldsymbol{e}_0$$

Substituting this expression into eq. 1.158, we have:

$$\delta_2 = -\mathrm{e}^{-t/\tau_0}\,E_0(\boldsymbol{e}\cdot\nabla\sigma)\int_0^t \mathrm{e}^{t/\tau_0}\sin\omega t\,\mathrm{d}t$$

This integral is well known:

$$\int \mathrm{e}^{t/\tau_0}\sin\omega t\,\mathrm{d}t = \frac{\mathrm{e}^{t/\tau_0}}{1/\tau_0^2+\omega^2}\left[\frac{1}{\tau_0}\sin\omega t-\omega\cos\omega t\right]$$

whence

$$\int_0^t \mathrm{e}^{t/\tau_0}\sin\omega t\,\mathrm{d}t = \frac{1}{1/\tau_0^2+\omega^2}\left[\omega+\mathrm{e}^{t/\tau_0}\left\{\frac{1}{\tau_0}\sin\omega t-\cos\omega t\right\}\right]$$

Therefore we have:

$$\delta_2(t) = -\frac{E_0}{1/\tau_0^2+\omega^2}\left[\omega\,\mathrm{e}^{-t/\tau_0}+\frac{1}{\tau_0}\sin\omega t-\omega\cos\omega t\right](\boldsymbol{e}_0\cdot\nabla\sigma)$$

In the quasistationary approximation when both the time of observation t and the period $T = 2\pi/\omega$ of the excitation are much greater than the relaxation time τ_0, we have:

$$\delta_2(t) = -\left[\tau_0\sin\omega t-\omega\tau_0^2\cos\omega t\right]E_0(\boldsymbol{e}_0\cdot\nabla\sigma)$$

Neglecting the second-order term and assuming that the field is not zero, that is, ωt is not a multiple of π, we finally obtain the expression for the volume charge density under quasistationary harmonic conditions:

$$\delta_2(t) = -\tau_0 E_0 \sin\omega t\,(\boldsymbol{e}_0 \cdot \nabla\sigma) = -\tau_0\,\boldsymbol{E}(t)\cdot \operatorname{grad}\sigma \tag{1.165}$$

We have so far only investigated the volume charge density. Let us next consider time-varying surface charges. Combining the following equations:

$$j_n^{(2)} - j_n^{(1)} = -\frac{\partial \Sigma}{\partial t} \text{ and } E_n^{(2)} - E_n^{(1)} = \frac{\Sigma}{\varepsilon_0}$$

we have:

$$\sigma_2 E_n^{(2)} - \sigma_1 E_n^{(1)} = \frac{1}{2}\left[(\sigma_2+\sigma_1)\left(E_n^{(2)} - E_n^{(1)}\right) + (\sigma_2-\sigma_1)\left(E_n^{(2)} - E_n^{(1)}\right)\right] = -\frac{\partial \Sigma}{\partial t}$$

or

$$\sigma^{av}\frac{\Sigma}{\varepsilon_0} + (\sigma_2-\sigma_1)E_n^{av} = -\frac{\partial \Sigma}{\partial t}$$

where $\sigma^{av} = (\sigma_1+\sigma_2)/2$.

Thus the equation for surface density is a differential equation of the first order similar to that for volume charge density:

$$\frac{\partial \Sigma}{\partial t} + \frac{1}{\tau_{0s}}\Sigma = (\sigma_1-\sigma_2)E_n^{av} \tag{1.166}$$

where $\tau_{0s} = \varepsilon_0/\sigma^{av}$ is the relaxation time for the surface charge.

In accord with eq. 1.154, the solution to eq. 1.166 can be written as:

$$\Sigma = \Sigma_0\,\mathrm{e}^{-t/\tau_{0s}} + \mathrm{e}^{-t/\tau_{0s}}(\sigma_1-\sigma_2)\int_0^t E_n^{av}(t)\,\mathrm{e}^{t/\tau_{0s}}\,\mathrm{d}t \tag{1.167}$$

or

$$\Sigma = \Sigma_1 + \Sigma_2$$

where:

$$\Sigma_1 = \Sigma_0\,\mathrm{e}^{-t/\tau_{0s}}$$

$$\Sigma_2 = (\sigma_1-\sigma_2)\,\mathrm{e}^{-t/\tau_{0s}}\int_0^t E^{av}(t)\,\mathrm{e}^{t/\tau_{0s}}\,\mathrm{d}t \tag{1.168}$$

Thus, two types of surface charges occur in this case. The first one, Σ_1, corresponds to the situation in which some charge with density Σ_0 is placed on the interface. In accord with eq. 1.168, such a charge decays exponentially with a time constant τ_{0s} that is controlled by the conductivity and dielectric constant of the medium. Inasmuch as the relaxation time τ_{0s} is usually very small with respect to measurement times, we will no further consider this type of charge and concentrate on the second type.

As was the case for volume charges, surface charges of the second kind arise as the consequence of the presence of an electromagnetic field only. Considering again that the relaxation time is very small, it is appropriate to expand the expression of Σ_2 in eq. 1.168 in a power series of τ_{0s}. Carrying out this expansion as indicated before, and discarding all the terms but the leading one, we obtain:

$$\Sigma_2(t) = \tau_{0s}(\sigma_1 - \sigma_2)E_n^{av}(t) \tag{1.169}$$

Replacing τ_{0s} by its expression from eq. 1.166, it is readily seen that at any given time, eq. 1.169 is identical to eq. 1.141 that described the surface distribution of charges due to a constant field. It can therefore be concluded that in the quasistationary approximation, the surface density of charges is locally controlled by the instantaneous value of the electric field. It is clear that the corresponding expression of the surface charge density follows directly from the differential eq. 1.166 provided that one neglects the term $\partial\Sigma/\partial t$ in comparison with the term Σ/τ_{0s}. This is equivalent to the condition:

$$\frac{\partial \Sigma}{\partial t} \ll \frac{1}{\tau_{0s}}\Sigma \tag{1.170}$$

In accord with eq. 1.141, $E_n^{av}(p)$ is the electric field contributed by all sources except the charge located in the immediate neighborhood of the point p. For this reason, the right-hand side of eq. 1.166 can be interpreted as the flux of the current density caused by the field of external charges only, through a closed surface with unit cross-sectional area as shown in Fig. 1.34.

The quantity Σ/τ_0 can be written as:

$$\frac{1}{\tau_{0s}}\Sigma(p) = \frac{\sigma_1 + \sigma_2}{2\varepsilon_0}\Sigma(p) = \sigma_1\frac{\Sigma(p)}{2\varepsilon_0} + \sigma_2\frac{\Sigma(p)}{2\varepsilon_0} \tag{1.171}$$

As was shown before, the term $\Sigma(p)/2\varepsilon_0$ indicates the magnitude of the normal component of the electric field caused by the charge at the point p. The term Σ/τ_{0s} therefore describes the flux of current density through the closed surface shown in Fig. 1.34 due to the elementary charge $\Sigma\,\mathrm{d}S$ only. Thus, in accord with eq. 1.166, the flux of current density caused by external sources is compensated by two fluxes, namely:

- the change in surface charge density with time, $\partial\Sigma(p)/\partial t$
- the flux caused by the electric field from the charge $\Sigma(p)$, that is, $\Sigma(p)/\tau_{0s}$.

In the quasistationary approximation, when condition 1.170 applies, eq. 1.166 can be rewritten as:

$$\sigma_1(E_n^{(1)}(p) + E_n^{av}(p)) = \sigma_2(E_n^{(2)}(p) + E_n^{av}(p))$$

where:

$$E_n^{(1)} = -\Sigma/2\varepsilon_0 \qquad E_n^{(2)} = \Sigma/2\varepsilon_0 \tag{1.172}$$

or

$$j_n^{(1)} = j_n^{(2)}$$

In the quasistationary approximation, the normal component of current density is a continuous function of position at an interface between media with different conductivities as was the case for direct currents. It is obvious that this result directly follows from eq. 1.133 by neglecting the term $\partial\Sigma/\partial t$. Our considerations should however make clear that condition:

$$\frac{\partial\Sigma}{\partial t} \to 0 \quad \text{as} \quad \frac{\tau_{0s}}{t} \to 0$$

does not necessarily mean the absence of surface charges, because in the quasistationary approximation the derivative $\partial\Sigma/\partial t$ needs only be small with respect to the flux Σ/τ_{0s}.

The postulate of conservation of the charge has allowed us to investigate in detail the distribution of charges in a conducting medium. It also serves as a basis for the introduction of the concept of displacement currents which plays a vital role in propagating electromagnetic fields.

Let us start from two equations, one of which is:

$$\text{curl}\, \boldsymbol{H} = \boldsymbol{j}$$

which is derived from Biot–Savart law, and the other one being the postulate of conservation of the charge written in differential form:

$$\text{div}\, \boldsymbol{j} = -\frac{\partial\delta}{\partial t}$$

It can be readily seen that these equations are contradictory inasmuch as:

$$\text{div}\,\text{curl}\, \boldsymbol{H} = \text{div}\, \boldsymbol{j} = 0$$

To solve this problem, it is necessary to add a term on the right-hand side of eq. 1.110, so that we obtain:

$$\text{curl}\, \boldsymbol{H} = \boldsymbol{j} + \boldsymbol{X} \tag{1.173}$$

where $\boldsymbol{X}$ is an undetermined quantity at this stage.

We can choose this quantity to satisfy eq. 1.132. Performing the divergence operation on eq. 1.173, we have:

$$0 = \operatorname{div} \boldsymbol{X} + \operatorname{div} \boldsymbol{j}$$

or

$$\operatorname{div} \boldsymbol{X} = \dot{\delta}$$

As is well known:

$$\operatorname{div} \boldsymbol{D} = \delta$$

where

$$\boldsymbol{D} = \varepsilon \boldsymbol{E}$$

is the electric displacement vector and $\varepsilon = \varepsilon_r \varepsilon_0$, while δ is the density of free charges.

Whence:

$$\operatorname{div} \boldsymbol{X} = \frac{\partial}{\partial t} \operatorname{div} \boldsymbol{D} = \operatorname{div} \frac{\partial \boldsymbol{D}}{\partial t} = \operatorname{div} \dot{\boldsymbol{D}}$$

One possible solution to the problem is the vector:

$$\boldsymbol{X} = \frac{\partial \boldsymbol{D}}{\partial t} = \dot{\boldsymbol{D}}$$

Substituting $\boldsymbol{X} = \dot{\boldsymbol{D}}$ on the right-hand side of eq. 1.173, we obtain the second Maxwell equation:

$$\operatorname{curl} \boldsymbol{H} = \boldsymbol{j} + \frac{\partial \boldsymbol{D}}{\partial t} \tag{1.174}$$

Numerous experiments have shown the appropriateness of selecting the vector $\boldsymbol{X}$ in this form. This quantity was called a displacement current. As follows from the second Maxwell equation, there are two sources for the magnetic field: conduction currents and displacement currents. Applying Stoke's theorem, we obtain the integral form of the second Maxwell equation:

$$\oint_L \boldsymbol{H} \cdot \mathrm{d}\boldsymbol{l} = \int_S (\boldsymbol{j} + \dot{\boldsymbol{D}})\, \mathrm{d}\boldsymbol{S} \tag{1.175}$$

Using an approach that was previously described, it is readily seen that displacement currents make no difference in the surface analog of this equation as derived for constant fields, that is, the tangential component of the magnetic field remains a continuous function at an interface:

$$H_t^{(2)} - H_t^{(1)} = 0 \tag{1.176}$$

Thus in the general case, we have three forms for the second Maxwell equation:

$$\begin{aligned} &\oint_L \boldsymbol{H} \cdot \mathrm{d}\boldsymbol{l} = \int_S (\boldsymbol{j} + \dot{\boldsymbol{D}})\,\mathrm{d}\boldsymbol{S} \\ &\operatorname{curl} \boldsymbol{H} = \boldsymbol{j} + \frac{\partial \boldsymbol{D}}{\partial t} \\ &H_t^{(2)} - H_t^{(1)} = 0 \end{aligned} \tag{1.177}$$

1.4. Faraday's Law and the First Maxwell Equation

Early investigators of electric and magnetic fields observed that when the magnetic induction vector $\boldsymbol{B}$ changes with time throughout a surface S bounded by a contour L, an electromotive force $\mathscr{E}$ exists along that contour with an intensity:

$$\mathscr{E} = -\frac{\partial \psi}{\partial t} \tag{1.178}$$

where ψ is the magnetic flux through surface S bounded by contour L (Fig. 1.37):

$$\psi = \int_S \boldsymbol{B} \cdot \mathrm{d}\boldsymbol{S}$$

where $\boldsymbol{B} = \mu \boldsymbol{H}$, $\mu = \mu_r \mu_0$, and $\partial/\partial t$ denotes the time derivative. The contour L can have any shape and need not be located solely within a conducting medium, it can in particular intersect media with various properties, including insulating ones.

As is well known, the electromotive force can also be presented as:

$$\mathscr{E} = \oint_L \boldsymbol{E} \cdot \mathrm{d}\boldsymbol{l} \tag{1.179}$$

where $\boldsymbol{E}$ is the electric field vector at each point along contour L. Equation 1.178 can therefore be rewritten as:

$$\oint_L \boldsymbol{E} \cdot \mathrm{d}\boldsymbol{l} = -\frac{\partial \psi}{\partial t} \tag{1.180}$$

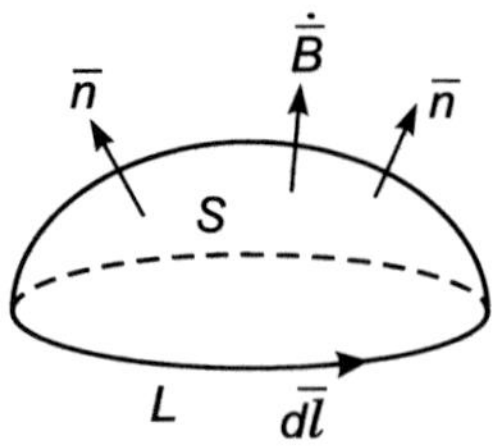

Figure 1.37. The time rate of change of magnetic induction flux throughout a surface S bounded by a contour L causes an electromotive force to exist along that contour.

This expression can be interpreted in a natural sense as follows: a change with time of the magnetic flux, ψ, gives rise to an electric field. This phenomenon was first observed and reported by Faraday and has been called electromagnetic induction. The relationship between the electric field and the rate of change of the magnetic flux, as described by eq. 1.180, is one of the most fundamental relationships in physics.

By convention, the electric field due to electromagnetic induction is called inductive electric field and noted $\boldsymbol{E}^{ind}$, emphasizing the origin of this particular component. One can rewrite eq. 1.180 in the following form:

$$\oint_L \boldsymbol{E}^{ind} \cdot \mathrm{d}\boldsymbol{l} = -\frac{\partial \psi}{\partial t} \tag{1.181}$$

It is a basic fact that a change in magnetic flux with time gives rise to a specific electromotive force. To determine the field however, additional information must be provided.

Up to this point, we have considered only one source for the electric field, namely electric charges. In addition to charges, a change in magnetic field with time provides a second mechanism for the development of an electric field. This fact is the fundamental basis of electromagnetic induction. The electric field can generally be attributed to two sources as well as the magnetic field. One can of course readily think of particular cases in which one of these sources does not exist as for example:

- A static field, which is a constant field in time arises from the presence of electrical charges only.

- An alternating electromagnetic field in which the current flow is tangential to interfaces between media of different conductivity so that the normal component of the electric field is zero and that charges do not arise. This happens for example when an induction probe is located on the axis of a borehole and the medium possesses cylindrical symmetry: in this case, the electric field has a pure inductive character.

Both sources of electric field, however, play an essential role in general cases, and this must be understood in order to solve significant interpretation problems which appear when charges arise on the borehole and other interfaces. In this respect, suppose that an electric field arises from both types of sources, namely:

- electric charges which vary with time but create at each instant a field $\boldsymbol{E}^c$ described by Coulomb's law
- a change in the flux of the magnetic field with time, $\partial\psi/\partial t$.

The total electric field can in this case be presented as the sum:

$$\boldsymbol{E} = \boldsymbol{E}^c + \boldsymbol{E}^{ind} \tag{1.182}$$

whence

$$\boldsymbol{E}^{ind} = \boldsymbol{E} - \boldsymbol{E}^c \tag{1.183}$$

Combining eqs. 1.181 and 1.183, we have:

$$\oint\limits_L \boldsymbol{E} \cdot \mathrm{d}\boldsymbol{l} - \oint\limits_L \boldsymbol{E}^c \cdot \mathrm{d}\boldsymbol{l} = -\frac{\partial\psi}{\partial t}$$

As was shown in the first section, the circulation of Coulomb's electric field is zero and therefore:

$$\oint\limits_L \boldsymbol{E}^{ind} \cdot \mathrm{d}\boldsymbol{l} = \oint\limits_L \boldsymbol{E} \cdot \mathrm{d}\boldsymbol{l} = -\frac{\partial\psi}{\partial t}$$

This result sometimes leads to a misunderstanding of the role played by charges in forming the electromagnetic field. This consideration actually merely shows that the electromotive force due to the Coulomb electric field is zero. But this conclusion cannot be extended to the electric field itself. The Coulomb electric field influences the distribution of the currents which in turn create an alternating magnetic field, therefore the inductive electric field does in general depend on the distribution of charges.

We will now write Faraday's law in various forms. Using first the definition of the magnetic flux:

$$\psi = \int\limits_S \boldsymbol{B} \cdot \mathrm{d}\boldsymbol{S}$$

we have:

$$\oint\limits_L \boldsymbol{E} \cdot \mathrm{d}\boldsymbol{l} = -\frac{\partial}{\partial t} \int\limits_S \boldsymbol{B} \cdot \mathrm{d}\boldsymbol{S}$$

We will not consider here the electromotive force induced by moving the integration path L, and thus the last equation can be rewritten as follows:

$$\oint_L \boldsymbol{E} \cdot \mathrm{d}\boldsymbol{l} = -\int_S \frac{\partial \boldsymbol{B}}{\partial t} \cdot \mathrm{d}S = -\int_S \dot{\boldsymbol{B}} \cdot \mathrm{d}\boldsymbol{S} \tag{1.184}$$

where $\dot{\boldsymbol{B}} = \partial \boldsymbol{B}/\partial t$.

This equation is an exact formulation of Faraday's law and is also considered to be the first Maxwell equation in integral form. In the left hand side, the vector $\mathrm{d}\boldsymbol{l}$ indicates the direction in which the integration is carried along contour L, while the vector $\mathrm{d}\boldsymbol{S}$ in the right hand side represents the direction normal to the surface S. A relationship has to be given between the orientation $\mathrm{d}\boldsymbol{S}$ in order to keep its physical meaning to Faraday's law. This relationship is in fact the well known right-hand rule, that is, an observer looking in the direction of $\mathrm{d}\boldsymbol{S}$ sees that the contour L is given a counter-clockwise orientation by $\mathrm{d}\boldsymbol{l}$. It is only when this is true that eq. 1.184 correctly describes the electromagnetic induction phenomenon.

Next, making use of Stoke's theorem, we obtain the differential form of the first Maxwell equation:

$$\oint_L \boldsymbol{E} \cdot \mathrm{d}\boldsymbol{l} = \int_S \operatorname{curl} \boldsymbol{E} \cdot \mathrm{d}\boldsymbol{S} = -\int_S \frac{\partial \boldsymbol{B}}{\partial t} \cdot \mathrm{d}\boldsymbol{S}$$

whence:

$$\operatorname{curl} \boldsymbol{E} = -\frac{\partial \boldsymbol{B}}{\partial t} \tag{1.185}$$

where the functions $\boldsymbol{E}$ and $\boldsymbol{B}$ are considered in the near vicinity of the same point.

Equations 1.184 and 1.185 both describe the same physical law, but the second form can be applied only at points where the electric field is a continuous function of the spatial variable.

Considering that in many problems, we must examine electromagnetic fields in media with discontinuous changes in properties, it is desirable to derive a surface analog of the first Maxwell equation. At interfaces between media with different conductivity for example, the normal component of the electric field is known to be a discontinuous function of the spatial variable. For this reason we will further proceed with eq. 1.184, evaluating it along the path shown in Fig. 1.38, so that we have:

$$E_t^{(2)} - E_t^{(1)} = 0$$

where t is an arbitrary direction tangent to the interface. In its general form, this equation is:

$$\boldsymbol{n} \times (\boldsymbol{E}_2 - \boldsymbol{E}_1) = 0 \tag{1.186}$$

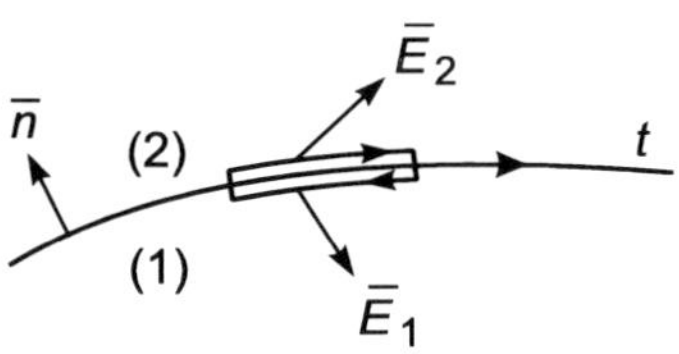

Figure 1.38. Evaluation of Faraday's law near an interface.

The right hand side is zero because the flux of the magnetic field goes to zero as the area enclosed by path L vanishes.

In accord with eq. 1.186, one can say that the tangential component of the time varying electric field is a continuous function of position as is the case for the electric field caused by charges only.

Thus we have obtained three different forms of the first Maxwell equation:

$$\oint_L \boldsymbol{E} \cdot d\boldsymbol{l} = -\int_S \dot{\boldsymbol{B}} \cdot d\boldsymbol{S}$$

$$\text{curl}\, \boldsymbol{E} = -\frac{\partial \boldsymbol{B}}{\partial t} \tag{1.187}$$

$$\boldsymbol{n} \times (\boldsymbol{E}_2 - \boldsymbol{E}_1) = 0$$

It should be emphasized that each of these equations describes the electromagnetic induction phenomenon.

We will now examine a few examples that demonstrate some features of electromagnetic induction when the field changes with time slowly enough. In that case, displacement currents can be neglected.

1.4.1. Example I: The Vortex Electric Field of a Solenoid

Suppose that a magnetic field arises as the consequence of an alternating current flowing in an infinitely long cylindrical solenoid as shown in Fig. 1.39. It is well known that the magnetic field is uniform and nonzero inside the solenoid, and zero outside. Inasmuch as both vectors $\boldsymbol{B}$ and $\partial \boldsymbol{B}/\partial t$ are directed along the z-axis, an induction (vortex) electric field develops in horizontal planes. Moreover, due to the axial symmetry, the vector lines of the electric field are circles with centers located on the axis of the solenoid. The electric field caused by the variation of the magnetic field has but one component E_ϕ which is a function of the radius r only. Making use of eq. 1.184 along any circle with radius r located in a horizontal plane and centered on the x-axis, we have:

$$\int_L \boldsymbol{E} \cdot d\boldsymbol{l} = E \cdot 2\pi r = -\frac{\partial \phi}{\partial t}$$

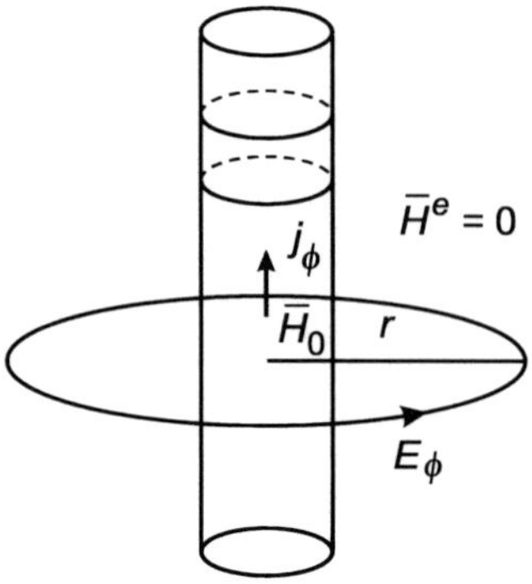

Figure 1.39. Excitation of a magnetic field by an alternating current flowing in an infinitely long cylindrical solenoid.

or

$$E_\phi = -\frac{1}{2\pi r}\frac{\partial \phi}{\partial t} \tag{1.188}$$

where $\partial\psi/\partial t$ is the rate of change of the magnetic flux within the area bounded by the circle of radius r. Suppose that the magnetic field varies with time as follows:

$$H = H_0\, f(t)$$

Then, in accord with eq. 1.188, the vortex field inside the solenoid ($r < a$) is:

$$E_\phi^i = -\frac{\pi r^2}{2\pi r} B_0\, f'(t) = -\frac{B_0}{2} r\, f'(t) \qquad r \leqslant a \tag{1.189}$$

That is, the electric field inside the solenoid increases linearly with the radius r.

Considering now horizontal circles located outside the solenoid, it is clear that the flux and its time-derivative do not depend on the radius of the circles, and that at any given time, we have:

$$\psi = B_0 \pi a^2\, f(t)$$
$$\frac{\partial \psi}{\partial t} = B_0 \pi a^2\, f'(t)$$

The voltage along any of these circles is therefore also independent of the radius and according to eq. 1.188, we can write:

$$E_\phi^e = -\frac{B_0}{2\pi r}\pi a^2\, f'(t) = -\frac{B_0 a^2}{2r}\, f'(t) \qquad r \geqslant a \tag{1.190}$$

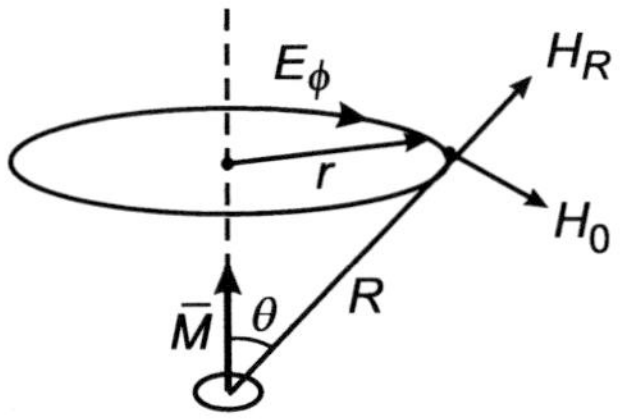

Figure 1.40. A magnetic dipole at the origin of a spherical system of coordinates and with its moment $M(t)$ directed along the z-axis.

That is, the vortex electric field outside the solenoid decreases inversely proportional to r.

This example clearly demonstrates that a vortex electric field can exist at points where the magnetic field is absent.

In the next several examples, we will also consider vortex electric fields caused by the change of the magnetic field with time.

1.4.2. Example II: The Vortex Electric Field of a Magnetic Dipole in a Free Space

In this example, we will consider a magnetic dipole of moment $M(t)$ directed along the z-axis at the origin of a spherical system of coordinates (Fig. 1.40). We will neglect the influence of displacement currents. Considering that in this approximation, the magnetic field is defined by the instantaneous intensity of the current in the dipole, one can calculate the field in the same manner as if it were a static magnetic field. In accord with eq. 1.135, we have the following equation for the alternating magnetic field caused by a magnetic dipole in free space:

$$H_R(t) = \frac{2M(t)}{4\pi R^3}\cos\theta \qquad H_\theta(t) = \frac{M(t)}{4\pi R^3}\sin\theta \qquad H_\phi = 0 \tag{1.191}$$

Inasmuch as the vector representing the magnetic field lies in a longitudinal plane and as a consequence of the axial symmetry, a vortex electric field, arising as a result of the change of this magnetic field with time, has but one component E_ϕ. The vector lines of the field are therefore circles centered on the z-axis.

Making use of eq. 1.184, we have:

$$E_\phi = -\frac{1}{2\pi r}\dot{\psi} \tag{1.192}$$

where ψ is the flux penetrating the area bounded by a circle with radius r (Fig.1.40). Inasmuch as the vector normal to this area is parallel to the z-axis, we have the following expressions for the flux ψ:

$$\psi = \int\limits_S \boldsymbol{B}\cdot \mathrm{d}\boldsymbol{S} = \int\limits_S B_z\,\mathrm{d}S = \int\limits_S B_z 2\pi r\,\mathrm{d}r \tag{1.193}$$

where B_z is the vertical component of the magnetic induction. As can be seen from Fig. 1.40 we have:

$$B_z = B_R \cos\theta - B_\theta \sin\theta$$

and considering eq. 1.191, we obtain:

$$B_z = \frac{\mu M}{4\pi R^3}(3\cos^2\theta - 1) \tag{1.194}$$

Substituting this result into eq. 1.193 and integrating, we obtain:

$$\dot{\psi} = \frac{\partial\psi}{\partial t} = \frac{1}{2}\mu\frac{M(t)}{R^3}r^2 \tag{1.195}$$

where $R = (r^2 + z^2)^{1/2}$. Therefore, we can write the expression for the inductive electric field as:

$$E_\phi = -\mu\frac{\dot{M}(t)}{4\pi R^2}\sin\theta \tag{1.196}$$

It should be expected that the electric field is zero on the z-axis ($\theta = 0$), since the flux through a surface bounded by a circle of vanishing radius vanishes as well. With increasing radius of the electric lines, there is always some critical radius r for which the magnetic field lines start to intersect the surface S twice and in opposite direction. For this reason, the magnetic flux and the corresponding vortex electric field gradually decrease with a further increase in the radius of the circle.

Thus, neglecting displacement currents, the electromagnetic field of an alternating magnetic dipole in free space is described as follows:

$$H_R = \frac{2M(t)}{4\pi R^3}\cos\theta \qquad H_\theta = \frac{M(t)}{4\pi R^3}\sin\theta \qquad E_\phi = -\mu\frac{\dot{M}(t)}{4\pi R^2}\sin\theta \tag{1.197}$$

It is an essential feature of the behavior of this field that along with a magnetic field at each point of space, there also is an electric field. One might suspect that if the medium has nonzero conductivity, this field will give rise to a current flow.

The field described by eq. 1.197 is caused by the currents flowing in the magnetic dipole only, and for this reason it is referred to as the *primary field*. Several examples of primary fields will be considered in this section.

Let us now briefly describe the electromagnetic field of a magnetic dipole when its moment varies with time in some simple ways.

Case 1

Suppose that the current in the dipole changes sinusoidally, that is:

$$M = M_0 \sin\omega t \tag{1.198}$$

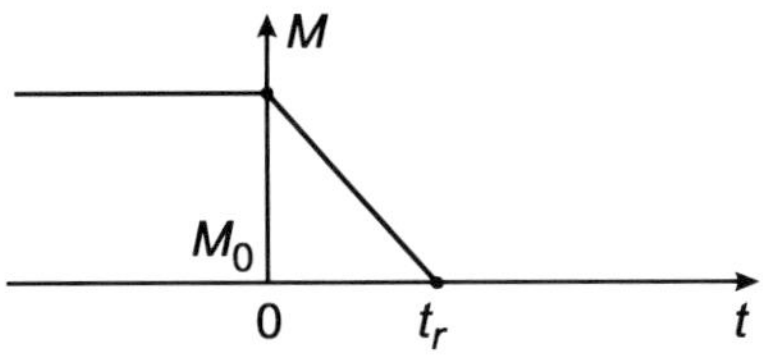

Figure 1.41. Variation of the magnetic moment with time as defined in eq. 1.200.

where M_0 is the magnitude of the moment, $\omega = 2\pi f = 2\pi/T$ is the angular frequency and T the period of oscillation. In accord with eqs. 1.197–1.198, we have the following expression for the magnetic field:

$$\begin{aligned} H_R &= \frac{2M_0}{4\pi R^3}\cos\theta \sin\omega t \\ H_\theta &= \frac{M_0}{4\pi R^3}\sin\theta\sin\omega t \end{aligned} \tag{1.199}$$

while the electric field can be written as:

$$E_\phi = \frac{\mu\omega M_0}{4\pi R^2}\sin(\omega t - \pi/2)\sin\theta$$

and one can therefore say that the primary electric field exhibits a phase shift of 90° with respect to the current in the dipole or the primary magnetic field as well. The last equation plays a basic role in the theory of induction logging as developed by H. Doll.

Case 2

Consider a dipole whose moment varies with time as follows (see Fig. 1.41):

$$M = \begin{cases} M_0 & t \leqslant 0 \\ M_0 - at & 0 \leqslant t \leqslant t_r \\ 0 & t \geqslant t_r \end{cases} \tag{1.200}$$

where $a = M_0/t_r$.

This relationship describes a primary field which is constant at negative times, then decreases linearly over the interval $0 \leqslant t \leqslant t_r$, and is exactly zero at times larger than t_r. A primary vortex electric field will exist only within the time interval over which the magnetic field varies, and in view of the linear dependence of the magnetic field on time during that range, the electric field will assume a constant value. Thus, we have:

$$H_R = \begin{cases} \dfrac{2M_0}{4\pi R^3}\cos\theta & t \leqslant 0 \\ \dfrac{2M(t)}{4\pi R^3}\cos\theta & 0 \leqslant t \leqslant t_r \\ 0 & t \geqslant t_r \end{cases}$$

$$H_\theta = \begin{cases} \dfrac{M_0}{4\pi R^3}\sin\theta & t \leqslant 0 \\ \dfrac{M(t)}{4\pi R^3}\sin\theta & 0 \leqslant t \leqslant t_r \\ 0 & t \geqslant t_r \end{cases} \tag{1.201}$$

$$E_\phi = \begin{cases} 0 & t \leqslant 0 \\ \dfrac{\mu a}{4\pi R^2}\sin\theta = \dfrac{\mu M_0}{4\pi R^2 t_r}\sin\theta & 0 \leqslant t \leqslant t_r \\ 0 & t \geqslant t_r \end{cases}$$

The curves given in Fig. 1.42 illustrate the behavior of all the components of the field with time.

1.4.3. Example III: The Inductive Electric Field due to the Magnetic Field of a Current Flowing in a Circular Loop

In induction logging, the receiver of an induction probe measures the field caused by the currents induced in the surrounding medium. As will be shown later, these currents in most cases flow along circles. This is why it is appropriate to explore some features of the electromagnetic field caused by a sole current ring.

In this case, we will assume that the source of the electromagnetic field is an alternating current $I(t)$ flowing in a circular loop of radius a as shown in Fig. 1.43. As is well known, the magnetic field of a constant current flowing in a loop can be expressed in terms of elliptic integrals. Inasmuch as displacement currents are ignored, the magnetic field is defined by the instantaneous value of the conduction current in the loop, that is:

$$\begin{aligned} B_r &= \frac{\mu I(t)}{\pi}\frac{z}{r\left[a+r)^2+z^2\right]^{1/2}}\left[-K+\frac{a^2+r^2+z^2}{(a-r)^2+z^2}E\right] \\ B_z &= \frac{\mu I(t)}{\pi}\frac{z}{\left[a+r)^2+z^2\right]^{1/2}}\left[K+\frac{a^2-r^2-z^2}{(a-r)^2+z^2}E\right] \\ B_\phi &= 0 \end{aligned} \tag{1.202}$$

where K and E are the elliptic integrals of first and second kind, respectively, which are described in most mathematical handbooks. Considering the axial symmetry as well as

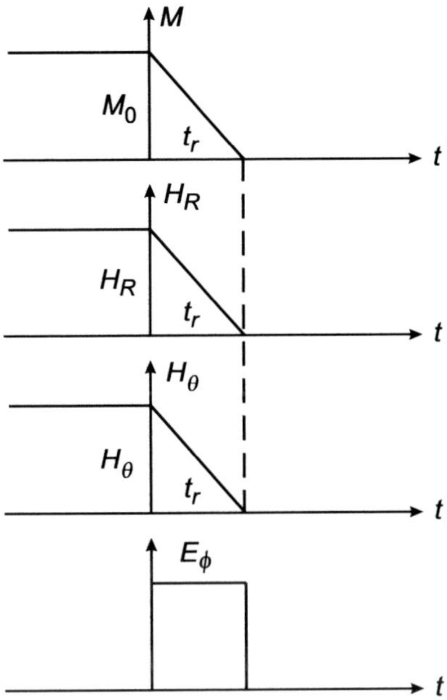

Figure 1.42. Variation with time of the components of the electromagnetic field caused by a magnetic dipole of moment varying according to eq. 1.200.

the fact that the magnetic field lines lie in longitudinal planes, the electric field that arises due to the variation of the magnetic field with time has but one component E_ϕ that is, in a cylindrical system of coordinates (r, ϕ, z):

$$\boldsymbol{E} = (0, E_\phi, 0)$$

To describe the vortex electric field, one can in principle make use of the first Maxwell equation in its integral form:

$$\oint_L \boldsymbol{E} \cdot \mathrm{d}\boldsymbol{l} = -\int_S \boldsymbol{B} \cdot \mathrm{d}\boldsymbol{S}$$

This approach, however, leads to the integration of elliptic functions, which is a somewhat cumbersome approach. Fortunately, there is a much more efficient way to derive an expression for the electric field. Considering that the potential $\boldsymbol{A}$ ($\boldsymbol{H} = \operatorname{curl} \boldsymbol{A}$) as well as the magnetic field $\boldsymbol{H}$ are defined by the instantaneous value of current in the source loop one can write:

$$A_\phi = \frac{I(t)}{4\pi} \oint_L \frac{\mathrm{d}l_\phi}{R}$$

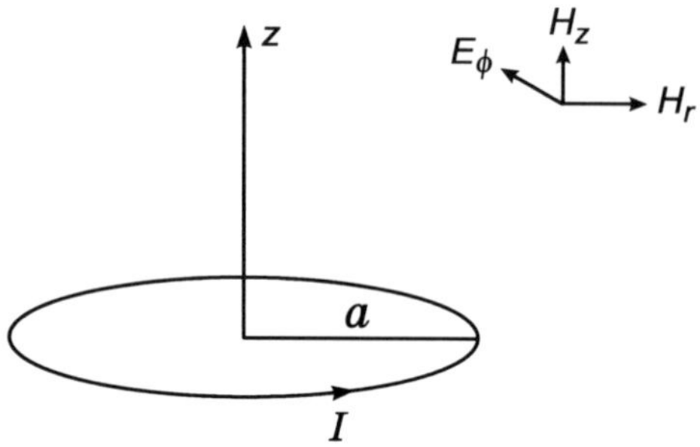

Figure 1.43. Components of the electromagnetic field caused by an alternating current flowing in a circular loop.

or, after some transformations, we have:

$$A_\phi = \frac{I(t)}{\pi k}\left(\frac{a}{r}\right)^{1/2}\left[(1-\frac{1}{2}k^2)K - E\right]$$

where $k^2 = 4ar/((a+r)^2+z^2)$.

From the first Maxwell equation:

$$\operatorname{curl} \boldsymbol{E} = -\frac{\partial \boldsymbol{B}}{\partial t}$$

follows that:

$$\operatorname{curl} \boldsymbol{E} = -\mu\frac{\partial}{\partial t}\operatorname{curl} \boldsymbol{A} = -\mu \operatorname{curl}\frac{\partial \boldsymbol{A}}{\partial t}$$

or

$$\operatorname{curl}\left(\boldsymbol{E} + \mu\frac{\partial \boldsymbol{A}}{\partial t}\right) = 0$$

From this we have:

$$E_\phi = -\mu\frac{\partial \boldsymbol{A}}{\partial t} - \operatorname{grad} U \qquad (1.203)$$

where U is a scalar potential. In accord with this last equation, we have the following expression for the component E_ϕ:

$$E_\phi = -\mu\frac{\partial A_\phi}{\partial t} - \frac{1}{r}\frac{\partial U}{\partial \phi}$$

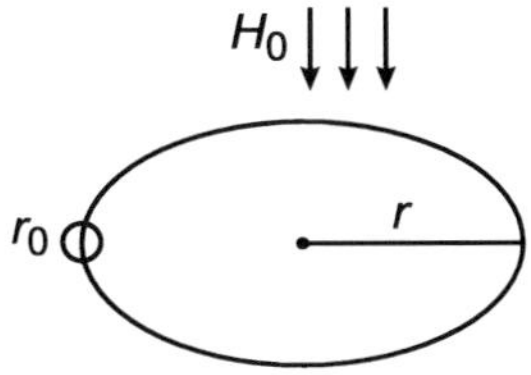

Figure 1.44. Induction of currents in a thin conducting ring placed in an alternating field.

It should be noted at this point that the vector potential $\boldsymbol{A}$, which we have used here, is identical at any instant to the vector potential caused by a constant current flow.

Taking into account the axial symmetry, the second term in the last expression vanishes so that we have:

$$E_\phi = -\mu \frac{\partial A_\phi}{\partial t} \tag{1.204}$$

This equation permits us to write down the expression for E_ϕ, that is:

$$E_\phi = -\mu \frac{\mathrm{d}I(t)}{\mathrm{d}t} \frac{1}{\pi k} \left(\frac{a}{r}\right)^{1/2} \left[\left(1 - \frac{1}{2}k^2\right) K - E\right] \tag{1.205}$$

In conclusion, it should be emphasized that the vector lines of the electric field are circles lying in horizontal planes with centers located on the z-axis. It is an easy matter to show that the electric field given in eq. 1.205 is practically identical to that of a magnetic dipole (eq. 1.197) when the distance of the observation point to the source is significantly larger than the radius of the loop.

1.4.4. Example IV: Induction of a Current in a Thin Conducting Ring Situated within a Primary Alternating Field

The induction process can be described as follows (Fig. 1.44). With a change of the primary magnetic field with time a primary vortex electric field arises. We will consider for simplicity that this field has only one component E_ϕ^0 which is tangential to the plane of the ring. According to Ohm's law, this field causes current to flow in the ring. This current in turn generates a so-called secondary electromagnetic field, and it should then be obvious that the density of the induced current is in fact determined by both primary and secondary fields. In accord with Ohm's law, we can write:

$$j_\phi = \sigma(E_\phi^0 + E_\phi^s) \tag{1.206}$$

where j_ϕ, is the current density in the ring, and σ is its conductivity. E_ϕ^0 is the primary electric field; E_ϕ^s is the secondary one.

We will now use Faraday's law (eq. 1.178) to find the current in the ring:

$$\mathscr{E} = -\frac{\partial \psi}{\partial t} \qquad (1.207)$$

The flux ψ through a surface bounded by the ring can be written as the sum:

$$\psi = \psi_0 + \psi_s \qquad (1.208)$$

where ψ_0 is the flux of the primary magnetic field and ψ_s is the one of the secondary field. Thus eq. 1.207 can be rewritten as:

$$\mathscr{E} = -\frac{\partial \psi_0}{\partial t} - \frac{\partial \psi}{\partial t} \qquad (1.209)$$

In this equation, only the term $\partial\psi_0/\partial t$ is known, while the electromotive force, $\mathscr{E}$, and the rate of change of the secondary magnetic flux, $\partial\psi_s/\partial t$, are unknowns. Our objective is to determine the current I flowing in the ring and we will therefore attempt to express both unknowns in the last equation in terms of this function. First of all, making use of Ohm's law in integral form, we have:

$$\mathscr{E} = RI \qquad (1.210)$$

where R is the resistance of the ring given by:

$$R = \rho\frac{l}{S} \qquad (1.211)$$

Here ρ is the resistivity of the ring, l is its circumference, and S its cross-sectional area. According to Biot–Savart law, it is clear that the magnetic flux ψ_s caused by the current flow in the ring is directly proportional to I, and can therefore be written as:

$$\psi_s = LI \qquad (1.212)$$

where L is a coefficient of proportionality known as the inductance of the ring. According to eq. 1.212, one could say that the inductance of the ring is the ratio of the magnetic flux through the ring to the current in the ring:

$$L = \frac{\psi_s}{I}$$

The inductance is controlled by the geometrical parameters of the ring. Its determination usually involves the solution of rather complex problems. In some special cases, however, this task is relatively easy and in particular the expression for the inductance of a thin ring in free space is known to be:

$$L = r\mu_0\left(\ln\frac{8r}{r_0} - 1.75\right) \qquad (1.213)$$

Inductance is measured in henries per meter in the m.k.s. system of units; r_0 is the cross-section radius.

If instead of one ring we have a n-turn coil, the inductance increases as the square of the number of turns:

$$L = r\mu_0 n^2 \left(\ln \frac{8r}{r_0} - 1.75 \right) \tag{1.214}$$

The simple form of the conductive volume (a thin uniform circular ring), and the assumption that the current density is uniform over the cross-section of the ring has allowed us to find a simple expression for the coefficient of the proportionality between the secondary magnetic flux and the secondary current intensity I. Substituting this result into eq. 1.209 yields a differential equation from which the current I can be determined:

$$L\frac{\mathrm{d}I}{\mathrm{d}t} + RI = -\frac{\partial \psi_0}{\partial t}$$

or

$$\frac{\mathrm{d}I}{\mathrm{d}t} + \frac{1}{\tau_0} I = f(t) \tag{1.215}$$

where

$$\tau_0 = L/R$$

and

$$f(t) = -\frac{1}{L}\frac{\partial \psi_0}{\partial t}$$

Making use of results obtained earlier in this chapter (eq. 1.154), we have the following solution for the induction current:

$$I = I_0 \mathrm{e}^{-t/\tau_0} - \mathrm{e}^{-t/\tau_0} \frac{1}{L} \int_0^t \mathrm{e}^{t/\tau_0} \frac{\partial \psi_0}{\partial t} \, \mathrm{d}t \tag{1.216}$$

We will now consider the behavior of the induced current in two specific cases.

Case 1

Let us assume that the primary magnetic field varies with time as shown in Fig. 1.42, so that we have the following expression for $\partial \psi_0 / \partial t$:

$$\frac{\partial \psi_0}{\partial t} = \begin{cases} 0 & t \leqslant 0 \\ -\dfrac{\psi_0}{t_r} & 0 \leqslant t \leqslant t_r \\ 0 & t \geqslant t_r \end{cases} \tag{1.217}$$

where t_r is called the *ramp time*. During the interval of the time over which the primary magnetic flux does not change with time ($t < 0$), there are no induced currents in the ring, that is:

$$I(t) = 0 \qquad t < 0$$

During the ramp time the primary flux changes with time, and the induced current is determined by the rate of change of the flux time, as well as by parameters of the ring (R and L). After the primary flux disappears, the behavior of the induced current is controlled by parameters R and L only. In this case, eq. 1.215 is simplified and we have:

$$\frac{\mathrm{d}I}{\mathrm{d}t} + \frac{I}{\tau_0} = 0 \tag{1.218}$$

A solution of this equation is:

$$I = C\,\mathrm{e}^{-t/\tau_0} \qquad t_r \geqslant t \tag{1.219}$$

The parameter τ_0 is commonly called time constant of the ring, inasmuch as it represents the rate at which the current decays in the absence of external sources.

In order to express the constant C, we will investigate the behavior of the induced current flow during the ramp time. In accord with eqs. 1.216 and 1.217, we have:

$$I(t) = I_0\,\mathrm{e}^{-t/\tau_0} + \mathrm{e}^{-t/\tau_0}\frac{\psi_0}{t_r L}\int\limits_0^t \mathrm{e}^{t/\tau_0}\,\mathrm{d}t = I_0\,\mathrm{e}^{-t/\tau_0} + \frac{\tau_0}{t_r}\frac{\psi_0}{L}(1 - \mathrm{e}^{-t/\tau_0}) \tag{1.220}$$

where I_0 represents the amplitude of the current at the instant $t = 0$. Inasmuch as at this instant there is no current in the ring, we have:

$$I(t) = \frac{\tau_0}{t_r}\frac{\psi_0}{L}(1 - \mathrm{e}^{-t/\tau_0}) \tag{1.221}$$

The constant C is readily found from eqs. 1.219 and 1.220. We have in fact:

$$I(t_r) = C\,\mathrm{e}^{-t_r/\tau_0} = \frac{\tau_0}{t_r}\frac{\psi_0}{L}(1 - \mathrm{e}^{-t_r/\tau_0})$$

Thus:

$$C = \frac{\tau_0}{t_r}\frac{\psi_0}{L}(\mathrm{e}^{t_r/\tau_0} - 1) \tag{1.222}$$

Therefore, we obtain the following expression for the current induced in the ring:

$$I(t) = \begin{cases} 0 & t \leqslant 0 \\ \dfrac{\tau_0}{t_r}\dfrac{\psi_0}{L}(1 - \mathrm{e}^{-t/\tau_0}) & 0 \leqslant t \leqslant t_r \\ \dfrac{\tau_0}{t_r}\dfrac{\psi_0}{L}(\mathrm{e}^{t_r/\tau_0} - 1)\mathrm{e}^{-t/\tau_0} & t \geqslant t_r \end{cases} \tag{1.223}$$

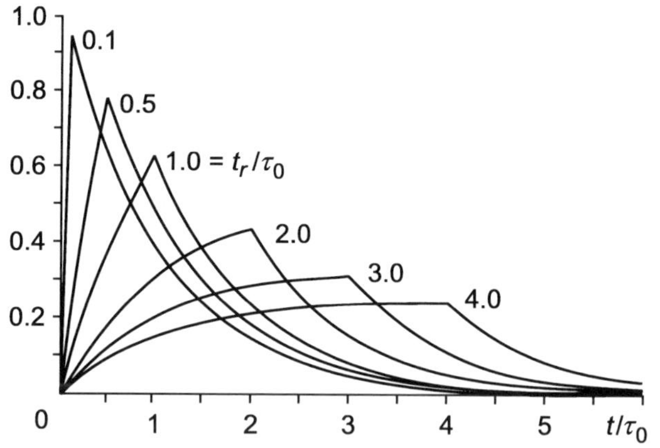

Figure 1.45. Behavior of induced currents as defined by eq. 1.223.

In accord with the last equations, the intensity of the induced current gradually increases during the ramp time, reaching a maximum at the instant $t = t_r$, and then starts to decrease exponentially.

Suppose that the ramp time t_r is much less than the time constant τ_0. Expanding then the exponential terms in eq. 1.223 in a power series and discarding all terms but the leading ones, we obtain:

$$I(t) = \begin{cases} 0 & t \leqslant 0 \\ \dfrac{t}{t_r}\dfrac{\psi_0}{L} & 0 \leqslant t \leqslant t_r \\ \dfrac{\psi_0}{L}\mathrm{e}^{-t/\tau_0} & t \geqslant t_r \quad \text{if } t_r \ll \tau_0 \end{cases} \tag{1.224}$$

On the contrary, if $t_r \gg \tau_0$, the induced current increases linearly at first, and then asymptotically approaches the following maximum value:

$$\frac{\tau_0}{t_r}\frac{\psi_0}{L}$$

after which it decreases exponentially as before.

The behavior of induced currents in both situations is shown in Fig. 1.45.

We will now investigate the induction of current in the ring when the primary current and magnetic flux change as a step function of time (Fig. 1.46). It is obvious that the behavior of induced currents in this case is described by eq. 1.224 as t_r approaches zero, that is:

$$I(t) = \frac{\psi_0}{L}\mathrm{e}^{-t/\tau_0} \tag{1.225}$$

Thus the initial value of the induced current does not depend on the resistance R of the ring, but on the primary flux and the inductance of the ring only.

Because there is always in practice a nonzero ramp time, the initial value of the current should be interpreted as being its value at the instant $t = t_r$, provided that t_r is much less than τ_0.

It is interesting to obtain the same result directly from eq. 1.215. Integrating both parts yields:

$$R\int_0^{t_r} I\,\mathrm{d}t + L\int_0^{t_r} \frac{\mathrm{d}I}{\mathrm{d}t}\,\mathrm{d}t = -\int_0^{t_r} \frac{\partial\psi_0}{\partial t}\,\mathrm{d}t$$

whence:

$$R\int_0^{t_r} I\,\mathrm{d}t + L\left\{I(t_r) - I(0)\right\} = \psi_0(0) - \psi_0(t_r) \qquad (1.226)$$

Inasmuch as we have the following initial conditions:

$$\psi_0(0) = \psi_0 \qquad I(0) = 0$$

and that at time $t = t_r$ the primary flux is zero, eq. 1.226 can be rewritten as:

$$R\int_0^{t_r} I\,\mathrm{d}t + LI(t_r) = \psi_0 \qquad (1.227)$$

The integrand $I\,\mathrm{d}t$ indicates the total quantity of charge passing through the ring during the time $\mathrm{d}t$. It is obvious that with decreasing ramp time, the quantity of charge tends to zero and in the limit, when the magnetic field changes as a step function, we have:

$$LI(t_r) = \psi_0 \qquad \text{as } t_r \to 0$$

that is, the initial current is:

$$I(0) = \frac{\psi_0}{L} \qquad \text{as } t_r \to 0 \qquad (1.228)$$

This is exactly the same result as obtained with eq. 1.225.

The analysis carried out earlier shows that the error caused by discarding the integral term decreases with decreasing ratio t_r/τ_0, that is, eq. 1.228 becomes more precise with increasing inductance or decreasing resistance of the ring.

Considering that for positive times, the current satisfies a homogeneous differential equation, we obtain again:

$$I(t) = \frac{\psi_0}{L}\,\mathrm{e}^{-t/\tau_0} \qquad (1.229)$$

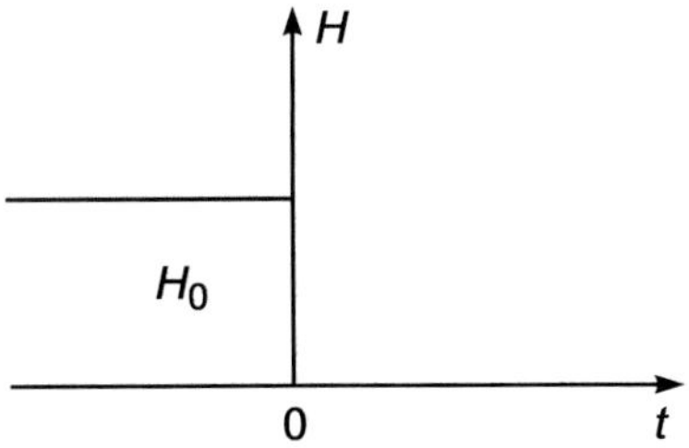

Figure 1.46. The variation of magnetic field with time for an ideal step function excitation.

Thus, at the initial instant, the current induced in the ring does not depend on the conductivity but is defined by the primary flux and the inductance (geometrical parameter) of the ring only.

The equality:

$$LI(0) = \psi_0 \tag{1.230}$$

is an essential feature of electromagnetic induction. The left-hand side in fact defines the magnetic flux of induced current through the ring at instant $t = 0$, when the primary flux disappears. Thus an induced current I arises with such magnitude that at the first instant its magnetic flux $LI(0)$ is exactly equal to the primary flux ψ_0. This result will later be generalized to include more complicated models of conducting media.

Case 2

Suppose that the primary magnetic field varies sinusoidally as:

$$H_0 \sin \omega t \tag{1.231}$$

where H_0 is the amplitude of the field, f is its frequency, ω is its angular frequency and T its period of oscillation. In contrast to the previous case, we consider here a steady field which is assumed to have been established far before the time of observation, and has been repeating itself periodically ever since. In order to find the current induced in the ring, we will make use of eq. 1.216. Since the primary flux is simply expressed as $\psi_0 \sin \omega t$, we have:

$$I(t) = I_0 \, \mathrm{e}^{-t/\tau_0} - \frac{\omega \psi_0}{L} \, \mathrm{e}^{-t/\tau_0} \int_0^t \mathrm{e}^{t/\tau_0} \cos \omega t \, \mathrm{d}t \tag{1.232}$$

Because:

$$\int e^{ax} \cos bx \, \mathrm{d}x = \frac{\mathrm{e}^{ax}}{a^2 + b^2} [a \cos bx + b \sin bx]$$

we obtain:

$$e^{-t/\tau_0}\int_0^t e^{t/\tau_0}\cos\omega t dt = \frac{1}{1/\tau_0^2+\omega^2}\left[\frac{1}{\tau_0}\cos\omega t + \omega\sin\omega t\right] - \frac{1}{\tau_0}\frac{e^{-t/\tau_0}}{1/\tau_0^2+\omega^2}$$

where $\tau_0 = L/R$.

Thus, the induced current is expressed as:

$$I(t) = I_0\, e^{-t/\tau_0} - \frac{\omega\psi_0 R}{R^2+\omega^2L^2}\cos\omega t - \frac{\omega^2\psi_0 L}{R^2+\omega^2L^2}\sin\omega t + \frac{\omega\psi_0 R}{R^2+\omega^2L^2}\, e^{-t/\tau_0}$$

Inasmuch as we are interested in the induced current for an established sinusoidal process, that is, for t much larger than τ_0, we have:

$$I(t) = -\frac{\omega\psi_0}{R^2+\omega^2L^2}\left[R\cos\omega t + \omega L\sin\omega t\right] \tag{1.233}$$

Let us introduce the following notations:

$$a = -\frac{\omega\psi_0 R}{R^2+\omega^2L^2} \quad \text{and} \quad b = -\frac{\omega\psi_0 L}{R^2+\omega^2L^2} \tag{1.234}$$

Correspondingly we have:

$$I(t) = a\cos\omega t + b\sin\omega t \tag{1.235}$$

that is, the induced current can be presented as the sum of two separate oscillations. One of them, $b\sin\omega t$, which changes synchronously with the primary magnetic field, is called the *inphase* component of the current and is expressed as follows:

$$\mathrm{In}\, I = b\sin\omega t$$

The other one, $a\cos\omega t$, representing an oscillation shifted by 90° with respect to the primary source, is called the *quadrature* component of the current and is written:

$$\mathrm{Q}\, I = a\cos\omega t$$

Equation 1.235 suggests that it is desirable to treat the induced current as the sum of an inphase and a quadrature component, the intensity of which is given by eq. 1.234.

It is important to note that in induction logging the quadrature component is usually measured.

We can write the inphase and quadrature components as:

$$a = A\sin\phi \qquad b = A\cos\phi \tag{1.236}$$

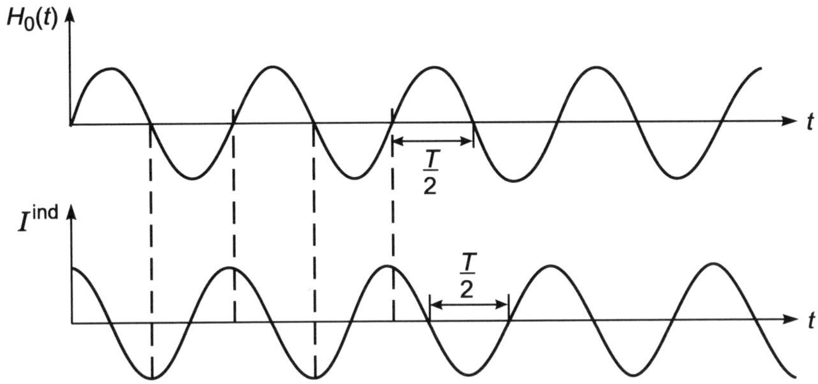

Figure 1.47. Sinusoidal variation of the primary field and of induced currents showing the existence of a phase shift between them.

so that we obtain:

$$I = A\left[\sin\phi\cos\omega t + \cos\phi\sin\omega t\right] = A\sin(\omega t + \phi) \tag{1.237}$$

Therefore, the induced current and the primary field are both sinusoidal functions, each having the same frequency ω, and being characterized in general by two parameters A and ϕ.

The parameter A is the amplitude of the secondary current, that is, this current oscillates and reaches a maximum value A each time when the argument $\omega t + \phi$ is an odd multiple of $\pi/2$.

The quantity ϕ indicates that there is a phase shift between the primary field and the induced current, that is, they oscillate asynchronously as shown in Fig. 1.47. In accord with eqs. 1.234 and 1.236, we have:

$$A = (a^2 + b^2)^{1/2} = \frac{\omega\psi_0}{(R^2 + \omega^2 L^2)^{1/2}} \tag{1.238}$$

and:

$$\tan\phi = a/b$$

or

$$\phi = \tan^{-1}\left(\frac{R}{\omega L}\right)$$

A representation of the induced current is shown in Fig. 1.48 in the form of inphase and quadrature components as well as amplitude and phase.

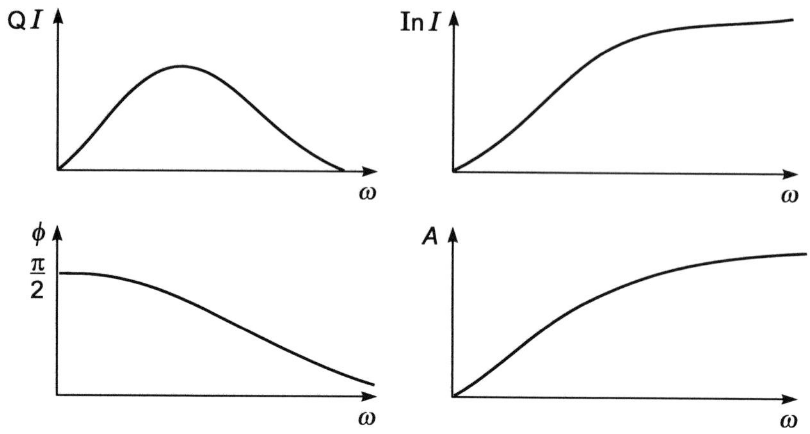

Figure 1.48. Curves representing the inphase and quadrature components of the induced current, as well as its amplitude and phase.

In spite of the simplicity of the model of the thin circular ring, the frequency response of the induced current contains some general features which are inherent to much more complicated cases as will be demonstrated in later chapters.

We will consider here only briefly the low-frequency part of the spectrum. When the frequency is sufficiently low it is possible to expand the expression in eq. 1.234 in a series of ω and discarding all terms but the leading ones, we obtain:

$$a = -\frac{\omega\psi_0}{R} \qquad \text{and} \qquad b = -\frac{\psi_0 L}{R^2}\omega^2 \tag{1.239}$$

or

$$\text{Q}\,I = -\frac{\psi_0\omega}{R}\cos\omega t \qquad \text{and} \qquad \text{In}\,I = -\frac{\psi_0 L}{R^2}\omega^2 \sin\omega t$$

From these expressions it is apparent that at low frequencies the quadrature component of the induced current is dominant and is directly proportional to the conductivity of the ring and to the frequency, while it does not depend on the inductance. This behavior can readily be explained as follows: if we neglect the flux caused by the induced current the total flux through the ring is the same as the primary one, that is ψ_0. As it changes with time, we have:

$$\frac{\partial\psi}{\partial t} = \frac{\partial\psi_0}{\partial t} = \omega\psi_0\cos\omega t$$

and therefore in accord with Ohm's law, we have:

$$\text{Q}\,I = -\frac{\omega\psi_0}{R}\cos\omega t$$

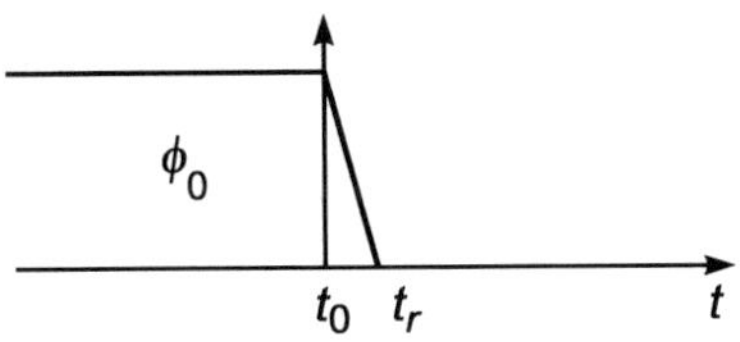

Figure 1.49. Example of a finite but short ramp time from the termination of a constant magnetic flux.

Thus at low frequencies the quadrature component is directly proportional to the primary field, frequency and conductivity. It is important to state that this behavior stands when more complicated conductors are considered. It is also appropriate to notice that eq. 1.239 for the quadrature component is very basic in the theory of induction logging as developed by H. Doll.

1.4.5. Example V: Behavior of the Electromagnetic Field at the Early Stage and High Frequencies in a Conducting Medium

Suppose that we have an arbitrarily oriented system of n conducting rings. The equation determining the intensity of the current induced in the kth ring can be written as:

$$R_k I_k = -\frac{\partial \psi_k}{\partial t} = -\frac{\partial \psi_{0k}}{\partial t} - \frac{\partial \psi_{sk}}{\partial t} \tag{1.240}$$

where R_k and I_k are the resistance and current in the kth ring, and ψ_{sk} are the magnetic fluxes of the primary and secondary fields, respectively. It is clear that the magnetic flux ψ_{sk} can be expressed as:

$$\psi_{sk} = M_{1k} I_1 + M_{2k} I_2 + \cdots + L_k I_k + \cdots + M_{nk} I_n$$

where L_k is the inductance of the kth ring, M_{ik} is the mutual inductance of the i-th and k-th rings, that is, the ratio of the magnetic flux through the k-th ring, to the current I_i in the i-th ring, that is:

$$\psi_{ik} = M_{ik} I_i$$

Equation 1.240 can correspondingly be rewritten as:

$$R_k I_k + L_k \frac{\partial I_k}{\partial t} + \sum_{\substack{p=1 \\ p \neq k}}^{n} M_{pk} \frac{\partial I_p}{\partial t} = \frac{\partial \psi_{0k}}{\partial t} \tag{1.241}$$

Assume now that the primary flux ψ_0 caused by external sources starts to change from a value ψ_0 to zero at the instant $t = t_0$, and that this change takes place in a very short time t_r (Fig. 1.49). Integrating eq. 1.241 with respect to time, we obtain:

$$R_k \int_{t_0}^{t_0+t_r} I_k \mathrm{d}t + L_k \int_{t_0}^{t_0+t_r} \frac{\partial I_k}{\partial t}\mathrm{d}t + \sum_{\substack{p=1\\p\neq k}}^{n} M_{pk} \int_{t_0}^{t_0+t_r} \frac{\partial I_p}{\partial t}\mathrm{d}t = \psi_{0k}(t_0)$$

since:

$$\psi_{0k}(t_0 + t_r) = 0$$

Taking into account that induced currents are absent at the first instant ($t = t_0$) and that the interval t_r is very short, this last equation can be approximated as follows:

$$L_k I_k(t_0 + t_r) + \sum_{\substack{p=1\\p\neq k}}^{n} M_{pk} I_p(t_0 + t_r) = \psi_{0k}(t_0)$$

Introducing the notation:

$$t_0^- = t_0 \qquad t_0^+ = t_0 + t_r$$

we have:

$$L_k I_k(t_0^+) + \sum_{\substack{p=1\\p\neq k}}^{n} M_{pk} I_p(t_0^+) = \psi_{0k}(t_0^-) \tag{1.242}$$

On the left-hand side of this expression, we have a representation of the magnetic flux through the kth ring caused by the currents induced in all the other rings just after switching, while on the right-hand side is the expression for the primary flux ψ_0 before switching. Thus we observe again a principal feature of electromagnetic induction when the primary flux changes as a step function:

$$\psi_0 = \begin{cases} \psi_0 & t \leqslant t_0 \\ 0 & t > t_0 \end{cases}$$

In fact at the very first instant, the currents induced in all the rings have such a magnitude that the magnetic flux they caused in any ring is exactly equal to the primary flux.

Now we are prepared to describe the asymptotic behavior of the field in a conducting medium. We will make no assumption on the uniformity of the medium or the locations of the primary sources. Let us suppose that the primary flux is instantaneously switched off at the instant $t = t_0$. At times smaller than t_0, the magnetic field has been constant so that no induced currents are present in the conductor before t_0. Correspondingly, the circulation of the magnetic field inside the medium along any arbitrary path is zero during that range of time:

$$\oint_L \boldsymbol{H}_0 \cdot \mathrm{d}\boldsymbol{l} = 0 \qquad t < t_0 \tag{1.243}$$

provided that the path of integration does not enclosed a current line of the primary sources.

A conducting medium can be presented as consisting of a system of current rings with arbitrary shapes and in this way one can apply results that were obtained earlier. Inasmuch as the flux through any ring at the instant $t = t_0$ remains the same as that for earlier times the magnetic field at any point in the conducting medium does not change either. This conclusion stems from the fact that for an arbitrary surface inside the conductor, we have:

$$\psi(t_0^+) = \psi_0$$

Thus immediately after disappearance of the primary flux ψ_0 we have:

$$H(t_0^+) = H_0 \tag{1.244}$$

Let us emphasize that this relationship does not exist outside the conductor.

From eqs. 1.243 and 1.244 it follows that the circulation of the magnetic field for any path inside a conductor is zero at the instant t_0 and therefore there are no induced currents:

$$\boldsymbol{j} = 0 \text{ and } \boldsymbol{E} = 0 \qquad \text{if } t = t_0^+ \tag{1.245}$$

There must, however, be some sources of the magnetic field which maintain the primary field when the source is switched off. These sources are induced surface currents which are situated close to the source of the primary field if this one is located outside the conductor. If the source is located within the conductor, as in induction logging, induced currents initially exist near the source only.

Induced currents, concentrated on the surface of a conductor or near the primary sources, decay with time as the electromagnetic energy is converted into heat and appears at various points in the medium. It is obvious that the decay of the field takes place more rapidly in a highly resistance medium, while it decreases slowly in a conductive medium.

Let us note that in solving many boundary problems related to the calculation of nonstationary fields, conditions 1.244 or 1.245 are extremely important and are usually referred to as *initial conditions.* They are in essence a modification of Faraday's law and therefore must be satisfied by any nonstationary field in a conducting medium.

Suppose now that the primary field changes as a sinusoid with relatively high frequency (Fig. 1.50a). Such a signal can be approximated qualitatively by a system of pulses of alternating sign (Fig. 1.50b).

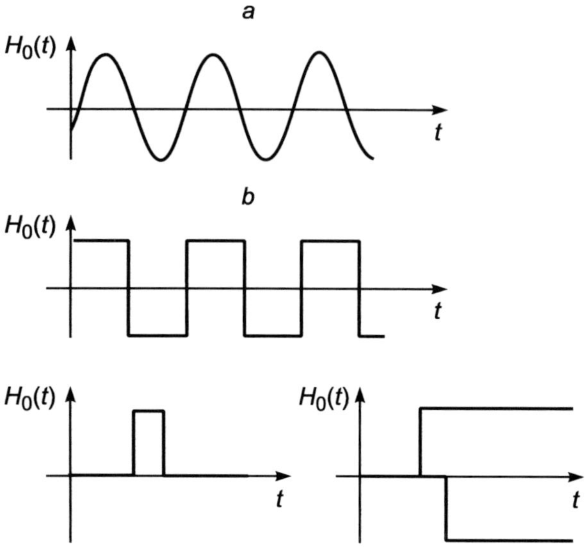

Figure 1.50. Presentation of a high-frequency oscillation as a system of step functions.

Then each pulse can be presented as the difference of two step functions arising with an interval of time of half the period of the oscillations.

It is clear that if the current changes in the primary source with relatively high frequency, the induced currents essentially remain either on the surface of the conductor or close to the source depending on whether the source is located outside or within the conductor. This explains why the high-frequency asymptote coincides with that of the early stage in transient electromagnetic fields.

1.5. Electromagnetic Field Equations

In previous sections, by making use of Gauss's and Stoke's theorems, we have developed the basic laws for the electromagnetic fields in the form of equations. In accord with these laws the electromagnetic field must satisfy the following set of equations:

$$\oint_L \boldsymbol{E} \cdot d\boldsymbol{l} = -\frac{\partial}{\partial t} \int_S \boldsymbol{B} \cdot d\boldsymbol{S} \tag{1.246}$$

$$\int_L \boldsymbol{H} \cdot d\boldsymbol{l} = \int_S \boldsymbol{j} \cdot d\boldsymbol{S} + \int_S \frac{\partial \boldsymbol{D}}{\partial t} \cdot d\boldsymbol{S} \tag{1.247}$$

$$\int_S \boldsymbol{D} \cdot d\boldsymbol{S} = e \tag{1.248}$$

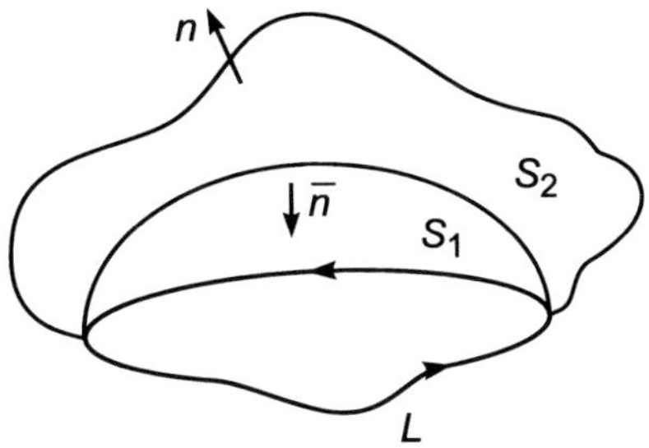

Figure 1.51. Two surfaces S_1, and S_2, bounded by a contour L.

$$\int_S \boldsymbol{B} \cdot \mathrm{d}\boldsymbol{S} = 0 \tag{1.249}$$

where $\boldsymbol{E}$ and $\boldsymbol{H}$ are electric and magnetic field vectors, $\boldsymbol{B}$ and $\boldsymbol{D}$ are magnetic and electric induction vectors, e is a charge in a volume surrounded by a surface S, and $\boldsymbol{j}$ is conduction current density. The various vectors are by a set of relationships known as the constitutive equations:

$$\boldsymbol{D} = \varepsilon \boldsymbol{E} \qquad \boldsymbol{B} = \mu \boldsymbol{H} \qquad \boldsymbol{j} = \sigma \boldsymbol{E}$$

where ε, μ and σ are the dielectric permeability, the magnetic permeability the electrical conductivity of a medium, respectively. The paths of integration L, can be arbitrarily situated, and in some cases they can cross the boundaries between media having different properties. Equations 1.246 through 1.249 are called the Maxwell's equations in integral form, and each one of them describes a specific physical law. For this reason, any distribution of an electromagnetic field must satisfy these equations. They define the field at any point in the medium, including points situated on interfaces. Maxwell's equations describe the field everywhere regardless of the nature of the change in electrical properties from one region to another.

The first equation (eq. 1.246) is in essence Faraday's law, while the second equation (eq. 1.247) follows from a combination of Ampere's law and the postulate of conservation of charge. The third equation (eq. 1.248) is obtained from Coulomb's law for a nonalternating electric field. However, it remains valid regardless of how quickly the field may change. In order to demonstrate this we will use the postulate of conservation of charge (eqs. 1.131–1.132):

$$\oint_S \boldsymbol{j} \cdot \mathrm{d}\boldsymbol{S} = -\dot{e} \quad \text{or} \quad \operatorname{div} \boldsymbol{j} = -\frac{\partial \delta}{\partial t} \tag{1.250}$$

Applying the second Maxwell equation (1.247), twice along contour L, once in one direction, and then in the opposite direction, and considering two surfaces (S_1 and S_2), bounded by the same contour L (Fig. 1.51), we have:

$$\oint_L \boldsymbol{H} \cdot \mathrm{d}\boldsymbol{l} = \int_{S_1} \boldsymbol{j} \cdot \mathrm{d}\boldsymbol{S} + \int_{S_1} \dot{\boldsymbol{D}} \cdot \mathrm{d}\boldsymbol{S}$$

$$\oint_L \boldsymbol{H} \cdot \mathrm{d}\boldsymbol{l} = \int_{S_2} \boldsymbol{j} \cdot \mathrm{d}\boldsymbol{S} + \int_{S_2} \dot{\boldsymbol{D}} \cdot \mathrm{d}\boldsymbol{S}$$

Adding the two equations and considering that the surfaces S_1 and S_2 form a closed surface, we obtain:

$$0 = \oint_S \boldsymbol{j} \cdot \mathrm{d}\boldsymbol{S} + \oint_S \frac{\partial \boldsymbol{D}}{\partial t} \cdot \mathrm{d}\boldsymbol{S}$$

and in accord with eq. 1.250:

$$\oint_S \frac{\partial \boldsymbol{D}}{\partial t} \cdot \mathrm{d}\boldsymbol{S} = \dot{e}$$

whence

$$\oint_S \boldsymbol{D} \cdot \mathrm{d}\boldsymbol{S} = e$$

By analogy, using the first Maxwell equation, we also have:

$$\oint_S \boldsymbol{B} \cdot \mathrm{d}\boldsymbol{S} = 0$$

The fourth Maxwell equation (eq. 1.249) represents the fact that the magnetic flux through an enclosed surface is zero. This consideration demonstrates that the field equations can also be written as other sets of equations:

$$\oint_L \boldsymbol{E} \cdot \mathrm{d}\boldsymbol{l} = -\frac{\partial}{\partial t} \int_S \boldsymbol{B} \cdot \mathrm{d}\boldsymbol{S}$$

$$\oint_L \boldsymbol{H} \cdot \mathrm{d}\boldsymbol{l} = \int_S \boldsymbol{j} \cdot \mathrm{d}\boldsymbol{S} + \frac{\partial}{\partial t} \int_S \boldsymbol{D} \cdot \mathrm{d}\boldsymbol{S} \tag{1.251}$$

$$\oint_S \boldsymbol{j} \cdot \mathrm{d}\boldsymbol{S} = -\frac{\partial e}{\partial t}$$

inasmuch as eq. 1.248 and 1.249 can be derived from the system given in eq. 1.251. However, we will use the basic system of equations (eqs. 1.246–1.249). It must be obvious that in any actual situation, the electromagnetic field has a finite value everywhere in space. However, in order to simplify the computation of fields, often some assumptions are made about the sources for the primary field. For example, in place of an actual source magnetic or electric dipoles may be considered. This type of approximation immediately leads to the existence of infinitely large values for the field at infinitesimal distances from such sources. Therefore, eqs. 1.246–1.249 cannot be applied in the immediate vicinity of such idealized sources. For this reason, some very small volume in which the source is situated is conceptionally surrounded by a surface on which the field almost coincides with that caused by the currents and charges of such a primary source. In other words, near the source the total field has to approach the primary field. One can say that this condition characterizes the type, intensity and location of a primary field source.

On the other hand, with an unlimited increase in distance from the source the field must decrease in a proper way. This condition at infinity must be taken into account in the full description of a field. Finally, there is one more condition which appears when a transient field is being considered. For example if the current or charges representing the source or the primary field change in the form of a step function at some moment $t = t_0$, eqs. 1.246 and 1.247 cannot be applied, since the derivatives with respect to time are not well defined at this moment. Therefore, at this instant, Maxwell's equations are replaced by an initial condition as described in section 1.4.

Thus, a full description of the electromagnetic field includes not only Maxwell's equations, as given by eq. 1.246–1.249, but also conditions that must be met near the primary source and at infinity, along with an initial condition. Thus, the following series of steps can be recognized in defining an electromagnetic field, making use of eqs. 1.246–1.249:

1. Determination of a set of functions, satisfying the system of integral equations.

2. Choice among these functions of those which satisfy the condition at infinity.

3. Choice among the remaining functions of those which satisfy the condition near the source.

4. Choice among the remaining functions of those which satisfy the initial condition, if a transient field is being considered.

From the physical point of view, it is apparent that a solution found in this way represents the electromagnetic field generated by the given distribution of sources. However, for the solution of a variety of problems, it is frequently preferable to apply differential equations. For this reason, we will consider a differential form of Maxwell's equations:

$$\begin{aligned} \operatorname{curl} \boldsymbol{E} &= -\frac{\partial \boldsymbol{B}}{\partial t} \qquad & \operatorname{div} \boldsymbol{D} &= \delta \\ \operatorname{curl} \boldsymbol{H} &= \sigma \boldsymbol{E} + \frac{\partial \boldsymbol{D}}{\partial t} \qquad & \operatorname{div} \boldsymbol{B} &= 0 \end{aligned} \tag{1.252}$$

here δ is a free charge.

In contrast to the integral form, given in eq. 1.251, all the vectors that enter into each of the equations in 1.252 are considered at a single point. The essential feature of Maxwell's equations, written in the differential form, is that they describe the field only at points where the first derivatives of the field exist, that is, where the divergence and curl have meaning. Thus, unlike Maxwell's equations in the integral form, the set of equations given in eqs. 1.252 can be applied for so-called well behaved points.

However, as is obvious, there can be points, lines, and surfaces where some components of the electromagnetic field are discontinuous functions of the spatial variables. For example, the normal component of the electric field is usually a discontinuous function of the spatial variables at an interface separating two media with different resistivity. As a consequence, we must make use of surface analogies of eqs. 1.252 at such interfaces. According to results obtained in previous sections these may be represented as a continuity of tangential components of electric and magnetic fields, that is:

$$\boldsymbol{n} \times (\boldsymbol{E}_2 - \boldsymbol{E}_1) = 0 \qquad \boldsymbol{n} \times (\boldsymbol{H}_2 - \boldsymbol{H}_1) = 0 \tag{1.253}$$

where $\boldsymbol{n}$ is a unit vector of the interface, $\boldsymbol{E}_2$ and $\boldsymbol{E}_1$, and $\boldsymbol{H}_2$ and $\boldsymbol{H}_1$ are electric and magnetic fields on either side of such an interface.

Thus, in essence, eqs. 1.253 are surface analogs to the corresponding Maxwell's equations given in differential form in eq. 1.252. Therefore, starting from the system of differential equations (eq. 1.252) the problem of defining the field consists of the following steps:

1. Determination of a set of functions satisfying the differential equations in 1.252.
2. The choice among these functions of those satisfying the condition at infinity.
3. The choice among the remaining functions of those having the given behavior near the source of the primary field.
4. A choice among still remaining functions of those that satisfy the boundary conditions given in eq. 1.253.
5. A choice among the remaining functions, $\boldsymbol{E}$ and $\boldsymbol{H}$, of those that satisfy the initial conditions, if a nonstationary field is being considered.

Taking into account, that:

$$\boldsymbol{D} = \varepsilon \boldsymbol{E} \qquad \boldsymbol{B} = \mu \boldsymbol{H} \qquad \boldsymbol{j} = \sigma \boldsymbol{E}$$

the system of equations in 1.252 usually contains two unknowns, namely the electric and magnetic field intensities. One can say that we have four differential equations in partial derivatives of the first order with respect to two unknown vectors, but more accurately, to six unknown components of the electromagnetic field.

Very frequently it is more convenient to derive equations in which the electric and magnetic fields are separated, than to make use of the set of equations in 1.252. Let us consider points in the medium where the parameters σ, μ and ε do not change:

$$\frac{\partial \sigma}{\partial l} = \frac{\partial \varepsilon}{\partial l} = \frac{\partial \mu}{\partial l} = 0$$

and where dl is an arbitrarily oriented displacement. As has been shown previously (section 1.3) electric charges are absent at such points, and therefore, Maxwell's equations take the form:

$$\begin{aligned} \operatorname{curl} \boldsymbol{E} &= -\mu \frac{\partial \boldsymbol{H}}{\partial t} \qquad & \operatorname{div} \boldsymbol{E} = 0 \\ \operatorname{curl} \boldsymbol{H} &= \sigma \boldsymbol{E} + \varepsilon \frac{\partial \boldsymbol{E}}{\partial t} \qquad & \operatorname{div} \boldsymbol{H} = 0 \end{aligned} \tag{1.254}$$

From the first Maxwell equation, we have:

$$\operatorname{curl}\operatorname{curl} \boldsymbol{E} = -\mu \frac{\partial}{\partial t} \operatorname{curl} \boldsymbol{H}$$

In making use of the vector identity:

$$\operatorname{curl}\operatorname{curl} \boldsymbol{E} = \operatorname{grad}\operatorname{div} \boldsymbol{E} - \nabla^2 \boldsymbol{E}$$

and of the second Maxwell equation we obtain:

$$\operatorname{grad}\operatorname{div} \boldsymbol{E} = -\nabla^2 \boldsymbol{E} = -\mu \frac{\partial}{\partial t} \left(\sigma \boldsymbol{E} + \varepsilon \frac{\partial \boldsymbol{E}}{\partial t} \right)$$

Taking into account the third Maxwell equation $\operatorname{div} \boldsymbol{E} = 0$, we have:

$$\nabla^2 \boldsymbol{E} - \mu\sigma \frac{\partial \boldsymbol{E}}{\partial t} - \mu\varepsilon \frac{\partial^2 \boldsymbol{E}}{\partial t^2} = 0 \tag{1.255}$$

where $\nabla^2 \boldsymbol{E} = \Delta \boldsymbol{E}$ is known as the Laplacian of the electric field.

Similarly, using the second equation in 1.364, we have:

$$\operatorname{curl}\operatorname{curl} \boldsymbol{H} = \operatorname{grad}\operatorname{div} \boldsymbol{H} - \nabla^2 \boldsymbol{H} = \sigma \operatorname{curl} \boldsymbol{E} + \varepsilon \frac{\partial}{\partial t} \operatorname{curl} \boldsymbol{E}$$

In making use of the first and fourth Maxwell equations, we obtain:

$$\nabla^2 \boldsymbol{H} - \sigma\mu \frac{\partial \boldsymbol{H}}{\partial t} - \mu\varepsilon \frac{\partial^2 \boldsymbol{H}}{\partial t^2} = 0 \tag{1.256}$$

Thus, for points in the medium, where the electric and magnetic properties do not vary spatially we have obtained equations involving only the electric or magnetic fields. The two equations are of identically the same form, being the second order in partial derivatives. They are sometimes known as telegraph equations for a conductive medium.

When these equations are used, the determination of the electromagnetic field can be done in almost the same sequence of steps as before:

1. Definition of various functions that satisfy eqs. 1.255 and 1.256.

2. The choice among these functions of those that satisfy the conditions at infinity.
3. The choice among those whose behavior near the source corresponds that for the primary field.
4. The choice among the remaining functions of those that satisfy surface conditions given in eq. 1.255.
5. The choice among the still remaining functions of those that satisfy initial condition if a nonstationary field is being considered.

Now let us consider some special cases.

Case 1

First of all, assume that an electromagnetic field does not change with time, that is, all the derivatives with respect to time are zero, and that:

$\boldsymbol{D} = \varepsilon_0 \boldsymbol{E}$ and $\boldsymbol{B} = \mu_0 \boldsymbol{H}$

Then with accord with eqs. 1.252 and 1.253 we have the following equations for well behaved points and for interfaces:

$$\begin{aligned} &\operatorname{curl} \boldsymbol{E} = 0 && \operatorname{curl} \boldsymbol{H} = \boldsymbol{j} \\ &\operatorname{div} \boldsymbol{E} = \delta/\varepsilon_0 && \operatorname{div} \boldsymbol{H} = 0 \end{aligned} \tag{1.257}$$

and

$$\begin{aligned} &\boldsymbol{n} \times (\boldsymbol{E}_2 - \boldsymbol{E}_1) = 0 && \boldsymbol{n} \times (\boldsymbol{H}_2 - \boldsymbol{H}_1) = 0 \\ &E_n^{(2)} - E_n^{(1)} = \Sigma/\varepsilon_0 && H_n^{(2)} - H_n^{(1)} = 0 \end{aligned} \tag{1.258}$$

where Σ is the surface density of charge. In this case of a constant field ($\partial/\partial t = 0$), the system is split into two parts as follows:

$$\operatorname{curl} \boldsymbol{E} = 0 \quad \operatorname{div} \boldsymbol{E} = \delta/\varepsilon_0 \tag{\#1}$$

and at interfaces:

$$\boldsymbol{n} \times (\boldsymbol{E}_2 - \boldsymbol{E}_1) = 0 \qquad E_n^{(2)} - E_n^{(1)} = \Sigma/\varepsilon_0$$

and

$$\operatorname{curl} \boldsymbol{H} = \boldsymbol{j} \quad \operatorname{div} \boldsymbol{H} = 0 \tag{\#2}$$

and at interfaces:

$$\boldsymbol{n} \times (\boldsymbol{H}_2 - \boldsymbol{H}_1) = 0 \quad H_n^{(2)} - H_n^{(1)} = 0$$

One part (#1) defines the electric field and clearly shows that the sole source of the field is electric charge, which can exist at points where the conductivity changes, such as at interfaces. It is clear that the electric field can be found without any knowledge of the magnetic field, and that the electric field is governed only by Coulomb's law. When the electric field has been determined, current density can be calculated using Ohm's law, $\boldsymbol{J} = \sigma \boldsymbol{E}$, and by making use of the second part (#2), the magnetic field can be found as well. It can also be calculated using the Biot–Savart law.

At this point we will consider a very important case, that of a quasistationary field, which is often also called a quasistatic field.

Case 2

Suppose that in the second equation describing the field (eq. 1.252) we can ignore the second term, which represents displacement current. Then, system 1.252 can be written as:

$$\begin{aligned} &\operatorname{curl} \boldsymbol{E} = -\mu \frac{\partial \boldsymbol{H}}{\partial t} \qquad \operatorname{curl} \boldsymbol{H} = \boldsymbol{j} \\ &\operatorname{div} \boldsymbol{E} = \delta/\varepsilon_0 \qquad \operatorname{div} \boldsymbol{H} = 0 \end{aligned} \tag{1.259}$$

and at interfaces as:

$$\begin{aligned} &\boldsymbol{n} \times (\boldsymbol{E}_2 - \boldsymbol{E}_1) = 0 \qquad \boldsymbol{n} \times (\boldsymbol{H}_2 - \boldsymbol{H}_1) = 0 \\ &E_n^{(2)} - E_n^{(1)} = \Sigma/\varepsilon_0 \qquad H_n^{(2)} - H_n^{(1)} = 0 \end{aligned} \tag{1.260}$$

(the surface density of current $i = 0$).

From these expressions it can be seen that the electric field has two sources: the first being volume and surface charges and the second being a change of the magnetic field with time. Therefore, the electric field can be represented as a sum:

$$\boldsymbol{E} = \boldsymbol{E}^c + \boldsymbol{E}^v$$

where $\boldsymbol{E}^c$ and $\boldsymbol{E}^v$ are caused by charges and a change of magnetic field with time, respectively. At the same time it is important to emphasize that there is a relation between these fields, since distribution of induced currents and charges depend on each other.

In contrast to the behavior of the electric field, the quasistationary magnetic field has but one source, conduction currents. Comparing the equations for the magnetic field (eq. 1.259) with those for a constant magnetic field (eq. 1.257) we see that they are precisely the same. This means that the magnetic field at any point in the medium is defined by the instantaneous values of current density throughout the conduction medium, and can be calculated using only the Biot–Savart law. One can say that the quasistationary approximation means that we neglect propagation time for electromagnetic energy, that is, it is assumed that the field travels instantaneously from the transmitter to the receiver.

In order to emphasize this, let us write down the field equations 1.255 and 1.256 for this approximation. Discarding terms involving displacement currents, we have:

$$\Delta \boldsymbol{E} = \sigma\mu\frac{\partial \boldsymbol{E}}{\partial t} \qquad \Delta \boldsymbol{H} = \sigma\mu\frac{\partial \boldsymbol{H}}{\partial t} \tag{1.261}$$

These equations are known as the diffusion equations, that is, they describe the penetration of energy, but do not take into account wave propagation. They can be used provided that the time at which the signal is recorded or the period of oscillations significantly exceeds the travel time for the field from the source to the observation point. It can be said that the quasistationary approximation is valid when conduction currents dominate over displacement currents in a conducting medium, and the arrival time for a signal in an insulator is much less than the time at which measurements are performed, or the period of the oscillations being observed. This assumption is equivalent to stating that the signal arrives instantly at all points where the field is to be measured.

It might be worth noting that even though the propagation effect is not considered in the quasistationary field approach this field contains some essential features of propagation.

Case 3

At this point we will examine a special case in which the electromagnetic field varies as a sinusoidal or cosinusoidal function of time. This leads to some important simplifications in the presentation of Maxwell's equations (through use of the so-called operator notation). Suppose that we have a sinusoidal oscillation:

$$M = M_0 \sin(\omega t + \phi) \tag{1.262}$$

where M_0 is the amplitude of the oscillation, ϕ is its phase, and ω is the angular frequency. Making use of Euler's formula:

$$\mathrm{e}^{\mathrm{i}(\omega t+\phi)} = \cos(\omega t + \phi) + \mathrm{i}\sin(\omega t + \phi)$$

we can write eq. 1.262 as the imaginary part of an exponential term:

$$M_0 \sin(\omega t + \phi) = \mathrm{Im}\, \hat{M}\mathrm{e}^{\mathrm{i}\omega t} \tag{1.263}$$

where $\hat{M} = M_0\,\mathrm{e}^{\mathrm{i}\phi}$. Therefore, we have:

$$\hat{M}\,\mathrm{e}^{\mathrm{i}\omega t} = M_0\,\mathrm{e}^{\mathrm{i}\phi}\,\mathrm{e}^{\mathrm{i}\omega t} = M_0\,\mathrm{e}^{\mathrm{i}(\omega t+\phi)}$$

Whence

$$\begin{aligned}\mathrm{Im}(\hat{M}\,\mathrm{e}^{\mathrm{i}\omega t}) &= \mathrm{Im}\left[M_0\,\mathrm{e}^{\mathrm{i}(\omega t+\phi)}\right] \\ &= \mathrm{Im}\left[M_0\left\{\cos(\omega t+\phi) + \mathrm{i}\sin(\omega t+\phi)\right\}\right] = M_0 \sin(\omega t + \phi)\end{aligned}$$

Similarly, a cosinusoidal oscillation can be represented by the real part of a complex function:

$$M_0 \cos(\omega t + \phi) = \operatorname{Re} \hat{M}\, \mathrm{e}^{\mathrm{i}\omega t} \tag{1.264}$$

where $\hat{M}$ is equal again to $M_0\, \mathrm{e}^{\mathrm{i}\phi}$.

Let us emphasize that the complex amplitude, $\hat{M}$, is defined by the real amplitude, M_0, and the phase, ϕ, of an oscillation.

Inasmuch as:

$$\hat{M}\, \mathrm{e}^{\mathrm{i}\omega t} = M_0 \cos(\omega t + \phi) + \mathrm{i}\, M_0 \sin(\omega t + \phi)$$

and since both terms on the right-hand side of this equality are solutions of Maxwell's equations, one can operate using only the function $\hat{M}\, \mathrm{e}^{\mathrm{i}\omega t}$. After finding a function that satisfies this system of equations one must then take either the imaginary or real part. The representation of a solution by the form $\hat{M}\, \mathrm{e}^{\mathrm{i}\omega t}$ has remarkable feature. In fact, it is actually the product of two functions, one being the complex amplitude, $\hat{M}$, which is a function of coordinates and the properties of the medium as well as of frequency, but which does not depend on time. The second multiplier, $\mathrm{e}^{\mathrm{i}\omega t}$, depends on time in a simple manner and, as is readily seen, after differentiation still remains an exponential. This fact permits us to write Maxwell's equations in a form which does not contain the argument t, and this essentially facilitates the solution. It is appropriate to note that the sinusoidal function, which is being considered, has been in effect for such a long time that there is no need to take into account an initial condition.

Thus, representing a field and charges in the form:

$$\boldsymbol{H} = \hat{\boldsymbol{H}}\, \mathrm{e}^{\mathrm{i}\omega t} \qquad \boldsymbol{E} = \hat{\boldsymbol{E}}\, \mathrm{e}^{\mathrm{i}\omega t} \qquad \delta = \hat{\delta}\, \mathrm{e}^{\mathrm{i}\omega t} \tag{1.265}$$

and substituting them in Maxwell's equations (eq. 1.252) we obtain:

$$\begin{aligned} \operatorname{curl} \hat{\boldsymbol{E}} &= -\mathrm{i}\omega\mu\boldsymbol{H} & \operatorname{div} \hat{\boldsymbol{E}} &= \delta \\ \operatorname{curl} \hat{\boldsymbol{H}} &= \sigma\hat{\boldsymbol{E}} + \mathrm{i}\omega\varepsilon\hat{\boldsymbol{E}} & \operatorname{div} \hat{\boldsymbol{H}} &= 0 \end{aligned} \tag{1.266}$$

inasmuch as:

$$\frac{\partial}{\partial t} \mathrm{e}^{\mathrm{i}\omega t} = \mathrm{i}\omega\, \mathrm{e}^{\mathrm{i}\omega t}$$

Similarly, we have the following for the eqs. 1.255 and 1.256:

$$\begin{aligned} \nabla^2 \hat{\boldsymbol{H}} - (\mathrm{i}\sigma\mu\omega - \omega^2\varepsilon\mu)\hat{\boldsymbol{H}} = 0 \\ \nabla^2 \hat{\boldsymbol{E}} - (\mathrm{i}\sigma\mu\omega - \omega^2\varepsilon\mu)\hat{\boldsymbol{E}} = 0 \end{aligned} \tag{1.267}$$

The quantity:

$$k^2 = \mathrm{i}\sigma\mu\omega - \omega^2\varepsilon\mu \tag{1.268}$$

is usually considered to be the square of a wave number, k. For quasistationary behavior of the field in which displacement currents are neglected, the wave number can be written as:

$$k = (\mathrm{i}\sigma\mu\omega)^{1/2} \tag{1.269}$$

Alternate forms are:

$$k = (\mathrm{i}\sigma\mu\omega)^{1/2} = \left(\frac{\sigma\mu\omega}{2}\right)^{1/2}(1+\mathrm{i}) = \frac{(1+\mathrm{i})}{h} \tag{1.270}$$

where h is the skin depth, defined as:

$$k = \left(\frac{2}{\sigma\mu\omega}\right)^{1/2} = \frac{10^3}{2\pi}\left(\frac{10\rho}{f}\right)^{1/2} \tag{1.271}$$

Here ρ is the resistivity in ohm·meters, f is the frequency in Hertz, and μ is the magnetic permeability, normally taken to be the value for free space, which is $4\pi \times 10^{-7}$ H/m.

Maxwell's equations can be written as follows for a harmonic quasistationary field behavior:

$$\begin{aligned} &\operatorname{curl} \boldsymbol{E} = -\mathrm{i}\omega\mu\boldsymbol{H} \qquad &&\operatorname{div} \boldsymbol{E} = \delta/\varepsilon_0 \\ &\operatorname{curl} \boldsymbol{H} = \sigma\boldsymbol{E} &&\operatorname{div} \boldsymbol{B} = 0 \end{aligned} \tag{1.272}$$

The *hat* notation previously used to indicate a complex amplitude has been omitted for simplicity.

By algebraic recombination of these four equations we have the Helmholtz equations:

$$\nabla^2\boldsymbol{H} - \mathrm{i}\sigma\mu\omega\boldsymbol{H} = 0 \qquad \nabla^2\boldsymbol{E} - \mathrm{i}\sigma\mu\omega\boldsymbol{E} = 0 \tag{1.273}$$

The system of equations in 1.272 is particularly simplified in the case in which a medium consists of parts within which conductivity is constant, that is, a piecewise uniform medium. In this case, electric charges can arise only at interfaces and within the uniform pieces the volume density of charge is zero. Therefore, in place of eqs. 1.272 within each volume we have:

$$\begin{aligned} &\operatorname{curl} \boldsymbol{E} = -\mathrm{i}\omega\mu\boldsymbol{H} \qquad &&\operatorname{div} \boldsymbol{E} = 0 \\ &\operatorname{curl} \boldsymbol{H} = \sigma\boldsymbol{E} &&\operatorname{div} \boldsymbol{H} = 0 \end{aligned}$$

The piecewise uniform medium is the most widely used model for a geoelectric section and it is appropriate here to formulate again the steps to use in determining the quasistationary harmonic field for this type of model.

The steps are:

1. Determination of solution functions that satisfy the systems of equations:

$$\begin{aligned} &\operatorname{curl} \boldsymbol{E} = -\mathrm{i}\omega\mu\boldsymbol{H} && \operatorname{div} \boldsymbol{E} = 0 \\ &\operatorname{curl} \boldsymbol{H} = \sigma\boldsymbol{E} && \operatorname{div} \boldsymbol{B} = 0 \end{aligned} \tag{1.274}$$

   or

$$\nabla^2\boldsymbol{H} - k^2\boldsymbol{H} = 0 \qquad \nabla^2\boldsymbol{E} - k^2\boldsymbol{E} = 0$$

   where $k^2 = \mathrm{i}\sigma\mu\omega$.

2. A choice among those functions of those which satisfy the source condition for the primary field.

3. The choice among the remaining functions of those which satisfy the condition at infinity.

4. The selection among the still remaining functions of those satisfying boundary conditions at each interface, that is, the continuity of tangential components of the electromagnetic field.

As has been pointed out previously, inasmuch as the field under investigation must be stationary, there is no initial condition to be met, and because of this the solution is made such simpler. It might also be noted that a solution of eq. 1.273 will be in the form of a set of complex amplitudes for the electric and magnetic fields. In accord with eq. 1.264 we obtain the amplitude M_0 and phase of an oscillation ϕ_0 in the basic form (eq. 1.262).

It should be apparent that when a solution has been obtained for harmonic fields a solution can also be derived for any arbitrary time dependence through the use of the Fourier transform. Most frequently the electric and magnetic field vectors cannot be completely described by using only a single spatial component. For this reason a solution can turn out to be very cumbersome. Some simplification can be obtained by making use of various auxiliary functions. There are two ways in which such auxiliary functions can be introduced. One approach follows from use of the third equation in the set 1.274:

$$\operatorname{div} \boldsymbol{E} = 0 \tag{1.275}$$

This is an approach which is commonly used when the field is energized using a non-grounded loop, as for example, a magnetic dipole. In this case only inductive excitation of the field takes place. In accord with eq. 1.275, the electric field can be defined as being a spatial derivation of the vector potential, $\boldsymbol{A}^*$:

$$\boldsymbol{E} = \operatorname{curl} \boldsymbol{A}^* \tag{1.276}$$

because the relationship:

$$\operatorname{div} \operatorname{curl} \boldsymbol{A}^* = 0$$

always applies for any vector.

The function $\boldsymbol{A}^*$ is called a vector potential of the magnetic type. It should be obvious that the same electric field can be described by an infinite number of different functions $\boldsymbol{A}^*$. For example, the gradient of any function can be added to some fixed potential $\boldsymbol{A}^*$ to provide the result:

$$\operatorname{curl}(\boldsymbol{A}^* + \operatorname{grad}\phi) = \operatorname{curl}\boldsymbol{A}^* + \operatorname{curl}\operatorname{grad}\phi$$

Taking into account another vector identity:

$$\operatorname{curl}\operatorname{grad}\phi = 0$$

we have:

$$\operatorname{curl}(\boldsymbol{A}^* + \operatorname{grad}\phi) = \operatorname{curl}\boldsymbol{A}^* = \boldsymbol{E}$$

This ambiguity in definition of $\boldsymbol{A}^*$ can be used to our advantage in simplifying equations when the vector potential is used, as well as to express both vectors of the field in terms of this single function.

To obtain a solution we substitute eq. 1.276 into the second equation of 1.274 so that we have:

$$\operatorname{curl}\boldsymbol{H} = \sigma\operatorname{curl}\boldsymbol{A}^* = \operatorname{curl}\sigma\boldsymbol{A}^*$$

since σ is considered to be a constant. This can all be written as:

$$\operatorname{curl}(\boldsymbol{H} - \sigma\boldsymbol{A}^*) = 0$$

whence:

$$\boldsymbol{H} - \sigma\boldsymbol{A}^* = \operatorname{grad}\phi \tag{1.277}$$

where ϕ is some scalar function. Just as is the case with the vector potential this function is ambiguous.

Substituting expressions for vector quantities $\boldsymbol{E}$ and $\boldsymbol{H}$ in terms of functions $\boldsymbol{A}^*$ and ϕ in the first equation of 1.274 we obtain:

$$\operatorname{curl}\operatorname{curl}\boldsymbol{A}^* = -\mathrm{i}\omega\mu(\sigma\boldsymbol{A}^* + \operatorname{grad}\phi)$$

Inasmuch as:

$$\operatorname{curl}\operatorname{curl}\boldsymbol{A}^* = \operatorname{grad}\operatorname{div}\boldsymbol{A}^* - \nabla^2\boldsymbol{A}^*$$

where ∇^2 is Laplacian, we have:

$$\operatorname{grad}\operatorname{div}\boldsymbol{A}^* - \nabla^2\boldsymbol{A}^* = -\mathrm{i}\sigma\mu\omega\boldsymbol{A}^* - \mathrm{i}\omega\mu\operatorname{grad}\phi \tag{1.278}$$

The expressions which have been obtained for the functions $\boldsymbol{A}^*$ and ϕ are quite complicated. In order to simplify this last equation, we now choose a pair of function $\boldsymbol{A}^*$ and ϕ, that satisfy the condition:

$$\operatorname{div}\boldsymbol{A}^* = -\mathrm{i}\omega\mu\phi \tag{1.279}$$

By using a gauge condition the differential equation becomes a Helmholtz equation for the vector potential $\boldsymbol{A}^*$:

$$\nabla^2\boldsymbol{A}^* - k^2\boldsymbol{A}^* = 0 \tag{1.280}$$

This is precisely the same equation for either the electric or the magnetic field. Making use of the condition 1.279 both vectors, comprising an electromagnetic field, are expressed in terms of a single vector potential quantity, $\boldsymbol{A}^*$. In accord with eqs. 1.276, 1.277 and 1.279 we have the following representations for the two vector quantities:

$$\begin{aligned}\boldsymbol{E} &= \operatorname{curl}\boldsymbol{A}^* \\ \boldsymbol{H} &= \sigma\boldsymbol{A}^* - \frac{1}{\mathrm{i}\omega\mu}\operatorname{grad}\operatorname{div}\boldsymbol{A}^*\end{aligned} \tag{1.281}$$

The behavior of the vector potential at interfaces follows from the required continuity for tangential components on the electric and magnetic fields at those boundaries. It is not particularly difficult to formulate conditions near the source of the primary field nor at infinity.

Let us examine another way for introducing the vector potential. From the fourth equation in system 1.274 we can represent the magnetic field as being:

$$\boldsymbol{H} = \operatorname{curl}\boldsymbol{A} \tag{1.282}$$

The function $\boldsymbol{A}$ is called a vector potential of the electrical type and this definition is normally used when the electromagnetic field is energized through the use of a grounded wire. Substituting this definition for the vector potential into the first equation of set 1.274 we have:

$$\operatorname{curl}\boldsymbol{E} = -\mathrm{i}\omega\mu\operatorname{curl}\boldsymbol{A}$$

or

$$\operatorname{curl}(\boldsymbol{E} + \mathrm{i}\omega\mu\boldsymbol{A}) = 0$$

whence:

$$\boldsymbol{E} + \mathrm{i}\omega\mu\boldsymbol{A} = \operatorname{grad} U$$

or

$$\boldsymbol{E} = -\mathrm{i}\omega\mu\boldsymbol{A} + \operatorname{grad} U \tag{1.283}$$

Substituting eqs. 1.282 and 1.283 into the second equation of the set 1.274, we have:

$$\operatorname{curl}\operatorname{curl}\boldsymbol{A} = -\mathrm{i}\sigma\mu\omega\boldsymbol{A} + \sigma\operatorname{grad} U$$

or

$$\operatorname{grad}\operatorname{div}\boldsymbol{A} - \nabla^2\boldsymbol{A} = -\mathrm{i}\sigma\mu\omega\boldsymbol{A} + \sigma\operatorname{grad} U$$

Considering that there are an infinite number of functions $\boldsymbol{A}$ and U, that will satisfy eqs. 1.282 and 1.283, we will seek a pair of them that simplifies the last equation. One such choice is:

$$U = \frac{1}{\sigma}\operatorname{div}\boldsymbol{A} \tag{1.284}$$

With this gauge condition we again obtain a Helmholtz equation for the electric vector potential:

$$\nabla^2\boldsymbol{A} - k^2\boldsymbol{A} = 0 \tag{1.285}$$

and both vectors, comprising the electromagnetic field, are expressed in terms of a single vector potential functions, $\boldsymbol{A}$:

$$\begin{aligned}\boldsymbol{H} &= \operatorname{curl}\boldsymbol{A}\\ \boldsymbol{E} &= -\mathrm{i}\omega\mu\boldsymbol{A} + \frac{1}{\sigma}\operatorname{grad}\operatorname{div}\boldsymbol{A}\end{aligned} \tag{1.286}$$

As in the previous case the behavior of this vector potential near the source and at infinity, as well as at interfaces, follows from the corresponding behavior of the electric and magnetic fields under these conditions.

In conclusion, let us review some of the results which are contained here. If one of the vector potentials is found, then the electric and magnetic fields can be determined by taking corresponding derivatives in accord with either eq. 1.281 or eq. 1.286. When an electromagnetic field is caused by both induced currents and charges it is necessary in most cases to make use of both vector potentials to determine the field, but there are some important exceptions. For example, the most important features of the theory of induction logging can be derived, making use of the vector potential of the magnetic type, $\boldsymbol{A}^*$. Just as is the case with harmonic fields solutions for vector potentials can be extended to the case in which the functions depend arbitrarily on time.

1.6. Relationships between Various Responses of the Electromagnetic Field

In this section we will explore some general relationships between the various responses of an electromagnetic field. First of all we will start from a relationship between the quadrature and inphase components of the field. For example, representing the complex amplitude of the electric field as being the sum of two components:

$$\boldsymbol{E} = \mathrm{In}\,\boldsymbol{E} + \mathrm{i}\,\mathrm{Q}\,\boldsymbol{E}$$

and substituting these into Helmholtz equation (1.273) we have:

$$\nabla^2(\mathrm{In}\,\boldsymbol{E} + \mathrm{i}\,\mathrm{Q}\,\boldsymbol{E}) - \mathrm{i}\sigma\mu\omega(\mathrm{In}\,\boldsymbol{E} + \mathrm{i}\,\mathrm{Q}\,\boldsymbol{E}) = 0$$

or

$$\begin{aligned} \nabla^2\,\mathrm{In}\,\boldsymbol{E} &= \mathrm{i}\sigma\mu\omega\;\mathrm{Q}\,\boldsymbol{E} \\ \nabla^2\,\mathrm{Q}\,\boldsymbol{E} &= \sigma\mu\omega\;\mathrm{In}\,\boldsymbol{E} \end{aligned} \tag{1.287}$$

Thus, there is a relationship between the inphase and quadrature components of the spectrum. Let us examine this in more detail. We will make use of a solution in the form $M\,\mathrm{e}^{\mathrm{i}\omega t}$, where M is a complex amplitude of the spectrum. In obtaining an actual sinusoidal solution one should take the imaginary part of this expression:

$$M_0 \sin(\omega t + \phi) = \mathrm{Q}\,M\,\mathrm{e}^{\mathrm{i}\omega t}$$

If the solution contains the complex amplitude term from the physical point of view this means that there is a phase shift and thus the field can be represented as being the sum of the quadrature (Q) and the inphase (In) components. We will have:

$$M = \mathrm{In}\,M + \mathrm{i}\;\mathrm{Q}\,M = M_0\cos\phi + \mathrm{i}\,M_0\sin\phi \tag{1.288}$$

where M_0 and ϕ are the amplitude and phase of an oscillation, respectively.

Using the conventional symbols for representing a complex variable we can write M as:

$$M(z) = U + \mathrm{i}\,V \tag{1.289}$$

where U and V are the real and imaginary parts of the function $M(z)$ and z is an argument defined as:

$$z = x + \mathrm{i}\,y$$

where x and y are coordinates on the complex plane z. Usually the complex amplitude M of an electromagnetic field is an analytic function of frequency, ω. If this is the case,

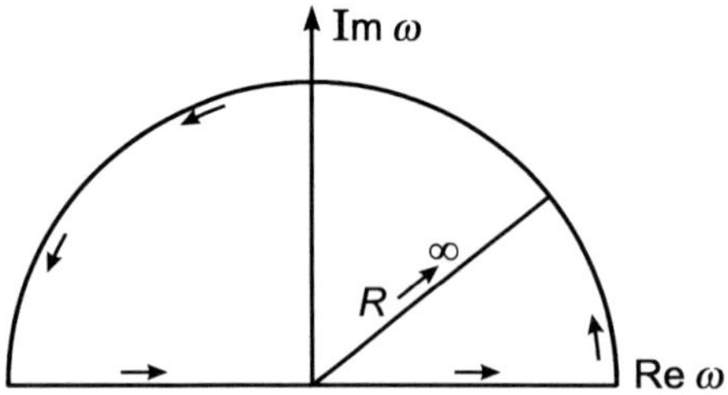

Figure 1.52. Path of integration consisting of a semi-circle with an infinitely large radius and centered on the x-axis.

a necessary and sufficient condition for analyticity of a function M the Cauchy–Reimann is:

$$\frac{\partial U}{\partial x} = \frac{\partial V}{\partial y} \qquad \frac{\partial U}{\partial y} = -\frac{\partial V}{\partial x} \tag{1.290}$$

The Cauchy–Reimann conditions express the relationship that exists between the real and imaginary parts of an analytic function in the complex plane in differential form. In our case the complex variable z is the frequency:

$$\omega = \mathrm{Re}\,\omega + \mathrm{i}\,\mathrm{Im}\,\omega$$

and we will seek a relationship between the quadrature and the inphase components of the field for real values of ω, because the electromagnetic field is observed only at real frequencies. For this purpose let us use the Cauchy formula which shows that if the function $M(z)$ is analytic within a contour C, as well as along this contour, and a is any point in the z-plane, then:

$$\oint_C \frac{M(z)}{z-a}\mathrm{d}z = 2\pi\mathrm{i}\,M(a)\begin{cases} 1 & \text{if } a \in C \\ 1/2 & \text{if } a \text{ is on } C \\ 0 & \text{if } a \notin C \end{cases} \tag{1.291}$$

The Cauchy formula permits us to evaluate $M(a)$ at any point within the contour C, when the values of $M(z)$ are known along this contour. This relationship is a consequence of the close connection which exists among all values of an analytic function on the complex plane z.

Let us consider a path consisting of a semi-circle with an infinitely large radius, centered on the x-axis. The internal area of the contour includes the upper half-plane as shown in Fig. 1.52. We will attempt to find a quadrature component for the function $M = U + \mathrm{i}V$ by assuming that the inphase component U is known along the z-axis or vice versa. Using the Cauchy formula, we have:

$$M(\xi) = \frac{1}{\mathrm{i}\pi}P\oint \frac{M(z)}{z-\xi}\,\mathrm{d}z \tag{1.292}$$

The point $\xi = \varepsilon + \mathrm{i}\,\eta$ lies along the path of integration and the symbol P indicates the principle value of the integral that is to be used. Inasmuch as the path of integration coincides with the x-axis ($\eta = 0$) we have:

$$M(\varepsilon, 0) = \frac{1}{\mathrm{i}\pi} P \int_{-\infty}^{\infty} \frac{M(x, 0)}{x - \varepsilon}\, \mathrm{d}x \tag{1.293}$$

In developing eq. 1.293 it has been assumed that the value for the integral along the semi-circular part of the path of integration vanishes as the radius increases without limit.

Because:

$$M(\varepsilon, 0) = U(\varepsilon, 0) + \mathrm{i}\,V(\varepsilon, 0)$$

and

$$M(x, 0) = U(x, 0) + \mathrm{i}\,V(x, 0)$$

we obtain:

$$\begin{aligned} U(\varepsilon, 0) &= \frac{1}{\pi} P \int_{-\infty}^{\infty} \frac{V(x, 0)}{x - \varepsilon}\, \mathrm{d}x \\ V(\varepsilon, 0) &= -\frac{1}{\pi} P \int_{-\infty}^{\infty} \frac{U(x, 0)}{x - \varepsilon}\, \mathrm{d}x \end{aligned} \tag{1.294}$$

The integrands in these expressions are characterized by a singularity which can readily be removed by making use of the identity:

$$P \int_{-\infty}^{\infty} \frac{\mathrm{d}x}{x - \varepsilon} = 0 \tag{1.295}$$

Now we can rewrite eq. 1.294 in the form:

$$U(\varepsilon, 0) = \frac{1}{\pi} \int_{-\infty}^{\infty} \frac{V(x, 0) - V(\varepsilon, 0)}{x - \varepsilon}\, \mathrm{d}x \tag{1.296}$$

$$V(\varepsilon, 0) = -\frac{1}{\pi} \int_{-\infty}^{\infty} \frac{U(x, 0) - U(\varepsilon, 0)}{x - \varepsilon}\, \mathrm{d}x \tag{1.297}$$

inasmuch as:

$$V(\varepsilon, 0) \int_{-\infty}^{\infty} \frac{\mathrm{d}x}{x - \varepsilon} = U(\varepsilon, 0) \int_{-\infty}^{\infty} \frac{\mathrm{d}x}{x - \varepsilon} = 0$$

It can readily be seen that the integrands in eqs. 1.296 and 1.297 do not have singularities, and these expressions establish the relationship between the real and imaginary parts of some analytic function.

Let us return to consideration of the complex amplitude of the field:

$$M(\omega) = \mathrm{In}\, M(\omega) + \mathrm{i}\,\mathrm{Q}\, M(\omega)$$

In accord with eqs. 1.296 and 1.297 the relationship between the quadrature and inphase components of the field are:

$$\mathrm{In}\, M(\omega_0) = \frac{1}{\pi} \int\limits_{-\infty}^{\infty} \frac{\mathrm{Q}\, M(\omega) - \mathrm{Q}\, M(\omega_0)}{\omega - \omega_0}\, \mathrm{d}\omega \tag{1.298}$$

$$\mathrm{Q}\, M(\omega_0) = -\frac{1}{\pi} \int\limits_{-\infty}^{\infty} \frac{\mathrm{In}\, M(\omega) - \mathrm{In}\, M(\omega_0)}{\omega - \omega_0}\, \mathrm{d}\omega \tag{1.299}$$

Thus, when the spectrum of one of the components is known, the other component describing a field can be calculated by making use of either eq. 1.298 or 1.299.

It is now a simple matter to find the relationship between the amplitude and the phase responses of a field components. Taking the logarithm of the complex amplitude M we have:

$$\ln M = \ln M_0 + \mathrm{i}\phi \tag{1.300}$$

From this equation we see that the relationship between the amplitude and phase responses is the same as that for the quadrature and the inphase components. For example, for the phase response we have:

$$\phi(\omega_0) = -\frac{1}{\pi} \int\limits_{-\infty}^{\infty} \frac{\ln M_0(\omega) - \ln M_0(\omega_0)}{\omega - \omega_0}\, \mathrm{d}\omega \tag{1.301}$$

For practical purposes it is preferable to express the right-hand side of eq. 1.301 in another form. After some algebraic operations, we obtain:

$$\phi(\omega_0) = \frac{1}{\pi} \int\limits_{-\infty}^{\infty} \frac{\mathrm{d}L}{\mathrm{d}u} \left| \ln \coth \frac{u}{2} \right| \mathrm{d}u \tag{1.302}$$

where:

$$L = \ln M_0 \qquad u = \ln(\omega/\omega_0)$$

It can be seen from eq. 1.302 that the phase response depends on the slope of the amplitude response curve when plotted on a logarithmic frequency scale. Inasmuch as the integration

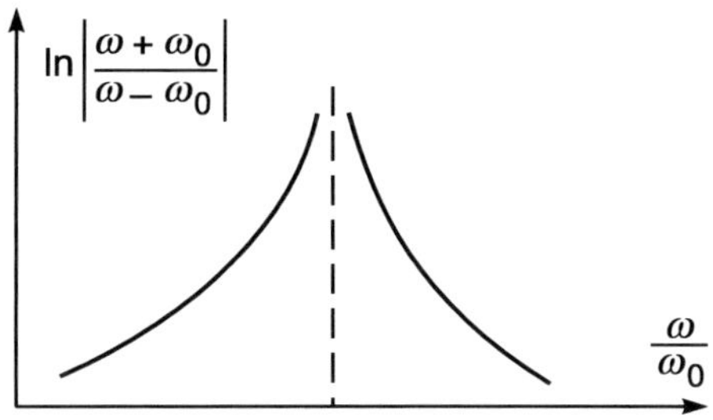

Figure 1.53. The kernel function (weighting factor) for the transform given in eq. 1.302.

is carried out over the entire frequency range, the phase at any particular frequency ω_0 depends on the slope of the amplitude response curve over the entire frequency spectrum. However, the relative importance of the slope over various portions of the spectrum is controlled by a weighting factor $|\ln\coth(u/2)|$ which can also be written as:

$$\left|\ln\frac{\omega+\omega_0}{\omega-\omega_0}\right|$$

The weighting factor is shown graphically in Fig. 1.53. It increases as the frequency is close to ω_0, and becomes logarithmically infinite at that point. Therefore, the slope of the amplitude response near the frequency for which the phase is to be calculated is much more important than the slope of the amplitude response curve at more distant frequencies.

It should be noted that calculation of the amplitude response from the phase can only be done with an accuracy of some constant. Equations 1.301 and 1.302 lead us to the following conclusions. First of all, measurement of the phase response does not provide additional information on the geoelectric section when the amplitude response is already known. However, it may well be that the shape of the phase response curve more clearly reflects some diagnostic features of this section than does the amplitude response curve.

It is important to stress that while there is in essence a unique relationship between the quadrature and the inphase responses, as well as between the amplitude and phase responses, this does not mean that there is a point by point relationship between them. In fact measuring both amplitude and phase at one or a few frequencies provides two types of information characterizing the geoelectric section in a different manner. The same conclusion can be derived for the quadrature and the inphase components.

We will now investigate the relationship between frequency domain and time domain responses. In most cases considered in this section a transient electromagnetic field is excited by a step function current in the source. Moreover the theory of the transient induction logging described in this monograph will be developed for this type of excitation. For this reason the relationship between frequency response and transient response corresponding to this single type of excitation will be our principal concern. The information we need is obtained through use of the Fourier transform which takes the well

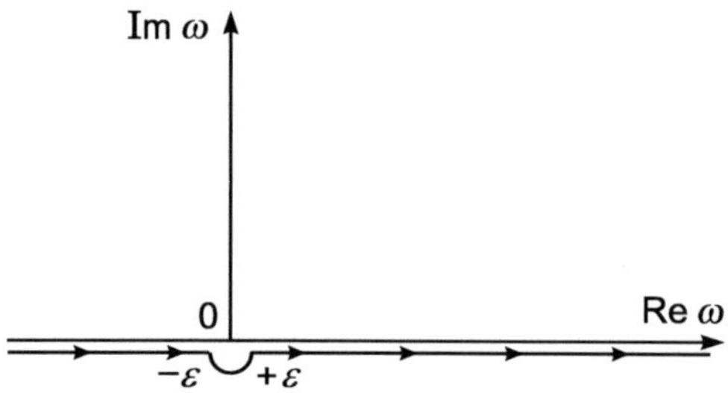

Figure 1.54. The path of integration for eq. 1.306.

known form:

$$F(t) = \frac{1}{2\pi} \int_{-\infty}^{\infty} F^*(\omega)\, e^{-i\omega t}\, d\omega \tag{1.303}$$

$$F^*(\omega) = \int_{-\infty}^{\infty} F(t)\, e^{i\omega t}\, dt$$

When the current in the source changes as a step function in time, the primary magnetic field accompanying this current does likewise:

$$H_0(t) = F_0(t) = \begin{cases} H_0 & t < 0 \\ 0 & t \geqslant 0 \end{cases} \tag{1.304}$$

According to eq. 1.303 the spectrum for the primary magnetic field is:

$$H_0(\omega) = F_0^*(\omega) = H_0 / i\omega \tag{1.305}$$

The amplitude of this spectrum decreases inversely proportionally to frequency while the phase remains constant.

Inasmuch as low frequencies prevail in the spectrum of the primary field when step function excitation is used application of this excitation is often preferable in practice. In accord with eq. 1.303 the primary magnetic field can be written as:

$$H_0(t) = \frac{H_0}{2\pi} \int_{-\infty}^{\infty} \frac{1}{i\omega} e^{-i\omega t}\, d\omega \tag{1.306}$$

where the path of integration is not permitted to pass through the point $\omega = 0$ (Fig. 1.54). Let us write the right-hand integral as a sum:

$$\frac{1}{2\pi i} \int_{-\infty}^{\infty} \frac{e^{-i\omega t}}{\omega}\, d\omega = \frac{1}{2\pi i} \int_{-\infty}^{-\varepsilon} \frac{e^{-i\omega t}}{\omega}\, d\omega + \frac{1}{2\pi i} \int_{-\varepsilon}^{+\varepsilon} \frac{e^{-i\omega t}}{\omega}\, d\omega + \frac{1}{2\pi i} \int_{+\varepsilon}^{\infty} \frac{e^{-i\omega t}}{\omega}\, d\omega$$

and integrate along a semi-circle around the origin whose radius tends to zero.

In calculating the middle integral we will introduce new variables, ρ and φ:

$$\omega = \rho\,\mathrm{e}^{\mathrm{i}\varphi}$$

Then, we have:

$$\mathrm{d}\omega = \mathrm{i}\,\rho\,\mathrm{e}^{\mathrm{i}\varphi}\,\mathrm{d}\varphi$$

and

$$\frac{1}{2\pi\mathrm{i}}\int\limits_{-\varepsilon}^{+\varepsilon}\frac{\mathrm{e}^{-\mathrm{i}\omega t}}{\omega}\,\mathrm{d}\omega = \frac{1}{2\pi\mathrm{i}}\int\limits_{\pi}^{2\pi}\frac{\mathrm{i}\,\rho\,\mathrm{e}^{\mathrm{i}\varphi}\,\mathrm{d}\varphi}{\rho\,\mathrm{e}^{\mathrm{i}\varphi}} = \frac{1}{2} \quad \text{if } \rho\to 0$$

Correspondingly, the second expression for the primary field when the variable of integration takes on only real values is:

$$H_0(t) = \frac{H_0}{2} + \frac{H_0}{2\pi}\int\limits_{-\infty}^{\infty}\frac{\mathrm{e}^{-\mathrm{i}\omega t}}{\omega}\,\mathrm{d}\omega \tag{1.307}$$

Now, making use of the principle of superposition, we obtain the following expression for a nonstationary field:

$$H(t) = \frac{H_0}{2\pi\mathrm{i}}\int\limits_{-\infty}^{\infty}\frac{H(\omega)}{\omega}\,\mathrm{e}^{-\mathrm{i}\omega t}\,\mathrm{d}\omega \tag{1.308}$$

or

$$H(t) = \frac{H_0}{2\pi} + \frac{H_0}{2\pi\mathrm{i}}\int\limits_{-\infty}^{\infty}\frac{H(\omega)}{\omega}\,\mathrm{e}^{-\mathrm{i}\omega t}\,\mathrm{d}\omega \tag{1.309}$$

where $H(\omega) = \mathrm{In}\,H(\omega) + \mathrm{i}\,\mathrm{Q}\,H(\omega)$ is the complex amplitude of the spectrum of the chosen component of the magnetic field, which is assumed to be known.

Let us write eq. 1.309 in the form:

$$\begin{aligned} H(t) = & \frac{H_0}{2} + \frac{H_0}{2\pi}\int\limits_{-\infty}^{\infty}\frac{\mathrm{Q}\,H(\omega)\cos\omega t - \mathrm{In}\,H(\omega)\sin\omega t}{\omega}\,\mathrm{d}\omega \\ & - \frac{\mathrm{i}}{2\pi}H_0\int\limits_{-\infty}^{\infty}\frac{\mathrm{Q}\,H(\omega)\sin\omega t + \mathrm{In}\,H(\omega)\cos\omega t}{\omega}\,\mathrm{d}\omega \end{aligned} \tag{1.310}$$

Inasmuch as

$$\begin{aligned} \text{In}\, H(\omega) &= \text{In}\, H(-\omega) \\ \text{Q}\, H(\omega) &= -\,\text{Q}\, H(-\omega) \end{aligned} \tag{1.311}$$

the second integral in 1.310 is zero, and therefore:

$$H(t) = \frac{H_0}{2} + \frac{H_0}{\pi} \int_0^{\infty} \frac{\text{Q}\, H(\omega) \cos \omega t - \text{In}\, H(\omega) \sin \omega t}{\omega} \, \mathrm{d}\omega \tag{1.312}$$

For negative times, $H(t) = H_0$, and by substituting $t = -t$:

$$H_0 = \frac{H_0}{2} + \frac{H_0}{\pi} \int_0^{\infty} \frac{\text{Q}\, H(\omega) \cos \omega t + \text{In}\, H(\omega) \sin \omega t}{\omega} \, \mathrm{d}\omega$$

or

$$0 = -\frac{H_0}{2} + \frac{H_0}{\pi} \int_0^{\infty} \frac{\text{Q}\, H(\omega) \cos \omega t + \text{In}\, H(\omega) \sin \omega t}{\omega} \, \mathrm{d}\omega \tag{1.313}$$

Let us take note that in these last two expressions time is taken as a positive quantity.
Combining eqs. 1.312 and 1.313, we obtain:

$$H(t) = \frac{2}{\pi} H_0 \int_0^{\infty} \frac{\text{Q}\, H(\omega)}{\omega} \cos \omega t \, \mathrm{d}\omega$$

and

$$H(t) = H_0 - \frac{2}{\pi} H_0 \int_0^{\infty} \frac{\text{In}\, H(\omega)}{\omega} \sin \omega t \, \mathrm{d}\omega \tag{1.314}$$

Correspondingly for derivatives with respect to time of the magnetic induction, B, we have:

$$\begin{aligned} \dot{B}(t) &= -\frac{2}{\pi} H_0 \int_0^{\infty} \text{Q}\, B(\omega) \sin \omega t \, \mathrm{d}\omega \\ \dot{B}(t) &= -\frac{2}{\pi} H_0 \int_0^{\infty} \text{In}\, B(\omega) \cos \omega t \, \mathrm{d}\omega \end{aligned} \tag{1.315}$$

Equations 1.312 and 1.315 permit us to calculate the transient response when either the quadrature or the inphase component of the frequency spectrum is known. It is often convenient to introduce new scale variables in these equations. Such a scale variable can be defined as:

$$y = \frac{1}{4\pi^2}\left(\frac{a}{h}\right)^2 = \frac{\sigma\mu a^2}{8\pi^2}\,\omega = \frac{\omega}{8\pi^2\alpha}$$

where h is skin depth, a is a linear dimension, for example, a radius of the borehole, and

$$\alpha = 1/\sigma\mu a^2$$

It should be obvious that:

$$\omega t = \left(\frac{\tau}{2\pi h}\right)^2 = \left(\frac{\tau}{a}\right)^2 y$$

where

$$\tau = 2\pi\left(\frac{2t}{\sigma\mu}\right)^{1/2} \quad \text{and} \quad t = \frac{1}{8\pi^2\alpha}\left(\frac{\tau}{a}\right)^2$$

Using these variables:

$$\begin{aligned}
\dot{B}(t) &= -16\pi\alpha H_0 \int\limits_0^\infty \operatorname{In} B\left(8\pi^2\alpha y\right) \cos\left[(\tau/a)^2 y\right]\, dy \\
\dot{B}(t) &= -16\pi\alpha H_0 \int\limits_0^\infty \operatorname{Q} B\left(8\pi^2\alpha y\right) \sin\left[(\tau/a)^2 y\right]\, dy
\end{aligned} \qquad (1.316)$$

Usually, because of the complexity of the expressions for the frequency spectrum, the only way to obtain numerical results from eqs. 1.312 to 1.316 is by numerical integration.

Up to this point we have examined the relationship between frequency and transient responses and have derived formulae for calculating the time-domain field for the case in which the primary field changes as a step function of time. In so doing we have assumed that the frequency spectrum for the field is known. However, in practice, the use of this type of excitation meets with some practical difficulties. For example, due to the inductance in a transmitter loop, the current cannot be terminated instantly, and because of this, in place of a step current behavior there is a gradual decrease of current in the transmitter. The time required for the current to vanish in the transmitter is usually called the ramp time. In order to investigate the effect caused by such behavior of the transmitter current, it is appropriate to use calculations of the transient field generated by a step function excitation, rather than refer to frequency domain fields by applying Fourier transform to the measurements. The further approach is based on the use of Duhamel's integral which is described below.

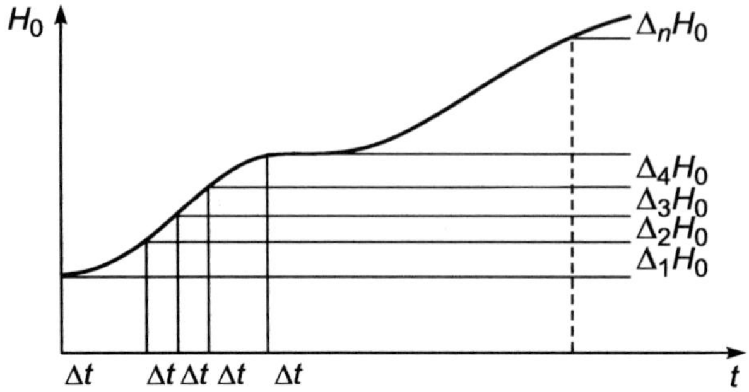

Figure 1.55. Presentation of the primary field for determination of Duhamel's integral.

Assume that a primary field varies with time like the function shown in Fig. 1.55. It should be clear that this function can be thought of as being the sum of step functions with the amplitude $\Delta H_0(\tau)$, where τ is the instant at which the excitation occurs. Also let us assume that the transient field caused by the unit step function is known and is described by the function $A^*(t-\tau)$. It is clear that a step function with the amplitude $\Delta H_0(\tau)$ generates a transient field given by $\Delta H_0(\tau)A^*(t-\tau)$.

Adding the actions of all such step functions occurring at various times we find the expressions for the total transient response for any component of the magnetic field:

$$H_i(t) = H_0(0)A_i^{*0}(t) + \sum_{t=0}^{\tau=t} \Delta H_0(\tau)A_i^*(t-\tau)$$

$$= H_0(0)A_i^{*0}(t) + \sum_{\tau=0}^{\tau=t} \frac{\partial H_0(\tau)}{\partial \tau} A_i^*(t-\tau)\,\Delta\tau$$

As can be seen from Fig. 1.55, the approximation:

$$\Delta H_0(\tau) \simeq \frac{\partial H_0(\tau)}{\partial \tau}\Delta\tau$$

becomes more accurate with a decrease of the interval $\Delta\tau$. In this expression $A_i^*(t-\tau)$ is the response of the medium to the ith component of the magnetic field H_i, when the unit current step function occurs at the instant $t=\tau$. For instance for inductive excitation of the field A_i^{*0} is identically zero, and therefore we have:

$$H_i(t) = \sum_{\tau=0}^{\tau=t} \frac{\partial H_0(\tau)}{\partial \tau} A_i^*(t-\tau)\Delta\tau$$

In the limiting case as $\Delta\tau$ approaches zero we obtain a convolution integral:

$$H_i(t) = \int\limits_0^t \frac{\partial H_0(\tau)}{\partial \tau} A_i^*(t-\tau)\,\mathrm{d}\tau \tag{1.317}$$

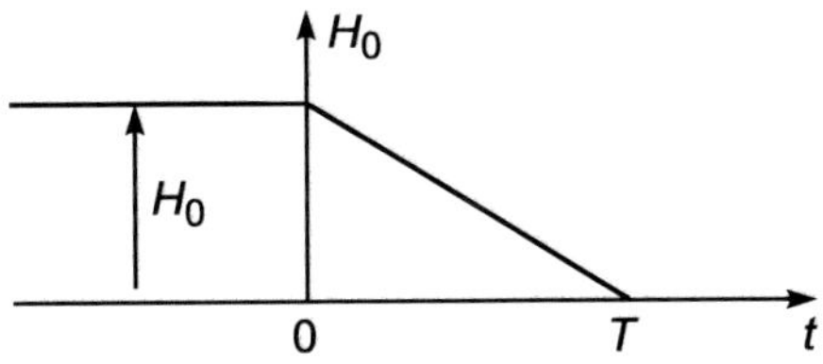

Figure 1.56. Time variation of the primary field described by eq. 1.319.

This integral is also called Duhamel's integral and it permits us to find the transient response for an arbitrary shape of a current excitation when the transient response of the medium for a step function excitation is already known.

Integrating by parts the right-hand side of eq. 1.317, we obtain:

$$\begin{aligned}\int_0^t \frac{\partial H_0(\tau)}{\partial \tau} A_i^*(t-\tau)\,\mathrm{d}\tau &= H_0(\tau)\,A_i^*(t-\tau)\Big|_0^t - \int_0^t H_0(\tau)\frac{\partial A_i^*(t-\tau)}{\partial \tau}\,\mathrm{d}\tau \\ &= H_0(t)A_i^*(0) - H_0(0)A_i^*(t) - \int_0^t H_0(\tau)\frac{\partial A_i^*(t-\tau)}{\partial \tau}\,\mathrm{d}\tau\end{aligned} \tag{1.318}$$

This is known as the second form of Duhamel's integral.

Similar expressions can be written for the electric field.

Let us now consider an example. Suppose that the behavior of the primary field described by the linear function shown in Fig. 1.56, that is:

$$H_0(\tau) = \begin{cases} H_0 & \tau < 0 \\ H_0(1-\tau/T) & 0 \leqslant \tau \leqslant T \\ 0 & \tau \geqslant T \end{cases} \tag{1.319}$$

where T is the ramp time.

Substituting eq. 1.319 into eq. 1.317 we obtain:

$$H_i(t) = -\frac{H_0}{T}\int_0^T A_i^*(t-\tau)\,\mathrm{d}\tau \qquad \text{if } t \geqslant T \tag{1.320}$$

In the limit as T approaches zero and applying the central limit theorem, we have:

$$H_i(t) = -A_i^*(t)H_0 = A_i^{-1}H_0$$

where A_i^{-1} is the transient field caused by unit step function where the current is turned off.

Chapter 2

ELECTROMAGNETIC FIELD OF THE MAGNETIC DIPOLE IN A UNIFORM CONDUCTING MEDIUM

We will start a development of the theory of induction logging assuming that an induction probe is placed in a uniform conducting medium and consists of two coils, as shown in Fig. 2.1. One coil is the source of the primary alternating field, while the other serves as a receiver measuring an electromotive force and therefore the magnetic field. Considering such a model we are not able to investigate an influence of a borehole, an invaded zone, finite thickness of a layer, eccentricity, efficiency of many coil probes and many other parameters, defining radial and vertical characteristics of the induction logging. All these questions will be analyzed in detail in the next chapters. At the same time it is appropriate to notice that very often signals, measured by an induction probe in real conditions, are close to those, which would be measured in a uniform medium with a resistivity of a formation.

Simplicity of this model allows us to investigate not only frequency responses of the magnetic field, measured by a receiver, but also a distribution of currents in a conducting medium. Such will help us to understand deeper physical principles of the induction logging as well as some approximate methods of calculation of fields which are widely used for interpretation.

In most cases dimensions of the transmitter coil are significantly smaller than a diameter of a borehole and distances to interfaces between layers. For this reason one can replace a coil with an alternating current by the magnetic dipole with the moment:

$$M = M_0\, \mathrm{e}^{-\mathrm{i}\omega t} \tag{2.1}$$

where $M_0 = SnI_0$ is the moment amplitude; I_0 is the current amplitude; $\omega = 2\pi f$ is angular frequency; f is the frequency; n is the number of turns; S is the area of one turn.

As was shown in the previous chapter the quasistationary electromagnetic field is described by equations:

$$\operatorname{curl} \boldsymbol{E} = -\frac{\partial \boldsymbol{B}}{\partial t} \tag{2.2}$$

$$\operatorname{curl} \boldsymbol{H} = \sigma \boldsymbol{E} \tag{2.3}$$

$$\operatorname{div} \boldsymbol{E} = 0 \tag{2.4}$$

$$\operatorname{div} \boldsymbol{H} = 0 \tag{2.5}$$

since a uniform medium is considered.

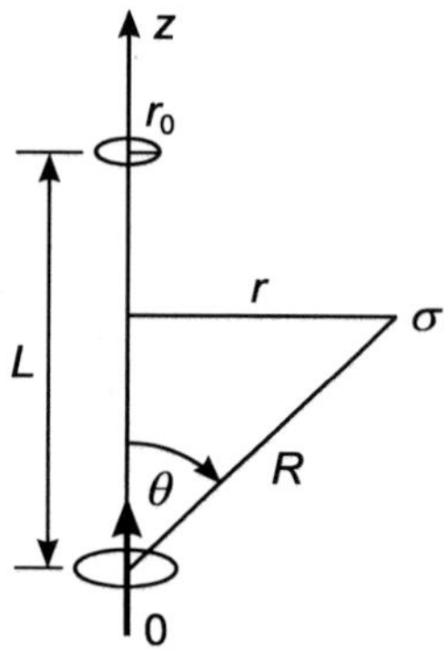

Figure 2.1. Two-coil induction probe in a uniform conducting medium.

In accord with eqs. 2.4 and 2.3 we have correspondingly:

$$\boldsymbol{E} = \operatorname{curl} \boldsymbol{A}^* \tag{2.6}$$
$$\boldsymbol{H} = \sigma \boldsymbol{A}^* - \operatorname{grad} U^* \tag{2.7}$$

where $\boldsymbol{A}^*$ and U^* are vector and scalar potentials, respectively.
Substituting expressions 2.6 and 2.7 into the first Maxwell equation, 2.2, we obtain:

$$\operatorname{curl}\operatorname{curl} \boldsymbol{A}^* = -\sigma\mu \frac{\partial \boldsymbol{A}^*}{\partial t} + \mu \operatorname{grad} \frac{\partial U^*}{\partial t}$$

Taking into account dependence on time, shown in eq. 2.1, and making use of the equality:

$$\operatorname{curl}\operatorname{curl} \boldsymbol{A}^* = \operatorname{grad}\operatorname{div} \boldsymbol{A}^* - \nabla^2 \boldsymbol{A}^*$$

we have:

$$\operatorname{grad}\operatorname{div} \boldsymbol{A}^* - \nabla^2 \boldsymbol{A}^* = \mathrm{i}\sigma\mu\omega \boldsymbol{A}^* - \mathrm{i}\mu\omega \operatorname{grad} U^* \tag{2.8}$$

where $\boldsymbol{A}^*$ and $\operatorname{grad} U^*$ are complex amplitudes of potentials.
Assuming, that:

$$\operatorname{div} \boldsymbol{A}^* = -\mathrm{i}\omega\mu\, U^* \tag{2.9}$$

we obtain from 2.8 the equation for the vector potential $\boldsymbol{A}^*$:

$$\nabla^2 \boldsymbol{A}^* + k^2 \boldsymbol{A}^* = 0 \tag{2.10}$$

where $k^2 = \mathrm{i}\sigma\mu\omega$.

Let us choose the spherical system of coordinates R, θ, ϕ and cylindrical system r, ϕ, with a common origin, where the dipole is placed and with axis z, $(\sin\phi = 0)$, coinciding with the direction of the dipole moment.

As was demonstrated in Chapter 1, the primary vortex electrical field, caused by a change with time of the primary magnetic field has only the component $E_\phi^{(0)}$. Correspondingly, one can expect that the secondary vortex electric field also has the same component. Making such an assumption and using the relation: $\boldsymbol{E} = \text{curl}\,\boldsymbol{A}^*$ it is appropriate to derive expressions for the electromagnetic field with the help of one component of the vector potential A_z^*, that is in the cylindrical system of coordinates:

$$\boldsymbol{A}^* = (0,\, 0,\, A_z^*) \tag{2.11}$$

Due to spherical symmetry one will look for a solution of the vector potential, $\boldsymbol{A}^*$ as a function depending on coordinate R only, that is:

$$A_z^* = A_z^*(R) \tag{2.12}$$

Respectively, it is more convenient to consider eq. 2.10 in a spherical coordinate system. As is well known, one can write eq. 2.10 in the form:

$$\frac{\partial^2 A_z^*}{\partial R^2} + \frac{2}{R}\frac{\partial A_z^*}{\partial R} + k^2 A_z^* = 0$$

or

$$\frac{\partial^2 (A_z^* R)}{\partial R^2} + k^2 (A_z^* R) = 0$$

Whence

$$A_z^* = C\,\frac{\mathrm{e}^{\mathrm{i}kR}}{R} + D\,\frac{\mathrm{e}^{-\mathrm{i}kR}}{R} \tag{2.13}$$

Inasmuch as $k^2 = \mathrm{i}\sigma\mu\omega$ we have:

$$k = (\sigma\mu\omega/2)^{1/2}\,(1 + \mathrm{i})$$

and

$$\mathrm{i}kR = (\sigma\mu\omega/2)^{1/2}\,(\mathrm{i} - 1)R \tag{2.14}$$

For this reason, the second term of the right-hand side of eq. 2.13 increases with increasing distance from the dipole.

Correspondingly, the vector potential can be described by only the first term of eq. 2.13 and we have:

$$A_z^* = C\,\frac{\mathrm{e}^{\mathrm{i}kR}}{R} \tag{2.15}$$

where C is an unknown coefficient. Whence:

$$\operatorname{div} \boldsymbol{A}^* = \frac{\partial A_z^*}{\partial z} = C\,\frac{\mathrm{e}^{\mathrm{i}kR}}{R^2}\,(\mathrm{i}kR - 1)\cos\theta$$

since

$$\frac{\partial A_z^*}{\partial \phi} = \frac{\partial A_z^*}{\partial \theta} = 0$$

and therefore in accord with eq. 2.9 we obtain:

$$\mathrm{i}\omega\mu\, U^* = C\,\frac{\mathrm{e}^{\mathrm{i}kR}}{R^2}\,(1 - \mathrm{i}kR)\cos\theta$$

Thus:

$$U^* = C\,\frac{\mathrm{e}^{\mathrm{i}kR}}{\mathrm{i}\omega\mu R^2}\,(1 - \mathrm{i}kR)\cos\theta \tag{2.16}$$

Making use of expressions for vector and scalar potentials one can define vectors of electric and magnetic fields by formulas 2.6 and 2.7. The electric field is located in planes perpendicular to the dipole axis and it has only the component E_ϕ.

In accord with eq. 2.6:

$$E_\phi = \operatorname{curl}_\phi \boldsymbol{A}^* = \frac{1}{R}\left(\frac{\partial}{\partial R}(RA_\theta^*) - \frac{\partial A_R^*}{\partial \theta}\right) \tag{2.17}$$

It is obvious that:

$$A_R^* = A_z^*\cos\theta \qquad A_\theta^* = -A_z^*\sin\theta$$

Substituting these expressions into eq. 2.17 after simple transformations we obtain:

$$E_\phi = C\frac{1}{R^2}\,\mathrm{e}^{\mathrm{i}kR}\,(1 - \mathrm{i}kR)\sin\theta \tag{2.18}$$

In particular, for a nonconducting medium letting $\sigma = 0$ we have:

$$E_\phi = C\frac{1}{R^2}\sin\theta$$

since $k = 0$.

It is obvious that this expression describes the primary vortex electrical field, caused by a change of the primary magnetic field with time, that is:

$$E_\phi^{(0)} = C\frac{1}{R^2}\sin\theta$$

On the other hand, as follows from the previous chapter (section 1.4), we have:

$$E_\phi^{(0)} = \frac{\mathrm{i}\omega\mu M}{4\pi R^2}\sin\theta$$

Comparing these last two expressions we define the unknown constant C:

$$C = \frac{\mathrm{i}\omega\mu M}{4\pi} \tag{2.19}$$

The vector of the magnetic field has two components:

$$\begin{aligned} H_\theta &= \sigma A_\theta^* - \operatorname{grad}_\theta U^* \\ H_R &= \sigma A_R^* - \operatorname{grad}_R U^* \end{aligned}$$

In accord with eq. 2.16 we have:

$$\begin{aligned} \operatorname{grad}_R U^* &= \frac{M}{4\pi}\,\mathrm{e}^{\mathrm{i}kR}\left(\frac{2\mathrm{i}k}{R^2} - \frac{2}{R^3} + \frac{k^2}{R}\right)\cos\theta \\ \operatorname{grad}_\theta U^* &= -\frac{M}{4\pi R^3}\,\mathrm{e}^{\mathrm{i}kR}\,(1 - \mathrm{i}kR)\sin\theta \end{aligned}$$

Therefore:

$$\begin{aligned} H_\theta &= \frac{M}{4\pi R^3}\,\mathrm{e}^{\mathrm{i}kR}\,(-k^2R^2 + 1 - \mathrm{i}kR)\sin\theta \\ H_R &= \frac{2M}{4\pi R^3}\,\mathrm{e}^{\mathrm{i}kR}\,(1 - \mathrm{i}kR)\cos\theta \end{aligned}$$

Finally, we obtain the following expressions for complex amplitudes of the electromagnetic field of the magnetic dipole:

$$E_\phi = \frac{\mathrm{i}\omega\mu M}{4\pi R^2}\,\mathrm{e}^{\mathrm{i}kR}\,(1 - \mathrm{i}kR)\sin\theta \tag{2.20}$$

$$H_R = \frac{2M}{4\pi R^3}\,\mathrm{e}^{\mathrm{i}kR}\,(1 - \mathrm{i}kR)\cos\theta \tag{2.21}$$

$$H_\theta = \frac{M}{4\pi R^3}\,\mathrm{e}^{\mathrm{i}kR}\,(1 - \mathrm{i}kR - k^2R^2)\sin\theta \tag{2.22}$$

Proceeding from these equations we will investigate the behavior of the electromotive force induced in the receiver of the two coil induction probe as well as the main features of the distribution of induced currents.

Suppose that a receiver coil is significantly smaller than the length of the induction probe. In other words, it will be assumed that all turns of the coil have the same area and they are located at some distance from the transmitter coil. It is clear that similar assumptions are applied to the transmitter coil since it is replaced by the magnetic dipole.

By definition the electromotive force in one turn of the receiver coil is defined as:

$$\mathscr{E} = \oint \boldsymbol{E} \cdot \mathrm{d}\boldsymbol{l}$$

Taking into account the axial symmetry the latter is significantly simplified and we have:

$$\mathscr{E} = E_\phi \oint \mathrm{d}l_\phi = E_\phi \, 2\pi r_0$$

where r_0 is the radius of the turn.

Making use of eq. 2.20 we obtain the following expression for the electromotive force in the coil receiver having n turns:

$$\mathscr{E} = E_\phi n 2\pi r_0 = \frac{\mathrm{i}\omega\mu M_T M_R}{2\pi R^3}\, \mathrm{e}^{\mathrm{i}kR}\,(1 - \mathrm{i}kR) \tag{2.23}$$

where $M_T = n_T S_T$ is the transmitter moment; $M_R = n_R S_R = n_R \pi r_0^2$ is the receiver moment; $\sin\theta = r_0/R$, $R = (L^2 + r^2)^{1/2}$.

Inasmuch as $r_0 \ll L$, eq. 2.23 can be rewritten as:

$$\mathscr{E} = \frac{\mathrm{i}\omega\mu M_T M_R}{2\pi L^3}\, \mathrm{e}^{\mathrm{i}kL}\,(1 - \mathrm{i}kL) \tag{2.24}$$

It is clear that the primary electromotive force induced in the receiver due to a change of the primary magnetic field is:

$$\mathscr{E}_0 = \frac{\mathrm{i}\omega\mu M_T M_R}{2\pi L^3} \tag{2.25}$$

Correspondingly, instead of eq. 2.24 we have:

$$\mathscr{E} = \mathscr{E}_0\, \mathrm{e}^{\mathrm{i}kL}\,(1 - \mathrm{i}kL) \tag{2.26}$$

Usually the primary electromotive force, which does not contain any information about the conductivity of the medium, is dominant.

It is appropriate here to derive eq. 2.26 proceeding from the magnetic field. In accord with Faraday's law the electromotive force is equal to:

$$\mathscr{E} = -\frac{\partial \psi}{\partial t}$$

If the area of the receiver coil of the induction probe is small with respect to its length (Fig. 2.1), one can assume that within this area the magnetic field is uniform, and it is directed perpendicular to the horizontal plane, i.e.:

$$H_R = H_z$$

Then

$$\psi = \pi r_0^2 n_R H_z$$

and making use of eq. 2.21 as well as the relation:

$$\frac{\partial \psi}{\partial t} = -\mathrm{i}\omega\psi$$

we again obtain eq. 2.26:

$$\mathscr{E} = \mathscr{E}_0 \, \mathrm{e}^{\mathrm{i}kL} \, (1 - \mathrm{i}kL)$$

Before we will begin to consider the full expression for the electromotive force, eq. 2.26, let us make some almost obvious comments concerning eq. 2.25:

- The presence of multiplier $\mathrm{i} = \sqrt{-1}$ indicates that the primary electromotive force, $\mathscr{E}_0$, is shifted in phase by 90° with respect to the transmitter current. It results from the fact that the electromotive force is proportional to the derivative of the primary magnetic field with time which changes as a cosinusoid when the current in the transmitter is a sinusoidal oscillation.
- The appearance of ω on the right-hand side of eq. 2.25 is also caused by derivation of the flux with time.
- The term $M_T/2\pi L^3$ describes the vertical component of the primary magnetic field within the receiver coil.
- The presence of μ results from the fact that the electromotive force is proportional to the rate of a change of the flux of the vector of the magnetic induction $\boldsymbol{B}$. Practically, in all cases which will be considered, the magnetic permeability will be taken to be equal to that of free space, i.e.:

  $$\mu = \mu_0 = 4\pi \times 10^{-7} \text{ H/m}$$

- The multiplier M_R reflects the fact that the flux through the receiver is directly proportional to its area and number of turns.

Now let us investigate the general case with the two-coil induction probe located in a uniform conducting medium. At the point of the receiver the magnetic field in accord with eq. 2.21 is described by an equation for its complex amplitude:

$$H_z = \frac{M_T}{2\pi L^3} \, (1 - \mathrm{i}kL) \, \mathrm{e}^{\mathrm{i}kL} \tag{2.27}$$

This field is caused by the current in the transmitter coil and induced currents in the conducting medium. Of course, all currents change as sinusoidal functions, but in general

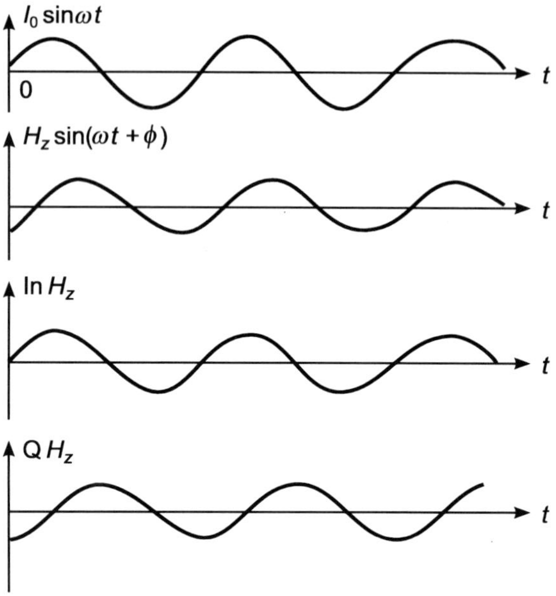

Figure 2.2. Presentation of the magnetic field behavior at the receiver.

they are shifted from each other in phase. It means that the magnetic field, H_z, can be presented as:

$$H_z = \frac{M_T}{2\pi L^3} A \sin(\omega t + \phi) \tag{2.28}$$

where $(M_T/2\pi L^3)A$ is the amplitude, while ϕ is the phase of the sinusoid describing the magnetic field.

It is essential that parameter ϕ characterizes the phase shift between the magnetic field at the receiver and the current in the transmitter (Fig. 2.2).

It is clear that in those cases, where the influence of induced currents is negligible or completely absent, the phase is equal to zero. In other words, the magnetic field, H_z, is equal to the primary one which varies synchronously with the transmitter current.

In conventional induction logging instead of amplitude and phase, quantities such as magnitudes of quadrature and sometimes inphase components are measured.

Making use of eq. 2.28 we have:

$$H_z = \frac{M_T}{2\pi L^3} A \cos\phi \sin\omega t + \frac{M_T}{2\pi L^3} A \sin\phi \cos\omega t$$

Inasmuch as the primary field:

$$H_z^{(0)} = \frac{M_T}{2\pi L^3}$$

is easily calculated, it is convenient to express the measured fields in units of the primary one. Then we have

$$h_z = H_z/H_z^{(0)} = A\cos\phi\sin\omega t + A\sin\phi\cos\omega t$$

or

$$h_z = \mathrm{In}\, h_z \sin\omega t + \mathrm{Q}\, h_z \cos\omega t \tag{2.29}$$

In accord with eq. 2.29 the field h_z is the sum of two oscillations: one of them synchronously changes with the primary magnetic field, while the other is shifted in phase by 90°. Correspondingly functions $\mathrm{In}\, h_z$ and $\mathrm{Q}\, h_z$ describe magnitudes of the inphase and quadrature components of the field expressed in units of the primary field.

The inphase component of the field, as follows from Biot–Savart law, is caused by the current in the transmitter and the inphase component of induced currents in a medium, while the quadrature component of the magnetic field is generated by the quadrature component of induced currents only. Therefore, one can write:

$$\mathrm{In}\, H_z = H_z^{(0)} + \mathrm{In}\, H_z^s \qquad \mathrm{Q}\, H_z = \mathrm{Q}\, H_z^s \tag{2.30}$$

or

$$\mathrm{In}\, h_z = 1 + \mathrm{In}\, h_z^s \qquad \mathrm{Q}\, h_z = \mathrm{Q}\, h_z^s \tag{2.31}$$

The index ^{s} means the secondary field caused by induced currents.

It is proper to emphasize that this consideration is based on the condition that the phase shift is defined with respect to the primary magnetic field. Let us make one more comment. The inphase component of the secondary magnetic field either coincides with the phase of the primary magnetic field or it is shifted by 180°.

An electric diagram illustrating these relations between the primary and secondary fields is shown in Fig. 2.3a.

As concerns the electromotive force, a definition of its quadrature and inphase components can be done in two ways. In fact, we can compare either a phase shift of the electromotive force with the current in the transmitter or with the primary electromotive force $\mathscr{E}_0$. In the future, the latter approach will be used and correspondingly one can write (Fig. 2.3b):

$$\mathrm{In}\, \mathscr{E} = |\mathscr{E}_0| + \mathrm{In}\, \mathscr{E}^s \qquad \mathrm{Q}\, \mathscr{E} = \mathrm{Q}\, \mathscr{E}^s \tag{2.32}$$

or

$$\mathrm{In}\, \mathscr{E} = 1 + \mathrm{In}\, \mathscr{E}^s \qquad \mathrm{Q}\, \mathscr{E} = \mathrm{Q}\, \mathscr{E}^s \tag{2.33}$$

In accord with eq. 2.25 we have:

$$|\mathscr{E}_0| = \frac{\omega\mu M_T M_R}{2\pi L^3} \tag{2.34}$$

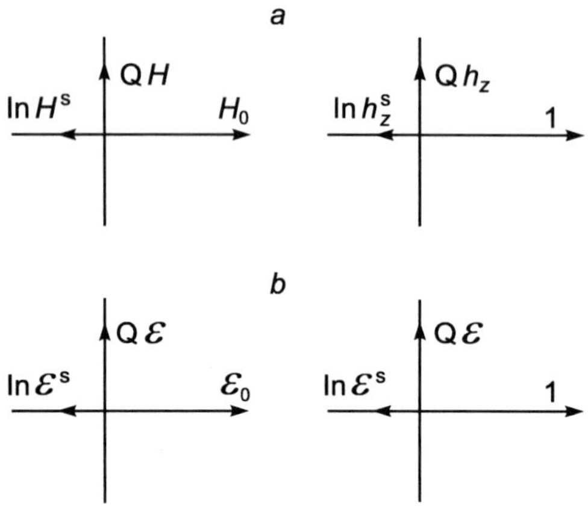

Figure 2.3. Electric diagram of the magnetic field and the electromotive force.

and $\mathscr{E}$ is the electromotive force expressed in units of the primary one, $|\mathscr{E}_0|$.

Comparing eqs. 2.26 and 2.27 we obtain:

$$\mathrm{In}\, h_z = \mathrm{In}\, \mathscr{E} \qquad \mathrm{Q}\, h_z = \mathrm{Q}\, \mathscr{E} \tag{2.35}$$

In other words, results of an analysis performed for the magnetic field, H_z, are directly applied for the electromotive force induced in a receiver coil as shown in Fig. 2.1 and vice versa.

Before we will start to investigate frequency responses of the field it is appropriate to notice the following. In accord with eqs. 2.21, 2.22 and 2.26 we have:

$$H_R = \frac{2M_T}{4\pi R^3} h_R \qquad H_\theta = \frac{M_T}{4\pi R^3} h_\theta \qquad \mathscr{E} = \frac{\mathrm{i}\omega\mu M_T M_R}{2\pi L^3} h_z$$

where:

$$h_R = \mathrm{e}^{\mathrm{i}kR}(1 - \mathrm{i}kR) \qquad h_\theta = \mathrm{e}^{\mathrm{i}kR}(1 - \mathrm{i}kR - k^2R^2) \qquad h_z = \mathrm{e}^{\mathrm{i}kL}(1 - \mathrm{i}kL) \tag{2.36}$$

Thus, equations for the field as well as for the electromotive force present themselves as the product of two terms. One of them depends on the moment of the dipole, the distance between the dipole and the receiver and in the case of the electromotive force, on the frequency and the magnetic permeability. The second term is a function of only one argument, namely kR, which can be presented as:

$$kR = \frac{R}{h}(1 + \mathrm{i}) = p(1 + \mathrm{i}) \tag{2.37}$$

where $h = (2/\sigma\mu\omega)^{1/2}$ is the skin depth.

Therefore, the electromagnetic field of the magnetic dipole expressed in the units of the primary field or the normalized electromotive force is defined by the parameter p:

$$p = \frac{R}{h} = \frac{2\pi R}{10^3(10\rho/f)^{1/2}} \tag{2.38}$$

In the case in which the field or the electromotive force is investigated in the receiver of the two-coil induction probe the distance R is replaced by the length of the probe, L, that is:

$$p = \frac{2\pi L}{10^3(10\rho/f)^{1/2}} \tag{2.39}$$

It is obvious that parameter p characterizes a distance, expressed in units of the skin depth, and this fact vividly demonstrates that an influence of induced currents in a surrounding medium is not defined by the value of frequency or resistivity and the distance between the dipole and an observation point, but it depends on the ratio between this distance and the skin depth only, that is the parameter p.

Furthermore, we will focus on the magnetic field and the electromotive force at the receiver of the two-coil induction probe. Substituting eq. 2.37 into 2.27 the magnetic field H_z can be represented as a sum of two components, namely the quadrature component which is shifted in phase by 90° with respect to the primary magnetic field, H_0, or the current in the source, and the inphase component which is shifted in phase by 0° or 180° with respect to the primary field, and we have:

$$\mathrm{Q}\,H_z = H_0\,\mathrm{Q}\,h_z \qquad \mathrm{In}\,H_z = H_0\,\mathrm{In}\,h_z \tag{2.40}$$

where:

$$\mathrm{Q}\,h_z = \mathrm{e}^{-p}\left[(1+p)\sin p - p\cos p\right] \tag{2.41}$$

$$\mathrm{In}\,h_z = \mathrm{e}^{-p}\left[(1+p)\cos p + p\sin p\right] \tag{2.42}$$

where $p = L/h$, i.e. the length of the induction probe measured in units of the skin depth.

Making use of eq. 2.35 we have for the electromotive force in the receiver:

$$\mathrm{Q}\,\mathscr{E} = |\mathscr{E}_0|\,\mathrm{Q}\,h_z \qquad \mathrm{In}\,\mathscr{E} = |\mathscr{E}_0|\,\mathrm{In}\,h_z \tag{2.43}$$

It is appropriate to emphasize again that according to the Biot–Savart law the quadrature component of the magnetic field arises from currents induced in a medium for which the phase is shifted by ±90° with respect to the current in the dipole source, while the inphase component is the algebraic sum of the primary and secondary fields. The inphase component of the secondary field is contributed by induction currents in the medium shifted by 180° or 0° with respect to the source current.

The understanding of this relation between induced currents and the measured field turns out to be extremely useful for explanation of the main features of field and electromotive force behavior.

First, let us consider the behavior of the field over the range of a small parameter p ($p \ll 1$).

Expanding the exponential e^{ikL} in the form of a power series:

$$e^{ikL} = \sum_{n=0}^{\infty} \frac{(ikL)^n}{n!}$$

and substituting this into eq. 2.27, after some simple algebra, we have:

$$h_z = 1 + \sum_{n=0}^{\infty} \frac{1-n}{n!}(ikL)^n = 1 + \sum_{n=0}^{\infty} \frac{1-n}{n!}\, 2^{n/2} p^n e^{i3\pi n/4} \tag{2.44}$$

It follows from this that for $p \ll 1$:

$$\mathrm{Q}\, h_z \simeq p^2 - \frac{2}{3}p^3$$

or

$$\mathrm{In}\, h_z \simeq 1 - \frac{2}{3}p^3$$

or

$$\begin{aligned} \mathrm{Q}\, H_z &\simeq \frac{M_T}{2\pi L^3}\left[\frac{\sigma\mu\omega L^2}{2} - \frac{2}{3}\left(\frac{\sigma\mu\omega L^2}{2}\right)^{3/2}\right] \\ \mathrm{In}\, H_z &\simeq \frac{M_T}{2\pi L^3}\left[1 - \frac{2}{3}\left(\frac{\sigma\mu\omega L^2}{2}\right)^{3/2}\right] \end{aligned} \tag{2.45}$$

Correspondingly, for both components of the electromotive force we have:

$$\begin{aligned} \mathrm{Q}\, \mathscr{E} &\simeq \frac{\omega\mu M_T M_R}{2\pi L^3}\left[\frac{\sigma\mu\omega L^2}{2} - \frac{2}{3}\left(\frac{\sigma\mu\omega L^2}{2}\right)^{3/2}\right] \\ \mathrm{In}\, \mathscr{E} &\simeq \frac{\omega\mu M_T M_R}{2\pi L^3}\left[1 - \frac{2}{3}\left(\frac{\sigma\mu\omega L^2}{2}\right)^{3/2}\right] \qquad \text{if } p \ll 1 \end{aligned} \tag{2.46}$$

Table 2.1 gives some idea about values of parameter p for various conductivities and resistivities for a probe length L of 1 m. As is seen from this table the parameter p is less than unity in a relatively resistive medium ($\rho > 5$ ohm·m) even at higher frequencies.

In accord with eqs. 2.45 and 2.46 one can make the following conclusions:

TABLE 2.1
Values of parameter p; $L = 1$ m

ρ, ohm·m \ f, kHz	1	10	20	50	100
100	0.0063	0.0200	0.028	0.045	0.063
50	0.0089	0.0282	0.039	0.063	0.089
10	0.0200	0.0630	0.089	0.140	0.200
5	0.0280	0.0890	0.125	0.198	0.279
1	0.0630	0.2001	0.280	0.450	0.630
0.1	0.2007	0.6304	0.890	1.401	2.002

1. At the range of very small parameter p the quadrature component of the field prevails over the inphase component of the secondary field and we have:

$$Q\,H_z \simeq \frac{M_T}{4\pi L}\,\sigma\mu\omega$$

or

$$Q\,H_z \simeq H_0\,p^2 \tag{2.47}$$

and

$$Q\,\mathscr{E} \simeq \frac{M_T M_R}{4\pi L}\mu^2\omega^2\sigma$$

or

$$\mathscr{E} \simeq |\mathscr{E}_0|\,p^2 \tag{2.48}$$

Thus in this range the quadrature component of the magnetic field is directly proportional to the conductivity and the frequency and inversely proportional to the probe length. As will be shown later this dependence on frequency and conductivity remains valid also for a nonuniform medium.

Equations 2.47–2.48 describe the field and the electromotive force with an error not exceeding 10%, provided that the parameter p is smaller than 0.1. In this case the quadrature component of the electromotive force, containing the information about the conductivity, constitutes only 1% or less of the primary electromotive force. For this reason cancellation of the latter in the induction probe is usually performed with a high accuracy.

2. The inphase component of the secondary field is very small with respect to the primary field:

$$\text{In}\,H_z^s \simeq \frac{M_T}{6\pi\sqrt{2}}(\sigma\mu\omega)^{3/2} \tag{2.49}$$

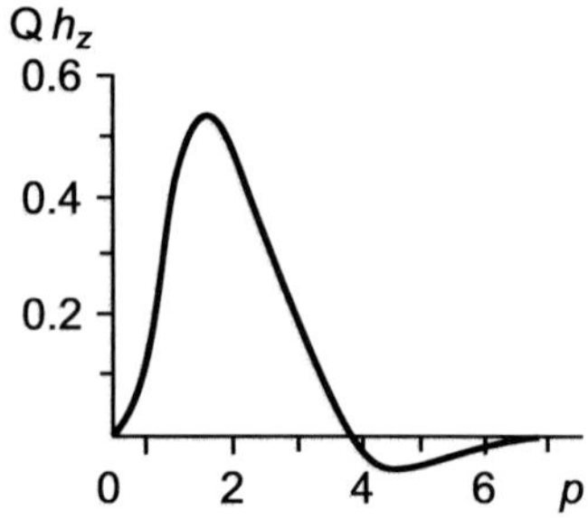

Figure 2.4. Frequency response of the quadrature component $\mathrm{Q}\,h_z$.

and

$$\mathrm{In}\,H_z^s \ll \mathrm{Q}\,H_z \ll H_0 \tag{2.50}$$

as well as

$$\mathrm{In}\,\mathscr{E}^s \ll \mathrm{Q}\,\mathscr{E} \ll |\mathscr{E}_0| \tag{2.51}$$

For this reason measuring the inphase component of the secondary field or $\mathrm{In}\,\mathscr{E}^s$ in the range of small parameters is a difficult task. At the same time this part of the field is more sensitive to a change of conductivity than the quadrature component. Also it is interesting to notice that the leading term of the inphase component of the secondary field does not depend on the probe length L. Later we will prove that all these features are inherent for a field in a nonuniform medium. As follows from eqs. 2.41 and 2.42, in the opposite case, i.e. within a range of large parameters of p, both components of the field H_z, as well as the electromotive force tend to zero, oscillating near this limit:

$$\begin{aligned} \mathrm{Q}\,H_z &= H_0\,\mathrm{e}^{-p}\,(\sin p - \cos p) \to 0 \\ \mathrm{In}\,H_z &= H_0\,\mathrm{e}^{-p}\,(\sin p + \cos p) \to 0 \quad \text{as } p \to \infty \end{aligned} \tag{2.52}$$

The latter means that at the range of very large parameters the inphase component of the secondary field approaches in magnitude the primary field, that is:

$$\mathrm{In}\,H_z \to -H_{0z} \quad \text{as } p \to \infty \tag{2.53}$$

Frequency responses of both components are presented in Figs. 2.4–2.5 and corresponding values of $\mathrm{Q}\,h_z$ and $\mathrm{In}\,h_z$ are given in Table 2.2. Also this table contains values of the amplitude and the phase of the secondary field and function σ_a/σ calculated in the following way:

$$A = \left[(\mathrm{In}\,h_z^s)^2 + (\mathrm{Q}\,h_z)^2\right]^{1/2} \qquad \phi = \arctan\frac{\mathrm{Q}\,h_z}{\mathrm{In}\,h_z} \qquad \frac{\sigma_a}{\sigma} = \frac{2}{\mu\omega\sigma L^2}\,\mathrm{Q}\,h_z \tag{2.54}$$

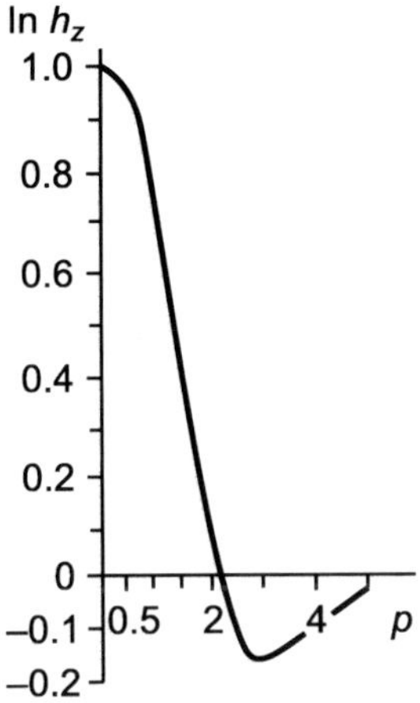

Figure 2.5. Behavior of frequency response of the inphase component of the secondary field In h_z.

The reason for the introduction of the function σ_a/σ will be explained in the next chapters.

As is seen from Figs. 2.4–2.5 and Table 2.2 with an increase of the parameter p the quadrature component also increases. At the beginning it is directly proportional to p^2, then it grows slower, reaches a maximum ($p = 1.6$) and with further increase of p decreases and oscillates as it approaches zero.

The behavior of the inphase component of the secondary field, In h_z^s, has a completely different character. First of all, at the beginning it increases in magnitude proportional to p^3. With further increase of parameter p the magnitude of In h_z^s reaches the primary field, then it becomes greater and at the range of large values of p it oscillates approaching the primary field. Unlike the quadrature component the function In h_z^s does not change sign, but it remains always negative. In other words, for all values of the parameter p the phase shift between the primary field and the inphase component of the secondary field is 180°.

In conclusion of this description of frequency responses of the field let us notice that a significant part of the ascending branch of the quadrature component is described by the approximate expression 2.45.

The behavior of the field as the parameter p varies can be explained in terms of the distribution of induced currents in the medium making use of eq. 2.20 and Ohm's law in its differential form:

$$\boldsymbol{J} = \sigma \boldsymbol{E}$$

We have the following expression for the current density at every point in a uniform full-space:

$$J_\phi = \frac{\mathrm{i}\sigma\mu\omega M_T}{4\pi R^2}\,(1 - \mathrm{i}kR)\,\mathrm{e}^{\mathrm{i}kR}\,\sin\theta \tag{2.55}$$

TABLE 2.2
Values of components of the field and function σ_a/σ

p	$Q\,h_z$	$\text{In}\,h_z^s$	A	ϕ	σ_a/σ
0.01	0.9933×10^{-4}	0.6616×10^{-6}	0.9933×10^{-4}	-0.1564×10^{1}	0.9933
	0.1403×10^{-3}	0.1111×10^{-5}	0.1403×10^{-3}	0.1562	0.9920
	0.1981	0.1865	0.1981	0.1561	0.9905
	0.2796	0.3131	0.2796	0.1559	0.9887
0.02	0.3946	0.5253	0.3947	0.1557	0.9866
	0.5567	0.8810×10^{-5}	0.5567	0.1554	0.9841
	0.7849×10^{-3}	0.1476×10^{-4}	0.7850×10^{-3}	0.1551	0.9811
	0.1106×10^{-2}	0.2473	0.1106×10^{-2}	0.1548	0.9775
0.04	0.1557	0.4140	0.1557	0.1545	0.9733
	0.2191	0.6922×10^{-4}	0.2192	0.1539	0.9683
	0.3079	0.1156×10^{-3}	0.3081	0.1533	0.9623
	0.4322	0.1929	0.4327	0.1526	0.9551
0.08	0.6059	0.5212	0.6067	0.1517	0.9467
	0.8477×10^{-2}	0.5341	0.8494×10^{-2}	0.1507	0.9366
	0.1183×10^{-1}	0.8859×10^{-3}	0.1187×10^{-1}	0.1496	0.9247
	0.1648	0.1465×10^{-2}	0.1654	0.1482	0.9106
0.16	0.2288	0.2416	0.2301	0.1465	0.8938
	0.3164	0.3970	0.3189	0.1445	0.8739
	0.4354	0.6491×10^{-2}	0.4402	0.1422	0.8505
	0.5958	0.1055×10^{-1}	0.6051	0.1395	0.8229
0.32	0.8094×10^{-1}	0.1704	0.8272×10^{-1}	0.1363	0.7904
	0.1089	0.2730	0.1123	0.1325	0.7525
	0.1451	0.4351	0.1514	0.1280	0.7085
	0.1905	0.6788×10^{-1}	0.2022	0.1228	0.6577
0.64	0.2457	0.1048	0.2671	0.1167	0.5999
	0.3099	0.1590	0.3484	0.1096	0.5351
	0.3799	0.2361	0.4473	0.1014×10^{1}	0.4638
	0.4488	0.3412	0.5638	0.9207	0.3674
1.28	0.5052	0.4772	0.6950	0.8138	0.3083
	0.5336	0.6414	0.8344	0.6939	0.2303
	0.5169	0.8212	0.9704	0.6617	0.1577
	0.4434	0.9923	0.1086×10^{1}	0.4202	0.9568×10^{-1}
2.56	0.3165	0.1121×10^{1}	0.1165	0.2751	0.4830
	0.1630	0.1177	0.1188	0.1373	0.1758×10^{-1}
	$+0.2903\times10^{-1}$	0.1154	0.1154	-0.2514×10^{1}	$+0.2215\times10^{-2}$
	-0.4275	0.1081	0.1082	$+0.3950$	-0.2306
5.12	-0.4570	0.1013×10^{1}	0.1014×10^{1}	0.4506	-0.1743×10^{-2}
	-0.1665×10^{-1}	0.9868	0.9870	0.1687×10^{-1}	-0.4493×10^{-3}
	$+0.1843\times10^{-2}$	0.9923	0.9920	-0.1857×10^{-2}	-0.3516×10^{-4}

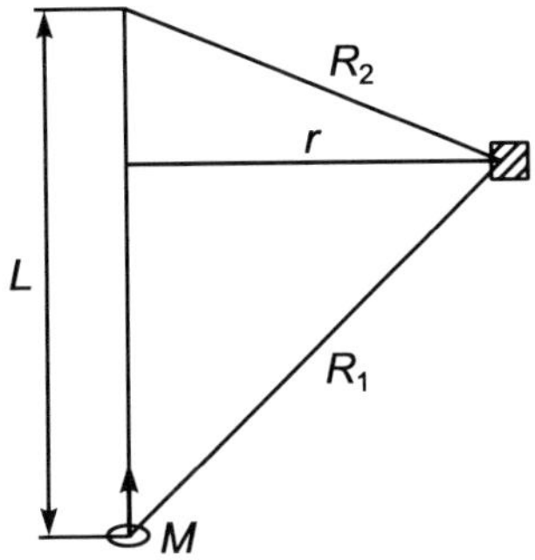

Figure 2.6. Definition of an elementary toroid in which induction current flows.

where R and θ are spherical coordinates of an observation point where the current density is calculated.

When the wave number is broken into its real and imaginary parts we can represent the current density as the sum of two components: one component is shifted by 90° relative to the current in the source dipole and is termed the quadrature component, and the other component is shifted in phase by 180° with respect to the current in the source and which is termed the inphase component. The expressions for the components of the current density are:

$$\mathrm{Q}\ J_\phi = \frac{\sigma\mu\omega}{4\pi}\frac{rM_T}{R^3}\,\mathrm{e}^{-p}\,[(1+p)\cos p + \sin p] \tag{2.56}$$

$$\mathrm{In}\ J_\phi = -\frac{\sigma\mu\omega}{4\pi}\frac{rM_T}{R^3}\,\mathrm{e}^{-p}\,[(1+p)\sin p - \cos p] \tag{2.57}$$

where $r/R = \sin\theta$ and $p = R/h$ is the distance from the dipole to the observation point expressed in units of the skin depth.

Equations 2.56 and 2.57 suggest that it is reasonable to imagine two components of the current density at every point which are distributed in an entirely different manner. In order to investigate their distribution let us first of all understand the physical meaning of the term: $(\sigma\mu\omega/4\pi)(rM_T/R^3)$ which is present in both expressions for the current density components. It is obvious that the current flow in the medium can theoretically be subdivided into currents flowing in a series of elemental toroids or within rings which have a common axis with that of the dipole and which lie in planes perpendicular to this axis as shown in Fig. 2.6.

Now let us define a current density induced in a very thin ring within a unit cross-section due to the primary magnetic field, $\boldsymbol{H}_0$ (see Fig. 2.6).

The magnetic flux piercing a ring with a radius r is:

$$\psi_0 = \int\limits_0^r \mu H_{0z} 2\pi r\,\mathrm{d}r \tag{2.58}$$

where H_{0z} is the z-component of the field due to current in the dipole source only.

The vertical component of the magnetic field caused by the vertical dipole in free space, as follows from Chapter 1, can be presented as:

$$H_{0z} = \frac{M_T}{4\pi R^3}\,(3\cos^2\theta - 1) \tag{2.59}$$

Substituting this expression for H_{0z} into eq. 2.58 and integrating we obtain the following expression for the flux, ψ_0, of the primary magnetic field enclosed by the toroid:

$$\psi_0 = \frac{1}{2}\,\mu\,\frac{M_T r^2}{R^3} \tag{2.60}$$

The magnetic flux ψ_0 and the EMF in the toroid are related through the formula:

$$\mathscr{E} = \mathscr{E}_0\,\mathrm{e}^{-\mathrm{i}\omega t} = -\frac{\partial}{\partial t}\,\psi_0\,\mathrm{e}^{-\mathrm{i}\omega t} = \mathrm{i}\omega\psi_0\,\mathrm{e}^{-\mathrm{i}\omega t} \tag{2.61}$$

On the other hand, the electromotive force is equal to the integral of the electric field along the closed path surrounding the magnetic flux:

$$\mathscr{E} = \oint \boldsymbol{E}_0\,\mathrm{d}\boldsymbol{l} = 2\pi r E_{0\phi} \tag{2.62}$$

because the field is characterized by axial symmetry. It follows from eqs. 2.60–2.62 that the electric field E_ϕ along a toroidal ring is:

$$E_{0\phi} = \frac{\mathrm{i}\omega\psi_0}{2\pi r} = \frac{\mathrm{i}\omega\mu}{4\pi}\,\frac{rM_T}{R^3} \tag{2.63}$$

It is clear the $E_{0\phi}$ is the primary vortex electrical field caused by a change of the primary magnetic field with time only.

From this we obtain the expression for the current density at any point of the toroid:

$$J_{0\phi} = \sigma E_{0\phi} = \frac{\mathrm{i}\sigma\mu\omega}{4\pi}\,\frac{rM_T}{R^3} \tag{2.64}$$

The physical meaning of this last expression should be obvious: it is the current density induced by the primary magnetic field of the dipole source alone. As can be seen from eq. 2.64 the current density $J_{0\phi}$ is shifted by 90° in phase with respect to the current of the source. If we could neglect any secondary effects caused by magnetic field accompanying induced currents in the medium the character of the current distribution could be defined precisely by eq. 2.64.

In this case the current density at any point in the medium depends on the distance from the source, and on the angle θ, that is, on geometric parameters only and it is directly proportional to the transmitter moment, conductivity, frequency and magnetic permeability. One can say that the actual distribution of the current density in a conducting medium could be described by $J_{0\phi}$, if the effect of interaction between currents, that

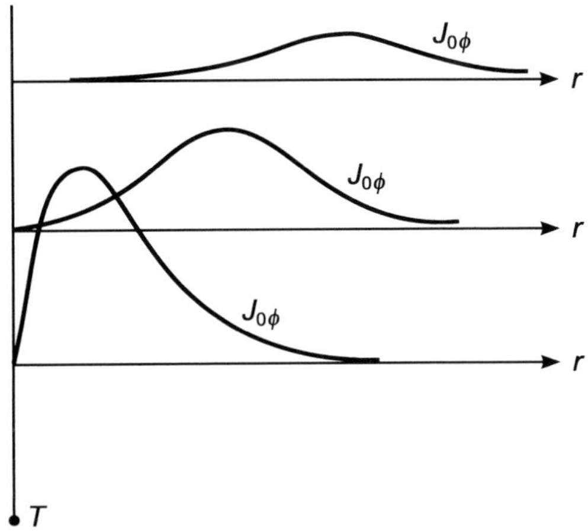

Figure 2.7. Curves showing the current density in planes perpendicular to the z-axis as a function of the distance r.

is, the skin effect, were to be negligible. Graphs of function $J_{0\phi}$ in planes perpendicular to the dipole axis are shown in Fig. 2.7. It can be seen that with increasing z the radius of the ring with maximum current also increases.

Let us define the quantity:

$$J_0 = \frac{\sigma\mu\omega M r}{4\pi R^3} = |J_{0\phi}|$$

and rewrite eqs. 2.56 and 2.57 in the form:

$$\mathrm{Q}\, J_\phi = J_0\, \mathrm{e}^{-p}\, [(1+p)\cos p + \sin p] \tag{2.65}$$

$$\mathrm{In}\, J_\phi = -J_0\, \mathrm{e}^{-p}\, [(1+p)\sin p - \cos p] \tag{2.66}$$

An analysis of these expressions permits us to explore how the actual current density J_ϕ differs from J_0 for various values of the parameter p and specifically for different distances from the source. Curves for the quadrature and inphase components of the current density normalized to J_0 are shown in Figs 2.8 and 2.9. For small values of parameter, p, the quadrature component of the current density is essentially the same as the current density J_0; that is, the interaction between currents is negligible in this case. With an increase of the parameter, p, the ratio $\mathrm{Q}\, J_\phi/J_0$ decreases, passes through zero, and for larger values of the parameter p approaches zero in an oscillating manner. Curves for the ratio of the inphase component of the actual current density to J_0 has a completely different character. For small values of p, the ratio of $\mathrm{In}\, J_\phi/J_0$ approaches zero, then increases to a maximum when the value of the parameter p is about 1.5 and for larger values of p, tends to zero again in an oscillating manner. Therefore, the actual distribution of J_0 is determined

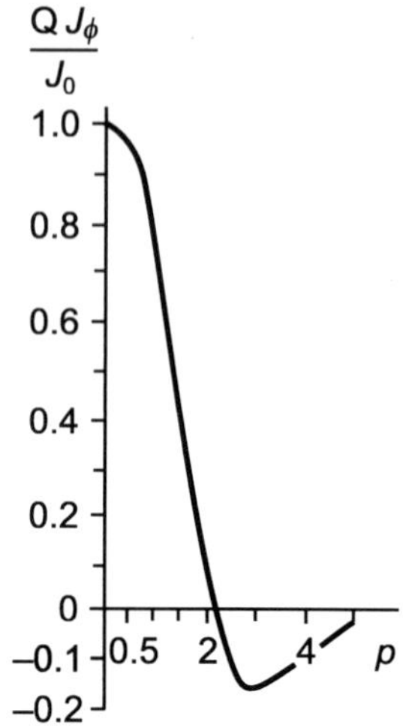

Figure 2.8. Graph of the ratio Q J_ϕ/J_0 as a function of the parameter p.

both by geometric factors and by interaction effect, known as skin effect. The last factor is taken into account in the case of a uniform full space by the parameter p.

Comparing curves in Figs 2.8 and 2.9 we can see that for small values of p the quadrature component of current density dominates, then for larger values of parameter p there is a range over which the inphase component is significantly larger in value. And finally for large values of p both components approach to zero in an oscillatory manner. The curve in Fig. 2.8 as well as that in Fig. 2.9 can be viewed from two points.

If the conductivity and frequency are held constant the curve shows a change in the quadrature component related to the current density J_0 when the distance of the observation point from the dipole source is increased. On the other hand, the observation point can be held fixed in the medium and the conductivity or the frequency can be varied over a wide range. This permits us to explain the main features of the behavior of the quadrature component of the magnetic field following from the distribution of the quadrature component of the current density.

As can be seen from Fig. 2.8 for relatively small values of p the current density Q J_ϕ does not practically differ from J_0. If the frequency is low enough and the medium possess a relatively high resistivity, the range of distances over which the actual current density is almost equal to J_0 can be so large, that the magnetic field at an observation point is entirely defined by currents in this area. In principle, according to Biot–Savart law all induced currents define the magnetic field measured in the receiver. However, in practice there is always a part of the medium where induced currents bring the main contribution to this field but the influence of the other part is negligible, taking into account the accuracy of measuring. It is natural that the size of this area and its position essentially depend on frequency, conductivity, length of the probe, as well as the component of the

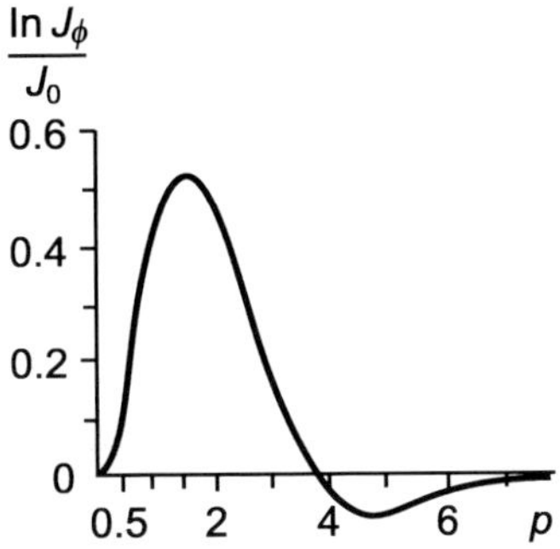

Figure 2.9. Graph of the ratio In J_ϕ/J_0 as a function of the parameter p.

field measured. It is clear that dimensions and a position of this zone, in fact, defines a depth of investigation of the method. In the case as the quadrature component of the magnetic field is measured, as follows from equation 2.65 and Fig. 2.8, this area is located around the induction probe. In accord with eq. 2.64 the current density Q J_0 quickly decreases with the distance from the dipole. The interaction between induced currents only emphasizes this tendency (see Fig. 2.8).

Now we are prepared to describe the main features of both components of the magnetic field, and therefore the electromotive force, as measured in the receiver.

As was pointed out for relatively low frequencies and high resistivities an area around the induction probe where the quadrature component of the current density, Q J_ϕ, practically coincides with $J_{0\phi}$, is sufficiently large. Because the current density is directly proportional to frequency (eq. 2.64), within the area, the magnetic field caused by these currents is also proportional to frequency. Over the range of the parameter p, when the dimensions of the area in which the actual current density is close to the current density J_0 is still significantly larger than the distance from the source to the measurement site, the quadrature component of the magnetic field would be also proportional to ω. With an increase in value of the parameter p (for example, as would be caused by an increase in frequency), the area in which the currents are essentially equal to J_0 becomes smaller, and therefore the external part of the medium in which the actual currents are less than J_0 begins to have an effect, resulting in the observation that the rate of growth of the magnetic field is reduced. With a further increase of frequency as a consequence of the rapid decrease of the quadrature component of the current density, the growth of the quadrature component of the magnetic field ceases and it begins to decrease.

By analogy the behavior of the inphase component of the magnetic field can also be explained with the use of the inphase component of currents. Here it is appropriate to notice the following. Unlike the previous case a zone of currents which gives the main contribution to the inphase component of the magnetic field is present in a confined zone, a position which essentially depends on conductivity and frequency. In particular, with a decrease of frequency is shifted far away from the induction probe and when it is located at sufficiently large distances, the measured magnetic field, In H_z^s, in accord with Biot–Savart law, is practically independent on the probe length (see eq. 2.49). In other words, a part of medium between the probe and this zone, as well as that part which is located outside of this zone, has no noticeable influence on the field measured by a receiver. As follows from Fig. 2.9, with a decrease of the parameter p (higher resistivity, lower frequency) the depth of investigation in measuring the inphase component of the secondary magnetic field increases.

In summary, let us enumerate some of the aspects of field behavior which are of practical interest. First, the quadrature component of the magnetic field is determined principally by currents flowing near the source and the observation site. With an increase in frequency the effect of interaction of currents from relatively far distant parts of the medium becomes less important. At the same time, this effect, i.e., attenuation or the skin effect, can become significant if the radius of the zone in which the quadrature component of the current density is practically the same as $J_{0\phi}$ becomes less than the distance from the transmitter to the receiver. It is important to emphasize here that in measuring the quadrature component of the field there is always a frequency ω_0 below which the depth of investigation is practically defined by only the length of the induction probe (a uniform medium). This conclusion remains also valid for a nonuniform medium, but in this general case, the depth of investigation is subjected to the influence of all geoelectric parameters.

The inphase component of the field can be more sensitive to effects from distant parts of the medium than the quadrature component. With decreasing frequency the depth of investigation, when the inphase component is being measured, gradually increases regardless of the distance between the source and the observation point and, in the general case, parameters of a geoelectric section. Here we observe the fundamental difference from the point of an influence of the frequency on the depth of investigation, depending on whether the quadrature and inphase components are measured at the range of small parameters p.

Now let us make some comments. One concerns the role of the induction probe length. With an increase in the distance between the source and the observation point, L, the induction number p also increases, and, therefore, the skin effect manifests itself more strongly. It is obvious that the position of the receiver of the induction probe does not change the character of the current distribution in the conducting medium. This fact can be explained as follows. When the separation between the source and the receiver is small, the depth of investigation, for example, measuring the quadrature component, is small as well, and the electromotive force, $Q\mathscr{E}$, in the receiver is defined primarily by currents which are approximately equal to J_0. In other words, the currents which reflect the skin depth for given values of σ and ω are situated at distances that exceed the range of investigation. As the separation between the source dipole and the observation point increases the depth of investigation of the probe also becomes greater and correspondingly the relative contribution on distance currents which have undergone the skin effect becomes more significant.

In this connection it is appropriate to emphasize that the distribution of the quadrature component of induced currents near the probe is not practically subjected to the skin effect and only distant currents are subjected to the influence of their interaction. At the same time, the inphase component of induced currents regardless of the distance from the dipole, is strongly influenced by the skin effect.

In conclusion, let us present eq. 2.54 in the explicit form, since some advantages of measuring the amplitude and the phase in the induction logging will be considered in detail. In accord with this equation we have:

$$\mathscr{E} = \mathscr{E}_0 \, A \, \mathrm{e}^{\mathrm{i}\phi} \tag{2.67}$$

here

$$A = \mathrm{e}^{-p} \left[(1+p)^2 + p^2\right]^{1/2} \tag{2.68}$$

$$\phi = p - \arctan \frac{p}{1+p} \tag{2.69}$$

Chapter 3

METHODS FOR THE SOLUTION OF DIRECT PROBLEMS OF INDUCTION LOGGING

The main task of the theory of induction logging is to determine the dependence of the quasistationary electromagnetic field, measured by a probe receiver, on the resistivity of a medium. Our investigations will naturally be based on Maxwell's equations. As was shown in Chapter 1 the problem of field determination can be formulated in the following way. All space can be represented as a sum of areas with constant parameters μ_i and σ_i, where μ_i is the magnetic permeability and σ_i is the conductivity of area D_i. Within every area D_i electric and magnetic fields satisfy Helmholtz equations:

$$\nabla^2 \boldsymbol{H}^i + k_i^2 \boldsymbol{H}^i = 0 \text{ and } \nabla^2 \boldsymbol{E}^i + k_i^2 \boldsymbol{E}^i = 0 \tag{3.1}$$

where $k^2 = \mathrm{i}\,\sigma_i \mu_i \omega$.

At interfaces of confined areas D_i, Helmholtz equations are replaced by boundary conditions which require continuity of tangential components of electric and magnetic fields. Near a source and at infinity a field must satisfy corresponding conditions which depend on the type of source and the parameters of the medium.

It is clear that we have formulated a boundary problem for harmonic fields which are used in conventional induction logging. The transition to a nonstationary field can be easily done applying Fourier integral.

In most cases which will be considered the electric field has only a component E_ϕ, which is tangential to any interface. For simplicity we will assume that in a chosen curvilinear system of coordinates α, β, ϕ, interfaces are characterized by coordinate α_i.

Then from the first Maxwell equation:

$$\operatorname{curl} \boldsymbol{E} = \mathrm{i}\omega\mu \boldsymbol{H}$$

we have:

$$H_\beta^i = \frac{1}{\mathrm{i}\omega\mu_i} \frac{1}{h_\alpha h_\phi} \frac{\partial (h_\phi E_\phi)}{\partial \alpha} \tag{3.2}$$

where h_α and h_ϕ are metric coefficients of the corresponding coordinate system.

Thus a solution of the direct problem of induction logging consists of determination of a function E_ϕ^i, which satisfies the Helmholtz equation inside every area and the boundary conditions:

$$E_\phi^k = E_\phi^l \text{ and } \frac{1}{\mu_k} \frac{\partial}{\partial \alpha}(h_\phi E_\phi^k) = \frac{1}{\mu_l} \frac{\partial}{\partial \alpha}(h E_\phi^l) \tag{3.3}$$

and which corresponds to the field behavior near a source and at infinity.

It is appropriate to notice that there will be cases when an electric field cannot be described by component E_ϕ only. In such cases, for example, when there is a displacement of an induction probe with respect to the well's axis, a more general approach to the solution of the boundary problem will be used.

In this chapter we will describe basic methods which have been mainly used in solving boundary problems of induction logging.

3.1. The Method of Separation of Variables

The Helmholtz equation:

$$\nabla^2 E^i_\phi \boldsymbol{\phi}_0 + k_i^2 E^i \boldsymbol{\phi}_0 = 0$$

is presented in partial derivatives of the second order. Here $\boldsymbol{\phi}_0$ is a unit vector tangential to the coordinate line ϕ. Let us assume that the electric field depends on coordinates α and β only (a more general case will be considered in a special chapter).

We will look for a solution of $E^i_\phi(\alpha, \beta)$ as a product of two functions, each of which depends only on one coordinate:

$$E^i_\phi = T_i(\alpha) P_i(\beta)$$

Substituting this expression into Helmholtz equation and by separating variables we obtain two normal differential equations of the second order. It is important to emphasize here that the method of separation of variables is applicable only for some orthogonal curvilinear coordinate systems.

Solutions of these two equations are functions of coordinates α and β, respectively, as well as of the wave number of the corresponding medium. Inasmuch as they are differential equations of the second order they have two independent solutions:

$$\begin{aligned} T^i_m &= A'_{mi} U_m(k_i, \alpha) + B'_{mi} V_m(k_i, \alpha) \\ P^i_m &= C'_{mi} \phi_m(k_i, \beta) + D'_{mi} \psi_m(k_i, \beta) \end{aligned} \tag{3.4}$$

here m is a constant of separation.

A general solution is presented by an integral (in some cases it can be a sum) with respect to m:

$$E^i_\phi = \int\limits_{-\infty}^{\infty} \left[A'_{mi} U_m(k_i, \alpha) + B'_{mi} V_m(k_i, \alpha)\right] \left[C'_{mi} \phi_m(k_i, \beta) + D'_{mi} \psi_m(k_i, \beta)\right] dm$$

Usually, functions T^i_m and P^i_m are complex functions which depend on complex arguments. As will be shown later, in many cases a solution can be constructed with the help of only one function of coordinate β, for instance, $\phi_m(k_i, \beta)$. Then the general solution can be written as:

$$E^i_\phi = \int\limits_{-\infty}^{\infty} [A_{mi}U_m(k_i, \alpha) + B_{mi}V_m(k_i, \alpha)]\, \phi_m(k_i, \beta)\, \mathrm{d}m \tag{3.5}$$

The unknown coefficients A_{mi} and B_{mi} are defined by the boundary conditions and the specific field behavior near the source and at infinity.

In accord with eqs. 3.3 and 3.5 we have the following equations for coefficients A_m and B_m:

$$\int\limits_{-\infty}^{\infty} [A_{mk}U_m(k_k, \alpha) + B_{mk}V_m(k_k, \alpha)]\, \phi_m(k_k, \beta)\mathrm{d}m$$

$$= \int\limits_{-\infty}^{\infty} [A_{ml}U_m(k_l, \alpha) + B_{ml}V_m(k_l, \alpha)]\, \phi_m(k_l, \beta)\, \mathrm{d}m \quad \text{if } \alpha = \alpha_{kl} \tag{3.6}$$

$$\frac{1}{\mu_k}\frac{\partial}{\partial\alpha}\int\limits_{\infty}^{\infty} h_\phi\, [A_{mk}U_m(k_k, \alpha) + B_{mk}V_{mk}(k_k, \alpha)]\, \phi_m(k_k, \beta)\, \mathrm{d}m$$

$$= \frac{1}{\mu_l}\frac{\partial}{\partial\alpha}\int\limits_{-\infty}^{\infty} h_\phi\, [A_{ml}U_m(k_l, \alpha) + B_{ml}V_m(k_l, \alpha)]\, \phi_m(k_l, \beta)\, \mathrm{d}m \quad \text{if } \alpha = \alpha_{kl} \tag{3.7}$$

where α_{lk} is the value of coordinate α, characterizing the interface between media with wave numbers k_l, and k_k, respectively.

In general, functions $\phi(k_k, \beta)$ and $\phi(k_l, \beta)$ are not orthogonal, inasmuch as they depend on the wave number k of a corresponding area of the medium. For this reason, in order to determine unknown coefficients A_m and B_m, it is necessary to present one of these as a function of the other. For example, having substituted this expansion of function $\phi_m(k_l, \beta)$ by functions $\phi_m(k_k, \beta)$ and making use of the orthogonality of functions $\phi_m(k_k, \beta)$, we obtain an infinite system of equations with an infinite number of unknowns.

In those cases in which function ϕ_m does not depend on the wave number k (this takes place when an interface coincides with a spherical one, the surface of circular cylinder or a plane), the determination of unknown coefficients is simplified to a great extent. In fact, instead of an infinite system we obtain for every harmonic, m, only two equations with four unknowns:

$$\begin{aligned} &A_{mk}U_m(k_k, \alpha) + B_{mk}V_m(k_k, \alpha) = A_{ml}U_m(k_l, \alpha) + B_{ml}V_m(k_l, \alpha) \\ &\frac{1}{\mu_k}\frac{\partial}{\partial\alpha}h_\phi\, [A_{mk}U_m(k_k, \alpha) + B_{mk}V_m(k_k, \alpha)] \\ &\qquad = \frac{1}{\mu_l}\frac{\partial}{\partial\alpha}h_\phi\, [A_{ml}U_m(k_l, \alpha) + B_{ml}V_m(k_l, \alpha)] \qquad \text{if } \alpha = \alpha_{kl} \end{aligned} \tag{3.8}$$

Making use of conditions near a source and at infinity this system, applied for every interface, completely defines the unknown coefficients A_m and B_m.

Thus the method of separation of variables in Helmholtz equation allows us in such cases to present a solution in an explicit form. In problems with cylindrical interfaces, as will be shown later, a solution can be written in the form of an improper integral containing Bessel functions of complex argument. In media with horizontal interfaces a field is also expressed through improper integrals, but in these cases the integrand is much simpler.

Both cases, when either cylindrical or horizontal interfaces are present, are of great practical interest in developing the theory of induction logging.

Solving problems with one and two cylindrical interfaces, or with one and two horizontal interfaces, by the method of separation of variables we obtain expressions which are convenient for analytical investigation as well as for programming.

However, with an increase of the number of interfaces expressions for the field become cumbersome and it is practically impossible to perform their analytical investigation. Significant difficulties also arise in programming these equations. It becomes specially noticeable when, instead of a piecewise uniform medium, a continuous change of resistivity is assumed. For example, such a behavior is observed in the invaded zone due to penetration of mud water into the formation.

When solving problems with a complicated resistivity distribution, either in a horizontal or vertical direction, it is often appropriate to apply the so-called method of shells. This approach has two merits, namely:

- a uniform algorithm of calculations which does not depend on the number of layers with different resistivity
- all formulae contain only functions with real arguments.

3.2. The Method of Shells

The idea of this method can be described in the following way. Let us imagine a set of surfaces $\alpha_i = const$. In such a way a conducting medium can be presented as a system of sufficiently thin layers. Then each layer can be replaced by an infinitesimally thin shell located at the middle part of this layer, provided that the longitudinal conductance, S, of the layer and that of the corresponding shell are the same. Here the longitudinal conductance of the layer is a product of its conductivity and thickness.

Therefore, instead of a continuous conducting medium, we obtain a system of thin conducting shells. In this case the exact boundary conditions can be replaced by approximate ones which do not require information about the field inside the shell. Correspondingly, we do not need to solve Helmholtz equation because outside and between the shells the field obeys Laplace equation. Thus our problem is reduced to the determination of a harmonic function satisfying approximate boundary conditions at the surface of shells.

However, determination of functions simultaneously satisfying boundary conditions at all shells requires a solution of a system of $2n$ equations with $2n$ unknowns. For this reason a method of reflections is suggested which can be used for the inductive excitation

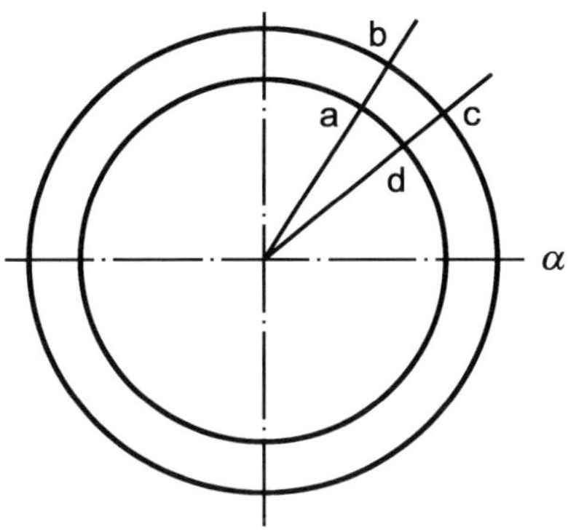

Figure 3.1. Illustration of the derivation of approximate boundary conditions.

of the field, and it allows us in a relatively simple manner to derive expressions for the field in the presence of a shell system, if some characteristics of the field for every shell are known.

3.2.1. Derivation of Approximate Boundary Conditions on a Shell Surface

Let us assume that in a curvilinear orthogonal system of coordinates α, β and ϕ, a shell surface coincides with one of the coordinate surfaces α. It is supposed that the electric field has a component E_ϕ only. Applying Ampere's law (the second Maxwell equation in the integral form) to the path '*abcd*', shown in Fig. 3.1 we obtain:

$$\oint_L \boldsymbol{H} \cdot \mathrm{d}\boldsymbol{l} = \int_Q \boldsymbol{j} \cdot \mathrm{d}\boldsymbol{Q}$$

or

$$(H^e_\beta - H^i_\beta)\,\mathrm{d}\beta = \boldsymbol{J}_\phi\,\mathrm{d}l_\alpha \cdot \mathrm{d}l_\beta \tag{3.9}$$

where: $\boldsymbol{H}$ is the magnetic field vector; $\mathrm{d}\boldsymbol{l}$ is the vector tangential to coordinate line β equal by magnitude $\mathrm{d}l_\beta$; H^e_β and H^i_β, are components of the magnetic field tangential to the surface outside and inside of it, respectively; $\mathrm{d}\boldsymbol{Q}$ is a vector surface element with a magnitude equal to $\mathrm{d}l_\alpha\,\mathrm{d}l_\beta$ (dl_α and $\mathrm{d}l_\beta$ are elementary displacements along coordinate lines α and β, respectively). These displacements along coordinate lines are expressed through coordinates as:

$$\mathrm{d}l_\alpha = h_\alpha\,\mathrm{d}\alpha \qquad \mathrm{d}l_\beta = h_\beta\,\mathrm{d}\beta \qquad \mathrm{d}l_\phi = h_\phi\,\mathrm{d}\phi$$

where h_α, h_β, h_ϕ are metric coefficients.

In problems considered here, the cylindrical system of coordinates is mainly used and correspondingly $h_\alpha = 1$. Taking into account this fact and making use of Ohm's law:

$$\boldsymbol{j} = \sigma \boldsymbol{E}$$

we present eq. 3.9 in the form:

$$H^e_\beta - H^i_\beta = \sigma\, \mathrm{d}\alpha\, E_\phi = S E_\phi \tag{3.10}$$

where $S = \sigma\, \mathrm{d}\alpha$ is the longitudinal conductance of the shell.

From the first Maxwell equation in differential form we have:

$$\boldsymbol{H} = \frac{1}{\mathrm{i}\omega\mu} \operatorname{curl} \boldsymbol{E}$$

As is well known the expression for curl in the orthogonal system of coordinates is written as:

$$\operatorname{curl} \boldsymbol{E} = \frac{1}{h_\alpha h_\beta h_\phi} \begin{vmatrix} h_\alpha \boldsymbol{I}_\alpha & h_\beta \boldsymbol{I}_\beta & h_\phi \boldsymbol{I}_\phi \\ \dfrac{\partial}{\partial \alpha} & \dfrac{\partial}{\partial \beta} & \dfrac{\partial}{\partial \phi} \\ h_\alpha E_\alpha & h_\beta E_\beta & h_\phi E_\phi \end{vmatrix}$$

where $\boldsymbol{I}_\alpha$, $\boldsymbol{I}_\beta$, $\boldsymbol{I}_\phi$ are unit vectors of coordinate system.

Inasmuch as $E_\alpha = E_\beta = 0$, we obtain:

$$\begin{aligned} H_\alpha &= -\frac{1}{\mathrm{i}\omega\mu} \frac{1}{h_\beta h_\phi} \frac{\partial (h_\phi E_\phi)}{\partial \beta} \\ H_\beta &= \frac{1}{\mathrm{i}\omega\mu} \frac{1}{h_\alpha h_\phi} \frac{\partial (h_\phi E_\phi)}{\partial \alpha} \end{aligned} \tag{3.11}$$

Substituting eq. 3.11 in eq. 3.10 we derive the first approximate boundary condition for the electric field at the shell surface:

$$\frac{\partial (h_\phi E^e_\phi)}{\partial \alpha} - \frac{\partial (h_\phi E^i_\phi)}{\partial \alpha} = \mathrm{i} n E_\phi \qquad \text{if } \alpha = \alpha_0 \tag{3.12}$$

where $n = \omega\mu S h_\phi$.

The second boundary condition requires continuity of the tangential component of the electric field:

$$E^e_\phi = E^i_\phi \qquad \text{if } \alpha = \alpha_0 \tag{3.13}$$

For this reason we can use the right-hand side of eq. 3.12 for either E^e_ϕ or E^i_ϕ.

The boundary conditions 3.12 and 3.13 sufficiently accurately describe the field near the shell surface provided that the value of the skin depth, $(2/\sigma\mu\omega)^{1/2}$, within an elementary layer is much greater than its thickness and the field slightly changes inside this layer along its normal. Correspondingly, as the conducting medium is presented as a system of elementary layers these conditions have to be met.

3.2.2. Calculation of the Electromagnetic Field Caused by Induced Currents in One Shell

As was mentioned above, the electric field has only one component E_ϕ, which satisfies Laplace equation $\nabla^2 \boldsymbol{E} = \nabla^2 E_\phi \boldsymbol{\phi}_0 = 0$ and depends on two coordinates α and β. Solving Laplace equation by the method of separation of variables we find:

$$E_\phi = \int\limits_0^\infty \left[A_m U_m(\alpha) + B_m V_m(\alpha)\right] \phi_m(\beta)\, \mathrm{d}m \tag{3.14}$$

where m is a separation constant. Functions $U_m(\alpha)$ and $V_m(\alpha)$ are radial functions (m-harmonic), while $\phi_m(\beta)$ is an angular one (in spherical coordinates $\alpha = R$, $\beta = \theta$; in cylindrical coordinates $\alpha = r$, $\beta = z$). One of the radial functions tends to zero with an increase of α, but the other one increases. For this reason the secondary field outside a shell is described by only one function, for example, $A_m U_m(\alpha)$ which decreases as the coordinate α increases. In the internal area one of the radial functions has usually a finite value, while the other one tends to infinity at some point or line. Correspondingly, the solution within the internal area of the shell is also described by one radial function which, for instance, has the form $B_m V_m(\alpha)$.

First we will assume that sources of the primary magnetic field are located outside the shell and their vortex electric field E_ϕ^0 can be presented as the integral of a product of radial and angular functions:

$$E_\phi^0 = \mathrm{i}\omega\mu \int\limits_0^\infty C_m^e V_m(\alpha) \phi_m(\beta)\, \mathrm{d}m \tag{3.15}$$

where C_m^e are known coefficients. We will look for a solution outside the shell as a sum of the primary field E_ϕ^0 and the field caused by induced currents in the shell:

$$E_\phi^e = \mathrm{i}\omega\mu \int\limits_0^\infty \left[C_m^e V_m(\alpha) + B_m^e U_m(\alpha)\right] \phi_m(\beta)\, \mathrm{d}m \tag{3.16}$$

The field inside of the shell can be presented as:

$$E_\phi^i = \mathrm{i}\omega\mu \int\limits_0^\infty A_m^e V_m(\alpha) \phi_m(\beta)\, \mathrm{d}m \tag{3.17}$$

Substituting eqs. 3.16 and 3.17 into boundary conditions 3.12 and 3.13 and making use of the orthogonality of angular functions $\phi_m(\beta)$, we obtain for the determination of coefficients A_m^e and B_m^e two equations with two unknowns:

$$\begin{aligned} &C_m^e \left(h_\phi' V_m + h_\phi V_m'\right) + B_m^e \left(h_\phi' U_m + h_\phi U_m'\right) - A_m^e \left(h_\phi' V_m + h_\phi V_m'\right) = \mathrm{i}n\, A_m^e V_m \\ &C_m^e V_m + B_m^e U_m = A_m^e V_m \qquad \text{if } \alpha = \alpha_0 \end{aligned} \tag{3.18}$$

Solving this system with respect to coefficients A^e_m and B^e_m, we obtain:

$$A^e_m = C^e_m \frac{U_m d_m - t_m V_m}{in\, U_m V_m + U_m d_m - t_m V_m} \tag{3.19}$$

$$B^e_m = -C^e_m \frac{in\, V^2_m}{in\, U_m V_m + U_m d_m - t_m V_m} \tag{3.20}$$

where

$$d_m = h'_\phi V_m + h_\phi V'_m \qquad t_m = h'_\phi U_m + h_\phi U'_m$$

Let us introduce notations:

$$P_{em} = \frac{U_m d_m - t_m V_m}{in\, U_m V_m + U_m d_m - t_m V_m} \tag{3.21}$$

$$W_{em} = -\frac{in\, V^2_m}{in\, U_m V_m + U_m d_m - t_m V_m} \tag{3.22}$$

The amplitude of function P_{em} alters with a change of n from zero to unity and characterizes the attenuation of the m-harmonic of the primary field in passing the shell. At the same time the structure of the field of a corresponding harmonic does not change. The magnitude of function W_{em} changes with an increase of n from zero to the value V_m/U_m and characterizes the intensity of reflection of the m-harmonic of the primary field from the shell for the external excitation. The physical meaning of coefficients P_{em} and W_{em} is almost obvious, specially if a change of parameter n is caused by a change of frequency ω. Under low frequencies very small currents are induced in the shell, and correspondingly a field within an internal area is close to the primary one, i.e. $P_{em} \simeq 1$, while the secondary field outside is small, $W_{em} \to 0$. On the other hand, for high frequencies the intensity of induced currents increases and correspondingly the secondary field outside increases, while within the internal area it tends to zero (effect of full screening).

Now let us consider the case, when sources of the primary field are located inside the shell (internal excitation). We will present expressions for the primary electric field in the form:

$$E^0_\phi = \mathrm{i}\omega\mu \int\limits_0^\infty C^i_m U_m(\alpha)\phi_m(\beta)\, \mathrm{d}m \tag{3.23}$$

We will look for a solution inside and outside the shell in the following forms, respectively:

$$E^i_\phi = \mathrm{i}\omega\mu \int\limits_0^\infty \left(C^i_m U_m(\alpha) + A^i_m V_m(\alpha)\right) \phi_m(\beta)\, \mathrm{d}m \tag{3.24}$$

$$E^e_\phi = \mathrm{i}\omega\mu \int\limits_0^\infty B^i_m U_m(\alpha)\phi_m(\beta)\, \mathrm{d}m \tag{3.25}$$

Substituting eqs. 3.24 and 3.25 into the boundary conditions we obtain the system:

$$B^i_m\left(h'_\phi U_m + h_\phi U'_m\right) + C^i_m\left(h'_\phi U_m + h_\phi U'_m\right) - A^i_m\left(h'_\phi V_m + h_\phi V_m + h_\phi V'_m\right) = \mathrm{i}n\, B^i_m U_m$$
$$C^i_m U_m + A^i_m V_m = B^i_m U_m \tag{3.26}$$

Solving this system with respect to A^i_m and B^i_m we have:

$$B^i_m = \frac{d_m U_m - t_m V_m}{\mathrm{i}n\, U_m V_m + U_m d_m - t_m V_m} C^i_m \tag{3.27}$$

$$A^i_m = -\frac{\mathrm{i}n\, U^2_m}{\mathrm{i}n\, U_m V_m + U_m d_m - t_m V_m} C^i_m \tag{3.28}$$

Let us introduce notations:

$$B^i_m = P_{im} C^i_m \qquad A^i_m = W_{im} C^i_m \tag{3.29}$$

where

$$W_{im} = -\frac{\mathrm{i}n\, U^2_m}{\mathrm{i}n\, U_m V_m + U_m d_m - t_m U_m} \tag{3.30}$$

Comparing eqs. 3.21 and 3.27 we see that:

$$P_{em} = P_{im} \tag{3.31}$$

It is essential to know that for the internal excitation, as well as for the external one, every harmonic of the primary field passing the shell is reduced but that its structure has not been changed. Therefore, the shell, characterized by coordinate $\alpha = const$, does not distort the structure of any harmonic, regardless of the location of the primary source with respect to the shell.

This feature takes place also for shells having a finite thickness. However, in the latter determination the field is related to a solution of Helmholtz equation.

3.2.3. The Field in a Presence of Two Confocal Shells

Let us consider some examples of the different location of field sources.

External excitation

We suppose that field sources are located outside both shells and, first of all, derive formulae for a field caused by one harmonic of the primary field only:

$$E^0_m = \mathrm{i}\omega\mu C^e_m V_m(\alpha)\phi_m(\beta) \tag{3.32}$$

For simplicity of transformations let $i\omega\mu C_m^{\prime e} = 1$. The process of forming the secondary field due to the m-harmonic can be presented in the following way. The field E_m^0 is partly reflected by the shell with the reflection coefficient W_{ie}:

$$W_{ie}U_m(\alpha)\phi_m(\beta) \tag{3.33}$$

The other part of the primary field passes the first shell without a change of its structure and it is equal to:

$$P_1 V_m(\alpha)\phi_m(\beta) \tag{3.34}$$

since $P_{1e} = P_{1i} = P_1$.

Arriving at the internal shell it is partly reflected (coefficient of reflection W_{2e}):

$$W_{2e}P_1 U_m(\alpha)\phi_m(\beta) \tag{3.35}$$

The remaining part passes through the internal shell, decreasing with a factor P_2:

$$P_1 P_2 V_m(\alpha)\phi_m(\beta) \tag{3.36}$$

Let us return to that part which is reflected from the internal shell (eq. 3.35). It arrives at the internal surface of the external shell and is partly reflected (the reflection coefficient W_{1i}):

$$P_1 W_{2e} W_{1i} V_m(\alpha)\phi_m(\beta)$$

The other part goes through the external shell and is reduced with a factor P_1:

$$P_1^2 W_{2e} U_m(\alpha)\phi_m(\beta)$$

This process continues until induced currents are established in both shells. Let us write down sums which describe fields outside, between, and inside shells:

Outside shells

$$V_m\phi_m + W_{1e}U_m\phi_m + P_1^2 W_{2e}U_m\phi_m + P_1^2 W_{2e}^2 W_{1i}U_m\phi_m + P_1^2 W_{2e}^3 W_{1i}^2 U_m\phi_m + \cdots$$

Between shells

$$P_1 V_m\phi_m + P_1 W_{2e}U_m\phi_m + P_1 W_{2e}W_{1i}V_m\phi_m + P_1 W_{2e}^2 W_{1i}U_m\phi_m + P_1 W_{2e}^2 W_{1i}^2 V_m\phi_m + \cdots$$

Inside shells

$$P_1 P_2 V_m\phi_m + P_1 P_2 W_{2e}W_{1i}V_m\phi_m + P_1 P_2 W_{2e}^2 W_{1i}^2 V_m\phi_m + \cdots$$

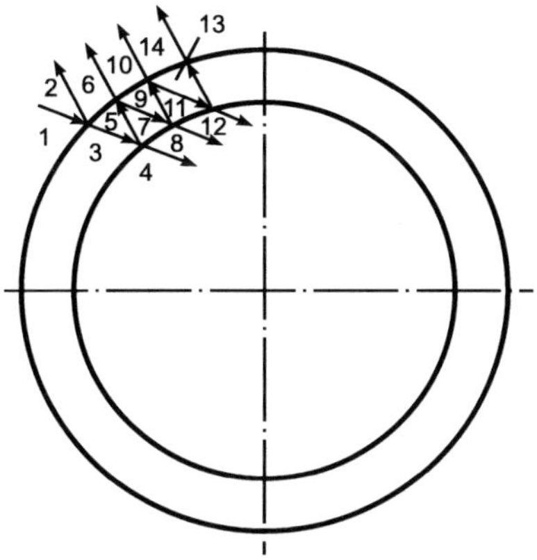

Figure 3.2. Illustration of formation of the field for external excitation.

For all these series summation is easy since they present geometric progression. For example, outside the shells we have:

$$V_m\phi_m + \left(W_{1e} + \frac{P_1^2 W_{2e}}{1 - W_{2e}W_{1i}}\right) U_m\phi_m \tag{3.37}$$

while inside shells we obtain:

$$\frac{P_1 P_2}{1 - W_{2e}W_{1i}} V_m\phi_m \tag{3.38}$$

The field-forming process when sources are located outside shells is illustrated in Fig. 3.2.

In accord with eqs. 3.37 and 3.38 in the case of external excitation of the field, two shells are equivalent to one shell which has the following expressions for coefficient of screening and reflections, respectively:

$$P_{em} = \frac{P_1 P_2}{1 - W_{2e}W_{1i}} \tag{3.39}$$

$$W_{em} = W_{1e} + \frac{P_1^2 W_{2e}}{1 - W_{2e}W_{1i}} \tag{3.40}$$

Internal excitation

Now we will assume that sources of the primary field are located inside shells and as before consider the case when the field is caused by the m-harmonic of the primary field:

$$E_m^0 = U_m(\alpha)\phi_m(\beta) \tag{3.41}$$

letting $i\omega\mu C^i = 1$.

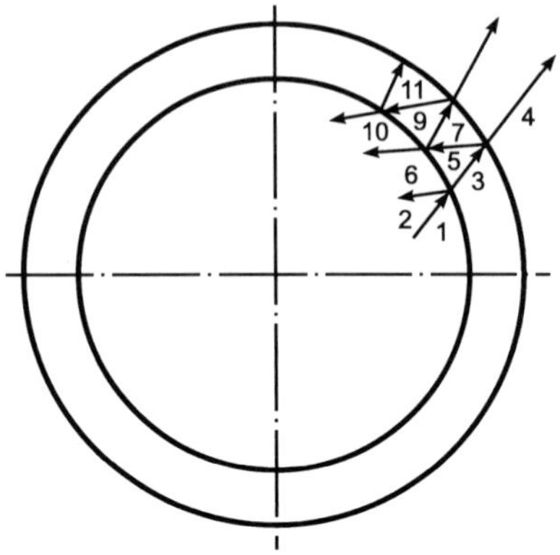

Figure 3.3. Illustration of process of arising the secondary field; sources of the primary field are located inside shells.

In this case the process of forming the secondary field caused by induced currents in both shells, in essence, does not differ from the previous case, when sources of the primary field were located outside. For this reason we will describe only the process of field formation illustrated by Fig. 3.3 and write down the corresponding sums:

Outside shells

$$\left[P_1P_2U_m + P_1P_2W_{1i}W_{2e}U_m + P_1P_2W_{1i}^2W_{2e}^2U_m + \cdots\right]\phi_m$$

Between shells

$$\left[P_2U_m + P_2W_{1i}V_m + P_2W_{1i}W_{2e}U_m + P_2W_{1i}^2W_{2e}V_m + P_2^2W_{1i}^3W_{2e}^2U_m + \cdots\right]\phi_m$$

Inside shells

$$\left[U_m + W_{2i}V_m + P_2^2W_{1i}^2V_m + P_2^2W_{1i}^2W_{2e}V_m + P_2^2W_{1e}^2V_m + \cdots\right]\phi_m$$

Performing summation of these expressions we obtain the following equations outside and inside shells:

$$\frac{P_1P_2}{1-W_{1i}W_{2e}}U_m(\alpha)\phi_m(\beta) \tag{3.42}$$

$$U_m\phi_m + \left(W_{2i} + \frac{P_2^2W_{1i}}{1-W_{1i}W_{2e}}\right)V_m(\beta)\phi_m(\beta) \tag{3.43}$$

Thus for the internal excitation, the field caused by induced currents in both shells is equivalent to that caused by currents in one shell with the screening coefficient:

$$P_{im} = P_{em} = \frac{P_1P_2}{1-W_{1i}W_{2e}}$$

and the reflection coefficient:

$$W_{im} = W_{2i} + \frac{P_2^2 W_{1i}}{1 - W_{1i}W_{2e}} \tag{3.44}$$

In general, when we have a system of confocal shells calculation of the secondary field caused by the m-harmonic of the primary field can be performed in the following way. First of all, three characteristics — W_{em}, W_{im} and P_m — of two shells, located closer to the source of the primary field than others are calculated. Then they are replaced by one shell, which is equivalent to them, and its characteristics are defined from eqs. 3.39, 3.40 and 3.44. After this the screening and reflection coefficients of the third shell are calculated and then the characteristics of the shell which is equivalent to all three shells are defined. This process continues until all shells are replaced by one which is equivalent to all of them. Calculations are made using eq. 3.39, 3.40 and 3.44, while characteristics of every shell are defined from eqs. 3.21, 3.22 and 3.30. As has been mentioned above these coefficients are expressed through well known functions with real arguments. The final expression for the field will be obtained after summation of the secondary fields, caused by all harmonics of the primary field. It is clear that components of the magnetic field are defined from Maxwell's equation:

$$\boldsymbol{H} = \frac{1}{\mathrm{i}\omega\mu}\,\mathrm{curl}\,\boldsymbol{E}$$

From this consideration it follows that in presenting a conducting medium as a system of confocal shells it is rather simple for us to calculate the field. Inasmuch as the conductivity, σ, is present only in parameter n (eq. 3.12) the calculation procedure does not depend on the distribution of medium conductivity. In other words, the field is calculated by the same formulae (eqs. 3.21, 3.22, 3.30, 3.39, 3.40, and 3.44) for a uniform medium, for a piecewise uniform medium, and for the general case of an arbitrary change of resistivity in the direction of coordinate α.

As an example, let us examine the application of the method of shells for the calculation of the field of a vertical magnetic dipole, located on the borehole axis when conductivity σ is a function of the distance from the axis only. In particular, we can imagine a medium with several uniform parts such as a borehole, invaded zone, formation.

Applying a system of confocal cylindrical surfaces, $r = const$, we theoretically divide the conducting medium in sufficiently thin cylindrical layers with a common axis coinciding with the borehole axis. In general, the layer thickness is defined by conditions, formulated at the beginning of this section, as well as by the character of the change of resistivity within this elementary layer. Then every layer is replaced by an infinitely thin shell located at the middle of the layer. The longitudinal conductance of the layer and the corresponding shell is the same. Thus instead of a continuous medium we obtain a system of thin coaxial cylindrical shells. As is well known, such replacement is possible, if induced currents do not intersect cylindrical surfaces $r = const$. This requirement is met as the source of the field is the vertical magnetic dipole, inasmuch as vector lines of the electric field are circles located in horizontal planes with centers on the borehole axis.

As in the case of a uniform medium the field can be described with the help of the z-component of the vector potential (A_z^*) only. This function satisfies Laplace equation between shells, which in cylindrical coordinates can be written as:

$$\frac{\partial^2 A_z^*}{\partial r^2} + \frac{1}{r}\frac{\partial A_z^*}{\partial r} + \frac{\partial^2 A_z^*}{\partial z^2} = 0 \tag{3.45}$$

since the field is independent of coordinate ϕ. After separation of variables in eq. 3.45 we obtain two normal differential equations of the second order:

$$Z'' + m^2 Z = 0 \tag{3.46}$$

$$R'' + \frac{1}{r}R' - m^2 R = 0 \tag{3.47}$$

where:

$$Z'' = \frac{\mathrm{d}^2 Z}{\mathrm{d}z^2} \qquad R' = \frac{\mathrm{d}R}{\mathrm{d}r} \qquad R'' = \frac{\mathrm{d}^2 R}{\mathrm{d}r^2}$$

Functions Z and R depend on coordinates z and r, respectively, and are related with component A_z^* as $A_z^* = R \cdot Z$; m is a separation variable.

Solutions of the first equation are harmonic functions $\sin mz$ and $\cos mz$. The second equation is the modified Bessel equation, solutions of which are functions $I_0(mr)$ and $K_0(mr)$. Inasmuch as the field is an even function with respect to coordinate z, it cannot contain $\sin mz$. For this reason, the general solution presents a combination of functions such as $K_0(mr)\cos mz$ and $I_0(mr)\cos mz$.

As follows from the previous chapter the vector potential of the magnetic dipole in a free space is described by the function:

$$\frac{\mathrm{i}\omega\mu M}{4\pi(r^2 + z^2)^{1/2}}$$

which can be presented through elementary solutions:

$$A_z^{*(0)} = \frac{\mathrm{i}\omega\mu M}{4\pi}\frac{2}{\pi}\int_0^\infty K_0(mr)\cos mz\,\mathrm{d}m \tag{3.48}$$

The electric field is related with the z-component of the vector potential as:

$$E_\phi = -\frac{\partial^2 A_z^*}{\partial r^2}$$

Therefore:

$$E_\phi^{(0)} = \frac{\mathrm{i}\omega\mu M}{4\pi}\frac{2}{\pi}\int_0^\infty mK_1(mr)\cos mz\,\mathrm{d}m \tag{3.49}$$

since

$$\frac{\partial}{\partial r}K_0(mr) = -mK_1(mr)$$

The primary excitation is presented as a sum (more precisely, the integral) of elementary excitations such as $mK_1(mr)\cos mz$. The penetration of these harmonics in a direction perpendicular to the borehole axis depends on the value of m. As follows from the behavior of the function $K_1(mr)$ with an increase of argument mr, for example due to m, a harmonic decays more rapidly. For this reason excitation of the most removed parts of a medium is realized by harmonics which are characterized by relatively small values of m.

The method of reflections, described above, is applied for every harmonic. The total electric field is expressed through the integral, and its integrand defines a reaction of a medium due to action of corresponding harmonic of the primary field. For this reason for electric field in the internal area we have the following expression:

$$E_\phi = \frac{i\omega\mu M}{4\pi}\frac{2}{\pi}\int\limits_0^\infty mW_i I_1(mr)\cos mz\, dm \tag{3.50}$$

where W_i is the function, characterizing an interaction of all shells.

Now using results given at the beginning of this section we will replace functions $U_m(\alpha)$ and $V_m(\alpha)$ by functions $mK_1(mr)$ and $mI_1(mr)$, respectively. Taking into account that in the cylindrical system of coordinates the metric coefficient $h_\phi = r$, we obtain expressions for reflection and screening coefficients, describing the field in the presence of one shell:

$$\begin{aligned}
W_{me} &= -\frac{in\, I_1^2(mr)}{1 + in\, I_1 K_1} \\
P_m &= \frac{1}{1 + inI_1K_1} \\
W_{mi} &= -\frac{in\, K_1^2(mr)}{1 + in\, I_1 K_1}
\end{aligned} \tag{3.51}$$

where $n = \omega\mu Sr$; $s = \sigma\Delta r$ is the longitudinal conductance of the shell; r is its radius; W_{me} is the reflection coefficient of the shell for the external excitation by harmonic $mK_1(mr)\cos mz$; W_{mi} is the reflection coefficient of the shell for the internal excitation by the same harmonic.

Coefficient P_m characterizes the decrease of the harmonic passing the shell. Interaction of shells is calculated by formulae 3.39, 3.40 and 3.44. Function W_i characterizes the interaction of all shells and it is the kernel function of the expression for the electric field in eq. 3.50.

As is known, the vertical component of the magnetic field is related with the electrical field in the following way:

$$H_z = -\frac{1}{i\omega\mu}\frac{1}{r}\frac{\partial(rE_\phi)}{\partial r}$$

Taking into account that:

$$\frac{1}{r}\frac{\partial}{\partial r} r I_1(mr) = \frac{1}{r}\left[I_1(mr) + mrI_1'(mr)\right] = mI_0(mr) \quad \text{and} \quad I_0(0) = 1$$

we obtain:

$$h_z = \frac{L^3}{\pi}\int_0^\infty m^2 W_i \cos mz \, dm \tag{3.52}$$

where h_z is the magnetic field at the borehole axis in units of the field in a free space, $2M/4\pi L^3$; L is the length of the two-coil induction probe.

Let us consider some features of the behavior of function $m^2 W_i$ for a given value of m. We will present function $m^2 W_{mi}$ in the form:

$$m^2 W_{mi} = \frac{m^2 n K_1^2}{\left[1 + n^2 (I_1 K_1)^2\right]^{1/2}} \exp\left[-i\left(\frac{\pi}{2} + \operatorname{arccoth} nI_1K_1\right)\right]$$

The product $I_1 K_1$ does not exceed unity. Therefore, for small values of n the phase of the secondary field slightly differs from 90°. In other words, induced currents in shells are shifted in phase by 90° with respect of the magnetic dipole current. With a decrease of a shell's radius such behavior is observed at higher frequencies and for more conductive shells. With an increase of a shell radius the argument, mr, increases, and a value of function $\omega\mu S r I_1(mr) K_1(mr)$ tends to the limit $\omega\mu S/2m$. If $\omega\mu S/2 \ll m$, the phase of induced currents in a shell is close to 90°.

However, harmonics with large values of m slightly penetrate into a medium. For this reason excitation of currents in shells with a relatively large radius is realized by harmonics having small values of m for which the inequality $\omega\mu S/2 \gg m$ is valid and correspondingly the phase of currents approaches to 180°.

This analysis is useful to define the role of various parts of the integrand in eq. 3.52 in calculating quadrature and inphase components of the field.

Thickness of shells is chosen from calculations and it depends on both parameters, $\omega\mu S r$ and m. For example, with an increase of m the shell thickness must be smaller. It is reasonable to choose a constant ratio between the shell thickness and its radius within a certain interval of a change of radius r. Numerical analysis shows that in practice this ratio changes from 0.02 to 0.05. The maximal radius of the shell, which is the most remote from the borehole axis, essentially depends on m. For larger values of mr the reflection coefficient for the internal excitation, W_{mi}, decreases as e^{-2mr}. Therefore, it is sufficient to satisfy the condition $r_{max} \leqslant 10/m$.

The minimal radius of the shell naturally coincides with the radius of nonconducting part of the induction probe. In those cases, when the argument mr is small ($mr \ll 1$), it is convenient to use approximate expressions for shell coefficients. For $x > 0$ functions $I_1(x)$ and $K_1(x)$ tend to $x/2$ and $1/x$ correspondingly and therefore instead of eq. 3.51

we have:

$$
\begin{aligned}
W_{me} &= -\frac{in\, m^2 r^2}{4(1+in/2)} \\
P_m &= \frac{1}{(1+in/2)} \\
W_{mi} &= \frac{-in}{m^2 r^2 (1+in/2)} \qquad \text{if } mr < 1
\end{aligned}
\tag{3.53}
$$

In conclusion let us notice that the method of shells can be considered as the algorithm of the calculation of the integrand in eq. 3.52 describing the magnetic field in the borehole.

3.3. The Method of Integral Equations

The analysis of the electromagnetic field of a vertical magnetic dipole located either on the axis of cylindrical interfaces (formations of an infinite thickness) or in a medium with horizontal interfaces only allows us to investigate the influence of the borehole and the invasion zone, as well as the effect caused by a finite thickness of the formation. For such models application of the separation of variables method is the most natural approach enabling us to present the field in a explicit form by known functions. It is a much more complicated problem when the vertical magnetic dipole is located on the borehole axis and the formation has a finite thickness. In this case the method of separation of variable cannot be used, since both cylindrical and horizontal interfaces are present and it is more appropriate to apply such numerical methods as integral equations or finite elements.

In fact, during the last 30 years the use of integral equations has allowed us to move significantly forward in the theory and interpretation of induction logging. This is the main reason why we will describe here only this numerical method. At the same time it is reasonable to point out that both methods have been used, provided that a model of the medium and a field have cylindrical symmetry with the common axis. Until now this restriction has not permitted us to investigate a field behavior in the case when the boundaries between a formation and a surrounding medium are not perpendicular to the borehole axis.

Now let us suppose that a vertical magnetic dipole is located on the borehole axis and the medium possesses axial symmetry (Fig. 3.4). In accord with the Biot–Savart law the current of the magnetic dipole creates the primary magnetic field and its change with time generates the primary vortex electric field. Due to the axial symmetry this electric field does not intersect boundaries between media with different conductivities. Because of this no electric charges develop and as a result of the existence of the vortex electric field currents arise at every point, of the conductive medium with a density given by:

$$
J_\phi = \sigma(E_{0\phi} + E_{s\phi}) \tag{3.54}
$$

where $E_{0\phi}$ is the primary vortex electric field strength; $E_{s\phi}$ is a secondary vortex electric field caused by the magnetic field from induced currents in a conductive medium; σ is the conductivity at a given point.

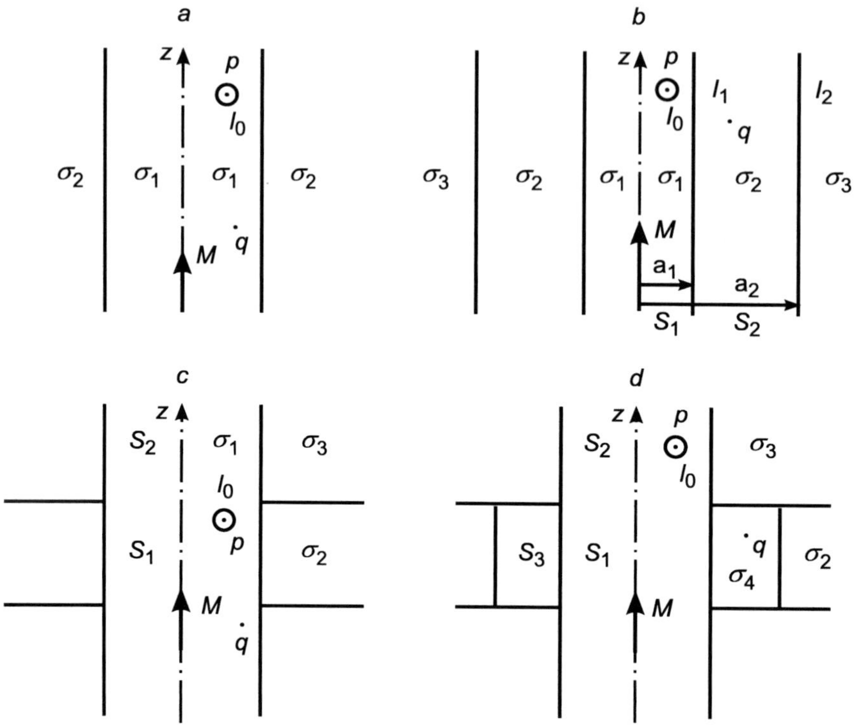

Figure 3.4. Models in induction logging with axial symmetry.

Inasmuch as electric charges are absent, the induced currents as well as the primary vortex electric field $E_{0\phi}$, have only an azimuthal component J_ϕ in the cylindrical system of coordinates r, ϕ, z (Fig. 3.4). It is obvious that interaction between current filaments does not change the direction of current flow in this case. Thus the total electric field is:

$$E_\phi = E_{0\phi} + E_{s\phi} \tag{3.55}$$

As was shown in Chapter 1 a circular current filament passing an elementary current tube at the point q creates the vortex electric field at point p (Fig. 3.4c) equal to:

$$\mathrm{i}\omega\mu G(p,q) J_\phi(q)\,\mathrm{d}S$$

where $\mathrm{d}S$ is the cross-sectional area of the tube and $G(p,q)$ is a function, which depends on geometric parameters and can be expressed through complete elliptical integrals, $J_\phi(q)$ is the current density at the point q.

Now applying the principle of superposition the total electric field can be written as:

$$E_\phi(p) = E_{0\phi}(p) + \mathrm{i}\omega\mu \int\limits_S G(p,q)\, J_\phi(q)\,\mathrm{d}S \tag{3.56}$$

where S is the half-plane described by equations:

$$r > 0 \qquad -\infty < z < \infty$$

Making use of Ohm's law:

$$J_\phi(q) = \sigma(q)E_\phi(q)$$

we obtain:

$$E_\phi(p) = E_{0\phi}(p) + \mathrm{i}\omega\mu \int_S \sigma(q)G(p,q)E_\phi(q)\,\mathrm{d}S \tag{3.57}$$

This is an integral equation of the Fredholm type of the second kind with respect to an unknown total field, E_ϕ. Replacing the elementary surface $\mathrm{d}S$ by $\mathrm{d}r\mathrm{d}z$ we have:

$$E_\phi(p) = E_{0\phi}(p) + \mathrm{i}\omega\mu \int_{-\infty}^{\infty} \mathrm{d}z \int_0^{\infty} \sigma(q)G(p,q)E_\phi(q)\,\mathrm{d}r \tag{3.58}$$

Taking into account eq. 3.55 the integral equation with respect to the secondary field has the following form:

$$E_{s\phi}(p) = F(p) + \mathrm{i}\omega\mu \int_{-\infty}^{\infty} \mathrm{d}z \int_0^{\infty} \sigma(q)G(p,q)E_{s\phi}(q)\,\mathrm{d}r \tag{3.59}$$

where

$$F(p) = \mathrm{i}\omega\mu \int_{-\infty}^{\infty} \mathrm{d}z \int_0^{\infty} \sigma(q)E_{0\phi}(q)G(p,q)\,\mathrm{d}r$$

is the known function.

One can conceptually replace the half-plane with a system of small cells within each of which the electric field is practically constant. In doing so the integral equation 3.59 can approximately be rewritten as:

$$E_{s\phi}(p) \simeq F(p) + \mathrm{i}\omega\mu \sum_{n=1}^{N} \sigma(q)G(p,q)E_{s\phi}(q)\Delta S$$

Having written this equation for every cell, we obtain a system of N linear equations with N unknown terms.

However this equation is not used in practice since the infinite limit with respect to distance r creates serious numerical problems for the determination of the electric field. Also it does not allow us to derive relatively simple asymptotic formulae, except for one special case corresponding to Doll's approximation when the skin effect is completely ignored.

In order to facilitate calculations of the field and obtain asymptotic formulae for the field, we will derive an integral equation as the area of integration with respect to distance r is limited.

First of all let us assume that the invasion zone is absent and proceeding from Green's formula we will obtain an integral equation for component E_ϕ in which integration is performed over the cross-section of the borehole only. At the beginning, suppose that the formation is uniform with conductivity σ_2 (Fig. 3.4a).

It is well known that the electric field satisfies Helmholtz equation:

$$\nabla^2 \boldsymbol{E} + k^2 \boldsymbol{E} = 0 \tag{3.60}$$

where $k^2 = \mathrm{i}\sigma\mu\omega$.

Let us represent the electric field as a sum:

$$\boldsymbol{E} = \boldsymbol{E}_0 + \boldsymbol{E}_1 \tag{3.61}$$

where $\boldsymbol{E}_0$ is the field in the uniform medium with conductivity of the formation, σ_2, consisting of the field caused by the dipole current in a free space and the field of eddy currents induced in the conducting medium. This field satisfies the following equation at all points in the medium:

$$\nabla^2 \boldsymbol{E}_0 + k_2^2 \boldsymbol{E}_0 = 0 \tag{3.62}$$

where $k_2^2 = \mathrm{i}\sigma_2\mu\omega$. The function $\boldsymbol{E}_1$ is the field due to the presence of the borehole with conductivity σ_1 and radius a.

In accord with eqs. 3.60 and 3.61 the electric field $\boldsymbol{E}_1$ is a solution of the equation:

$$\nabla^2 \boldsymbol{E}_1 = -k^2 \boldsymbol{E}_1 - k^2 \boldsymbol{E}_0 - \nabla^2 \boldsymbol{E}_0 \tag{3.63}$$

Taking into account eq. 3.63 we have for the formation and the borehole, respectively:

$$\nabla^2 \boldsymbol{E}_1 = -k_2^2 \boldsymbol{E}_1 \qquad \text{if } r > a \tag{3.64}$$

$$\nabla^2 \boldsymbol{E}_1 = -k_1^2 \boldsymbol{E}_1 + (k_2^2 - k_1^2) \boldsymbol{E}_0 \qquad \text{if } r < a \tag{3.65}$$

where $k_1^2 = \mathrm{i}\sigma_1\mu\omega$ and σ_1 is the borehole conductivity.

It is obvious that:

$$\boldsymbol{E}_0 = E_0 \boldsymbol{I}_\phi \qquad \boldsymbol{E}_1 = E_1 \boldsymbol{I}_\phi \tag{3.66}$$

We will introduce a function $\boldsymbol{G} = G\boldsymbol{I}_\phi$, which is continuous with the first derivative over all space and satisfies the equation

$$\nabla^2 \boldsymbol{G} + k_2^2 \boldsymbol{G} = 0 \tag{3.67}$$

except a point p at which the field $\boldsymbol{E}_1$ is defined. At this point the function $\boldsymbol{G} = G\boldsymbol{I}_\phi$ has a logarithmic singularity.

Now we will consider the expression:

$$\boldsymbol{G}\nabla^2\boldsymbol{E}_1 - \boldsymbol{E}_1\nabla^2\boldsymbol{G}$$

Inasmuch as:

$$\nabla^2\boldsymbol{E}_1 = \nabla^2 E_1\boldsymbol{I}_\phi = E_1\nabla^2\boldsymbol{I}_\phi + \boldsymbol{I}_\phi\nabla^2 E_1$$

and

$$\nabla^2\boldsymbol{G} = \nabla^2 G\boldsymbol{I}_\phi = G\nabla^2\boldsymbol{I}_\phi + \boldsymbol{I}_\phi\nabla^2 G$$

we have:

$$\begin{aligned}\boldsymbol{G}\nabla^2\boldsymbol{E}_1 - \boldsymbol{E}_1\nabla^2\boldsymbol{G} &= E_1\boldsymbol{G}\nabla^2\boldsymbol{I}_\phi + G\nabla^2 E_1 - G\boldsymbol{E}_1\nabla^2\boldsymbol{I}_\phi - E_1\nabla^2 G \\ &= G\nabla^2 E_1 - E_1\nabla^2 G\end{aligned} \tag{3.68}$$

It is also appropriate to make the following comment. In practice the magnetic dipole presents itself as a small horizontal loop with its center located on the borehole axis. In approaching this loop the primary electric field tends to infinity as a logarithmic function, while the electric field caused by induced currents does not have a singularity. Correspondingly, the electric field $\boldsymbol{E}_1$ is a continuous function everywhere in the borehole as well as in the formation. Taking into account the two-dimensionality of the model we will use a two-dimensional form of Green's formula. Let us assume that point p is located inside the borehole. Then for the borehole and the formation parts of the space we have:

$$\int\limits_{S_i}(G\nabla^2 E_1 - E_1\nabla^2 G)\,\mathrm{d}S = \int\limits_{l_0}\left(G\frac{\partial E_1}{\partial n_+} - E_1\frac{\partial G}{\partial n_+}\right)\,\mathrm{d}l + \int\limits_{l}\left(G\frac{\partial E_1}{\partial n_+} - E_1\frac{\partial G}{\partial n_+}\right)\,\mathrm{d}l \tag{3.69}$$

and

$$\int\limits_{S_e}(G\nabla^2 E_1 - E_1\nabla^2 G)\,\mathrm{d}S = \int\limits_{l}\left(G\frac{\partial E_1}{\partial n_-} - E_1\frac{\partial G}{\partial n_-}\right)\,\mathrm{d}l \tag{3.70}$$

where $n_+ = r$, $n_- = -r$; l is the straight line on the borehole surface which is parallel to z-axis; l_0 is a contour around point p.

From eqs. 3.64 to 3.67 we have for the formation:

$$G\nabla^2 E_1 - E_1\nabla^2 G = 0$$

and in the borehole:

$$G\nabla^2 E_1 - E_1\nabla^2 G = (k_2^2 - k_1^2)E_0 G + (k_2^2 - k_1^2)E_1 G$$

Near the point p the field E_1 is bounded but the function G increases without limit as $\ln r$, where r is the radius for the circumference l_0. Therefore the value for the integral along the contour l_0 as $r \to 0$ tends to the value $-2\pi E_1(p)$. Combining eqs. 3.69 and 3.70 and making use of the continuity of the tangential components of the electric and magnetic fields the integrals along line l vanish. Respectively, we obtain an integral equation, which includes a surface integral only over a restricted area in the radial direction corresponding to half the cross-section of the borehole:

$$E_1(p) = \frac{k_1^2 - k_2^2}{2\pi} \int\limits_{S_i} E_0(q)(k_2, p, q)\,\mathrm{d}S + \frac{k_1^2 - k_2^2}{2\pi} \int\limits_{S_i} E_1(q)G(k_2, p, q)\,\mathrm{d}S \tag{3.71}$$

The function $G(k_2, p, q)$ describes with accuracy of a constant the electric field in a uniform medium with conductivity σ_2, generated by a circular current filament. As will be shown in the next chapter it can be expressed through the proper integral from the elementary function.

In addition we will remember that the electromotive force induced in the receiver coil is defined as:

$$\mathscr{E} = 2\pi r_0 n E$$

where r_0 and n are radius and number of turns of this coil, respectively.

Now we will suppose that there is an invasion zone and the formation has an infinite thickness (Fig. 3.4b). In this case Green's function satisfies equation:

$$\nabla^2 \boldsymbol{G} + k_3^2 \boldsymbol{G} = 0 \tag{3.72}$$

and it has a logarithmic singularity at point p.

In accord with eq. 3.65 we have for the field $\boldsymbol{E}_1$:

$$\begin{aligned} \nabla^2 \boldsymbol{E}_1 &= -k_1^2 \boldsymbol{E}_1 + (k_3^2 - k_1^2)\boldsymbol{E}_0 \qquad 0 < r < a_1 \\ \nabla^2 \boldsymbol{E}_1 &= -k_2^2 \boldsymbol{E}_1 + (k_3^2 - k_2^2)\boldsymbol{E}_0 \qquad a_1 < r < a_2 \\ \nabla^2 \boldsymbol{E}_1 &= -k_3^2 \boldsymbol{E}_1 \end{aligned} \tag{3.73}$$

Applying Green's formulas for borehole, invasion zone and formation we have, respectively:

$$\int\limits_{S_1} (G\nabla^2 E_1 - E_1 \nabla^2 G)\,\mathrm{d}S = -2\pi E_1(p) + \int\limits_{l_1} \left(G\frac{\partial E_1}{\partial r} - E_1 \frac{\partial G}{\partial r} \right) \mathrm{d}l \tag{3.74}$$

$$\int\limits_{S_2} (G\nabla^2 E_1 - E_1 \nabla^2 G)\,\mathrm{d}S = \int\limits_{l_1} \left(-G\frac{\partial E_1}{\partial r} + E_1 \frac{\partial G}{\partial r} \right) \mathrm{d}l + \int\limits_{l_2} \left(G\frac{\partial E_1}{\partial r} - E_1 \frac{\partial G}{\partial r} \right) \mathrm{d}l \tag{3.75}$$

$$\int\limits_{S_3} (G\nabla^2 E_1 - E_1 \nabla^2 G)\,\mathrm{d}S = \int\limits_{l_2} \left(-G\frac{\partial E_1}{\partial r} + E_1 \frac{\partial G}{\partial r} \right) \mathrm{d}l \tag{3.76}$$

where l_1 and l_2 are contours defining boundaries between the borehole and the invasion zone and the latter with the formation, while S_1, S_2 and S_3 are their cross-sections, respectively.

Now taking into account eqs. 3.72–3.73 and performing summation of eqs. 3.74–3.76 we obtain the integral equation, which contains two surface integrals over areas corresponding to cross-sections of the borehole and the invasion zone:

$$\begin{aligned} E_1(p) &= \frac{k_1^2 - k_3^2}{2\pi} \int\limits_{S_1} E_0(q) G(k_3, p, q)\, \mathrm{d}S + \frac{k_2^2 - k_3^2}{2\pi} \int\limits_{S_2} E_0(q) G(k_3, p, q)\, \mathrm{d}S \\ &+ \frac{k_1^2 - k_3^2}{2\pi} \int\limits_{S_1} E_1(q) G(k_3, p, q)\, \mathrm{d}S + \frac{k_2^2 - k_3^2}{2\pi} \int\limits_{S_2} E_1(q) G(k_3, p, q)\, \mathrm{d}S \end{aligned} \tag{3.77}$$

It is obvious that integral equations 3.71 and 3.77 coincide with each other, if $k_1 = k_2$ or $k_2 = k_3$.

We have illustrated derivation of the integral equation in two cases when the solution of the boundary problems can be obtained in the explicit form, making use of the method of separation of variables. In both cases the same Green's function corresponding to a uniform medium with the formation conductivity has been used.

Let us notice that unlike eq. 3.59 the integral equation 3.77 allows one to obtain directly very simple and sufficiently accurate formulae for the field which will be described in the next paragraph of this chapter. In order to derive the integral equation for the case when the formation has a finite thickness (Fig. 3.4c), we will introduce a new Green's function which satisfies the following conditions:

- It is a solution of equations:

$$\nabla^2 \boldsymbol{G} + k_2^2 \boldsymbol{G} = 0 \qquad \nabla^2 \boldsymbol{G} + k_3^2 \boldsymbol{G} = 0 \tag{3.78}$$

  in a horizontally layered medium when the formation and the surrounding medium are characterized by wave numbers k_2 and k_3, respectively.

- Function $\boldsymbol{G} = G\boldsymbol{I}_\phi$ and its first derivative with respect to coordinate z, $\partial G/\partial z$, are continuous functions at interfaces between the formation and the surrounding medium.

From a physical point of view, the function $\boldsymbol{G}$ presents itself with accuracy of a multiplier of the electrical field of a circular filament passing through the point p in the horizontally layered medium. As will be shown in the next chapters this function can be expressed through the proper integral.

We will present the total electric field, $\boldsymbol{E}$, as before in the form of the sum:

$$\boldsymbol{E} = \boldsymbol{E}_0 + \boldsymbol{E}_1 \tag{3.79}$$

where $\boldsymbol{E}_0$ is the electric field of the magnetic dipole in the horizontally layered medium, i.e.:

$$\nabla^2 \boldsymbol{E}_0 = -k_2^2 \boldsymbol{E}_0 \tag{3.80}$$

in the formation, and

$$\nabla^2 \boldsymbol{E}_0 = -k_3^2 \boldsymbol{E}_0 \tag{3.81}$$

in the surrounding medium provided that the borehole is absent.

It is clear that:

$$\boldsymbol{G} = G\boldsymbol{I}_\phi \qquad \boldsymbol{E}_0 = E_0\boldsymbol{I}_\phi \qquad \boldsymbol{E}_1 = E_1\boldsymbol{I}_\phi$$

Taking into account eq. 3.63 as well as eqs. 3.80–3.81 we have in the surrounding medium:

$$\nabla^2 \boldsymbol{E}_1 = -k_3^2 \boldsymbol{E}_1 \tag{3.82}$$

in the formation:

$$\nabla^2 \boldsymbol{E}_1 = -k_2^2 \boldsymbol{E}_1 \tag{3.83}$$

in the part of the borehole, located against the surrounding medium:

$$\nabla^2 \boldsymbol{E}_1 = (k_3^2 - k_1^2)\boldsymbol{E}_0 - k_1^2 \boldsymbol{E}_1 \tag{3.84}$$

and finally in the part of the borehole located against the formation:

$$\nabla^2 \boldsymbol{E}_1 = (k_2^2 - k_1^2)\boldsymbol{E}_0 - k_1^2 \boldsymbol{E}_1 \tag{3.85}$$

Correspondingly, function $G\nabla^2 E_1 - E_1\nabla^2 G$ is equal to zero within the surrounding medium and the formation while it is equal to:

$$(k_3^2 - k_1^2)E_0 G + (k_3^2 - k_1^2)E_1 G$$

in the part of the borehole, located against the surrounding medium, and

$$(k_2^2 - k_1^2)E_0 G + (k_2^2 - k_1^2)E_1 G$$

in the part of the borehole, located against the formation.

Now applying Green's formula we obtain the integral equation with respect to the electric field, E_1:

$$E_1(p) = F_1(p) + \frac{k_3^2 - k_1^2}{2\pi} \int\limits_{S_2} E_1(q)G(p,q)\,\mathrm{d}S + \frac{k_2^2 - k_1^2}{2\pi} \int\limits_{S_1} E_1(q)G(p,q)\,\mathrm{d}S \tag{3.86}$$

where

$$F_1(p) = \frac{k_3^2 - k_1^2}{2\pi} \int\limits_{S_2} E_0(q) G(p,q)\, \mathrm{d}S + \frac{k_2^2 - k_1^2}{2\pi} \int\limits_{S_1} E_0(q) G(p,q)\, \mathrm{d}S \tag{3.87}$$

is a known function.

It is clear that the half cross-section of the borehole S is equal to:

$$S = S_1 + S_2$$

The integral equation 3.86 allows us to determine the electric field $E_1(p)$ and therefore the total electric field:

$$E = E_0 + E_1$$

which creates the electromotive force in the receiver coil.

Generalizing this result in the case, when there is an invasion zone within the formation (Fig. 3.4d), we have:

$$\begin{aligned} E_1(p) = F_2 + \frac{k_3^2 - k_1^2}{2\pi} \int\limits_{S_2} E_1(q) G(p,q)\, \mathrm{d}S \\ + \frac{k_3^2 - k_1^2}{2\pi} \int\limits_{S_1} E_1(q) G(p,q)\, \mathrm{d}S + \frac{k_4^2 - k_1^2}{2\pi} \int\limits_{S_3} E_0(q) G(p,q)\, \mathrm{d}S \end{aligned} \tag{3.88}$$

where S_3 is the half cross-section of the invasion zone and $k_4^2 = \mathrm{i}\sigma_4\mu\omega$:

$$\begin{aligned} F_2 = \frac{k_3^2 - k_1^2}{2\pi} \int\limits_{S_2} E_0(q) G(p,q)\, \mathrm{d}S \\ + \frac{k_2^2 - k_1^2}{2\pi} \int\limits_{S_1} E_0(q) G(p,q)\, \mathrm{d}S + \frac{k_4^2 - k_2^2}{2\pi} \int\limits_{S_3} E_0(q) G(p,q)\, \mathrm{d}S \end{aligned} \tag{3.89}$$

Further simplification of numerical problems is related with derivation of integral equations with respect to tangential components of the electromagnetic field in which integration is performed along the line l, characterizing the borehole surface.

In order to eliminate the surface integral, we will choose Green function, which obeys the following conditions:

- Inside the borehole function $\boldsymbol{G}$ satisfies Helmholtz equation:

  $$\nabla^2 \boldsymbol{G}_1 + k_1^2 \boldsymbol{G}_1 = 0 \tag{3.90}$$

  and it has a logarithmic singularity at point p.

- Outside the borehole we have for the formation and the surrounding medium, respectively:

$$\nabla^2 \boldsymbol{G}_2 + k_2^2 \boldsymbol{G}_2 = 0 \qquad\qquad \nabla^2 \boldsymbol{G}_2 + k_3^2 \boldsymbol{G}_2 = 0 \tag{3.91}$$

  Function $\boldsymbol{G}_2$ is continuous along with its first derivative with respect to coordinate z at horizontal interfaces (Fig. 3.4c) and it does not have singularities.

It is clear that the total electric field $\boldsymbol{E} = E\boldsymbol{I}_\phi$ satisfies the following equations:

$$\begin{aligned} &\nabla^2 \boldsymbol{E} + k_1^2 \boldsymbol{E} = 0 \quad \text{in the borehole} \\ &\nabla^2 \boldsymbol{E} + k_2^2 \boldsymbol{E} = 0 \quad \text{in the formation} \\ &\nabla^2 \boldsymbol{E} + k_3^2 \boldsymbol{E} = 0 \quad \text{in the surrounding medium} \end{aligned} \tag{3.92}$$

and it is a continuous function at interfaces of the medium.

From the first Maxwell equation $\boldsymbol{E} = \mathrm{i}\omega\mu\boldsymbol{H}$, we have:

$$H_z = \frac{1}{\mathrm{i}\omega\mu r}\frac{\partial(rE_\phi)}{\partial r} = \frac{1}{\mathrm{i}\omega\mu}\left(\frac{\partial E_\phi}{\partial r} + \frac{1}{r}E_\phi\right)$$

Inasmuch as both components H_z and E_ϕ are continuous functions, the first derivative $\partial E_\phi/\partial r$ is also continuous at the borehole surface and at interfaces between the formation and the surrounding medium.

Applying Green's formula outside the borehole and taking into account eqs. 3.68 and 3.91 we obtain:

$$\int\limits_l \left(E\frac{\partial G_2}{\partial r} - G_2\frac{\partial E}{\partial r}\right) \mathrm{d}l = 0 \tag{3.93}$$

where l is the straight line along the borehole surface.

Now we will apply Green's formula for functions E and G_1 inside the borehole. Then we have:

$$\int\limits_l \left(E\frac{\partial G_1}{\partial r} - G_1\frac{\partial E}{\partial r}\right) \mathrm{d}l + \int\limits_{l_1} \left(E\frac{\partial G_1}{\partial n} - G_1\frac{\partial E}{\partial n}\right) \mathrm{d}l + \int\limits_{l_2} \left(E\frac{\partial G_1}{\partial n} - G_1\frac{\partial E}{\partial n}\right) \mathrm{d}l = 0 \tag{3.94}$$

where l_1 is the contour, surrounding the observation point p, while l_2 is the contour around the current ring representing the source of the primary field. The value of integral around point p is equal to $-2\pi E(p)$, since function G_1 behaves as $\ln r$ near this point.

In approaching the source of the primary field, E tends to that caused by the primary magnetic field only. Therefore:

$$E = E_0^0 + E_1 \to \frac{\mathrm{i}\omega\mu I}{2}\ln r$$

and

$$\frac{\partial E}{\partial r} \rightarrow \frac{\partial E_0^0}{\partial r} \rightarrow \frac{\mathrm{i}\omega\mu I}{2\pi r}$$

Correspondingly, the integral around contour l_2 is equal to:

$$-G_1 \int\limits_{l_2} \frac{\partial E}{\partial n} \mathrm{d}l = G_1 \int\limits_{l_2} \frac{\partial E}{\partial r} \mathrm{d}l = \mathrm{i}\omega\mu I G_1$$

Since the electric field of the source in a uniform medium with conductivity σ_1 is equal to:

$$E_0 = \frac{\mathrm{i}\omega\mu I}{2\pi} G_1$$

we have $2\pi E_0(p)$ for the integral along contour l_2.

Correspondingly, instead of eq. 3.94 we obtain:

$$-2\pi(E - E_0) + \int\limits_{l} \left(E\frac{\partial G_1}{\partial r} - G_1\frac{\partial E}{\partial r} \right) \mathrm{d}l = 0$$

or

$$E(p) = E_0(p) + \frac{1}{2\pi} \int\limits_{l} \left(E\frac{\partial G_1}{\partial r} - G_1\frac{\partial E}{\partial r} \right) \mathrm{d}l \tag{3.95}$$

If point p is located on the contour l, it is an integral equation with two unknowns, E and $\partial E/\partial r$. The latter is expressed through tangential component of the magnetic field, H_z.

Subtracting eq. 3.93 from 3.95 we obtain:

$$E(p) = E_0(p) + \frac{1}{2\pi} \int\limits_{l} \left(E\frac{\partial G^*}{\partial r_q} - G^*\frac{\partial E}{\partial r_q} \right) \mathrm{d}l,$$

where $G^* = G_1 - G_2$ and $\partial/\partial r_q$ means the derivative at point q. The last operation permits us to reduce the order of the singularity.

Taking the normal derivative at the point p we have:

$$\frac{\partial E}{\partial r_p} = \frac{\partial E_0}{\partial r_p} + \frac{1}{2\pi} \int\limits_{l} \left(E\frac{\partial^2 G^*}{\partial r_p \partial r_q} - \frac{\partial G^*}{\partial R_p}\frac{\partial E}{\partial r_p} \right) \mathrm{d}l \tag{3.96}$$

Thus eqs. 3.95 and 3.96 form a system of two integral equations in terms of the electric field E and its first derivative with respect to r, $\partial E/\partial r$.

When these functions are found along the contour l, we can determine the electric field inside the borehole by making use of the computational formula 3.95. This approach has been used for the investigation of radial and vertical responses of induction probes when the formation has a finite thickness.

3.4. Approximate Methods of Field Calculation in Induction Logging

In this section we will describe two methods of field calculation which have played an essential role in developing such aspects of induction logging as:

- theory
- interpretation
- invention of principle of *focusing* of the induction probes
- choice of optimal parameters of multi-coil induction probes
- choice of field frequency and understanding of the influence of the skin effect on radial responses of induction probes.

3.4.1. Doll's Theory of Induction Logging

In 1949 Henry Doll developed an approximate theory of induction logging. The basis of this theory is the assumption that for a sufficiently resistive medium and at relatively low frequencies one can neglect interaction of induced currents. For this reason the phase of these currents is $90°$, regardless of the distance from the transmitter coil, and the measured signal is a sum of elementary signals created by currents in various parts of the medium, which depend on the conductivity of the corresponding part of the medium only.

As follows from the analysis of the field of a magnetic dipole in a uniform medium (Chapter 2), such behavior of the field and induced currents is closer to reality with decreasing frequency as well as conductivity.

The range of frequencies and resistivities of a nonuniform medium and also geometric parameters, when this theory remains valid will be established by comparison with results of calculations, making use of the exact solution.

Application of Doll's theory permits us to derive simple expressions for the quadrature component of the magnetic field in a medium with horizontal and cylindrical interfaces, and in many cases this theory allows us to evaluate with sufficient accuracy the influence of currents, induced in the borehole and in the invasion zone as well as in other parts of the medium. It is appropriate to consider formulae based on this theory as asymptotical ones, which are valid for large values of the skin depth with respect to such parameters as:

- borehole radius
- invasion zone radius
- formation thickness
- distance from the magnetic dipole to the observation point.

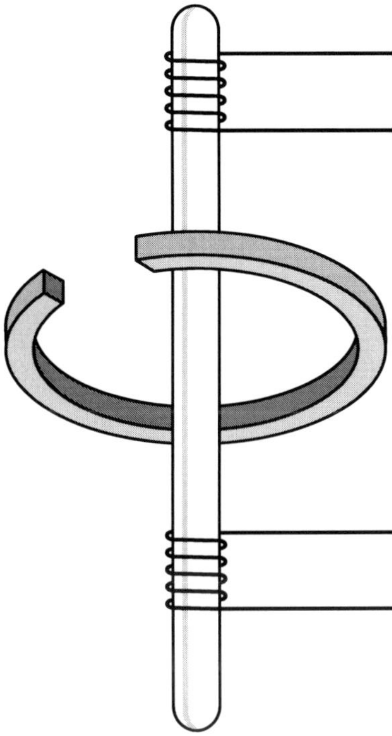

Figure 3.5. Position of an elementary ring with respect to the induction probe.

Let us consider a part of the medium formed by two horizontal planes and two coaxial cylindrical surfaces having a common axis with the borehole (Fig. 3.5). This part presents itself as a horizontal ring filled by a uniform medium. Its cross-section, dS, is almost square and, one will assume that this area equals unity (d$S = 1$). It is essential that the dimensions of the cross-section are small with respect to the ring's radius. Doll called this part of the medium an elementary unit ring.

Now we will find a signal at the receiver of a two-coil induction probe caused by an induced current from this ring. As was shown in Chapter 2 the current induced in the elementary unit ring is:

$$I = \frac{\mathrm{i}\sigma\mu\omega}{4\pi}\frac{rM_T}{R_1^3} \qquad \text{if } \mathrm{d}S = 1 \tag{3.97}$$

where σ is the ring conductivity, and R_1 is the distance from the transmitter coil to the ring (Fig. 3.6).

The current in the elementary ring with radius r generates the secondary magnetic field which has only the vertical component at its axis:

$$H_z = \frac{I}{2}\frac{r^2}{R_2^3} \tag{3.98}$$

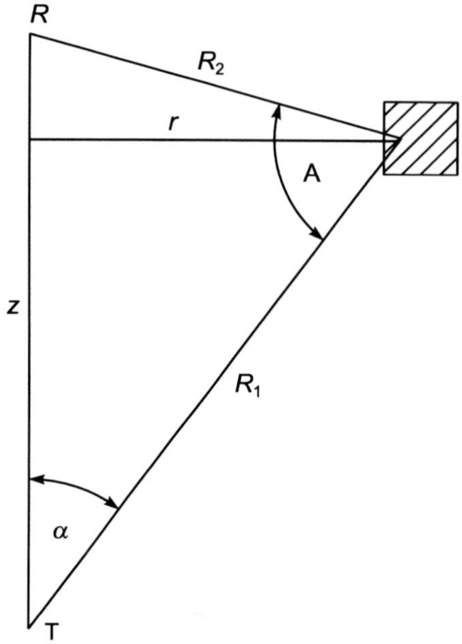

Figure 3.6. Illustration of eq. 3.97.

where R_2 is the distance from points of the ring to the receiver coil.

Respectively, the flux of the magnetic induction vector piercing the receiver is:

$$\phi = \frac{\mu}{2}\frac{Ir^2}{R_3^2}S_2n_2$$

where S_2 and n_2 are area and number of turns of the receiver, respectively.

For the electromotive force in the receiver, arising due to a change of the magnetic field with time, we have:

$$\mathscr{E} = \mathrm{i}\omega\phi = \frac{\mathrm{i}\omega\mu}{2}\frac{Ir^2}{R_2^3}S_2n_2 \tag{3.99}$$

Substituting eq. 3.97 into 3.99 we finally obtain the expression for the electromotive force in the receiver coil:

$$\mathscr{E} = -\frac{\pi}{2}f^2\mu^2\sigma I_0S_1S_2n_1n_2\frac{r^3}{R_1^3R_2^3} \tag{3.100}$$

where I_0 is the current in the transmitter coil; S_1 and n_1 are the area and number of turns of the transmitter coil; f is the field frequency.

Let us present this equation in the form:

$$\mathscr{E} = K_0 \sigma g_0 \tag{3.101}$$

where

$$K_0 = -\frac{\pi}{2} f^2 \mu^2 S_1 S_2 n_1 n_2$$

is the probe coefficient, and

$$g_0 = \frac{r^3}{R_1^3 R_2^3} \tag{3.102}$$

is the function depending on the radius and location of the ring as well as on the probe length, L. Doll called this function by the geometric factor of the elementary ring or the elementary geometric factor. Thus, the signal generated by the current in the elementary ring of a medium, is directly proportional to its conductivity and geometric factor of the ring. Now we will present the function g_0 in cylindrical coordinates, r, z with the origin at the middle of the induction probe. Inasmuch as:

$$R_1 = \left[r^2 + (L/2 + z)^2\right]^{1/2} \qquad R_2 = \left[r^2 + (L/2 - z)^2\right]^{1/2}$$

we have for the function g_0:

$$g_0 = \frac{r^3}{[r^2 + (L/2 + z)^2]^{3/2}\,[r^2 + (L/2 - z)^2]^{3/2}} \tag{3.103}$$

Instead of function g_0, we introduce, following Doll, a new function q:

$$q = \frac{L}{2} g_0 = \frac{L}{2} \frac{r^3}{[r^2 + (L/2 + z)^2]^{3/2}\,[r^2 + (L/2 - z)^2]^{3/2}} \tag{3.104}$$

At the same time the probe coefficient K_0 is multiplied by $2/L$. In this case, as will be shown later, the geometric factor of all space is equal to unity.

Knowing the induced current in an elementary ring of a medium, one can calculate a signal caused by currents in a whole space. In fact, making use of the principle of superposition and neglecting interaction of induced currents the electromotive force is equal to the sum of the signals from all elementary rings, i.e.:

$$\mathscr{E} = K \int\limits_S \sigma q \, \mathrm{d}S \tag{3.105}$$

where $K = (2/L)K_0$ and $\mathrm{d}S$ is the cross-section of the elementary ring.

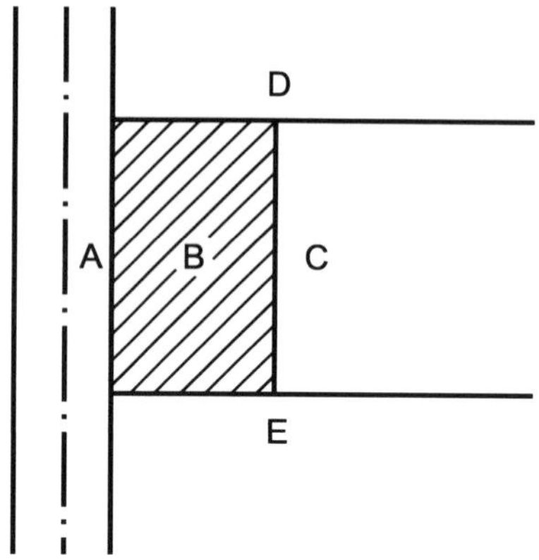

Figure 3.7. Illustration of eq. 3.106.

In a general case conductivity can be a continuous function of coordinates of a point. If a medium is uniform we have:

$$\mathscr{E} = K\sigma \int\limits_S q \, \mathrm{d}S = K\sigma \int\limits_r \mathrm{d}r \int\limits_z \mathrm{d}z$$

Inasmuch as radii of elementary rings change from 0 to ∞ but coordinate z varies from $-\infty$ to ∞, we can write for a uniform medium:

$$\mathscr{E} = K\sigma \int\limits_0^\infty \mathrm{d}r \int\limits_{-\infty}^{\infty} q \, \mathrm{d}z$$

For illustration we will consider a nonuniform medium with a conductivity distribution as shown in Fig. 3.7. Then, taking into account axial symmetry, it is natural to denote parts of the medium with various conductivity by capital letters A, B, C, D, E and so on. Contribution of every uniform part of the medium to the total signal is proportional to the product of the corresponding conductivity and geometric factor of this part. The latter can be presented as a sum of geometric factors of elementary rings over the area of the considered part of the medium.

For example, if conductivities of parts A, B, C, D, and E are equal to σ_A, σ_B, σ_C, σ_D, and σ_E, respectively, the total electromotive force is:

$$\mathscr{E} = K\left(\sigma_A \iint\limits_A q \, \mathrm{d}S + \sigma_B \iint\limits_B q \, \mathrm{d}S + \sigma_C \iint\limits_C q \, \mathrm{d}S + \sigma_D \iint\limits_D q \, \mathrm{d}S + \sigma_E \iint\limits_E q \, \mathrm{d}S\right) \tag{3.106}$$

where $\iint_A q\,\mathrm{d}S$, $\iint_B q\,\mathrm{d}S$ and so on are geometric factors of corresponding parts of the medium.

Introducing notations:

$$G_A = \iint\limits_A q\,\mathrm{d}S \qquad G_B = \iint\limits_B q\,\mathrm{d}S \qquad G_C = \iint\limits_C q\,\mathrm{d}S \qquad \text{and so on}$$

we obtain the following expression for the magnitude of the electromotive force:

$$\mathscr{E} = K\left(\sigma_A G_A + \sigma_B G_B + \sigma_C G_C + \sigma_D G_D + \sigma_E G_E\right) \tag{3.107}$$

Inasmuch as for whole space the geometric factor is equal to unity:

$$\sigma_A G_A + \sigma_B G_B + \sigma_C G_C + \sigma_D G_D + \sigma_E G_E + \cdots = 1 \tag{3.108}$$

the ratio $\mathscr{E}/K$ is equal to the conductivity of this medium. In the general case of a nonuniform medium this ratio is called the apparent conductivity, σ_a. Therefore, we have:

$$\sigma_a = \frac{\mathscr{E}}{K} = \sigma_A G_A + \sigma_B G_B + \sigma_C G_C + \sigma_D G_D + \sigma_E G_E \tag{3.109}$$

From this equation it follows that to some extent the conductivity and the geometric factor have a similar influence. For instance, a part of the medium with high conductivity and with relatively small dimensions can contribute the same signal as an area with lower conductivity but with greater dimensions.

Now we will investigate the behavior of the elementary geometric factor g in detail. In accord with eq. 3.104 it is very simple to show that the geometric factor q depends on the angle under which both coils of the induction probe are seen from points of the corresponding ring, and it is equal to:

$$q = \frac{\sin^3 A}{2L^2} \tag{3.110}$$

In fact, as follows from Fig. 3.6:

$$\frac{\sin A}{L} = \frac{\sin\alpha}{R_2} \qquad \sin\alpha = \frac{r}{R_1} \qquad \frac{\sin A}{L} = \frac{r}{R_1 R_2}$$

and therefore:

$$q = \frac{L}{2}\frac{r^3}{R_1^3 R_2^3} = \frac{L}{2}\frac{\sin^3 A}{L^3} = \frac{\sin^3 A}{2L^2}$$

In other words, for the given probe the elementary geometric factor is completely defined by the angle under which the probe is seen from points of the elementary ring. Thus, all elementary rings have the same geometric factor, if the probe is seen under the same angle from the ring's points. Consequently, they contribute the same signal, provided that all these rings have the same conductivity. From this consideration it is obvious that the geometric configuration of a section of elementary rings with the same geometric factor in a vertical plane are circles, passing through transmitter and receiver coils of the induction probe (Fig. 3.6). Elementary rings for which $\sin A = 1$ have maximal geometric factor, equal $1/2L^2$. Cross-sections of these rings are located on the circle with radius equal to $L/2$.

Making use of the concept of a elementary geometric factor it is not difficult to define a signal caused by currents in various parts of the conducting medium. In particular, if a medium is uniform, signals caused by its different parts depend on corresponding geometric factors only. In developing the theory of induction logging in media with either cylindrical or horizontal interfaces, Doll also introduced useful concepts of geometric factors of an elementary cylinder and an elementary horizontal layer which will be considered later.

Finally, let us make the following comments:

- According to Doll's theory induced currents arise due to the primary vortex electric field only:

  $$E_\phi^0 = \frac{i\omega\mu M_T r}{R_1^3}$$

  In other words, interaction between currents (skin effect) is neglected, and, respectively, every element of a medium manifests itself independently, regardless of the resistivity of neighboring parts of the medium. It also means that this theory does not allow us to evaluate the inphase component of the secondary magnetic field.

- As was shown in Chapter 2, near the source the quadrature component of the current is practically defined by the primary vortex electric field. Correspondingly, if the area, where such behavior takes place, is greater than the depth of investigation of the given probe, Doll's theory describes the field behavior with a sufficient accuracy.

- From a mathematical point of view Doll's theory can be considered as the first approximation of the integral equation 3.56, when the thickness of the skin layer tends to infinity.

Now we will describe also simple, but more accurate method of field calculation.

3.4.2. The Approximate Theory of Induction Logging, Taking into Account the Skin Effect in the External Area

The analysis of the field of a magnetic dipole in a uniform conducting medium (Chapter 2) has clearly demonstrated that with an increase of the distance from the source the quadrature component of induced currents becomes smaller with respect to that corresponding to Doll's theory. Moreover, comparison of the vertical component of the magnetic field on the borehole axis, calculated from the exact solution and making use of Doll's formulae, confirms this conclusion about the distribution of the quadrature component of currents. In fact, values of the actual turn out to be smaller than those calculated from Doll's theory.

Now we will describe a method which under certain conditions takes correctly into account the skin effect, i.e. interaction of currents in a conducting medium. The idea of this method is very simple. Let us present all current space around the induction probe as a sum of two areas, namely:

- the internal area, where the induction probe is located
- the external area.

For simplicity we will suppose that the conductivity of the external area is constant. Later this assumption will be omitted.

We will suppose that two conditions are valid:

- Induced currents in the internal area which contribute to a signal in the receiver are shifted in phase by 90°, and their density depends on geometric parameters and the conductivity at a given point. In other words, interaction between currents induced within this area is practically absent, that is the primary vortex electric field generates induced currents only.
- The density of vortex currents in the external area does not depend on the resistivity distribution within the internal area, that is, interaction of two currents, located in both areas, can be neglected.

The second condition emphasizes the fact that the skin effect manifests itself, first of all, at relatively large distances from the source. Proceeding from these two assumptions we will derive sufficiently simple expressions for the quadrature component of the magnetic field.

This component of the magnetic field on the borehole axis can be presented as a sum of two magnetic fields caused by currents in the internal and external areas:

$$\mathrm{Q}\,H_z = \mathrm{Q}\,H_z^i + \mathrm{Q}\,H_z^e$$

or

$$\mathrm{Q}\,h_z = \mathrm{Q}\,h_z^i + \mathrm{Q}\,h_z^e \tag{3.111}$$

where $\mathrm{Q}\,h_z$ is the vertical component of the magnetic field, related to the field in a free space, H_z^0:

$$H_z^0 = \frac{M_T}{2\pi L^3}$$

$Q\,h_z^i$ and $Q\,h_z^e$ are quadrature components of the magnetic field, caused by currents within the internal and external areas, respectively.

Making use of results obtained in Chapter 2 and formulae of Doll's theory, the first condition allows us to present the magnetic field due to currents in the internal area as:

$$Q\,h_z^i = \frac{\omega\mu L^2}{2}\sigma_a^i \tag{3.112}$$

where μ is the magnetic permeability of the free space equals to $4\pi \times 10^{-7}$ H/m; ω is the angular frequency; L is the length of a two coil induction probe.

In accord with eq. 3.109 the apparent conductivity σ_a^i of the internal area, related with an actual conductivity distribution as:

$$\sigma_a^i = \sigma_A G_A + \sigma_B G_B + \sigma_C G_C + \cdots + \sigma_F G_F \tag{3.113}$$

where G_A, G_B, G_C and G_F are geometric factors of areas with conductivities σ_A, σ_B, σ_C and σ_F, respectively.

For instance let us suppose that the conductivities of an internal area and an external area are equal to each other. In this case we have a uniform medium, and the field can be presented in the form:

$$Q\,h_z^{un} = Q\,h_{ze}^i + Q\,h_z^e \tag{3.114}$$

inasmuch as the field of currents in the external area, in accord with the second condition, does not depend on the conductivity distribution within the internal area. Here $Q\,h_z^{un}$ is the quadrature component of the magnetic field in a uniform medium with conductivity of the external area σ_e. The function $Q\,h_{ze}^i$ is the quadrature component of the magnetic field due to currents in the internal area when it has conductivity σ_e. As follows from the first condition this part of the field can be expressed through the geometric factor of the internal area G_i:

$$Q\,h_{ze}^i = \frac{\omega\mu L^2}{2}\sigma_e G_i \tag{3.115}$$

Therefore, for the quadrature component of the magnetic field, caused by currents in the external area we have:

$$Q\,h_z^e = Q\,h_z^{un} - Q\,h_{ze}^i = Q\,h_z^{un} - \frac{\omega\mu L^2}{2}\sigma_e G_i \tag{3.116}$$

Correspondingly, for the total quadrature component of the field we obtain:

$$Q\,h_z = \frac{\omega\mu L^2}{2}\sigma_a^i + Q\,h_z^{un} - \frac{\omega\mu L^2}{2}\sigma_e G_i = Q\,h_z^{un} + \frac{\omega\mu L^2}{2}\left[\sigma_a^i - \sigma_e G_i\right] \tag{3.117}$$

Thus for a field determination on the borehole axis it is sufficient to know the geometric factors of corresponding parts of the internal area and the field of the magnetic dipole in a uniform medium with conductivity of the external area σ_e. In particular, the conductivity of the internal area can vary as a continuous function. An expression of the field in a uniform medium and its values are given in Chapter 2.

Calculation of geometrical factors, i.e. integrals of type $\int_S q\, dS$ is a rather simple numerical problem, and for the most important cases it is already performed in detail.

Let us notice that the first term in the right-hand side of eq. 3.117 is subjected to the influence of the skin effect in the same manner, as it takes place in uniform medium with conductivity σ_e.

Now we will show that with a decrease of parameter $p = L/h$ (h is the skin depth), eq. 3.117 approaches to that derived from Doll's theory.

As was demonstrated in the previous chapter, the quadrature component of the magnetic field in a uniform medium can be expressed as:

$$\frac{\mathrm{Q}\, H_z^{un}}{H_z^0} = \mathrm{Q}\, h_z = \frac{\sigma_e \omega \mu L^2}{2} \qquad \text{if } \frac{L}{h} = L\left(\frac{\sigma_e \mu \omega}{2}\right)^{1/2} \ll 1 \tag{3.118}$$

Substituting eq. 3.118 into eq. 3.117 we obtain:

$$\mathrm{Q}\, h_z = \frac{\omega \mu L^2}{2}\left(\sigma_e(1 - G_i) + \sigma_A G_A + \sigma_B G_B + \cdots + \sigma_F G_F\right)$$

or

$$\mathrm{Q}\, h_z = \frac{\omega \mu L^2}{2}\left(\sigma_e G_e + \sigma_A G_A + \sigma_B G_B + \cdots + \sigma_F G_F\right) \tag{3.119}$$

where G_e is the geometric factor of the external area.

This equation for $\mathrm{Q}\, h_z$, coincides with that for the magnetic field in a uniform medium derived by Doll.

Making use of the relation between the apparent conductivity and the field, introduced by Doll, we have:

$$\sigma_a = \frac{2}{\omega \mu L^2}\mathrm{Q}\, h_z = \sigma_a^{un} + \sigma_a^i - \sigma_e G_i$$

or

$$\frac{\sigma_a}{\sigma_e} = \frac{\sigma_a^{un}}{\sigma_e} + \frac{\sigma_a^i}{\sigma_e} - G_i = \frac{\sigma_a^{un}}{\sigma_e} + \frac{\sigma_A}{\sigma_e}G_A + \frac{\sigma_B}{\sigma_e}G_B + \cdots + \frac{\sigma_F}{\sigma_e}G_F - G_i \tag{3.120}$$

Values of function σ_a^{un}/σ_e are given in Chapter 2.

It is obvious that with a decrease of internal area dimensions and an increase of medium resistivity this method of field calculation will be valid for higher frequencies.

Comparison with results of calculations using the exact solution will allow us later to characterize boundaries of application of this method.

Now we will consider two more examples when the external medium is uniform.

As follows from results obtained above, we have for the field on the borehole axis, provided that the formation has infinite thickness (two layered-medium with one cylindrical interface):

$$Q\,h_z = Q\,h_z^{un} + \frac{\omega\mu L^2}{2}(\sigma_1 - \sigma_2)G_1(\alpha) \qquad (3.121)$$

and for the apparent conductivity:

$$\frac{\sigma_a}{\sigma_2} = \frac{\sigma_a^{un}}{\sigma_2} + \left(\frac{\sigma_1}{\sigma_2} - 1\right) G_1(\alpha) \qquad (3.122)$$

where:
σ_1 is the borehole conductivity;
σ_2 is the formation conductivity;
$G_1(\alpha)$ is the geometric factor of the borehole;
α is the ratio of the length, L, of the induction probe to the borehole radius, a_1: $\alpha = L/a_1$;
$Q\,h_z^{un}$ and σ_a^{un} are the quadrature component of the magnetic field and the apparent conductivity in a uniform medium with the formation resistivity, respectively.

As will be demonstrated in the next chapter the geometric factor of the borehole can be expressed through the integral:

$$G_1(\alpha) = 1 - \frac{2\alpha}{\pi} \int_0^\infty \frac{m}{2} \left[2K_0(m)K_1(m) - m\left(K_1^2 - K_0^2\right)\right] \cos m\alpha \, dm$$

where $K_0(m)$, $K_1(m)$ are modified Bessel functions of the second type. If the internal area includes both the borehole and the invasion zone, expressions for the quadrature component, $Q\,h_z$, and the apparent conductivity, σ_a, have the form:

$$Q\,h_z = Q\,h_z^{un}(\sigma_3) + \frac{\omega\mu L^2}{2}(\sigma_1 - \sigma_3)G_1(\alpha) + \frac{\omega\mu L^2}{2}(\sigma_2 - \sigma_3)G_2(\alpha) \qquad (3.123)$$

and

$$\frac{\sigma_a}{\sigma_3} = \frac{\sigma_a^{un}}{\sigma_3} + \left(\frac{\sigma_1}{\sigma_3} - 1\right) G_1(\alpha) + \left(\frac{\sigma_2}{\sigma_3} - 1\right) G_2(\alpha) \qquad (3.124)$$

where:
σ_1, σ_2 and σ_3 are conductivities of the borehole, the invasion zone and the formation, respectively;
$G_1(\alpha)$ and $G_2(\alpha)$ are geometric factors of the borehole and the invasion zone:

$$G_2(\alpha) = G_1\left(\frac{\alpha}{\beta}\right) - G_1(\alpha) \qquad \beta = a_2/a_1$$

a_2 and a_1 are radii of the invasion zone and the borehole, correspondingly.

Later we will demonstrate that this method of the field calculation permits us to obtain with a sufficient accuracy values of the quadrature component of the field in most practical cases where induction logging is applied.

Thus for the field calculation in media with cylindrical interfaces two functions should be known, namely:

- the quadrature component of the magnetic field of the dipole in a uniform medium with the formation conductivity
- the geometric factor of the cylinder, $G_1(\alpha)$.

A similar approach can, in principle, be used in a medium with horizontal interfaces when the formation has a finite thickness.

Until now we have assumed that the external area is a uniform one. This restriction allows us to obtain very simple formulae for the field in a media with cylindrical interfaces.

However, in a medium with both cylindrical and horizontal interfaces, when the formation has a finite thickness, the external area is not uniform, anymore.

In order to derive formulae for such a case, let us present the field in a medium with two horizontal interfaces, i.e. as the formation has the finite thickness and the borehole is absent, $\mathrm{Q}\,h_0$, as a sum:

$$\mathrm{Q}\,h_0 = \mathrm{Q}\,h_1 + \mathrm{Q}\,h_e \tag{3.125}$$

where:
$\mathrm{Q}\,h_1$ is the quadrature component of the magnetic field, caused by induced currents in the vertical cylinder with the radius of the borehole;
$\mathrm{Q}\,h_e$ is the quadrature component of the magnetic field, caused by induced currents outside the borehole;
σ_2 and σ_3 are conductivities of the formation and the surrounding medium, respectively.

Whence:

$$\mathrm{Q}\,h_e = \mathrm{Q}\,h_0 - \mathrm{Q}\,h_1 \tag{3.126}$$

The magnetic field $\mathrm{Q}\,h_1$ can be expressed through geometric factors in the following way:

$$\mathrm{Q}\,h_1 = \frac{\omega\mu L^2}{2}\left(\sigma_2 G_1^A + \sigma_3 G_1^B\right) \tag{3.127}$$

where G_1^A is a geometric factor of the borehole part (A), located against the formation, while G_1^B is the geometric factor of the rest part of the borehole (B), that is:

$$G_1^A = \int\limits_A q\,\mathrm{d}S \qquad G_1^B = \int\limits_B q\,\mathrm{d}S \tag{3.128}$$

It is obvious that the geometric factor of the borehole, G_1, can be presented as a sum:

$$G_1(\alpha) = G_1^A + G_1^B \tag{3.129}$$

Thus, the quadrature component of the magnetic field, caused by currents induced outside the borehole, can be written as

$$Q\,h_e = Q\,h_0 - \frac{\omega\mu L^2}{2}\left(\sigma_2 G_1^A + \sigma_3 G_1^B\right) \tag{3.130}$$

Taking into account the magnetic field caused by currents in the borehole with conductivity σ_1 we obtain for the total quadrature component on the borehole axis the following expression:

$$\begin{aligned} Q\,h &= Q\,h_0 - \frac{\omega\mu L^2}{2}\left(\sigma_2 G_1^A + \sigma_3 G_1^B\right) + \frac{\omega\mu L^2}{2}\sigma_1 G_1 \\ &= Q\,h_0 + \frac{\omega\mu L^2}{2}\left(\sigma_1 G_1 - \sigma_2 G_1^A - \sigma_3 G_1^B\right) \\ &= Q\,h_0 + \frac{\omega\mu L^2}{2}\left[(\sigma_1 - \sigma_3)G_1 + (\sigma_3 - \sigma_2)G_1^A\right] \end{aligned} \tag{3.131}$$

The magnetic field of the vertical magnetic dipole, h_0, in a horizontal layered medium is expressed in the explicit form. For example, if the induction probe is located symmetrically with respect to the formation boundary we have:

$$h_0 = h_z^{un}(\sigma_2) + L^3 \int\limits_0^\infty \frac{m^3}{m_2^2} K_{12}\,\mathrm{e}^{-2m_2 H_1} \frac{\mathrm{e}^{m_2 H_1} + K_{12}\coth m_2 L}{1 - K_{12}^2\,\mathrm{e}^{-2m_2 H_1}}\,dm \qquad \text{if } H_1 \geqslant L$$

and

$$h_0 = 2L^3 \int\limits_0^\infty \frac{m^3 m_2\,\mathrm{e}^{-H_1(m_2 - m_3)}}{(m_2 + m_3)^2(1 - K_{12}^2\,\mathrm{e}^{-2m_2 H})}\,dm \qquad \text{if } H_1 \leqslant L$$

where $h_z^{un}(\sigma_2)$ is the magnetic field of the magnetic dipole in a uniform medium with conductivity of the formation:

$$m_2 = (m^2 - k_2^2)^{1/2} \qquad m_3 = (m^2 - k_3^2)^{1/2} \qquad K_{12} = (m_2 - m_3)/(m_2 + m_3)$$

H_1 is the formation thickness; L is the probe length.

In the more complicated case, when there is an invasion into the formation, we have:

$$Q\,h = Q\,h_0 + \frac{\omega\mu L^2}{2}\left[(\sigma_2 - \sigma_3)G_2^A + (\sigma_4 - \sigma_3)G_1^A + (\sigma_1 - \sigma_4)G_1\right] \tag{3.132}$$

where:
$Q\,h_0$ is the quadrature component of the magnetic field when the borehole and the invasion zone are absent;
σ_1, σ_2, σ_3 and σ_4 are conductivities of the borehole, the invasion zone, the formation and the surrounding medium, respectively;
G_2^A is the geometric factor of the invasion zone;
G_1 is the geometric factor of the borehole;
G_1^A is the geometric factor of the part of the borehole, located against the invasion zone.

Function G_2^A is expressed through G_1^A in the same manner as the geometric factor of the invasion zone. G_2 is related with the geometric factor of the borehole G_1.

Therefore, determination of the magnetic field on the borehole axis when the formation has a finite thickness, consists of calculation of the field in a horizontally layered medium and geometric factors of vertical cylinders with a finite height which are coaxial to the borehole.

It is obvious that equations for the field corresponding a medium with one or two coaxial cylindrical interfaces can be obtained, as particular cases, from eq. 3.132.

Let us notice that in the case of a medium with two coaxial cylindrical interfaces one can derive a field equation, which is valid for higher frequencies and conductivities of the borehole and the invasion zone. However, in this case the field on the borehole axis in a medium with one cylindrical interface has to be known. From calculation of the field in this medium we can obtain values of the field on the borehole axis as its radius is equal to that of the invasion zone of the given model. Then, having replaced the central part of the invasion zone by a medium with the borehole resistivity, we obtain a three-layered medium, and correspondingly the quadrature component of the magnetic field is defined from equation:

$$Q\,h = Q\,h_0 + \frac{\omega\mu L^2}{2}(\sigma_1 - \sigma_2)G_1(\alpha) \tag{3.133}$$

where:
$Q\,h_0$ is the quadrature component on the borehole axis, the radius of which is equal to that of the invasion zone, a_2;
σ_1 and σ_2 are conductivities of the borehole and the invasion zone;
$\alpha = L/a_1$; L is the probe length.

We have described different aspects of this method for calculation of the quadrature component of the magnetic field in various models. In conclusion, it is appropriate to make several comments.

1. Formulae, obtained in this section, directly follow from the integration equation 3.88. In fact, eq. 3.89 is its first approximation and, if both conditions about current distribution are valid, leads to the same results as eq. 3.312.

2. This method was suggested almost 35 years ago, and it was very useful in developing the interpretation of induction logging, the determination of frequencies and geoelectric parameters of a section, where focusing induction probes are effective.

3. Until now this method has been considered from a point of calculation of the quadrature component of the magnetic field, which is mainly measured in induction logging. At the same time it allows us to obtain some information about the inphase component of the field also.

Inasmuch as it was assumed that the skin effect manifests itself in an external area only, i.e. interaction between currents within the internal area is negligible, we can think that the inphase component is caused by currents within the external area only. Proceeding from this consideration we can rewrite some of eqs. 3.111–3.133 in a more general form. Validity of this step also follows from analysis of the integral equation 3.88 and its first approximation 3.89.

In the simplest case, when there is one cylindrical interface only, instead of eq. 3.121 we have:

$$h = h^{un}(\sigma_2) + \frac{i\omega\mu L^2}{2}(\sigma_1 - \sigma_2)G_1(\alpha) \tag{3.134}$$

It is obvious that only the first term contains an inphase component of the magnetic field, which coincides with that in a uniform medium with conductivity of the formation, σ_2. In other words, within a range of relatively small parameters L/h induced currents in the borehole do not influence the inphase component. Similar results are obtained when an invasion zone is present. In accord with eq. 3.123 we have:

$$h = h_0(\sigma_3) + \frac{i\omega\mu L^2}{2}(\sigma_1 - \sigma_3)G_1(\alpha) + \frac{i\omega\mu L^2}{2}(\sigma_2 - \sigma_3)G_2(\alpha) \tag{3.135}$$

Again the inphase component of the magnetic field in the borehole is not practically subjected to the influence of induced currents in the borehole and in the invasion zone, and it coincides with the inphase component in a uniform medium with the formation conductivity, σ_3. In this approximation induced currents in the borehole and in the invasion zone contribute to the quadrature component of the field. This consideration clearly shows that the inphase component of the magnetic field has a different sensitivity to geoelectric parameters of a medium than the quadrature component, and therefore they are characterized by different depths of investigation. It is clear that the analysis of the current distribution in a uniform medium, performed in Chapter 2, is in complete agreement with these results.

Understanding this feature of field behavior is important for the further development of the interpretation of induction logging. Moreover, some of the induction probes, currently used in practice, are based on measuring both components of the magnetic field.

Let us consider one more case when the formation has a finite thickness. Then according to eq. 3.132 we have:

$$h = h_0 + \frac{i\omega\mu L^2}{2}\left[(\sigma_2 - \sigma_3)G_2^A + (\sigma_4 - \sigma_3)G_1^A + (\sigma_1 - \sigma_4)G_1\right] \tag{3.136}$$

Consequently the inphase component of the field is the same as that of horizontally layered medium, i.e. it is defined by the conductivity and thickness of the formation as well as the conductivity of the surrounding medium. Later it will be demonstrated that with a decrease of frequency the inphase component approaches that for a uniform medium with conductivity σ_4, i.e. medium, surrounding the formation.

4. In the next chapters, considering models which are of practical interest for induction logging, we will establish the conditions, when this method of field calculation can be applied with sufficient accuracy.

Chapter 4

ELECTROMAGNETIC FIELD OF A VERTICAL MAGNETIC DIPOLE ON THE AXIS OF A BOREHOLE

In this chapter we will derive an expression for the vertical component of the magnetic field on the axis of a borehole when the source of the primary field is a vertical magnetic dipole and the formation has an infinite thickness. Special attention will be paid to the analysis of frequency responses of quadrature and inphase components of the field, including their asymptotic behavior. The influence of various parameters of a geoelectric section will also be investigated. Such questions as the influence of finite dimensions of coils, displacement of the induction probe with respect to the borehole axis, the role of magnetic permeability and dielectric constant will be studied.

4.1. Formulation of the Boundary Problem

In formulating this boundary problem we will suppose that:

- All media surrounding the induction probe are uniform and isotropic.
- The electrical properties of the medium do not change in the direction parallel to the borehole axis. Practically it means that the top and bottom of the bed, against which the probe is located, are significantly distant from it.
- The space filled by the borehole mud has the shape of an infinitely long circular cylinder.
- An intermediate medium located between the borehole and the bed presents itself as a system of coaxial cylindrical layers, the axis of which coincides with the borehole axis.
- The transmitter and receiver coils, forming the induction probe, are located on the borehole axis, and they can be considered as dipoles because their dimensions are usually small with respect to the induction probe length and the borehole radius. At the same time the influence of the finite dimensions of these coils as well as the eccentricity will be studied.

Thus, the boundary problem is formulated in the following way. The medium is separated by a set of $n-1$ coaxial cylindrical surfaces with radii a_1, a_2, $a_3, \ldots, a_{n-1}$ into n

parts filled with uniform and isotropic media having conductivity σ_m. We will assume that the magnetic permeability and the dielectric constant are constants and equal to those in a free space, that is:

$$\mu_m = \mu_0 = 4\pi \times 10^{-7} \text{ H/m} \qquad \varepsilon_m = \varepsilon_0 = \frac{1}{36\pi} \times 10^{-9} \text{ F/m}$$

Later we will consider a more general case.

The vertical magnetic dipole is located on the borehole axis, and its moment is a sinusoidal function of time. Due to a change of the primary magnetic field with time a primary electrical field arises which has only an azimuthal component $E_\phi^{(0)}$, as was shown above. Correspondingly, induced currents arise in the conducting medium which also have only the azimuthal component since interaction between currents does not change their direction, and electrical charges do not arise at interfaces. Therefore, the sources of the secondary field are induced currents located in horizontal planes which have only components J_ϕ, and their vector lines are circles with centers on the borehole axis. In other words, we can say that the geometry of currents is the same as that for a uniform medium (Chapter 2).

The system of Maxwell's equations for the quasistationary field $\boldsymbol{E}e^{i\omega t}$ and $\boldsymbol{H}e^{i\omega t}$ is:

$$\text{curl}\,\boldsymbol{E} = -i\omega\mu\boldsymbol{H} \qquad (4.1)$$

$$\text{curl}\,\boldsymbol{H} = \sigma\boldsymbol{E} \qquad (4.2)$$

$$\text{div}\,\boldsymbol{E} = 0 \qquad (4.3)$$

$$\text{div}\,\boldsymbol{H} = 0 \qquad (4.4)$$

Unlike in Chapter 2 we will use here another dependence on time: $e^{i\omega t}$ instead of $e^{-i\omega t}$. In accord with eq. 4.3 the complex amplitude of the electric field can he presented as:

$$\boldsymbol{E} = -i\omega\mu\,\text{curl}\,\boldsymbol{A}^* \qquad (4.5)$$

and substituting eq. 4.5 into eq. 4.1 we obtain:

$$\boldsymbol{H} = \text{curl}\,\text{curl}\,\boldsymbol{A}^* = \text{grad}\,\text{div}\,\boldsymbol{A}^* - \nabla^2\boldsymbol{A}^* \qquad (4.6)$$

From eq. 4.2 we have:

$$\text{curl}\,\boldsymbol{H} = -i\sigma\mu\omega\,\text{curl}\,\boldsymbol{A}^* = k^2\,\text{curl}\,\boldsymbol{A}^*$$

or

$$\boldsymbol{H} = k^2\boldsymbol{A}^* - \text{grad}\,U \qquad (4.7)$$

where $\boldsymbol{A}^*$ is a vector potential while U is a scalar potential.

Applying the gauge condition:

$$U = -\,\text{div}\,\boldsymbol{A}^* \qquad (4.8)$$

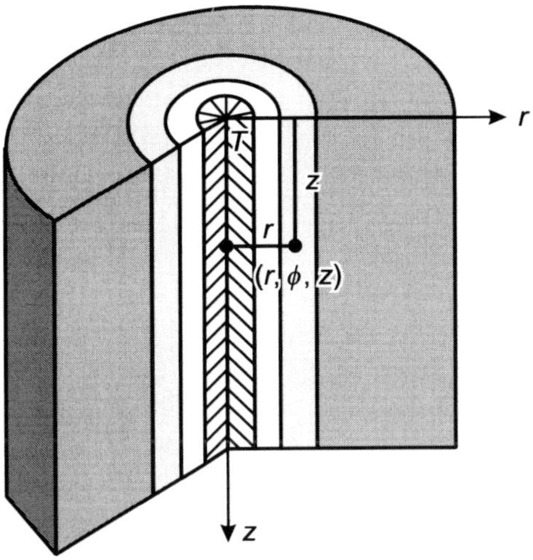

Figure 4.1. A model of a medium with cylindrical interfaces.

we obtain for the vector potential Helmholtz equation:

$$\nabla^2 \boldsymbol{A}^* + k^2 \boldsymbol{A}^* = 0 \tag{4.9}$$

and the components of the electromagnetic field are only expressed through function $\boldsymbol{A}^*$:

$$\boldsymbol{E} = -\mathrm{i}\omega\mu \,\mathrm{curl}\, \boldsymbol{A}^*$$
$$\boldsymbol{H} = k^2 \boldsymbol{A}^* + \mathrm{grad}\,\mathrm{div}\, \boldsymbol{A}^*$$

where $k^2 = -\mathrm{i}\sigma\mu\omega$. The wave number $k = (\sigma\mu\omega/2)^{1/2}(1-\mathrm{i}) = (1-\mathrm{i})/h$; $h = (2/\sigma\mu\omega)^{1/2}$. Sometimes the following notations will be used: $\lambda = 2\pi h$ and $k^2 = -\mathrm{i}\chi$, $\chi = \sigma\mu\omega$, where λ is called the wavelength even if the effect of propagation is not observed in the quasistationary approximation.

4.2. Derivation of the Formula for the Vertical Component of the Magnetic Field

Let us choose the cylindrical system of coordinates $(r,\ \phi,\ z)$ and the vertical magnetic dipole is placed at the origin of this system (Fig. 4.1). The moment of the magnetic dipole is oriented along the z-axis. We will look for a solution using only the z-component of the vector-potential, A_z^*. As follows from Maxwell's equations the vector potential must satisfy several conditions:

- Function A_z^* is a solution of Helmholtz equation in every part of the medium:

$$\nabla^2 A_z^* + k^2 A_z^* = 0 \qquad \text{if } r^2 + z^2 \neq 0$$

In the cylindrical system of coordinates this equation can be written in the form:

$$\frac{1}{r}\frac{\partial}{\partial r}\left(r\frac{\partial A_z^*}{\partial r}\right)+\frac{1}{r^2}\frac{\partial^2 A_z^*}{\partial \phi^2}+\frac{\partial^2 A_z^*}{\partial z^2}+k^2A_z^*=0 \tag{4.10}$$

- Near the origin of coordinates system the function A_z^* tends to the vector potential of the magnetic dipole in a uniform medium, that is:

$$A_z^{(0)}=\frac{M\mathrm{e}^{-ik_1R}}{4\pi R} \qquad \text{if } R=(r^2+z^2)^{1/2}\to 0$$

here M is the dipole moment.

- In passing the interface $r=a_m$, tangential components of the electric and magnetic field are continuous functions. The electrical field has only the component E_ϕ, but the magnetic field is characterized by two components: H_r and H_z, and they are related with the vector potential as:

$$E_\phi=-\mathrm{i}\omega\mu\frac{\partial Aza}{\partial r} \qquad H_r=\frac{\partial^2 A_z^*}{\partial r\partial z} \qquad H_z=k^2A_z^*+\frac{\partial^2 A_z^*}{\partial z^2} \tag{4.11}$$

Therefore, boundary conditions for the vector potential A_z^* at the interface can be written in the form:

$$\begin{aligned}\mu_m\frac{\partial A_{z,\,m}^*}{\partial r}&=\mu_{m+1}\frac{\partial A_{z,\,m+1}^*}{\partial r}\\ k_m^2A_{z,\,m}^*+\frac{\partial^2 A_{z,\,m}^*}{\partial z^2}&=k_{m+1}^2A_{z,\,m+1}^*+\frac{\partial^2 A_{z,\,m+1}^*}{\partial z^2}\end{aligned} \tag{4.12}$$

- With an increase of the distance from the magnetic dipole the function A_z^* tends to zero.

- Due to the axial symmetry of the field the vector potential and all components of the field do not depend on the ϕ coordinate, that is $A_z^*=A_z^*(r,\,z)$.

- The vector potential and all components of the field do not depend on the sign of the z-coordinate due to the symmetry with respect to the plane passing through the source and which is perpendicular to the z-axis, that is:

$$A_z^*(r,\,z)=A_z^*(r,\,-z)$$

We will look for a solution of Helmholtz equation as a product of two functions:

$$A_z^*=T(r)\Phi(z)$$

Substituting this expression for A_z^* into eq. 4.10 we obtain instead of Helmholtz equation two normal differential equations of the second order:

$$\frac{\mathrm{d}^2\Phi(z)}{\mathrm{d}z^2}+\lambda^2\Phi(z)=0$$

$$\frac{d^2T(y)}{dz^2} + \frac{1}{y}\frac{dT(y)}{dy} - T(y) = 0$$

where $y = (\lambda^2 + i\chi)^{1/2} r$, λ is the variable of separation.

Solutions of the first equation are functions $\cos \lambda z$ and $\sin \lambda z$, while solutions of the second equation are modified Bessel functions of zero order, $I_0\,(y_m)$ and $K_0\,(y_m)$, where $y_m = (\lambda^2 + i\chi_m)^{1/2} r = \lambda_m r$, and $\lambda_m = (\lambda^2 + i\chi)^{1/2}$. Taking into account the symmetry of the field with respect to the plane $z = 0$, the expression for the vector potential within the borehole can be written as:

$$A_1^* = \frac{M}{4\pi}\left(\frac{e^{-ik_1R}}{R} + \frac{2}{\pi}\int_0^\infty C_1 I_0(\lambda_1 r)\cos\lambda z\, d\lambda\right) \tag{4.13}$$

since function $K_0(\lambda r)$ tends to infinity as $r \to 0$.

It is well known that the primary excitation, e^{-ik_1R}/R, can be presented in the form:

$$\frac{e^{-ik_1R}}{R} = \frac{2}{\pi}\int_0^\infty K_0(\lambda_1 r)\cos\lambda z\, d\lambda$$

Thus:

$$A_{z,1}^* = \frac{M}{2\pi^2}\int_0^\infty [C_1 I_0(\lambda_1 r) + D_1 K_0(\lambda_1 r)]\cos\lambda z\, d\lambda$$

where $D_1 = 1$.

In a general case:

$$A_{z,m}^* = \frac{M}{2\pi^2}\int_0^\infty [C_m I_0(\lambda_m r) + D_m K_0(\lambda_m r)]\cos\lambda z\, d\lambda \tag{4.14}$$

where $m = 1, 2, \ldots, n$.

The right-hand side of the expression for the vector potential contains $2n$ unknown coefficients C_m and D_m. The first boundary condition allows us to obtain $n-1$ equations for their determination. In accord with eq. 4.12 they have the following form:

$$\begin{aligned}\mu_m[\lambda_m C_m I_1(\lambda_m a_m) - \lambda_m D_m K_1(\lambda_m a_m)] = \mu_{m+1}[&\lambda_{m+1} C_{m+1} I_1(\lambda_{m+1} a_m) \\ &- \lambda_{m+1} D_{m+1} K_1(\lambda_{m+1} a_m)]\end{aligned}$$

From the second boundary condition we also obtain $n - 1$ equations as:

$$\lambda_m^2[C_m I_0(\lambda_m a_m) + D_m K_0(\lambda_m a_m)] = \lambda_{m+1}^2[C_{m+1} I_0(\lambda_{m+1} a_m) + D_m K_0(\lambda_{m+1} a_m)]$$

Two additional equations which are necessary for a solution are obtained from the condition near the source and at infinity. Inside the borehole where the dipole is located the coefficient D_1 is equal to unity, while in the external medium the field must decrease with an increase of the distance r and therefore $C_n = 0$ inasmuch as function $I_0(\lambda_n r)$ unlimitedly grows as $r \to \infty$.

Thus we have $2n-2$ unknown coefficients:

$$C_1,\ C_2,\ C_3,\ C_4,\ C_5,\ \ldots,\ C_{n-1} \qquad D_1,\ D_2,\ D_3,\ D_4,\ D_5,\ \ldots,\ D_{n-1},\ D_n$$

for the determination of which the following system of $2n-2$ equations has to be solved:

$$\mu_1\lambda_1 I_1(\lambda_1 a_1)C_1 - \mu_1\lambda_1 K_1(\lambda_1 a_1) - \mu_2\lambda_2 I_1(\lambda_2 a_1)C_2 + \mu_2\lambda_2 K_1(\lambda_2 a_1)D_2 = 0$$
$$\mu_2\lambda_2 I_1(\lambda_2 a_2)C_2 - \mu_2\lambda_2 K_1(\lambda_2 a_2)D_2 - \mu_3\lambda_3 I_1(\lambda_3 a_2)C_3 + \mu_3\lambda_3 K_1(\lambda_3 a_3)D_3 = 0$$

$$\cdots\cdots\cdots\cdots\cdots\cdots\cdots\cdots\cdots\cdots\cdots\cdots$$

$$\mu_{n-2}\lambda_{n-2} I_1(\lambda_{n-2}a_{n-2})C_{n-2} - \mu_{n-2}\lambda_{n-2} K_1(\lambda_{n-2}a_{n-2})D_{n-2}$$
$$- \mu_{n-1}\lambda_{n-1} I_1(\lambda_{n-1}a_{n-1})C_{n-1} + \mu_{n-1}\lambda_{n-1} K_1(\lambda_{n-1}a_{n-1})D_{n-1} = 0$$
$$\mu_{n-1}\lambda_{n-1} I_1(\lambda_{n-1}a_{n-1})C_{n-1} - \mu_{n-1}\lambda_{n-1} K_1(\lambda_{n-1}a_{n-1})D_{n-1}$$
$$+ \mu_n\lambda_n K_1(\lambda_n a_{n-1})D_n = 0$$

and

$$\lambda_1^2 I_0(\lambda_1 a_1)C_1 + \lambda_1^2 K_0(\lambda_1 a_1) - \lambda_2^2 I_0(\lambda_2 a_1)C_2 - \lambda_2^2 K_0(\lambda_2 a_1)D_2 = 0$$
$$\lambda_2^2 I_0(\lambda_2 a_2)C_2 + \lambda_2^2 K_0(\lambda_2 a_2) - \lambda_3^2 I_0(\lambda_3 a_2)C_3 - \lambda_3^2 K_0(\lambda_3 a_2)D_3 = 0$$

$$\cdots\cdots\cdots\cdots\cdots\cdots\cdots\cdots\cdots\cdots\cdots\cdots$$

$$\lambda_{n-2}^2 I_0(\lambda_{n-2}a_{n-2})C_{n-2} + \lambda_{n-2}^2 K_0(\lambda_{n-2}a_{n-2})D_{n-2} - \lambda_{n-1}^2 I_0(\lambda_{n-1}a_{n-2})C_{n-1}$$
$$- \lambda_{n-1}^2 K_0(\lambda_{n-1}a_{n-2}) = 0$$
$$\lambda_{n-1}^2 I_0(\lambda_{n-1}a_{n-1})C_{n-1} + \lambda_{n-1}^2 K_0(\lambda_{n-1}a_{n-1})D_{n-1} - \lambda^2 K_0(\lambda_n a_{n-1})D_n = 0$$

Solving this system with respect to C_1 we obtain:

$$C_1 = \Delta_1/\Delta$$

where Δ is determinant of the system:

$$\begin{aligned} \Delta &= \mu_1\lambda_1 I_1(\lambda_1 a_1)\delta_1 - \lambda_1^2 I_0(\lambda_1 a_1)\delta_2 \\ \Delta_1 &= \mu_1\lambda_1 K_1(\lambda_1 a_1)\delta_1 - \lambda_1^2 K_0(\lambda_1 a_1)\delta_2 \end{aligned} \tag{4.15}$$

Now let us consider a three-layered medium which consists of three parts, namely the external one with the formation resistivity, ρ_3, the internal one with resistivity ρ_1 (borehole) and the intermediate part with resistivity ρ_2 (invasion zone). In this case we have the following expressions for δ_1, δ_2, Δ and Δ_1 in eqs. 4.15:

$$\delta_1 = -\lambda_2^2 I_0(\lambda_2 a_1)[\mu_2\lambda_2\lambda_3^2 K_1(\lambda_2 a_2)K_0(\lambda_3 a_2) - \mu_3\lambda_3\lambda_2^2 K_0(\lambda_2 a_2)K_1(\lambda_3 a_2)]$$
$$+ \lambda_2^2 K_0(\lambda_2 a_1)[-\mu_2\lambda_2\lambda_3^2 I_1(\lambda_2 a_2)K_0(\lambda_3 a_2) - \mu_3\lambda_3\lambda_2^2 I_0(\lambda_2 a_2)K_1(\lambda_3 a_2)]$$

$$\delta_2 = -\mu_2\lambda_2 I_1(\lambda_2 a_1)[\mu_2\lambda_2\lambda_3^2 K_1(\lambda_2 a_2)K_0(\lambda_3 a_2) - \mu_3\lambda_3\lambda_2^2 K_0(\lambda_2 a_2)K_1(\lambda_3 a_2)]$$
$$- \mu_2\lambda_2^2 K_1(\lambda_2 a_1)[-\mu_2\lambda_2\lambda_3^2 I_1(\lambda_2 a_2)K_0(\lambda_3 a_2) - \mu_3\lambda_3\lambda_2^2 I_0(\lambda_2 a_2)K_1(\lambda_3 a_2)]$$

$$\Delta = [-\mu_1\lambda_1\lambda_2^2 I_0(\lambda_2 a_1)I_1(\lambda_1 a_1) + \mu_2\lambda_2\lambda_1^2 I_0(\lambda_1 a_1)I_1(\lambda_2 a_2)][\mu_2\lambda_2\lambda_3^2 K_1(\lambda_2 a_2)K_0(\lambda_3 a_2)$$
$$- \mu_3\lambda_3\lambda_2^2 K_0(\lambda_2 a_2)K_1(\lambda_3 a_2)] + [\mu_1\lambda_1\lambda_2^2 I_1(\lambda_1 a_1)K_0(\lambda_2 a_1)$$
$$+ \mu_2\lambda_2\lambda_1^2 I_0(\lambda_1 a_1)K_1(\lambda_2 a_1)][-\mu_2\lambda_2\lambda_3^2 I_1(\lambda_2 a_2)K_0(\lambda_2 a_2)$$
$$- \mu_3\lambda_3\lambda_2^2 I_0(\lambda_2 a_2)K_1(\lambda_3 a_2)]$$

$$\Delta_1 = [-\mu_1\lambda_1\lambda_2^2 I_0(\lambda_2 a_1)K_1(\lambda_1 a_1) - \mu_2\lambda_2\lambda_1^2 K_0(\lambda_1 a_1)I_1(\lambda_3 a_1)][\mu_2\lambda_2\lambda_3^2 K_1(\lambda_2 a_2)K_0(\lambda_3 a_2)$$
$$- \mu_3\lambda_3\lambda_2^2 K_0(\lambda_2 a_1)K_1(\lambda_3 a_2)] + [\mu_1\lambda_1\lambda_2^2 K_0(\lambda_2 a_1)K_1(\lambda_1 a_1)$$
$$- \mu_2\lambda_2\lambda_1^2 K_0(\lambda_1 a_1)K_1(\lambda_2 a_1)][-\mu_2\lambda_2\lambda_3^2 I_1(\lambda_2 a_2)K_0(\lambda_3 a_2)$$
$$- \mu_3\lambda_3\lambda_2^2 I_0(\lambda_2 a_2)K_1(\lambda_3 a_2)]$$

Inasmuch as we will consider mainly a medium which has a uniform magnetic permeability, μ, it is convenient to present function C_1 in the form:

$$C_1 = \Delta_1'/\Delta'$$

$$\Delta_1' = [-\lambda_2 I_0(\lambda_2 a_1)K_1(\lambda_1 a_1) - \lambda_1 K_0(\lambda_1 a_1)I_1(\lambda_2 a_1)][\lambda_3 K_1(\lambda_2 a_2)K_0(\lambda_3 a_2)$$
$$- \lambda_2 K_0(\lambda_2 a_2)K_1(\lambda_3 a_2)] + [\lambda_2 K_0(\lambda_2 a_1)K_1(\lambda_1 a_1) - \lambda_1 K_0(\lambda_1 a_1)K_1(\lambda_2 a_1)] \quad (4.16)$$
$$\times [-\lambda_3 I_1(\lambda_2 a_2)K_0(\lambda_3 a_2) - \lambda_2 I_0(\lambda_2 a_2)K_1(\lambda_3 a_2)]$$

$$\Delta' = [-\lambda_2 I_0(\lambda_2 a_1)I_1(\lambda_1 a_1) + \lambda_1 I_0(\lambda_1 a_1)I_1(\lambda_2 a_1)][\lambda_3 K_1(\lambda_2 a_2)K_0(\lambda_3 a_2)$$
$$- \lambda_2 K_0(\lambda_2 a_2)K_1(\lambda_3 a_2)] + [\lambda_2 I_1(\lambda_1 a_1)K_0(\lambda_2 a_1) + \lambda_1 I_0(\lambda_1 a_1)K_1(\lambda_2 a_1)] \quad (4.17)$$
$$\times [-\lambda_3 I_1(\lambda_2 a_2)K_0(\lambda_3 a_2) - \lambda_2 I_0(\lambda_2 a_2)K_1(\lambda_3 a_2)]$$

$$\lambda_1 = (\lambda^2 + \mathrm{i}s_1\chi_3)^{1/2} \qquad \lambda_2 = (\lambda^2 + \mathrm{i}s_2\chi_3)^{1/2} \qquad \lambda_3 = (\lambda^2 + \mathrm{i}\chi_3)^{1/2}$$
$$\chi_3 = \sigma_3\mu\omega$$

where:
σ_3 is the formation conductivity;
$s_1 = \sigma_1/\sigma_3$, $s_2 = \sigma_2/\sigma_3$;
a_1 and a_2 are the radii of the borehole and the invasion zone, respectively.

For Bessel functions the following equations are known:

$$I_0(x) = \sum_{k=0}^{\infty} \frac{1}{(k!)^2}\left(\frac{x}{2}\right)^{2k}$$
$$I_1(x) = \sum_{k=0}^{\infty} \frac{1}{k!(k+1)!}\left(\frac{x}{2}\right)^{2k+1}$$
$$K_0(x) = -\left(\ln\frac{x}{2} + C\right) I_0(x) + \sum_{k=1}^{\infty} \frac{1}{(k!)^2}\left(\frac{x}{2}\right)^{2k} \sum_{m=1}^{k} \frac{1}{m} \quad (4.18)$$

$$K_1(x) = \frac{1}{x} + \left(\ln\frac{x}{2} + C\right) I_1(x) - \sum_{k=1}^{\infty} \frac{1}{k!(k+1)!} \left(\frac{x}{2}\right)^{2k+1} \left[\sum_{m=1}^{k} \frac{1}{m} + \frac{1}{2(k+1)}\right]$$

where C is Euler's constant equal to 0.57721566...

For large argument values, the calculation of Bessel functions can be performed by asymptotical formulae:

$$\begin{aligned}
I_0(x) &= \frac{e^x}{(2\pi x)^{1/2}} \left[1 + \sum_{k=1}^{\infty} \frac{(2k-1)!!}{k!(8x)^k}\right] \\
I_1(x) &= \frac{e^x}{(2\pi x)^{1/2}} \left[1 - \sum_{k=1}^{\infty} \frac{(2k-3)!!(2k+1)!!}{k!(8x)^k}\right] \\
K_0(x) &= e^{-x} \left(\frac{\pi}{2x}\right)^{1/2} \left[1 + \sum_{k=1}^{\infty} \frac{(-1)^k\{(2k-1)!!\}^2}{k!(8x)^k}\right] \\
K_1(x) &= e^{-x} \left(\frac{\pi}{2x}\right)^{1/2} \left[1 - \sum_{k=1}^{\infty} \frac{(-1)^k(2k-3)!!(2k+1)!!}{k!(8x)^k}\right]
\end{aligned} \tag{4.19}$$

Let $\mu_2 = \mu_3 \neq \mu_1$ and $\lambda_2 = \lambda_3$, then in accord with eq. 4.15 we have for a two-layered medium:

$$C_1 = \frac{\Delta_1}{\Delta} = \frac{\mu_1\lambda_2 K_0(\lambda_2 a_1) K_1(\lambda_1 a_1) - \mu_2\lambda_1 K_0(\lambda_1 a_1) K_1(\lambda_2 a_1)}{\mu_1\lambda_2 K_0(\lambda_2 a_1) I_1(\lambda_1 a_1) + \mu_2\lambda_1 I_0(\lambda_1 a_1) K_1(\lambda_2 a_1)} \tag{4.20}$$

It is obvious that a similar equation will be obtained if $\mu_1 = \mu_2 \neq \mu_3$ and $\lambda_1 = \lambda_2 \neq \lambda_3$.

According to eq. 4.11 the components of the electromagnetic field in the first medium are:

$$E_\phi = E_{0\phi} - \frac{i\omega\mu}{4\pi}\frac{2M}{\pi} \int_0^\infty \lambda_1 C_1 I_1(\lambda_1 r) \cos\lambda z \, d\lambda \tag{4.21}$$

$$H_z = H_{0z} - \frac{M}{4\pi}\frac{2}{\pi} \int_0^\infty \lambda_1^2 C_1 I_0(\lambda_1 r) \cos\lambda z \, d\lambda \tag{4.22}$$

$$H_r = H_{0r} - \frac{M}{4\pi}\frac{2}{\pi} \int_0^\infty \lambda\lambda_1 C_1 I_1(\lambda_1 r) \sin\lambda z \, d\lambda \tag{4.23}$$

where $E_{0\phi}$, H_{0z}, H_{or} are components of the field in a uniform medium with conductivity σ_1.

In particular, on the borehole axis, we have:

$$H_r = E_\phi = 0$$

and

$$H_z = H_{0z} - \frac{M}{4\pi}\frac{2}{\pi}\int_0^\infty \lambda_1^2 C_1 \cos\lambda z \, \mathrm{d}\lambda$$

The primary magnetic field along the z-axis is equal to $M/2\pi L^3$, and correspondingly the expression for the vertical component of the magnetic field presented in units of the primary field is:

$$h_z = \frac{H_z}{H_z^{(0)}} = h_z^{(0)} - \frac{L^3}{\pi}\int_0^\infty \lambda_1^2 C_1 \cos\lambda L \, \mathrm{d}\lambda \tag{4.24}$$

where the function $h_z^{(0)}$ was described in detail in Chapter 2, and L is the length of a two-coil induction probe.

Thus the magnetic field due to induced currents in a conducting medium can be expressed through the improper integral, while the integrand consists of a product of two terms: the complex function $\lambda_1^2 C_1$ and the oscillating multiplier $\cos\lambda L$.

Now let us investigate a behavior of function $\lambda_1^2 C_1$ for different values of argument λ, when $\mu_1 = \mu_2 = \mu_3 = 4\pi \times 10^{-7}$ H/m.

First we will consider a two-layered medium (the invasion zone is absent).

In accord with eq. 4.20:

$$\lambda_1^2 C_1 = \lambda_1^2 \frac{\lambda_2 K_0(\lambda_2 a) K_1(\lambda_1 a) - \lambda_1 K_1(\lambda_1 a) K_0(\lambda_1 a)}{\lambda_2 K_0(\lambda_2 a) I_1(\lambda_1 a) + \lambda_1 K_1(\lambda_2 a) I_0(\lambda_1 a)} \tag{4.25}$$

where:

$$\lambda_1 = (\lambda^2 + \mathrm{i}\chi_1)^{1/2}$$
$$\lambda_2 = (\lambda^2 + \mathrm{i}\chi_2)^{1/2}$$
$$\chi_1 = \sigma_1 \mu\omega$$
$$\chi_2 = \sigma_2 \mu\omega$$

and:

$$\lambda_1 = (\lambda^4 + \chi_1^2)^{1/4} \, \mathrm{e}^{\mathrm{i}\arctan(\chi_1/\lambda^2)/2}$$
$$\lambda_2 = (\lambda^4 + \chi_2^2)^{1/4} \, \mathrm{e}^{\mathrm{i}\arctan(\chi_2/\lambda^2)/2}$$

In general, the magnitude of the argument changes from zero to infinity, and its phase alters from $-\pi/4$ for $\lambda = 0$ to 0 as λ tends to infinity.

In two special cases, the expression for $\lambda_1^2 C_1$, (eq. 4.20) is slightly simplified:

- The borehole is nonconductive, that is $\sigma_1 = 0$. Then, $\lambda_1 = \lambda$ and

$$\lambda_1^2 C_1 = \lambda^2 \frac{\lambda_2 K_0(\lambda_2 a) K_1(\lambda a) - \lambda K_1(\lambda_2 a) K_0(\lambda a)}{\lambda_2 K_0(\lambda_2 a) I_1(\lambda a) + \lambda K_1(\lambda_2 a) I_0(\lambda a)} \tag{4.26}$$

- The formation around the borehole is an insulator, that is $\sigma_2 = 0$. Then we have:

$$\lambda_1^2 C_1 = \lambda_1^2 \frac{\lambda K_0(\lambda a) K_1(\lambda_1 a) - \lambda_1 K_1(\lambda a) K_0(\lambda_1 a)}{\lambda K_0(\lambda a) I_1(\lambda_1 a) + \lambda_1 K_1(\lambda a) I_0(\lambda_1 a)} \tag{4.27}$$

Let $\lambda \to 0$, then in accord with eq. 4.25 we have:

$$\lambda_1^2 C_1 \to \mathrm{i}\chi_1 \frac{\sqrt{\mathrm{i}\chi_2} K_0(\sqrt{\mathrm{i}\chi_2} a) K_1(\sqrt{\mathrm{i}\chi_1} a) - \sqrt{\mathrm{i}\chi_1} K_1(\sqrt{\mathrm{i}\chi_2} a) K_0(\sqrt{\mathrm{i}\chi_1} a)}{\sqrt{\mathrm{i}\chi_2} K_0(\sqrt{\mathrm{i}\chi_2} a) I_1(\sqrt{\mathrm{i}\chi_1} a) + \sqrt{\mathrm{i}\chi_1} K_1(\sqrt{\mathrm{i}\chi_2} a) I_0(\sqrt{\mathrm{i}\chi_1} a)} \tag{4.28}$$

where $a = a_1$. Thus function $\lambda_1^2 C_1$ has a finite value for $\lambda = 0$, provided parameters χ_1 and χ_2 are not equal to zero.

As follows from eq. 4.18 for small values of λ, Bessel functions can be replaced by approximate formulae:

$$I_0(\lambda a) \to 1 \qquad I_1(\lambda a) \to \lambda a/2 \qquad K_0(\lambda a) \to -(\ln(\lambda a/2) + C) \qquad K_1(\lambda a) \to 1/\lambda a$$

Substituting these expressions into eqs. 4.26 and 4.27 we obtain:

$$\begin{aligned} \lambda_1^2 C_1 &= \lambda^2 \frac{\sqrt{\mathrm{i}\chi_2} K_0(\sqrt{\mathrm{i}\chi_2} a)/\lambda a + \lambda K_1(\sqrt{\mathrm{i}\chi_2} a)\,(\ln(\lambda a/2) + C)}{\sqrt{\mathrm{i}\chi_2} K_0(\sqrt{\mathrm{i}\chi_2} a)(\lambda a/2) + \lambda K_1(\sqrt{\mathrm{i}\chi_2} a)} \\ &\to \frac{1}{a} \frac{\sqrt{\mathrm{i}\chi_2} K_0(\sqrt{\mathrm{i}\chi_2} a)}{\sqrt{\mathrm{i}\chi_2}\, K_0(\sqrt{\mathrm{i}\chi_2} a)(a/2) + K_1(\sqrt{\mathrm{i}\chi_2} a)} \qquad \text{if } \lambda_1 = \lambda \end{aligned} \tag{4.29}$$

and

$$\begin{aligned} \lambda_1^2 C_1 &= \mathrm{i}\chi_1 \frac{-\lambda(\ln(\lambda a/2) + C) K_1(\sqrt{\mathrm{i}\chi_1} a) - \lambda_1 K_0(\sqrt{\mathrm{i}\chi_1} a)/\lambda a}{-\lambda(\ln(\lambda a/2) + C) I_1(\sqrt{\mathrm{i}\chi_1} a) + \lambda_1 I_0(\sqrt{\mathrm{i}\chi_1} a)/\lambda a} \\ &\to -\mathrm{i}\chi_1 \frac{K_0(\sqrt{\mathrm{i}\chi_1} a)}{I_0(\sqrt{\mathrm{i}\chi_1} a)} \qquad \text{if } \lambda_2 = \lambda \end{aligned} \tag{4.30}$$

Now we will consider the behavior of function $\lambda_1 C_1$ (eq. 4.25) for large values of λ. As is known for $x \to \infty$, we have:

$$I_0(x) \approx \frac{\mathrm{e}^x}{\sqrt{2\pi x}} \left(1 + \frac{0.125}{x}\right) \qquad I_1(x) \approx \frac{\mathrm{e}^x}{\sqrt{2\pi x}} \left(1 - \frac{0.375}{x}\right)$$

$$K_0(x) \approx \mathrm{e}^{-x} \sqrt{\frac{\pi}{2x}} \left(1 - \frac{0.125}{x}\right) \qquad K_1(x) \approx \mathrm{e}^{-x} \sqrt{\frac{\pi}{2x}} \left(1 + \frac{0.375}{x}\right)$$

Substituting these expressions into eq. 4.25, after simple algebra we obtain:

$$\lambda_1^2 C_1 \to \pi \lambda_1^2 e^{-2\lambda_1 a} \frac{\lambda_2 - \lambda_1}{\lambda_2 + \lambda_1} \left(1 + \frac{0.750}{\lambda a}\right) \to 0 \tag{4.31}$$

Therefore, the real and imaginary parts of function $\lambda_1^2 C_1$ decrease very rapidly when $\lambda a \gg 1$. The latter allows us to evaluate an upper limit of integration.

Now we will investigate function $\lambda_1^2 C_1$ in the case of a three-layered medium:

$$m_1 = -\lambda_2 I_0(\lambda_2 a_1) K_1(\lambda_1 a_1) - \lambda_1 K_0(\lambda_1 a_1) I_1(\lambda_2 a_1) \tag{4.32}$$

$$n_1 = \lambda_3 K_1(\lambda_2 a_2) K_0(\lambda_3 a_2) - \lambda_2 K_0(\lambda_2 a_2) K_1(\lambda_3 a_2) \tag{4.33}$$

$$m_2 = \lambda_2 K_0(\lambda_2 a_1) K_1(\lambda_1 a_1) - \lambda_1 K_0(\lambda_1 a_1) K_1(\lambda_2 a_1) \tag{4.34}$$

$$n_2 = -\lambda_3 I_1(\lambda_2 a_2) K_0(\lambda_3 a_2) - \lambda_2 I_0(\lambda_2 a_2) K_1(\lambda_3 a_2) \tag{4.35}$$

$$m_3 = -\lambda_2 I_0(\lambda_2 a_1) I_1(\lambda_1 a_1) + \lambda_1 I_0(\lambda_1 a_1) I_1(\lambda_2 a_1) \tag{4.36}$$

$$n_3 = \lambda_2 I_1(\lambda_1 a_1) K_0(\lambda_2 a_1) + \lambda_1 I_0(\lambda_1 a_1) K_1(\lambda_2 a_1) \tag{4.37}$$

$$\lambda_1^2 C_1 = \lambda_1^2 \frac{m_1 n_1 + m_2 n_2}{m_3 n_1 + n_2 n_3} \tag{4.38}$$

If $\lambda_2 = \lambda_3 \neq \lambda_1$, function $n_1 = 0$ and $\lambda_1^2 C_1 = m_2/n_3$, that is we obtain the formula for a two-layered medium (the interface with radius $r = a_1$). If $\lambda_1 = \lambda_2 \neq \lambda_3$ we have:

$$m_1 = -\lambda_2 [I_0(\lambda_2 a_1) K_1(\lambda_2 a_1) + K_0(\lambda_2 a_1) I_1(\lambda_2 a_1)] = -1/a_1$$

$$m_2 = 0 \qquad m_3 = 0 \qquad n_3 = 1/a_1$$

$$\lambda_1^2 C_1 = \lambda_2^2 \frac{m_1 n_1}{n_2 n_3} = -\lambda_2^2 \frac{n_1}{n_2} = \lambda_2^2 \frac{\lambda_3 K_1(\lambda_2 a_2) K_0(\lambda_3 a_2) - \lambda_2 K_0(\lambda_2 a_2) K_1(\lambda_3 a_2)}{\lambda_3 I_1(\lambda_2 a_2) K_0(\lambda_3 a_2) + \lambda_2 I_0(\lambda_2 a_2) K_1(\lambda_3 a_2)}$$

that corresponds to a two-layered medium, as the interface radius is equal to a_2.

For three special cases, expressions of $\lambda_1^2 C_1$ are somewhat simplified.

Nonconducting borehole ($\sigma_1 = 0$, $\sigma_2 \neq 0$, $\sigma_3 \neq 0$, $\lambda_1 = \lambda$)

$$\begin{aligned}
m_1 &= -\lambda_2 I_0(\lambda_2 a_1) K_1(\lambda a_1) - \lambda K_0(\lambda a_1) I_1(\lambda_2 a_1) \\
m_2 &= \lambda_2 K_0(\lambda_2 a_1) K_1(\lambda a_1) - \lambda K_0(\lambda a_1) K_1(\lambda_2 a_1) \\
m_3 &= -\lambda_2 I_0(\lambda_2 a_1) I_1(\lambda a_1) + \lambda I_0(\lambda a_1) I_1(\lambda_2 a_1) \\
n_3 &= \lambda_2 I_1(\lambda a_1) K_0(\lambda_2 a_1) + \lambda I_0(\lambda a_1) K_1(\lambda_2 a_1)
\end{aligned} \tag{4.39}$$

Nonconducting intermediate zone ($\sigma_2 = 0$, $\sigma_1 \neq 0$, $\sigma_3 \neq 0$, $\lambda_2 = \lambda$)

$$\begin{aligned}
m_1 &= -\lambda I_0(\lambda a_1) K_1(\lambda_1 a_1) - \lambda_1 K_0(\lambda_1 a_1) I_0(\lambda a_1) \\
n_1 &= \lambda_3 K_1(\lambda a_2) K_0(\lambda_3 a_2) - \lambda K_0(\lambda a_2) K_1(\lambda_3 a_2) \\
m_2 &= \lambda K_0(\lambda a_1) K_1(\lambda_1 a_1) - \lambda_1 K_0(\lambda_1 a_1) K_1(\lambda a_1) \\
n_2 &= -\lambda_3 I_1(\lambda a_2) K_0(\lambda_3 a_2) - \lambda I_0(\lambda a_2) I_1(\lambda a_1) \\
m_3 &= -\lambda I_0(\lambda a_1) I_1(\lambda_1 a_1) + \lambda_1 I_0(\lambda_1 a_1) I_1(\lambda a_1) \\
n_3 &= \lambda I_1(\lambda_1 a_1) K_0(\lambda a_1) + \lambda_1 I_0(\lambda_1 a_1) K_1(\lambda a_1)
\end{aligned} \tag{4.40}$$

Nonconducting bed ($\sigma_1 \neq 0$, $\sigma_2 \neq 0$, $\sigma_3 = 0$, $\lambda_3 = \lambda$)

$$n_1 = \lambda K_1(\lambda_2 a_2) K_0(\lambda a_2) - \lambda_2 K_0(\lambda_2 a_2) K_1(\lambda a_2)$$
$$n_2 = -\lambda I_1(\lambda_2 a_2) K_0(\lambda a_2) - \lambda_2 I_0(\lambda_2 a_2) K_1(\lambda a_2)$$

If the conductivities of all three media are not zero it is clear that function $\lambda_1^2 C_1$ tends to the finite limit as $\lambda \to 0$. Now consider these three special cases:

Case 1

If $\sigma_1 = 0$ and $\lambda \to 0$ we have:

$$m_1 \to \frac{\lambda_2 I_0(\lambda_2 a_1)}{\lambda a_1} \qquad m_2 \to \frac{\lambda_2}{\lambda a_1} K_0(\lambda_2 a_1)$$
$$m_3 \to \lambda\left(-\frac{\lambda_2 a_1}{2} I_0(\lambda_2 a_1) + I_1(\lambda_2 a_1)\right) = -\lambda \frac{\lambda_2 a_1}{2} I_2(\lambda_2 a_1)$$
$$n_3 \to \lambda\left(-\frac{\lambda_2 a_1}{2} K_0(\lambda_2 a_1) + K_1(\lambda_2 a_1)\right) = -\lambda \frac{\lambda_2 a_1}{2} K_2(\lambda_2 a_1)$$

Functions n_1 and n_2 also have finite values and therefore:

$$\lambda_1^2 C_1 = \lambda^2 \frac{(-\lambda_2 I_0(\lambda_2 a_1) n_1 + \lambda_2 K_0(\lambda_2 a_1) n_2)/\lambda a_1}{\lambda \lambda_2 a_1 (I_2(\lambda_2 a_1) n_1 + K_2(\lambda_2 a_1) n_2)/2} = \frac{2}{a_1^2} \frac{(I_0(\lambda_2 a_1) n_1 + K_0(\lambda_2 a_1) n_2)}{(-I_2(\lambda_2 a_1) n_1 + K_2(\lambda_2 a_1) n_2)}$$

Case 2

If $\sigma_2 = 0$ and λ goes to zero we have:

$$m_1 = -\lambda\left(K_1(\lambda_1 a_1) + \frac{\lambda_1 a_1}{2} K_0(\lambda_1 a_1)\right) = -\lambda \frac{\lambda_1 a_1}{2} K_2(\lambda_1 a_1)$$
$$n_1 = \frac{\lambda_3}{\lambda a_2} K_0(\lambda_3 a_2) \qquad m_2 = -\frac{\lambda_1}{\lambda a_1} K_0(\lambda_1 a_1)$$
$$n_2 = -\lambda\left(\frac{\lambda_3 a_2}{2} K_0(\lambda_3 a_2) + K_1(\lambda_3 a_2)\right) = -\lambda \frac{\lambda_3 a_2}{2} K_2(\lambda_3 a_2)$$
$$m_3 = -\lambda\left(I_1(\lambda_1 a_1) - \frac{\lambda_1 a_1}{2} I_0(\lambda_1 a_1)\right) = \lambda \frac{\lambda_1 a_1}{2} I_2(\lambda_1 a_1)$$
$$n_3 = \frac{\lambda_1}{\lambda a_1} I_0(\lambda_1 a_1)$$

Thus:

$$\lambda_1^2 C_1 = \lambda_1^2 \frac{(a_2/a_1) K_0(\lambda_1 a_1) K_2(\lambda_3 a_2) - (a_1/a_2) K_0(\lambda_3 a_2) K_2(\lambda_1 a_1)}{(a_1/a_2) K_0(\lambda_3 a_2) I_2(\lambda_1 a_1) - (a_2/a_1) I_0(\lambda_1 a_1) K_2(\lambda_3 a_2)}$$

Case 3

If $\sigma_3 = 0$ and $\lambda \to 0$ then:

$$n_1 = -\frac{\lambda_2}{\lambda a_2} K_0(\lambda_2 a_2) \qquad n_2 = -\frac{1}{a_2} I_0(\lambda_2 a_2) \qquad \lambda_1 C_1 \to \lambda_1 \frac{m_1}{m_2}$$

Now making use of the asymptotic behavior of Bessel functions, we will study the behavior of function $\lambda_1^2 C_1$, when the variable of integration increases unlimitedly ($\lambda \to \infty$).

In accord with eq. 4.14 we have

$$m_1 = -\frac{e^{(\lambda_2-\lambda_1)a_1}}{2a_1\sqrt{\lambda_1\lambda_2}}(\lambda_1+\lambda_2) \qquad n_2 = -\frac{e^{(\lambda_2-\lambda_3)a_2}}{2a_2\sqrt{\lambda_1\lambda_2}}(\lambda_1+\lambda_2)$$

$$n_1 = \frac{e^{-(\lambda_1+\lambda_2)a_2}}{2a_2\sqrt{\lambda_2\lambda_3}}(\lambda_3-\lambda_2) \qquad m_3 = \frac{e^{(\lambda_1+\lambda_2)a_1}}{2a_1\sqrt{\lambda_2\lambda_3}}(\lambda_3-\lambda_2)$$

$$m_2 = -\frac{e^{-(\lambda_1+\lambda_2)a_1}}{2a_1\sqrt{\lambda_1\lambda_2}}(\lambda_2-\lambda_1) \qquad n_3 = \frac{e^{(\lambda_1-\lambda_2)a_1}}{2a_1\sqrt{\lambda_1\lambda_2}}(\lambda_1+\lambda_2)$$

Therefore, as $\lambda \to \infty$

$$m_1 n_1 \to 0 \qquad m_2 n_3 \to 0 \qquad n_2 n_3 \to -\frac{1}{a_1 a_2}$$

The product $m_3 n_1$ also tends to zero, since $a_2 > a_1$. An unlimited increase of function m_3, as $\lambda \to \infty$, should be taken into account when the corresponding computer program code is prepared. Thus with increasing λ, function $\lambda_1^2 C_1$ decreases exponentially.

Now let us consider the integral on the right-hand side of eq. 4.24 from the following point of view. It can be interpreted as an infinite sum of cylindrical harmonics with complex amplitude $\lambda_1^2 C_1$:

$$\lambda_1^2 C_1 \cos \lambda L$$

In this expression the variable of separation λ plays the role of spatial frequency. It is clear that with an increase in λ, the corresponding harmonic changes rapidly. Now we will investigate how various spatial harmonics are sensitive to different parts of the medium provided that the electromagnetic frequency ω, is the same.

As follows from eqs. 4.32–4.37, function n_1 decreases more rapidly than others when spatial frequency λ increases. Correspondingly, we can neglect terms containing n_1 in eq. 4.38 when λ is sufficiently large. Therefore, instead of eq. 4.38 we have:

$$\lambda_1^2 C_1 = \lambda_1^2 m_2 / n_3$$

i. e. the expression for a two-layered medium when the external layer has resistivity ρ_2. In other words, harmonics with higher spatial frequencies have smaller depths of penetration.

After the analysis of the behavior of function $\lambda_1^2 C_1$ we will describe a method to calculate the integral on the right-hand side of eq. 4.24. As was pointed out above the integrand contains an oscillating multiplier $\cos \lambda L$, which complicates the numerical integration, specially for large ratios of L/a_1. Application of Simpson's well known method with a uniform step consists of a replacement of the whole integrand $\lambda_1^2 C_1 \, \cos \lambda L$ with a polynomial $a_1 + a_1\lambda + a_2\lambda^2$.

The presence of the oscillating function $\cos \lambda L$ implies that only for very small steps of integration we can achieve a sufficient accuracy of calculations. For this reason integration with a uniform step requires a large amount of computer time. It is much more efficient to perform the integration with a nonuniform step which is a modification of Fillon's method. The idea of this is that the interval of integration is presented as a sum of elements of different length. Inasmuch as with an increase of λ the integrand decreases, it is natural to increase also the length of these elements. Within each element $[\lambda_i \div \lambda_{i+2}]$ the non-oscillating part of the integrand $\lambda_1^2 C_1$ is replaced by the polynomial:

$$\phi(\lambda) = \lambda_1^2 C_1 = a_1 + a_1\lambda + a_2\lambda^2$$

Therefore, the integral over this element is:

$$I_i = \int_{\lambda_i}^{\lambda_{i+2}} \lambda_1^2 C_1 \, \cos \lambda L \, \mathrm{d}\lambda = \int_{\lambda_i}^{\lambda_{i+2}} (a_0 + a_1\lambda + a_2\lambda_2^2) \, \cos \lambda L \, \mathrm{d}\lambda$$

$$= a_0 \int_{\lambda_i}^{\lambda_{i+2}} \cos \lambda L \, \mathrm{d}\lambda + a_1 \int_{\lambda_i}^{\lambda_{i+2}} \lambda \cos \lambda L \, \mathrm{d}\lambda + a_2 \int_{\lambda_i}^{\lambda_{i+2}} \lambda^2 \cos \lambda L \, \mathrm{d}\lambda$$

Integrals on the right-hand side, as it is well known, can be expressed by elementary functions:

$$\int_{\lambda_i}^{\lambda_{i+2}} \cos \lambda L \, \mathrm{d}\lambda = \frac{\sin \lambda_{i+2} L - \sin \lambda_i L}{L}$$

$$\int_{\lambda_i}^{\lambda_{i+2}} \lambda \cos \lambda L \, \mathrm{d}\lambda = \left(\frac{\cos \lambda L}{L^2} + \frac{\lambda \sin \lambda L}{L} \right) \Bigg|_{\lambda_i}^{\lambda_{i+2}}$$

$$\int_{\lambda_i}^{\lambda_{i+2}} \lambda^2 \cos \lambda L \, \mathrm{d}\lambda = \left\{ \frac{2\lambda}{L^2} \cos \lambda L + \left(\frac{\lambda^2}{L} - \frac{2}{L^2} \right) \sin \lambda L \right\} \Bigg|_{\lambda_i}^{\lambda_{i+2}}$$

Coefficients a_0, a_1 and a_2 are defined from the system:

$$a_0 + a_1\lambda_i + a_2\lambda_i^2 = \phi(\lambda_i)$$
$$a_0 + a_1\lambda_{i+1} + a_2\lambda_{i+1}^2 = \phi(\lambda_{i+1})$$
$$a_0 + a_1\lambda_{i+1} + a_2\lambda_{i+2}^2 = \phi(\lambda_{i+2})$$

The final result presents a sum of integrals I_i over the whole integration interval. Numerical analysis as well as eq. 4.31 show that the maximal value of λ does not exceed 160 if $a_1 = 0.1$ m.

The integration interval is usually divided in two parts, namely:

- the initial one: $10^{-8} \leqslant \lambda \leqslant 10^{-1}$
- the remaining one: $10^{-1} < \lambda < 160$.

From comparison of results of calculation within the external part the step of integration is usually chosen in the following way: $\lambda_{i+1} = \sqrt[8]{2}\lambda_i$, sometimes it is replaced by a smaller step: $\lambda_{i+1} = \sqrt[16]{2}\lambda_i$. The value of the ratio a_1/λ_1 ($\lambda_1 = 2\pi h_1$) is changed with the step $\sqrt{2}$ from 10^{-4} to 1, that allows us to obtain the total spectrum practically for all geoelectric sections of interest. The ratio of the length, L, of a two-coil induction probe to the borehole radius, a_1, is altered from 2 to 30 with the uniform step:

$$2,\ 4,\ 6,\ 8,\ 10, \ldots, 28,\ 30$$

For a two-layered medium the following σ_2/σ_1 ratios were considered:

$$\frac{1}{128}, \frac{1}{64}, \frac{1}{32}, \frac{1}{16}, \frac{1}{8}, \frac{1}{4}, \frac{1}{2}, 2,\ 4,\ 6,\ 16,\ 32$$

For a three-layered medium (the invaded zone is present) calculations were performed for the following parameters:

$$\frac{a_2}{a_1} = 2,\ 4,\ 8,\ 16 \qquad \frac{\sigma_3}{\sigma_1} = \frac{1}{128}, \frac{1}{64}, \frac{1}{32}, \frac{1}{16}, \frac{1}{8}, \frac{1}{4}, \frac{1}{2}, 1 \qquad \frac{\rho_2}{\rho_1} = 4,\ 8,\ 16,\ 32,\ 64,\ 128$$

Results of calculations are presented as a spectrum of four quantities, such as:

- the quadrature component of the magnetic field: $\mathrm{Q}\, h_z$
- the inphase component of the secondary field: $\mathrm{In}\, h_z - 1$
- the amplitude of the secondary field: $A = ((\mathrm{In}\, h_z - 1)^2 + (\mathrm{Q}\, h_z)^2)^{1/2}$.
- the function $\sigma_a/\sigma_1 = (2/\sigma_1\mu\omega L^2)\, \mathrm{Q}\, h_z$, where σ_a is the apparent conductivity function introduced by H. Doll.

Before we investigate the spectra of these functions it is appropriate to investigate their asymptotical behavior.

As follows from eq. 4.24 the field h_z, expressed in units of the primary field, depends on the following parameters (three-layered medium):

$$\frac{a_1}{h_1}, \frac{L}{a_1}, \frac{a_2}{a_1}, \frac{\rho_2}{\rho_1}, \frac{\rho_3}{\rho_1}$$

First of all, let us investigate the range of small parameters a_i/h_i and correspondingly L/h_1, where a_i is a radius of any interface while h_i is the skin depth in any part of the medium.

4.3. The Quadrature Component of the Magnetic Field at the Range of Very Small Model Parameters

By definition the range of small parameters a_i/h_i and L/h_i corresponds to conditions when the skin depth in every uniform part of the medium is much greater than its geometric parameters, such as the radius of the borehole and the invaded zone, the length of two-coil probe, that is:

$$h_i \gg L \qquad h_i \gg a_i \qquad h_i \gg a_2$$

where $i = 1, 2, 3$.

This relationship can take place due to either:

- the relatively low frequencies of the field, or
- the sufficiently resistive medium, or
- the probe length is relatively small, i.e. measurements are performed near the field source.

There is one common feature in all these cases, namely the strong influence of the primary electric field. In other words, in the limit we can neglect the interaction of induced currents and consider that the current density at every point of a medium is defined by the primary electric field only. In accord with the results described in Chapter 3 (Doll's theory), we have:

$$\boldsymbol{J}_\phi = \sigma \boldsymbol{E}_\phi^{(0)} = \frac{\mathrm{i}\,\sigma\mu\omega Mr}{4\pi R^3} = -\frac{k^2 Mr}{4\pi R^3} = -\frac{\mathrm{i}\,\chi Mr}{4\pi R^3}$$

and therefore the magnetic field measured in this approximation has to be proportional to k^2 as follows from Biot–Savart law. Correspondingly, in order to derive formulae for this approximation it is necessary to expand the right-hand side of eq. 4.24 in a series and discard all terms but the first one which is proportional to k^2.

In accord with results obtained in Chapter 2 the first term, $h_z^{(0)}$, can be presented as:

$$h_z^{(0)} \simeq \frac{\mathrm{i}\,\sigma_1\mu\omega L^2}{2} \qquad \text{if } L/h_1 \ll 1 \tag{4.41}$$

Now we have to find the leading term of the expansion of the integral in eq. 4.24. Let us start from the simplest case of a two-layered medium (the invasion zone is absent):

$$\lambda_1^2 C_1 = \lambda_1^2 \frac{\lambda_2 K_0(\lambda_2 a_1) K_1(\lambda_1 a_1) - \lambda_1 K_1(\lambda_2 a_1) K_0(\lambda_1 a_1)}{\lambda K_0(\lambda_2 a_1) I_1(\lambda_1 a_1) + \lambda_1 K_1(\lambda_2 a_1) I_0(\lambda_1 a_1)} \tag{4.42}$$

It is obvious that:

$$\begin{aligned} \lambda_1 &= (\lambda^2 + k_1^2)^{1/2} = \lambda\left(1 + \frac{k_1^2}{\lambda^2}\right) \approx \lambda + \frac{1}{2}\frac{k_1^2}{\lambda} \\ \lambda_2 &= (\lambda^2 + k_2^2)^{1/2} = \lambda\left(1 + \frac{k_2^2}{\lambda^2}\right) \approx \lambda + \frac{1}{2}\frac{k_2^2}{\lambda} \end{aligned} \tag{4.43}$$

and

$$I_0(\lambda a_1) = I_0\left(\lambda a + \frac{1}{2}\frac{k_1^2 a_1}{\lambda}\right) = I_0(\lambda a_1) + \frac{1}{2}\frac{k_1^2 a_1}{\lambda} I_0'(\lambda a_1)$$

where:

$$\begin{aligned}
I_0'(\lambda a_1) &= \frac{\partial I_0(\lambda a_1)}{\partial(\lambda a_1)} \\
I_1(\lambda_1 a_1) &= I_1\left(\lambda a_1 + \frac{1}{2}\frac{k_1^2 a_1}{\lambda}\right) = I_1(\lambda a_1) + \frac{1}{2}\frac{k_1^2 a_1}{\lambda} I_1'(\lambda a_1) \\
K_0(\lambda_1 a_1) &= K_0\left(\lambda a_1 + \frac{1}{2}\frac{k_1^2 a_1}{\lambda}\right) = K_0(\lambda a_1) + \frac{1}{2}\frac{k_1^2 a_1}{\lambda} K_0'(\lambda a_1) \\
K_1(\lambda_1 a_1) &= K_1\left(\lambda a_1 + \frac{1}{2}\frac{k_1^2 a_1}{\lambda}\right) = K_1(\lambda a_1) + \frac{1}{2}\frac{k_1^2 a_1}{\lambda} K_1'(\lambda a_1) \\
K_0(\lambda_2 a_1) &= K_0\left(\lambda a_1 + \frac{1}{2}\frac{k_2^2 a_1}{\lambda}\right) = K_0(\lambda a_1) + \frac{1}{2}\frac{k_2^2 a_1}{\lambda} K_0'(\lambda a_1) \\
K_1(\lambda_2 a_1) &= K_1\left(\lambda a_1 + \frac{1}{2}\frac{k_2^2 a_1}{\lambda}\right) = K_1(\lambda a_1) + \frac{1}{2}\frac{k_2^2 a_1}{\lambda} K_1'(\lambda a_1)
\end{aligned} \tag{4.44}$$

Substituting eqs. 4.43 and 4.44 into eq. 4.42 and making use of the recurrence relations of Bessel functions:

$$\begin{aligned}
&I_0'(x) = -I_1(x) \qquad K_0'(x) = -K_1(x) \\
&I_{v-1}(x) - I_{v+1}(x) = \frac{2v}{x} I_v(x) \\
&K_{v-1}(x) - K_{v+1}(x) = -\frac{2v}{x} K_v(x) \\
&I_{v-1}(x) + I_{v+1}(x) = 2I_v'(x) \\
&K_{v-1}(x) + K_{v+1}(x) = -2K_v'(x)
\end{aligned}$$

After simple transformations we obtain:

$$\lambda_1^2 C_1 = \mathrm{i}(\chi_2 - \chi_1)\frac{\lambda a_1}{2}\left[2K_0(\lambda a_1)K_1(\lambda a_1) - \lambda a_1(K_1^2 - K_0^2)\right] \tag{4.46}$$

where a_1 is the borehole radius.

Thus the quadrature component of the magnetic field, expressed in units of the primary field, is:

$$\mathrm{Q}\,h_z = \frac{\sigma_1\mu\omega L^2}{2} + \frac{L^3}{\pi}(s_1 - 1)\chi_1 \int_0^\infty \frac{\lambda a_1}{2}\left[2K_0K_1 - \lambda a_1(K_1^2 - K_0^2)\right]\cos\lambda L\,\mathrm{d}\lambda \tag{4.47}$$

where $s_1 = \sigma_2/\sigma_1$.

Let us introduce notations:

$$m = \lambda a_1 \qquad \alpha = L/a_1 \tag{4.48}$$

Then eq. 4.47 can be rewritten as:

$$Q\,h_z = \frac{\omega\mu L^2}{2}\left(\sigma_1 + (\sigma_2 - \sigma_1)\frac{2\alpha}{\pi}\int\limits_0^\infty \frac{m}{2}\left[2K_0(m)K_1(m) - m(K_1^2 - K_0^2)\right]\cos m\alpha\, \mathrm{d}m\right)$$

or

$$Q\,h_z = \frac{\omega\mu L^2}{2}(\sigma_1 G_1 - \sigma_2 G_2) \tag{4.49}$$

where

$$G_2 = \frac{2\alpha}{\pi}\int\limits_0^\infty \frac{m}{2}\left[2K_0(m)K_1(m) - m(K_1^2 - K_0^2)\right]\cos m\alpha\, \mathrm{d}m \tag{4.50}$$

and

$$G_1 = 1 - G_2 \tag{4.51}$$

Functions G_2 and G_1 are the geometric factors of the formation and the borehole, respectively.

It is essential that G_2 and G_1 depend on only one parameter $\alpha = L/a_1$, characterizing the length of the two-coil probe, and in accord with eq. 4.51 their sum is equal to unit

$$G_1 + G_2 = 1 \tag{4.52}$$

Applying the same approach for a three-layered medium (eq. 4.38) we obtain the following expression for the quadrature component:

$$\begin{aligned} Q\,h_z = \frac{\omega\mu L^2}{2}\Bigg\{&\sigma_1 + (\sigma_2 - \sigma_1)\frac{2\alpha}{\pi}\int\limits_0^\infty \frac{m}{2}\left[2K_0(m)K_1(m) - m(K_1^2 - K_0^2)\right]\cos m\alpha\, \mathrm{d}m \\ &+ (\sigma_3 - \sigma_2)\frac{2\alpha}{\beta\pi}\int\limits_0^\infty \frac{m}{2}\left[2K_0K_1 - m(K_1^2 - K_0^2)\right]\cos\left(\frac{\alpha}{\beta}m\right)\mathrm{d}m\Bigg\} \end{aligned} \tag{4.53}$$

where $\alpha = L/a_1$, $\beta = a_2/a_1$.

Thus the latter can be written in the form:

$$h_z = \frac{\mu\omega L^2}{2}(\sigma_1 G_1 + \sigma_2 G_2 + \sigma_3 G_3) \tag{4.54}$$

where:

$$G_1 = 1 - \frac{2\alpha}{\pi} \int_0^\infty \frac{m}{2} \left[2K_0K_1 - m(K_1^2 - K_0^2)\right] \cos m\alpha \, dm$$

$$\begin{aligned} G_2 = & \frac{2\alpha}{\pi} \int_0^\infty \frac{m}{2} \left[2K_0K_1 - m(K_1^2 - K_0^2)\right] \cos m\alpha \, dm \\ & - \frac{2\alpha}{\beta\pi} \int_0^\infty \frac{m}{2} \left[2K_0K_1 - m(K_1^2 - K_0^2)\right] \cos\left(\frac{\alpha}{\beta} m\right) dm \end{aligned} \tag{4.55}$$

$$G_3 = \frac{2\alpha}{\beta\pi} \int_0^\infty \frac{m}{2} \left[2K_0K_1 - m(K_1^2 - K_0^2)\right] \cos\left(\frac{\alpha}{\beta} m\right) dm$$

These functions, G_1, G_2, and G_3 are the geometric factors of the borehole, the invasion zone and the formation, respectively.

Let us introduce the function:

$$G(x) = \frac{2x}{\pi} \int_0^\infty \frac{m}{2} \left[2K_0K_1 - m(K_1^2 - K_0^2)\right] \cos mx \, dm \tag{4.56}$$

which is naturally called the geometric factor of the external medium or formation with resistivity ρ_3, or the geometric factor of the bed. It is clear that other geometric factors can be expressed through this function. Indeed in accord with eq. 4.55 we have:

$$\begin{aligned} G_1(\alpha) \quad &= 1 - G(\alpha) \\ G_2(\alpha, \beta) &= G(\alpha) - G(\alpha/\beta) \\ G_3(\alpha, \beta) &= G(\alpha/\beta) \end{aligned} \tag{4.57}$$

that is, the geometric factors of the borehole and invasion zone are related with function G in a very simple manner.

As follows from eqs. 4.57 the sum of geometric factors, as in the case of two-layered medium, is equal to unity:

$$G_1 + G_2 + G_3 = 1 \tag{4.58}$$

Thus the geometric factor of each cylindrical layer can be described by the function G or $1 - G$. The latter, that is $1 - G$, is the geometric factor of the cylinder. If the radius of the cylinder coincides with that of the borehole function, $1 - G$ is the geometric factor of the borehole.

Therefore, due to the absence of interaction of induced currents in the range of small parameters the quadrature component of the field in a medium with coaxial cylindrical interfaces can be described only with the help of function G or $1-G$. In a general case of n-layered medium we have:

$$Q\,h_z = \frac{\omega\mu L^2}{2}\sum_{i=1}^{N-1}\sigma_i G_i \tag{4.59}$$

where G_1 is the geometric factor of the borehole equal to:

$$G_1 = 1 - \frac{2\alpha}{\pi}\int_0^\infty \frac{m}{2}\left[2K_0K_1 - m(K_1^2 - K_0^2)\right]\cos m\alpha\, dm$$

where $\alpha = L/a_1$

$$G_i = \frac{2\alpha_{i-1}}{\pi}\int_0^\infty \frac{m}{2}\left[2K_0K_1 - m(K_1^2 - K_0^2)\right]\cos m\alpha_{i-1}\, dm$$
$$- \frac{2\alpha_i}{\pi}\int_0^\infty \frac{m}{2}\left[2K_0K_1 - m(K_1^2 - K_0^2)\right]\cos m\alpha_i\, dm$$

where $\alpha_{i-1} = L/a_{i-1}$, $\alpha_i = L/a_i$; a_{i-1} and a_i are the radii of the internal and external interfaces of the i-th cylindrical layer.

Finally:

$$G = G_{N-1} = \frac{2\alpha_{i-1}}{\pi}\int_0^\infty \frac{m}{2}\left[2K_0K_1 - m(K_1^2 - K_0^2)\right]\cos m\alpha_{i-1}\, dm$$

is the geometric factor of the formation, that is the bed, and $\alpha_{i-1} = L/a_{i-1}$.

It is clear that:

$$\sum_{i=1}^{N-1} G_i = 1 \tag{4.60}$$

Inasmuch as all geometric factors of all cylindrical layers are expressed through either function G or $1-G$, let us describe their behavior in detail.

Thus:

$$G_1(\alpha) = 1 - G = 1 - \frac{2\alpha}{\pi}\int_0^\infty A(m)\cos(\alpha m)\, dm, \tag{4.61}$$

where

$$A(m) = \frac{m}{2}\left[2K_0K_1 - m(K_1^2 - K_0^2)\right] \tag{4.62}$$

We will consider the dependence of this function on m. For sufficiently large values of m we have:

$$K_0(m) \approx \mathrm{e}^{-m}\left(\frac{\pi}{2m}\right)^{1/2}\left(1-\frac{0.125}{m}\right)$$
$$K_1(m) \approx \mathrm{e}^{-m}\left(\frac{\pi}{2m}\right)^{1/2}\left(1+\frac{0.375}{m}\right)$$

For this reason when $m \to \infty$:

$$A(m) \to \frac{3}{4}\pi \mathrm{e}^{-2m} \to 0$$

In the opposite case as $m \to 0$:

$$K_0(m) \to -\left(\ln\frac{m}{2}+C\right) \qquad K_1(m) \to \frac{1}{m}$$

Substituting these values into eq. 4.62 we obtain:

$$A(m) \to K_0(m) \to -\left(\ln\frac{m}{2}+C\right) \qquad \text{as } m \to 0 \tag{4.63}$$

that is, the integrand has a logarithmic singularity as m tends to zero.

In order to remove this singularity we will make use of the following equation:

$$\frac{1}{(1+\alpha^2)^{1/2}} = \frac{2}{\pi}\int_0^\infty K_0(m)\cos m\alpha \, dm$$

Then the function G_1 can be presented in the form:

$$G_1(\alpha) = 1 - \frac{2}{\pi}\int_0^\infty A(m)\cos m\alpha \, dm = 1 - \frac{\alpha}{(1+\alpha^2)^{1/2}} + \frac{2}{\pi}\int_0^\infty [K_0(m)-A(m)]\cos m\alpha \, dm \tag{4.64}$$

In accord with eq. 4.63 the integrand in eq. 4.64 does not have singularities, and its calculation presents a relatively simple task. Values of function G_1 are given in Table 4.1. Corresponding values of the geometric factor $G = 1 - G_1$ are presented in Table 4.2.

Let us investigate the behavior of the integral at the right-hand side of eq. 4.64 when parameter α increases. In this case due to oscillating character of the integrand the value of the integral:

$$\int_0^\infty \Phi(m)\cos m\alpha \, dm$$

TABLE 4.1
Values of function $G_1(\alpha)$

α	$G_1(\alpha)$	α	$G_1(\alpha)$	α	$G_1(\alpha)$	α	$G_1(\alpha)$	α	$G_1(\alpha)$
0.2	0.8829	3.6	0.08149	7.0	0.02158	10.4	0.009599	13.8	0.005390
0.4	0.7700	3.8	0.07348	7.2	0.02037	10.6	0.009232	14.0	0.005235
0.6	0.6650	4.0	0.06653	7.4	0.01926	10.8	0.008887	14.2	0.005085
0.8	0.5701	4.2	0.06047	7.6	0.01824	11.0	0.008559	14.4	0.004943
1.0	0.4866	4.4	0.05517	7.8	0.01729	11.2	0.008250	14.6	0.004806
1.2	0.4146	4.6	0.05051	8.0	0.01642	11.4	0.007957	14.8	0.004675
1.4	0.3534	4.8	0.04640	8.2	0.01561	11.6	0.007679	15.0	0.004549
1.6	0.3019	5.0	0.04276	8.4	0.01486	11.8	0.007416	15.2	0.004428
1.8	0.2588	5.2	0.03951	8.6	0.01416	12.0	0.007166	15.4	0.004312
2.0	0.2229	5.4	0.03661	8.8	0.01351	12.2	0.006929	15.6	0.004200
2.2	0.1929	5.6	0.03401	9.0	0.01290	12.4	0.006702	15.8	0.004003
2.4	0.1679	5.8	0.03167	9.2	0.01233	12.6	0.006487	16.0	0.003890
2.6	0.1469	6.0	0.02956	9.4	0.01180	12.8	0.006282	17.0	0.003530
2.8	0.1292	6.2	0.02765	9.6	0.01130	13.0	0.006087	18.0	0.003140
3.2	0.1143	6.4	0.02592	9.8	0.01084	13.2	0.005901	19.0	0.002820
3.2	0.1016	6.6	0.02434	10.0	0.01046	13.4	0.005722	20.0	0.002540
3.4	0.0907	6.8	0.02290	10.2	0.00998	13.6	0.005553		

where

$$\Phi(m) = K_0(m) - \frac{m}{2}\left[2K_0K_1 - m(K_1^2 - K_0^2)\right]$$

is defined by the behavior of the function $\Phi(m)$ and its derivatives near zero. In fact, integrating by parts we obtain:

$$\begin{aligned}
\int_0^\infty \Phi(m)\cos m\alpha\,\mathrm{d}m &= \frac{1}{\alpha}\int_0^\infty \Phi(m)\,\mathrm{d}(\sin\alpha m) \\
&= \frac{1}{\alpha}\Phi\sin\alpha m\Big|_0^\infty - \frac{1}{\alpha}\int_0^\infty \Phi'(m)\sin\alpha m\,\mathrm{d}m = \frac{1}{\alpha}\Phi\sin\alpha m\Big|_0^\infty + \frac{1}{\alpha^2}\int_0^\infty \Phi'(m)\,\mathrm{d}(\cos m\alpha) \\
&= \frac{1}{\alpha}\Phi\sin\alpha m\Big|_0^\infty + \frac{1}{\alpha^2}\Phi'(m)\cos\alpha m\Big|_0^\infty - \frac{1}{\alpha^2}\int_0^\infty \Phi''(m)\cos m\alpha\,\mathrm{d}m
\end{aligned} \tag{4.65}$$

For large values of m function $\Phi(m)$ and its derivatives tend to zero and therefore instead of eq. 4.65 we have:

$$\int_0^\infty \Phi(m)\cos m\alpha\,\mathrm{d}m = \frac{1}{\alpha}\,\Phi(0) + \frac{1}{\alpha^2}\Phi'(0) - \frac{1}{\alpha^2}\int_0^\infty \Phi''(m)\cos m\alpha\,\mathrm{d}m \tag{4.66}$$

TABLE 4.2
Values of function $G = 1 - G_1(a)$

α	$G(\alpha)$	α	$G(\alpha)$	α	$G(\alpha)$	α	$G(\alpha)$	α	$G(\alpha)$
0.2	0.1170	3.6	0.9185	7.0	0.9784	10.4	0.9904	13.8	0.9946
0.4	0.2299	3.8	0.9265	7.0	0.9796	10.6	0.9907	14.0	0.9947
0.6	0.3349	4.0	0.9334	7.4	0.9807	10.8	0.9911	14.2	0.9949
0.8	0.4298	4.2	0.9395	7.6	0.9817	11.0	0.9914	14.4	0.9950
1.0	0.5133	4.4	0.9448	7.8	0.9827	11.2	0.9817	14.6	0.9951
1.2	0.5853	4.6	0.9448	8.0	0.9835	11.4	0.9920	14.8	0.9953
1.4	0.6465	4.8	0.9535	8.2	0.9843	11.6	0.9923	15.0	0.9954
1.6	0.6980	5.0	0.9572	8.4	0.9851	11.8	0.9925	15.2	0.9955
1.8	0.7411	5.2	0.9604	8.6	0.9858	12.0	0.9928	15.4	0.9956
2.0	0.7770	5.4	0.9633	8.8	0.9864	12.2	0.9930	15.6	0.9957
2.2	0.8070	5.6	0.9659	9.0	0.9870	12.4	0.9932	15.8	0.9959
2.4	0.8320	5.8	0.9683	9.2	0.9876	12.6	0.9935	16.0	0.9961
2.6	0.8530	6.0	0.9704	9.4	0.9881	12.8	0.9937		
2.8	0.8707	6.2	0.9723	9.6	0.9886	13.0	0.9939		
3.0	0.8856	6.4	0.9740	9.8	0.9891	13.2	0.9940		
3.2	0.8983	6.6	0.9756	10.0	0.9895	13.4	0.9942		
3.4	0.9092	6.8	0.9770	10.2	0.9900	13.6	0.9944		

For small values of m ($m \to 0$) we have:

$$K_0(m) \backsimeq -\ln m - \frac{m^2}{4}\ln m + \frac{m^2}{4} - C + \dots$$
$$K_1(m) \backsimeq \frac{1}{m} + \frac{m}{2}\ln m - \frac{m}{4} + \dots$$

Substituting these expressions into $\Phi(m)$ we obtain:

$$\begin{aligned} \Phi(m) &\approx \frac{1}{2} + \frac{1}{4}m^2 \ln m \\ \Phi'(m) &\approx \frac{m}{2}\ln m \\ \Phi''(m) &\approx \frac{1}{2}\ln m \qquad \text{as } m \to 0 \end{aligned}$$

Thus:

$$\int_0^\infty \Phi(m)\cos m\alpha \, dm \to -\frac{1}{2\alpha^2}\int_0^\infty \ln m \cos m\alpha \, dm \approx \frac{1}{2\alpha^2}\int_0^\infty K_0(m)\cos m\alpha \, dm \to \frac{1}{2\alpha^2}\frac{\pi}{2\alpha} \tag{4.67}$$

Substituting this expression for the integral into eq. 4.64 we obtain:

$$G_1(\alpha) \simeq 1 - \frac{1}{\left(1 + \dfrac{1}{\alpha^2}\right)^{1/2}} + \frac{1}{2\alpha^2} = \frac{1}{\alpha^2} \qquad \text{if } \alpha \gg 1 \tag{4.68}$$

Therefore, for large values of α the geometric factor of the cylinder, in particular of the borehole, decreases inversely proportional to α^2, that is:

$$G_1(\alpha) \simeq \frac{1}{\alpha^2} = \frac{a_1^2}{L^2} \qquad \text{if } \alpha \gg 1 \tag{4.69}$$

Let us notice that this behavior of the geometric factor of the borehole defines a radial characteristics of the simplest *focusing* induction probe consisting of three coils.

Comparison with calculated results (Table 4.1) shows that eq. 4.69 describes with sufficient accuracy the value of function G_1 if the ratio $L/a_1 > 4$ ($\alpha > 4$). Using the same approach for obtaining asymptotical expressions for function $G_1(\alpha)$ we obtain the following terms of an expansion. For example, a more accurate expression of the geometric factor $G_1(\alpha)$ for large values of α has the form:

$$G_1 \simeq \frac{1}{\alpha^2} + \frac{3\ln 2\alpha - 4.25}{\alpha^4} \tag{4.70}$$

In the opposite case of small values of α, function $G_1(\alpha)$ approaches to unity as:

$$G_1(\alpha) = 1 - 0.5862\alpha \tag{4.71}$$

In other words, for small values of parameter α the geometric factor of the borehole is close to unity, and with an increase of α, this function, $G_1(\alpha)$, decreases inversely proportional to α.

Behavior of the geometric factor $G_1(\alpha)$ is shown in Fig. 4.2.

It is appropriate to notice here that in the range of small parameters the interaction of induced currents is negligible. We can obtain an expression for the geometric factor of the borehole by integrating the geometric factor of an elementary ring over a cylinder. Applying this approach H. Doll derived the following expression:

$$G_1(\alpha) = 1 - \frac{1}{(1 + 4/\alpha^2)^{1/2}} \left(E(k) + \frac{2}{\alpha^2}[E(k) - K(k)]\right) \tag{4.72}$$

here E and K are elliptical integrals of the first and second kind and $k = \alpha(\alpha^2 + 4)^{1/2}$. It is clear that function G_1 or $1 - G_1$ allows us to investigate radial characteristics of a two-coil induction probe in the range of small parameters L/h_i, and a_i/h_i, in a medium with cylindrical interfaces only, that is as we cannot take into account a finite thickness of a formation.

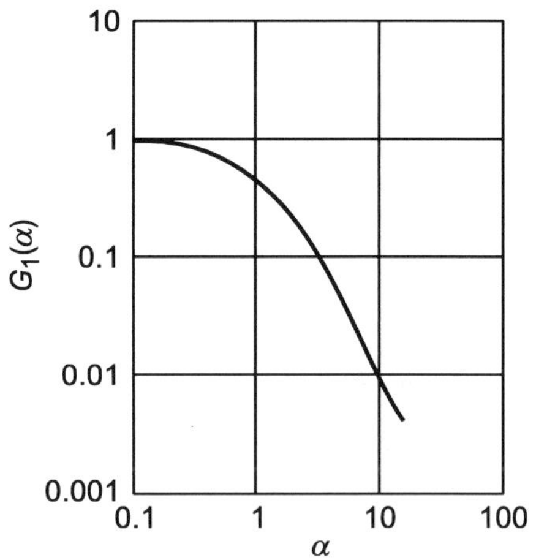

Figure 4.2. Geometric factor of the borehole, $G_1(\alpha)$.

Before we study this subject it is appropriate to make three comments:

- The theory of induction logging developed by H. Doll is based on the concept of geometric factors.
- In many cases the parameters of geoelectric section defining the field behavior are very small. For example, if the length of the induction probe is 1 m, the frequency of the field is 20×10^3 Hz and the resistivity of the medium is 5 ohm·m, we have $p = L/h \approx 0.12$. In a more resistive medium, parameter p is even smaller for this frequency. Correspondingly, for a relatively resistive medium this theory of induction logging provides sufficiently accurate information about radial characteristics of this method.
- It is obvious that with increasing conductivity or frequency the influence of the skin effect becomes more prominent, and this method of calculation of the field becomes less accurate, and it is necessary to use either the exact solution or a better approximation.

After these comments we will consider radial characteristics of induction logging proceeding from the geometric factor G_1.

First of all we will introduce the concept of the apparent conductivity which, as well as apparent resistivity, is widely used in all electrical methods.

Let us define the apparent conductivity in the following way:

$$\frac{\sigma_a}{\sigma_1} = \frac{\mathrm{Q}\, H_z}{\mathrm{Q}\, H_z^{un}(\sigma_1)}$$

or

$$\frac{\sigma_a}{\sigma_1} = \frac{Q\,\mathscr{E}}{Q\,\mathscr{E}^{un}(\sigma_1)} \tag{4.73}$$

where σ_1 is the conductivity of the borehole mud. $Q\,H_z$ and $Q\,\mathscr{E}$ are the magnetic field and the electromotive force measured by the receiver coil on the borehole axis while $Q\,H_z^{un}$ and $\mathscr{E}^{un}(\sigma_1)$ are the magnetic field and the electromotive force, measured by the receiver coil when the induction probe is located in a uniform medium with conductivity σ_1.

In accord with eq. 4.73 the ratio σ_a/σ_1 shows how the field or the electromotive force, measured on the borehole axis, differs from the same quantities in a uniform medium with conductivity σ_1. In other words, this ratio characterizes the influence of the formation surrounding the borehole.

It is appropriate to present equations 4.73 in another form, namely:

$$\sigma_a = \frac{\sigma_1}{Q\,H_z^{un}(\sigma_1)}\,Q\,H_z \qquad \text{and} \qquad \sigma_a = \frac{\sigma_1}{Q\,\mathscr{E}^{un}(\sigma_1)}\,Q\,\mathscr{E} \tag{4.74}$$

or

$$\sigma_a = K_H\,Q\,H_z \qquad \text{and} \qquad \sigma_a = K_{\mathscr{E}}\,Q\,\mathscr{E} \tag{4.75}$$

where K_H and $K_{\mathscr{E}}$ are coefficients of a two-coil induction probe measuring the magnetic field and the electromotive force, respectively.

Inasmuch as the range of small parameters is considered:

$$Q\,H^{un}(\sigma_1) = \frac{\mu\omega M_T}{4\pi L}\sigma_1 \qquad Q\,\mathscr{E}^{un} = \frac{\mu^2\omega^2 M_T M_R}{4\pi L}\sigma_1$$

Therefore, for coefficients of the probe we have:

$$K_H = \frac{4\pi L}{\mu\omega M_T} \qquad K_{\mathscr{E}} = \frac{4\pi L}{\mu^2\omega^2 M_T M_R} \tag{4.76}$$

It is essential that coefficients of the probe do not depend on the conductivity, and they are defined by parameters of the induction probe (length, moments of the coils) and the frequency.

Making use of eq. 4.59 we have the following expression for the apparent conductivity:

$$\sigma_a = \sum_{i=1}^{N-1} \sigma_i G_i \tag{4.76}$$

In particular, for a two-layered medium when the invasion zone is absent we have:

$$\sigma_a = \sigma_1 G_1 + \sigma_2 G_2 \tag{4.77}$$

and for a three-layered medium:

$$\sigma_a = \sigma_1 G_1 + \sigma_2 G_2 + \sigma_3 G_3 \tag{4.78}$$

These last three equations allow us to investigate the radial responses of a two-coil induction probe in detail, in other words, to evaluate the influence of the borehole, the invasion zone and the formation as a function of the induction probe length, L, for various geoelectric parameters.

4.4. Radial Characteristics of a Two-coil Induction Probe at the Range of Small Parameters

In order to analyze the depth of investigation of a two-coil induction probe let us consider curves describing the dependence of the function σ_a/σ_1 on the length of the probe expressed in units of the borehole radius, L/a_1. We will call these curves induction lateral sounding curves (ILS-2). The number 2 means that two-coil induction probes are considered. Such terminology can be explained in the following way. Suppose that measurements are simultaneously performed along the borehole axis with induction probes of different lengths. As follows from the behavior of the geometric factor $G_1(\alpha)$ a relatively short induction probe mainly provides information about the resistivity of the borehole mud. A signal measured with a longer probe contains more information about the resistivity of the invasion zone. In other words, the latter has a greater depth of investigation. Performing measurements by even longer probes we obtain more information about the invasion zone and the resistivity of the formation. In principle we can always apply such a long induction probe that the measured signal will be a function of the formation conductivity only, i.e. induced currents within the borehole and the invasion zone will not practically affect the signal.

Thus, measurements with a system of induction probes of different lengths allow us in principle to define parameters of a geoelectric section in the radial direction, that is to perform the lateral soundings. Correspondingly, the curve of the apparent conductivity for probes with various lengths presents the result of such soundings.

It is appropriate to distinguish three main cases. They are:

- the invasion zone is absent
- the resistivity of the invasion zone has an intermediate value between the resistivity of the borehole and that of the formation, that is: $\sigma_1 > \sigma_2 > \sigma_3$
- the resistivity of the invasion zone exceeds both the borehole and the formation resistivities, that is: $\sigma_1 > \sigma_2 < \sigma_3$.

First of all let us consider a two-layered model (the invasion zone is absent). The family of two-layered curves of apparent conductivity is shown in Fig. 4.3. Along the x- and y-axes σ_a/σ_1 and L/a_1 are plotted, respectively. The curve index is the ratio σ_2/σ_1.

As is seen from Fig. 4.3 every curve has two asymptotes, presented by straight lines, which are parallel to the axis of abscissa. The one to the right corresponds to the conductivity of the formation, σ_2, while the one on the left is described by the equation: $\sigma_a = \sigma_1$. The latter is a common asymptote for all curves since with a decrease of the probe length the geometric factor of the borehole tends to unity and in the limit, regardless of conductivity of the formation, the quadrature component of the field is defined by induced currents in the borehole only. With an increase of the ratio L/a_1 the apparent conductivity, σ_a/σ_1, gradually changes and for large values of L/a_1 approaches to the asymptote on the right: $\sigma_a = \sigma_2$. It is essential to emphasize that apparent conductivity curves do not intersect their asymptote.

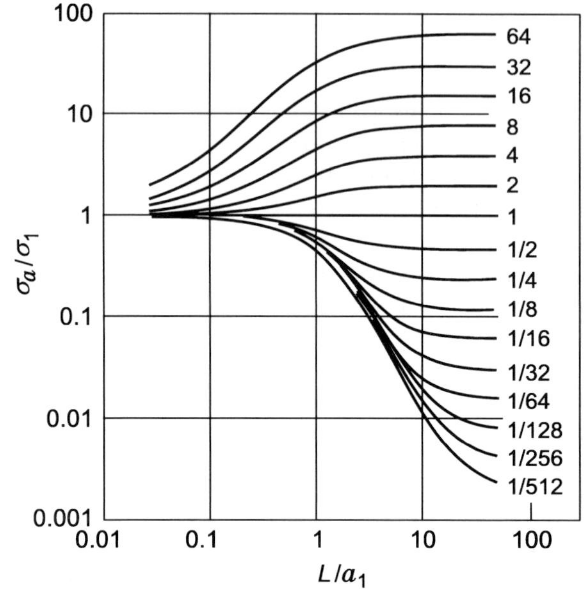

Figure 4.3. Two-layered curves of ILS-2. Index of curves σ_2/σ_1.

Curves with index $\sigma_2/\sigma_1 > 1$ relatively quickly approach to the asymptote on the right and practically for $L/a_1 \geqslant 5$ the value of the apparent conductivity is almost equal to the conductivity of the formation, σ_2. With a decrease of the curve index, that is with an increase of the resistivity of the formation or conductivity of the borehole, the approach to the asymptote to the right takes place for greater values L/a_1. In accord with eqs. 4.60 and 4.69 we have the following presentation for apparent conductivity σ_a which is valid for relatively large values of $\alpha = L/a_1$:

$$\frac{\sigma_a}{\sigma_1} \simeq \frac{\sigma_2}{\sigma_1}\left(1 - \frac{1}{\alpha^2}\right) + \frac{1}{\alpha^2} \qquad \text{if } \alpha \gg 1 \tag{4.79}$$

In particular if the inequality:

$$\frac{\sigma_2}{\sigma_1} \gg \frac{1}{\alpha^2}$$

takes place within a certain range of lengths of the induction probe we have:

$$\sigma_a = \sigma_1 \frac{1}{\alpha^2}$$

that is the apparent conductivity changes inversely proportional to L^2.

Now let us consider the second and third cases when there is an invasion zone between the borehole and the formation. Corresponding apparent conductivity curves are presented in Figs. 4.4–4.11. As is seen from the curves in the second case ($\rho_1 < \rho_2 < \rho_3$) three-layered curves are similar to two-layered ones, but differ from them by a slower rate of change with an increase of $\alpha = L/a_1$. If the penetration of the borehole solution into the formation is relatively large the right-hand branch of a three-layered curve tends to that of a two-layered one with index σ_3/σ_2, but the left-hand branch almost coincides with the two-layered curve with index σ_2/σ_1.

As it concerns the intermediate part of the curve, it is more lifted than a two-layered curve with the same index. Since $\sigma_1 > \sigma_2 > \sigma_3$, the presence of the invasion zone usually causes a significant increase of the apparent conductivity with respect to a two-layered model and therefore it requires the application of longer probes for the determination of the conductivity of a formation, σ_3.

In the third case ($\sigma_1 > \sigma_2 < \sigma_3$) for a relatively small probe length the apparent conductivity curves have a more gentle slope than that of two-layered curves, provided the conductivity of the bed is significantly greater than the conductivity of the invasion zone.

For relatively small values of σ_3/σ_2, the curves have a minimum. It is obvious that the influence of penetration of the borehole solution on the apparent conductivity increases with an increase of the radius of the invasion zone and its conductivity with respect to σ_3.

It is appropriate to notice here that in logging methods based on application of direct current the direction of current lines depends on the relation between resistivities of borehole, invasion zone, and bed. At the same time in induction logging, when the source of the field is a vertical magnetic dipole located on the borehole axis, current lines, regardless of the distribution of resistivity in a radial direction, present themselves as circles located in horizontal planes with their centers on the borehole axis, and they do not intersect the cylindrical surface between media with different conductivity.

It is useful here to define the quadrature component of the magnetic field caused by induced currents within a thin cylindrical shell. In accord with eq. 4.57 we have:

$$Q\,h_z = \frac{\mu\omega L^2}{2}\sigma\left[G_1\left(\frac{L}{r_2}\right) - G_1\left(\frac{L}{r_1}\right)\right] \tag{4.80}$$

where r_1 and r_2 are the internal and external radii of the shell. With a decrease of the shell's thickness the difference in geometric factors in eq. 4.80 can be approximately replaced by a product of the thickness, Δr, and the derivative of the geometric factor, G_1 with respect to r. Then we have:

$$Q\,h_z \simeq \frac{\mu\omega L^2}{2} Sq \tag{4.81}$$

where S is the longitudinal conductance of the shell equal to $\sigma\Delta r$, and q is the geometric factor of the shell which is:

$$q(r) = -\frac{L}{r^2}G_1'\left(\frac{L}{r}\right)$$

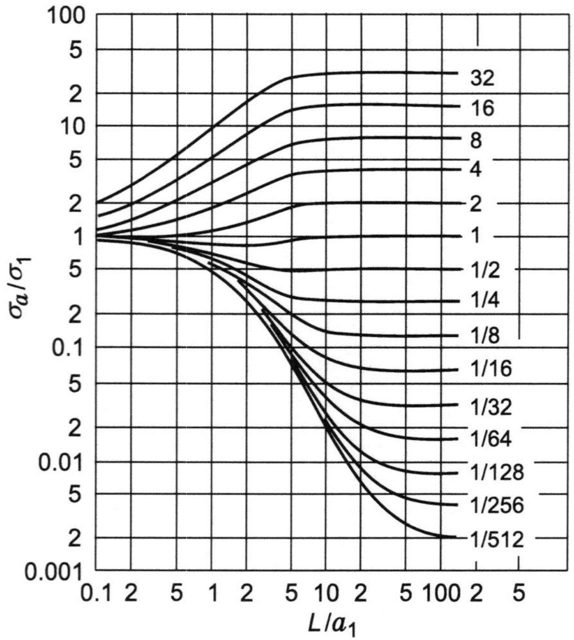

Figure 4.4. Three-layered curves of σ_a/σ_1 ($a_2/a_1 = 2$, $\sigma_2/\sigma_1 = 1/4$). Curve index σ_3/σ_1.

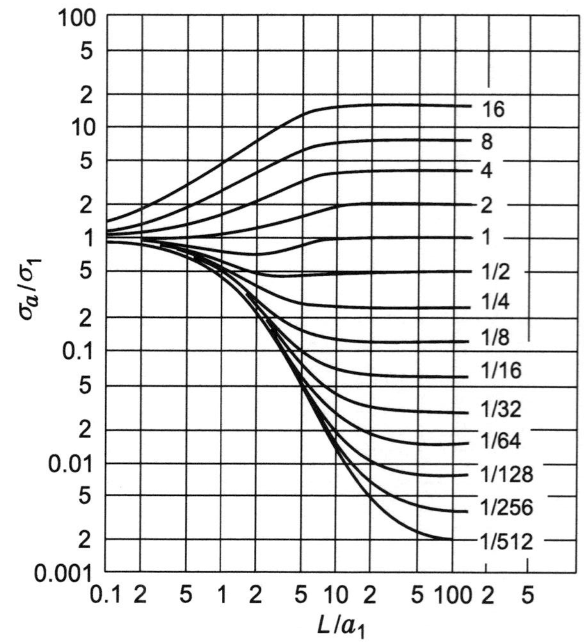

Figure 4.5. Three-layered curves of σ_a/σ_1 ($a_2/a_1 = 2$, $\sigma_2/\sigma_1 = 1/64$). Curve index σ_3/σ_1.

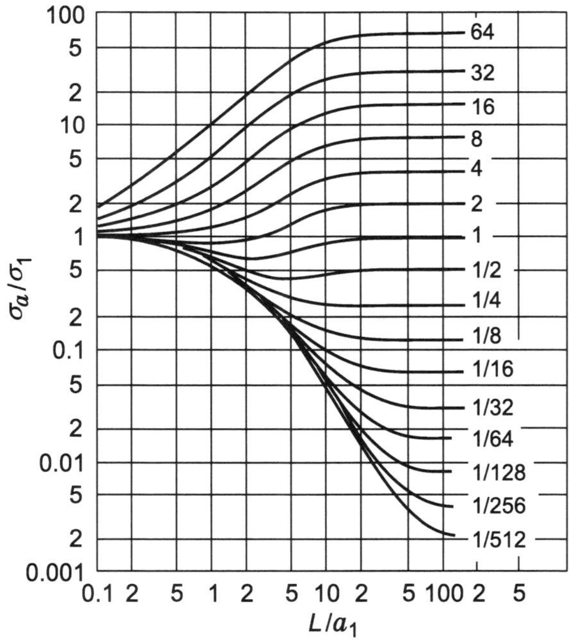

Figure 4.6. Three-layered curves of σ_a/σ_1 ($a_2/a_1 = 4$, $\sigma_2/\sigma_1 = 1/4$). Curve index σ_3/σ_1.

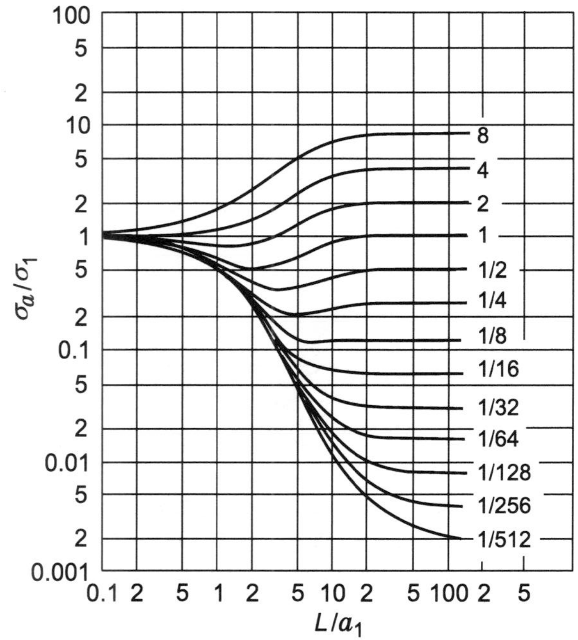

Figure 4.7. Three-layered curves of σ_a/σ_1 ($a_2/a_1 = 4$, $\sigma_2/\sigma_1 = 1/128$). Curve index σ_3/σ_1.

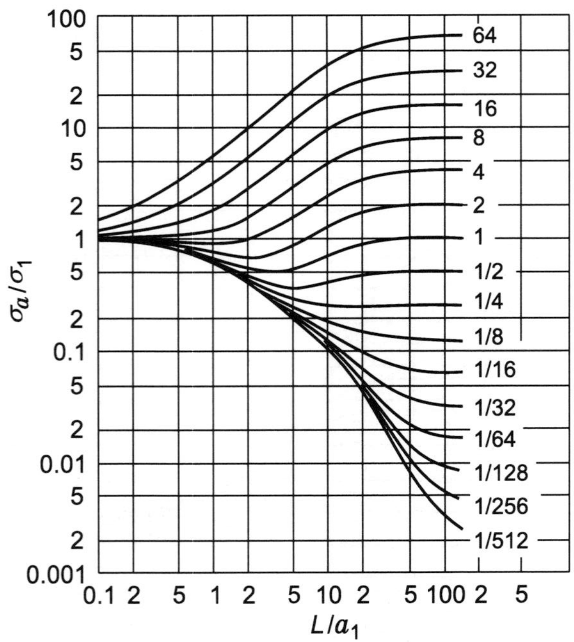

Figure 4.8. Three-layered curves of σ_a/σ_1 ($a_2/a_1 = 8$, $\sigma_2/\sigma_1 = 1/4$). Curve index σ_3/σ_1.

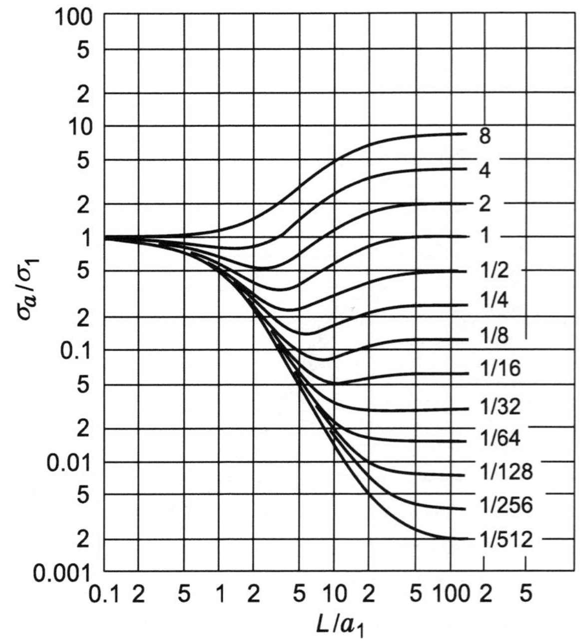

Figure 4.9. Three-layered curves of σ_a/σ_1 ($a_2/a_1 = 8$, $\sigma_2/\sigma_1 = 1/128$). Curve index σ_3/σ_1.

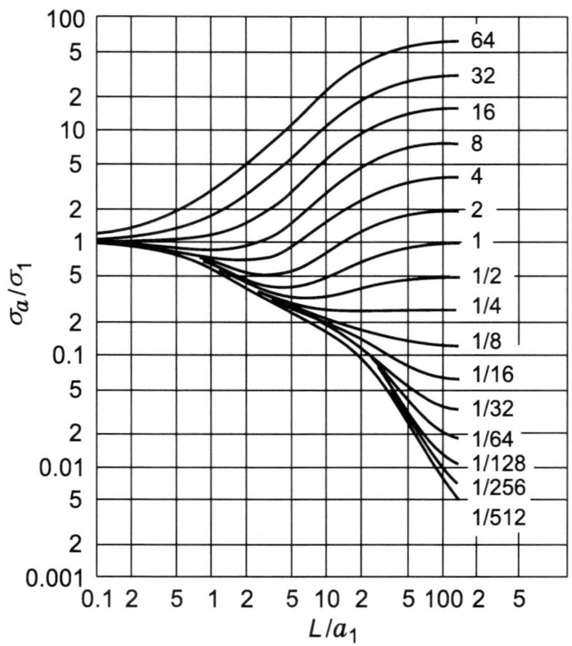

Figure 4.10. Three-layered curves of σ_a/σ_1 ($a_2/a_1 = 16$, $\sigma_2/\sigma_1 = 1/4$). Curve index σ_3/σ_1.

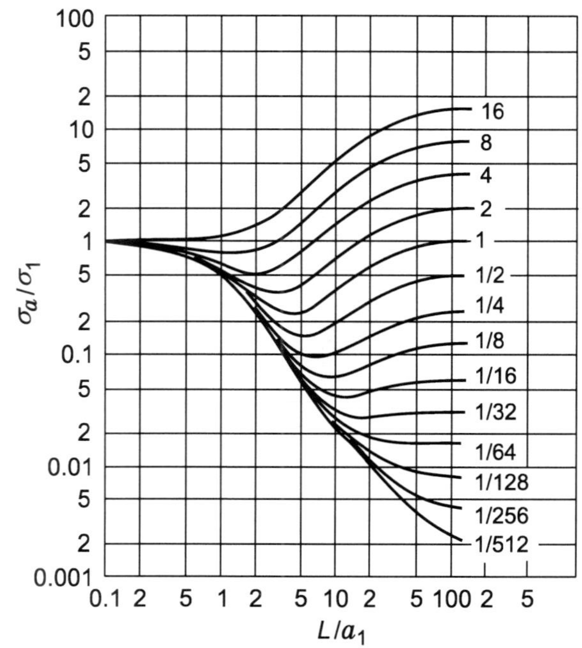

Figure 4.11. Three-layered curves of σ_a/σ_1 ($a_2/a_1 = 16$, $\sigma_2/\sigma_1 = 1/64$). Curve index σ_3/σ_1.

where r is the average radius of the shell.

Within a certain range of a change of conductivity, σ, and thickness, Δr, of shells, induced currents within them generate practically the same field if the longitudinal conductance, $S = \sigma \Delta r$, remains constant. With an increase of the radius r, that is moving away from the borehole axis, this equivalence by S can be applied to shells with greater thickness.

In the range of small parameters the principle of equivalence is only defined by geometric factors related with the distribution of currents within the corresponding cylindrical layer.

Consideration of three-layered curves of ILS-2 allows us also to introduce a concept of an approximate equivalence of right-hand and left-hand branches of three-layered curves and two-layered curves with some specially chosen parameters.

Let us suppose that there is slight penetration of a borehole solution into a formation, and we investigate the right-hand branch of a three-layered curve. If the probe length is significantly greater than the radius of the borehole and that of the invasion zone, instead of eq. 4.78 we can write:

$$\sigma_a = \frac{\sigma_1}{\alpha^2} + \frac{\sigma_2}{\alpha^2}\left(\beta^2 - 1\right) + \sigma_3 G_3\left(\frac{\alpha}{\beta}\right) \tag{4.82}$$

or

$$\sigma_a = \frac{\sigma_1\left[1 + \frac{\sigma_2}{\sigma_1}\left(\beta^2 - 1\right)\right]}{\beta^2\alpha^2} + \sigma_3 G_3\left(\frac{\alpha}{\beta}\right) \tag{4.83}$$

where $\beta = a_2/a_1$. Therefore the right-hand branch of a three-layered curve is equivalent to that of a two-layered one with the following parameters:

- The radius of a fictitious borehole is equal to that of the invasion zone.
- The conductivity of a fictitious borehole solution is related with the parameters of the three-layered model as:

$$\sigma_a = \frac{\sigma_1 + \sigma_2\left(\beta^2 - 1\right)}{\beta} \tag{4.84}$$

- The conductivities of the bed in both cases are the same.

For deep penetration of the filtrate into the formation eq. 4.84 can be applied only for relatively long induction probes. For this reason instead of eq. 4.82 we can write:

$$\sigma_a = \frac{\sigma_1}{\alpha^2} + \sigma_2\left[G_1\left(\frac{\alpha}{\beta}\right) - \frac{1}{\alpha^2}\right] + \sigma_3 G_3\left(\frac{\alpha}{\beta}\right)$$

or

$$\sigma_a \simeq \sigma_2 G_1\left(\frac{\alpha}{\beta}\right) + \sigma_3 G_3\left(\frac{\alpha}{\beta}\right) \qquad \text{if } L/a_1 \gg 1 \tag{4.85}$$

that is, we obtain the expression of σ_a for a two-layered medium with borehole radius a_2 and conductivity σ_2. With an increase of the radius of the invasion zone and the resistivity of the borehole solution eq. 4.85 describes the right-hand branch of a three-layered curve with higher accuracy.

When the length of the induction probe tends to zero one can use the approximate expression of function $G(\alpha)$ as $\alpha \to 0$. Then instead of eq. 4.85 we have:

$$\sigma_a = \sigma_1 G_2(\alpha) + \sigma_2 G_3(\alpha) \tag{4.86}$$

Therefore, the left-hand branch of a three-layered curve approximately corresponds to a two-layered curve provided the radius of the invasion zone is sufficiently large.

Considering radial responses of two-coil induction probes in the range of small parameters it is natural to investigate cases when the resistivity within the intermediate zone changes as a continuous function. In accord with eq. 4.59 we have:

$$Q\, h_z = \frac{\omega\mu L^2}{2}\sigma_a$$

where

$$\sigma_a = \sigma_1 G_1(\alpha) + \sum_{i=1}^{N} \sigma_i q_i + \sigma_3 G_3\left(\frac{\alpha}{\beta}\right) \tag{4.87}$$

and $\beta = a_2/a_1$; q_i is the geometric factor of a thin cylindrical layer with conductivity σ_i.

As will be shown in Chapter 6, the function q_i can be presented in the form:

$$q(\alpha) = \frac{\Delta r}{r}\frac{1}{\alpha^2}G(\alpha)$$

Function $G(\alpha)$ is tabulated and its values are presented in Table 4.6. If an intermediate (invasion) zone is divided uniformly by a set of cylindrical shells, eq. 4.87 can be rewritten in the form:

$$\sigma_a = \sigma_1 G_1(\alpha) + \frac{\Delta r}{r}\sum_{i=1}^{N} \sigma_i \frac{1}{\alpha_i} C(\alpha_i) + \sigma_3 G_3(\alpha_N) \tag{4.88}$$

Examples of apparent conductivity curves for various change of resistivity within an intermediate zone are presented in Fig. 4.12.

The first pair of curves (A) corresponds to the condition: $\rho_1 < \rho_2 < \rho_3$. In one case the resistivity of the invasion zone is constant: $\rho_2 = 8\rho_1$, while in the other case the resistivity ρ_2 gradually increases from that of the borehole to that of the formation.

The second pair of curves (B) corresponds to condition: $\rho_1 < \rho_2 > \rho_3$. Again one curve describes the behavior of the apparent conductivity as ρ_2 is constant, while in the second case the resistivity of the invasion zone decreases.

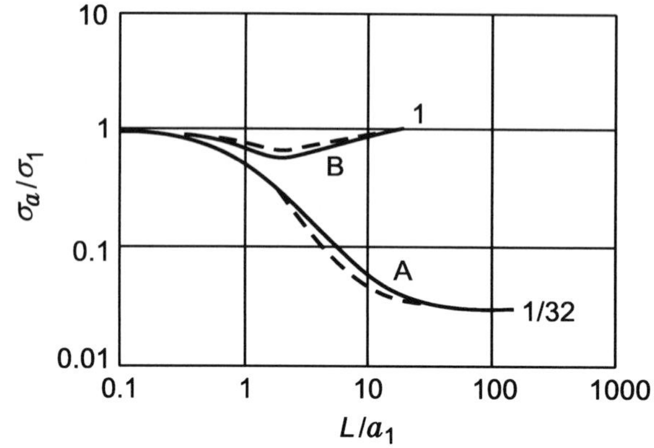

Figure 4.12. Apparent conductivity curves: (A) $\sigma_2/\sigma_1 = 1/8$, $\sigma_3/\sigma_1 = 1/32$, $a_2/a_1 = 4$ is solid line, $\rho_2 = 0.31\mathrm{e}^{1.16\alpha}$ is dotted line; (B) $\sigma_2/\sigma_1 = 1/8$, $\sigma_3/\sigma_1 = 1$, $a_2/a_1 = 4$ is solid line, $\rho_2 = 16\mathrm{e}^{-0.69\alpha}$ is dotted line.

4.5. Influence of the Skin-effect in the Formation on the Radial Characteristics of a Two-coil Induction Probe

The theory of small parameters developed by H. Doll does not take into account the skin effect. For this reason values of the quadrature component of the field calculated by this method are always greater than actual values of the field, Q H_z, measured on the borehole axis. It becomes specially noticeable when results of calculations based on Doll's formulae and the exact solution in beds with a finite thickness are compared with each other.

First we will assume that the skin effect manifests itself only in a bed. Then for the quadrature component of the field and the apparent conductivity for two- and three-layered media we have (see eqs. 3.123–3.127):

$$\begin{aligned} \mathrm{Q}\, h_z &= \mathrm{Q}\, h_z^{un} + \frac{\omega\mu L^2}{2}(\sigma_1 - \sigma_2)G_1(\alpha) \\ \frac{\sigma_a}{\sigma_2} &= \frac{\sigma_a^{un}}{\sigma_2} + \left(\frac{\sigma_1}{\sigma_2} - 1\right) G_1(\alpha) \end{aligned} \tag{4.89}$$

and:

$$\begin{aligned} \mathrm{Q}\, h_z &= \mathrm{Q}\, h_z^{un} + \frac{\omega\mu L^2}{2}(\sigma_1 - \sigma_3)G_1(\alpha) + \frac{\omega\mu L^2}{2}(\sigma_2 - \sigma_3)G_2(\alpha) \\ \frac{\sigma_a}{\sigma_3} &= \frac{\sigma_a^{un}}{\sigma_3} + \left(\frac{\sigma_1}{\sigma_3} - 1\right) G_1(\alpha) + \left(\frac{\sigma_2}{\sigma_3} - 1\right) G_2(\alpha) \end{aligned} \tag{4.90}$$

where Q h_z^{un} is the quadrature component of the field in a uniform medium with conductivity σ_3, and:

$$\sigma_a^{un} = \frac{2}{\mu\omega L^2}\mathrm{Q}\, h_z^{un} \tag{4.91}$$

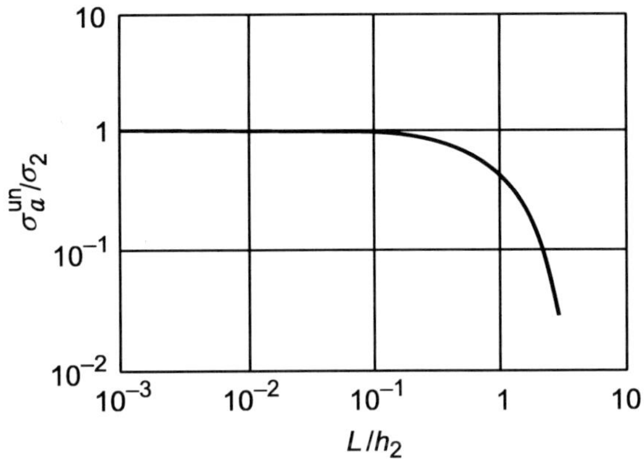

Figure 4.13. Behavior of function σ_a^{un}/σ_2.

Formulae 4.89–4.90 are valid if one can neglect the skin effect in the borehole and the invasion zone while it displays in the same way as in a uniform medium with a resistivity of the bed (formation).

Comparison with results of calculations based on the exact solution allows us to define the limits of application of this approximate theory in models with cylindrical interfaces. With errors not exceeding 10% we can use these formulae, provided:

$$\max\left(\frac{a_1}{h_1}, \frac{a_2}{h_2}\right) < 0.3 \qquad \text{and} \qquad \frac{L}{h_3} < 1.5 \tag{4.91a}$$

It is important to emphasize that the analysis of calculations based on the exact solution and this theory allows us to establish the character of distribution of induced currents in a medium that is of great practical interest for the development and application of multi-coil *focusing* induction probes.

In accord with eq. 4.77 for the apparent conductivity, σ_a, in the range of very small parameters we have:

$$\frac{\sigma_a}{\sigma_2} = 1 + \left(\frac{\sigma_1}{\sigma_2} - 1\right) G_1(\alpha) \tag{4.92}$$

The latter differs from the second equation of 4.89 by the fact that the first term of the right-hand side is unity instead of function σ_a^{un}/σ_2. As was shown in Chapter 2 this function decreases from unity to zero as parameter $(\sigma\mu\omega/2)^{1/2}L$ increases (Fig. 4.13).

It is clear that if the resistivity of the borehole mud is greater than that of the bed and the probe length, L, exceeds several times the radius a_1, the main signal is defined by induced currents in the surrounding medium. For this reason the decrease of the signal on the borehole axis due to the skin effect is almost the same as in a uniform medium with conductivity σ_2.

In a more general case, when the conductivity of the borehole is greater than that of the bed $\sigma_1 \gg \sigma_2$ the influence of the skin effect often manifests itself in a lesser degree than in the previous case. With an increase of the probe length, the effect caused by the interaction of currents in the formation becomes more noticeable. It is related with two factors. First of all, the influence of the currents in the borehole is reduced since function $G_1(\alpha)$ decreases. Second, the relative contribution of the quadrature component of the currents in removed parts of the medium (which are smaller than those in Doll's theory) increases.

The ratio:

$$\frac{\sigma_a^{un}/\sigma_2 + (\sigma_1/\sigma_2 - 1)\, G_1(\alpha)}{1 + (\sigma_1/\sigma_2 - 1)\, G_1(\alpha)}$$

characterizes the decrease of a signal in a two-layered medium due to the skin effect in a bed.

Let us consider one example illustrating to some extent condition 4.91a.

Suppose that the borehole radius a_1 is equal to 0.1 m. Then it is a simple matter to obtain maximal values of the product $\sigma_2 f$, which corresponds to the condition 4.91a:

$$\begin{aligned} f_{max} &< 5 \times 10^5 \frac{\rho_2}{L^2} \\ f_{max} &< 5 \times 10^6 \rho_1 \end{aligned} \tag{4.93}$$

It is obvious that frequencies used in conventional induction logging ($200 < f < 600$ kHz) satisfy these conditions.

Relations 4.93 are of great practical interest inasmuch as they define a range of resistivities and frequencies for which induced currents in the borehole, contributing to a signal, are shifted in phase by 90°, and the skin effect in the bed manifests itself in the same manner as in a uniform medium with conductivity σ_1.

When conditions 4.91a are valid an increase of the borehole radius and its conductivity lead to a reduction of the influence of the internal skin effect and application of the theory of small parameters (Doll's theory) becomes more grounded.

However, with a further increase of the borehole radius or its conductivity, unequalities 4.91a become invalid, and for the calculation of the field it is necessary to apply the exact solution.

In accord with eqs. 3.78 and 4.90, expressions for the apparent conductivities for three-layered medium within the range of small parameters and in the case when the skin effect manifests itself only in the bed are:

$$\frac{\sigma_a}{\sigma_3} = 1 + \left(\frac{\sigma_1}{\sigma_3} - \frac{\sigma_2}{\sigma_3}\right) G_1(\alpha) + \left(\frac{\sigma_2}{\sigma_3} - 1\right) G_1\left(\frac{\alpha}{\beta}\right) \tag{4.94}$$

and

$$\frac{\sigma_a}{\sigma_3} = \frac{\sigma_a^{un}}{\sigma_3} + \left(\frac{\sigma_1}{\sigma_3} - \frac{\sigma_2}{\sigma_3}\right) G_1(\alpha) + \left(\frac{\sigma_2}{\sigma_3} - 1\right) G_1\left(\frac{\alpha}{\beta}\right) \tag{4.95}$$

where $\beta = a_1/a_2$.

As in the case of the two-layered medium we have instead of unity in eq. 4.95 the term σ_a^{un}/σ_3, which takes into account the skin effect in the bed and allows us in every given case to define the range where eq. 4.94 can be applied.

It is obvious that with an increase of the probe length the depth of investigation also increases, that is, the probe becomes more sensitive to removed parts of a medium. For this reason the skin effect display is more noticeable for longer probes regardless of the character of the resistivity distribution within the borehole and the invasion zone.

With an increase of resistivity, ρ_2, ($\rho_2 > \rho_3$) the contribution of induced currents within the invasion zone in a measured signal decreases, and correspondingly the influence of the skin effect in the bed becomes stronger.

In those cases when the resistivity of the invasion zone has an intermediate value: $\rho_1 < \rho_2 < \rho_3$, the relative contribution of currents in the bed decreases and correspondingly eq. 4.94, derived from the assumption that the skin effect is absent, provides more accurate results.

From a physical point of view it is clear that, with an increase of the radius, a_2, of the invasion zone, the upper limit of frequencies and conductivities, when the approximate theory taking into account the skin effect in the bed is valid, decreases.

Conditions 4.91a and numerous results of calculations based on the exact and approximate solutions show that — if the radius of the invasion zone does not exceed 0.5–0.6 m for the most typical values of resistivity of the borehole, the invasion zone and the bed and frequencies used in conventional induction probes — we can apply eq. 4.95 for the determination of the apparent conductivity, σ_a. The ratio:

$$\frac{\dfrac{\sigma_a^{un}}{\sigma_3} + \left(\dfrac{\sigma_1}{\sigma_3} - \dfrac{\sigma_2}{\sigma_3}\right) G_1(\alpha) + \left(\dfrac{\sigma_2}{\sigma_3} - 1\right) G_1\left(\dfrac{\alpha}{\beta}\right)}{1 + \left(\dfrac{\sigma_1}{\sigma_3} - \dfrac{\sigma_2}{\sigma_3}\right) G_1(\alpha) + \left(\dfrac{\sigma_2}{\sigma_3} - 1\right) G_1\left(\dfrac{\alpha}{\beta}\right)}$$

allows, as in the case of a two-layered medium, for given parameters of geoelectric section to define errors related with the application of the theory of small parameters.

In accord with this theory suggested by H. Doll the influence of frequency is defined from:

$$\frac{Q\mathscr{E}}{\mathscr{E}_0} = \frac{\omega\mu L^2}{2}\sigma_a \tag{4.96}$$

where $Q\mathscr{E}$ is the quadrature component of the electromotive force, induced by a magnetic field of currents in a medium, $\mathscr{E}_0$ is the primary electromotive force, L is the probe length, and σ_a is the apparent conductivity, which does not depend on frequency. For this reason within a range of small parameters, an increase in the frequency results in an increase of the relative contribution of the useful signal with respect to the primary field, but the relation between signals caused by induced currents in different parts of a medium does not change. In other words, the depth of investigation is not reduced.

Here it is appropriate to notice that in many cases the theory of small parameters provides practically correct values of the quadrature component of the field in spite of the fact that the skin effect is not taken into account. From a physical point of view it means that the considered induction probe is not sensitive to those parts of a medium where the skin effect manifests itself.

With an increase in frequency the influence of the skin effect in parts of a medium located relatively close to a borehole becomes stronger and correspondingly Doll's theory cannot be applied. An increasing frequency, first of all, leads to the fact that the signal becomes more sensitive to the skin effect in an external area, i.e. in the bed. For this reason, function σ_a is defined by the right-hand part of eq. 4.95 and therefore it depends on the frequency. From the analysis of the field of the magnetic dipole in a uniform medium (Chapter 2) follows, that with an increase in frequency the influence of relatively remote parts of the medium decreases. If the conductivities of the borehole and invasion zone essentially exceed that of the formation and the skin effect manifest itself practically in the latter, it cannot affect the value of the apparent conductivity, σ_a. For this reason the relative contribution of the secondary field increases with frequency almost in the same manner as follows from Doll's theory. If the internal areas (borehole and invasion zone) are less conductive than the bed, the ratio of the secondary signal to the primary one increases slower than the frequency does.

Tables of functions $G_1(\alpha)$ and σ_3^{un}/σ_3 allow us to define for every given case the influence of the frequency on the magnitude of the signal, provided the skin effect manifests itself in the bed only.

Now let us consider several examples.

4.5.1. Example I: Two-layered Medium (Invasion Zone is Absent)

Case 1

The resistivity of a borehole is relatively high ($\sigma_1 \ll \sigma_2$). In accord with eq. 4.89:

$$\frac{Q\mathscr{E}}{\mathscr{E}_0} = \frac{\sigma_2 \omega \mu L^2}{2} \left[\frac{\sigma_a^{un}}{\sigma_2} - G_1 \left(\frac{L}{a_1} \right) \right] \tag{4.97}$$

When the probe length, L, exceeds several times the borehole radius, a_1, function $G_1(\alpha)$ is usually much smaller than σ_a^{un}/σ_2. In fact, eq. 4.97 is valid when conditions 4.93 are met. In this case function σ_a^{un}/σ_2 decreases from unity to 0.3 if the probe length, L, is equal to 1 m. Correspondingly, function $G_1(\alpha)$ does not exceed 0.01, if $a_1 = 0.1$ m. For this reason we have:

$$\frac{Q\mathscr{E}}{\mathscr{E}_0} \simeq \frac{\sigma_2 \omega \mu L^2}{2} \frac{\sigma_a^{un}}{\sigma_2} \qquad \text{if } \sigma_1 \ll \sigma_2 \tag{4.98}$$

As is seen from Table 2.2 an increase of conductivity of the bed, σ_2, or a frequency of more than 100 times (change of parameter $\sigma_2 \mu \omega$ from 0.01 to 256 for $L = 1$ m) leads to a decrease of the function σ_a^{un}/σ_2 almost of three times. For example if the frequency equals 60 kHz a change of the parameter $\sigma_2 \mu \omega$ from 0.01 to 0.64 corresponds to a change of resistivity, ρ_2, from 48 to 0.8 ohm·m, and the ratio $Q\mathscr{E}/\mathscr{E}_0$ increases almost 30 times.

Case 2

The borehole is more conductive ($\sigma_1 > \sigma_2$). In this case we have:

$$\frac{Q\mathscr{E}}{\mathscr{E}_0} = \frac{\sigma_2\mu\omega L^2}{2}\left[\frac{\sigma_a^{un}}{\sigma_2} + \left(\frac{\sigma_1}{\sigma_2} - 1\right) G_1(\alpha)\right] \simeq \frac{\sigma_2\mu\omega L^2}{2}\left(\frac{\sigma_a^{un}}{\sigma_2} + \frac{\sigma_1}{\sigma_2} G_1(\alpha)\right)$$

For instance, if $\alpha = L/a_1 = 10$ and $\sigma_1/\sigma_2 = 31$ we have:

$$\frac{Q\mathscr{E}}{\mathscr{E}_0} = \frac{\sigma_2\mu\omega L^2}{2}\left(\frac{\sigma_a^{un}}{\sigma_2} + 0.3\right) \tag{4.99}$$

Calculations demonstrate that the influence of frequency and conductivity of a formation on the magnitude of the ratio $Q\mathscr{E}/\mathscr{E}_0$ is practically the same as in the previous case. At the range of small values of parameter $\sigma_2\mu\omega$ the relative contribution of currents induced in the bed constitutes about 80% while for a value of $\sigma_2\mu\omega = 0.64$ the contribution of the formation is equal to 70% but the ratio $Q\mathscr{E}/\mathscr{E}_0$ essentially increases. For this reason with an increase of the frequency the depth of investigation of a two-layered medium by a two-coil induction probe does not change until the signal from the formation is greater or at least comparable with that caused by induced currents in the borehole. Also the natural limitation of a further increase of frequency is related with a nonunique interpretation, inasmuch as the spectrum of the quadrature component has a maximum.

There is another factor defining an upper limit of frequency. It is dictated by the fact that the efficiency of focusing multi-coil induction probes takes place provided that currents in the borehole have to be shifted in phase by 90°.

As was mentioned above, eqs. 4.89 are valid if:

- Induced currents in the borehole which make a contribution to the measured signal are shifted in phase by 90° with respect to the current in the dipole.

- The skin effect in the bed manifests itself in the same manner as in a uniform medium with the resistivity of the bed.

As follows from a comparison of the exact solution with conditions 4.91a, with an increase of the probe length the accuracy of the calculations in eqs. 4.89 decreases. It is related with the fact that, first of all, the second condition of 4.91a becomes incorrect. It means that induced currents in the bed begin to distribute in a different manner than in a uniform medium. For shorter probes, as the influence of a more conductive borehole is significant, the discrepancy between results of calculation based on approximate and exact solutions become smaller. Therefore, one can think that the first condition (the phase shift by 90° of currents in the borehole, which define the signal in a receiver) is valid for larger values of parameter a_1/h_2 and in particular, of frequency. For example, if $\rho_2/\rho_1 = 16$ and $L/a_1 = 3$, an error does not exceed 3%, if $a_1/h < 0.12$. This fact has a certain practical interest since the application of a short focusing probe using a high frequency can be useful for the determination of the resistivity of an invasion zone when the effect of induced currents in the borehole is compensated.

4.5.2. Example II: Three-Layered Medium (Invasion Zone is Present)

In accord with eq. 4.90 we have:

$$\frac{Q\mathscr{E}}{\mathscr{E}_0} = \frac{\sigma_2\mu\omega L^2}{2}\left[\frac{\sigma_a^{un}}{\sigma_2} + \left(\frac{\sigma_1}{\sigma_3} - \frac{\sigma_2}{\sigma_3}\right) G_1(\alpha) + \left(\frac{\sigma_2}{\sigma_3} - 1\right) G_1\left(\frac{\alpha}{\beta}\right)\right] \tag{4.100}$$

where $\beta = a_2/a_1$.

As has been shown above, the range of frequencies and conductivities, when this equation can be applied, becomes smaller with an increase of the radius of the invasion zone, a_2. The influence of the frequency on the ratio $Q\mathscr{E}/\mathscr{E}_0$ and the distribution character of induced currents essentially depends on the resistivity and the radius of the invasion zone. The deeper the penetration of the borehole filtrate and the smaller the resistivity ρ_2, the smaller the influence of the skin effect in the bed on the value of $Q\mathscr{E}/\mathscr{E}_0$. However, the relative contribution of currents in the bed also decreases.

With an increase of resistivity of the borehole solution and the invasion zone, provided that radius a_2 is relatively small, an increase of the frequency also results in a decrease of the relative contribution from the bed. However, in this case the signal from the formation decreases slower in spite of the fact that the skin effect manifests itself stronger.

In the range of very small parameters (Doll's approximation) in a uniform medium with an increase of the probe length the ratio $Q\mathscr{E}/\mathscr{E}_0$ also increases though the magnitude of the secondary field decreases inversely proportional to the probe length. In a nonuniform medium the relation between signal and probe length becomes more complicated. For instance in a two-layered medium with a more conductive borehole the secondary field can decrease more rapidly for relatively short probes than in a uniform medium but with an increase of the probe length it starts to decrease as $1/L$. It is obvious that a decrease of the rate of change of the relative anomaly of the secondary field, $Q\mathscr{E}/\mathscr{E}_0$ with an increase of the probe length begins when the induction probe starts to feel parts of a medium where the influence of the skin effect is significant.

The area of application of the approximate theory, taking into account the skin effect in an external medium, can be enlarged. Along with a consideration of the field in a piecewise uniform medium (resistivity of the invasion zone is constant) it is possible to consider a field in models, when the resistivity is a continuous function. For example, for an arbitrary change of resistivity within the invasion zone expressions for the field and the apparent conductivity are:

$$Q\,h_z = Q\,h_z^{un} + \frac{\omega\mu L^2}{2}(\sigma_1 - \sigma_3)G_1(\alpha) + \frac{\omega\mu L^2}{2}\int_{a_1}^{a_2} [\sigma(r) - \sigma_3]q(r)\,dr \tag{4.101}$$

$$\frac{\sigma_a}{\sigma_3} = \frac{\sigma_a}{\sigma_3} + \left(\frac{\sigma_1}{\sigma_3} - 1\right) G_1(\alpha) + \int_{a_1}^{a_2} \left(\frac{\sigma(r)}{\sigma_3} - 1\right) q(r)\,dr \tag{4.102}$$

where $q(r)$ is the geometric factor of a thin cylindrical layer with radius r.

4.6. Asymptotic Behavior of the Magnetic Field in the Borehole in the Range of Small Parameters

In the previous sections we have investigated the behavior of the leading term of the quadrature component of the magnetic field which is directly proportional to the frequency ω. In such an approximation it was assumed that the inphase component of the field is equal to zero. Now we will consider this range of small parameters ($(L/h)_{max} < 1$ and $(a/h)_{max} < 1$) in more detail. In other words, the next terms of the series describing the quadrature and inphase components of the magnetic field will be derived. As was shown in Chapter 2 the magnetic field, $H_z^{(0)}$, on the axis of the vertical magnetic dipole in a uniform medium can be presented as:

$$H_z^{(0)} = \frac{M}{2\pi L^3}\left(1 + \sum_{n=2}^{\infty} \frac{1-n}{n!}(\mathrm{i}kL)^n\right)$$

and discarding all terms except the two first ones we have:

$$H_z^{(0)} = \frac{M}{2\pi L^3}\left(1 + \frac{k^2L^2}{2} + \frac{1}{2}\mathrm{i}(kL)^3 + \ldots\right) \tag{4.103}$$

It is clear that the second term of this series is the leading term of the quadrature component since $k^2 = \mathrm{i}\sigma\mu\omega$, while the last term:

$$\frac{M}{2\pi L^3}\frac{1}{3}\mathrm{i}(kL)^3 = \frac{M}{6\pi}\mathrm{i}k^3 \tag{4.104}$$

defines the leading term of the inphase component as well as the second term of a series describing the quadrature component. It is essential that this term (eq. 4.104) does not depend on coordinates of an observation point.

In order to obtain an expansion of the magnetic field on the borehole axis in the range of small parameters, and in particular to derive a term proportional to k^3, one will apply several ways.

First of all, let us use of the expression of the field, H_z, derived in section 4.2 (eq. 4.22):

$$H_z = H_z^{(0)} - \frac{M}{4\pi}\frac{2}{\pi}\int_0^{\infty} \lambda_1^2 C_1 \cos\lambda z \, \mathrm{d}\lambda \tag{4.105}$$

where $H_z^{(0)}$ is the field in a uniform medium with conductivity σ_1. Function C_1 is given by eqs. 4.16–4.17 and L is the two-coil induction probe length. Let us notice that according to eq. 4.11:

$$H_z = k^2 A_z^* + \frac{\partial^2 A_z^*}{\partial z^2}$$

and the equality:

$$\frac{e^{-ikR}}{R} = \frac{2}{\pi}\int_0^\infty K_0(\lambda_1 r)\cos\lambda r \,d\lambda \tag{4.106}$$

we have:

$$H_z^{(0)} = -\frac{M}{4\pi}\frac{2}{\pi}\int_0^\infty \lambda_1^2 K_0(\lambda_1 r)\cos\lambda r \,d\lambda \tag{4.107}$$

where $\lambda_1 = (\lambda^2 - k^2)^{1/2}$ and $R = (r^2 + z^2)^{1/2}$.

Now we will present the integral in the right-hand side of eq. 4.105 as a sum of two integrals, such as:

$$\int_0^\infty \lambda_1^2 C_1 \cos\lambda z \,d\lambda = \int_0^{\lambda_0} \lambda_1^2 C_1 \cos\lambda z \,d\lambda + \int_{\lambda_0}^\infty \lambda_1^2 C_1 \cos\lambda z \,d\lambda \tag{4.108}$$

where λ_0 is very small arbitrary number $\lambda_0 \ll 1$.

First suppose that the invasion zone is absent. Then in accord with eq. 4.20 we have:

$$C_1 = \frac{\lambda_2 K_0(\lambda_2 a_1)K_1(\lambda_1 a_1) - \lambda_1 K_0(\lambda_1 a_1)K_1(\lambda_2 a_1)}{\lambda_2 K_0(\lambda_2 a_1)I_1(\lambda_1 a_1) + \lambda_1 I_0(\lambda_1 a_1)K_1(\lambda_2 a_1)} \tag{4.109}$$

where a_1 is the borehole radius; $\lambda_1 = (\lambda^2 - k_1^2)^{1/2}$, $\lambda_2 = (\lambda^2 - k_2^2)^{1/2}$.

Function C_1 depends on both parameters λ_1 and λ_2. As soon as the value of λ is greater than the magnitude of wave numbers these radicals can be expanded in series with respect to ratio k^2/λ^2. Correspondingly, function C_1 also can be presented as a series:

$$C_1 = \sum_{n=1}^\infty a_n \left(\frac{k_1}{\lambda}\right)^{2n} \tag{4.110}$$

where a_n are coefficients which depend on geoelectric parameters of the medium and the probe length.

Inasmuch as the external integral:

$$\int_{\lambda_0}^\infty \lambda_1^2 C_1 \cos\lambda z \,d\lambda$$

does not contain point $\lambda = 0$, we can replace function C_1 by the series (eq. 4.110), and then we obtain:

$$\int_{\lambda_0}^\infty \lambda_1^2 C_1 \cos\lambda z \,d\lambda = \sum_{n=1}^\infty b_n(k)^{2n} \tag{4.111}$$

Therefore, the series describing the external integral consists of only even powers of wave number, k, that is whole powers of ω. This fact allows us to conclude that odd powers of k, in particular k^3, can be derived only from expansion of the internal integral, provided

$$k \to 0 \qquad \text{and} \qquad \lambda \to 0 \tag{4.112}$$

In other words, information about terms of a series proportional to odd powers of k, that is fractional powers of ω, is contained in function $\lambda_1^2 C_1$ corresponding to the initial part of integration.

Taking into account the behavior of modified Bessel functions for small arguments:

$$I_0(x) \sim 1 \quad I_1(x) \sim \frac{x}{2} \quad K_0(x) \sim -\ln x \quad K_1(x) \sim \frac{1}{x}$$

function C_1 can be presented as:

$$C_1 \approx \frac{\lambda_2 K_0(\lambda_2 a_1) K_1(\lambda_1 a_1) - \lambda_1 K_0(\lambda_1 a_1) K_1(\lambda_2 a_1)}{\lambda_1 K_1(\lambda_2 a_1)}$$

or

$$C_1 \simeq \frac{\lambda_2}{\lambda_1} \frac{K_1(\lambda_1 a_1)}{K_1(\lambda_2 a_1)} K_0(\lambda_2 a_1) - K_0(\lambda_1 a_1) \tag{4.113}$$

Replacing ratio $K_1(\lambda_1 a_1)/K_1(\lambda_2 a_1)$ by its asymptotic value we finally have:

$$C_1 \simeq \frac{\lambda_2^2}{\lambda_1^2} K_0(\lambda_2 a_1) - K_0(\lambda_1 a_1)$$

and

$$\lambda_1^2 C_1 \simeq \lambda_2^2 K_0(\lambda_2 a_1) - \lambda_1^2 K_0(\lambda_1 a_1) \tag{4.114}$$

Thus, the internal integral can be presented as:

$$\int_0^{\lambda_0} \lambda_1^2 C_1 \cos\lambda z \, \mathrm{d}\lambda \simeq \int_0^{\lambda_0} \lambda_2^2 K_0(\lambda_2 a_1) \cos\lambda z \, \mathrm{d}\lambda - \int_0^{\lambda_0} \lambda_1^2 K_0(\lambda_1 a_1) \cos\lambda z \, \mathrm{d}\lambda \tag{4.115}$$

Comparing eqs. 4.114 with eq. 4.107 and keeping in mind that we are interested in odd powers of k, we can write the following equality:

$$\frac{M}{4\pi} \frac{2}{\pi} \int_0^{\infty} \lambda_1^2 C_1 \cos\lambda z \, \mathrm{d}\lambda = -H_z^0(k_2 R) + H_z^0(k_1 R) \tag{4.116}$$

where $H_z^0(k_2 R)$ and $H_z(k_1 R)$ are the magnetic fields in a uniform medium with resistivity of a bed and a borehole, respectively, at points with coordinates z and a, that is on the borehole surface; $R = (z^2 + a^2)^{1/2}$.

Substituting eq. 4.116 into eq. 4.105 we have:

$$H_z = H_z^0(k_1 z) + H_z^0(k_2 R) - H_z^0(k_1 R) \tag{4.117}$$

It is appropriate to emphasize again that the latter equality is valid only in the range of small parameters for terms of a series proportional to odd powers of wave number, k.

Now in order to facilitate further algebra let us make use of eq. 4.11 to write a similar equation for the vector potential A_z^*. Then we have:

$$A_z^*(z) = A_z^{(0)}(k_1 z) + A_z^{(0)}(k_2 R) - A_z^0(k_1 R)$$

or applying eq. 4.106 we finally obtain:

$$A_z^*(z) = \frac{M}{4\pi}\left(\frac{\mathrm{e}^{\mathrm{i}k_1 z}}{z} + \frac{\mathrm{e}^{\mathrm{i}k_2 R}}{R} - \frac{\mathrm{e}^{\mathrm{i}k_1 R}}{R}\right) \tag{4.118}$$

where A_z^* is the vector potential on the borehole axis.

Inasmuch as the magnetic field, H_z, is related with the vector potential A_z^* by the equation:

$$H_z = k^3 A_z^* + \frac{\partial^2 A_z^*}{\partial z^2}$$

and we are interested in the leading term of the inphase component, proportional to k^3, only, let us expand the right-hand part of eq. 4.118 in a series and collect terms which give a contribution to this part of the field. Doing so we obtain for the second term of a series describing the magnetic field on the borehole axis:

$$\mathrm{i}\frac{M}{6\pi}k_2^3 \tag{4.119}$$

Comparing with eq. 4.104 we see that the leading term of the inphase component of the field H_z coincides with that in a uniform medium with conductivity of a formation:

$$\operatorname{In} H_z \to \operatorname{In} H_z^{un}(k_2 L) \tag{4.120}$$

This result does not depend on the ratio of conductivities as well as the probe length. In other words, in the range of small parameters, the borehole becomes transparent when the inphase component of the secondary field is measured.

Now let us demonstrate that the same result is valid for a three-layered medium. We will proceed from equations 4.32–4.38 assuming that condition 4.112 is valid. Then functions m_1, n_1, m_2, n_2, and n_3 can be essentially simplified. Taking into account the behavior of modified Bessel functions I_0, I_1, and K_0 for small arguments, we have:

$$m_1 \simeq \lambda_2 K_1(\lambda_1 a_1) = -\frac{\lambda_2}{\lambda_1 a_1} \qquad n_1 \simeq \frac{\lambda_3}{\lambda_2 a_2} K_0(\lambda_3 a_2) - \frac{2}{\lambda_3 a_2} K_0(\lambda_2 a_2)$$

$$m_2 \simeq \frac{\lambda_2}{\lambda_1 a_1} K_0(\lambda_3 a_2) - \frac{\lambda_1}{\lambda_2 a_1} K_0(\lambda_1 a_1) \qquad n_2 \simeq -\frac{\lambda_2}{\lambda_3 a_2}$$

$$m_3 \simeq -\frac{\lambda_1 \lambda_2 a_1}{2} + \frac{\lambda_1 \lambda_2 a_1}{2} \to 0 \qquad n_3 \simeq \frac{\lambda_1}{\lambda_2 a_1}$$

Whence, for small values of λ and k, we obtain:

$$C_1 \approx \frac{m_1 n_1 + m_2 n_2}{n_2 n_3} = \frac{m_1 n_1}{n_2 n_3} + \frac{m_2}{n_2}$$

Inasmuch as:

$$n_2 n_3 = -\frac{\lambda_1}{\lambda_3 a_1 a_2} \qquad m_1 n_1 = -\frac{\lambda_3}{\lambda_1 a_1 a_2} K_0(\lambda_3 a_2) + \frac{\lambda_2^2}{\lambda_1 \lambda_2 a_1 a_2} K_0(\lambda_2 a_2)$$

$$\frac{m_1 n_1}{n_2 n_3} = \frac{\lambda_3^2}{\lambda_1^2} K_0(\lambda_3 a_2) - \frac{\lambda_2^2}{\lambda_1^2} K_0(\lambda_2 a_2)$$

$$\frac{m_2}{n_3} = \frac{\lambda_2^2}{\lambda_1^2} K_0(\lambda_2 a1) - K_0(\lambda_1 a_1)$$

we have the following expression for function $\lambda_1^2 C_1$:

$$\lambda_1^2 C_1 \approx \lambda_3^2 K_0(\lambda_3 a_2) - \lambda_2^2 K_0(\lambda_2 a_2) + \lambda_2^2 K_0(\lambda_2 a_1) - \lambda_1^2 K_0(\lambda_1 a_1) \tag{4.121}$$

Thus, the internal integral has the form:

$$\int\limits_0^{\lambda_0} \lambda_1^2 C_1 \cos \lambda z \, \mathrm{d}\lambda = \int\limits_0^{\lambda_0} \lambda_3^2 K_0(\lambda_2 a_2) \cos \lambda z \, \mathrm{d}\lambda - \int\limits_0^{\lambda_0} \lambda_2^2 K_0(\lambda_2 a_2) \cos \lambda z \, \mathrm{d}\lambda$$
$$+ \int\limits_0^{\lambda_0} \lambda_2^2 K_0(\lambda_2 a_1) \cos \lambda z \, \mathrm{d}\lambda - \int\limits_0^{\lambda_0} \lambda_1^2 K_0(\lambda_1 a_1) \cos \lambda z \, \mathrm{d}\lambda \tag{4.122}$$

In accord with eq. 4.107 and taking into account that our purpose is to determine the term proportional to k^3, we can write the following expression for the field on the borehole axis:

$$H_z = H_z^{(0)}(k_1 z) + H_z^{(0)}(k_3 R_2) - H_z^{(0)}(k_3 R_2) + H_z^{(0)}(k_2 R_1) - H_z^{(0)}(k_1 R_1) \tag{4.123}$$

where $R_1 = (z^2 + a_1^2)^{1/2}$, $R_2 = (z^2 + a_2^2)^{1/2}$, or for the vector-potential A^* we obtain:

$$A_z^*(z) = \frac{M}{4\pi} \left(\frac{\mathrm{e}^{\mathrm{i}k_1 z}}{z} + \frac{\mathrm{e}^{\mathrm{i}k_3 R_2}}{R_2} - \frac{\mathrm{e}^{\mathrm{i}k_2 R_2}}{R_2} + \frac{\mathrm{e}^{\mathrm{i}k_2 R_1}}{R_2} - \frac{\mathrm{e}^{\mathrm{i}k_1 R_1}}{R_1} \right)$$

Repeating calculations performed for a two-layered medium we again derive the same term proportional to k^3, that is:

$$\mathrm{i}\frac{M}{6\pi} k_3^2 \tag{4.124}$$

As in the previous case in the range of small parameters the inphase component of the field tends to that corresponding to a uniform medium with the conductivity of the formation, σ_3.

Now, taking into account the results derived in the previous section for the leading term of the quadrature component, we have for the magnetic field on the borehole axis the following approximation:

$$H_z \simeq \frac{M}{4\pi}\left(\frac{1}{L}\sum_{i=1}^{3} k_i^2 G_i + \frac{2}{3}\mathrm{i}k_3^3\right) \tag{4.125}$$

where $k_i^2 = \mathrm{i}\sigma_i\mu\omega$ and L is the probe length.

For the quadrature and inphase components of the field we have:

$$\begin{aligned} \mathrm{Q}\,H_z &\simeq \frac{M}{4\pi}\left(\frac{\mu\omega}{L}\sum_{i=1}^{3}\sigma_i G_i - \frac{\sqrt{2}}{3}(\sigma_3\mu\omega)^{3/2}\right) \\ \mathrm{In}\,H_z^s &= -\frac{M}{4\pi}\frac{\sqrt{2}}{3}(\sigma_3\mu\omega)^{3/2} \end{aligned} \tag{4.126}$$

where σ_i and G_i, are conductivity and geometric factor of the corresponding part of a medium such as borehole, invasion zone and formation.

This result can be easily generalized and it can be applied, for instance, to the case in which the resistivity of the invasion zone changes arbitrarily in the radial direction.

Let us remember that $\mathrm{In}\,H_z^s$ is the secondary field which is shifted in phase by 180° with respect to the primary field.

From eq. 4.126 it follows that the leading terms of the quadrature and inphase components are related with parameters of a geoelectric section in a completely different manner and therefore, in general, have a different depth of investigation. This question will be considered later in detail.

It is appropriate to notice that formulae 4.126 can be derived in a much simpler way proceeding from the approximate theory of induction logging described in the previous section. According to eq. 4.90 we can write:

$$H_z = H_z^{(0)} h_z = H_z^{(0)}\left(h_z^{un}(\sigma_3) + \frac{\mathrm{i}\mu\omega L^2}{2}\sum_{i=1}^{2}(\sigma_i - \sigma_3)G_i\right) \tag{4.127}$$

or

$$H_z = \frac{M}{2\pi L^3} h_z^{un}(\sigma_3) + \frac{\mathrm{i}\mu\omega M}{4\pi L}\sum_{i=1}^{2}(\sigma_i - \sigma_3)G_i \tag{4.128}$$

where $(M/2\pi L^3)h_z^{un}$ is the vertical component of the magnetic field in a uniform medium with conductivity σ_3.

Expanding the right-hand side of eq. 4.128 in a series and discarding all terms but the first ones we again obtain formulae 4.126:

$$\mathrm{Q}\,H_z \approx \frac{M}{4\pi}\left(\frac{\mu\omega}{L}\sum_{i=1}^{3}\sigma_i G_i - \frac{\sqrt{2}}{3}(\sigma_3\mu\omega)^{3/2} + \ldots\right)$$

$$\mathrm{In}\,H_z^s \approx -\frac{M}{4\pi}\frac{\sqrt{2}}{3}(\sigma_3\mu\omega)^{3/2}$$

Let us emphasize that the second approach of obtaining the leading terms of these series will be also used for more complicated cases, in particular, when the bed has a finite thickness.

As follows from eqs. 4.126 the second term of the quadrature component and the leading term of the inphase component of the magnetic field, H_z, do not depend on the probe length, nor on the geoelectrical parameters of the borehole and the invasion zone. Therefore, regardless of the separation of the coils measuring these quantities we can essentially increase the depth of investigation on the induction probe.

Now it is appropriate to notice the following. We have derived only two terms of the series describing the quadrature component of the field and the leading term for the inphase component. In order to obtain subsequent terms of both series it is necessary to perform much more cumbersome transformations. There are at least two approaches allowing us to solve this problem. The first one is based on expansion of the internal and external integrals in eq. 4.100 into a series with respect to k. The second method uses the integral equation with respect to the electric field described in Chapter 3. Expanding the integrand in eq. 3.77 in a series with respect to k and making use of the method of subsequent approximation we can obtain the following terms of a series in the range of small parameters.

In general this series has the form:

$$H_z = H_z^{(0)}\left(\sum_{n=1}^{\infty} a_{1n}k^{2n} + \sum_{n=1}^{\infty} a_{2n}k^{2n+1} + \ln k\sum_{n=1}^{\infty} a_{3n}k^{n}\right) \tag{4.129}$$

In the next section we will give expressions for some of these coefficients when the probe length, L, is several times larger than the radius of the borehole or that of the invasion zone.

At the same time it is important to define coefficients of terms containing odd powers of k for arbitrary probe length, since as is well known, these terms are responsible for the late stage of the transient field. Applying the first approach and using only the internal integral of eq. 4.108 we have for field h_z:

Two-layered medium

$$f_3k_1^3 + f_5k_1^5 + f_7k_1^7 + l_7k_1^7\ln k \tag{4.130}$$

where:

$$f_3 = \frac{\alpha^3 s^{3/2}}{3} \qquad f_5 = f_3\left(\frac{\alpha^3 s}{10} - \frac{1-s}{2}\right)$$

$$f_7 = f_3\left[\frac{\alpha^4 s^2}{280} - \frac{\alpha^2 s(1-s)}{20} + \frac{5}{32}(1-s)^2 - \frac{s(1-s)}{10}\left(C - \frac{77}{60} + \frac{\ln s}{2}\right)\right] \tag{4.131}$$

$$l_7 = -f_3\frac{s}{10}(1-s)$$

where C is Euler's constant, $s = \sigma_2/\sigma_1$, and $\alpha = L/a_1$.

Three-layered medium

$$\Phi_3 k_1^3 + \Phi_5 k_1^5 \tag{4.132}$$

where:

$$\Phi_3 = \frac{1}{3}\alpha^3 s_1^{3/2} \qquad \Phi_5 = \Phi_3\left[\frac{\alpha^3 s_1}{10} - \frac{s_{12}}{2}\right] \qquad s_1 = \sigma_3/\sigma_1 \qquad s_2 = \sigma_2/\sigma_1$$
$$s_{12} = 1 - s_2 + (s_2 - s_1)\beta^2 \qquad \beta = a_2/a_1 \tag{4.133}$$

4.7. Behavior of the Field on the Borehole Axis in the Near and Far Zones

In induction logging the length of the probe usually exceeds the borehole radius and sometimes this ratio, L/a_1 reaches ten and more. Correspondingly, it is appropriate to investigate the behavior of the field when the parameter $\alpha = L/a_1$ is large. As will be shown in this section, within this range of parameters:

$$L/a_1 \gg 1 \qquad L/a_2 \gg 1 \tag{4.134}$$

the field possesses some features which can be used for increasing the depth of investigation. We will assume that conditions 4.134 define the range of parameters α characterizing the behavior of the field in the far zone.

On the other hand, the near zone corresponds to conditions when the probe length, L, is smaller than the borehole diameter, d, that is:

$$L < d_1 \tag{4.135}$$

Since this case is hardly a very practical one, the field behavior will be considered only briefly for two extreme situations, namely when parameter a_1/h_1 is either very large or very small.

In both cases we will proceed from eq. 4.24. Introducing a new variable, $m = \lambda a_1$, this equation can be presented in the form:

$$h_z = h_z^{(0)} - \frac{\alpha^3}{\pi}\int_0^\infty m_1^2 C_1 \cos m\alpha \, dm \tag{4.136}$$

where:

$$m_1 = (m^2 - k_1^2 a_1^2)^{1/2} \qquad m_2 = (m^2 - k_2^2 a_1^2)^{1/2} \qquad m_3 = (m^2 - k_3^2 a_1^2)^{1/2}$$
$$h_z^{(0)} = \mathrm{e}^{\mathrm{i}k_1 L}(1 - \mathrm{i}k_1 L)$$
$$\alpha = L/a_1$$

As follows from eq. 4.126 in the range of small parameters we have:

$$\mathrm{Q}\, H_z \simeq \frac{M}{4\pi}\frac{\mu\omega\sigma_1}{L} \qquad H_z^s \simeq -\frac{M}{4\pi}\frac{\sqrt{2}}{3}(\sigma_2\mu\omega)^{3/2} \tag{4.137}$$

since $G_1 \to 1$, $G_2 \to 0$.

Therefore with a decrease of the probe length, L, the quadrature component of the magnetic field tends to that in a uniform medium with the conductivity of the borehole. At the same time, regardless of how small the distance is between the transmitter and the receiver, the inphase component of the secondary field coincides with that corresponding to a uniform medium with the conductivity of the formation.

Now suppose that the small separation L is fixed and consider a field behavior at the high-frequency spectrum as:

$$k \to \infty \tag{4.138}$$

First assume that the formation is the ideal conductor, i. e., $k_2 = \infty$. Then in accord with eq. 4.136 we have:

$$h_z = h_z^{(0)} - \frac{\alpha^3}{\pi}\int\limits_0^\infty m_1^2 \frac{K_1(m_1)}{I_1(m_1)} \cos m\alpha \, dm$$

or using the asymptotic formulae for Bessel functions:

$$h_z = h_z^{(0)} - \alpha^3 \int\limits_0^\infty m_1 \mathrm{e}^{-2m_1} \cos m\alpha \, dm \quad \text{if } k_1 \to \infty$$

It is obvious that the latter integral is proportional to $\mathrm{e}^{2\mathrm{i}k_1 a_1}$. Therefore, with an increase of frequency, induced currents concentrate near the source, and the field is defined by the term, $h_z^{(0)}$, which is proportional to $\mathrm{e}^{\mathrm{i}k_1 L}$, provided that:

$$L < 2a_1 \tag{4.139}$$

Now let us explore the opposite case when the external medium is an insulator, that $k_2 = 0$ and $m_2 = m$. Then for the integral of eq. 4.136 we obtain:

$$\alpha^3 \int\limits_0^\infty m_1^2 \mathrm{e}^{2m_1} \frac{mK_0(m) - m_1 K_1(m)}{mK_0(m) + m_1 K_1(m)} \cos m\alpha \, dm$$

As $k_1 \to \infty$ and $\alpha \to 0$ the magnitude of the fraction of the integrand is also close to unit, and, respectively, the secondary field is proportional to $e^{2ik_1a_1}$, as in the previous case. We can show that this result remains valid in the general case of a finite conductivity of the formation. Thus, in the near zone ($L < 2a_1$) the field tends to that of a uniform medium with the borehole resistivity when the parameter a_1/h_1 increases.

For obtaining asymptotic formulae describing the field behavior in the far zone ($\alpha \gg 1$) we will take into account the distribution of the singularities of the integrand $m_1^2C_1$ on the complex plane of m.

In accord with eq. 4.136, the variable of integration, m, has only real values: $0 \leqslant m \leqslant \infty$ a relative probe length plays the role of multiplier of the argument of oscillating term: $\cos m\alpha$. For small values of parameter $|ka|$, function $m_1^2C_1$ rapidly decreases with an increase of m. Moreover, due to the presence of oscillating factor $\cos m\alpha$, the contribution of the integrand corresponding to large values of m becomes small. For this reason the integral:

$$\int_0^\infty m_1^2 C_1 \cos m\alpha \, dm \tag{4.140}$$

is defined by the behavior of integrand $m_1^2C_1$ near $m \to 0$. This consideration allowed us to derive asymptotical formulae for geometric factors of borehole, invasion zone and formation.

With an increase of wave number $|k|$ the integrand $m_1^2C_1$ begins to decrease slower with increasing variable m: for $m < |ka|$ it does not practically change. Correspondingly, the fact that the amount of oscillations increase as $\alpha \to \infty$ does not imply that a value of the integral is mainly defined by the integrand at the initial part of integration. For this reason, in order to obtain asymptotical formulae for the far zone, it is necessary to perform preliminary transformations of the expression for the field (eq. 4.136).

Inasmuch as the integrand of eq. 4.140 is an even function with respect to m the integral can be written in the form:

$$\begin{aligned} I &= \frac{\alpha^3}{2\pi}\int_0^\infty m_1^2 C_1 \cos m\alpha \, dm = \frac{\alpha^3}{2\pi}\int_0^\infty m_1^2 C_1 e^{im\alpha}\, dm + \frac{\alpha^3}{2\pi}\int_0^\infty m_1^2 C_1 e^{-im\alpha}\, dm \\ &= \frac{\alpha^3}{2\pi}\int_{-\infty}^\infty m_1^2 C_1 e^{im\alpha}\, dm \end{aligned} \tag{4.141}$$

We will make use of Cauchy's theorem, according to which the integral value of an analytical function does not change under deformation of an integration contour if it does not intersect singularities on the complex plane of variable m. It is clear that deforming the contour of integration in the upper half-plane ($\operatorname{Im} m > 0$) exponent $e^{im\alpha}$ with an increase of $\operatorname{Im} m$ tends to zero.

The integrand of eq. 4.140 in general use has two types of singularities namely branch points and poles. First we will consider a two-layered medium when the invasion zone is absent. Analysis of zeroes of a determinant of function C_1, as well as calculations shows

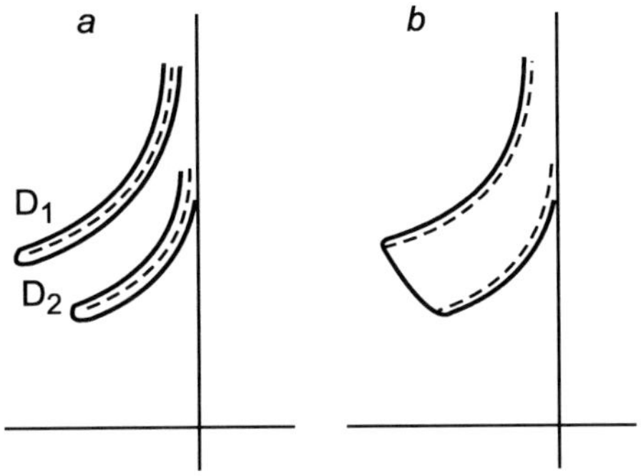

Figure 4.14. Contour integration for derivation of eq. 4.149.

that poles are absent in upper half plane if $\lambda_1\alpha_1 > 8$, $(\rho_2 > \rho_1)$. At the same time in upper half-plane there are two branch points: $\mathrm{i}k_1$ and $\mathrm{i}k_2$. Because of the presence of radicals m_1 and m_2 let us consider the integrand of eq. 4.141 on four leaves of a surface which are connected along cross-section lines. From condition $\mathrm{Re}\, m_j > 0$, which arises as a result of a solution of the boundary problem, it follows that these cross-sections must distinguish areas where the real part of radicals m_1 and m_2 is positive. For this reason equations of cross-section lines are:

$$\mathrm{Re}\, m_1 = 0 \qquad \mathrm{Re}\, m_2 = 0 \tag{4.142}$$

Letting $m = x + \mathrm{i}y$ and $k_j^2 a_1^2 = \mathrm{i}\chi_j^2$, we obtain:

$$\frac{\sqrt{(x^2-y^2)^2 + (2xy + \chi_j^2)^2} + (x^2 - y^2)}{2} = 0$$

This equality is valid provided:

$y = -\chi_j^2/2x$ and $|y| > |x|$

Thus the contour of integration along the real axis of m is replaced by that along both sides of two cross-sections, where $\mathrm{Re}\, m_1 = 0$ and $\mathrm{Re}\, m_2 = 0$, respectively, and within an area surrounding the real axis of m and these contours singularities are absent (Fig. 4.14).

Therefore,

$$\oint m_1^2 C_1 \mathrm{e}^{\mathrm{i}m\alpha}\, \mathrm{d}m = 1$$

or

$$\int_{-\infty}^{\infty} m_1^2 C_1 \mathrm{e}^{\mathrm{i}m\alpha}\, \mathrm{d}m = \int_{D_1} m_1^2 C_1 \mathrm{e}^{\mathrm{i}m\alpha}\, \mathrm{d}m + \int_{D_2} m_1^2 C_1 \mathrm{e}^{\mathrm{i}m\alpha}\, \mathrm{d}m \tag{4.143}$$

inasmuch as the integral along of a semi-circle of infinitely long radius is equal to zero.

Integrating along the cross-section D_1, where $\operatorname{Re} m_1 = 0$ we can introduce new variable of integration $m_1 = it$. Here t is the parameter of the cross-section which alters from 0 to ∞ on the right side of the cross-section and from $-\infty$ to 0 on its left side since passing around the branch point the radical m_1 changes sign.

The variable of integration of m along contour D_1 can be presented as:

$$m = (m_1^2 - \mathrm{i}\chi_1^2)^{1/2} = (-t^2 - \mathrm{i}\chi_1^2)^{1/2} = \mathrm{i}(t^2 + \mathrm{i}\chi_1^2)^{1/2}$$

and correspondingly:

$$\mathrm{d}m = \frac{it\,\mathrm{d}t}{(t^2 + \mathrm{i}\chi_1^2)^{1/2}} \qquad m_2 = [-t^2 + \mathrm{i}(\chi_2^2 - \chi_1^2)]^{1/2}$$

Thus, for the integral along the cross-section D_1 we have the following expression:

$$\int_0^\infty (-t^2) \left[\frac{m_2 K_0(m_2) K_1(\mathrm{i}t) - \mathrm{i}t K_0(-\mathrm{i}t) K_1(m_2)}{m_2 K_0(m_2) I_1(\mathrm{i}t) + \mathrm{i}t K_1(m_2) I_0(\mathrm{i}t)} - \frac{m_2 K_0(m_2) K_1(\mathrm{i}t) + \mathrm{i}t K_0(-\mathrm{i}t) K_1(m_2)}{m_2 K_0(m_2) I_1(\mathrm{i}t) - \mathrm{i}t K_1(m_2) I_0(-\mathrm{i}t)} \right] \times \frac{\mathrm{i}t\mathrm{e}^{-\alpha(t^2+\mathrm{i}\chi_1^2)^{1/2}}}{(t^2 + \mathrm{i}\chi_1^2)^{1/2}}\,\mathrm{d}t \tag{4.144}$$

Making use of relations:

$$\begin{aligned} I_0(-\mathrm{i}t) &= I_0(\mathrm{i}t) \qquad K_0(-\mathrm{i}t) = K_0(\mathrm{i}t) + \mathrm{i}\pi I_0(\mathrm{i}t) \\ I_1(-\mathrm{i}t) &= -I_1(\mathrm{i}t) \quad K_1(-\mathrm{i}t) = -K_1(\mathrm{i}t) + \mathrm{i}\pi I_1(\mathrm{i}t) \end{aligned} \tag{4.145}$$

we will transform eq. 4.144.

Then the second term of parentheses of this equation can be rewritten as:

$$\begin{aligned} &\frac{m_2 K_0(m_2)[-K_1(\mathrm{i}t) + \mathrm{i}\pi I_1(\mathrm{i}t)] + \mathrm{i}t K_1(m_2)[K_0(\mathrm{i}t) + \mathrm{i}\pi I_0(\mathrm{i}t)]}{-m_2 K_0(m_2) I_1(\mathrm{i}t) - \mathrm{i}t K_1(m_2) I_0(\mathrm{i}t)} \\ &\quad = \frac{m_2 K_0(m_2) K_1(\mathrm{i}t) - \mathrm{i}t K_1(m_2) K_0(\mathrm{i}t)}{M_2 K_0(m_2) I_1(\mathrm{i}t) + \mathrm{i}t K_1(m_2) I_0(\mathrm{i}t)} - \mathrm{i}\pi \end{aligned} \tag{4.146}$$

The first term in eq. 4.146 is equal to the first term in parentheses of eq. 4.144. For this reason the integral along the cross-section D_1 is equal to:

$$\pi \int_0^\infty \frac{t^3 \mathrm{e}^{-\alpha(t^2+\mathrm{i}\chi_1^2)^{1/2}}}{(t^2 + \mathrm{i}\chi_1^2)^{1/2}}\,\mathrm{d}t$$

This integral being multiplied by α/π presents the field of the magnetic dipole, h_z, in a uniform medium with conductivity σ_1.

In accord with eqs. 4.136 and 4.146 the field on the borehole axis is expressed only through the integral along the cross-section D_2 ($\operatorname{Re} m_2 = 0$). Making replacement of variables $m_2 = \mathrm{i}t$, we have:

$$m_2 = \mathrm{i}(t^2 + \mathrm{i}\chi_2^2)^{1/2} \qquad \mathrm{d}m = \frac{\mathrm{i}t\,\mathrm{d}t}{(t^2 + \mathrm{i}\chi_2^2)^{1/2}} \qquad m_1 = \left[-t^2 + \mathrm{i}\left(\chi_1^2 - \chi_2^2\right)\right]^{1/2}$$

Correspondingly, the integral along path D_2 can be written as:

$$\int_0^\infty m_1^2 \left[\frac{\mathrm{i}tK_0(\mathrm{i}t)K_1(m_1) - m_1K_0(m_1)K_1(\mathrm{i}t)}{\mathrm{i}tK_0(\mathrm{i}t)I_1(m_1) + m_1I_0(m_1)K_1(\mathrm{i}t)} + \frac{\mathrm{i}tK_0(-\mathrm{i}t)K_1(m_1) + m_1K_0(m_1)K_1(-\mathrm{i}t)}{-\mathrm{i}tK_0(-\mathrm{i}t)I_1(m_1) + m_1I_0(m_1)K_1(-\mathrm{i}t)}\right]$$
$$\times \frac{\mathrm{i}t\mathrm{e}^{-\alpha(t^2+\mathrm{i}\chi_2^2)^{1/2}}}{(t^2 + \mathrm{i}\chi_2^2)^{1/2}}\,\mathrm{d}t \tag{4.147}$$

Making use of relations 4.145 and presenting this integrand as one fraction we obtain for the numerator of the square parentheses of eq. 4.147 the following expression:

$$m_1\mathrm{i}t[I_0(m_1)K_1(m_1) + I_1(m_1)K_0(m_1)][K_0(-\mathrm{i}t)K_1(\mathrm{i}t) + K_0(\mathrm{i}t)K_1(-\mathrm{i}t)]$$

Inasmuch as

$$I_1(x)K_0(x) + I_0(x)K_1(x) = 1/x$$

the latter is equal to $\mathrm{i}\pi$.

Therefore, after corresponding transformations the field h_z on the borehole axis is expressed through the integral along right-hand side of the cross-section D_2:

$$h_z = \frac{\alpha^3}{2}\int_0^\infty \frac{m_1^2 t\,\mathrm{e}^{-\alpha(t^2+\mathrm{i}\chi_2^2)^{1/2}}}{\zeta_1\zeta_2(t^2 + \mathrm{i}\chi_2^2)^{1/2}}\,\mathrm{d}t$$
$$\zeta_1 = \mathrm{i}tK_0(\mathrm{i}t)I_1(m_1) + m_1K_1(\mathrm{i}t)I_0(m_1)$$
$$\zeta_2 = -\mathrm{i}tK_0(-\mathrm{i}t)I_1(m_1) + m_1K_1(-\mathrm{i}t)I_0(m_1) \tag{4.148}$$

The integrand can be presented as a product of two functions:

$$F(m_1, t)\,t^3\,\frac{\mathrm{e}^{-\alpha(t^2+\mathrm{i}\chi_2^2)^{1/2}}}{(t^2 + \mathrm{i}\chi_2^2)^{1/2}}$$

At the initial part of integration function F slightly depends on parameter t. The second multiplier $\mathrm{e}^{-\alpha(t^2+\mathrm{i}\chi_2^2)^{1/2}}/(t^2+\mathrm{i}\chi_2^2)^{1/2}$ is the integrand of the Sommerfeld integral describing the field in a uniform medium with the resistivity of the formation.

If the wave length in the formation is much greater than the borehole radius:

$$\chi_2^2 = 8\pi^2 a_1^2/\lambda_2^2 \ll 1$$

then for sufficiently large values of α we can let $t = 0$ in expression of $F(m_1, t)$ and take this function out of the integral. Thus we obtain the asymptotical equation for the field in the far zone:

$$h_z = \frac{1}{I_0^2(a_1\sqrt{k_1^2 - k_2^2})} e^{-k_2 L}(1 + k_2 L) \qquad \text{if } \frac{L}{a_1} \gg 1 \text{ and } \frac{\lambda_i}{a_1} > 1 \tag{4.149}$$

Making use of expansion of Bessel function $I_0(x)$ in a series by x and discarding all terms except the first two: $I_0(x) \simeq 1 + x^2/4$ we have, instead of eq. 4.149:

$$h_z \simeq h_z^{(0)}(k_2 L) - \frac{1}{2}(k_1^2 - k_2^2)a_1 h_z^{(0)}(k_2 L)$$

If the inphase component of the secondary field is small and therefore In $h_z \simeq 1$,we obtain:

$$h_z \simeq -\frac{(k_1^2 - k_2^2)a_1^2}{2} + h_z^{(0)}(k_2 L) \qquad \text{if } \alpha > 1 \tag{4.150}$$

which for small parameters corresponding to Doll's approximation coincides with eq. 4.89.

Expression 4.149 allows us to easily obtain the expansion at the range of small parameters $k_1 L$ and $k_2 L$. In fact, expanding the right-hand part of this equation in the series we have:

$$h_z \simeq 1 + f_2 k_1^2 + f_3 k_1^3 + f_4 k_1^4 + f_5 k_1^5 \tag{4.151}$$

where:

$$f_2 = -\frac{1}{2}[\alpha^2 s + (1 - s)] \qquad f_3 = \frac{\alpha^3 s^{3/2}}{3}$$
$$f_4 = \frac{1}{4}\left[-\frac{\alpha^4 s^2}{2} + (1 - s)^2 s + \frac{5}{8}(1 - s)^2\right] \tag{4.152}$$
$$f_5 = f_3\left(\frac{\alpha^3 s}{10} - \frac{1 - s}{2}\right) \qquad \alpha = \frac{L}{a_1} \qquad s = \frac{\sigma_2}{\sigma_1}$$

As follows from eq. 4.149 in the range of large parameters: $|ka| \gg 1$, for example at the high frequency spectrum, the field, h_z, tends to zero.

It means that induced currents concentrate near the source, and the secondary field is almost equal by magnitude to the primary field but it has the opposite sign. However, at the far zone unlike the near one the influence of the formation resistivity remains regardless of the frequency.

Let us notice that eq. 4.149 is valid not only for a quasistationary field but also in a more general case when the influence of displacement currents is significant. This fact is of great practical interest for the realization of dielectric logging.

If resistivities of borehole and formation are essentially different from each other eq. 4.149 can be replaced by an approximate equation:

$$h_z \simeq \frac{1}{I_0^2(k_1 a_1) h_z^{(0)}(k_2 L)} \qquad \text{for } |k_1| \gg |k_2| \tag{4.153}$$

Comparison of the results of calculation using exact formulae and the asymptotic eq. 4.149 shows that, if the skin depth in the borehole is greater than its radius and $\rho_2 > \rho_1$ the error in determination of the amplitude and the phase of the field h_z by eq. 4.149 does not exceed 5%, provided that parameter α satisfies the condition:

$$\alpha = L/a_1 > 4 \tag{4.154}$$

Let us present the complex amplitude of the field in the form:

$$h_z = A(ka_1)\,\mathrm{e}^{\mathrm{i}\phi(ka_1)}\,A_z^0(k_2L)\,\mathrm{e}^{\mathrm{i}\phi(k_2L)} \tag{4.155}$$

Therefore, the ratio of amplitudes as well as the difference of the phases of the magnetic field measured by two probes do not depend on electrical properties and the borehole radius ($\alpha \gg 1$).

Now we will derive expressions for the vertical component of the field on the borehole axis when there is an invasion zone and measurements are performed at the far zone. Taking into account that the integrand in eq. 4.136 is an even function we will consider integration along whole axis m and, applying Cauchy's theorem, the contour of integration then will be deformed in the upper part of the complex plane of m without intersection of singularities on this plane.

Singularities of the integrand of function C_1 are in general poles and branch points. Numerical analysis shows that for relatively large values of wavelength $\lambda = 2\pi h$ poles are absent in the upper half-plane where only three branch points are located, namely $\mathrm{i}k_1$, $\mathrm{i}k_2$, and $\mathrm{i}k_3$.

Because of the presence of radicals we will consider the integrand in eq. 4.136 on eight leaves surface of radicals whose leaves are connected along the cross-section lines $\mathrm{Re}\, m_j = 0$ $(j = 1, 2, 3)$.

Let us present the integrand in the form:

$$m_1^2 C_1 = m_1^2 \frac{l_1 n_1 + l_2 n_2}{l_3 n_1 + n_2 n_3} \tag{4.156}$$

where:

$$l_1 = -m_2 I_0(m_2) K_1(m_1) - m_1 K_0(m_1) I_1(m_2)$$

$$n_1 = m_3 K_1(m_2\beta) K_0(m_3\beta) - m_2 K_0(m_2\beta) K_1(m_3\beta)$$

$$l_2 = m_2 K_0(m_2) K_1(m_1) - m_1 K_0(m_1) K_1(m_2)$$

$$n_2 = -m_3 I_1(m_2\beta) K_0(m_3\beta) - m_2 I_0(m_2\beta) K_1(m_3\beta)$$

$$l_3 = -m_2 I_0(m_2) I_1(m_1) + m_1 I_0(m_1) I_1(m_2)$$

$$n_3 = m_2 I_1(m_1) K_0(m_2) + m_1 I_0(m_1) K_1(m_2)$$

$$m_1 = (m^2 + k_1^2 a_1^2)^{1/2} \qquad m_2 = (m^2 + k_2^2 a_1^2)^{1/2} \qquad m_3 = (m^2 + k_3^2 a_1^2)^{1/2}$$

$$\beta = a_2/a_1$$

Again letting $m = x + \mathrm{i}y$ and $k_j^2 a_1^2 = \mathrm{i}\chi_j^2$, we will obtain equation of the cross-section: $y = -\chi_j^2/2x$ and $|y| > |x|$. Thus, instead of the real axis of m the contour of integration consists of contours along sides of three cross-sections where $\operatorname{Re} m_j = 0$, and within the area bounded by the real axis and this contour singularities are absent.

In the case of a two-layered medium the integral along the cross-section $\operatorname{Re} m_1$, turned out to be equal to that describing the field of the magnetic dipole in a uniform medium with resistivity ρ_1. In the case of a medium with two and three cylindrical interfaces corresponding transformations become much more cumbersome. However we can show that for a medium with $n-1$ interfaces, integrals along sides of cross-sections $\operatorname{Re} m_j = 0$ $(j = 2, \ldots, n-1)$ are equal to zero, and therefore the integral remains along the cross-section $\operatorname{Re} m_j = 0$ only. For this reason the expression for the field in a three-layered medium has the form:

$$h_z = \frac{\alpha^3}{2} \int_0^\infty \frac{m_1^2 t\, \mathrm{e}^{-\alpha(t^2+k_3^2 a_1^2)^{1/2}}}{(l_2 n_1 + n_1 n_2)_-(l_2 n_1 + n_2 n_3)_+} \frac{\mathrm{d}t}{(t^2 + k_3^2 a_1^2)^{1/2}} \tag{4.157}$$

where radicals m_1 and m_2 are expressed through the variable of the integration t in the following way:

$$m_1 = \left[-t^2 + (k_1^2 - k_3^2) a_1^2\right]^{1/2} \qquad m_2 = \left[-t^2 + (k_2^2 - k_3^2) a_1^2\right]^{1/2}$$

and in expressions $(l_2 n_2 - n_1 n_2)_\pm$ m_3 is equal to $\pm \mathrm{i}t$, respectively.

Let us distinguish within function 4.157 expression:

$$\frac{t^3 \mathrm{e}^{-\alpha(t^2+k_3^2 a_1^2)^{1/2}}}{(t^2 + k_3^2 a_1^2)^{1/2}}$$

corresponding to the integrand of the Sommerfeld integral for a uniform medium with the conductivity of the formation. The other multiplier $F(m_1, m_2, m_3)$ as a function of t changes relatively slowly and therefore, for large values of α, letting $t = 0$ in function F we can take out the integral coefficient $F(m_1, m_2, 0)$. After integration we obtain:

$$h_z = F(m_1, m_2, 0) h_z^{(0)}$$

where $m_1 = (k_1^2 - m_2^2)^{1/2} a_1$ and $m_2 = (k_2^2 - k_3^2)^{1/2} a_1$.

Thus, the expression for the field in the far zone has the form:

$$\begin{aligned} h_z &= \frac{m_1 - m_3}{m_2 - m_3} \frac{\mathrm{e}^{-k_3 L}(1 + k_3 L)}{[I_0(\bar{m}_2\beta)\eta_1 + K_0(\bar{m}_2\beta)\eta_2]^2} \qquad \text{if } \frac{L}{a} \gg 1 \\ \eta_1 &= \bar{m}_2 K_0(\bar{m}_2) I_1(\bar{m}_1) + \bar{m}_1 I_0(\bar{m}_1) K_1(\bar{m}_2) \\ \eta_2 &= \bar{m}_2 I_0(\bar{m}_2\beta) I_1(\bar{m}_2) - \bar{m}_1 I_0(\bar{m}_2) I_1(\bar{m}_1) \end{aligned} \tag{4.158}$$

If the conductivities of invasion zone and formation are the same then:

$$\bar{m}_1 = (k_1^2 - k_2^2) a_1^2 \qquad \bar{m}_2 = 0$$

and we will obtain the known expression for a two-layered medium at the far zone:

$$h_z = \frac{1}{I_0^2\left(a_1\sqrt{k_1^2 - k_2^2}\right)} h_z^{(0)}(k_2 L)$$

By analogy, when $\sigma_1 = \sigma_2$, $\bar{m}_1 = \bar{m}_3$, $l_3 = 0$, $n_3 = 1$:

$$h_z = \frac{1}{I_0^2\left(a_2\sqrt{k_1^2 - k_3^2}\right)} h_z^{(0)}(k_3 L)$$

i.e. the field in a two-layered medium as the borehole radius equals to a_2.

The asymptotical formula 4.158, as in the case of a two-layered medium, satisfactory describes a field, h_z, at the far zone provided that the minimal skin depth is greater than the radius of the invasion zone.

Letting the maximal value of parameter $|ka_2 < 1|$ and performing the corresponding expansion of function F we obtain the following expression for the field at the far zone:

$$h_z = \frac{1}{I_0^2\left(a_1\sqrt{k_1^2 - k_2^2}\right) I_0^2\left(a_2\sqrt{k_2^2 - k_3^2}\right)} h_z^{(0)}(k_3 L) \tag{4.159}$$

In accord with eqs. 4.158–4.159 the ratio of amplitudes and difference of phases of fields measured by two coil probes of different length do not depend on parameters of the borehole and the invasion zone at the far zone. This behavior of the field is used in high-frequency induction logging.

In such the case soundings are based on measurements of the quadrature and inphase components with probes of different lengths, while the ratio of amplitudes and phase difference are calculated. However, there are some exceptions and the equipment measures directly one or both of these last parameters.

4.8. Frequency Responses of the Magnetic Field of the Vertical Magnetic Dipole on the Borehole Axis

Until now we have considered the behavior of the field in the range of small and large parameters (low- and high-frequency parts of the spectrum) as well as in the near and far zones. Now let us consider the main features of frequency responses of the vertical component of the magnetic field on the borehole axis. Results of numerical modeling presented in this section are based on calculations of the field, h_z, by eq. 4.24 for models of a medium with one and two cylindrical interfaces.

It is appropriate to describe the field behavior in the following order:

1. If the field excitation is realized by vertical magnetic dipole sources of the secondary field are induced currents vector lines of which are located in horizontal planes and they present themselves as circles with centers on the borehole axis.

2. In cylindrical system of coordinates with the z-axis oriented along the borehole axis and the dipole located in its origin the electric field has only one component E_ϕ, but the magnetic field has two components, H_r and H_z. On the borehole axis the electrical field, E_ϕ, and the component of the magnetic field, H_r, are both equal to zero. In other words, the magnetic field is directed along the borehole axis.

3. As is well known, the current density of induced currents, J_ϕ, at any point can be presented as a sum of two components, namely the inphase and quadrature ones. The inphase and quadrature components of induced currents are shifted in phase by 180° (or 0°) and 90° with respect to the dipole current. Distribution of these components, In J_ϕ and Q J_ϕ, is essentially different. The quadrature component of the current is dominant near the source and rapidly decreases with an increase of the distance from the dipole. With an increase of frequency and conductivity of the bed, dimensions of the area where the quadrature component prevails become smaller.

4. In a wide range of frequencies and conductivities of borehole, invasion zone and bed, the quadrature component prevails near an induction probe, and the skin effect in the bed manifests itself in the same manner as in a uniform medium with the resistivity of the bed.

5. Near the source the quadrature component of the current density is directly proportional to the frequency, but with an increase of distance its behavior is strongly subjected to an influence of the skin effect.

6. Near the dipole the inphase component of the current density is significantly less than the quadrature one, and with an increase of the distance it increases, reaches a maximum, and then rapidly approaches zero.

7. The quadrature and the inphase components of the magnetic field on the borehole axis are defined by the distribution of the quadrature and the inphase components of current density, respectively. It follows directly from Biot–Savart law.

Examples of a spectrum of the vertical component of the magnetic field, expressed in units of the primary field, as well as frequency responses of apparent conductivity curves, σ_a/σ_3, are presented in Fig. 4.15–4.41. Function σ_a/σ_3 is related with the field by equation:

$$\frac{\sigma_a}{\sigma_3} = \frac{2}{\sigma\mu\omega L^2}\,\mathrm{Q}\,h_z$$

where $\mathrm{Q}\,h_z$ is the quadrature component of the field expressed in units of the primary field.

The ratio a_1/λ_1 is plotted along the abscissa, where $\lambda_1 = 2\pi h_1 = 2\pi(2/\sigma_1\mu\omega)^{1/2}$ (h_1 is the thickness of the skin layer in the borehole). The index of the curves is ρ_3/ρ_1. In the case of a two-layered medium $\rho_2 = \rho_3$. For three-layered medium every set of curves has index:

$$\frac{\rho_2}{\rho_1},\quad \frac{a_2}{a_1} \text{ and } \frac{L}{a_1}$$

In accord with the definition of the apparent conductivity function σ_a in a uniform medium is equal to its conductivity only in the case when $\mathrm{Q}\,h_z = \sigma\mu\omega L^2/2$, i.e. the interaction between currents, defining a signal in a measuring coil of an induction probe, is negligible.

8. The quadrature component of the magnetic field, regardless of the resistivity of the medium, firstly increases directly proportional to frequency, reaches a maximum, and then tends to zero. The oscillating character of the behavior of $Q\,h_z$ at the right-oriented part of the response ($\lambda_1/a_1 \to 0$) is not shown since a logarithmical scale is used.

9. The left-hand asymptote of the frequency response of the quadrature component presents a straight line with a slope equal to 63°30' with respect to the abscissa axis. This part of the response corresponds to the case, when the signal is caused by induced currents, the intensity of which is defined by only the primary magnetic flux and the resistivity of a medium. As was mentioned above, the area of distribution of induced currents, shifted in phase by 90° with respect to the dipole current and caused by only the primary magnetic field, increases with a decrease of the frequency and an increase of the resistivity of the medium, specially the resistivity of the bed. At the same time with an increase of the probe length the depth of investigation increases and, correspondingly, the influence of the part of the medium which is closer to the induction probe decreases. For this reason the deviation of the frequency response of the quadrature component of the field from its left-hand asymptote begins earlier for longer probes.

10. We will call that part of the frequency response of the quadrature component, $Q\,h_z$, which practically coincides with its left-hand asymptote *Doll's range.* Within this range the quadrature component is significantly greater than the inphase one. In section 4.3 we described in detail the magnetic field and the apparent conductivity on the borehole axis as functions of geometric factors and resistivity distribution corresponding to *Doll's range.*

With an increase of frequency or conductivity of a medium the frequency response of $Q\,h_z$ is located lower its left-hand asymptote.

11. In a two-layered medium (the invasion zone is absent), when the resistivity of the borehole exceeds that of the formation ($\rho_1 > \rho_2$) departure from *Doll's range* commences practically for the same values of parameter L/a_1 as in a uniform medium with conductivity σ_2.

12. If the conductivity of the borehole exceeds that of the formation, $\sigma_1 > \sigma_2$, and the skin depth in the borehole is significantly larger than its radius, the behavior of the field corresponding to *Doll's range* manifests itself for greater values of parameter L/h_2 than in a uniform medium with resistivity ρ_2. It is explained by the fact that induced currents in the borehole and defining a signal are mainly shifted in phase by 90°, and they increase directly proportional to frequency and conductivity.

13. Similar features are observed for a three-layered medium: with an increase of the conductivity of the borehole and the invasion zone, as well as its radius a_2 (within certain limits) the behavior of the field corresponding to *Doll's range* takes place for larger values of L/h_3 than in a uniform medium with conductivity σ_3.

14. With an increase of parameter a_1/λ_1 the frequency response of the quadrature component $Q\,h_z$ departures from its left-hand asymptote and within a certain range of parameter a_1/λ_1 the skin effect is practically absent in the borehole and in the invasion

zone, but in the bed it manifests itself in the same manner as in a uniform medium with the bed resistivity. From a practical point of view this is the most important range of the frequency response for conventional tools of induction logging. The main features of the field behavior within this range have been described in detail in section 4.5.

15. Frequency responses of the quadrature component, $Q\,h_z$, for a two-layered medium has one maximum which to some extent increases with an increase of resistivity of the borehole. The position of the maximum is mainly defined by the resistivity of the formation. For example, an increase of the borehole conductivity of more than 100 times only slightly shifts the maximum to a range of lower frequencies. In some cases when the invasion zones is relatively large we can observe two maxima.

16. With a further increase of frequency the influence of induced currents in the bed becomes smaller, and the frequency responses in a three-layered medium almost coincide with those for a two-layered medium when the resistivity outside the borehole is equal to that of the invasion zone, ρ_2.

17. At the right-hand part of the frequency response of the quadrature component it relatively quickly decreases with frequency. Within *Doll's range* the influence of the borehole is defined by geometric factors and distribution of the medium's resistivity. With an increase of the frequency, the relative contribution of currents induced in the borehole increases, inasmuch as the current density in the bed grows slower than in the borehole (skin effect). At the range of very high frequencies, when the skin depth in the borehole and its radius are comparable, the influence of the bed conductivity on the quadrature component strongly increases (far zone). A similar effect takes place in the far zone as there is the invasion zone.

18. In a wide range of frequencies the influence of a more resistive borehole is not significant and the frequency response of the field, $Q\,h_z$, practically coincides with that corresponding to a uniform medium with the resistivity of the bed, if the skin depth h_1 is several times larger than the borehole radius.

19. In measuring only the quadrature component of the field, $Q\,h_z$, at one frequency it is possible to perform nonunique interpretation since the frequency response of this component has a maximum. This fact has to be taken into account choosing a frequency.

20. A choice of frequency or frequencies for induction logging cannot be done by using the results of calculation of the field in a medium with only cylindrical interfaces. However, these data allow us to investigate radial characteristics of two-coil induction probes, as well as other types of probes, consisting of several coils. In particular, comparison of calculations based on exact and approximate methods permits us to establish a range of frequencies and resistivities of a medium where it is reasonable to apply so-called *focusing* multi-coil probes.

21. The frequency responses of the inphase component of the secondary field, $\mathrm{In}\,h_z$, essentially differ from those for the quadrature one. At the range of small values of parameter a_1/λ_1 (low frequencies, high resistivity) function $\mathrm{In}\,h_z$ tends to zero as $\omega^{3/2}$, and the ratio of the inphase component of the secondary field to the quadrature one rapidly decreases. In this range of frequencies and resistivities the inphase component of induced currents within the borehole and the invasion zone is usually very small. For this

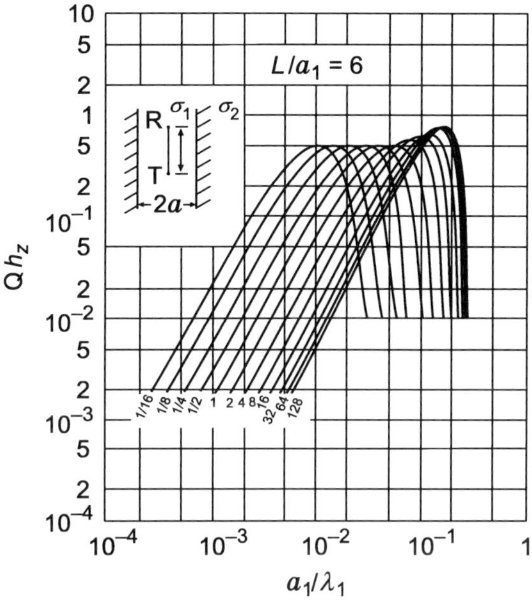

Figure 4.15. Function Q h_z.

reason it is practically proportional to the conductivity of the formation only. This very important feature of the inphase component, In h_z, at the range of small parameters is often used to increase the depth of investigation of induction logging.

With an increase of the ratio a_1/λ_1 the inphase component of the secondary magnetic field increases and then becomes greater than the quadrature component, but when the skin depth in the borehole is smaller than its radius, function In h_z approaches to -1, that is, all induced currents concentrate near the dipole. Thus, at the right-hand part of a frequency response of the secondary field the inphase component prevails.

22. In the far zone the influence of the borehole and the invasion zone does not depend on the length of the induction probe. Such a behavior of the field presents a certain practical interest, inasmuch as it allows us to increase the depth of investigation significantly, measuring the ratio of amplitudes and differences of phases by three-coil induction probes.

4.9. Influence of Finite Dimensions of Induction Probe Coils

As is well known, measurements of relatively small signals in induction logging require application of coils, the dimensions of which are comparable with the borehole radius and sometimes with the probe length. For this reason it is appropriate to investigate the influence of the dimensions of transmitter and receiver coils of the induction probe on the field behavior, and here we will describe the results of the calculations corresponding to

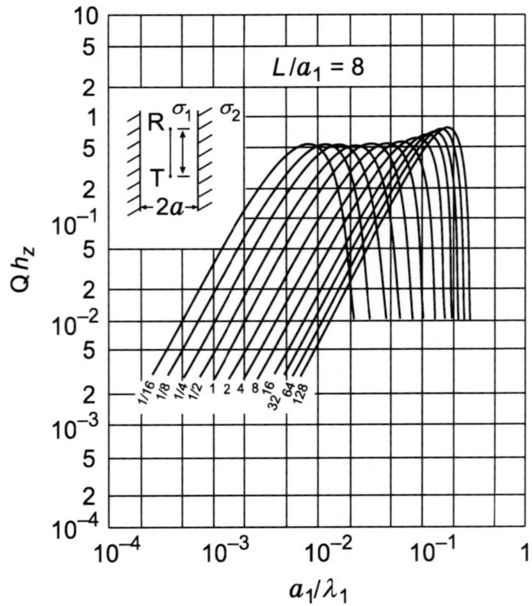

Figure 4.16. Function $\mathrm{Q}\,h_z$.

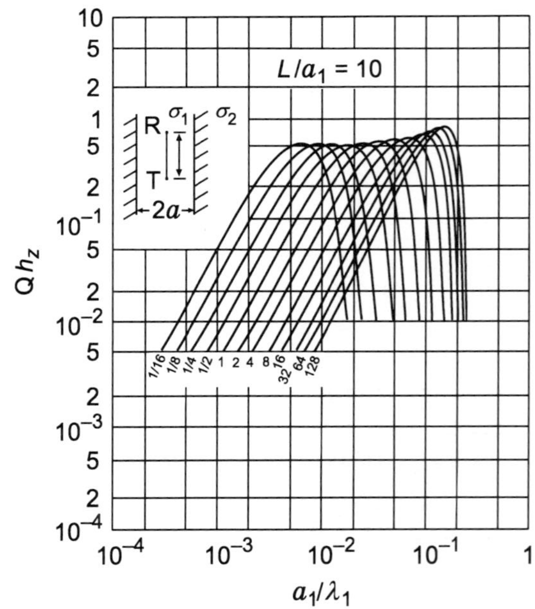

Figure 4.17. Function $\mathrm{Q}\,h_z$.

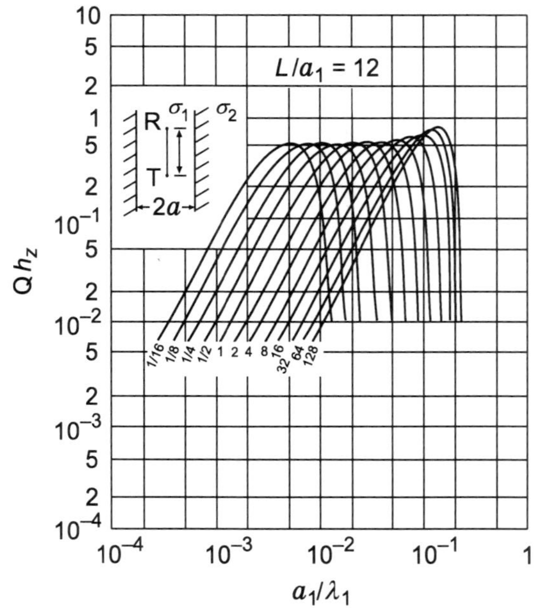

Figure 4.18. Function Q h_z.

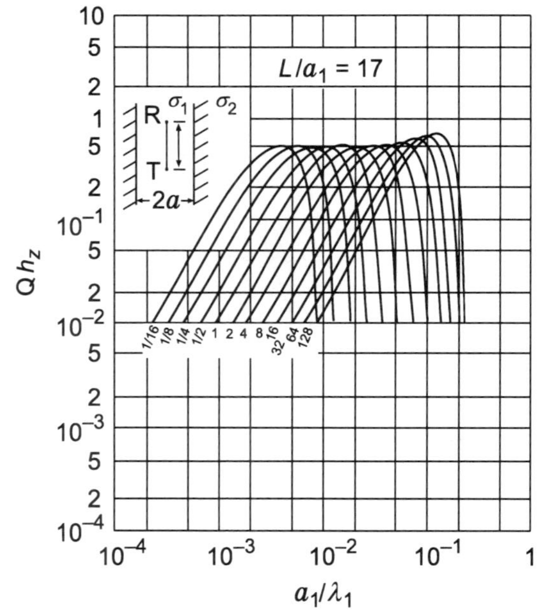

Figure 4.19. Function Q h_z.

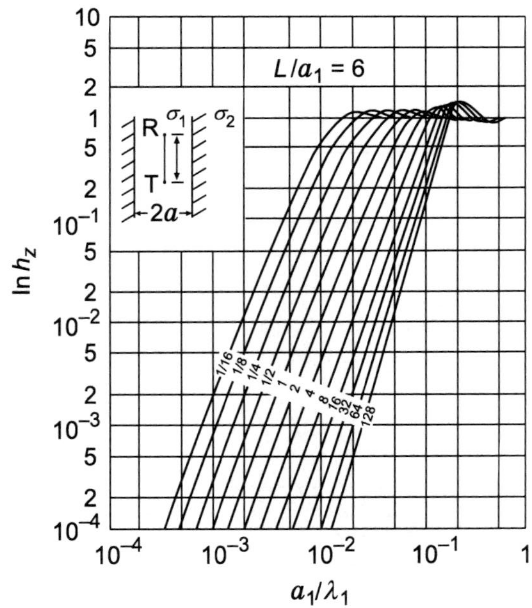

Figure 4.20. Function In h_z.

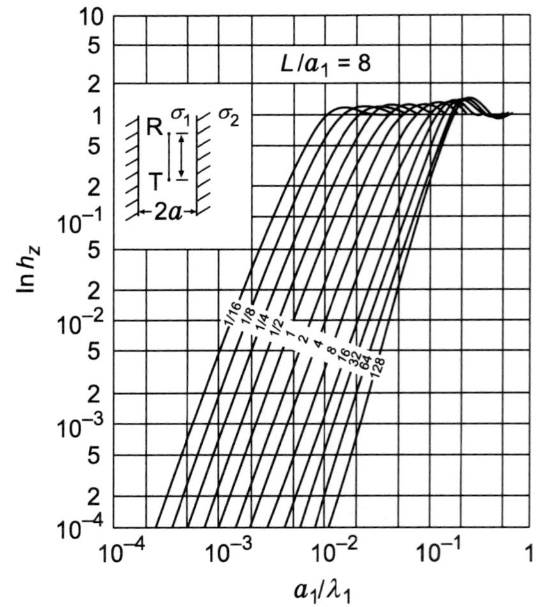

Figure 4.21. Function In h_z.

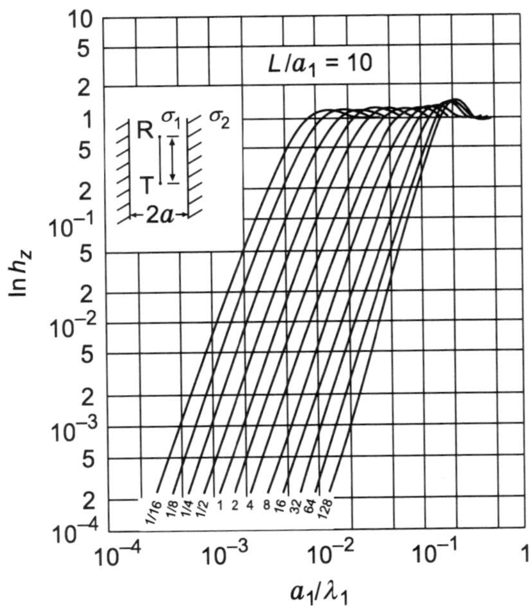

Figure 4.22. Function In h_z.

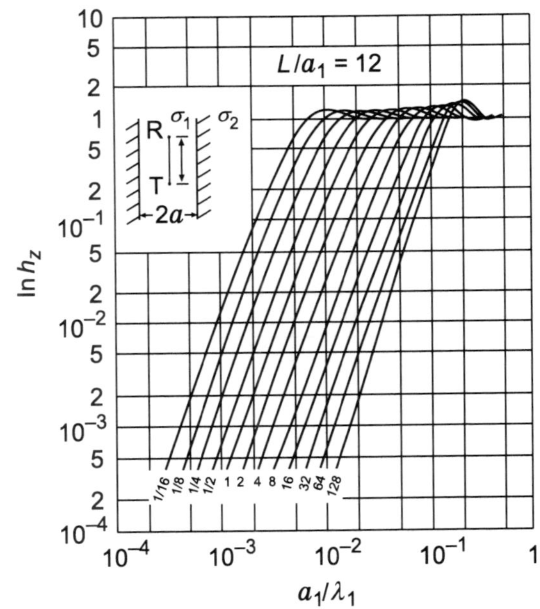

Figure 4.23. Function In h_z.

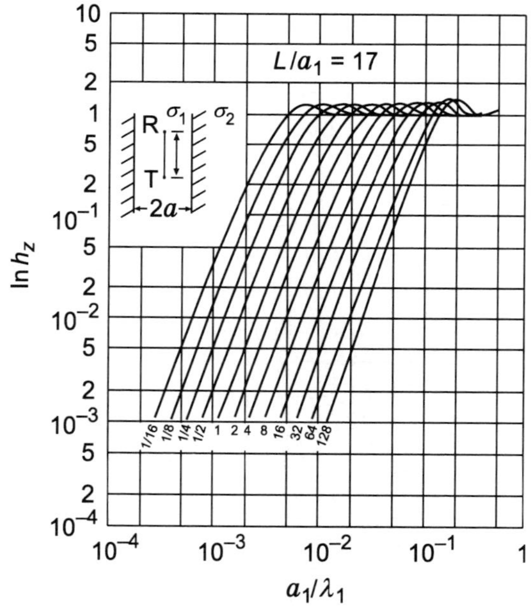

Figure 4.24. Function In h_z.

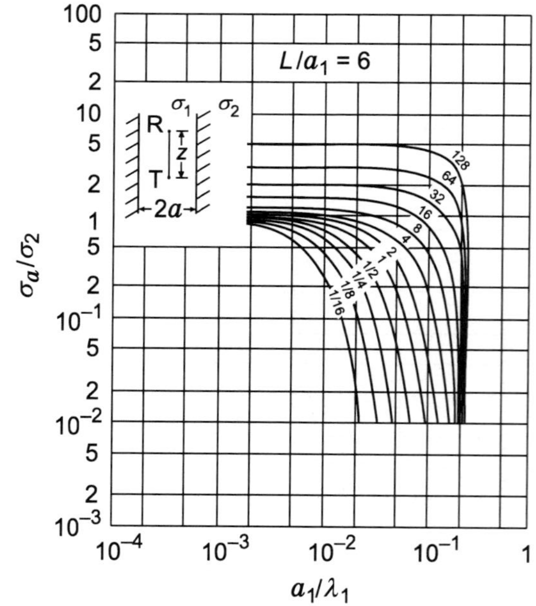

Figure 4.25. Function σ_a/σ_2.

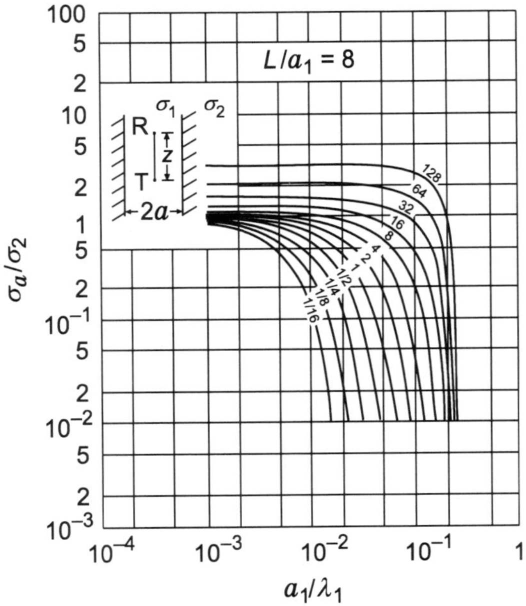

Figure 4.26. Function σ_a/σ_2.

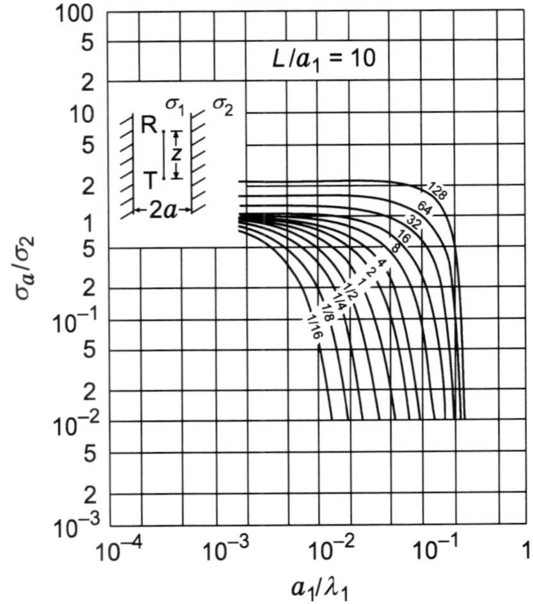

Figure 4.27. Function σ_a/σ_2.

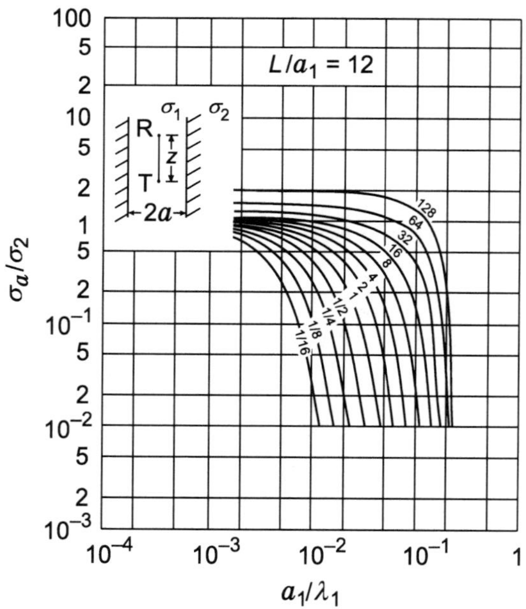

Figure 4.28. Function σ_a/σ_2.

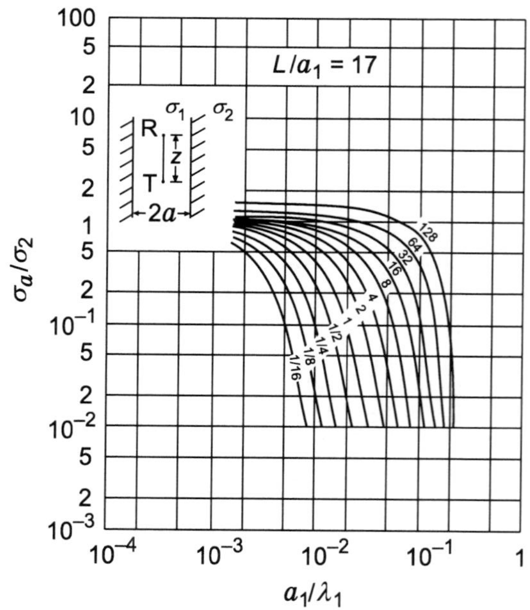

Figure 4.29. Function σ_a/σ_2.

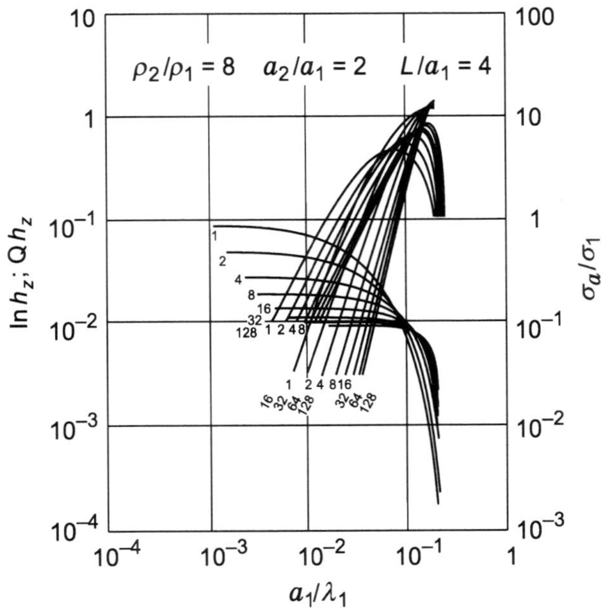

Figure 4.30. Functions $\text{In}\, h_z$, $\text{Q}\, h_z$, and σ_a/σ_1 (three-layered medium).

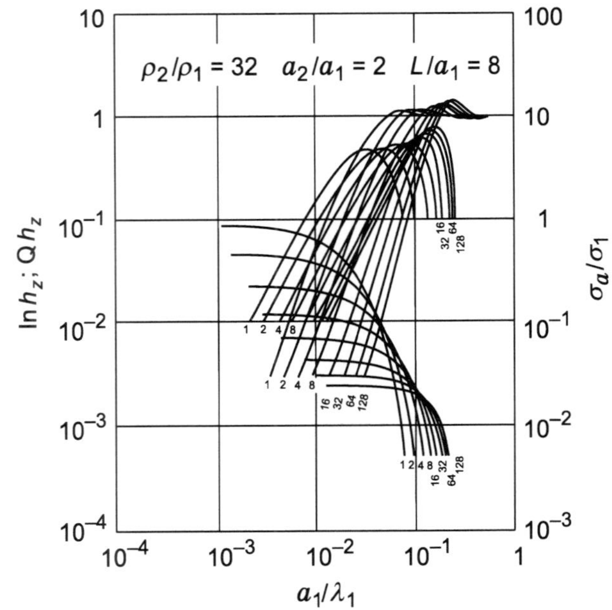

Figure 4.31. Functions $\text{In}\, h_z$, $\text{Q}\, h_z$, and σ_a/σ_1 (three-layered medium).

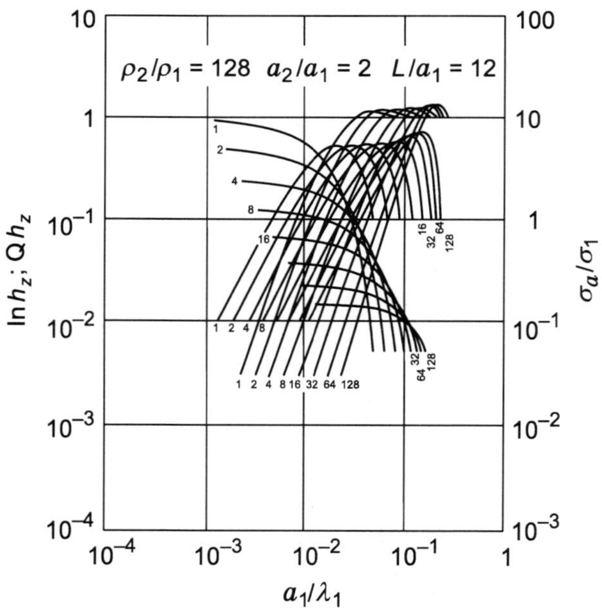

Figure 4.32. Functions In h_z, Q h_z, and σ_a/σ_1 (three-layered medium).

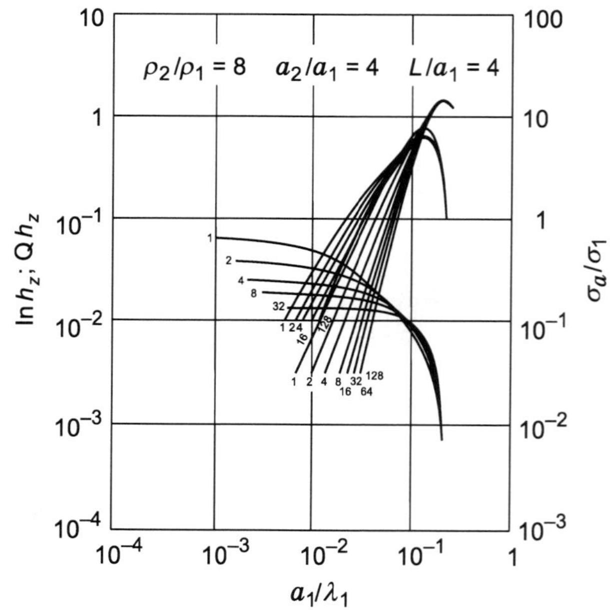

Figure 4.33. Functions In h_z, Q h_z, and σ_a/σ_1 (three-layered medium).

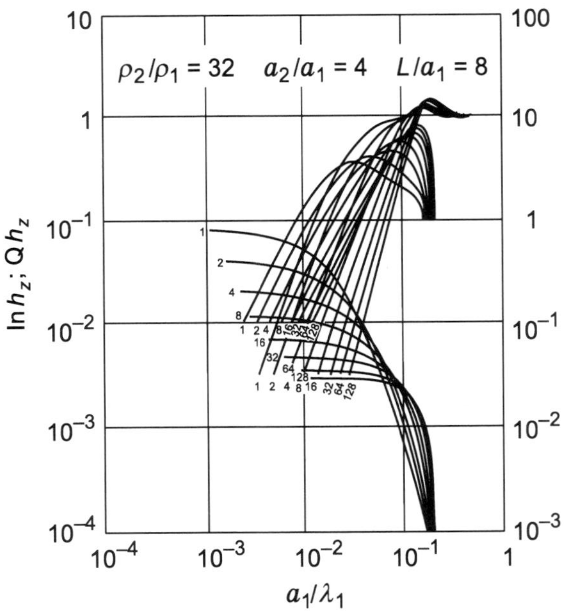

Figure 4.34. Functions $\text{In}\, h_z$, $\text{Q}\, h_z$, and σ_a/σ_1 (three-layered medium).

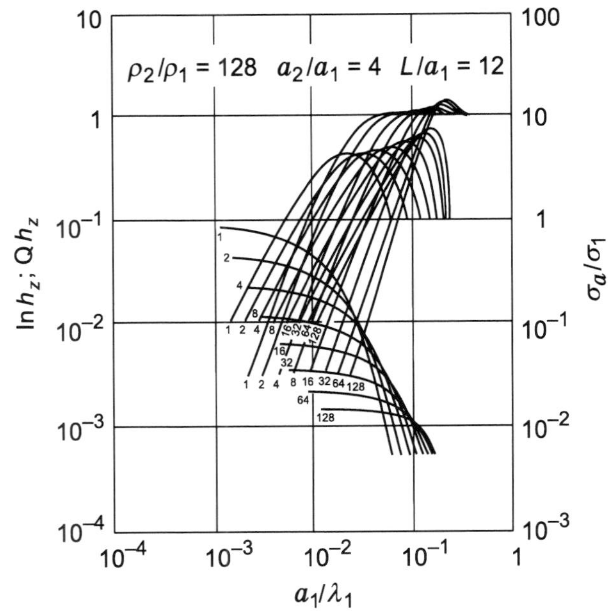

Figure 4.35. Functions $\text{In}\, h_z$, $\text{Q}\, h_z$, and σ_a/σ_1 (three-layered medium).

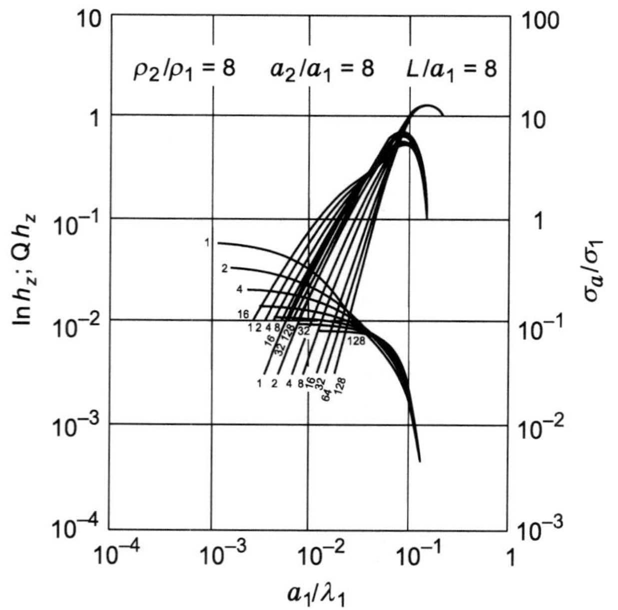

Figure 4.36. Functions $\ln h_z$, $Q\,h_z$, and σ_a/σ_1 (three-layered medium).

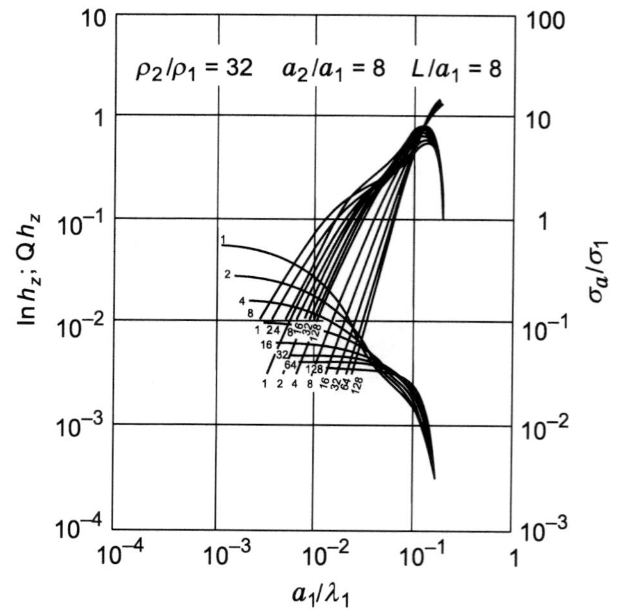

Figure 4.37. Functions $\ln h_z$, $Q\,h_z$, and σ_a/σ_1 (three-layered medium).

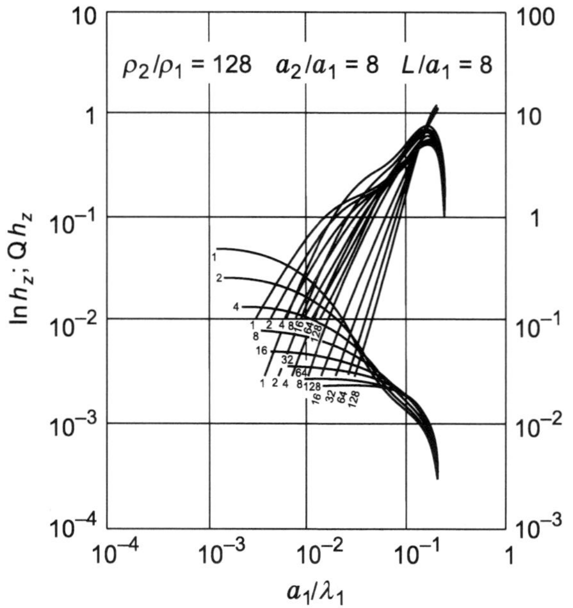

Figure 4.38. Functions In h_z, Q h_z, and σ_a/σ_1 (three-layered medium).

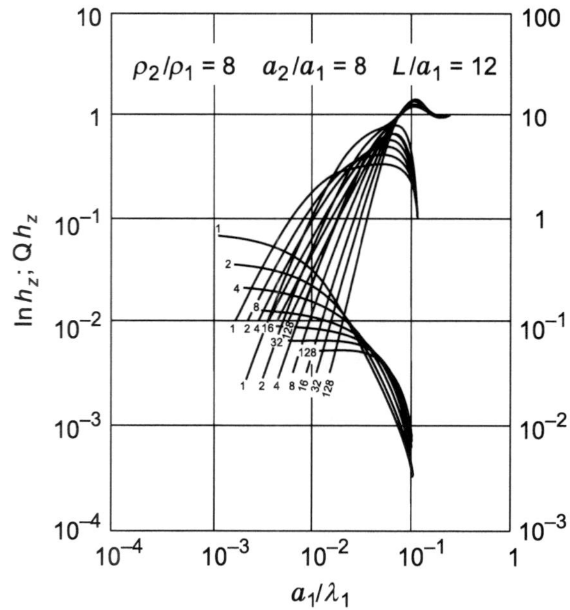

Figure 4.39. Functions In h_z, Q h_z, and σ_a/σ_1 (three-layered medium).

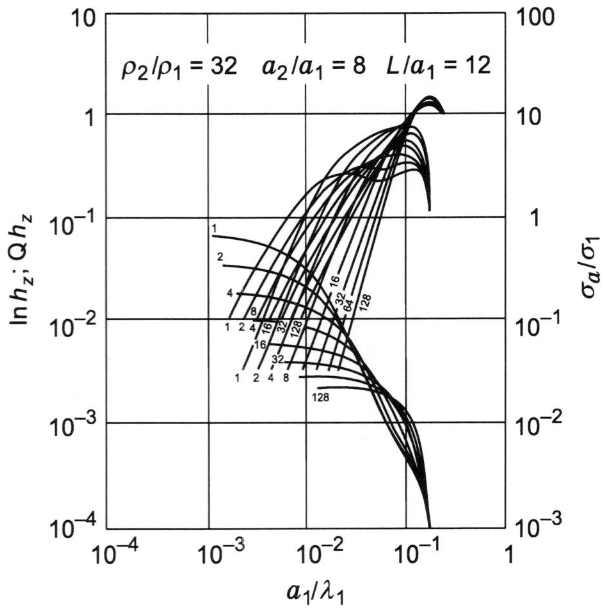

Figure 4.40. Functions $\ln h_z$, $Q h_z$, and σ_a/σ_1 (three-layered medium).

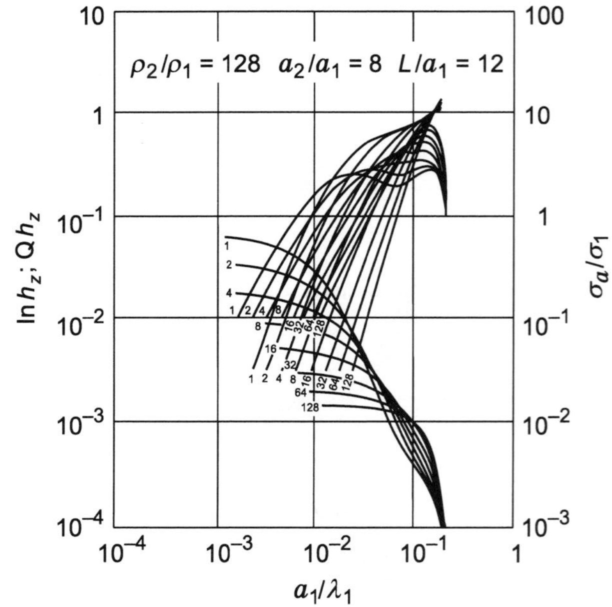

Figure 4.41. Functions $\ln h_z$, $Q h_z$, and σ_a/σ_1 (three-layered medium).

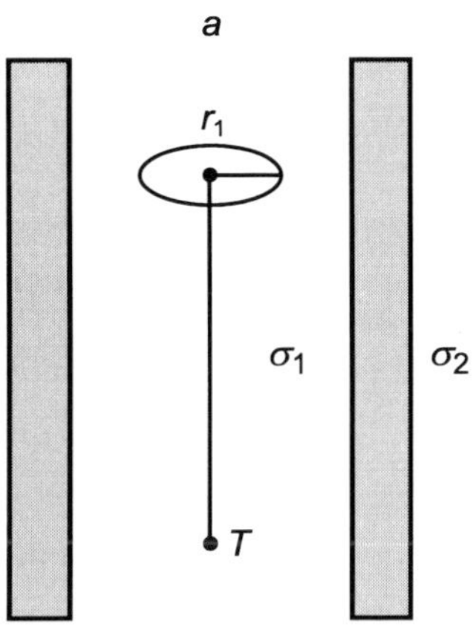

Figure 4.42a. Transmitter is the magnetic dipole, the receiver is the horizontal ring with radius r_1.

the following cases:

- The transmitter and receiver are both coils with one ring, placed on a nonconducting base of the probe.
- The transmitter is a coil with one ring, the receiver is a single-layered coil or vice versa. Radii of both the transmitter and the receiver are the same and equal to the radius of a nonconducting base of the probe.
- The transmitter and receiver are both single-layered coils, the radius of which coincides with that of the nonconducting base of the probe.
- One-coil induction probe.

Inasmuch as the influence of finite dimensions of coils is specially noticeable with relatively short probes, as the skin effect manifests itself weaker, the main attention is paid to the range of small parameters. As was shown above, in this case we can neglect the interaction of currents in the borehole and in the invasion zone, while the skin effect in the formation is displayed in the same manner as in a uniform medium with the formation's resistivity. Correspondingly, the field in the borehole can be expressed through a field in a uniform medium and geometric factors. For this reason we will investigate mainly the influence of dimensions of coils on geometric factors of the borehole and the bed.

Let us start from the simplest case when the field source is the vertical magnetic dipole but the receiver is the ring with radius r_1 located in a horizontal plane (Fig. 4.42a). As was shown in section two of this chapter the electrical field in the borehole can be written in the form:

$$E = E_0 + \frac{i\omega\mu}{4\pi}\frac{2}{\pi}\int_0^\infty \lambda_1^2 C_1 I_1(\lambda_1 r)\cos\lambda z\, d\lambda \tag{4.160}$$

where E_0 is the electrical field of the magnetic dipole in a uniform medium with conductivity σ_1:

$$E_0 = \frac{i\omega\mu M}{4\pi}\frac{e^{ik_1R}}{R^2}(1 - ik_1R)\sin\theta \tag{4.161}$$

Expanding function e^{ik_1R} in a series:

$$e^{ik_1R} \simeq 1 + ik_1R - \frac{k_1^2R^2}{2}$$

we will find an expression for E_0 in the range of small parameters:

$$E_0 \simeq \frac{i\omega\mu M}{4\pi}\frac{1}{R^2}\sin\theta + \frac{i\omega\mu M}{4\pi}\frac{k_1^2}{2}\sin\theta$$

where $k_1^2 = i\sigma_1\mu\omega$. The first term defines the vortex electric field of the magnetic dipole in free space while the second one describes the vortex electrical field of currents induced in a medium.

Making use of the approach considered in detail in section 4.3, the secondary electrical field in the borehole (two-layered medium) in the range of very small parameters can be presented as:

$$E = \frac{i\omega\mu M}{4\pi}\frac{k_1^2}{2} + \frac{i\omega\mu M}{4\pi}\frac{2}{\pi}(\sigma_2 - \sigma_1)i\omega\mu\int_0^\infty \frac{m}{2}\left[2K_0K_1 - m(K_1^2 - K_0^2)\right]\frac{I_1(\beta m)}{m}\cos m\alpha\, dm \tag{4.162}$$

where $\alpha = L/a_1$, $\beta = r_1/a_1$.

Since the medium is uniform, the secondary electrical field is:

$$E_0^s = \frac{i\omega\mu M}{4}\frac{i\sigma\mu\omega}{2}\sin\theta$$

an expression for the apparent conductivity can be written as:

$$\sigma_a = -\frac{8\pi E}{M\omega^2\mu^2\sin\theta}$$

In particular for a two-layered medium we have:

$$\frac{\sigma_a}{\sigma_1} = 1 + \frac{4}{\pi\sin\theta}\left(\frac{\sigma_2}{\sigma_1} - 1\right)\int_0^\infty \frac{m}{2}\left[2K_0K_1 - m(K_1^2 - K_0^2)\right]\frac{I_1(\beta m)}{m}\cos m\alpha\, dm \tag{4.163}$$

We will call the function:

$$G_2^*(\alpha,\beta) = \frac{4}{\pi \sin\theta} \int_0^\infty \frac{m}{2} \left[2K_0K_1 - m(K_1^2 - K_0^2)\right] \frac{I_1(\beta m)}{m} \cos m\alpha \, dm \tag{4.164}$$

the geometric factor of a formation.

Let us notice that:

$$\sin\theta = \frac{r_1}{(r_1^2 + L^2)^{1/2}} = \frac{1}{\left[1 + (\alpha/\beta)^2\right]^{1/2}}$$

If the ring radius, r_1, tends to zero, $(\beta \to 0)$ then

$$\frac{I_1(\beta m)}{m} \to \frac{\beta}{2} \text{ and } \sin\theta \to \frac{\beta}{\alpha}$$

Therefore, we have:

$$G_2^* \to \frac{2\alpha}{\pi} \int_0^\infty \frac{m}{2} \left[2K_0K_1 - m(K_1^2 - K_0^2)\right] \cos m\alpha \, dm = G_2$$

i.e. it approaches value of the geometric factor, G_2, for the infinitely small coils of the induction probe.

In accord with eq. 4.163 we have:

$$G_1^*(\alpha,\beta) = 1 - \frac{4\left[1 + (\alpha/\beta)^2\right]^{1/2}}{\pi} \int_0^\infty \frac{m}{2} \left[2K_0K_1 - m(K_1^2 - K_0^2)\right] \frac{I_1(\beta m)}{m} \cos m\alpha \, dm \tag{4.165}$$

where $G_1^*(\alpha,\beta)$ is the geometric factor of the borehole when the receiver is the ring of radius r_1.

As follows from eq. 4.165 for small values of variable m, the integrand tends to function $(\beta/2)K_0(m)$. For this reason applying methodics of calculation of the function $G_1(\alpha)$ we will rewrite eq. 4.165 in the form:

$$\begin{aligned} G^*(\alpha,\beta) &= 1 - \frac{(\beta^2+\alpha^2)^{1/2}}{(1+\alpha^2)^{1/2}} + \frac{(\beta^2+\alpha^2)^{1/2}}{(1+\alpha^2)^{1/2}} - \frac{4}{\pi}\left[1 + (\alpha/\beta)^2\right]^{1/2} \int_0^\infty \phi(m)\frac{I_1(\beta m)}{m} \, dm \\ &= 1 - \frac{(\beta^2+\alpha^2)^{1/2}}{(1+\alpha^2)^{1/2}} + \frac{4}{\pi}\left[1 + (\alpha/\beta)^2\right]^{1/2} \int_0^\infty \left[\frac{\beta}{2}K_0(m) - \phi(m)\frac{I_1(\beta m)}{m}\right] \\ &\quad \times \cos m\alpha \, dm \end{aligned} \tag{4.166}$$

TABLE 4.3
Values of function G_1^*

α \ β	0.025	0.050	0.1	0.2	0.3	0.4	0.5
1	0.4865	0.4860	0.4840	0.4763	0.4636	0.4463	0.4247
3	0.1142	0.1142	0.1139	0.1128	0.1109	0.1083	0.1051
5	0.0427	0.0427	0.0426	0.0442	0.0417	0.0408	0.0398
7	0.0215	0.0215	0.0215	0.2130	0.0210	0.0207	0.0202
9	0.0129	0.0128	0.0128	0.0127	0.0126	0.0123	0.0120
11	0.0086	0.0086	0.0085	0.0085	0.0084	0.0082	0.0080
13	0.0061	0.0061	0.0061	0.0060	0.0060	0.0058	0.0057
15	0.0045	0.0045	0.0045	0.0045	0.0044	0.0044	0.0043
17	0.0035	0.0035	0.0035	0.0035	0.0034	0.0034	0.0033

where

$$\phi(m) = \frac{m}{2}\left[2K_0K_1 - m(K_1^2 - K_0^2)\right]$$

and the equality:

$$\frac{1}{(1+\alpha^2)^{1/2}} = \frac{2}{\pi}\int_0^\infty K_0(m)\cos m\alpha \, dm$$

is used.

Unlike the previous case the integrand in eq. 4.166 does not have singularities, and this expression is convenient for calculations. Some results of numerical integration are presented in Table 4.3. As these data show, corrections are usually small and do not exceed 10% even when the radius of the receiver ring is equal to half of the borehole radius. With an increase of the radius the geometric factor, G_1^*, decreases, and it is specially noticeable for relatively small probes. It is obvious that with an increase of the probe the influence of the ring radius on the geometric factor G_1^* decreases.

In order to investigate the effect caused by finite dimensions of a transmitter coil let us first derive formulae for the vector potential of the electrical type corresponding to a current element. As is well known, complex amplitudes of the field are described by Maxwell's equations:

$$\operatorname{curl} \boldsymbol{E} = -i\omega\mu\boldsymbol{H} \tag{4.167}$$

$$\operatorname{curl} \boldsymbol{H} = \sigma\boldsymbol{E} \tag{4.168}$$

$$\operatorname{div} \boldsymbol{E} = 0 \tag{4.169}$$

$$\operatorname{div} \boldsymbol{H} = 0 \tag{4.170}$$

From eq. 4.170 we have:

$$\boldsymbol{H} = \operatorname{curl} \boldsymbol{A} \tag{4.171}$$

where $\boldsymbol{A}$ is the vector potential of the electrical type. The equation for potential $\boldsymbol{A}$ is found by substituting eq. 4.171 into eq. 4.168:

$$\operatorname{curl}\operatorname{curl}\boldsymbol{A} = \operatorname{grad}\operatorname{div}\boldsymbol{A} - \nabla^2\boldsymbol{A} = \sigma\boldsymbol{E}$$

On the other hand, from eqs. 4.167 and 4.171 we have:

$$\boldsymbol{E} = -\mathrm{i}\omega\mu\boldsymbol{A} - \operatorname{grad} U$$

or

$$\operatorname{grad}\operatorname{div}\boldsymbol{A} - \nabla^2\boldsymbol{A} = -\mathrm{i}\sigma\mu\omega\boldsymbol{A} - \sigma\operatorname{grad} U$$

The latter can be replaced by two equations:

$$\nabla^2\boldsymbol{A} + k^2\boldsymbol{A} = 0 \qquad (4.172)$$
$$\operatorname{div}\boldsymbol{A} = \sigma U \qquad (4.173)$$

where $k^2 = -\mathrm{i}\sigma\mu\omega$.

Direct substitution shows that function $(I\mathrm{d}\boldsymbol{l}/4\pi R)\mathrm{e}^{\mathrm{i}kR}$ is a solution of eq. 4.172 when the source is a current element $\mathrm{d}\boldsymbol{l}$:

$$\boldsymbol{A} = \frac{I\mathrm{d}\boldsymbol{l}}{4\pi R}\,\mathrm{e}^{-\mathrm{i}kR}$$

where I is the current and R is the distance from the element $\mathrm{d}\boldsymbol{l}$ to an observation point.

Now we can derive an expression for the vector potential of a current ring in a conducting medium.

First of all we will present a vector potential at an arbitrary point P as a sum of vector potentials caused by all elements of the current ring (Fig. 4.42b). The vector potential of the current element, $\mathrm{d}\boldsymbol{l}$, at the point P is written as:

$$\mathrm{d}\boldsymbol{A} = \frac{\mathrm{e}^{-\mathrm{i}kR(P,P_1)}I\mathrm{d}\boldsymbol{l}}{4\pi R(P,P_1)} \qquad (4.174)$$

If the current ring with radius, r_1, is located in a horizontal plane, the vector potential of the ring has only component A_ϕ. As can be easily seen from Fig. 4.42b, the radial component, A_r, is equal to zero.

Component $\mathrm{d}A_\phi$ at point P is:

$$\mathrm{d}A_\phi = \frac{\mathrm{e}^{-\mathrm{i}kR}}{R}\frac{I\cos\alpha\,\mathrm{d}l}{4\pi} = \frac{Ir_1}{4\pi}\frac{\mathrm{e}^{\mathrm{i}kR}}{R}\cos\alpha\,\mathrm{d}\alpha \qquad (4.175)$$

We will make use of the known integral presentation of function $\mathrm{e}^{-\mathrm{i}kR}/R$:

$$\frac{\mathrm{e}^{-\mathrm{i}kR}}{R} = \frac{2}{\pi}\int_0^\infty K_0\left[(\lambda^2 + \mathrm{i}\chi)^{1/2}d\right]\cos\lambda z\,\mathrm{d}\lambda \qquad (4.176)$$

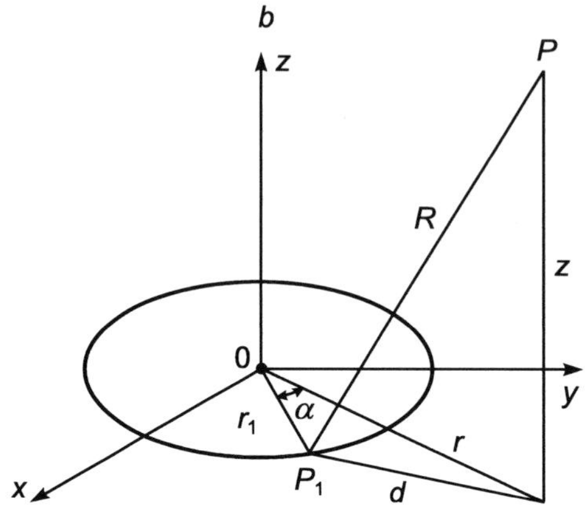

Figure 4.42b. Current ring in a conducting medium.

where $\chi = \sigma\mu\omega$, $R = (z^2 + d^2)^{1/2}$.

Substituting eq. 4.176 into eq. 4.175 and integrating along the ring we obtain an expression for A_ϕ:

$$A_\phi = \frac{Ir_1}{\pi}\frac{2}{\pi}\int_0^\infty \cos\lambda z \, \mathrm{d}\lambda \int_0^{2\pi} K_0[(\lambda^2 + \mathrm{i}\chi)^{1/2} d] \cos\alpha \, \mathrm{d}\alpha \tag{4.177}$$

In accord with the addition theorem of modified Bessel functions we have:

$$K_0(dv) = \begin{cases} K_0(rv)I_0(r_1 v) + 2\sum\limits_{m=1}^{\infty} K_m(rv)I_m(r_1 v)\cos m\alpha & \text{if } r \geqslant r_1 \\ K_0(r_1 v)I_0(rv) + 2\sum\limits_{m=1}^{\infty} K_m(r_1 v)I_m(rv)\cos m\alpha & \text{if } r \leqslant r_1 \end{cases} \tag{4.178}$$

Replacing function $K_0(dv)$ in eq. 4.177 by the right-hand side of equality 4.178 and applying orthogonality of trigonometric functions:

$$\int_0^{2\pi} \cos m\alpha \cos\alpha \, \mathrm{d}\alpha = \begin{cases} 0 & \text{if } m \neq 1 \\ \pi & \text{if } m = 1 \end{cases}$$

we obtain the integral presentation for the vector potential of the electrical type when the source is the current loop:

$$A_\phi = \frac{4Ir_1}{4\pi}\int_0^\infty I_1(r_1 v)K_1(rv)\cos\lambda z \mathrm{d}\lambda \qquad \text{if } r \geqslant r_1 \tag{4.179}$$

or

$$A_\phi = \frac{4Ir_1}{4\pi} \int_0^\infty K_1(r_1 v) I_1(rv) \cos \lambda z \mathrm{d}\lambda \qquad \text{if } r \leqslant r_1 \tag{4.180}$$

where $v = (\lambda^2 + \mathrm{i}\chi)^{1/2}$.

Inasmuch as the vector potential has only component A_ϕ which does not depend on ϕ, $\operatorname{div} \boldsymbol{A} = 0$ and, according to eq. 4.174, the scalar potential U is zero. For this reason the electrical field of the current ring in a uniform medium can be written in the form:

$$E_\phi = -\frac{\mathrm{i}\omega\mu}{4\pi} 4Ir_1 \int_0^\infty I_1(r_1 v) K_1(rv) \cos \lambda z \, \mathrm{d}\lambda \qquad \text{if } r \geqslant r_1 \tag{4.181}$$

or

$$E_\phi = -\frac{\mathrm{i}\omega\mu}{4\pi} 4Ir_1 \int_0^\infty K_1(r_1 v) I_1(rv) \cos \lambda z \, \mathrm{d}\lambda \qquad \text{if } r \leqslant r_1 \tag{4.182}$$

Now we are prepared to consider the field of a current ring in a nonuniform medium.

4.10. Electrical Field of a Current Ring in a Medium with Cylindrical Interfaces

The total current electrical field can be presented as a sum of two terms. The first one is the field in a uniform medium with the conductivity of the borehole, while the second one is caused by the nonuniformity of the medium, that is:

$$E_\phi = -\frac{\mathrm{i}\omega\mu}{4\pi} 4Ir_1 \left(\int_0^\infty I_1(r_1 v) K_1(rv) \cos \lambda z \, \mathrm{d}\lambda - \int_0^\infty C_1(\lambda) I_1(r_1 v) K_1(rv) \cos \lambda z \, \mathrm{d}\lambda \right)$$

$$\text{if } r \leqslant r_1 \tag{4.183}$$

Taking into account the axial symmetry for determination of the electromotive force in a measuring ring, it is sufficient to multiply the electric field E_ϕ by the ring length:

$$\mathscr{E} = 2\pi r E_\phi$$

Function $C_1(\lambda)$ in eq. 4.183 is defined from boundary conditions, and it does not depend on the type of source of the vortex electrical field provided that the axial symmetry is held.

For this reason we make use of results derived in the third section of this chapter, and then we have:

$$C_1(\lambda) = (\chi_2 - \chi_1)\frac{a_1}{2\lambda}\left[2K_0(\lambda a_1)K_1(\lambda a_1) - \lambda a_1(K_1^2 - K_0^2)\right] \\ + (\chi_3 - \chi_2)\frac{a_2}{2\lambda}\left[2K_0(\lambda a_2)K_1(\lambda a_2) - \lambda a_2(K_1^2 - K_0^2)\right] \tag{4.184}$$

where $\chi_1 = \sigma_1\mu\omega$, $\chi_2 = \sigma_2\mu\omega$, $\chi_3 = \sigma_3\mu\omega$; σ_1, σ_2 and σ_3 are conductivities of internal, middle and external parts of the medium, respectively; a_1 and a_2 are radii of cylindrical interfaces.

Let us suppose that transmitter and receiver rings of the introduction probe are placed on the nonconducting base of the probe with radius a_1, that is $r_1 = a_1$ and $\sigma_1 = 0$. Then expressions for components of the electrical field in media with one cylindrical interface ($r = a_1$) and two interfaces ($r = a_1$ and $r = a_2$) can be written, respectively:

$$E_2 = -\frac{i\omega\mu}{4\pi}4Ir_1\chi_2\int\limits_0^\infty \frac{r_1}{2}\left[2K_0(\lambda r_1)K_1(\lambda r_1) - \lambda r_1(K_1^2 - K_0^2)\right] I_1^2(\lambda r_1)\cos\lambda z\,d\lambda \tag{4.185}$$

$$E_3 = E_2 - \frac{i\omega\mu}{4\pi}4Ir_1(\chi_3 - \chi_2)\int\limits_0^\infty \frac{a_2}{2\pi}\left[2K_0(\lambda a_2)K_1(a_2) - \lambda a_2(K_1^2 - K_0^2)\right] I_1^2(\lambda r_1)\cos\lambda z\,d\lambda \tag{4.186}$$

Both equations describe the electrical field on the surface of the nonconducting base of the induction probe.

Now we will introduce the apparent conductivity from the equation:

$$\frac{\sigma_a}{\sigma_2} = \frac{E_3}{E_2} \tag{4.187}$$

which shows how the field E_3 in a medium with cylindrical interface $r = a_2$ differs from field E_2 on the surface of the probe base surrounded by a uniform medium with conductivity σ_2.

Thus, the expression for the apparent conductivity when the invasion zone is absent can be presented in the form:

$$\sigma_a = \sigma_2 + (\sigma_3 - \sigma_2)\,\frac{\displaystyle\int\limits_0^\infty \frac{a_2}{2\lambda}\left[2K_0(\lambda a_2)K_1(\lambda a_2) - \lambda a_2(K_1^2 - K_0^2)\right] I_1^2(\lambda r_1)\cos\lambda z\,d\lambda}{\displaystyle\int\limits_0^\infty \frac{a_1}{2\lambda}\left[2K_0(\lambda a_1)K_1(\lambda a_1) - \lambda a_1(K_1^2 - K_0^2)\right] I_1^2(\lambda r_1)\cos\lambda z\,d\lambda} \tag{4.188}$$

We will introduce in the integrals of the latter, new variables $m = \lambda a_2$ and $m = \lambda a_1$, respectively, and notations

$$\beta = \frac{a_2}{a_1} = \frac{a_2}{r_1} \qquad \alpha_2 = \frac{L}{a_2} \qquad \alpha_1 = \frac{L}{a_1} \qquad v = \frac{a_1}{a_2} = \frac{1}{\beta}$$

L is the probe length. Then we have:

$$\sigma_a = \sigma_2 + (\sigma_3 - \sigma_2)\,\frac{\int\limits_0^\infty \phi(m)\dfrac{I_1^2(vm)}{m^2}\cos m\alpha_2\,dm}{\int\limits_0^\infty \phi(m)\dfrac{I_1^2(vm)}{m^2}\cos m\alpha_1\,dm} \tag{4.189}$$

Let us call the function:

$$\beta\,\frac{\int\limits_0^\infty \phi(m)\dfrac{I_1^2(vm)}{m^2}\cos m\alpha_2\,dm}{\int\limits_0^\infty \phi(m)\dfrac{I_1^2(vm)}{m^2}\cos m\alpha_1\,dm} \tag{4.190}$$

the geometric factor of the formation, $G_2^*(\alpha,\beta)$, where:

$$\phi(m) = (m/2)\left[2K_0(m)K_1(m) - m(K_1^2 - K_0^2)\right]$$

First of all assume that the radius of the ring r_1 along with that of a nonconducting base of the probe tend to zero. Then the following conditions take place:

$$v = \frac{r_1}{a_2} \to 0 \qquad \alpha_1 = \frac{L}{a_1} \to \infty$$

From the first condition it follows that instead of $I_1^2(vm)/m^2$ we can write $v^2/4$ and correspondingly the numerator of eq. 4.190 takes the form:

$$\frac{\beta v^2}{4}\int\limits_0^\infty \phi(m)\cos m\alpha_2\,dm$$

As follows from the second condition, the integral of the denominator in eq. 4.190 is defined by the behavior of function $\phi(m)I_1^2(m)/m^2$ for small values of m. Inasmuch as $\phi(m) \to K_0(m)$ when $m \to 0$ we have:

$$\int\limits_0^\infty \phi(m)\frac{I_1^2(m)}{m^2}\cos m\alpha_1\,dm \to \frac{1}{4}\int\limits_0^\infty K_0(m)\cos m\alpha_1\,dm$$

$$\to \frac{\pi}{8}\frac{1}{(1+\alpha_1^2)^{1/2}} \to \frac{\pi}{8\alpha_1} \qquad \text{if } \alpha \to \infty$$

For the geometric factor of the formation, $G_2^*(\alpha,\beta)$, we obtain:

$$G_2^*(\alpha,\beta) = \frac{2\beta v^2\alpha_1}{\pi}\int\limits_0^\infty \phi(m)\cos m\alpha_2\,dm = \frac{2\alpha_2}{\pi}\int\limits_0^\infty \phi(m)\cos(\alpha_2 m)\,dm$$

which coincides with function $G_2(\alpha)$ corresponding to the case when the coils of the induction probe are infinitely small.

Making use of eq. 4.189 we obtain again a known relation between the geometric factors of a formation and a borehole:

$$G_1^*(\alpha,\beta) = 1 - G_2^*(\alpha,\beta)$$

In accord with eq. 4.190, the calculation of geometric factor $G_2^*(\alpha,\beta)$ is defined by numerical integration. The integrand in the numerator:

$$I = \int\limits_0^\infty \phi(m)\frac{I_1^2(vm)}{m^2}\cos(\alpha_2 m)\,dm$$

has a singularity as $m \to 0$, since:

$$\phi(m)\frac{I_1^2(vm)}{m^2} = \frac{v^2}{4}K_0(m)$$

For this reason let us present the integral in the form:

$$I = \frac{v^2}{4}\int\limits_0^\infty K_0(m)\cos m\alpha_2\,dm + \int\limits_0^\infty \left(\phi(m)\frac{I_1^2(vm)}{m^2} - \frac{v^2}{4}K_0(m)\right)\cos m\alpha_2\,dm$$

$$= \frac{v^2\pi}{8}\frac{1}{(1+\alpha^2)^{1/2}} + \int\limits_0^\infty F(m)\cos m\alpha_2\,dm$$

where

$$F(m) = \phi(m)\frac{I_1^2(vm)}{m^2} - \frac{v^2}{4}K_0(m)$$

Function $F(m)$ does not have a singularity as $m \to 0$.

Similar transformations have been performed with the integral of the denominator in eq. 4.190.

Examples of values of geometrical factor $G_1^*(\alpha,\beta)$ are given in Table 4.4 for various values of parameters α and β.

Analysis of these data shows that a change of function $G_1^*(\alpha,\beta)$, caused by a change of parameter β, is mainly related with a change of diameter of the nonconducting base of the probe if its length exceeds the borehole diameter. For this reason the value of the geometric factor of the borehole $G_1^*(\alpha,\beta)$ can be derived with a sufficient accuracy from relation:

$$G_1^*(\alpha,\beta) = G^*(\alpha) - G_1^*(\beta\alpha)$$

TABLE 4.4
Values of function $G_1^*(\alpha,\beta)$

α \ β	1	$\sqrt[4]{2}$	$\sqrt{2}$	$2/\sqrt[4]{2}$	2	4	8	16
1	0	0.10479	0.19174	0.26179	0.31723	0.43796	0.47424	0.48259
$\sqrt{2}$	0	0.07889	0.14350	0.19516	0.23547	0.31875	0.34800	0.34754
2	0	0.05373	0.09579	0.13037	0.15563	0.20560	0.21364	0.22189
$2\sqrt{2}$	0	0.03316	0.05843	0.07756	0.09159	0.11806	0.12483	0.12640
4	0	0.01843	0.03206	0.04211	0.04919	0.06224	0.06548	0.06630
$4\sqrt{2}$	0	0.00957	0.01654	0.02154	0.02492	0.03133	0.03288	0.03314
8	0	0.00490	0.00820	0.01058	0.01238	0.01544	0.01621	0.01643
$8\sqrt{2}$	0	0.00231	0.00407	0.00530	0.00616	0.00765	0.00789	0.00810

where function $G_1^*(\alpha)$ and $G_1^*(\beta\alpha)$ correspond to infinitely small coils. For shorter probes the relation between the radius of the borehole and the induction probe begins to play a more essential role.

The influence of the radius of the transmitter and the receiver loops, as well as the presence of a nonconducting base, on the geometric factor of the formation $G_2^*(\alpha,\beta)$ is very small. As was demonstrated above, for large values of the probe length ($\alpha \gg 1$), integrals in eq. 4.190 are mainly defined by the behavior of the integrand for small values of m. It is easily seen that the singularity of these functions near zero ($m \to 0$) has the same character as in the case of infinitely small coils. Therefore, with an increase of the probe length, the geometric factor of the borehole, regardless of the ratio r_1/a_1, decreases inversely proportional to α^2.

In accord with eq. 4.187, values of geometric factor $G_1^*(\alpha,\beta)$ allow us to define the ratio E_3/E_2. For the determination of the electrical field in a nonuniform medium E_3 it is necessary to calculate the field, E_2, corresponding to a uniform medium with the conductivity of the borehole. Expression for this field can be written in the form:

$$E_2 = -\frac{i\omega\mu}{4\pi} 4Ir_1\sigma_1\mu\omega \int_0^\infty \phi(m)\frac{I_1(m)}{m}\cos m\alpha_1\, dm \tag{4.191}$$

Usually, the induction probe length is significantly greater than the diameter of the nonconducting base, that is $\alpha \gg 1$, and as was established above:

$$|E_2| \to \frac{\sigma_1\mu^2\omega^2}{4\pi}\frac{I\pi r_1^3}{2L}$$

Correspondingly, for the electromotive force we have:

$$|\mathscr{E}_2| = 2\pi r_1|E_2| = \frac{\sigma_1\mu^2\omega^2}{4\pi}\cdot\frac{M_T M_R}{L} \tag{4.192}$$

where $M_T = \pi r_1^2 I$ and $M_R = \pi r_1^2$ are moments of the probe coils.

Equation 4.192 can be presented in the known form:

$$\frac{\mathscr{E}_2}{\mathscr{E}_0} = \frac{\sigma_1 \mu^2 \omega^2 L^2}{2}$$

where $\mathscr{E}_0$ is the electromotive force in a free space.

Asymptotic presentation of field E_2 and, respectively, the electromotive force $\mathscr{E}_2$ can be used with sufficient accuracy if $L/r_1 \geqslant 4$.

Now let us consider a more complicated case.

4.10.1. Electrical Field of a Single-layer Coil in a Medium with Cylindrical Interfaces

Suppose that n turns correspond to a unit of length of a transmitter coil and consider a uniform medium first. Then in accord with eqs. 4.179 and 4.180, the vector potential of the coil with height $\mathrm{d}z_0$ can be presented as:

$$\mathrm{d}A_\phi = \begin{cases} 4In_T r_1 \,\mathrm{d}z_0 \displaystyle\int_0^\infty I_1(r_1 v) K_1(rv) \cos\lambda(z - z_0)\,\mathrm{d}\lambda & \text{if } r \geqslant r_1 \\ 4In_T r_1 \,\mathrm{d}z_0 \displaystyle\int_0^\infty I_1(rv) K_1(r_1 v) \cos\lambda(z - z_0)\,\mathrm{d}\lambda & \text{if } r \leqslant r_1 \end{cases}$$

Here I is the current; r_1 is the coil radius; z is the distance from middle of the coil to the measuring ring; z_0 is the distance from the coil middle to the center of the ring with thickness $\mathrm{d}z_0$; $v = (\lambda^2 + \mathrm{i}\chi)^{1/2}$, $\chi = \sigma\mu\omega$.

The vector potential generated by the current in the whole coil is obtained by integration with respect to z_0:

$$A_\phi = \begin{cases} \dfrac{4In_T r_1}{4\pi} \displaystyle\int_0^\infty I_1(r_1 v) K_1(rv)\,\mathrm{d}\lambda \int_{-l/2}^{l/2} \cos\lambda(z - z_0)\,\mathrm{d}z_0 & \text{if } r \geqslant r_1 \\ \dfrac{4In_T r_1}{4\pi} \displaystyle\int_0^\infty I_1(rv) K_1(r_1 v)\,\mathrm{d}\lambda \int_{-l/2}^{l/2} \cos\lambda(z - z_0)\,\mathrm{d}z_0 & \text{if } r \leqslant r_1 \end{cases} \tag{4.193}$$

where l is the coil length.

Integral by z_0 is equal to $2\cos\lambda z \sin(\lambda l/2)$, and therefore we have:

$$A_\phi = \begin{cases} \dfrac{8In_T r_1}{4\pi} \displaystyle\int_0^\infty I_1(r_1 v) \frac{\sin(\lambda l/2)}{\lambda} K_1(rv) \cos\lambda z\,\mathrm{d}\lambda & \text{if } r \geqslant r_1 \\ \dfrac{8In_T r_1}{4\pi} \displaystyle\int_0^\infty K_1(r_1 v) \frac{\sin(\lambda l/2)}{\lambda} I_1(rv) \cos\lambda z\,\mathrm{d}\lambda & \text{if } r \leqslant r_1 \end{cases} \tag{4.194}$$

With a decrease of the coil length, l, ratio $(l/2\lambda)/\lambda$ can be in limit replaced by coefficient $l/2$ and we obtain the expression for the vector potential due to the current ring derived above ($ln_T = 1$). As is known, the electrical field is related with the vector potential by the relation: $E_\phi = -\mathrm{i}\omega\mu A_\phi$, and therefore we have:

$$E_\phi = \begin{cases} -\dfrac{\mathrm{i}\omega\mu 8 I n_T r_1}{4\pi} \displaystyle\int_0^\infty I_1(r_1 v)\frac{\sin(\lambda l/2)}{\lambda} K_1(rv)\cos\lambda z\, \mathrm{d}\lambda & \text{if } r \geqslant r_1 \\ -\dfrac{\mathrm{i}\omega\mu 8 I n_T r_1}{4\pi} \displaystyle\int_0^\infty K_1(r_1 v)\frac{\sin(\lambda l/2)}{\lambda} I_1(rv)\cos\lambda z\, \mathrm{d}\lambda & \text{if } r \leqslant r_1 \end{cases} \tag{4.195}$$

Now we consider a medium with cylindrical interfaces. In accord with eq. 4.195 and making use of results derived in previous sections the electric field can be written in the form:

$$E_\phi = -\frac{\mathrm{i}\omega\mu 8 I n_T r_1}{4\pi} \times \left(\int_0^\infty K_1(r_1 v)\frac{\sin(\lambda l/2)}{\lambda} I_1(rv)\cos\lambda z\, \mathrm{d}\lambda - \int_0^\infty C_1(\lambda) I_1(r_1 v)\frac{\sin(\lambda l/2)}{\lambda} I_1(rv)\cos\lambda z\, \mathrm{d}\lambda \right) \tag{4.196}$$

if $a_1 \geqslant r \leqslant r_1$.

Expression of function $C_1(\lambda)$ for the range of small parameters was derived above. We will suppose that the single-layered transmitter coil and the measuring ring are placed on a nonconducting base, and they have the same radius. Then for the component of the electrical field E_1 when the surrounding medium has resistivity ρ_1 and field E_2, as there is also an interface between media with resistivities ρ_1 and ρ_2, we obtain, respectively:

$$E_1 = -\frac{\mathrm{i}\omega\mu 8 I n_T r_1}{4\pi}\chi_1 \int_0^\infty \frac{r_1}{2\lambda}\left[2K_0K_1 - \lambda r_1(K_1^2 - K_0^2)\right]\frac{\sin(\lambda l/2)}{\lambda} I_1^2(\lambda r_1)\cos\lambda z\, \mathrm{d}\lambda \tag{4.197}$$

and

$$E_2 = E_1 - \frac{\mathrm{i}\omega\mu 8 I n_T r_1}{4\pi}(\chi_2 - \chi_1) \times \int_0^\infty \frac{a_1}{2\lambda}\left[2K_0K_1 - \lambda a(K_1^2 - K_0^2)\right]\frac{\sin(\lambda l/2)}{\lambda} I_1^2(\lambda r_1)\cos\lambda z\, \mathrm{d}\lambda \tag{4.198}$$

Both equations take into account that the central part of the borehole is occupied by a nonconducting base of the probe. At the same time in all formulae, the finite length of this base is not considered. From a physical point of view we can assume that the influence of a magnetic field caused by induced currents within the central part of the borehole, being a continuation of the probe base, should be very small. A further evaluation of the contribution of currents in these parts of the borehole will be done for infinitely small coils.

Making use of the relation:

$$\frac{\sigma_a}{\sigma_1} = \frac{E_2}{E_1}$$

we obtain the following expression for the apparent conductivity as the transmitter is a single-layered coil and the receiver is the ring:

$$\sigma_a = \sigma_1 + (\sigma_2 - \sigma_1)\beta^2 \frac{\int\limits_0^\infty \phi(m)\frac{\sin(ms_2)}{m}\frac{I_1^2(vm)}{m^2}\cos m\alpha_2\, dm}{\int\limits_0^\infty \phi(m)\frac{\sin(ms_1)}{m}\frac{I_1^2(vm)}{m^2}\cos m\alpha_1\, dm} \tag{4.199}$$

where:
σ_1 and σ_2 are conductivities of the borehole and the formation, respectively;
$\beta = a_1/r_1$, a_1 is the borehole radius;
r_1 is the radius of the coil and the nonconducting base;
$v = r_1/a_1$ is the coil radius expressed in units of the borehole radius;
$s_2 = l/2a_1$ and $s_1 = l/2r_1$ are ratios of the coil length to the borehole and coil diameters, respectively;
$\alpha_2 = L/a_1$, $\alpha_1 = L/r_1$; L is the probe length.

If the coil length tends to zero ($s_2 \to 0$, $s_1 \to 0$) then instead of eq. 4.199 we will obtain an expression for the apparent conductivity when both the transmitter and the receiver are rings. In fact, replacing $\sin(ms)/m$ by s we have:

$$\sigma_a = \sigma_1 + (\sigma_2 - \sigma_1)\beta^2 \frac{s_2}{s_1} \frac{\int\limits_0^\infty \phi(m)\frac{I_1^2(vm)}{m^2}\cos m\alpha_2\, dm}{\int\limits_0^\infty \phi(m)\frac{I_1^2(m)}{m^2}\cos m\alpha_1\, dm}$$

$$= \sigma_1 + (\sigma_2 - \sigma_1)\beta \frac{\int\limits_0^\infty \phi(m)\frac{I_1^2(vm)}{m^2}\cos m\alpha_2\, dm}{\int\limits_0^\infty \phi(m)\frac{I_1^2(m)}{m^2}\cos m\alpha_1\, dm}$$

Let us call the function:

$$\beta^2 \frac{\int\limits_0^\infty \phi(m) \frac{\sin s_2 m}{m} \frac{I_1^2(vm)}{m^2} \cos m\alpha_2 \, dm}{\int\limits_0^\infty \phi(m) \frac{\sin s_1 m}{m} \frac{I_1^2(m)}{m^2} \cos m\alpha_1 \, dm}$$

the geometric factor of the formation, G_2^*, when the transmitter is an one-layered coil and the receiver is the ring or vice versa. As follows from eq. 4.199, the geometric factors of the formation and the borehole are related as:

$$G_1^* = 1 - G_2^*$$

It is obvious that the geometric factor of the invasion zone is related with function G_1^* in the same manner as it takes place for very small coils.

The geometric factor of the borehole depends on three parameters α_2, β, and s_2, since $\alpha_1 = \beta\alpha_2$ and $s_1 = \beta s_2$. Results of calculation of function G_1^* illustrating its behavior are presented in Table 4.5. As is seen from this table and follows from physical consideration with an increase of the probe diameter the influence of the borehole decreases, and this effect in some cases can be significant.

A change of the geometric factor G_1^* for probes, the length of which exceeds the borehole diameter, is mainly related to a change of diameter of the nonconducting base. For this reason a value of the geometric factor of the borehole, $G_1^*(\alpha, \beta, s_2)$, with sufficient accuracy can be obtained from equation:

$$G_1^*(\alpha, \beta, s_2) = G_1^*(\alpha, s_2) - G_1^*(\alpha\beta, s_2)$$

where on the right-hand side functions $G_1^*(\alpha, s_2)$ and $G_1^*(\alpha\beta, s_2)$ correspond to geometric factors of the borehole when coils are very thin $(1/\beta \to 0)$.

4.10.2. Both Transmitter and Receiver of the Induction Probe are Single-layer Coils

If the amount of turns of the measuring coil with radius r_m per unit length is equal to n_m then for electromotive force in the coil with length dz (as follows from eq. 4.195) we have:

$$\mathrm{d}\mathscr{E} = -\frac{\mathrm{i}\omega\mu}{4\pi} M_{T_m} \, \mathrm{d}z \int\limits_0^\infty \frac{I_1(r_T v)\sin(\lambda l/2)}{\lambda} K_1(r_m v) \cos\lambda z \, \mathrm{d}\lambda \quad (r_m > r_T)$$

$$\mathrm{d}\mathscr{E} = -\frac{\mathrm{i}\omega\mu}{4\pi} M_{T_m} \, \mathrm{d}z \int\limits_0^\infty \frac{K_1(r_T v)\sin(\lambda l/2)}{\lambda} I_1(r_m v) \cos\lambda z \, \mathrm{d}\lambda \quad (r_m < r_T)$$

$$M_{T_m} = 16\pi I n_T n_m r_m r_T$$

TABLE 4.5
Values of function G_1^*

$s_2 = 0.2$

$\alpha \backslash \beta$	1	$\sqrt[4]{2}$	$2/\sqrt[4]{2}$	2	4	8	16
1	0	0.10565	0.23377	0.31990	0.44305	0.48076	0.49063
$\sqrt{2}$	0	0.07939	0.19679	0.23753	0.32224	0.34593	0.35180
2	0	0.05415	0.13124	0.15688	0.20743	0.22060	0.22405
$2\sqrt{2}$	0	0.03325	0.08802	0.09218	0.11884	0.12560	0.12703
4	0	0.01853	0.04222	0.04937	0.06249	0.06575	0.06648
$4\sqrt{2}$	0	0.00965	0.02150	0.02501	0.03136	0.03290	0.03324
8	0	0.00482	0.01067	0.01239	0.01547	0.01620	0.01360
$8\sqrt{2}$	0	0.00239	0.00527	0.00611	0.00761	0.00795	0.00804
16	0	0.00118	0.00261	0.00301	0.00373	0.00390	0.00393

$s_2 = 0.4$

$\alpha \backslash \beta$	1	$\sqrt[4]{2}$	$2/\sqrt[4]{2}$	2	4	8	16
1	0	0.10783	0.26969	0.32674	0.45882	0.50000	0.51242
$\sqrt{2}$	0	0.08101	0.20147	0.24362	0.33270	0.35820	0.36462
2	0	0.05512	0.13404	0.16047	0.21298	0.22688	0.23035
$2\sqrt{2}$	0	0.03370	0.07928	0.09376	0.12108	0.12803	0.12975
4	0	0.01872	0.04266	0.04991	0.06322	0.06661	0.06762
$4\sqrt{2}$	0	0.00969	0.02164	0.02517	0.03156	0.03303	0.03327
8	0	0.00487	0.01071	0.01243	0.01552	0.01621	0.01630
$8\sqrt{2}$	0	0.00240	0.00528	0.00612	0.00761	0.00793	0.00802
16	0	0.00119	0.00261	0.00301	0.00373	0.00389	0.00393

The total electromotive force in the measuring coil is defined from equations:

$$\begin{aligned}
\mathscr{E} &= -\frac{i\omega\mu}{4\pi} M_{T_m} \int_0^\infty \frac{I_1(r_T v)\sin(\lambda l/2)}{\lambda} K_1(r_m v) \int_{z_0-b/2}^{z_0+b/2} \cos\lambda z \, d\lambda \, dz \\
&= -\frac{i\omega\mu}{4\pi} 2M_{T_m} \int_0^\infty \frac{I_1(r_T v)\sin(\lambda l/2)}{\lambda} K_1(r_m v)\sin(\lambda b/2)\cos\lambda z_0 \, d\lambda \qquad \text{if } r_m \geqslant r_T
\end{aligned} \tag{4.200}$$

$$\mathscr{E} = -\frac{i\omega\mu}{4\pi} 2M_{T_m} \int_0^\infty \frac{K_1(r_T v)\sin(\lambda l/2)}{\lambda} I_1(r_m v)\sin(\lambda b/2)\cos\lambda z_0 \, d\lambda \qquad \text{if } r_m \leqslant r_T$$

where:
b is the length of the measuring coil;
z_0 is the distance between the centers of the transmitter and receiver coils.

TABLE 4.5
(Continued)

$s_2 = 0.8$

α \ β	1	$\sqrt[4]{2}$	$\sqrt{2}$	$2/\sqrt[4]{2}$	2
1	0	0.11650	0.21265	0.29147	0.35608
$\sqrt{2}$	0	0.08767	0.16073	0.22029	0.26815
2	0	0.05918	0.10741	0.14570	0.17544
$2\sqrt{2}$	0	0.03564	0.06345	0.08459	0.10036
4	0	0.01944	0.03391	0.04449	0.05214
$4\sqrt{2}$	0	0.01992	0.01707	0.02218	0.02580
8	0	0.00491	0.00839	0.01086	0.01261

$s_2 = 1.6$

α \ β	1	$\sqrt[4]{2}$	$\sqrt{2}$	$2/\sqrt[4]{2}$	2
1	0	0.12476	0.22674	0.31144	0.38257
$\sqrt{2}$	0	0.11113	0.20307	0.28044	0.34637
2	0	0.07742	0.14293	0.19758	0.24281
$2\sqrt{2}$	0	0.04457	0.08081	0.10958	0.13193
4	0	0.02272	0.04003	0.05295	0.06242
$4\sqrt{2}$	0	0.01090	0.01880	0.02448	0.02853
8	0	0.00516	0.00883	0.01143	0.01327

If linear dimensions of probe coils coincide ($l = b, r_m = r_T = r_1$) then:

$$\mathscr{E} = -\frac{32 i\omega\mu}{4\pi} I n_T n_m r_1^2 \int_0^\infty \frac{I_1(r_1 v)\sin^2(\lambda l/2)}{\lambda^2} K_1(r_1 v)\cos\lambda z_0 \, d\lambda \tag{4.201}$$

Equations 4.200 and 4.201 are derived provided that the coils are located in a uniform medium. Now assume that single-layered coils are placed on a nonconducting base of the probe which is located on the borehole axis. Then expressions for the apparent conductivity and geometric factors are derived in a similar manner as they were obtained in the previous case. For example, when the invasion zone is absent we have:

$$\sigma_a = \sigma_1 + (\sigma_2 - \sigma_1)\beta^3 \frac{\int_0^\infty \phi(m) \frac{\sin^2(ms_2)}{m^2} \frac{I_1^2(vm)}{m^2} \cos m\alpha_2 \, dm}{\int_0^\infty \phi(m) \frac{\sin^2(ms_1)}{m^2} \frac{I_1^2(vm)}{m^2} \cos m\alpha_1 \, dm} \tag{4.202}$$

where:

$\beta = a_1/r_1 \qquad v = 1/\beta$

TABLE 4.5
(Continued)

$s_2 = 2.0$

α \ β	1	$\sqrt[4]{2}$	$\sqrt{2}$	$2/\sqrt[4]{2}$	2
1	0	0.11919	0.21739	0.29917	0.36793
$\sqrt{2}$	0	0.11361	0.20782	0.28715	0.35473
2	0	0.09190	0.16959	0.23582	0.29273
$2\sqrt{2}$	0	0.05257	0.09653	0.13249	0.16136
4	0	0.02562	0.04551	0.06061	0.07185
$4\sqrt{2}$	0	0.01172	0.02028	0.02647	0.03090
8	0	0.00537	0.00920	0.01192	0.01385

$s_2 = 4.0$

α \ β	1	$\sqrt[4]{2}$	$\sqrt{2}$	$2/\sqrt[4]{2}$	2
1	0	0.09239	0.17057	0.23725	0.29457
$\sqrt{2}$	0	0.09282	0.17144	0.23856	0.29628
2	0	0.09338	0.17272	0.24060	0.29910
$2\sqrt{2}$	0	0.09252	0.17175	0.24002	0.29977
4	0	0.06963	0.13046	0.18398	0.23139
$4\sqrt{2}$	0	0.02294	0.04121	0.05540	0.06620
8	0	0.00764	0.01318	0.01716	0.02000

$$s_2 = l/2a_1 \qquad s_1 = l/2r$$
$$\alpha_2 = L/a_1 \qquad \alpha_1 = L/r_1$$

$L = z_0$ is the probe length.

Thus the geometric factor of the borehole can be presented in this case as:

$$G_1^* = 1 - \beta^3 \frac{\int\limits_0^\infty \phi(m) \frac{\sin^2(ms_2)}{m^2} \frac{I_1^2(vm)}{m^2} \cos m\alpha_2 \, dm}{\int\limits_0^\infty \phi(m) \frac{\sin^2(ms_1)}{m^2} \frac{I_1^2(vm)}{m^2} \cos m\alpha_1 \, dm} \tag{4.203}$$

and it depends on three parameters: α_2, β and s_2. Some values of the function G_1^* are given in Table 4.6.

Taking into account the behavior of the integrands in eq. 4.203 it is easy to demonstrate that with an increase of the probe length, function G_1^* decreases inversely proportional to α_2^2, regardless of the coil dimensions.

It is natural that with an increase of the diameter of the nonconducting base of the probe, the geometric factor G_1 decreases and that it is more noticeable for shorter induction probes.

TABLE 4.6
Values of function G_1^*

$s_2 = 0.2$

α \ β	1	$\sqrt[4]{2}$	$\sqrt{2}$	$2/\sqrt[4]{2}$	2	4	8	16
1	0	0.10636	0.19448	0.26574	0.36225	0.44817	0.48753	0.49797
$\sqrt{2}$	0	0.07993	0.14569	0.19835	0.23957	0.32574	0.35000	0.35673
2	0	0.05447	0.09812	0.01322	0.15808	0.20928	0.22274	0.22612
$2\sqrt{2}$	0	0.03339	0.05911	0.07844	0.09271	0.11952	0.12643	0.12817
4	0	0.01856	0.03235	0.04235	0.04955	0.05273	0.06603	0.06670
$4\sqrt{2}$	0	0.00966	0.01660	0.02156	0.02507	0.03141	0.03296	0.03325
8	0	0.00484	0.00827	0.01069	0.01240	0.01548	0.01840	0.01635
$8\sqrt{2}$	0	0.00239	0.00409	0.00528	0.00612	0.00751	0.00794	0.00803
16	0	0.00118	0.00202	0.00260	0.00301	0.00373	0.00389	0.00395

$s_2 = 0.4$

α \ β	1	$\sqrt[4]{2}$	$\sqrt{2}$	$2/\sqrt[4]{2}$	2	4	8	16
1	0	0.11078	0.20250	0.27718	0.33748	0.47926	0.53231	0.54937
$\sqrt{2}$	0	0.08324	0.15216	0.20782	0.25194	0.34766	0.37657	0.38413
2	0	0.05648	0.10208	0.13793	0.16548	0.22090	0.23580	0.23956
$2\sqrt{2}$	0	0.03435	0.06095	0.08105	0.09600	0.12424	0.13148	0.13331
4	0	0.01896	0.03301	0.04326	0.05060	0.06420	0.06760	0.06843
$4\sqrt{2}$	0	0.00977	0.01681	0.02182	0.02540	0.03185	0.33500	0.03399
8	0	0.00487	0.00832	0.10760	0.01250	0.01561	0.01633	0.01637
$8\sqrt{2}$	0	0.00240	0.00410	0.00529	0.00613	0.00762	0.00789	0.00789
16	0	0.00119	0.00202	0.00260	0.00302	0.00372	0.00386	0.00389

With an increase of the probe length the correction due to the presence of a nonconducting base tends to a constant which does not depend on the probe length, and it is equal to the square of the probe diameter/borehole diameter ratio:

$$\frac{G_1(\alpha, s, \beta)}{G_1(\alpha, s, \infty)} \simeq 1 - \frac{1}{\beta^2}$$

For example, if the probe diameter is equal to the borehole radius this correction factor exceeds 25%.

As is seen from data in Table 4.6, the geometric factor of the borehole G_1^* also increases with an increase of the coil length. If its length does not exceed the borehole diameter its influence is not significant, and it is measured in several percentages. Only for very short probes, the length of which is comparable with the borehole diameter ($L \simeq 2a_1 \div 4a_1$), coil size can have an essential influence.

Therefore, the main factors defining the value of the geometric factor of the borehole for conventional induction probes are length and diameter of the nonconducting base.

TABLE 4.6
(Continued)

$s_2 = 0.8$

α \ β	1	$\sqrt[4]{2}$	$\sqrt{2}$	$2/\sqrt[4]{2}$	2
1	0	0.12094	0.22005	0.30183	0.36982
$\sqrt{2}$	0	0.09644	0.17684	0.24355	0.29890
2	0	0.06530	0.11947	0.16349	0.19870
$2\sqrt{2}$	0	0.38600	0.06921	0.09290	0.11086
4	0	0.02053	0.03593	0.04728	0.05553
$4\sqrt{2}$	0	0.01024	0.01765	0.02294	0.02971
8	0	0.00499	0.00854	0.11047	0.01282
$8\sqrt{2}$	0	0.00244	0.00416	0.00538	0.00624
16	0	0.00120	0.00204	0.00264	0.00306

$s_2 = 4.0$

α \ β	1	$\sqrt[4]{2}$	$\sqrt{2}$	$2/\sqrt[4]{2}$	2
1	0	0.08821	0.16355	0.22839	0.28458
$\sqrt{2}$	0	0.08626	0.16006	0.22366	0.27886
2	0	0.08318	0.15457	0.21629	0.27002
$2\sqrt{2}$	0	0.07839	0.14608	0.20494	0.25619
4	0	0.07056	0.13216	0.18632	0.23425
$4\sqrt{2}$	0	0.05528	0.10449	0.14862	0.18846
8	0	0.02055	0.03815	0.05311	0.06578

In this section it is also appropriate to consider an induction probe which consists of only one coil. Such a probe can have a certain interest in measuring resistivity on the invasion zone.

First let us assume that the induction probe is located in a uniform medium and consists of one ring. Then, according to eq. 4.181 or eq. 4.182, the electromotive force induced in this ring is defined in the following way:

$$\mathscr{E} = 2\pi r_1 E_\phi = -\frac{i\omega\mu}{4\pi} 8\pi I r_1^2 \int\limits_0^\infty I_1(r_1 v) K_1(r_1 v)\, d\lambda \tag{4.204}$$

The integral in this equation diverges. In fact, for $\lambda \to \infty$ parameter $v = (\lambda^2 + i\chi)^{1/2}$ also tends to infinity as λ. For this reason, making use of asymptotic presentation of Bessel functions:

$$I_1(x) \to \frac{e^x}{\sqrt{2\pi x}} \qquad K_1(x) \to \frac{e^{-x}\sqrt{\pi}}{\sqrt{2x}}$$

we have:

$$I_1(r_1 v) K_1(r_1 v) \xrightarrow{\lambda \to \infty} I_1(r_1 \lambda) K_1(r_1 \lambda) \to \frac{1}{2\lambda r_1}$$

When the probe length, L, is not equal to zero (two-coil probe) the integrand contains the oscillating multiplier $\cos \lambda L$ which provides convergence of integrals in eqs. 4.181 and 4.182.

The electromotive force induced in one ring probe is caused by a change with time of the magnetic field of the current in the ring located in a free space as well as by the magnetic field of currents induced in a conducting medium. Thus:

$$\mathscr{E} = \mathscr{E}_0 + \mathscr{E}_1$$

where:

$$\mathscr{E}_0 = -\frac{i\omega\mu}{4\pi} 8\pi I r_1^2 \int_0^\infty I_1(r_1\lambda) K_1(r_1\lambda)\, d\lambda \tag{4.205}$$

is the electromotive force in a free space, and

$$\mathscr{E}_1 = -\frac{i\omega\mu}{4\pi} 8\pi I r^2 \int_0^\infty [I_1(r_1 v) K_1(r_1 v) - I_1(r_1\lambda) K_1(r_1\lambda)]\, d\lambda \tag{4.206}$$

The right-hand side of eq. 4.205 is equal to infinity. It results from the fact that the ring is assumed to be infinitely thin and correspondingly the current density and the magnetic field near its surface are infinitely large. Therefore:

$$\mathscr{E}_0 = -\frac{\partial \phi_0}{\partial t} = -\frac{\partial}{\partial t} \int_S \boldsymbol{B}_0 \cdot d\boldsymbol{S}$$

also tends to infinity.

The inductance of the ring is related with electromotive force by relation:

$$\mathscr{E} = -L \frac{\partial I}{\partial t}$$

whence

$$L_0 = 2r_1^2 \mu \int_0^\infty I_1(\lambda r_1) K_1(\lambda r_1)\, d\lambda \to \infty$$

As is well known, the inductance, L_0, is defined from equation:

$$L_0 = r_1 \mu \left(\ln \frac{8r_1}{r_0} - 1.76 \right) \tag{4.207}$$

where r_1 is the ring radius and r_0 is the radius of its cross-section.

In accord with eq. 4.207, the self-inductance L_0 tends to infinity as $r_0 \to 0$. Of course, under real conditions the cross-section radius, r_0, is not zero and correspondingly the self-inductance, L_0, has a finite value.

As is seen from eq. 4.206 the integrand does not have singularities and rapidly decreases when λ increases. In fact, for $\lambda \to \infty$ we have:

$$I_1(r_1 v)K_1(r_1 v) - I_1(r_1\lambda)K_1(r_1 \xrightarrow{\lambda\to\infty} \frac{1}{2(\lambda^2 + \mathrm{i}\chi)^{1/2} r_1} - \frac{1}{2\lambda r_1} \simeq -\frac{\mathrm{i}\chi}{4\lambda^3 r_1}$$

In a general case the resistance, introduced by currents in a medium into the ring, is a complex one, that is, it has inphase and quadrature components. In accord with eq. 4.206 we have:

$$Z = \frac{\mathscr{E}_1}{I} = \mathrm{i}\omega\mu 2 r_1^2 \int\limits_0^\infty [I_1(r_1 v)K_1(r_1 v) - I_1(r_1\lambda)K_1(r_1\lambda)]\, \mathrm{d}\lambda \tag{4.208}$$

As in the case of the dipole excitation the quadrature component of the current density prevails near the ring, that is in its vicinity induced currents are mainly shifted in phase by 90°. For this reason, the electromotive force induced by these currents in the ring is in phase with the source current. Therefore, parts of the medium which are close to the ring mainly introduce the active resistance. At the same time with an increase of the distance from the source the inphase component of currents increases, and correspondingly, these parts of the medium begin to contribute the reactive resistance.

If the frequency of the field and the conductivity of the medium are sufficiently low the active resistance introduced by induced currents prevails and its expression can be obtained in the following manner.

Let us present the integrand in eq. 4.208:

$$I_1(r_1 v)K_1(r_1 v) - I_1(r_1\lambda)K_1(r_1\lambda)$$

in a series by power of parameter $\mathrm{i}\chi$ and discard all terms except the first one. It is obvious that:

$$\begin{aligned} I_1(r_1\sqrt{\lambda^2+\mathrm{i}\chi})K_1(r_1\sqrt{\lambda^2+\mathrm{i}\chi}) &\simeq I_1(r_1\lambda)K_1(r_1\lambda) \\ &+ \frac{\mathrm{i}\chi r_1}{2\lambda}\left[I_1'(r_1\lambda)K_1(r_1\lambda) + I_1(r_1\lambda)K_1'(r_1\lambda)\right] \end{aligned}$$

Therefore we have:

$$R = -\omega\mu 2 r_1^2 \chi \int\limits_0^\infty \frac{r_1}{2\lambda}\left[I_1'(r_1\lambda)K_1(r_1\lambda) + I_1(r_1\lambda)K_1'(r_1\lambda)\right]\, \mathrm{d}\lambda$$

We will introduce a new variable: $m = r_1\lambda$, then:

$$R = \omega^2\mu^2\sigma 2 r_1^3 \int\limits_0^\infty \frac{1}{2m}\left[I_1'(r_1\lambda)K_1(r_1\lambda) + I_1(r_1\lambda)K_1'(r_1\lambda)\right]\, \mathrm{d}\lambda \tag{4.209}$$

Making use of the recurrence formulae for modified Bessel's functions:

$$I_1'(m) = I_0(m) - \frac{1}{m} I_1(m) \qquad K_1'(m) = K_0(m) - \frac{1}{m} K_1(m)$$

we obtain a final expression for the active resistance introduced by induced currents at the range of small parameters:

$$R = \omega^2 \mu^2 \sigma 2 r_1^3 \int_0^\infty \frac{1}{2m} \left(I_0(m) K_1(m) - I_1(m) K_0(m) - \frac{2}{m} I_1(m) K_1(m) \right) \mathrm{d}m \tag{4.210}$$

The integrand of the latter has singularity, as $m \to 0$. In fact we have for $m \to 0$:

$$I_0(m) \to 1 \qquad I_1(m) \to \frac{m}{2} \qquad K_1(m) \to \frac{1}{m} \qquad K_0(m) \to -\ln m$$

and

$$\left(I_0(m) K_1(m) - I_1(m) K_0(m) - \frac{2}{m} I_1(m) K_1(m) \right) \frac{1}{2m} \to \left(\frac{1}{m} - \frac{m}{2} K_0(m) - \frac{1}{m} \right) \frac{1}{2m}$$
$$\to -\frac{1}{4} K_0(m) \to \infty$$

For this reason it is appropriate to present eq. 4.210 in the form:

$$R = -\omega^2 \mu^2 \sigma 2 r_1^3 \left[\int_0^\infty \left\{ \frac{1}{4} K_0(m) - \frac{1}{2m} \left(I_0 K_1 - I_1 K_0 - \frac{2}{m} I_1 K_1 \right) \right\} \mathrm{d}m \right.$$
$$\left. - \frac{1}{4} \int_0^\infty K_0(m)\, \mathrm{d}m \right]$$
$$= -\omega^2 \mu^2 \sigma 2 r_1 \left[\int_0^\infty \left\{ \frac{1}{4} K_0(m) - \frac{1}{2m} \left(I_0 K_1 - I_1 K_0 - \frac{2}{m} I_1 K_1 \right) \right\} \mathrm{d}m - \frac{\pi}{8} \right]$$

since

$$\int_0^\infty K_0(m)\, \mathrm{d}m = \frac{\pi}{2}$$

Now we will suppose that the ring is placed on the nonconducting base of the induction probe located in a medium with conductivity σ_1. In accord with eq. 4.183 the electromotive force introduced into this ring is:

$$\mathscr{E} = -\frac{\mathrm{i}\omega\mu}{4\pi} 8\pi I r_1^2 \left(\int_0^\infty I_1(r_1) K_1(r_1 \lambda)\, \mathrm{d}\lambda - \int_0^\infty C_1(\lambda) I_1^2(r_\lambda \lambda)\, \mathrm{d}\lambda \right) \tag{4.211}$$

The second term defines the electromotive force caused by currents induced in a conducting medium, and correspondingly, at the range of small parameters, expressions for the electromotive force and the active resistance can be written as:

$$
\begin{aligned}
\mathscr{E}_1 &= -2Ir_1^3 a_1 \mu\omega\chi_1 \int_0^\infty \frac{m}{2}\left[2K_0K_1 - m(K_1^2 - K_0^2)\right]\frac{I_1^2(m)}{m^2}\,dm \\
R_1 &= -2r_1^3 a_1 \mu\omega\chi_1 \int_0^\infty \frac{m}{2}\left[2K_0K_1 - m(K_1^2 - K_0^2)\right]\frac{I_1^2(m)}{m^2}\,dm
\end{aligned}
\tag{4.212}
$$

Suppose there is a second cylindrical interface between two media with conductivities σ_1 and σ_2. Then, in accord with eq. 4.186, we have:

$$
\mathscr{E} = \mathscr{E}_1 - 2Ir_1^2 a_1 \mu\omega(\chi_3 - \chi_1)\int_0^\infty \frac{m}{2}\left[2K_0K_1 - m(K_1^2 - K_0^2)\right]\frac{I_1^2(mv)}{m^2}\,dm
$$

$$
R = R_1 - 2r_1^2 a_1 \mu\omega(\chi_3 - \chi_1)\int_0^\infty \frac{m}{2}\left[2K_0K_1 - m(K_1^2 - K_0^2)\right]\frac{I_1^2(mv)}{m^2}\,dm
$$

For an evaluation of the influence of a formation it is reasonable to introduce the apparent conductivity as:

$$
\frac{\sigma_a}{\sigma_1} = \frac{\mathscr{E}}{\mathscr{E}_1} \quad \text{or} \quad \frac{\sigma_a}{\sigma_1} = \frac{R}{R_1}
$$

It is obvious that:

$$
\frac{\sigma_a}{\sigma_1} = 1 + \left(\frac{\sigma_2}{\sigma_1} - 1\right)\beta^3 \frac{\int_0^\infty \frac{m}{2}\left[2K_0K_1 - m(K_1^2 - K_0^2)\right]\frac{I_1^2(mv)}{m^2}\,dm}{\int_0^\infty \frac{m}{2}\left[2K_0K_1 - m(K_1^2 - K_0^2)\right]\frac{I_1^2(m)}{m^2}\,dm} = 1 + \left(\frac{\sigma_2}{\sigma_1}\right)G_2
$$

where G_2 is the geometrical factor of the formation given by eq. 4.190, as $\alpha = 0$.

Finally, we will consider the last case where a single-layered coil presenting the induction probe is placed on the nonconducting base. Making use of results obtained above we have the following expressions for the electromotive force and active resistance introduced into a coil by induced currents in a medium with conductivity σ_1:

$$
\begin{aligned}
\mathscr{E}_1 &= -8Ir_1^5 n^2 \mu\omega\chi_1 \int_0^\infty \frac{m}{2}\left[2K_0K_2 - m(K_1^2 - K_0^2)\right]\frac{I_1^2(m)}{m^2}\frac{\sin^2 ms_1}{m^2}\,dm \\
R_1 &= -8r_1^5 n^2 \mu\omega\chi_1 \int_0^\infty \frac{m}{2}\left[2K_0K_2 - m(K_1^2 - K_0^2)\right]\frac{I_1^2(m)}{m^2}\frac{\sin^2 ms_1}{m^2}\,dm
\end{aligned}
\tag{4.213}
$$

If the probe is located on the borehole axis we have:

$$\mathscr{E} = \mathscr{E}_1 - 8Ir_1^2 n^2 a_1^3 \mu\omega(\chi_2 - \chi_1) \int_0^\infty \frac{m}{2}\left[2K_0K_1 - m(K_1^2 - K_0^2)\right] \frac{I_1^2(vm)}{m^2} \frac{\sin^2 ms_2}{m^2}\,\mathrm{d}m$$

$$R = R_1 - 8r_1^2 n^2 a_1^3 \mu\omega(\chi_2 - \chi_1) \int_0^\infty \frac{m}{2}\left[2K_0K_2 - m(K_1^2 - K_0^2)\right] \frac{I_1^2(vm)}{m^2} \frac{\sin^2 ms_2}{m^2}\,\mathrm{d}m \tag{4.214}$$

and finally:

$$\frac{\sigma_a}{\sigma_1} = 1 + \left(\frac{\sigma_2}{\sigma_1} - 1\right)\beta^3 \frac{\int_0^\infty \frac{m}{2}\left[2K_0K_1 - m(K_1^2 - K_0^2)\right] \frac{I_1^2(vm)}{m^2} \frac{\sin^2 ms_2}{m^2}\,\mathrm{d}m}{\int_0^\infty \frac{m}{2}\left[2K_0K_1 - m(K_1^2 - K_0^2)\right] \frac{I_1^2(m)}{m^2} \frac{\sin^2 ms_1}{m^2}\,\mathrm{d}m} \tag{4.215}$$

$$= 1 + \left(\frac{\sigma_2}{\sigma_1} - 1\right) G_2$$

where G_2 is the geometrical factor of the probe having a single-layered coil; r_1 is the coil radius; a_1 is the borehole radius; l is the coil length; $\beta = a_1/r_1$, $v = 1/\beta$, $s_2 = l/2a_1$, $s_1 = l/2r_1$.

For illustration, let us consider an induction probe which consists of two single-layered coils of different length and the same radius connected with each other in opposite directions and placed one inside the other (Fig. 4.43).

Then, for the electromotive force and active resistance in such the probe caused by inducted currents we have:

$$\mathscr{E}_1 = -8Ir_1^5\omega\mu\chi_1[n_1^2 I_1 - n_2^2 I_2]$$

$$R_1 = -8r_1^5\omega\mu\chi_1[n_1^2 I_1 - n_2^2 I_2] \tag{4.216}$$

$$I_1 = \int_0^\infty \frac{m}{2}\left[2K_0K_1 - m(K_1^2 - K_0^2)\right] \frac{I_1^2(m)}{m^2} \frac{\sin^2 ms_1}{m^2}\,\mathrm{d}m$$

$$I_2 = \int_0^\infty \frac{m}{2}\left[2K_0K_1 - m(K_2^2 - K_0^2)\right] \frac{I_1^2(m)}{m^2} \frac{\sin^2 ms_1'}{m^2}\,\mathrm{d}m$$

where:
σ_1 is the conductivity of a surrounding medium;
n_1 and n_2 are number of turns per unit of every coil length;
$s_1 = l/2r_1$ and $s_1' = b/2r_1$.

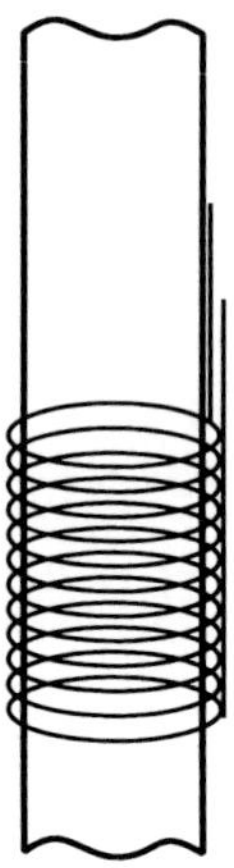

Figure 4.43. An induction probe with two transmitter-receiver coils.

In a medium with a cylindrical interface between the borehole and the formation we have:

$$\begin{aligned} \mathscr{E} &= \mathscr{E}_1 - 8Ir_1^2 a_1^3 \omega\mu(\chi_2 - \chi_1)\left[n_1^2 I_1' - n_2^2 I_2'\right] \\ R &= R_1 - 8r_1^2 a_1^3 \omega\mu(\chi_2 - \chi_1)\left[n_1^2 I_1' - n_2^2 I_2'\right] \end{aligned} \tag{4.217}$$

where:

$$I_1' = \int_0^\infty \frac{m}{2}\left[2K_0K_1 - m(K_1^2 - K_0^2)\right] \frac{I_1(vm)}{m^2} \frac{\sin^2 ms_2}{m^2}\, dm$$

$$I_2' = \int_0^\infty \frac{m}{2}\left[2K_0K_1 - m(K_1^2 - K_0^2)\right] \frac{I_1^2(vm)}{m^2} \frac{\sin^2 ms_2'}{m^2}\, dm$$

$$s_2 = l/2a_1 \qquad s_2' = b/2a_1$$

Correspondingly, the expression for the apparent conductivity is:

$$\begin{aligned} \frac{\sigma_a}{\sigma_1} &= 1 + \left(\frac{\sigma_2}{\sigma_1} - 1\right)\beta^3 \frac{n_1^2 I_1' - n_2^2 I_2'}{n_1^2 I_1 - n_2^2 I_2} = \beta^3 \frac{\dfrac{I_1'}{I_1} - \left(\dfrac{n_2}{n_1}\right)^2 \dfrac{I_2}{I_1}\dfrac{I_2'}{I_2}}{1 - \left(\dfrac{n_2}{n_1}\right)^2 \dfrac{I_2}{I_1}} \\ &= 1 + \left(\frac{\sigma_2}{\sigma_1} - 1\right) \frac{G_{12}^* - t^2 G_{22}^*}{1 - t^2} \end{aligned} \tag{4.218}$$

where G_{12}^* and G_{22}^* are geometric factors of formation for every coil. The latter can be written in the form:

$$\sigma_a = \sigma_1 G_1^* + \sigma_2 G_2^*$$

TABLE 4.7
Values of function G_1^*

$a/r_1 = 2.0$		$a/r_1 = 4.0$		$a/r_1 = 8.0$		
$l/2a_1$	G_1^*	$l/2a_1$	G_1^*	$l/2a_1$	G_1^*	I
0.2	0.5274	0.10	0.7665	0.050	0.8836	0.03979
0.3	0.5114	0.15	0.7591	0.075	0.8796	0.08663
0.4	0.5010	0.20	0.7509	0.100	0.8755	0.1489
0.5	0.4877	0.25	0.7424	0.125	0.8711	0.2242
0.6	0.4749	0.30	0.7339	0.150	0.8665	0.3115
0.7	0.4628	0.35	0.7253	0.175	0.8619	0.4095
0.8	0.4513	0.40	0.7169	0.200	0.8752	0.5169
0.9	0.4406	0.45	0.7086	0.225	0.8516	0.6330
1.0	0.4306	0.50	0.7003	0.250	0.8480	0.7569
1.1	0.4212	0.55	0.6926	0.275	0.8434	0.8879
1.2	0.4124	0.60	0.6849	0.300	0.8389	1.026
1.3	0.4043	0.65	0.6775	0.326	0.8345	1.170
1.4	0.3968	0.70	0.6702	0.350	0.8301	1.319
1.5	0.3895	0.75	0.6632	0.375	0.8257	1.474
1.6	0.3827	0.80	0.6565	0.400	0.8215	1.634
1.7	0.3764	0.85	0.6500	0.425	0.8173	1.797
1.8	0.3074	0.90	0.6436	0.450	0.8131	1.968
1.9	0.3468	0.95	0.6375	0.475	0.8091	2.141
2.0	0.3595	1.02	0.6316	0.500	0.8051	2.319
2.1	0.3545	1.05	0.6259	0.525	0.8011	2.500

where geometric factors of the borehole and formation are:

$$G_1^* = 1 - \frac{G_{12}^* - t^2 G_{22}^*}{1 - t^2} \qquad \text{or} \qquad G_1^* = \frac{G_{11}^* - t^2 G_{21}^*}{1 - t^2}$$

and:

$$G_2^* = \frac{G_{12}^* - t^2 G_{22}^*}{1 - t^2}$$

$$t = \frac{n_2}{n_1}\left(\frac{I_2}{I_1}\right)^{1/2}$$

Results of calculation of the geometric factor of the borehole, G_1^*, as well as integrals of types I_1, I_2, I_1', I_2', (I) are presented in Table 4.7.

In conclusion of this section let us notice that applying the approximate theory, taking into account the skin effect in the formation, these results can be used beyond the range of small parameters.

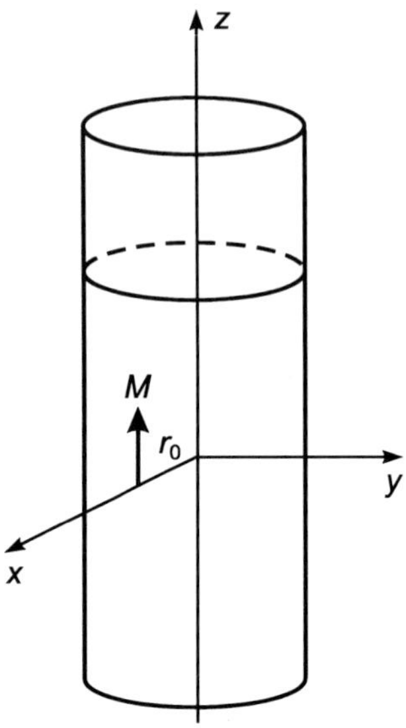

Figure 4.44. The vertical magnetic dipole displaced from the borehole axis.

4.11. Radial Responses of Two-coil Induction Probes Displaced with Respect to the Borehole Axis

Until now we have considered radial and frequency responses of two-coil induction probes located on the borehole axis. However, under real conditions if special centering devices are not used induction probes are usually displaced with respect to the axis, and for this reason it is essential to investigate the influence of this displacement on the radial responses of induction probes.

Let us introduce a Cartesian and cylindrical system of coordinates with common origin located at the borehole axis which coincides with the z-axis (Fig. 4.44). The radius of the borehole is a. Let us assume that the vertical magnetic dipole with moment $M = M_0 e^{i\omega t}$ is placed on the x-axis at distance r_0 from the origin. Unlike the previous model, when the vertical dipole is located on the borehole axis, in this case the primary vortex electrical field intersects a surface between the borehole and the formation. Correspondingly, electrical changes arise at this interface, and they provide continuity of the normal component of the current density. Therefore, current lines do not have a circular shape, located in horizontal planes, and possess a much more complicated form. For this reason, the quasistationary electromagnetic field in a cylindrical system of coordinates has all components: E_r, E_ϕ, E_z, H_r, H_ϕ, H_z which are related by Maxwell's equations:

$$\begin{aligned} &\operatorname{curl} \boldsymbol{E} = -i\omega\mu \boldsymbol{H} \qquad && \operatorname{div} \boldsymbol{E} = 0 \\ &\operatorname{curl} \boldsymbol{H} = \sigma \boldsymbol{E} && \operatorname{div} \boldsymbol{H} = 0 \end{aligned} \tag{4.219}$$

Let us present the electromagnetic field outside and inside the borehole as a sum of two terms:

$$\boldsymbol{E} = \boldsymbol{E}^{(1)} + \boldsymbol{E}^{(2)} \qquad \boldsymbol{H} = \boldsymbol{H}^{(1)} + \boldsymbol{H}^{(2)}$$

and each of them can be described with the help of the vertical component of the vector potential of, respectively, the magnetic and electrical type only:

$$\boldsymbol{E}^{(1)} = -\mathrm{i}\omega\mu \,\mathrm{curl}\, \boldsymbol{A}^* \qquad \boldsymbol{H}^{(2)} = \mathrm{curl}\, \boldsymbol{A} \tag{4.220}$$

where $\boldsymbol{A}^* = (0, 0, A^*)$, $\boldsymbol{A} = (0, 0, A)$. Then, in accord with eq. 4.219, we have:

$$\begin{aligned} \boldsymbol{H}^{(1)} &= k^2 \boldsymbol{A}^* - \mathrm{grad}\, U^* \\ \boldsymbol{E}^{(2)} &= -\mathrm{i}\omega\mu \boldsymbol{A} - \mathrm{grad}\, U \qquad k^2 = \mathrm{i}\sigma\mu\omega \end{aligned} \tag{4.221}$$

Choosing gauge conditions:

$$U^* = -\,\mathrm{div}\, \boldsymbol{A}^* \qquad \text{and} \qquad \sigma U = -\,\mathrm{div}\, \boldsymbol{A} \tag{4.222}$$

we obtain equations for both potentials:

$$\nabla^2 \boldsymbol{A}^* + k^2 \boldsymbol{A}^* = 0 \qquad \text{and} \qquad \nabla^2 \boldsymbol{A} + k^2 \boldsymbol{A} = 0$$

Taking into account eqs. 4.220–4.221 we will derive the following expressions for field components of magnetic and electrical type in a cylindrical system of coordinates:

$$\begin{aligned} E_r^{(1)} &= -\mathrm{i}\omega\mu \frac{1}{r}\frac{\partial A^*}{\partial \phi} & H_r^{(1)} &= \frac{\partial^2 A^*}{\partial r \partial z} \\ E_\phi^{(1)} &= -\mathrm{i}\omega\mu \frac{\partial A^*}{\partial r} & H_\phi^{(1)} &= \frac{1}{r}\frac{\partial^2 A^*}{\partial \phi \partial z} \\ E_z^{(1)} &= 0 & H_z^{(1)} &= k^2 A^* + \frac{\partial^2 A^*}{\partial z^2} \end{aligned} \tag{4.223}$$

$$\begin{aligned} E_r^{(2)} &= \frac{1}{\sigma}\frac{\partial^2 A}{\partial r \partial z} & H_r^{(2)} &= \frac{1}{r}\frac{\partial A}{\partial \phi} \\ E_\phi^{(2)} &= \frac{1}{\sigma r}\frac{\partial^2 A}{\partial \phi \partial z} & H_\phi^{(2)} &= -\frac{\partial A}{\partial r} \\ E_z^{(2)} &= \frac{1}{\sigma}\left(k^2 A + \frac{\partial^2 A}{\partial z^2}\right) & H_z^{(2)} &= 0 \end{aligned} \tag{4.224}$$

As follows from these equations vertical components of electrical and magnetic fields are absent in oscillations of magnetic and electrical types, respectively:

$$E_z^{(1)} = 0 \qquad H_z^{(2)} = 0 \tag{4.225}$$

In accord with eqs. 4.223 and 4.224, electrical and magnetic fields at any point are related with potentials A^* and A by formulae:

$$
\begin{aligned}
E_r &= -\mathrm{i}\omega\mu \frac{1}{r}\frac{\partial A^*}{\partial \phi} + \frac{1}{\sigma}\frac{\partial^2 A}{\partial r \partial z} \\
E_\phi &= \mathrm{i}\omega\mu \frac{1}{r}\frac{\partial A^*}{\partial r} + \frac{1}{\sigma r}\frac{\partial^2 A}{\partial \phi \partial z} \\
E_z &= \frac{1}{\sigma}\left(k^2 A + \frac{\partial^2 A}{\partial z^2}\right) \\
H_r &= \frac{\partial^2 A^*}{\partial \phi \partial z} + \frac{1}{r}\frac{\partial A}{\partial r} \\
H_\phi &= \frac{1}{r}\frac{\partial^2 A^*}{\partial \phi \partial z} - \frac{\partial A}{\partial r} \\
H_z &= k^2 A^* + \frac{\partial^2 A^*}{\partial z^2}
\end{aligned}
\tag{4.226}
$$

From continuity of tangential components of the field at boundary $r = a$ we will obtain conditions for vector-potentials A^* and A:

$$
\begin{aligned}
&\frac{1}{\sigma_1}\left(k_1^2 A_1 + \frac{\partial^2 A_1}{\partial z^2}\right) = \frac{1}{\sigma_2}\left(k_2^2 A_2 + \frac{\partial^2 A_2}{\partial z^2}\right) \\
&\frac{1}{\sigma_1}\left(\frac{1}{a}\frac{\partial^2 A_1}{\partial \phi \partial z} - k_1^2 \frac{\partial A_1^*}{\partial r}\right) = \frac{1}{\sigma_2}\left(\frac{1}{a}\frac{\partial^2 A_2}{\partial \phi \partial z} - k_2^2 \frac{\partial A_2^*}{\partial r}\right) \\
&k_1^2 A_1^* + \frac{\partial^2 A_1^*}{\partial z^2} = k_2^2 A_2^* + \frac{\partial^2 A_2^*}{\partial z^2} \\
&\frac{1}{a}\frac{\partial^2 A_1^*}{\partial \phi \partial z} - k_1^2 \frac{\partial A_1}{\partial r} = \frac{1}{a}\frac{\partial^2 A_2^*}{\partial \phi \partial z} - k_1^2 \frac{\partial A_2}{\partial r}
\end{aligned}
\tag{4.227}
$$

where σ_1, A_1^*, A_1 and σ_2, A_2^*, A_2 are conductivities and potentials of the borehole and the formation, respectively.

As was shown in Chapter 3, the electromagnetic field of the vertical magnetic dipole in a uniform medium can be described with the aid of the vertical component of the potential of the magnetic type only:

$$
A_0^* = \frac{M}{4\pi} \cdot \frac{2}{\pi}\frac{\mathrm{e}^{-\mathrm{i}k_1 R_1}}{R_1} = \frac{M}{4\pi}\frac{2}{\pi}\int_0^\infty K_0(\lambda_1 R)\cos\lambda z \, \mathrm{d}\lambda \tag{4.228}
$$

where $R_1 = |\boldsymbol{R} - \boldsymbol{r_0}|$, $\boldsymbol{r_0} = (r_0, 0, 0)$, $\lambda_1 = (\lambda^2 + k^2)^{1/2}$, $R = (r^2 + r_0^2 - 2rr_0\cos\phi)^{1/2}$; $\boldsymbol{R}$ and (r, ϕ) are radius vector and cylindrical coordinates of an observation point, respectively.

Let us present the primary potential A_0^* as a sum of angular harmonics. Making use of the addition theorem:

$$K_0(\lambda_1 R) = I_0(\lambda_1 r_0)K_0(\lambda_1 r) + 2\sum_{n=1}^{\infty} I_n(\lambda_1 r_0)K_n(\lambda_1 r)\cos n\phi$$

$$= 2\sum_{n=0}^{\infty}{}' I_n(\lambda_1 r_0)K_n(\lambda_1 r)\cos n\phi \qquad \text{if } r > r_0$$

we have:

$$A_0^* = \sum_{n=0}^{\infty}{}' A^* = \frac{M}{\pi^2}\sum_{n=0}^{\infty}{}' \cos n\phi \int_0^{\infty} I_n(\lambda_1 r_0)K_n(\lambda_1 r)\cos\lambda z \, \mathrm{d}\lambda \tag{4.229}$$

where superscript "$'$" means that the null harmonic is multiplied by coefficient 1/2.

Every angular harmonic A_{0n}^* of the primary field generates a corresponding harmonic A_n^*, describing the secondary field of the magnetic type. Taking into account the condition at infinity expressions for potentials A_1^* and A_2^* inside and outside the borehole are:

$$A_1^* = A_0 + \frac{M}{\pi^2}\sum_{n=0}^{\infty}{}' \cos n\phi \int_0^{\infty} c_n I_n(\lambda_1 r)K_n(\lambda_1 r)\, \mathrm{d}\lambda$$

$$A_2^* = \frac{M}{\pi^2}\sum_{n=0}^{\infty}{}' \cos(n\phi) \int_0^{\infty} d_n K_n(\lambda_2 r)\cos\lambda z \, \mathrm{d}\lambda \tag{4.230}$$

$$\lambda_1 = (\lambda^2 + k_1^2)^{1/2} \qquad \lambda_2 = (\lambda^2 + k_2^2)^{1/2}$$

In a uniform medium the magnetic dipole potential of the electrical type, A_0, is equal to zero. For this reason for the determination of this potential, A, we will make use of boundary conditions 4.227.

It is not difficult to see that if potentials of electrical type are presented in the form:

$$A_1 = \frac{M}{\pi^2}\sum_{n=0}^{\infty}{}' \sin(n\phi) \int_0^{\infty} a_n I_n(\lambda_1 r)\sin\lambda z \, \mathrm{d}\lambda$$

$$A_2 = \frac{M}{\pi^2}\sum_{n=0}^{\infty}{}' \sin(n\phi) \int_0^{\infty} b_n K_n(\lambda_2 r)\sin\lambda z \, \mathrm{d}\lambda \tag{4.231}$$

The unknown coefficients a_n, b_n, c_n, and d_n are defined from a linear system of four equations:

$$\frac{\lambda_1}{\sigma_1} a_n I_n(\lambda_1 a) = \frac{\lambda_2}{\sigma_2} b_n K_n(\lambda_2 a)$$

$$\frac{\lambda_1}{\sigma_1 a} n a_n I_n(\lambda_1 a) + \frac{k_1^2}{\sigma_1}\lambda_1 I_n(\lambda_1 r_0)K_n'(\lambda_1 a) - \frac{k_1^2}{\sigma_1}\lambda_1 c_n I_n'(\lambda_1 a)$$

$$= \frac{\lambda}{\sigma_2 a} n b_n K_n(\lambda_2 a) + \frac{k_2^2}{\sigma_2}\lambda_2 d_n K_n'(\lambda_2 a) \tag{4.232}$$

$$\lambda_1^2\left[I_n(\lambda_1 r_0)K_n(\lambda_1 a) + c_n I_n(\lambda_1 a)\right] = \lambda_2^2 d_n K_n(\lambda_2 a)$$
$$a_n\lambda_1 I_n'(\lambda_1 a) - \frac{\lambda}{a}nc_n I_n(\lambda_1 a) - \frac{\lambda}{a}nI_n(\lambda_1 r_0)K_n(\lambda_1 a) = b_n\lambda_2 K_n'(\lambda_2 a) - \frac{\lambda}{a}nd_n K_n(\lambda_2 a)$$

The electromotive force induced in a measuring coil of the probe is defined by the vertical component of magnetic field, H_z, which in accord with eq. 4.226 is expressed through the potential of the magnetic type, A_1^*, only. At the point with cylindrical coordinates $r = r_0$, $\phi = 0$ and $z = L$ for the magnetic field we have:

$$H_z = H_{0z} - \frac{M}{\pi^2}\sum_{n=0}^{\infty}{}' \lambda_1^2 c_n I_n(\lambda_1 r_0)\cos\lambda z \, \mathrm{d}z \tag{4.233a}$$

or

$$h_z = h_{0z} - \frac{2}{\pi}\alpha^3\sum_{n=0}^{\infty}{}' \int_0^{\infty} m_1^2 c_n I_n\left(m_1\frac{r_0}{a}\right)\cos m\alpha \, \mathrm{d}m \tag{4.233b}$$

where:

$$h_z = H_z/H_{0z} \qquad H_{0z} = M/2\pi L^3 \qquad h_{0z} = \mathrm{e}^{-\mathrm{i}kL}(1 + \mathrm{i}kL)$$
$$\alpha = L/a \qquad m = \lambda a \qquad c_n = \Delta_c/\Delta$$

$$\Delta_c = \left\{-(1-s)^2\frac{k_1^2 a_1^2}{m_2^4}m^2 n^2 - m_1^2\left[\frac{I_n'(m)}{I_n(m)} - s\frac{m_1}{m_2}\frac{K_n'(m_2)}{K_n(m_2)}\right]\left[\frac{K_n'(m_1)}{K_n(m_1)} - \frac{m_1}{m_2}\frac{K_n'(m_2)}{K_n(m_2)}\right]\right\} \times \frac{K_n(m_1)}{I_n(m_1)}I_n\left(m_1\frac{r_0}{a}\right) \tag{4.234}$$

$$\Delta = (1-s)^2\frac{k_1^2 a^2}{m_2^4}m^2 n^2 + m_1\left[\frac{I_n'(m)}{I_n(m)} - s\frac{m_1}{m_2}\frac{K_n'(m_2)}{K_n(m_2)}\right]\left[\frac{K_n'(m_1)}{K_n(m_1)} - \frac{m_1}{m_2}\frac{K_n'(m_2)}{K_n(m_2)}\right] \tag{4.235}$$
$$s = \sigma_2/\sigma_1$$

Let us notice that current lines corresponding to oscillations of the magnetic type are located in horizontal planes perpendicular to the borehole axis. Current lines of zero harmonic of the secondary field are circles with centers located on the borehole axis, and therefore they do not intersect the boundary between the borehole and the formation. Current lines of oscillations of the electrical type have a much more complicated form but their distribution has such a character that the field component $H_z^{(2)}$ is equal to zero everywhere.

Thus, in accord with eq. 4.233a field H_z consists of two terms, namely, field H_{0z} in a uniform medium with conductivity of the borehole, and the secondary field presented by the sum of angular harmonics, each of which is expressed through an improper integral. It is reasonable to notice that the integrand of zero harmonic differs from a corresponding function when the vertical magnetic dipole is located on the borehole axis by multiplier

TABLE 4.8
Values of function G_1^*

$\mathrm{Q}\, h_z \times 10^2$

ε \ a/h_1	0.125	0.025	0.05	0.1	0.2	0.4	0.8	1.6
0	0.111	0.435	1.68	6.19	20.7	52.4	41.8	−4.19
0.2	0.110	0.433	1.67	6.15	20.5	52.0	42.2	−4.66
0.5	0.107	0.421	1.62	5.95	19.8	49.8	44.2	−6.43

$\mathrm{In}\, h_z^s \times 10^4$

ε \ a/h_1	0.125	0.025	0.05	0.1	0.2	0.4	0.8	1.6
0	−0.200	−1.57	−12.2	915	636	3680	12110	−9510
0.2	−0.200	−1.57	−12.2	−913	633	3650	−12000	−9560
0.5	−0.200	−1.57	−12.1	−902	619	3500	−11400	−9900

$I_0(mr_0/a)$. In the case as the induction probe is not displaced with respect to the borehole axis ($r_0 = 0$) amplitudes of angular harmonics in eq. 4.233 are equal to zero for $n > 0$, and we obtain the known expression for the vertical component of the magnetic field.

For illustration values of quadrature and inphase components of h_z for various displacements $\varepsilon = r_0/a$, as the two-coil induction probe is located parallel to the borehole axis are given in Table 4.8. The ratio of conductivities $\sigma_2/\sigma_1 = 1/16$.

As follows from numerical analysis, in this case five angular harmonics describe the field with high accuracy for all considered values of a/h_1 where h_1 is the skin depth in the borehole. It is appropriate to notice that the influence of displacement on inphase and quadrature component of the field increases with an increase of frequency. At the same time within this range of frequencies the inphase component is less sensitive to displacement than is the quadrature component. For example, even if $a/h_1 = 1.6$ we have: $\mathrm{In}\, h_z^s(\varepsilon = 0.5)/\,\mathrm{In}\, h_z^s(\varepsilon = 0) = 1.04$, while $\mathrm{Q}\, h_z(\varepsilon = 0.5)/\,\mathrm{Q}\, h_z(\varepsilon = 0) = 1.51$. It is explained by the fact that within a wide range of frequencies the density of charges arising at the interface between the borehole and the formation is shifted in phase by 90° with respect to the current in the transmitter. Correspondingly, we can expect that the quadrature component of the field for a two-coil probe will be mainly subjected to the influence of eccentricity.

Now let us investigate in detail the low-frequency part of the spectrum when the skin depth in a more conductive medium is much greater than the length, L, of the two-coil induction probe:

$$|kL| \ll 1 \qquad L/a > 1 \tag{4.236}$$

Expanding integrand c_n in eq. 4.233b in a series by powers of $k_1^2a^2/m^2$ and discarding all terms except the first one we obtain:

$$c_n = -(1-s)\frac{k_1^2a^2}{m^2}I_n(m\varepsilon)$$

$$\times\left\{\frac{(1-s)n^2}{P_n(m)}K_n^2(m)-\left[\frac{m^2}{2}\left(K_{n-1}^2(m)-K_n^2(m)+\frac{2(n-1)}{m}K_nK_{n-1}\right)-nK_n^2(m)\right]\right\} \tag{4.237}$$

where

$$P_n(m)=m\left[\frac{I_n'(m)}{I_n(m)}-s\frac{K_n'(m)}{K_n(m)}\right]$$

Substituting eq. 4.237 into eq. 4.234 we have:

$$Q\,h_z=\frac{\sigma_1\omega\mu L^2}{2}G_1^*(\alpha,\varepsilon,s)+\frac{\sigma_2\omega\mu L^2}{2}G_2^*(\alpha,\varepsilon,s) \tag{4.238}$$

where:

$$G_2^*=\frac{2\alpha}{\pi}\sum_{n=0}^{\infty}{}' A_n(m)\cos m\alpha\, dm \tag{4.239}$$

$$A_n(m)=\left\{\frac{(1-s)n}{P_n(m)}K_n^2-\left[\frac{m^2}{2}\left(K_{n-1}^2-K_n^2+\frac{2(n-1)}{m}K_nK_{n-1}\right)\right]I_n(m\varepsilon)\right\} \tag{4.240}$$

and

$$G_1^*(\alpha,\varepsilon,s)+G_2^*(\alpha,\varepsilon,s)=1 \tag{4.241}$$

Inasmuch as functions G_1^* and G_2^* depend on geometric parameters and as well as on the ratio of conductivities σ_2 and σ_1 they are not in essence geometric factors. Function G_1^* for various values of α, ε and s is given in Table 4.9. As is seen from this table the eccentricity, ε, and ratio of conductivities, s, make an influence on function G_1^* regardless of the length of the two-coil induction probe. This influence is specially essential for relatively short probes and small conductivity of the formation. The length of the probe is usually several times larger than the borehole radius, a, and for this reason analysis of the behavior of function G_1^* for such probes is of great practical interest.

As was illustrated above the value of integrals $\int_0^\infty A_n(m)\cos m\alpha\, dm$, as $\alpha\to\infty$, is defined by behavior of functions $A_n(m)$ within the initial part of integration ($m\to 0$). In accord with eq. 4.240 for the zero-harmonic we have:

$$A_0(m)=\frac{m}{2}\left[2K_0(m)K_1(m)-m(K_1^2-K_0^2)\right]I_0^2(m\varepsilon)$$

Inasmuch as:

$$\begin{aligned}
I_0(m)\ &\simeq 1+\frac{m^2}{4}\\
K_0(m)&\simeq -\ln m-\frac{m^2}{4}\ln m+\frac{m^2}{4}\\
K_1(m)&\simeq \frac{1}{m}+\frac{m}{2}\ln m-\frac{m^2}{4}\qquad \text{as } m\to 0
\end{aligned}$$

TABLE 4.9
Values of function $G_1^*(\alpha, \varepsilon, s)\times 10^2$

	$s = 1/128$				$s = 1/2$			$s = 4$		
α \ ε	0	0.2	0.5	0.75	0.2	0.5	0.75	0.2	0.5	0.75
1	48.7	47.0	37.4	20.4	47.5	41.1	30.3	48.2	45.9	43.6
2	22.3	21.2	15.3	6.75	21.7	18.4	13.8	22.4	23.0	24.6
4	6.65	6.32	4.60	2.18	6.50	5.72	4.60	6.88	8.04	9.75
6	2.96	2.82	2.14	1.15	2.90	2.60	2.15	3.09	3.78	4.78
8	1.64	1.57	1.21	0.687	1.61	1.45	1.22	1.72	2.13	2.73
10	1.04	0.998	0.776	0.488	1.02	0.924	0.778	1.09	1.35	1.74
12	0.717	0.688	0.537	0.315	0.704	0.637	0.540	0.751	0.932	1.20
14	0.523	0.502	0.393	0.232	0.514	0.466	0.396	0.549	0.680	0.875
16	0.389	0.383	0.300	0.180	0.392	0.355	0.304	0.418	0.517	0.665
18	0.314	0.302	0.236	0.141	0.309	0.280	0.238	0.329	0.406	0.522
20	0.254	0.244	0.191	0.114	0.249	0.226	0.193	0.266	0.327	0.420

then

$$A_0(m) \simeq K_0(m) - \phi_0(m) \tag{4.242}$$

where

$$\phi_0(m) = \frac{1}{2}\left[1 + 2\left(\frac{r_0}{a}\right)^2\right]\frac{m^2}{4}\ln m$$

By analogy, for the first harmonic we obtain, as $m \to 0$:

$$A_1(m) \simeq \frac{1}{4}\left(\frac{r_0}{a}\right)^2\left(\frac{2}{1+s}\frac{2+3s-s^2}{(1+s)^2}m^2\ln m\right) \tag{4.243}$$

Making use of the Sommerfeld integral:

$$\int\limits_0^\infty K_0(m)\cos m\alpha\, dm = \frac{\pi}{2(1+\alpha^2)^{1/2}} \simeq \frac{\pi}{2\alpha}\left(1 - \frac{1}{2\alpha^2}\right) \qquad \text{if } \alpha \gg 1$$

and integrating by parts $\int_0^\infty A(m)\cos m\alpha\, dm$, we will obtain an asymptotic presentation for functions G^*_{10} and G^*_{11} corresponding to zero and the first harmonics as $\alpha \to \infty$:

$$G^*_{10} = \left[1 + \left(\frac{r_0}{a}\right)^2\right]\frac{1}{\alpha^2} \tag{4.244}$$

and

$$G^* = -\left(\frac{r_0}{a}\right)^2\frac{(2+3s-s^2)}{(1+s)^2}\frac{1}{\alpha^2} \tag{4.245}$$

We can show that functions $A_n(m)$ for higher order of harmonics $(n > 1)$ at the initial part of the integration have the form $T_n + M_n m^2$, where T_n and M_n are constants. Correspondingly, functions $G^*_{1n}(\alpha, \varepsilon, s)$ with an increase of the probe length decrease more rapidly than $1/\alpha^2$. Thus, adding eqs. 4.244 and 4.245 we obtain the leading term of the asymptotical expression of G^*_1:

$$G^*(\alpha, \varepsilon, s) \simeq \left[1 + \left(\frac{r_0}{a}\right)^2\frac{(s-1)(2s+1)}{(s+1)^2}\right]\frac{1}{\alpha^2} \tag{4.246}$$

Comparison with results of calculations of function $G^*_1(\alpha, \varepsilon, s)$, given in Table 4.9, shows that the error of determination of this function by eq. 4.246 does not exceed 5% if the probe length is at least six times greater than the borehole radius even for $\varepsilon = 0.75$. When the two-coil induction probe is located on the borehole axis, current lines do not intersect the surface of the borehole and they have a circular form. Correspondingly, we obtain the known expression for function G:

$$G(\alpha) \simeq \frac{1}{\alpha^2} \quad \text{if } \alpha \gg 1 \text{ and } r_0 = 0$$

If the borehole is more conductive than the formation, electrical charges arising at the boundary $r = a$ create a field which decreases the effect caused by the primary vortex electrical field. Respectively, in accord with eq. 4.246 function G_1^* turns out to be smaller than G_1. In the limited case, as $s = \sigma_2/\sigma_1 \to 0$ we have for function $G_1^*(\alpha, \varepsilon, s)$:

$$G_1^*(\alpha, \varepsilon) \simeq \frac{1}{\alpha^2}(1 - \varepsilon^2) \tag{4.247}$$

In the case where the formation is more conductive than the borehole at those points of the interface where the primary electrical field is directed into external or internal areas, negative or positive charges arise correspondingly, and their electrical field results in an increase of the current density within the borehole. For this reason function $G_1^*(\alpha, \varepsilon, s)$ becomes greater than $G_1(\alpha)$. In particular if $s \gg 1$ we have:

$$G_1^* \simeq \frac{1}{\alpha^2}(1 + 2\varepsilon^2) \tag{4.248}$$

For intermediate values of s: $0 < s < \infty$ the coefficient in front of ε^2 gradually increases from -1 to 2, and it is equal to zero when $s = 1$ (a uniform medium). In real conditions the maximal value of ε does not exceed 0.70-0.75 when the induction probe touches the borehole surface. In accord with eq. 4.246 the second term characterizing the influence of the probe displacement, is directly proportional to ε^2 and, in particular, for a more conductive borehole function $G_1^*(\alpha, \varepsilon, s)$ can decrease almost two times ($\alpha \gg 1$). This consideration also shows that relatively small displacements of the probe leads to an insignificant change of the geometric factor.

Within a certain range of frequencies and resistivities we can neglect the interaction between currents in the borehole, while in the formation the skin effect manifests itself in the same manner as in a uniform medium but the current density in the borehole and surface charges are directly proportional to frequency. For these conditions the field can be presented in the form:

$$h_z \simeq \frac{k_1^2 L^2}{2}(1 - s)G^*(\alpha, \varepsilon, s) + h_{0z}(k_2 L) \tag{4.249}$$

Calculations based on data given in Table 4.8 demonstrate the validity of this relation. Therefore, the inphase component of the field as well as the term of the quadrature component proportional to $\omega^{3/2}$ are in this range of frequencies defined only by the conductivity of the formation and, correspondingly, do not depend on the position of the induction probe with respect to the borehole axis regardless of its length.

4.12. The Influence of Magnetic Permeability and Dielectric Constant in Induction Logging

In principal the quasistationary electromagnetic fields applied in induction logging depend on conductivity and magnetic permeability. For this reason an investigation of the influence of the magnetic permeability can be of great practical interest, specially for two cases:

- A borehole solution contains magnetic elements which are used to increase the weight of this solution.
- The formation has a higher magnetic permeability. For example, such conditions arise when induction logging is applied in mining geophysics.

Results of analysis of the influence of the magnetic permeability described in this section are based on results of calculations of the magnetic field on the borehole, when an invasion zone is absent.

Let us first consider some features of the influence of μ in a uniform medium.

As was shown in Chapter 2 the electromagnetic field of the magnetic dipole can be presented as:

$$\begin{aligned} E_\phi &= \frac{i\omega\mu M}{4\pi R^2} e^{ikR}(1 - ikR)\sin\theta \\ H_R &= \frac{2M}{4\pi R^3} e^{ikR}(1 - ikR)\cos\theta \\ H_\theta &= \frac{M}{4\pi R^3} e^{ikR}(1 - ikR - k^2R^2)\sin\theta \end{aligned} \tag{4.250}$$

where R and θ are spherical coordinates of a point; k is wave number:

$$k = (\sigma\mu\omega/2)^{1/2}(1 + i)$$

In a nonconducting magnetic medium instead of eq. 4.250 we have:

$$E_{0\phi} = \frac{i\omega\mu M}{4\pi R^2}\sin\theta \qquad H_{0R} = \frac{2M}{4\pi R^3}\cos\theta \qquad H_{0\theta} = \frac{M}{4\pi R^3}\sin\theta \tag{4.251}$$

Thus in a uniform non-conducting medium the magnetic field does not depend on μ while the electric field related with the electromotive force as

$$\mathscr{E} = \oint \boldsymbol{E}\, d\boldsymbol{l}$$

is directly proportional to the magnetic permeability.

If a medium is conductive, magnetic permeability, μ, makes an influence on the magnetic field inasmuch as intensity of the skin effect is characterized by parameter $p = (\sigma\mu\omega)^{1/2}$. With an increase of the magnetic permeability, the interaction between currents in accord with Faraday's law increases and curves of distribution of induced currents, shown in Figs. 2.4–2.5 for both quadrature and inphase components, are shifted to the range of lower frequencies.

In accord with eq. 4.250, when the parameter p is small we have for components of the magnetic field:

$$Q\, H_z = p^2 H_{0z} = \frac{\sigma\mu\omega L^2}{2} H_{0z} \tag{4.252}$$

$$\text{In}\, H_z = (1 - \frac{2}{3}p^3)H_{0z} = H_{0z} - \frac{2}{3}p^3 H_{0z} \tag{4.253}$$

and for vertical component of vector of the magnetic induction $\boldsymbol{B}$ ($\boldsymbol{B} = \mu \boldsymbol{H}$), we obtain:

$$Q\, B_z = \frac{\sigma\mu^2\omega L^2}{2} H_{0z} \qquad \text{In}\, B_z = B_{0z} - \frac{2}{3} p^3 B_{0z} \tag{4.254}$$

where L is the probe length. Functions H_{0z} and B_{0z} are vertical components of the magnetic field and magnetic induction in a nonconducting medium. For this reason the electromotive force generated by the quadrature component of induced currents is directly proportional to the square of magnetic permeability while the inphase component of the electromotive force changes as a linear function of μ. For very small values of p we can neglect the second term in eq. 4.253 and therefore $\text{In}\, B_z = B_{0z} = \mu H_{0z}$. In other words, in this approximation the inphase component of the electromotive force is a function of μ only, and it does not practically depend on conductivity.

It is obvious that for given frequency, conductivity, and magnetic permeability this behavior takes place for shorter probes with higher accuracy. Correspondingly, in order to reduce the influence of conductivity, it is appropriate to apply relatively short induction probes, in particular, single-coil arrays. However, in this case we can expect an increase of the influence of borehole radius, caverns, and eccentricity. As is well known, the relation between magnetic permeability and susceptibility in Gauss system is:

$$\mu = 1 + \chi_0 \qquad \chi_0 = 4\pi \times 10^{-6} \chi$$

Let us assume that the magnetic susceptibility of a formation is small and $\chi_0 \ll 1$. Then, substituting expression for μ into eq. 4.254 we obtain:

$$\begin{aligned} Q\, B_z &\simeq \frac{\sigma\omega L^2}{2}(1 + 2\chi_0) H_{0z} \\ \text{In}\, B_z &\simeq H_{0z} + \chi_0 H_{0z} - \frac{2}{3}(1+\chi_0) H_{0z} \left[\frac{\omega\sigma(1+\chi_0)L^2}{2}\right]^{3/2} \\ &\simeq H_{0z} + \chi_0 H_{0z} - \frac{2}{3}\left(1 + \frac{5}{2}\chi_0\right) H_{0z} \left(\frac{\sigma\omega}{2} L^2\right)^{3/2} \end{aligned}$$

The primary field, H_{0z}, is usually compensated, and correspondingly the secondary inphase component:

$$\chi_0 H_{0z} - \frac{2}{3}\left(\frac{\sigma\omega L^2}{2}\right)^{3/2} H_{0z}$$

is measured.

From comparison of eqs. 4.252–4.254 it is seen that the quadrature component of the electromotive force is less influenced by the magnetic susceptibility than the inphase component. For this reason at the range of very small parameters, when the susceptibility is relatively small, the conductivity of a formation is defined by the quadrature component of the electromotive force while measurements of the inphase component allow us to determine the magnetic susceptibility. If parameter $p = (\sigma\mu\omega/2)^{1/2}$ is not small and the value of χ_0 is comparable with unity, determination of conductivity and magnetic permeability is usually a sufficiently complicated procedure inasmuch as both quadrature and inphase components of the electromotive force are influenced by σ and μ. It is clear that in measuring the amplitude of the secondary field the influence of the magnetic permeability increases.

Now let us consider the case when a two-coil induction probe is located on the borehole axis.

As was shown in section 4.3, the expression for the vertical component of the magnetic field on the axis of the borehole has the form:

$$h_z = h_{0z} - \frac{L^3}{\pi} \int\limits_0^\infty \lambda_1^2 C_1 \cos \lambda L \, d\lambda \tag{4.255}$$

where h_z is the magnetic field expressed in units of the primary field; h_{0z} is the magnetic field in a uniform medium with conductivity and magnetic permeability of the borehole.

Function C_1, defined from boundary conditions at interface $r = a$, is:

$$C_1 = \frac{\mu_1 \lambda_2 K_0(\lambda_2 a) K_1(\lambda_1 a) - \mu_2 \lambda_1 K_0(\lambda_1 a) K_1(\lambda_2 a)}{\mu_1 \lambda_2 K_0(\lambda_2 a) I_1(\lambda_1 a) + \mu_2 \lambda_1 I_0(\lambda_1 a) K_1(\lambda_2 a)} \tag{4.256}$$

where μ_1 and μ_2 are magnetic permeability of the borehole and the formation; a is the borehole radius; L is the probe length and:

$$\lambda_1 = (\lambda_+^2 i\chi_1)^{1/2} \qquad \lambda_2 = (\lambda_+^2 i\chi_2)^{1/2}$$

$$\chi_1 = \sigma_1 \mu_1 \omega \qquad \chi_2 = \sigma_2 \mu_2 \omega$$

Methodics of calculation of this integral was described above.

We will consider the range of small parameters when the quadrature component of the field is directly proportional to frequency. Then we can write:

$$Q\, h_z = \frac{\mu^2 \omega L^2}{2} [\sigma_1 G_1(\alpha, s) + \sigma_2 G_2(\alpha, s)] \tag{4.257}$$

Functions $G_1(\alpha, s)$ and $G_2(\alpha, s)$ depend on both geometric factor $\alpha = L/a_1$ and the ratio of magnetic permeabilities, $s = \mu_2/\mu_1$.

Unlike a medium which has a uniform magnetic permeability, in this case redistribution of the primary magnetic flux is a function of μ, and at the range of small parameters a density of induced currents is directly proportional to the flux of this field and a conductivity at a given point.

Analytical expressions for functions $G_1(\alpha, s)$ and $G_2(\alpha, s)$ can be obtained in the same manner as those for geometric factors $G_1(\alpha)$ and $G_2(\alpha)$, namely expanding the right-hand side of eq. 4.255 in a series by powers of ω and defining a coefficient which is directly proportional to frequency. Results of calculations of these functions are presented in Table 4.10 and Fig. 4.45.

TABLE 4.10
Values of functions $G_1(\alpha, s)$, $G_2(\alpha, s)$

$G_1(\alpha, s)$

α \ s	0.99	0.90	0.80	0.70	0.60	0.50
3.0	0.1160	0.1340	0.1610	0.1990	0.2500	0.0326
4.0	0.0675	0.0805	0.0099	0.1260	0.1670	0.2280
5.0	0.0436	0.0525	0.0662	0.0860	0.1160	0.1640
6.0	0.0302	0.0366	0.0466	0.0615	0.0837	0.1210
7.0	0.0220	0.0270	0.0345	0.0454	0.0627	0.0916
8.0	0.0168	0.0204	0.0262	0.0350	0.0484	0.0710
9.0	0.0131	0.0161	0.0206	0.0275	0.0382	0.0568
10.0	0.0106	0.0129	0.0168	0.0222	0.0307	0.0455

$G_2(\alpha, s)$

α \ s	0.99	0.90	0.80	0.70	0.60	0.50
3.0	0.870	0.973	1.098	1.259	1.476	1.777
4.0	0.930	1.023	1.160	1.337	1.574	1.910
5.0	0.944	1.045	1.186	1.370	1.619	1.972
6.0	0.950	1.054	1.197	1.383	1.632	1.996
7.0	0.955	1.071	1.200	1.386	1.639	2.000
8.0	0.956	1.068	1.200	1.384	1.636	1.000
9.0	0.954	1.056	1.197	1.379	1.629	1.986
10.0	0.953	1.056	1.192	1.374	1.621	1.973

TABLE 4.11
Values of function $F(\alpha, s)$

α \ s	0.90	0.90	0.80	0.70	0.60	0.50	0.10
3.0	0.986	1.107	1.259	1.458	1.726	2.103	0.985
4.0	0.998	1.104	1.260	1.463	1.741	2.138	0.982
5.0	0.988	1.098	1.252	1.456	1.735	2.136	0.976
6.0	0.980	1.091	1.244	1.445	1.716	2.116	0.972
7.0	0.977	1.098	1.234	1.431	1.702	2.092	0.966
8.0	0.973	1.088	1.226	1.419	1.684	2.071	0.962
9.0	0.967	1.072	1.218	1.407	1.667	2.043	0.958
10.0	0.964	1.069	1.209	1.396	1.652	1.019	0.952

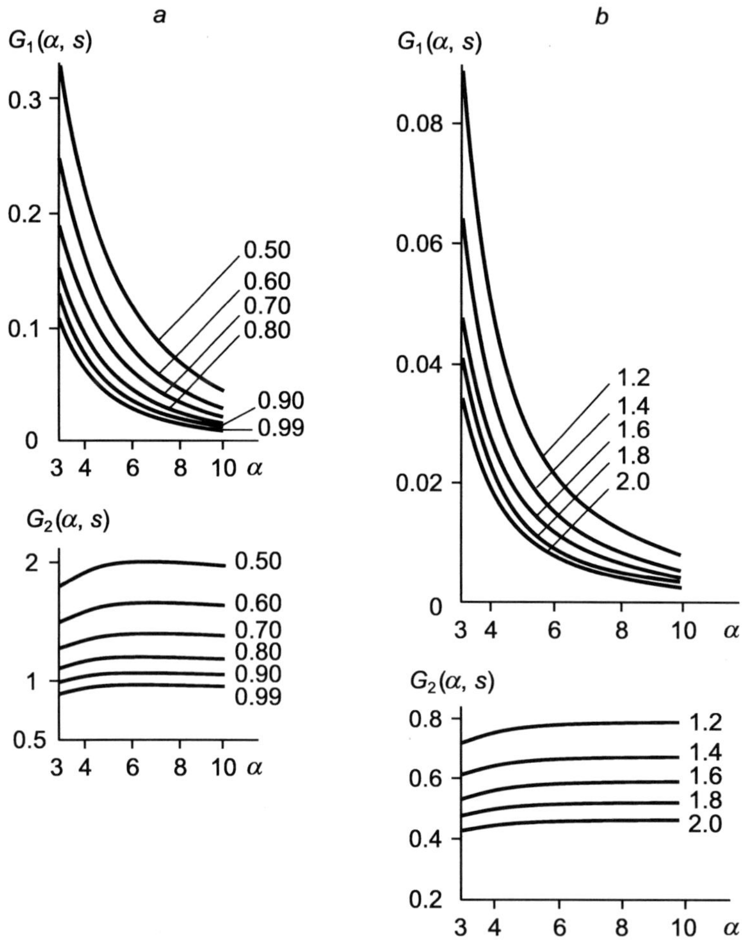

Figure 4.45. (a) Curves of functions $G_1(\alpha, s)$, $G_2(\alpha, s)$; (b) curves of functions $G_1(\alpha, s)$, $G_2(\alpha, s)$.

TABLE 4.12
Values of function $G_1^n(\alpha, s)$

α \ s	0.99	0.90	0.80	0.70	0.60	0.50
3.0	0.1180	0.1210	0.1280	0.1370	0.1450	0.1550
4.0	0.0680	0.0728	0.0790	0.0862	0.0960	0.1070
5.0	0.0442	0.0478	0.0529	0.0590	0.0668	0.0768
6.0	0.0308	0.0335	0.0375	0.0426	0.0488	0.0572
7.0	0.0225	0.0246	0.0280	0.0318	0.0368	0.0438
8.0	0.0172	0.0187	0.0214	0.0247	0.0287	0.0342
9.0	0.0135	0.0151	0.0169	0.0195	0.0228	0.0278
10.0	0.0110	0.0121	0.0139	0.0159	0.0186	0.0224

It is appropriate to notice that a sum of these functions is not generally equal to unity, since the magnetic permeabilities of borehole and formation are not equal to each other. As an example, function:

$$F(\alpha, s) = G_1(\alpha, s) + G_2(\alpha, s)$$

is presented in Table 4.11, and it does not practically depend on the probe length, and it turns out to be in essence of a function of the ratio of magnetic permeabilities, s, only.

In accord with eq. 4.257 in a medium which has a uniform conductivity we have:

$$\mathrm{Q}\, h_{0z} = \frac{\mu_2 \omega L^2}{2} \sigma_2 F \tag{4.258}$$

Inasmuch as function F is not equal to unit it is reasonable to normalize functions $G_1(\alpha, s)$ and $G_2(\alpha, s)$ in such a way that their sum would be equal to unit. For this purpose we will divide these functions by F. Then we obtain:

$$\mathrm{Q}\, h_z = \frac{\mu_2 \omega L^2 F}{2} [\sigma_1 G_1^n + \sigma_2 G_2^n] \tag{4.259}$$

where:

$$G_1^n = G_n/F \qquad G_2^n = G_n/F \tag{4.260}$$

Values of normalized functions $G_1^n(\alpha, s)$ and $G_2^n(\alpha, s)$ are given in Tables 4.12 and 4.13.

As follows from this table, as well as from physical point of view, with an increase of magnetic permeability of the borehole the magnetic field within the borehole as well as function G_1^n increase.

It is important to notice that in the case when magnetic permeabilities of the borehole and the formation are different function $G_1^n(\alpha, s)$ for large values of α behaves in the same manner as geometrical factor $G_1(\alpha)$, namely:

$$G_1^n(\alpha, s) \to k(s)/\alpha^2 \qquad \text{if } \alpha \gg 1 \tag{4.261}$$

TABLE 4.13
Values of function $G_2^n(\alpha, s)$

α \ s	0.99	0.90	0.80	0.70	0.60	0.50
3.00	0.882	0.879	0.872	0.863	0.855	0.845
4.0	0.932	0.927	0.921	0.914	0.904	0.893
5.0	0.956	0.952	0.947	0.941	0.933	0.923
6.0	0.969	0.966	0.962	0.957	0.951	0.943
7.0	0.977	0.975	0.972	0.968	0.963	0.956
8.0	0.983	0.981	0.979	0.975	0.971	0.966
9.0	0.986	0.985	0.983	0.980	0.970	0.972
10.0	0.989	0.988	0.986	0.984	0.981	0.978

TABLE 4.14
Values of function $k(s)$

s	1	0.9	0.8	0.7	0.6	0.5
$k(s)$	1.0	1.2	1.4	1.6	1.85	2.24

Values of $k(s)$ are given in Table 4.14.

Calculations show that asymptotical behavior of function G_1^n commences practically from the same values of α regardless of the ratio of magnetic permeabilities s.

With an increase of the probe length the influence of magnetic permeability on function G_1^n increases until a certain limit. Values of $\eta_s = G_1^n(s,\alpha)/G_1^n(0.99,\alpha)$ are given in Table 4.15.

In Chapter 7 we will consider in detail multi-coil induction probes which essentially allow a decrease on influence on induced currents in the borehole as well as in the invasion zone. At that time questions related to magnetic permeability will not be investigated more. For this reason let us here briefly demonstrate that multi-coil probes can be applied in order to reduce the influence of induced currents in a conductive and magnetic medium of the borehole (Fig. 4.46). As the first example we will consider a three-coil induction probe consisting of one generator and two measuring coils (Fig. 4.46a). The distance between the later is significantly less than that between the transmitter and receiver coils. Moments of receiver coils are chosen in such a way that the electromotive force in a free space, $\mathscr{E}_0^{(0)}$, is equal to zero.

In a uniform medium with conductivity σ_2, according to eq. 4.258, for a two-coil probe we have:

$$\mathscr{E}_0 = \frac{\sigma_2 \mu_2 \omega L^2}{2} \mathscr{E}_{01}^{(0)} F(s) \tag{4.262}$$

where $\mathscr{E}_{01}^{(0)}$ is the electromotive force in the receiver when the probe is located in a uniform nonconducting medium with magnetic permeability μ_1; $F(s)$ is the sum of functions G_1 and G_2.

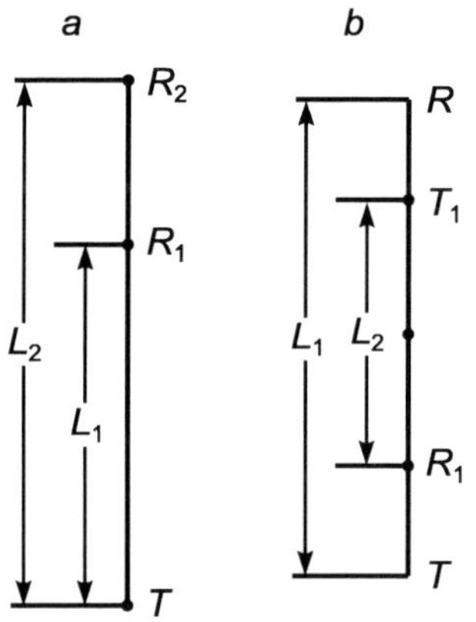

Figure 4.46. Three- and four-coil induction probes.

TABLE 4.15
Values of function $\eta_s = G_1^n(s,\alpha)/G_1^n(0.99,\alpha)$

η_s \ α	3.0	4.0	5.0	6.0	7.0	8.0	9.0	10.0
$\eta_{0.9}$	1.02	1.07	1.08	1.09	1.09	1.09	1.12	1.10
$\eta_{0.8}$	1.08	1.16	1.20	1.22	1.24	1.24	1.25	1.26
$\eta_{0.7}$	1.16	1.27	1.33	1.38	1.42	1.43	1.44	1.45
$\eta_{0.6}$	1.23	1.41	1.51	1.59	1.64	1.67	1.69	1.69
$\eta_{0.5}$	1.31	1.57	1.73	1.86	1.95	1.99	2.06	2.04

Therefore, for the electromotive force induced in two receivers of a three-coil probe we obtain:

$$\mathscr{E}_0 = \frac{\mu_2 \omega \mathscr{E}_{01}^{(0)}}{2}\left(L_1^2 F_1 - L_2^2 F_2\right)\sigma_2 = \frac{\omega \mu_2 L_1^2 F(s)}{2}\mathscr{E}_{01}^{(0)}(1-t^2)\sigma_2 \tag{4.263}$$

where $t = L_2/L_1$ and $F(s) = F_1(s) = F_2(s)$. Correspondingly, at the borehole axis the electromotive force is:

$$\mathscr{E} = \frac{\mu_2 \omega L_1^2 F(s)}{2}\left(\sigma_a^{(1)} - t^2 \sigma_a^{(2)}\right) \tag{4.264}$$

where $\sigma_a^{(1)}$ and $\sigma_a^{(2)}$ are apparent conductivities for a two-coil induction probe with lengths L_1 and L_2, respectively. For this reason an expression for apparent conductivity defined with the three-coil probe can be written as:

$$\frac{\sigma_1}{\sigma_2} = \frac{1}{1-t^2}\left(\frac{\sigma_a^{(1)}}{\sigma_2}L_1 - t^2\frac{\sigma_a^{(2)}}{\sigma_2}L_2\right)$$

TABLE 4.16
Values of function In h_z^s

η_s \ α	0.9	0.8	0.7	0.6	0.5
3.0	0.0728	0.159	0.263	0.391	0.550
4.0	0.100	0.222	0.374	0.568	0.822
5.0	0.114	0.255	0.435	0.671	0.991
6.0	0.120	0.271	0.466	0.726	1.088
7.0	0.122	0.278	0.480	0.753	1.139
8.0	0.123	0.280	0.485	0.765	1.164
9.0	0.123	0.280	0.485	0.768	1.173
10.0	0.122	0.279	0.484	0.766	1.172

or

$$\frac{\sigma_a}{\sigma_2} = \frac{1}{1-t^2}\left\{(N-1)\left[G_1^n(\alpha_1, s) - t^2 G_1^n(\alpha_2, s)\right] + (1-t^2)\right\} \tag{4.265}$$

where $N = \sigma_1/\sigma_2$.

If lengths of probes, L_1 and L_2, are significantly greater than the borehole radius ($\alpha \gg 1$) function $G_1^n(\alpha)$ tends to $k(s)/\alpha^2$ and the first term at the right-hand side of eq. 4.265:

$$(N-1)\left[G_1^n(\alpha_1, s) - t^2 G_1^n(\alpha_2, s)\right]$$

defining the signal caused by currents in a borehole, tends to zero.

Calculations of the quadrature component of the field and correspondingly function σ_a/σ_2, based on the exact solution, confirm that this type of induction probe can be efficiently used at the range of small parameters ($L/h_2 < 0.2 \div 0.3$).

In accord with eq. 4.263, a coefficient of the three-coil probe is a function of parameter s, in particular, of magnetic permeability of the borehole. For this reason in order to calculate the apparent conductivity, σ_a, it is necessary to define parameter s. It can be done by measuring the inphase component of the field, since within a wide range of frequencies and conductivities this component practically depends on parameter s only. Values of the inphase component of the magnetic field expressed in units of the primary field in a free space for a two-coil induction probe with various lengths, are given in Table 4.16.

We assume that, for most cases which are of great practical interest in induction logging, data presented in this table coincide with the magnetic field of a direct current. Therefore, by measuring the inphase component with a three-coil induction probe or with a probe of two coils, parameter s and, respectively, coefficient of the probe are defined. This enables us to calculate, the apparent conductivity by making use of quadrature component data.

As a second example of a probe which simultaneously measures conductivity and magnetic permeability and reduces the influence of induced currents in the borehole, we will consider a four-coil symmetrical probe (Fig. 4.46b). In accord with eq. 4.262 electromotive force induced in receivers of this probe, when it is located in a uniformly conductive medium, can be written as:

$$\mathscr{E}_0 = \frac{\omega\mu_2\sigma_2 L_1^2}{2}\mathscr{E}_{01}^{(0)}\left(1 - \frac{2c}{p} + \frac{c^2}{1-2p}\right)F(s)$$

where $p = L_2/L_1$ and c is the ratio of coil turns of the basic probe with length L_1 to those for the probe with length L_2.

Correspondingly, on the borehole axis we have:

$$\mathscr{E}_0 = \frac{\mu_2\sigma_2 L_1^2}{2}\mathscr{E}_{01}^{(0)}\left(\sigma_a^{(1)} - \frac{2c}{p}\sigma_a^{(2)} + \frac{c^2}{1-2p}\sigma_a^{(3)}\right)F(s)$$

where $\sigma_a^{(1)}, \sigma_a^{(2)}, \sigma_a^{(3)}$ are apparent conductivities for two coil probes with lengths L_1 and L_2 and $(L_1 - L_2)/2$, respectively.

For this reason an expression for the apparent conductivity of the four-coil induction probe is:

$$\frac{\sigma_a}{\sigma_2} = \frac{1}{\left(1 - \frac{2c}{p} + \frac{c^2}{1-2p}\right)}\left(\frac{\sigma_1}{\sigma_2}G_1^{(4)} + G_2^{(4)}\right) \tag{4.266}$$

Here:

$$\begin{aligned} G_1^{(4)} &= G_1^m(\alpha) - \frac{2c}{p}G_1^m(p\alpha) + \frac{c^2}{1-2p}G_1^m[(1-2p)\alpha] \\ G_2^{(4)} &= G_2^m(\alpha) - \frac{2c}{p}G_2^m(p\alpha) + \frac{c^2}{1-2p}G_2^m[(1-2p)\alpha] \end{aligned} \tag{4.267}$$

$\alpha = L_1/a_1$. G_1^n and G_2^m are functions given in Table 4.13.

The parameters of the probe (α, p and c) which allow us to reduce the influence of the borehole are defined from condition:

$$G_1^{(4)}(\alpha, p, c) = 0 \tag{4.268}$$

For the inphase component of electromotive force, expressed in units of that in a nonconducting medium with magnetic permeability μ, we have:

$$\frac{\operatorname{In}\mathscr{E}}{\mathscr{E}_{01}} = \operatorname{In} h_z(\alpha) - \frac{2c}{p}\operatorname{In} h_z(p\alpha) + \frac{c^2}{1-2p}\operatorname{In} h_z[(1-2p)\alpha] \tag{4.269}$$

Unlike the three-coil induction probe (Fig. 4.46a), simultaneous measuring of σ and μ, with the four-coil probe has two shortcomings:

- A four-coil probe includes relatively short two-coil probes, $(L_1/2 - L_2/2)$ which are usually subjected more strongly to the influence of a borehole radius, and its change can lead to significant errors in the determination of s.

- With a change of position, z, the magnetic permeability of the borehole can change, and correspondingly, the condition of compensation of currents in the borehole (eq. 4.268) becomes invalid. For large values of σ_1/σ_2 it can result in significant errors in the determination of the formation conductivity.

Finally, it is appropriate to make the following comment. In accordance with equations 4.258–4.259 and the analysis of functions $G_1(\alpha, s)$, $G_2(\alpha, s)$, the influence of the magnetic permeability on the quadrature component measured by the induction probe can be practically neglected if $\chi < 10^3$.

In conclusion of this section we will briefly consider the influence of dielectric constant, ε. In accord with Maxwell's equation:

$$\operatorname{curl} \boldsymbol{H} = \sigma \boldsymbol{E} + \frac{\partial \boldsymbol{D}}{\partial t} \tag{4.270}$$

the magnetic field is defined by current of conductivity ($\sigma \boldsymbol{E}$(and displacement currents ($\partial \boldsymbol{D}/\partial t$).

For a harmonic field: $E\mathrm{e}^{-\mathrm{i}\omega t}$, $H\mathrm{e}^{-\mathrm{i}\omega t}$, instead of eq. 4.270 we have:

$$\operatorname{curl} \boldsymbol{H} = \sigma \boldsymbol{E} - \mathrm{i}\omega\varepsilon \boldsymbol{E} \tag{4.271}$$

since $\boldsymbol{D} = \varepsilon \boldsymbol{E}$.

The theory of induction logging is based on the assumption that displacement currents are much smaller than conductive ones, and correspondingly we can neglect the term $\varepsilon\omega\boldsymbol{E}$ in eq. 4.271. In other words, it is assumed that:

$$\alpha = \frac{\omega\varepsilon}{\sigma} \ll 1 \tag{4.272}$$

Let us consider one numerical example. Let $f = 3\times10^5$ Hz, $\rho = 200$ ohm·m, and $\varepsilon = 20\,\varepsilon_0$. Then we have $\alpha \simeq 0.007$. Thus, even in a relatively resistive medium and making use of a sufficiently high frequency, parameter α is still very small.

Now let us suppose that the two-coil probe is located in a uniform medium and product kL is less than unity. Then, making use of results obtained in Chapter 2 we have for the vertical component, h_z:

$$h_z \simeq 1 + \frac{k^2L^2}{2} + \frac{\mathrm{i}k^3L^3}{3} \tag{4.273}$$

where $k^2 = \mathrm{i}\sigma\mu\omega(1 - \mathrm{i}\alpha)$. Whence:

$$\mathrm{Q}\, h_z \simeq \frac{\mu\omega\sigma L^2}{2} \qquad \mathrm{In}\, h_z^s \simeq -\frac{\sqrt{2}}{6}(\mu\omega\sigma)^{3/2}L^3 + \frac{1}{2}\frac{\mu\varepsilon}{2}L^2\omega^2 \tag{4.274}$$

i.e. with a decrease of frequency at the range of small parameters both components of the field tend to those corresponding to the quasistationary field.

For evaluation of the influence of the dielectric constant when the probe is located on the borehole axis we can make use of the approximate method taking into account the skin effect in the formation. Then, by analogy with the quasistationary case we have:

$$h_z \simeq H_{0z}(k_2L) + \frac{k_1^2 - k_2^2}{2}L^2G_1(\alpha) \tag{4.275}$$

i.e. the influence of dielectric constant is practically the same as in a uniform medium.

Chapter 5

QUASISTATIONARY MAGNETIC FIELD OF A VERTICAL MAGNETIC DIPOLE IN A FORMATION WITH A FINITE THICKNESS

In this chapter we will consider vertical responses of two-coil induction probes located arbitrarily with respect to interfaces between a bed and a surrounding medium. Special attention will be paid to the influence of frequency, ratio of conductivities and geometric factors such as formation thickness probe length and probe position. It is appropriate to notice that analyses performed in this chapter are used for investigation of vertical responses of multi-coil induction probes.

5.1. Derivation of Formulae for the Vertical Component of the Magnetic Field of a Vertical Magnetic Dipole

Suppose that there are two parallel interfaces which divide a space into three parts as shown in Fig. 5.1. The vertical magnetic dipole is placed at the origin of the cylindrical system of coordinates and its moment is oriented along the z-axis.

Let us assume that the magnetic permeability of the medium is equal to $4\pi \times 10^{-7}$ H/m.

As is well known, the quasistationary field is described by Maxwell's equations:

$$\begin{aligned} \operatorname{curl} \boldsymbol{E} &= \frac{\partial \boldsymbol{B}}{\partial t} \qquad \operatorname{div} \boldsymbol{E} = 0 \\ \operatorname{curl} \boldsymbol{H} &= \sigma \boldsymbol{E} \qquad \operatorname{div} \boldsymbol{H} = 0 \end{aligned} \tag{5.1}$$

The current in the dipole changes with time as function $e^{-i\omega t}$, and therefore electrical and magnetic fields change in the same manner. For this reason Maxwell's equations for complex amplitudes of the field can be presented in the form:

$$\begin{aligned} \operatorname{curl} \boldsymbol{E} &= i\omega\mu \boldsymbol{H} \qquad \operatorname{div} \boldsymbol{E} = 0 \\ \operatorname{curl} \boldsymbol{H} &= \sigma \boldsymbol{E} \qquad \operatorname{div} \boldsymbol{H} = 0 \end{aligned} \tag{5.2}$$

As follows from the third Maxwell equation: $\operatorname{div} \boldsymbol{E} = 0$, the electrical field can be expressed through a vector potential of the electrical type $\boldsymbol{A}^*$:

$$\boldsymbol{E} = i\omega\mu \operatorname{curl} \boldsymbol{A}^* \tag{5.3}$$

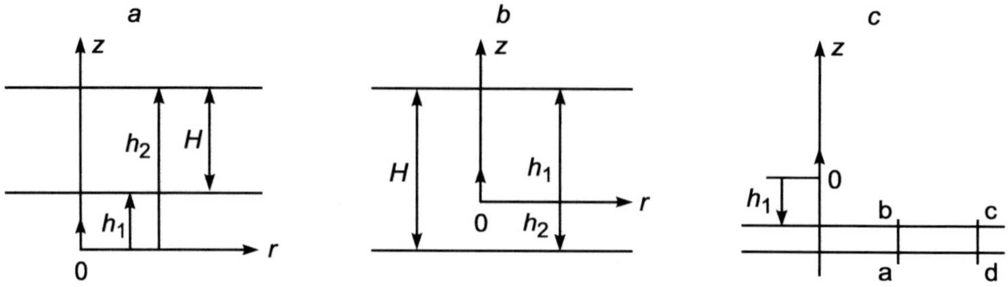

Figure 5.1. Various positions of the magnetic dipole with respect to interfaces.

Applying the same approach as in Chapter 4 we find:

$$\boldsymbol{H} = k^2\boldsymbol{A}^* + \text{grad}\,\text{div}\,\boldsymbol{A}^* \tag{5.4}$$

and

$$\nabla^2\boldsymbol{A}^* + k^2\boldsymbol{A}^* = 0 \tag{5.5}$$

where $k^2 = \mathrm{i}\sigma\mu\omega$.

Taking into account the axial symmetry and correspondingly the absence of surface electrical charges we will look for a solution with the help of the vertical component of the vector potential A_z^* only:

$$\boldsymbol{A}^* = (0, 0, A_z^*) \tag{5.6}$$

It is clear that function A_z^* in this case depends on two coordinates, r and z :

$$A_z^* = A_z^*(r,z) \qquad r = (x^2+y^2)^{1/2}$$

Making use of eqs. 5.3–5.4 we obtain the following expressions for the field components:

$$\begin{aligned} &H_r = \frac{\partial^2 A_z^*}{\partial r \partial z} \qquad H_z = k^2 A_z^* + \frac{\partial^2 A_z^*}{\partial z^2} \qquad H_\phi = 0 \\ &E_r = E_z = 0 \qquad E_\phi = -\mathrm{i}\omega\mu\frac{\partial A_z^*}{\partial r} \end{aligned} \tag{5.7}$$

Due to the continuity of tangential components of the field, boundary conditions for the vector potential at interfaces are:

$$A_{iz}^* = A_{kz}^* \qquad \frac{\partial A_{iz}^*}{\partial z} = \frac{\partial A_{kz}^*}{\partial z} \qquad \text{if } z = h_{ik} \tag{5.8}$$

Near the origin of the coordinate system the field tends to that of a magnetic dipole in a uniform medium, and therefore for the vector potential we have:

$$A_z^* \to \frac{M}{4\pi}\frac{\mathrm{e}^{\mathrm{i}kR}}{R} \quad \text{as } R \to 0$$

where M is the dipole moment and $R = (r^2 + z^2)^{1/2}$.

At infinity ($R \to \infty$) the field and correspondingly the vector potential vanish.

Thus, in order to find the field it is necessary to solve equations:

$$\begin{aligned} \nabla^2 A_{1z}^* + k_1^2 A_{1z}^* &= 0 \\ \nabla^2 A_{2z}^* + k_2^2 A_{2z}^* &= 0 \\ \nabla^2 A_{3z}^* + k_3^2 A_{3z}^* &= 0 \end{aligned} \tag{5.9}$$

and provide conditions 5.8 as well as a corresponding behavior of the field near the source and at infinity. Here k_1, k_2, and k_3 are wave numbers of every part of the medium.

First, let us consider particular solutions of equation:

$$\nabla^2 A_z^* + k^2 A_z^* = 0$$

If the vector potential, A_z^* depends on distance R only, the latter has the form:

$$\frac{\partial^2 A_z^*}{\partial R^2} + \frac{1}{R}\frac{\partial A_z^*}{\partial R} + k^2 A_z^* = 0$$

or

$$\frac{\partial^2 (R A_z^*)}{\partial R^2} + k^2 (R A_z^*) = 0$$

Whence

$$R A_z^* = A\mathrm{e}^{\mathrm{i}kR} + B\mathrm{e}^{-\mathrm{i}kR}$$

We will assume in this chapter that the wave number k has a positive imaginary part, i.e. $\mathrm{Re}\,\mathrm{i}k < 0$, here Re is the real part of the complex number, $\mathrm{i}k$. Function A_z^* tends to zero as $R \to \infty$, letting $B = 0$ and $A = M/4\pi$, for this reason we obtain the known expression for the vector potential of a magnetic dipole in a uniform medium:

$$A_z^* = \frac{M}{4\pi}\frac{\mathrm{e}^{\mathrm{i}kR}}{R} \tag{5.10}$$

Now, let us find a solution of eq. 5.5 in cylindrical coordinates (r, z), since the field does not depend on coordinate ϕ. Correspondingly, eq. 5.5 can be written in the form:

$$\frac{\partial^2 A_z^*}{\partial r^2} + \frac{1}{r}\frac{\partial A_z^*}{\partial r} + \frac{\partial^2 A_z^*}{\partial z^2} + k^2 A_z^* = 0$$

Letting $A_z^* = U(r)V(z)$ and applying the method of separation of variables we obtain two normal differential equations:

$$\frac{\partial^2 U}{\partial r^2} + \frac{1}{r}\frac{\partial U}{\partial r} + \lambda^2 U = 0$$

$$\frac{\partial^2 U}{\partial z^2} - (\lambda^2 - k^2)V = 0$$

where λ is the separation constant.

The first equation is called the Bessel equation and its solutions are Bessel functions of the first and second kind: $J_0(\lambda r)$ and $Y_0(\lambda r)$:

$$U(r) = AJ_0(\lambda r) + BY_0(\lambda r)$$

Function $Y_0(\lambda r)$ tends to infinity as $r \to 0$, and therefore it cannot describe a field.

The second equation has a solution:

$$U(z) = C\mathrm{e}^{-(\lambda^2-k^2)^{1/2}} + D\mathrm{e}^{(\lambda^2-k^2)^{1/2}}$$

Correspondingly, the general solution of eq. 5.5 can be presented in the form:

$$A_z^*(r,z) = \int_0^x \left[N_1\mathrm{e}^{(\lambda^2-k^2)^{1/2}z} + N_2\mathrm{e}^{-(\lambda^2-k^2)^{1/2}z}\right] J_0(\lambda r)\,\mathrm{d}\lambda \tag{5.11}$$

We will choose the sign of radical: $(\lambda^2 - k^2)^{1/2}$ in such a way that its real part is positive, i.e:

$$\mathrm{Re}(\lambda^2 - k^2)^{1/2} > 0 \tag{5.12}$$

We will present the field in a medium where the dipole is located as a sum:

$$A_{1z}^* = \frac{M}{4\pi}\frac{\mathrm{e}^{ikR}}{R} + A_{1z}^{*s} \tag{5.13}$$

where A_{1z}^{*s} is the vector potential of the secondary field.

As is known, the vector potential of the magnetic dipole can be expressed by a Sommerfeld integral:

$$\frac{M}{4\pi}\frac{\mathrm{e}^{ikR}}{R} = \frac{M}{4\pi}\int_0^x \frac{\lambda\mathrm{e}^{-(\lambda^2-k^2)^{1/2}|z|}}{(\lambda^2-k^2)^{1/2}} J_0(\lambda r)\,\mathrm{d}\lambda \tag{5.14}$$

Now we are prepared to derive formulae for the vector potential for various positions of the dipole with respect to the interfaces.

Let us introduce the following notations for different parts of a medium:

(1) stands for the medium where the dipole is placed;

(2) stands for the medium occupied by the bed;

(3) stands for the medium located at the opposite side of the bed.

5.1.1. The Field of the Magnetic Dipole Located outside the Bed

In accord with eq. 5.11 and taking into account the condition at infinity, expressions for the vector potential in every part of the medium can be written in the form (Fig. 5.1a):

$$A_{1z}^* = \frac{M}{4\pi}\int_0^\infty \left[\frac{\lambda}{\lambda_1}e^{\lambda_1|z|} + D_1 e^{\lambda_1 z}\right] J_0(\lambda r)\, d\lambda \qquad \text{if } z \leqslant h_1 \tag{5.15}$$

$$A_{2z}^* = \frac{M}{4\pi}\int_0^\infty \left[D_2 e^{\lambda_2 z} + D_3 e^{\lambda_3 z}\right] J_0(\lambda r)\, d\lambda \qquad \text{if } h_1 \leqslant z \leqslant h_2 \tag{5.16}$$

$$A_{3z}^* = \frac{M}{4\pi}\int_0^\infty D_4 e^{-\lambda_1 z} J_0(\lambda r)\, d\lambda \qquad \text{if } z \leqslant h_2 \tag{5.17}$$

where h_1 is the distance from the dipole to the nearest interface; $h_2 = h_1 + H$; H is the bed thickness; $\lambda_1 = (\lambda^2 - k_1^2)^{1/2}$, $\lambda_2 = (\lambda^2 - k_2^2)^{1/2}$.

In accord with boundary conditions 5.8 we obtain a system of linear equations with respect to D_1, D_2, D_3, and D_4:

$$\begin{aligned}
&\frac{\lambda}{\lambda_1}e^{-\lambda_1 h_1} + D_1 e^{\lambda_1 h_1} &&= D_2 e^{\lambda_2 h_1} + D_3 e^{-\lambda_2 h_1}\\
&-\lambda e^{-\lambda_1 h_1} + \lambda_1 D_1 e^{\lambda_1 h_1} &&= \lambda_2 D_2 e^{\lambda_2 h_1} - \lambda_2 D_3 e^{-\lambda_2 h_1}\\
&D_2 e^{\lambda_2 h_1} + D_3 e^{-\lambda_2 h_2} &&= D_4 e^{\lambda_1 h_2}\\
&\lambda_2 D_2 e^{-\lambda_2 h_2} - \lambda_2 D_3 e^{-\lambda_2 h_2} &&= -\lambda_1 D_4 e^{-\lambda_1 h_2}
\end{aligned} \tag{5.18}$$

Solving this system we have:

$$D_1 = -\frac{\lambda K_{12} e^{2\lambda_1 h_1}\left(1 - e^{-2\lambda_2 H}\right)}{\lambda_1\left(1 - K_{12}^2 e^{-2\lambda_2 H}\right)} \tag{5.19}$$

$$D_2 = \frac{2\lambda K_{12} e^{-(\lambda_1+\lambda_2)h_1} e^{-2\lambda_2 H}}{(\lambda_1 + \lambda_2)\left(1 - K_{12}^2 e^{-2\lambda_2 H}\right)} \tag{5.20}$$

$$D_3 = \frac{2\lambda e^{-(\lambda_1-\lambda_2)h_1}}{(\lambda_1 + \lambda_2)\left(1 - K_{12}^2 e^{-2\lambda_2 H}\right)} \tag{5.21}$$

$$D_4 = \frac{4\lambda_1\lambda_2 e^{-(\lambda_1-\lambda_2)H}}{(\lambda_1 + \lambda_2)^2\left(1 - K_{12}^2 e^{-2\lambda_1 H}\right)} \tag{5.22}$$

Substituting these expressions for the coefficients into eqs. 5.15–5.17 we have:

$$A_{1z}^* = \frac{M}{4\pi}\int_0^\infty \frac{\lambda}{\lambda_1}\left[e^{\lambda_1|z|} - \frac{K_{12}(1 - e^{-2\lambda_2 H})e^{-\lambda_1(2h_1 - z)}}{1 - K_{12}^2 e^{-2\lambda_2 H}}\right] J_0(\lambda r)\, d\lambda \tag{5.23}$$

$$A^*_{2z} = \frac{M}{4\pi} \int\limits_0^\infty \frac{2\lambda e^{-\lambda_1 h_1} e^{-\lambda_2(z-h_1)} \left(1 + K_{12} e^{2\lambda_2(z-h_1-H)}\right)}{(\lambda_1+\lambda_2)(1-K_{12}^2 e^{-2\lambda_2 H})} J_0(\lambda r)\, d\lambda \tag{5.24}$$

$$A^*_{3z} = \frac{M}{4\pi} \int\limits_0^\infty \frac{4\lambda\lambda_2 e^{-(\lambda_2-\lambda_1)H} e^{-\lambda_1 z}}{(\lambda_1+\lambda_2)^2 (1-K_{12}^2 e^{-2\lambda_2 H})} J_0(\lambda r)\, d\lambda \tag{5.25}$$

Parameter z in eqs. 5.14–5.25 is the length of the two-coil induction probe.

Usually in induction logging the vertical component of the magnetic field is measured on the borehole axis ($r = 0$). In accord with eqs. 5.23–5.25 we obtain equations for this component expressed in units of the primary field:

$$h_z^{(1)} = h_z^{(0)} - \frac{1}{2} \int\limits_0^\infty \frac{m^3}{m_1} \frac{K_{12}(1-e^{-2m_2\alpha})e^{-m_1(2\beta-1)}}{1-K_{12}^2 e^{-2m_2\alpha}}\, dm \qquad \text{if } \beta \geqslant 1 \tag{5.26}$$

$$h_z^{(2)} = \int\limits_0^\infty \frac{m^3 e^{-m_1\beta} e^{-m_2(1-\beta)} \left(1 + K_{12} e^{2m_2(1-\beta-\alpha)}\right)}{(m_1+m_2)(1-K_{12}^2 e^{-2m_2\alpha})}\, dm \qquad \text{if } 1 \geqslant \beta \geqslant 1-\alpha \tag{5.27}$$

$$h_z^{(3)} = \int\limits_0^\infty \frac{2m^3 m_2 e^{-(m_2-m_1)\alpha} e^{-m_1}}{(m_1+m_2)^2 (1-K_{12}^2 e^{-2m_2})}\, dm \qquad \text{if } \beta \leqslant 1 \tag{5.28}$$

where $m = \lambda z$, $h_z = H_z/(2M/4\pi z^3)$, $m_1 = (m^2 - k_1^2 z^2)^{1/2}$, $m_2 = (m^2 - k_2^2 z^2)^{1/2}$, $K_{12} = (m_1 - m_1)/(m_2 + m_1)$, $\alpha = H/z$, $\beta = h_1/z$; H is the bed thickness; $h_z^{(0)}$ is the vertical component of the magnetic field of the magnetic dipole in a uniform medium with conductivity σ_1.

The latter equation (eq. 5.28) corresponds to the case when the layer is located between the dipole and the observation point and, as it follows from this formula, the field does not depend on the position of the layer with respect to the probe coils.

5.1.2. The Field of the Magnetic Dipole Located within the Bed

Unlike the previous case the dipole is located within the bed (Fig. 5.1b), and for this reason expressions for the vector potential can be written in the form:

$$A^*_{1z} = \frac{M}{4\pi} \int\limits_0^\infty D_1 e^{\lambda_1 z} J_0(\lambda r)\, d\lambda \qquad \text{if } z \leqslant h_2 \tag{5.29}$$

$$A^*_{2z} = \frac{M}{4\pi} \int\limits_0^\infty \left(\frac{\lambda}{\lambda_2} e^{-\lambda_2 |z|} + D_2 e^{\lambda_2 z} + D_3 e^{-\lambda_2 z} \right) J_0(\lambda r)\, d\lambda \qquad \text{if } h_2 \leqslant z \leqslant h_1 \tag{5.30}$$

$$A^*_{3z} = \frac{M}{4\pi} \int\limits_0^\infty D_4 e^{-\lambda_1 z} J_0(\lambda r)\, d\lambda \qquad \text{if } z \geqslant h_1 \tag{5.31}$$

where h_1 is the distance from the dipole to the upper interface of the bed, $h_2 = H - h_1$; H is the bed thickness.

By making use of the boundary conditions to determine the unknown coefficients we have the following system:

$$\begin{aligned}
D_1\mathrm{e}^{-\lambda_1 h_2} &= \frac{\lambda}{\lambda_2}\mathrm{e}^{-\lambda_2 h_2} + D_2\mathrm{e}^{-\lambda_2 h_2} + D_3\mathrm{e}^{\lambda_2 h_2} \\
\lambda_1 D_1\mathrm{e}^{-\lambda_1 h_2} &= \lambda\mathrm{e}^{-\lambda_2 h_2} + \lambda_2 D_2\mathrm{e}^{-\lambda_2 h_2} - \lambda_2 D_3\mathrm{e}^{\lambda_2 h_2} \\
D_4\mathrm{e}^{-\lambda_1 h_1} &= \frac{\lambda}{\lambda_2}\mathrm{e}^{-\lambda_2 h_1} + D_2\mathrm{e}^{\lambda_2 h_1} + D_3\mathrm{e}^{-\lambda_2 h_1} \\
-\lambda_1 D_4\mathrm{e}^{-\lambda_1 h_1} &= -\lambda\mathrm{e}^{-\lambda_2 h_1} + \lambda_2 D_2\mathrm{e}^{\lambda_2 h_1} - \lambda_2 D_3\mathrm{e}^{-\lambda_2 h_1}
\end{aligned} \tag{5.32}$$

In this case we will only consider the field inside the bed inasmuch as expressions for the field outside the bed can be derived from eq. 5.27.

Solving the system 5.32 we find:

$$D_2 = \frac{\lambda K_{12}\mathrm{e}^{-2\lambda_2 h_1}\left(1 + K_{12}\mathrm{e}^{-2\lambda_2 h_2}\right)}{\lambda_2\left(1 - K_{12}^2\mathrm{e}^{-2\lambda_2 H}\right)} \tag{5.33}$$

$$D_3 = \frac{\lambda K_{12}\mathrm{e}^{-2\lambda_2 h_2}\left(1 + K_{12}\mathrm{e}^{-2\lambda_2 h_1}\right)}{\lambda_2\left(1 - K_{12}^2\mathrm{e}^{-2\lambda_2 H}\right)} \tag{5.34}$$

Substituting these expressions into eq. 5.30 we obtain:

$$A_{2z}^* = \frac{M}{4\pi}\int_0^\infty \left[\frac{\lambda}{\lambda_2}\mathrm{e}^{-\lambda_2|z|} + \frac{\lambda K_{12}\left[\mathrm{e}^{-\lambda_2(2h_1 - z)} + \mathrm{e}^{-\lambda_2(2h_2+z)} + 2K_{12}\mathrm{e}^{-\lambda_2 H}\cosh\lambda_2 z\right]}{\lambda_2\left(1 - K_{12}^2\mathrm{e}^{-2\lambda_2 H}\right)}\right] \times J_0(\lambda r)\,\mathrm{d}\lambda \tag{5.35}$$

In accord with eqs. 5.7 and 5.35, the expression for the vertical component of the magnetic field on the dipole axis related to the primary field is:

$$h_z = h_z^0 + \frac{1}{2}\int_0^\infty \frac{m^3 K_{12}\left[\mathrm{e}^{-(1+2\beta)m_2} + \mathrm{e}^{-(2\alpha-2\beta-1)m_2} + 2K_{12}\mathrm{e}^{-2\alpha m_2}\cosh m_2\right]}{m_2\left(1 - K_{12}^2\mathrm{e}^{-2\alpha m_2}\right)}\,dm \tag{5.36}$$

where $\alpha = H/z$, $\beta = h_2/z$.

If coils of the probe are located symmetrically with respect to interfaces, that is, $2\beta = \alpha - 1$, the latter equation can be presented as:

$$h_z = h_z^0 + \int_0^\infty \frac{m^3 K_{12}\mathrm{e}^{-2\alpha m_2}}{m_2}\,\frac{\mathrm{e}^{\alpha m_2} + K_{12}\cosh m_2}{1 - K_{12}^2\mathrm{e}^{-2\alpha m_2}}\,\mathrm{d}m \tag{5.37}$$

5.1.3. The Field of the Vertical Magnetic Dipole in the Presence of a Thin Conducting Plane

If the probe length is significantly greater than the bed thickness (Fig. 5.1c) and its conductivity essentially exceeds that of a surrounding medium and, finally, the skin depth inside the bed is much greater than its thickness, the bed can be replaced by a thin conducting plane with conductance S, equal to product of conductivity and thickness of the bed. Instead of the exact boundary conditions we can make use of two approximate conditions which do not require any information about the field inside the bed. The first boundary condition is continuity of the tangential component of the electrical field:

$$E_{1\phi} = E_{2\phi} \tag{5.38}$$

Circulation of the magnetic field along contour *abcd* is equal to the current piercing this contour (Fig. 5.1c). For this reason:

$$\oint \boldsymbol{H}\,\mathrm{d}\boldsymbol{l} = H_{1r}\,\mathrm{d}r - H_{2r}\,\mathrm{d}r = \sigma\,\mathrm{d}r\,\mathrm{d}h\,E_\phi \qquad \text{as } \mathrm{d}h \to 0$$

or

$$H_{1r} - H_{2r} = SE_\phi \tag{5.39}$$

where S is the conductance of the thin layer.

In accord with eqs. 5.7, 5.38 and 5.39, the boundary conditions for the vector potential have the form:

$$A^*_{1z} = A^*_{2z} \qquad \frac{\partial A^*_{1z}}{\partial z} - \frac{\partial A^*_{2z}}{\partial z} = -\mathrm{i}\omega\mu S A^*_{2z} \qquad \text{as } z = -h_1 \tag{5.40}$$

For function A^*_z outside of the conducting surface we have:

$$\begin{aligned} A^*_{1z} &= \frac{M}{4\pi}\int_0^\infty \left[\frac{\lambda}{\lambda_1}\mathrm{e}^{-\lambda_1|z|} + D_1\mathrm{e}^{-\lambda_1 z}\right] J_0(\lambda r)\,\mathrm{d}\lambda \qquad \text{if } z \leqslant h_1 \\ A^*_{2z} &= \frac{M}{4\pi}\int_0^\infty D_2\mathrm{e}^{-\lambda_2 z} J_0(\lambda r)\,\mathrm{d}\lambda \qquad \text{if } z \geqslant h_1 \end{aligned} \tag{5.41}$$

Substituting these expressions into eqs. 5.40 we obtain the system for determination of D_1 and D_2:

$$\begin{aligned} -D_1\mathrm{e}^{\lambda_1 h_1} + D_2\mathrm{e}^{-\lambda_1 h_1} &= \frac{\lambda}{\lambda_1}\mathrm{e}^{-\lambda_1 h_1} \\ \lambda_1 D_1\mathrm{e}^{\lambda_1 h_1} + (\lambda_1 - \mathrm{i}\omega\mu S)D_2\mathrm{e}^{-\lambda_1 h_1} &= \lambda\mathrm{e}^{-\lambda_1 h_1} \end{aligned}$$

Solving this system we have:

$$D_1 = \frac{\lambda K_s^2\mathrm{e}^{-2\lambda_1 h_1}}{\lambda_1(2\lambda_1 - K_s^2)} \qquad D_2 = \frac{2\lambda}{2\lambda_1 - K_s^2} \tag{5.42}$$

here $k_s^2 = \mathrm{i}\omega\mu S$.

Therefore:

$$A^*_{1z} = \frac{M}{4\pi}\int_0^\infty \left[\frac{\lambda}{\lambda_1}\mathrm{e}^{-\lambda_1|z|} + \frac{\lambda K_s^2 \mathrm{e}^{-2\lambda_1 h_1}}{\lambda_1(2\lambda_1 - K_s^2)}\right] J_0(\lambda r)\,\mathrm{d}\lambda$$

$$A^*_{2z} = \frac{M}{4\pi}\int_0^\infty \frac{2\lambda}{2\lambda_1 - K_s^2}\mathrm{e}^{\lambda_1 z} J_0(\lambda r)\,\mathrm{d}\lambda$$

Correspondingly, for the vertical component of the magnetic dipole along its axis we have:

$$\begin{aligned} h_z^{(1)} &= h_z^{(0)} + \frac{n_s}{2}\int_0^\infty \frac{m^3 \mathrm{e}^{m(1+2\alpha)}}{m_1(2m_1 - n_s)}\,dm \\ h_z^{(2)} &= \int_0^\infty \frac{m^3 \mathrm{e}^{m_1}}{2m_1 - n_s}\,dm \end{aligned} \qquad (5.43)$$

where $m = \lambda z$, $m_1 = (m^2 - k_1^2 z_1^2)^{1/2}$, $\alpha = h_1/z$, $n_s = \mathrm{i}\omega\mu S z$.

Formulae derived in this section have allowed us to investigate the vertical characteristics of two-coil induction probes for various frequencies and conductivities of medium as well as to describe curves of profiling.

First let us consider the range of small parameters when skin depth in every medium is greater than probe length and bed thickness.

5.2. The Vertical Responses of the Two-coil Induction Probe in the Range of Small Parameters

Suppose that the frequency and the conductivity of a medium are so low that both parameters $n_1 = \sigma_1\mu\omega z^2$ and $n_2 = \sigma_2\mu\omega z^2$ are much less than unity. In this case the theory of induction logging in a medium with horizontal interfaces can be easily developed. There are at least two approaches for the solution of this problem. One of them is based on obtaining asymptotic formulae proceeding from the exact solution developed in the first section. The second approach makes use of the concept of geometric factor of thin layer suggested by H. Doll. We will describe here the theory of the two-coil induction probe applying geometrical factor presentation.

5.2.1. Geometric Factor of an Elementary Layer

In accord with H. Doll a layer with a thickness which is much less than the probe length and is equal to unity, will be called the elementary layer. The geometric factor of such an elementary layer can be found by performing a summation of geometric factors of all elementary rings located at the same height z with respect to the origin and forming this layer. The radius of the elementary rings changes from zero to infinity. For this reason an expression of the geometric factor of the elementary layer, G_z, has the form:

$$G_z = \int_0^\infty q\,\mathrm{d}r \tag{5.44}$$

where q is the geometric factor of the elementary ring.

Making use of eq. 3.104 we have:

$$G_z = \frac{L}{2}\int_0^\infty \frac{r^3\,\mathrm{d}r}{[r^2+(L/2+z)^2]^{3/2}\,[r^2+(L/2-z)^2]^{3/2}}$$

where L is the probe length.

Introducing notations: $L/2+z=m$, $L/2-z=n$, and $r^3=x$, we obtain:

$$G_z = \frac{L}{2}\int_0^\infty \frac{x\,\mathrm{d}x}{[x^2+(m^2+n^2)x+m^2n^2]^{3/2}}$$

or (5.45)

$$G_z = \frac{L}{2}\int_0^\infty \frac{x\,\mathrm{d}x}{(x^2+bx+c)^{3/2}}$$

where $m^2+n^2=b$, $m^2n^2=c$.

The integral in eq. 5.45 is tabulated and it is equal to:

$$\frac{Lc}{(4c-b^2)c^{1/2}} - \frac{Lb}{2(4c-b^2)}$$

or

$$G_z = \frac{L}{2(m+n)^2}$$

There are two possible cases such as:

- $m > 0 \quad n > 0$
- $m < 0 \quad n < 0.$

In the first case the elementary layer is located between the coils of the probe ($L/2 > z$ and $z > -L/2$). Then its geometric factor, G_z, is:

$$G_z = 1/2L \tag{5.46}$$

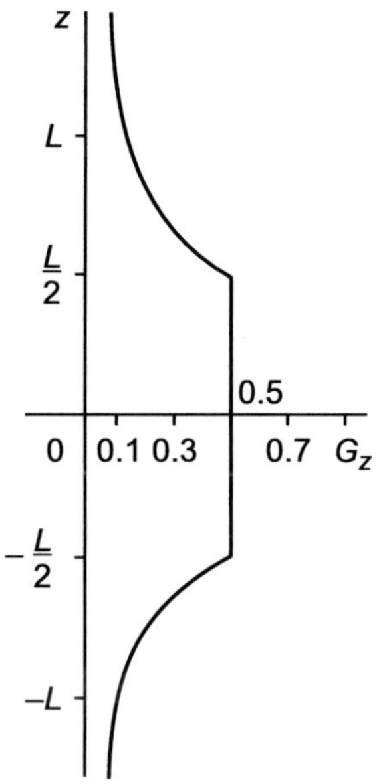

Figure 5.2. Geometric factor of the elementary layer G_z as a function of z.

In the second case the layer is located outside the probe ($z > L/2$, $z < -L/2$). Then we have:

$$G_z = L/8z^2 \tag{5.47}$$

As follows from eqs. 5.46–5.47 the geometric factor of the whole space is equal to unity. In fact we obtain:

$$G = \frac{L}{8}\int\limits_{L/2}^{\infty}\frac{dz}{z^2} + \frac{L}{2L} + \frac{L}{8}\int\limits_{L/2}^{\infty}\frac{dz}{z^2} = \frac{L}{8}\frac{2}{L} + \frac{1}{2} + \frac{L}{8}\frac{2}{L} = 1 \tag{5.48}$$

According to eqs. 5.46–5.47 the geometric factors of elementary layers located outside the probe decrease inversely proportional to z^2 while geometric factors of elementary layers located inside the interval between coils of the probe are equal to each other regardless of z. A curve illustrating the behavior of geometric factor G_z as a function of z is shown in Fig. 5.2.

Values of geometric factor of elementary layers and the distance, z, between them and the middle of a two-coil induction probe are plotted along horizontal and vertical axis, respectively.

Inasmuch as this dependence of function G_z on z reflects sensitivity of the probe to induced currents in elementary layers, H. Doll called this function G_z the vertical response of a two-coil induction probe.

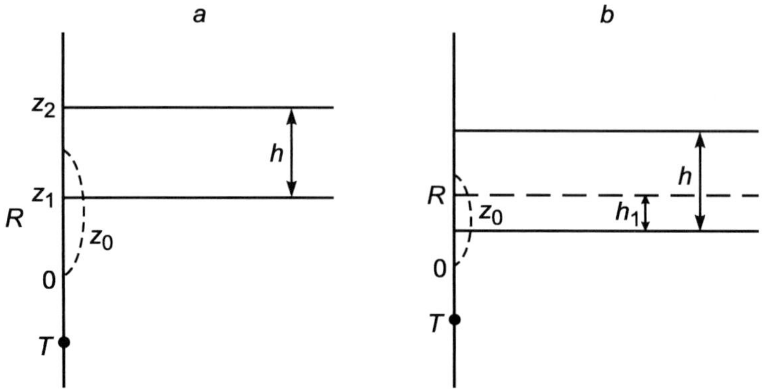

Figure 5.3. Position of the probe with respect to the bed.

5.2.2. Geometric Factor of a Layer with a Finite Thickness

Proceeding from the expression for the geometric factor of an elementary layer it is not difficult to find geometric factors of a layer with a finite thickness. For this purpose it is necessary to present the layer as a sum of elementary ones and perform summation of their geometric factors. Let us consider several positions of the two-coil probe with respect to the bed.

The probe is located outside the bed of finite thickness

In order to derive the geometric factor of this bed (Fig. 5.3a) we have to integrate function $q = L/8z^2$ by z within the interval from z_1 to z_2 where z_1 and z_2 are coordinates of the bed boundaries. Then we have:

$$G_b = \frac{L}{8}\int\limits_{z_1}^{z_2} \frac{dz}{z^2} = \frac{L}{8}\left(\frac{1}{z_1} - \frac{1}{z_2}\right) \tag{5.49}$$

Assuming that the coordinate origin is placed at the middle of the bed and taking into account that $z_1 = z_0 - L/2$ and $z_2 = z_0 + L/2$, instead of eq. 5.49 we have:

$$G_b = \frac{LH}{2}\frac{1}{z_0^2 - (H/2)^2} \tag{5.50}$$

where H is the bed thickness; z_0 is the distance from the middle of the bed to the center of a two-coil probe.

This equation is applied if the upper coil of the probe does not intersect the low boundary of the bed, i.e. it is valid if $z_1 \geqslant L/2$ or $z_0 \geqslant L/2 + H/2$.

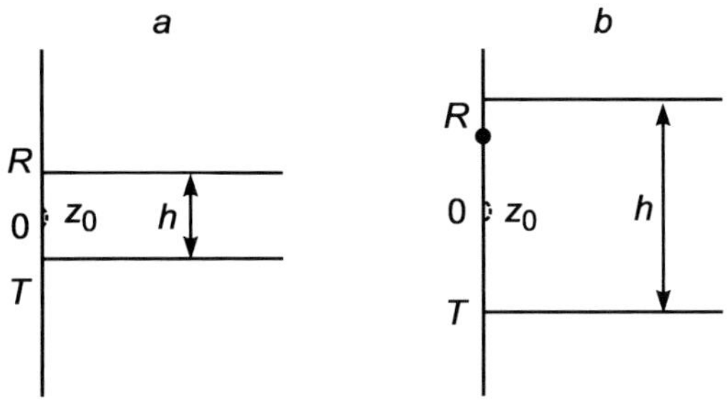

Figure 5.4. Position of the probe with respect to the bed.

One coil of the probe is located inside the bed while the other is outside of it

For deriving the geometric factor of the bed with thickness H for such a position of the probe (Fig. 5.36) we have to add geometric factors of parts of the bed located outside and inside the probe. In accord with eq. 5.47 the first of them, G_1, is:

$$G_1 = \frac{L}{8}\left(\frac{1}{L/2} - \frac{1}{z_0 + H/2}\right) = \frac{1}{4} - \frac{H}{8(z_0 + H/2)}$$

The geometric factor of that part of the bed which is located inside the probe and has thickness h_1 is calculated as:

$$G_2 = \frac{1}{2L}h_1 = \frac{1}{2L}\left(\frac{L}{2} - z_0 + \frac{H}{2}\right)$$

since $h_1 = L/2 - (z_0 - H/2)$. Therefore, for the geometric factor of the bed we have:

$$G_b = G_1 + G_2 = \frac{1}{2} - \frac{1}{2L}(z_0 - H/2) - \frac{L}{8(z_0 + H/2)} \tag{5.51}$$

This formula is applied until the upper receiver of the probe is located within the bed if its thickness is smaller than the probe length, $H < L$, i.e. when $z_2 > L/2$ or $L/2 - H/2 \leqslant z_0 \leqslant L/2 + H/2$. In the case, when the bed thickness is greater than the probe length $(H > L)$ this formula can be used until the lower coil does not intersect the lower boundary of the bed, i.e. when $z_1 \geqslant -L/2$ or $z_0 \geqslant H/2 - L/2$.

The probe is located against the bed

There are two possible variants (Fig. 5.4):

- The probe length exceeds the bed thickness $(H < L)$.
- The thickness of the bed is greater than the length of the probe $(H > L)$.

It is obvious that for the first variant we have:

$$G_b = H/2L \tag{5.52}$$

At the same time for the second variant we have:

$$\begin{aligned} G_b &= \frac{L}{2L} + \frac{L}{8}\left(\frac{1}{L/2} - \frac{1}{z_0 + H/2}\right) + \frac{L}{8}\left(\frac{1}{L/2} - \frac{1}{H/2 - z_0}\right) \\ &= \frac{1}{2} + \frac{1}{4} - \frac{L}{8(z_0 + H/2)} + \frac{1}{4} + \frac{L}{8(z_0 + H/2)} = 1 + \frac{LH}{8\left[z_0 - (H/2)^2\right]} \end{aligned} \tag{5.53}$$

These equations can be applied provided that:

$$0 \leqslant z_0 < L/2 - H/2 \qquad \text{if } H < L$$

and

$$0 \leqslant z_0 < H/2 - L/2 \qquad \text{if } H > L$$

Formulae derived for geometric factors allow us to determine the apparent conductivity for a two-coil induction probe located in a medium with two horizontal interfaces. As was shown in Chapter 4 we have:

$$\sigma_a = \sigma_1 G_1 + \sigma_2 G_2 \tag{5.54}$$

induction probe in a medium with one interface. where σ_1 and σ_2 are the conductivities of the bed and the surrounding medium while G_1 and G_2 are their geometric factors. By definition the sum of these factors is equal to unity, i.e:

$$G_2 = 1 - G_1$$

Before we will investigate the apparent conductivity in the presence of a bed having a finite thickness let us consider the influence of one horizontal interface.

If the probe is located in a medium with conductivity σ_2 (Fig. 5.5a) then in accord with eq. 5.49 the geometric factors of both half-spaces are:

$$G_1 = \frac{L}{8z_0} \quad G_2 = \frac{L}{8}\left(\frac{1}{L/2} - \frac{1}{z_0}\right) + \frac{L}{2L} + \frac{1}{8}\frac{1}{L/2}$$

and

$$\sigma_a = \sigma_2 - (\sigma_2 - \sigma_1)\frac{L}{8z_0}$$

This formula is applied provided $z_0 \geqslant L/2$, i.e. the coil of the probe does not intersect the interface.

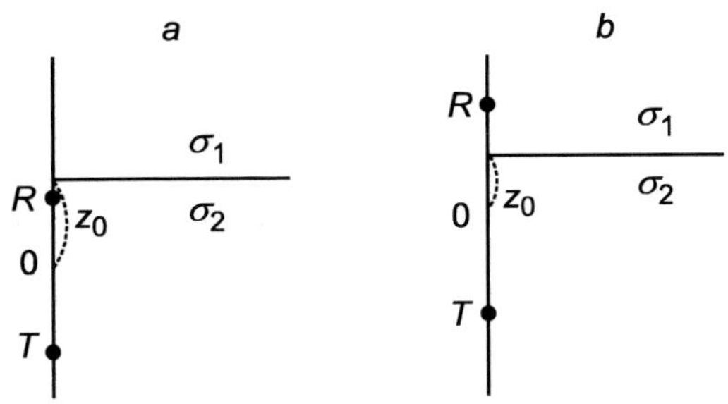

Figure 5.5. The two-coil induction probe in a medium with one interface.

In the case when coils of the induction probe are located in different parts of the medium (Fig. 5.5b) the geometric factors are:

$$G_1 = \frac{1}{8}\frac{1}{L/2} + \frac{1}{2L}\left(\frac{L}{2} - z_0\right) = \frac{1}{2} - \frac{z_0}{2L}$$

$$G_2 = \frac{1}{2L}\left(\frac{L}{2} + z_0\right) + \frac{L}{8}\frac{1}{L/2} = \frac{1}{2} + \frac{z_0}{2L}$$

Correspondingly, for the function σ_a we have:

$$\sigma_a = \frac{1}{2}(\sigma_1 + \sigma_2) + (\sigma_1 - \sigma_2)\frac{z_0}{2L}$$

In that case when the probe center is in a medium with conductivity σ_2, i.e. $z_0 < 0$, formulae derived above remain valid provided that conductivities σ_1 for various and σ_2 are changed by roles, namely:

$$\sigma_a = \frac{1}{2}(\sigma_1 + \sigma_2) + (\sigma_1 - \sigma_2)\frac{z_0}{2L} \qquad -L/2 < z_0 < 0$$

$$\sigma_a = \sigma_1 - (\sigma_1 - \sigma_2)\frac{L}{8z_0} \qquad -\infty < z_0 < -L/2$$

Apparent conductivity curves for various positions of the probe with respect to the interface are shown in Fig. 5.6.

Let us notice that the value of the apparent conductivity is equal to the mean value of both conductivities when the probe center is located at the interface.

Now we will investigate apparent conductivity curves in the presence of a bed. Inasmuch as function σ_a is symmetrical with respect to the middle of the bed we will restrict ourselves to considering this function when z_0 is positive.

In deriving formulae for apparent conductivity for various positions of the probe we will make use of the equations of geometric factors of a bed with finite thickness.

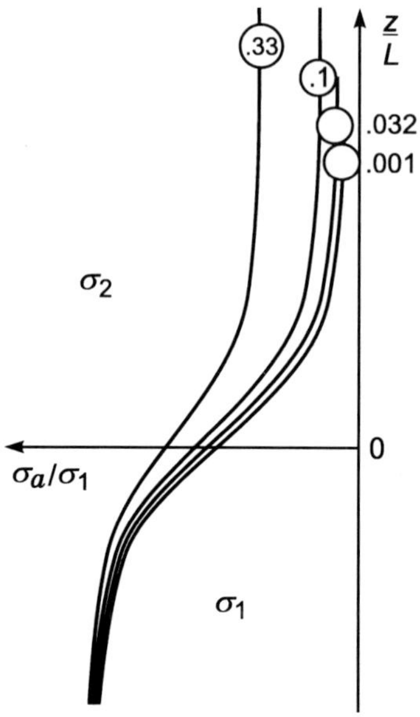

Figure 5.6. Curves of the apparent conductivity for various positions of the probe with respect to the interface.

The probe is located outside the bed

The expression for the apparent conductivity is:

$$\sigma_a = \sigma_1 G_1 + \sigma_2 G_2$$

The geometric factor of bed, G_1, for a given position with respect to the probe is:

$$G_1 = \frac{LH}{8\left[z_0^2 - (H/2)^2\right]}$$

The geometric factor of the surrounding medium is presented as a sum:

$$G_2 = G_{12} + G_{22}$$

where G_{12} is the geometric factor of that part of the surrounding medium which is located above the bed, and G_{22} is the geometric factor of the surrounding medium located beneath the bed. Correspondingly:

$$G_2 = \frac{L}{8(z_0 + H/2)} - \frac{L}{8(z_0 - H/2)} + 1$$

Substituting expressions derived for geometric factors we have:

$$\sigma_a = \sigma_2 + (\sigma_2 - \sigma_1)\frac{LH}{8\left[z_0^2 - (H/2)^2\right]} \tag{5.55}$$

This formula is applied if the upper coil does not intersect the low boundary of the bed, i.e. if $z_0 \geqslant H/2 + L/2$.

One of the coils is located within the bed

In this case we have:

$$\sigma_a = \sigma_2(G_{12} + G_{22}) + \sigma_1 G_1$$

here:

$$G_1 = \frac{1}{2} - \frac{z_0 - H/2}{2L} - \frac{L}{8(z_0 + H/2)}$$

$$G_{12} = \frac{L}{8(z_0 + H/2)} \qquad G_{22} = \frac{1}{4} + \frac{1}{2L}\left(\frac{L}{2} + z_0 - \frac{H}{2}\right)$$

and

$$\sigma_a = \sigma_1\left[\frac{1}{2} - \frac{z_0 - H/2}{2L} - \frac{L}{8(z_0 + H/2)}\right] + \sigma_2\left[\frac{L}{8(z_0 + H/2)} + \frac{1}{2} + \frac{z_0 - H/2}{2L}\right]$$

or

$$\sigma_a = \frac{\sigma_1 + \sigma_2}{2} + \frac{\sigma_2 - \sigma_1}{2L}(z_0 - H/2) + \frac{(\sigma_2 - \sigma_1)L}{8(z_0 + H/2)} \tag{5.56}$$

This formula is applied if:

$$\frac{L}{2} - \frac{H}{2} \leqslant z_0 \leqslant \frac{L}{2} + \frac{H}{2} \qquad \text{if } H \leqslant L$$

and

$$z_0 \geqslant \frac{H}{2} - \frac{L}{2} \qquad \text{if } H \geqslant L$$

The bed is located between the probe coils or the probe is located within the interval of the bed

If the bed is located between the probe coils ($H < L$) we have:

$$\sigma_a = \sigma_1 G_1 + \sigma_2 G_2$$

where:

$$G_1 = H/2L$$
$$G_2 = G_{12} + G_{22}$$
$$G_{12} = \frac{1}{4} + \frac{1}{2L}\left[\frac{L}{2} - z_0 - \frac{H}{2}\right] \qquad G_{22} = \frac{1}{4} + \frac{1}{2L}\left[\frac{L}{2} + z_0 - \frac{H}{2}\right]$$

and

$$\sigma_a = \sigma_1 \frac{H}{2L} + \sigma_2 \left[1 + \frac{1}{2L}\left(z_0 - \frac{H}{2}\right) - \frac{1}{2L}\left(z_0 + \frac{H}{2}\right)\right] = \sigma_2 + (\sigma_1 - \sigma_2)\frac{H}{2L} \tag{5.57}$$

This equation is applied if $0 \leqslant z_0 \leqslant L/2 - h/2$, as $H < L$.

If the probe is located within the interval of the bed ($H > L$) we have:

$$\sigma_a = \sigma_2 G_2 + \sigma_1 G_1$$

and

$$G_2 = G_{12} + G_{22}$$

where:

$$G_{12} = \frac{L}{8\left(z_0 + \dfrac{H}{2}\right)} \qquad G_{22} = \frac{L}{8\left(\dfrac{H}{2} - z_0\right)}$$
$$G_1 = 1 + \frac{LH}{8\left[z_0^2 - \left(\dfrac{H}{2}\right)^2\right]}$$

and

$$\sigma_a = \sigma_1 + (\sigma_1 - \sigma_2)\frac{LH}{8\left[z_0^2 - \left(\dfrac{H}{2}\right)^2\right]} \tag{5.58}$$

This formula is applied for $0 < z_0 \leqslant H/2 - L/2$.

Now let us introduce new variables:

$$z_0 = \eta L \qquad H = \xi L$$

Then the formulae have the form:

(1) $\sigma_a = \sigma_2 + (\sigma_1 - \sigma_2)\dfrac{\xi}{8\left[\eta^2 - (\xi/2)^2\right]}$

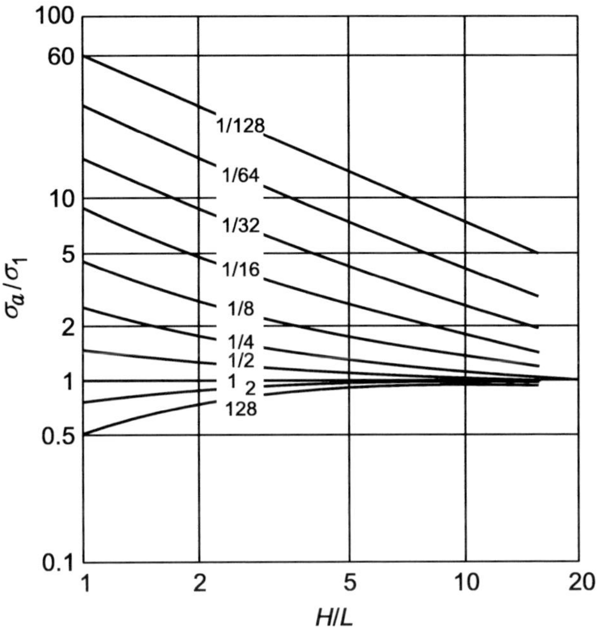

Figure 5.7. Curves of function σ_a/σ_1 for various H/L $(H/L \geqslant 1)$. Curve index σ_1/σ_2.

$$\text{if } \eta \geqslant 0.5 + \frac{\xi}{2}$$

$$(2)\ \sigma_a = \frac{1}{2}\sigma_1 + \frac{1}{2}\sigma_2 + \frac{\sigma_2 - \sigma_1}{2}\left(\eta - \frac{\xi}{2}\right) + \frac{\sigma_2 - \sigma_1}{8(\eta + \xi/2)}$$

$$\text{if } \ \eta < 0.5 + \frac{\xi}{2} \quad \text{as } \ H > L, \text{ i.e. } \xi < 1 \text{ and for}$$

$$\eta \geqslant 0.5 - \frac{\xi}{2} \quad \text{as } \ H > L;\ \xi > 1,\ \eta \geqslant \frac{\xi}{2} - 0.5$$

$$(3a)\ \sigma_a = \sigma_2 + (\sigma_1 - \sigma_2)\frac{\xi}{2};\ \xi < 1$$

$$\text{if } 0 < \eta < 0.5 - \frac{\xi}{2}$$

$$(3b)\ \sigma_a = \sigma_1 + \frac{\sigma_1 - \sigma_2}{8}\frac{\xi}{\eta^2 - (\xi/2)^2}$$

$$\text{if } \xi > 1,\ 0 < \eta \leqslant \frac{\xi}{2} - 0.5$$

Curves showing the dependence of function σ_a/σ_1 on the ratio of bed thickness to probe length, if $H/L \geqslant 1$, are presented in Fig. 5.7. The center of two-coil probe coincides with

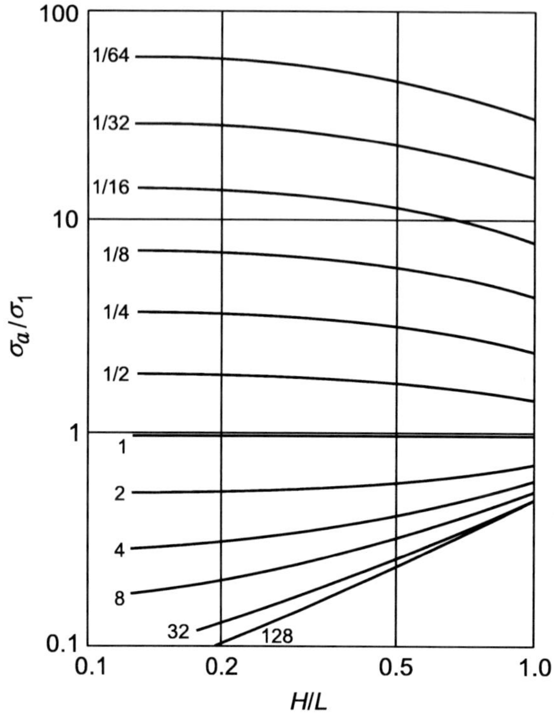

Figure 5.8. Curves of ratio σ_a/σ_1 as a function of H/L ($H/L \leqslant 1$). Curve index σ_1/σ_2.

the middle of the bed. Calculations have been made using the equation:

$$\frac{\sigma_a}{\sigma_1} = 1 + \left(\frac{\sigma_2}{\sigma_1} - 1\right)\frac{1}{2\xi} \tag{5.59}$$

As is seen from these curves with an increase of the conductivity of the surrounding medium and a decrease of the bed thickness the influence of the surrounding medium becomes greater. If the resistivity of the bed is significantly greater than that of the surrounding medium ($\sigma_1 \ll \sigma_2$), the apparent conductivity approaches the conductivity of the bed provided that its thickness is many times larger than the probe length ($H \gg L$). In other words, in such cases the vertical characteristic of the two-coil induction probe is essentially worse than the corresponding response of the normal probe.

If the bed conductivity is greater than that of the surrounding medium for most typical values of σ_1/σ_2 the influence of the surrounding medium becomes insignificant when $H/L > 4$.

Curves of function σ_a/σ_1, when the bed thickness is smaller than the probe length, are given in Fig. 5.8.

Formulae for calculation have the form:

$$\frac{\sigma_a}{\sigma_1} = \frac{\sigma_2}{\sigma_1} + \left(1 - \frac{\sigma_2}{\sigma_1}\right)\frac{H}{2L} \qquad (5.60)$$

or

$$\frac{\sigma_a}{\sigma_2} = 1 + \left(\frac{\sigma_1}{\sigma_2} - 1\right)\frac{H}{2L} \qquad (5.61)$$

In accord with eq. 5.61 if the bed resistivity is higher than that of the surrounding medium and its thickness is less than $0.2L$ such a bed cannot be practically noticed on curves of induction logging. This fact can be interpreted as an advantage of induction logging with respect to electrical logging where screening often makes interpretation sufficiently complicated.

In contrary, thin low resistive layers have essential influence on curves of induction logging. In fact with an increase of ratio σ_1/σ_2 the value of the apparent conductivity, σ_a, tends to a constant equal to $S_1/2L$ (S_1 is the longitudinal conductance of the bed: $\sigma_1 H$), and it can turn out to be much greater than σ_2. This fact is well seen from curves with index $\sigma_1/\sigma_2 > 1$ (Fig. 5.8). Let us assume that the probe is located against a system of thin layers. Then an expression for apparent conductivity, σ_a, can be obviously presented as:

$$\sigma_a = \sigma_2\left(1 - \sum_{i=1}^{n}\frac{h_i}{2L}\right) + \sigma_1\sum_{i=1}^{n}\frac{h_i}{2L} = \sigma_2\left(1 - \frac{H}{2L}\right) + \frac{\sigma_1 H}{2L} \qquad (5.62)$$

where h_i is the thickness of i-layer; n is number of layers, and $H = \sum_{i=1}^{n} h_i$.

Thus, in the range of small parameters a group of thin beds located against the probe is equivalent to one bed having the same conductivity and with a thickness equal to the sum of thicknesses of all beds. This principle of equivalence by S can be easily generalized for the more common case when conductivities and thicknesses of layers are different.

5.3. The Theory of the Two-coil Induction Probe in Beds with a Finite Thickness

In the previous section we have considered vertical responses of the two-coil induction probe in the range of small parameters as the skin effect could be neglected. Now, a more general case will be investigated proceeding from the results of calculation by the exact formulae derived in the first section.

We will assume that the two-coil probe is located symmetrically with respect to the boundary interfaces. Then, according to eqs. 5.28–5.37, the vertical component of the magnetic field on the z-axis, expressed in units of the primary field, is defined by three parameters:

- ratio of the probe length, L, to the thickness of skin depth, h_1, in the bed: L/h_1

- ratio of conductivities of the bed and the surrounding medium: σ_1/σ_2
- ratio of the bed thickness, H, to the probe length: H/L.

We will investigate frequency responses of quadrature and inphase components of the field measured by the receiver coil of the induction probe. Examples of the responses are presented in Figs. 5.9–5.22. Analysis of results of calculations allows us to outline the main features of field behavior, such as:

1. For small values of parameter L/h_1 (low frequency, high resistivity) the inphase component of the secondary field is much smaller than the quadrature component: $\operatorname{In} h_z \ll \operatorname{Q} h_z$. With an increase of parameter L/h_1 the inphase component, $\operatorname{In} h_z^s$, increases and oscillating approaches unity.

Comparing responses of quadrature and inphase components we can see that in the range of small parameters induced currents in the surrounding medium have an influence on the inphase component, $\operatorname{In} h_z^s$, which is much stronger than that on the quadrature component, $\operatorname{Q} h_z$. In the limit, as parameter L/h_1 tends to zero, the inphase component of the magnetic field approaches to that of a uniform medium with the conductivity of surrounding medium, σ_2:

$$\operatorname{In} h_z \to \operatorname{In} h_z^0(\sigma_2) \quad \text{as } L/h_1 \to 0$$

It is essential that this result does not depend on ratio of the bed thickness to the probe length, H/L, as well as the ratio of conductivities. In other words, with a decrease of parameters L/h_1 the bed becomes transparent for the inphase component regardless of how the probe length is small. It means that within this range of parameters L/h_1, the vertical response of the inphase component is much worse than that of the quadrature one.

2. In the range of small parameters the quadrature component of the field is directly proportional to frequency and conductivity. Such behavior of the quadrature component is inherent to *Doll's domain*, which therefore represents the left-hand asymptote of frequency response of function: $\operatorname{Q} h_z(L/h_1)$. With an increase of parameter L/h_1 the quadrature component increases, reaches a maximum and then oscillating goes to zero. Thus at the left part of the frequency response of the secondary field the quadrature component prevails while at the right part the inphase component $\operatorname{In} h_z^s$ is dominant.

It is appropriate to notice that the left-hand part of the frequency response of $h_z(L/h_1)$ is of a great practical interest because for frequencies used in conventional induction logging and the most typical geoelectrical sections parameter L/h_1 is usually less than unity.

In accord with eq. 5.37 the expression for vertical component of the magnetic field along the dipole axis is:

$$h_z = h_z^0 + \int_0^\infty \frac{m^3 K_{12} e^{-2\alpha m_2}}{m_2} \frac{e^{\alpha m_2} + K_{12} \cosh m_2}{1 - K_{12}^2 e^{-2\alpha m_2}} \, dm$$

where h_z^0 is the field in a uniform medium with a bed conductivity expressed in units of the primary field:

$$m_2 = (m^2 - \mathrm{i}n_2)^{1/2} \qquad m_1 = (m_2 - \mathrm{i}n_1)^{1/2} \qquad K_{12} = (m_2 - m_1)/(m_2 + m_1)$$
$$n_1 = \sigma_1\mu\omega L^2 \qquad n_2 = \sigma_2\mu\omega L^2 = Nn_1N = \sigma_2/\sigma_1 \qquad \alpha = H/L.$$

Let us consider an integral at the right-hand part of the formula for h_z as a function of n_1 and find its approximate value when $n_1 \rightarrow 0$. Expanding the integrand in a series by small parameters n_1 and considering only the first term we obtain for the integrand:

$$-\frac{1}{4}(N-1)n_1\mathrm{e}^{-\alpha m_1}$$

Correspondingly, the integral becomes equal to:

$$-\frac{\mathrm{i}}{4\alpha}(N-1)\sigma_2\mu\omega L^2$$

Thus, the field at the range of small parameter n_1 is:

$$h_z = \frac{\mathrm{i}\sigma_1\mu\omega L^2}{2} - \frac{\mathrm{i}}{4\alpha}\left(\frac{\sigma_1}{\sigma_2} - 1\right)\sigma_2\mu\omega L^2$$

Making use of relation between the quadrature component of field h_z and the apparent conductivity:

$$\sigma_a = \frac{2}{\mu\omega L^2}\,\mathrm{Q}\,h_z \tag{5.63}$$

we obtain an expression for σ_a, as the two-coil induction probe is located symmetrically within the bed boundaries and $L < H$:

$$\sigma_a = \sigma_1 + \frac{1}{2\alpha}(\sigma_2 - \sigma_1) \qquad \text{if } \alpha \geqslant 1 \tag{5.64}$$

The latter completely coincides with eq. 5.59.

Therefore, Doll's theory is in fact the theory of very small parameters, which characterize the linear dimensions of a model, expressed in units of the skin depth. For example, with a decrease of the probe length parameter L/h_1 decreases also. From a physical point of view this means that the influence of induced currents near the dipole, which are shifted in phase by 90° and do not interact with each other, increases.

Within *Doll's domain* there is a simple relation between the field, $\mathrm{Q}\,h_z$, or σ_a, and the parameters of a medium. For example when the probe is located symmetrically within the bed (eq. 5.64) the value of σ_a/σ_1 depends on two parameters: σ_2/σ_1 and H/L only, and it does not depend on absolute values of conductivity and frequency. For this reason the interpretation of induction logging within *Doll's domain* is in essence the same as in resistivity logging based on direct currents.

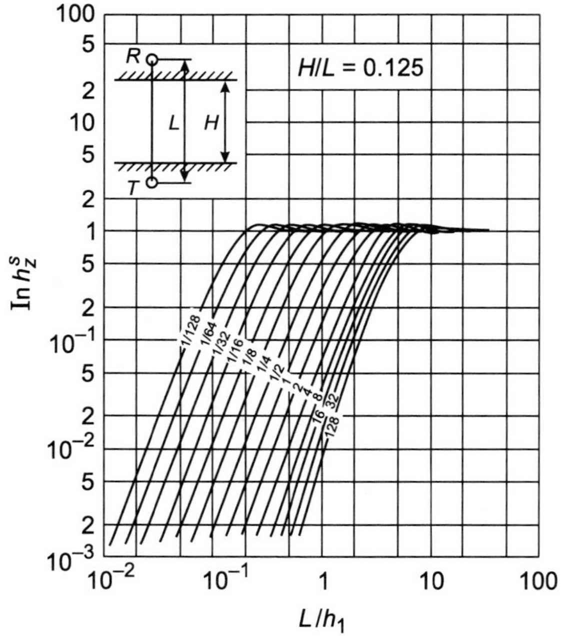

Figure 5.9. Frequency responses of $\ln h_z^s$. Curve index σ_1/σ_2.

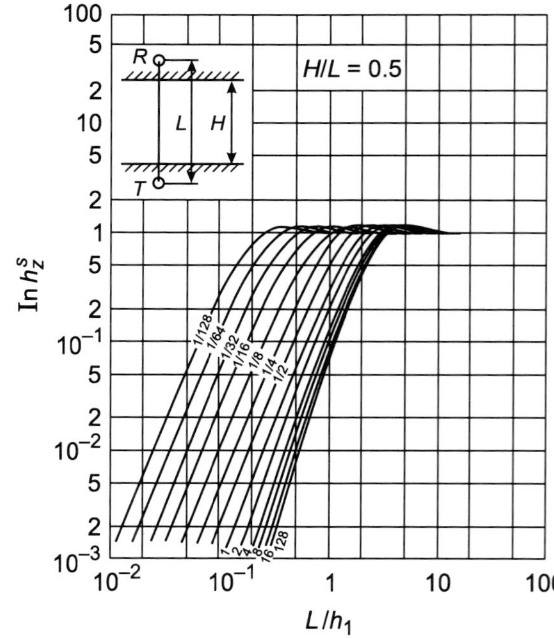

Figure 5.10. Frequency responses of $\ln h_z^s$. Curve index σ_1/σ_2.

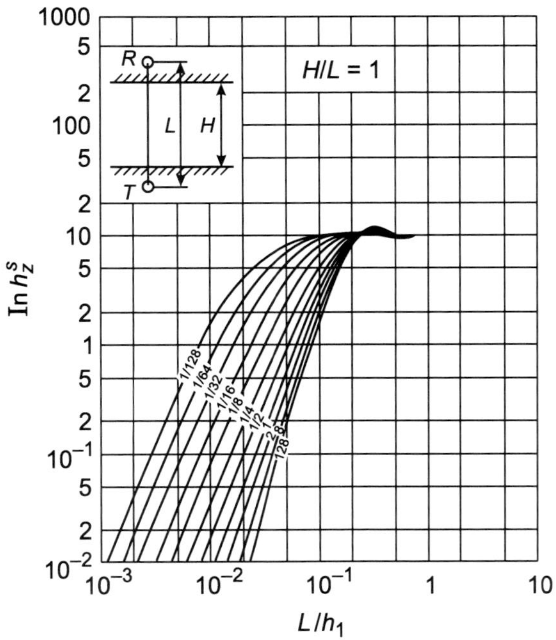

Figure 5.11. Frequency responses of $\mathrm{In}\, h_z^s$. Curve index σ_1/σ_2.

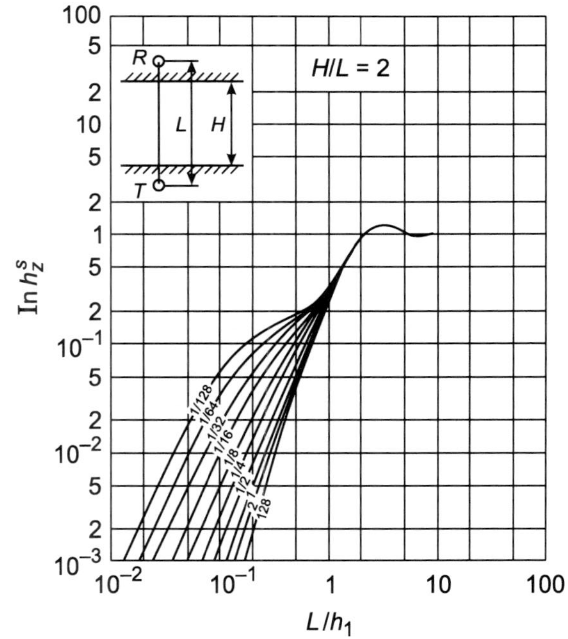

Figure 5.12. Frequency responses of $\mathrm{In}\, h_z^s$. Curve index σ_1/σ_2.

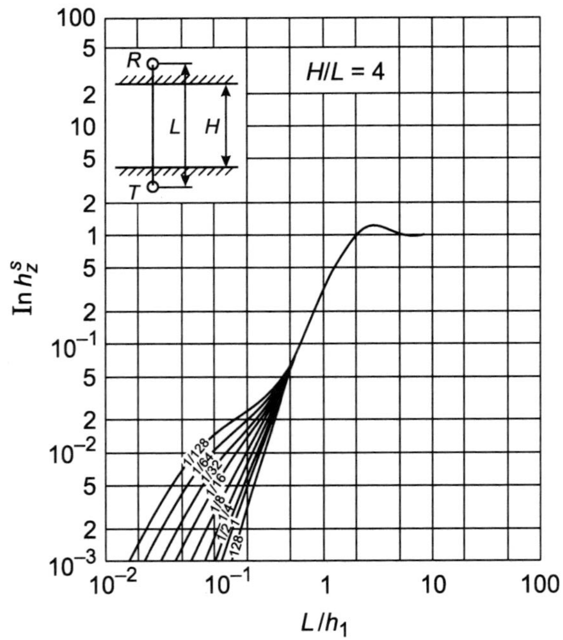

Figure 5.13. Frequency responses of In h_z^s. Curve index σ_1/σ_2.

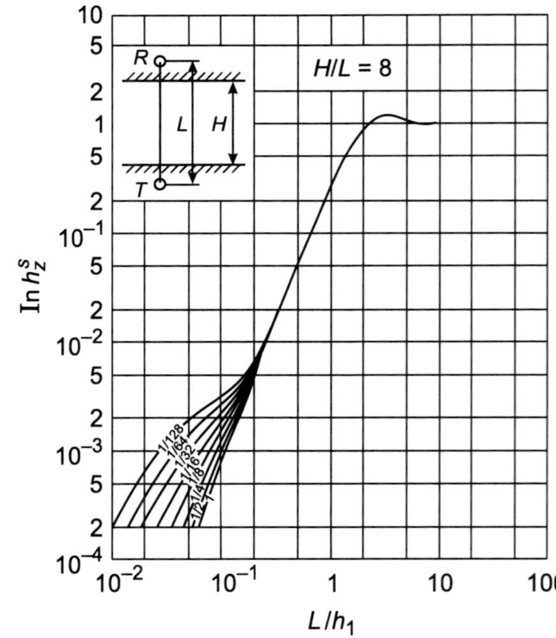

Figure 5.14. Frequency responses of In h_z^s. Curve index σ_1/σ_2.

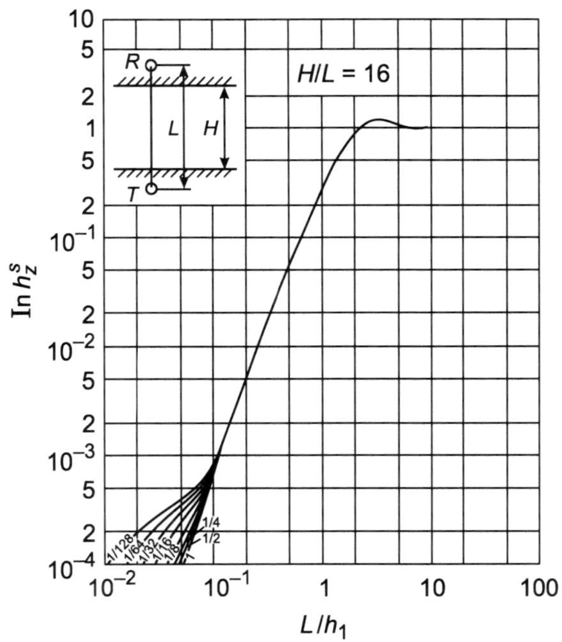

Figure 5.15. Frequency responses of In h_z^s. Curve index σ_1/σ_2.

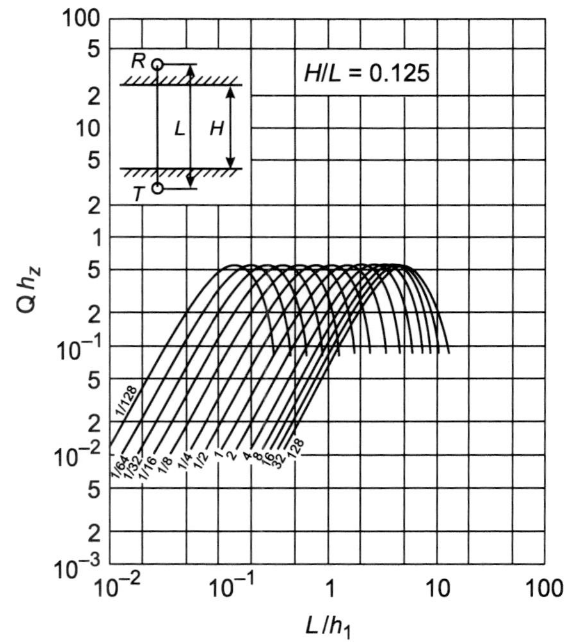

Figure 5.16. Frequency responses of Q h_z^s. Curve index σ_1/σ_2.

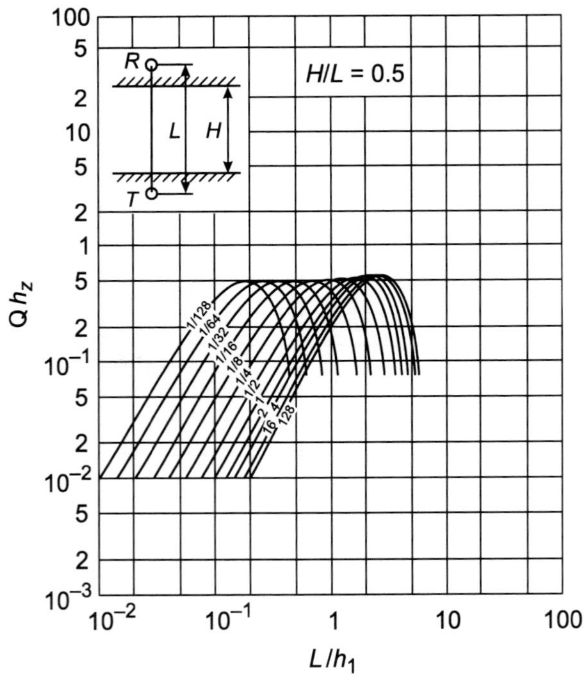

Figure 5.17. Frequency responses of $\mathrm{Q}\,h_z^s$. Curve index σ_1/σ_2.

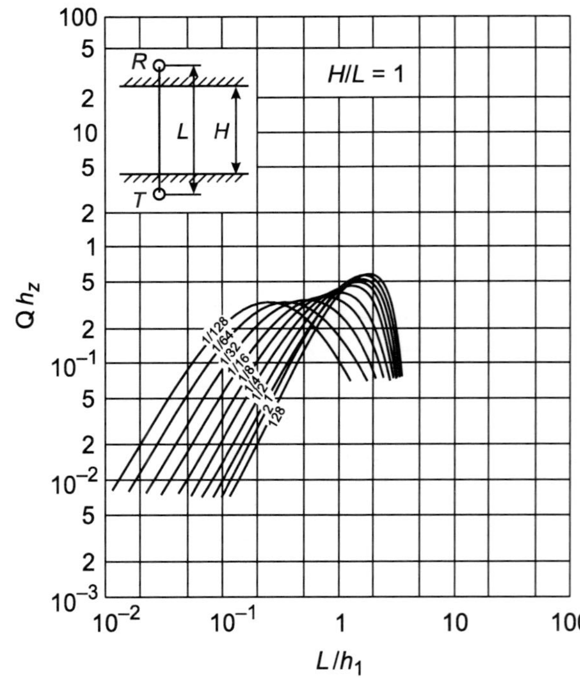

Figure 5.18. Frequency responses of $\mathrm{Q}\,h_z^s$. Curve index σ_1/σ_2.

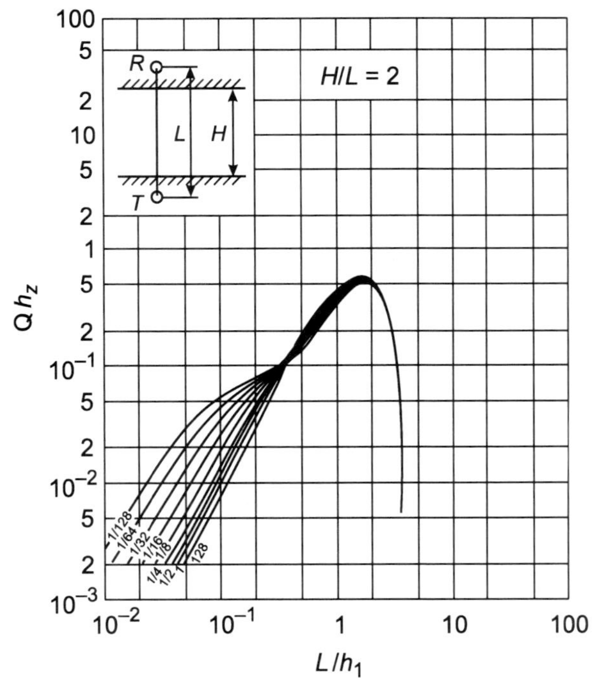

Figure 5.19. Frequency responses of $Q\,h_z^s$. Curve index σ_1/σ_2.

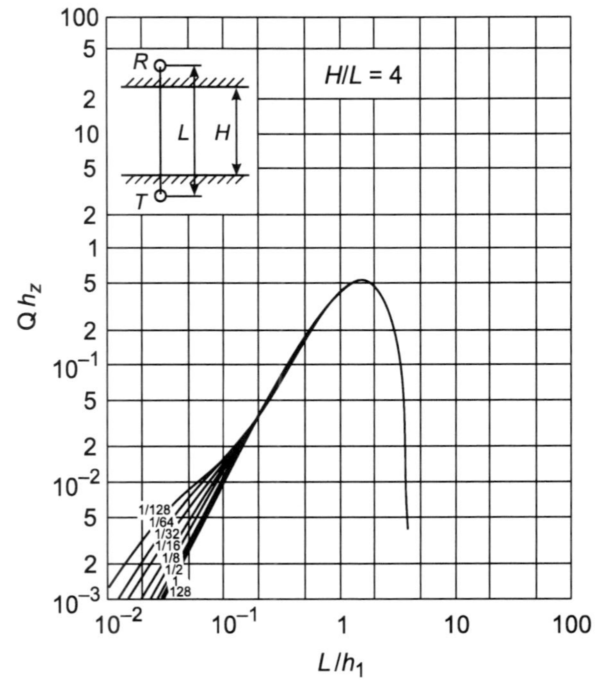

Figure 5.20. Frequency responses of $Q\,h_z^s$. Curve index σ_1/σ_2.

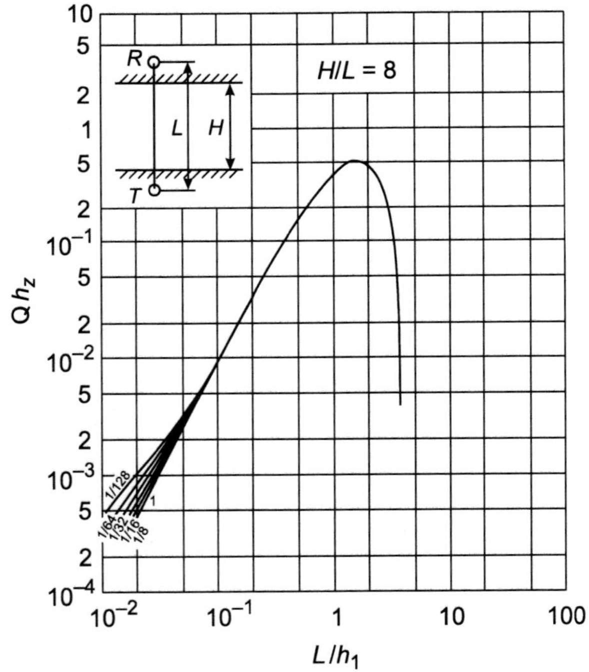

Figure 5.21. Frequency responses of $Q\,h_z^s$. Curve index σ_1/σ_2.

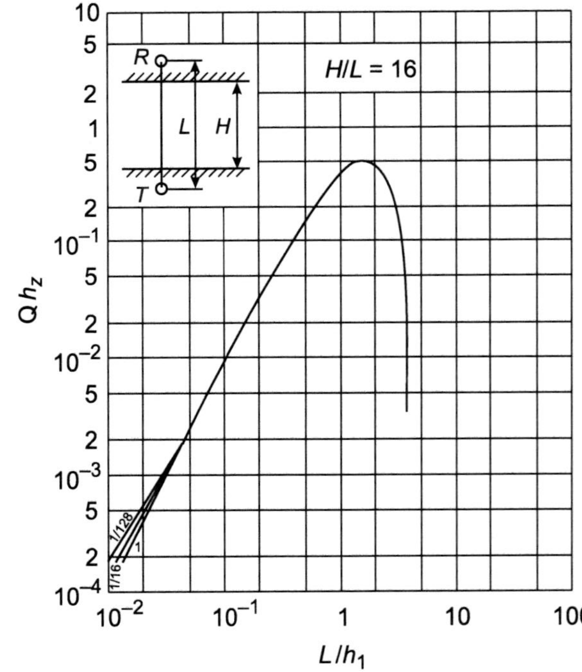

Figure 5.22. Frequency responses of $Q\,h_z^s$. Curve index σ_1/σ_2.

As was demonstrated in the previous section eq. 5.64 was derived from Doll's theory, based on the concept of the geometric factor, and strictly speaking it is valid only if all currents, regardless of the distance from the dipole, are shifted in phase by 90°, i.e. interaction between them is absent.

Inasmuch as this equation for the apparent conductivity, σ_a, has also been obtained from the exact solution, provided parameter L/h_1 is small, physical principles of Doll's theory can be interpreted in the following manner. The signal measured in the receiver coil of the induction probe is mainly defined by currents induced in the area within which the probe is located. For this reason eq. 5.64 is valid if currents within this area are shifted in phase by 90° and are proportional to frequency and conductivity regardless of the behavior of currents induced outside of this area.

Analysis of the field behavior in a uniform medium as well as in a medium with coaxially cylindrical interfaces demonstrated that:

- Near the dipole induced currents, which are proportional to frequency and shifted in phase by 90°, prevail.
- With an increase of the distance from the dipole the influence of the inphase component increases.
- The depth of investigation of the probe becomes greater with an increase of the probe length.

Therefore, we can expect that with increases of the bed thickness, the probe length and the ratio of conductivities of the surrounding medium and the bed (σ_2/σ_1), application of formulae of Doll's theory will lead to greater errors.

Now let us consider frequency responses of apparent conductivity when the two-coil probe is located symmetrically with respect to the bed boundaries (Figs. 5.23–5.31). Along the axes values L/h_1 and function σ_a/σ_1 are plotted, respectively. Index of curves is σ_1/σ_2.

As is well known, for electromagnetic fields apparent conductivity can be introduced in different ways. In accord with eq. 5.63 apparent conductivity is equal to conductivity in the uniform medium only within *Doll's domain* while outside this range they differ from each other. Such a way of introducing the apparent conductivity is justified by the fact that very often in the practice of conventional induction logging a field behavior either corresponds to that for *Doll's domain* or is sufficiently close to it.

As is seen from Figs. 5.23–5.31, all curves of the apparent conductivity at the left-hand part, i.e. within the range of small parameters, are parallel to the axis of abscissa that corresponds to *Doll's domain* but with an increase of σ_2/σ_1 the influence of the skin effect manifests earlier. This behavior is in complete agreement with our understanding of the distribution of quadrature component of induced currents in a conducting medium. In fact, with an increase of the distance from the dipole this component becomes smaller than that according to Doll's theory, and since with an increase of conductivity of the surrounding medium the role of this part of the medium also increases, deviation between results of calculation by exact and approximate solutions also increases. Practically this

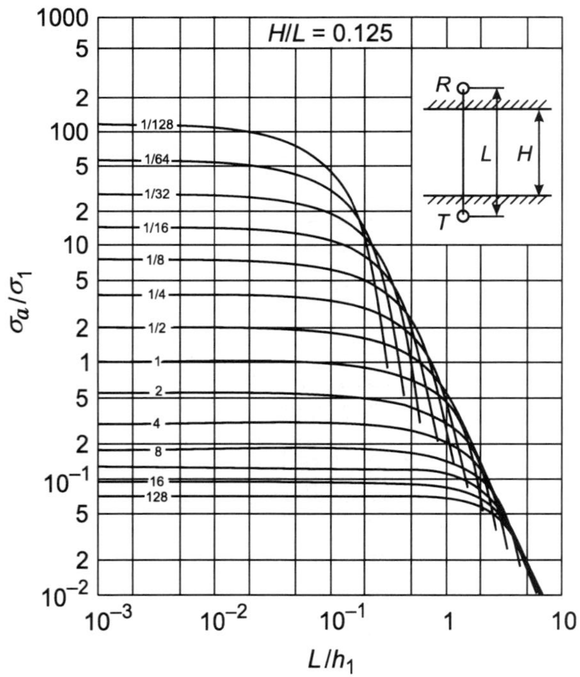

Figure 5.23. Apparent conductivity function. Curve index σ_1/σ_2.

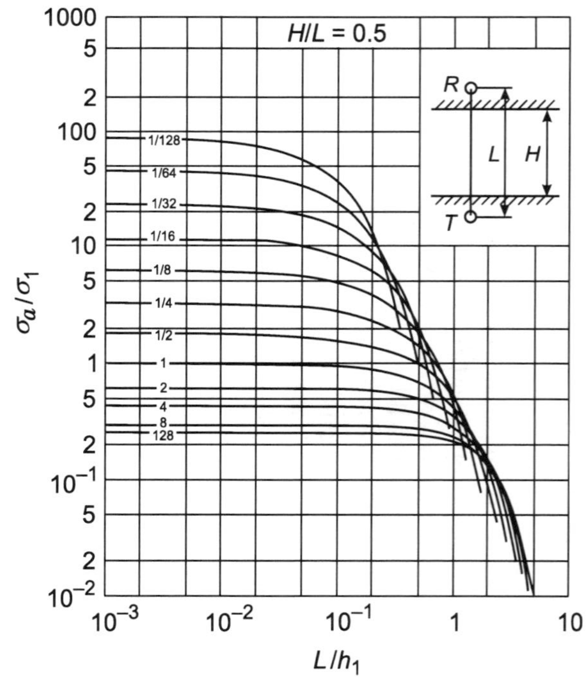

Figure 5.24. Apparent conductivity function. Curve index σ_1/σ_2.

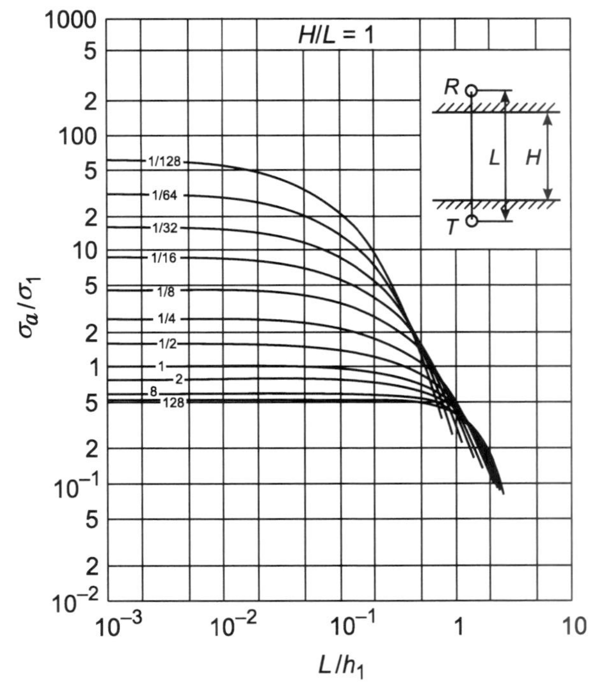

Figure 5.25. Apparent conductivity function. Curve index σ_1/σ_2.

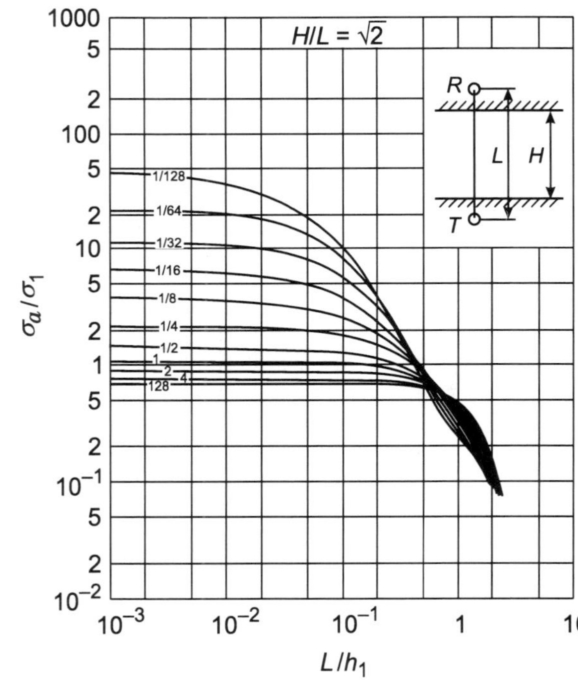

Figure 5.26. Apparent conductivity function. Curve index σ_1/σ_2.

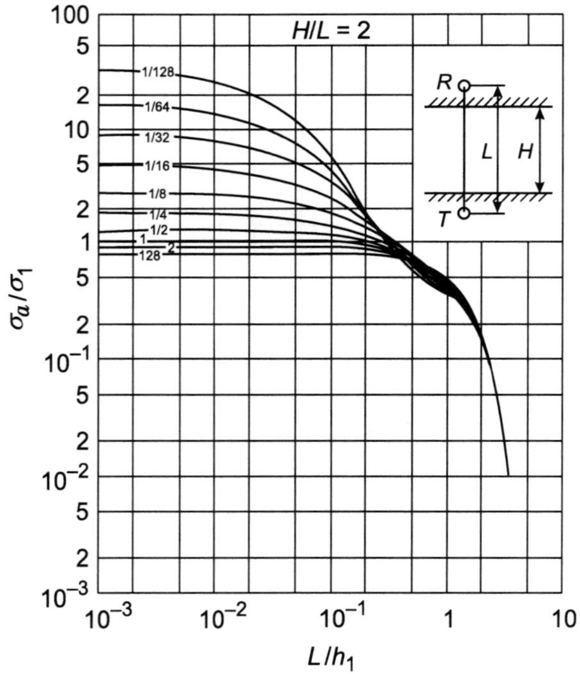

Figure 5.27. Apparent conductivity function. Curve index σ_1/σ_2.

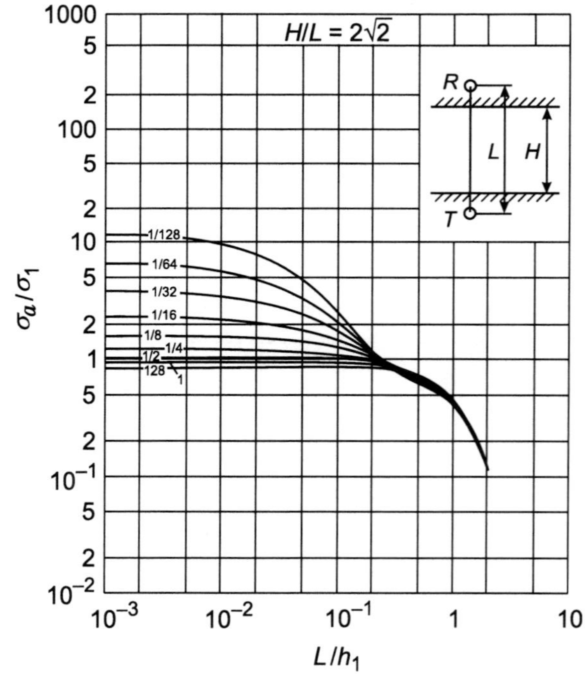

Figure 5.28. Apparent conductivity function. Curve index σ_1/σ_2.

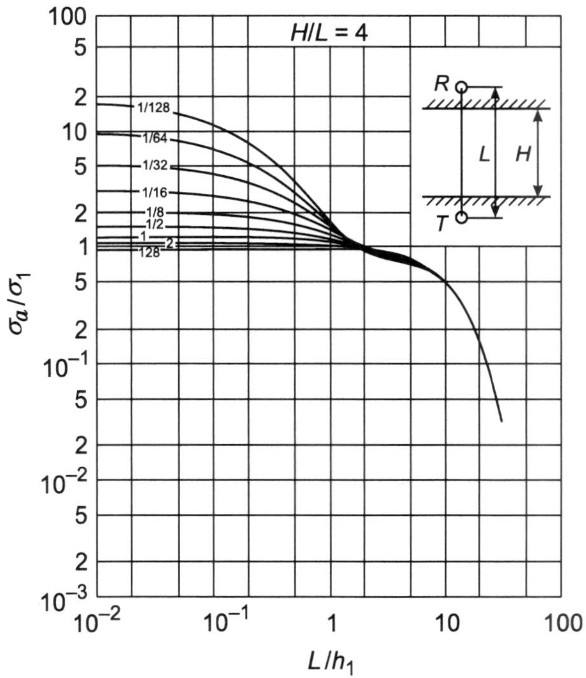

Figure 5.29. Apparent conductivity function. Curve index σ_1/σ_2.

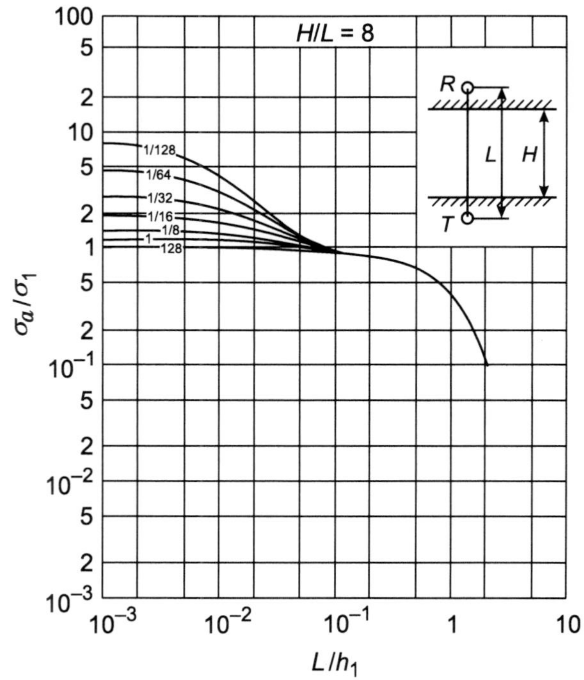

Figure 5.30. Apparent conductivity function. Curve index σ_1/σ_2.

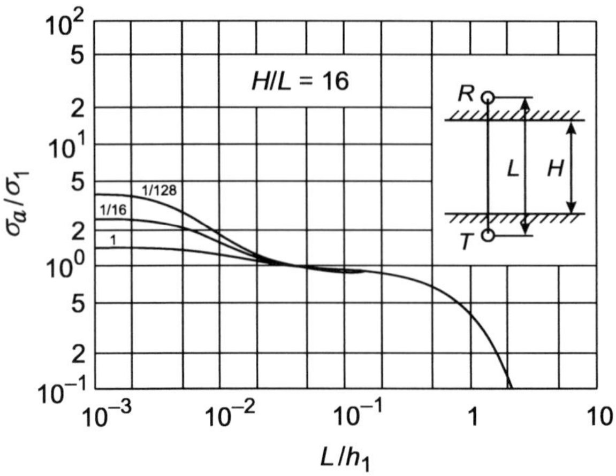

Figure 5.31. Apparent conductivity function. Curve index σ_1/σ_2.

TABLE 5.1
Maximal values of parameter L/h_1

σ_1/σ_2 \ α	1.0	$\sqrt{2}$	2.0	$2\sqrt{2}$	4.0	8.0
1/128	0.0085	0.0065	0.0048	0.0034	0.0027	0.0020
1.0	0.15	0.15	0.15	0.15	0.15	0.15
128	0.67	0.55	0.44	0.36	0.28	0.15

means that the influence of the surrounding medium on the value of σ_a with an increase of its conductivity becomes less than what follows from Doll's theory.

With an increase of the bed thickness, deviation from the left-hand asymptote is also observed at smaller frequencies specially as the ratio of conductivities σ_1/σ_2 decreases. For illustration the maximal values of parameter L/h_1, when the left-hand asymptote of curves σ_a/σ_1 is still practically observed, are given in Table 5.1.

It is convenient to relate values of this parameter with resistivity and frequency. As an example Table 5.2 contains some values of parameter L/h_1 for various frequencies and resistivities within the range 1–100 ohm·m for a two-coil induction probe with length 1 m.

In conventional induction logging the equipment is based on the use of frequencies within the range of 20–60 kHz. From comparison of curves of apparent conductivity (Figs. 5.26–5.31) and data given in Table 5.2 it is seen that for a frequency of 20 kHz and $\sigma_1 > \sigma_2$ corrections due to the internal skin effect are small and only for a relatively large bed thickness ($\alpha > 8$) and its low resistivity ($\rho_1 \simeq 1$ ohm·m) these corrections reach 10–20%. If the conductivity of the bed is smaller than that of the surrounding medium the influence of the skin effect can be significant.

For instance, if $f = 20$ kHz, $\rho_1 = 20$ ohm·m, $\rho_2 = 2.5$ ohm·m, $L = 1$ m and $H = 2$ m,

TABLE 5.2
Values of parameter L/h_1

ρ, ohm·m \ f, Hz	2×10^4	4×10^4	6×10^4	8×10^4	1×10^5	2×10^5
1	0.280	0.396	0.485	0.560	0.625	0.885
5	0.125	0.177	0.216	0.250	0.280	0.396
10	0.089	0.125	0.154	0.178	0.198	0.280
20	0.063	0.089	0.109	0.126	0.141	0.200
40	0.045	0.064	0.078	0.090	0.101	0.142
60	0.036	0.051	0.062	0.072	0.081	0.114
100	0.028	0.040	0.049	0.056	0.063	0.089

TABLE 5.3
Values of parameter L/h_1

α	$\sqrt{2}$	2	$2\sqrt{2}$	4.0	8.0	16.0
L/h_1	2.00	1.20	0.70	0.40	0.10	0.04

the value of apparent conductivity, σ_a/σ_1 is equal to 2.0 instead of 2.8 as it follows from eq. 5.64.

This influence of the internal skin effect manifests itself to a greater extent when higher frequencies are used, for example 60 kHz. Also, comparison of curves in Figs. 5.26–5.31 with the same index shows that with an increase of the probe length the influence of the skin effect becomes stronger. This is related to the fact that with an increase of the separation between coils, the sensitivity to more remote parts of the medium also increases.

These examples, as well as a more detailed analysis, demonstrate that calculations based on the exact solution which takes into account the skin effect are necessary in order to investigate the vertical responses of the introduction probe.

Consideration of the field on the borehole axis in a medium with cylindrical interfaces (borehole, invasion zone, formation) also shows that the skin effect has to be taken into account. However, as a rule, its influence on the radial responses is less than that on vertical ones.

Due to the internal skin effect with an increase of frequency induced currents concentrate more near the source, and correspondingly the influence of the surrounding medium decreases. It is vividly seen from curves of the apparent conductivity as the bed thickness exceeds the probe length (Figs. 5.26–5.31). Therefore, we can select such high frequency for which the influence of induced currents in the surrounding medium will already be practically negligible.

For illustration, values of parameter L/h_1 for which apparent conductivity curves, presented in Figs. 5.26–5.31, merge in one curve corresponding to a uniform medium with conductivity σ_1 ($\sigma_2 < 130\sigma_1$) are given in Table 5.3.

An increase of frequency is specially desirable when beds have relatively small thickness, ($L \simeq H$), and higher resistivity than that of the surrounding medium, $\sigma_1 \ll \sigma_2$. However, in such cases it is necessary to use very high frequencies, sometimes reaching dozens of MHz in order to eliminate the influence of the surrounding medium.

Significant technical difficulties related with measuring field components at such high frequencies, deterioration of radial response of the induction probe specially when invasion zone has intermediate resistivity ($\rho_1 < \rho_2 < \rho_3$), increase of the influence of currents in the borehole, they all essentially reduce principal possibilities to use frequencies in order to eliminate completely the effect of induced currents in the surrounding medium.

At the same time it is appropriate to emphasize that an increase of frequency within certain limits, when the radial response does not practically change, can essentially improve the vertical response of the probe. Curves of the apparent conductivity shown in Figs. 5.32–5.36 vividly demonstrate this tendency. In fact with an increase of parameter L/h_1 the influence of a more conductive surrounding medium becomes smaller and it is specially noticeable when the bed thickness increases.

Now let us consider the main features of behavior of function σ_a/σ_1 in those cases when the bed thickness is smaller than the probe length, and the bed is located between probe coils (Figs. 5.23–5.24).

As was shown above, in this case the field does not depend on the position of the bed inside the probe, and it can be presented in the form:

$$h_z = 2\int_0^\infty \frac{m^3 m_2\, \mathrm{e}^{-(m_2-m_1)\alpha}\, \mathrm{e}^{-m_1}}{(m_1+m_2)^2\,(1-K_{12}^2\,\mathrm{e}^{-2\alpha m_2})}\, dm \qquad \text{if } \alpha < 1$$

where $m_1 = (m^2 - k_1^2 L^2)^{1/2}$, $m_2 = (m^2 - k_2^2 L^2)^{1/2}$, and $K_{12} = (m_2 - m_1)/(m_2 + m_1)$. Comparing the curves of apparent conductivity for both cases ($\alpha > 1$ and $\alpha < 1$) we can see that in the latter case the asymptote on the left usually takes place for larger values of parameter L/h_1. Asymptotic presentation for function σ_a/σ_1, as $L/h_1 \to 0$, can be derived in the same manner as eq. 5.64 was obtained. Omitting intermediate operations we have:

$$\frac{\sigma_a}{\sigma_1} = \frac{\sigma_2}{\sigma_1} - \left(\frac{\sigma_2}{\sigma_1} - 1\right)\frac{\alpha}{2} \qquad \text{if } \alpha < 1 \tag{5.65}$$

This formula coincides with expressions derived with the help of the geometric factor in the previous section and with a reasonable accuracy describes the apparent conductivity σ_a/σ_1 for thin layers.

At the same time it is possible to derive a simple expression for the apparent conductivity which is valid for larger values of parameter L/h_1 proceeding from the method which takes into account the skin effect in the external area (Chapter 3). Having assumed that within the bed induced currents are shifted in phase by 90° and interaction between currents in the bed and the surrounding medium is absent we obtain:

$$\frac{\sigma_a}{\sigma_1} = \frac{\sigma_a^{un}}{\sigma_1} - \left(\frac{\sigma_2}{\sigma_1} - 1\right)\frac{\alpha}{2} \tag{5.66}$$

where σ_a^{un} is apparent conductivity in a uniform medium with conductivity σ_2. It is obvious that within *Doll's domain* this value coincides with σ_2.

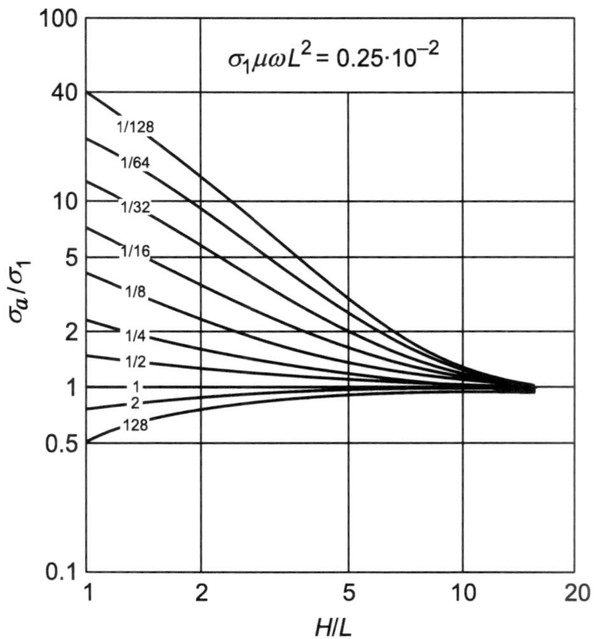

Figure 5.32. Dependence of apparent conductivity upon parameter α. Curve index σ_1/σ_2.

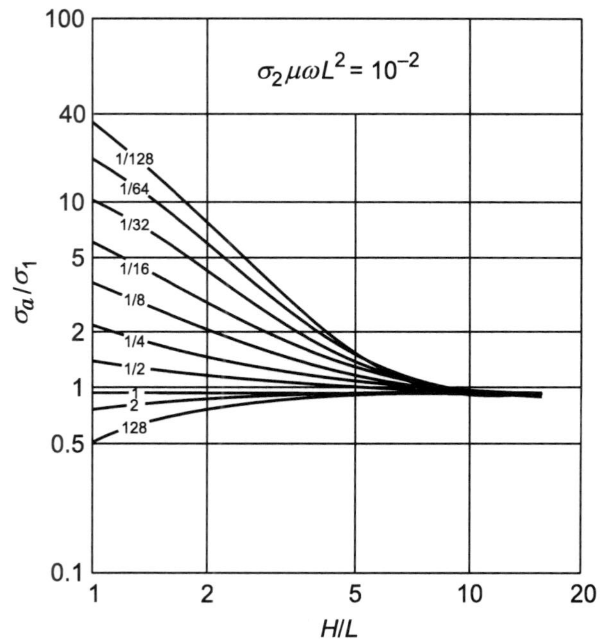

Figure 5.33. Dependence of apparent conductivity upon parameter α. Curve index σ_1/σ_2.

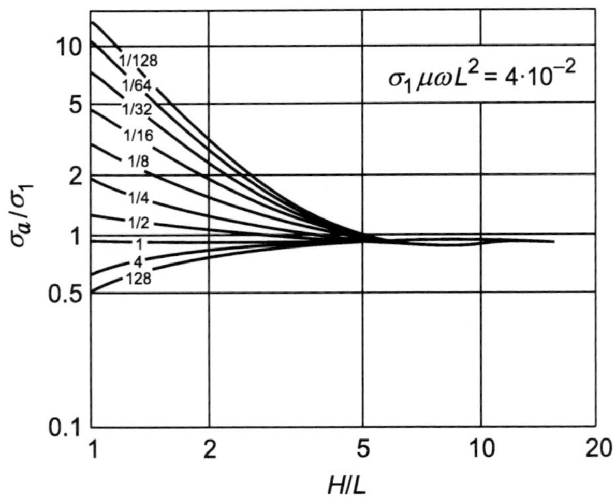

Figure 5.34. Dependence of apparent conductivity upon parameter α. Curve index σ_1/σ_2.

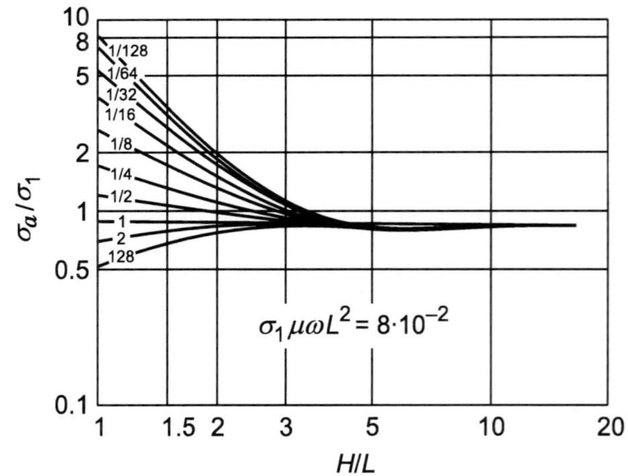

Figure 5.35. Dependence of apparent conductivity upon parameter α. Curve index σ_1/σ_2.

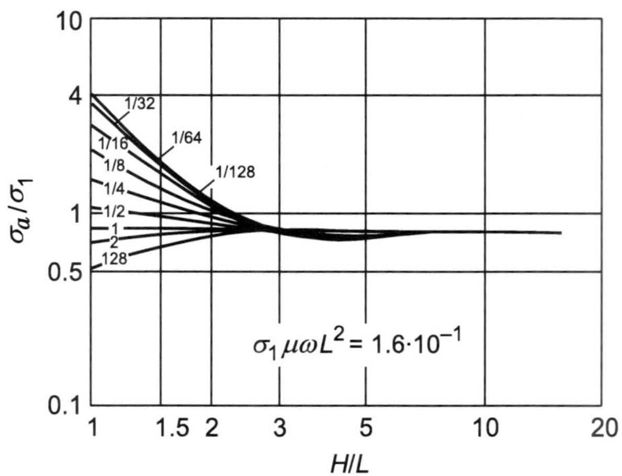

Figure 5.36. Dependence of apparent conductivity upon parameter α. Curve index σ_1/σ_2.

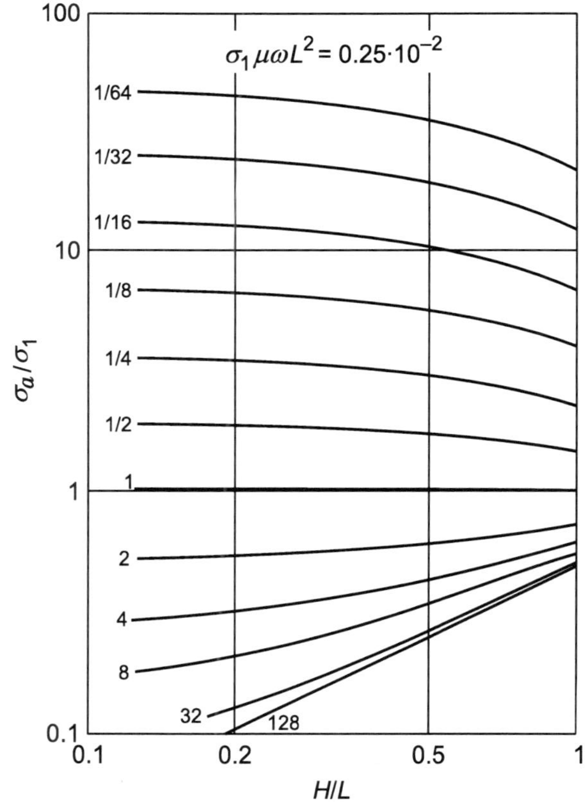

Figure 5.37. Dependence of apparent conductivity upon parameter α. Curve index σ_1/σ_2.

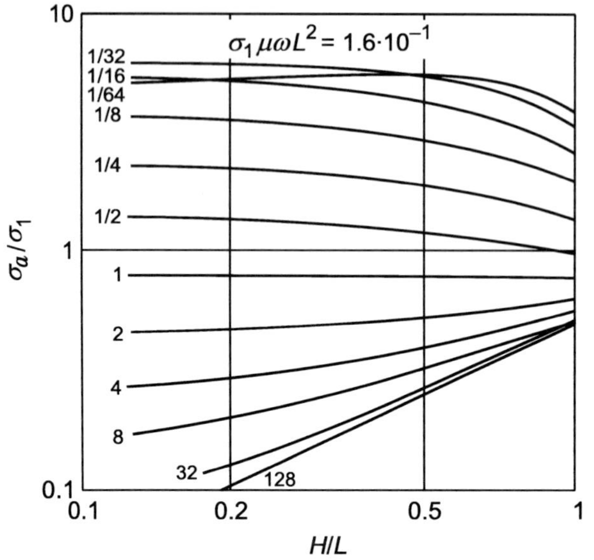

Figure 5.38. Dependence of apparent conductivity upon parameter α. Curve index σ_1/σ_2.

Equation 5.66 is valid for larger values of parameter L/h_1 than eq. 5.65, and this fact becomes more noticeable for relatively resistive beds.

Analysis of curves of apparent conductivity, σ_a/σ_1, (Figs. 5.37–5.38) shows that:

- Thin beds with resistivity greater than that of the surrounding medium are hardly noticeable at the range of small parameter L/h_1. For example, if $\alpha \leqslant 0.3$ and $\sigma_1/\sigma_2 \leqslant 1/8$ the influence of the bed does not exceed 5–10%.

- Thin conducting layers are sufficiently clearly distinguished by induction logging. For example, for small values of L/h_1, when $\alpha \simeq 0.3$ and $\sigma_1/\sigma_2 = 8$ the influence of the bed reaches 50%.

5.4. Curves of Profiling with a Two-coil Induction Probe in a Medium with Two Horizontal Interfaces (a Bed with Finite Thickness)

As was shown in the first section the signal and the apparent conductivity depend on the following parameters:

- Ratio between the probe length, L, and the skin depth in a medium, for example, in a surrounding one: L/h_2.

- Ratio of the bed thickness to the probe length, H/L.

- Ratio of conductivities of the bed and the surrounding medium, σ_1/σ_2.
- Position of the probe with respect to the bed, which can be characterized by the distance between the middle of the bed and the probe center, expressed in units of the probe length.

Of course, the electromotive force also depends on the moments of transmitter and receiver as well as on probe length and frequency.

Considering curves of profiling it is reasonable to distinguish four of the most typical positions of the probe with respect to interfaces of the bed. Formulae for field calculation are different from each other:

The probe is located outside the bed

$$h_z = h_z^{(0)} - \frac{1}{2}\int\limits_0^\infty \frac{m^3}{m_1}\frac{K_{12}(1-\mathrm{e}^{-2m_2\alpha})\mathrm{e}^{-m_2(2\beta_1-1)}}{1-K_{12}^2\mathrm{e}^{-2m_2\alpha}}\,dm \qquad \text{if } \beta_1 = h_1/L \geqslant 1 \qquad \alpha = H/L$$

One of the probe coils is located within the bed

$$h_z = \int\limits_0^\infty \frac{m^3\mathrm{e}^{-m_1\beta_2}\mathrm{e}^{-m_2(1-\beta_2)}\left(1+K_{12}\mathrm{e}^{2m_2(1-\beta-\alpha)}\right)}{(m_1+m_2)\left(1-K_{12}^2\mathrm{e}^{-2m_2\alpha}\right)}\,dm \qquad \text{if } \beta \leqslant 1$$

The probe is located within the bed

$$h_z = h_z^{(0)}(\sigma_1) + \frac{1}{2}\int\limits_0^\infty \frac{m^3K_{12}\left(\mathrm{e}^{-(1+2\beta_3)m_2}+\mathrm{e}^{-(2\alpha-2\beta_3-1)m_2}+2K_{12}\mathrm{e}^{-2\alpha m_2}\cosh m_2\right)}{m_2\left(1-K_{12}^2\mathrm{e}^{-2m_2}\right)}\,dm$$

$$\text{if } H > L$$

The bed is located between probe coils

$$h_z = 2\int\limits_0^\infty \frac{m^3m_2\mathrm{e}^{-(m_2-m_1)\alpha}\mathrm{e}^{-m_1}}{(m_1+m_2)^2\left(1-K_{12}^2\mathrm{e}^{-2m_2\alpha}\right)}\,dm \quad \text{if } H < L$$

As before we will use the following relation between the quadrature component of the field and the apparent conductivity:

$$\frac{\sigma_a}{\sigma_1} = \frac{2}{\sigma_1\mu\omega L^2}\,\mathrm{Q}\,h_z$$

Let us consider the influence of the main factors pointed out above on the shape of the profiling curves (Figs. 5.39–5.56). Curves in every figure correspond to certain values of

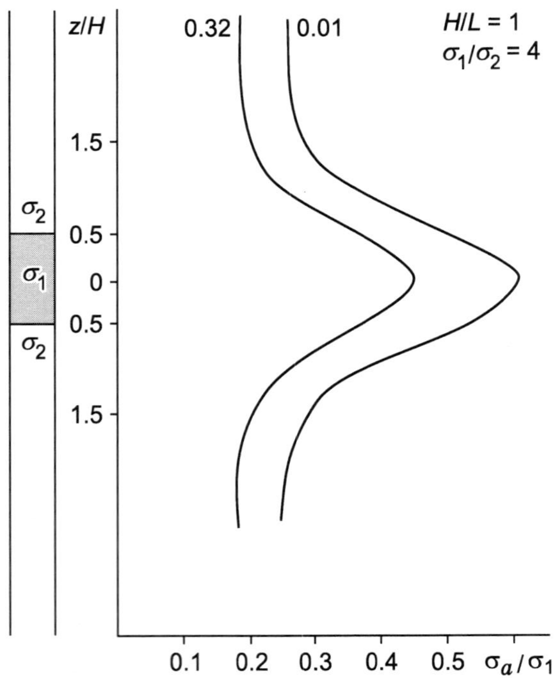

Figure 5.39. Profiling curve. Curve index $\sigma_2 \mu \omega L^2$.

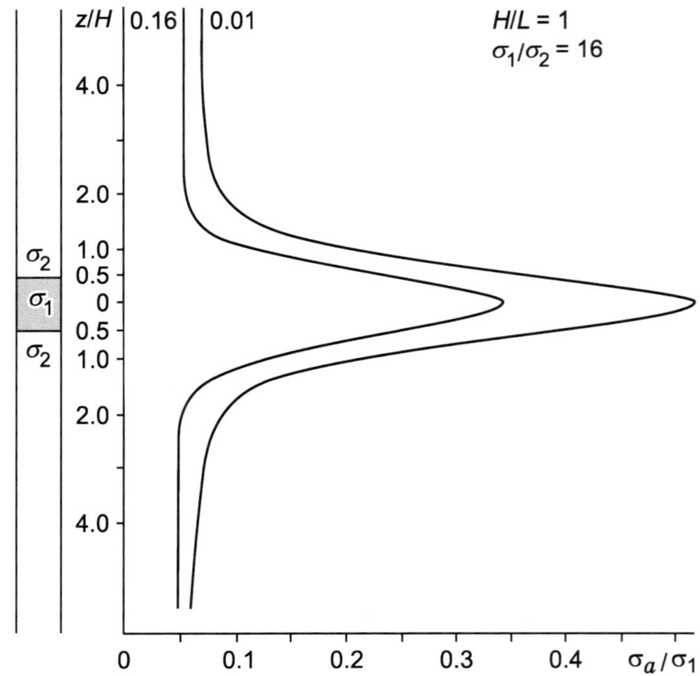

Figure 5.40. Profiling curve. Curve index $\sigma_2 \mu \omega L^2$.

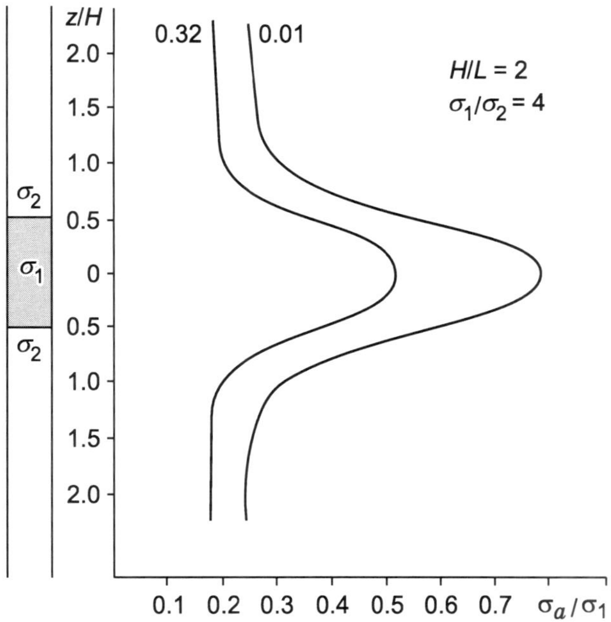

Figure 5.41. Profiling curve. Curve index $\sigma_2\mu\omega L^2$.

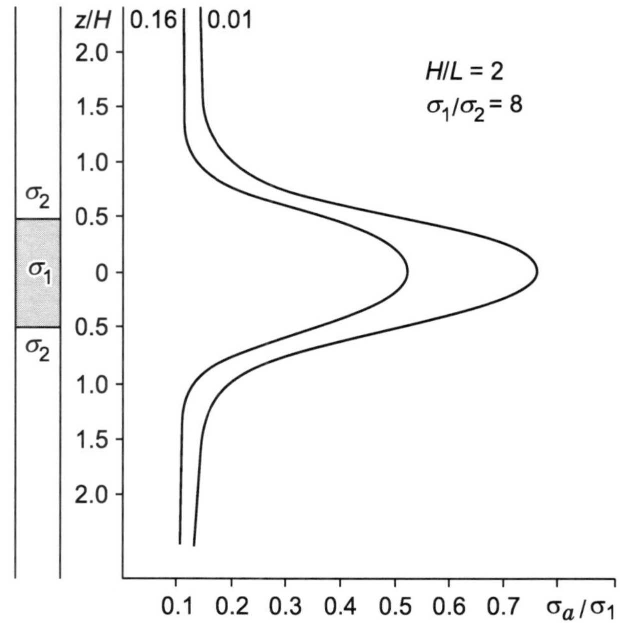

Figure 5.42. Profiling curve. Curve index $\sigma_2\mu\omega L^2$.

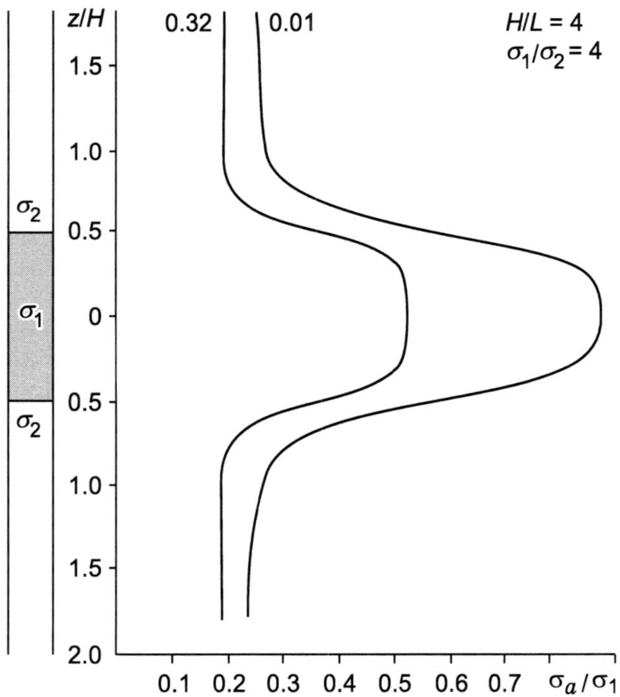

Figure 5.43. Profiling curve. Curve index $\sigma_2\mu\omega L^2$.

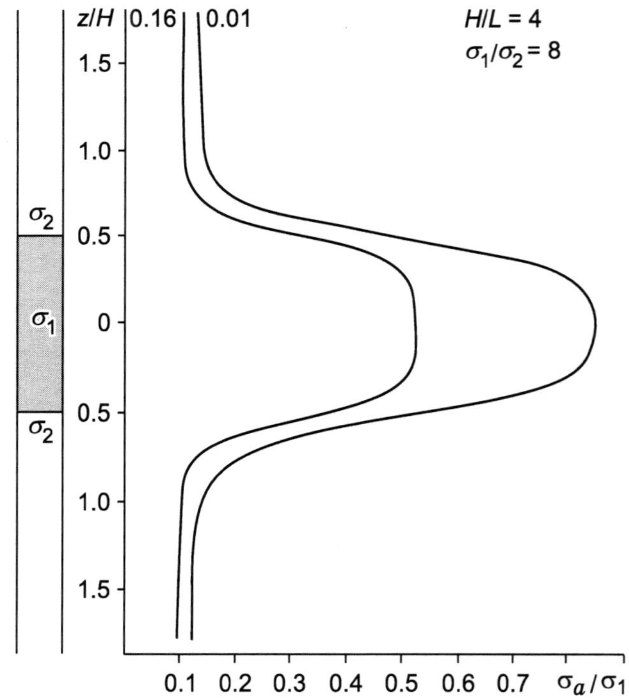

Figure 5.44. Profiling curve. Curve index $\sigma_2\mu\omega L^2$.

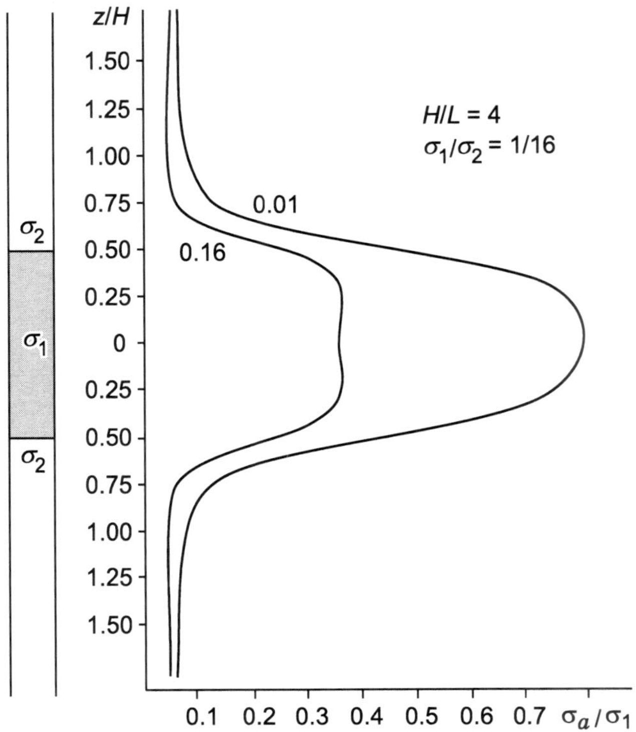

Figure 5.45. Profiling curve. Curve index $\sigma_2\mu\omega L^2$.

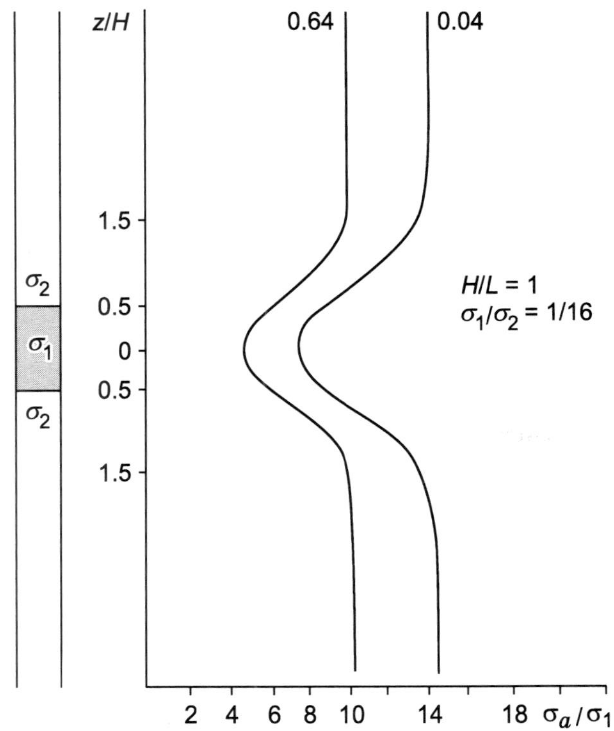

Figure 5.46. Profiling curve. Curve index $\sigma_2\mu\omega L^2$.

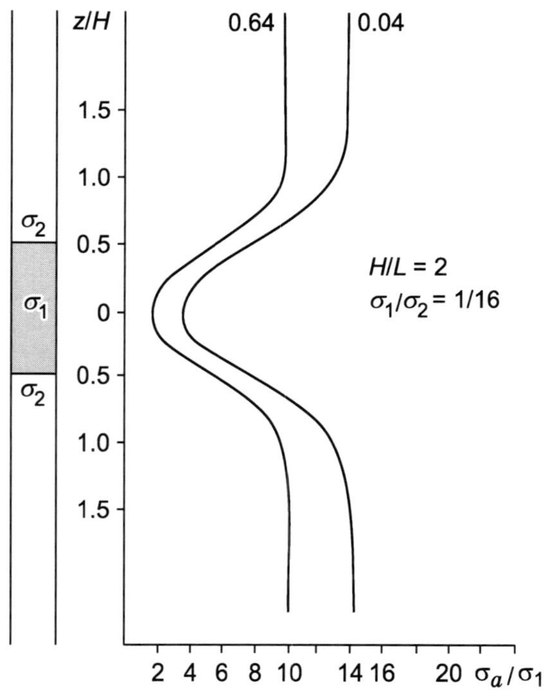

Figure 5.47. Profiling curve. Curve index $\sigma_2\mu\omega L^2$.

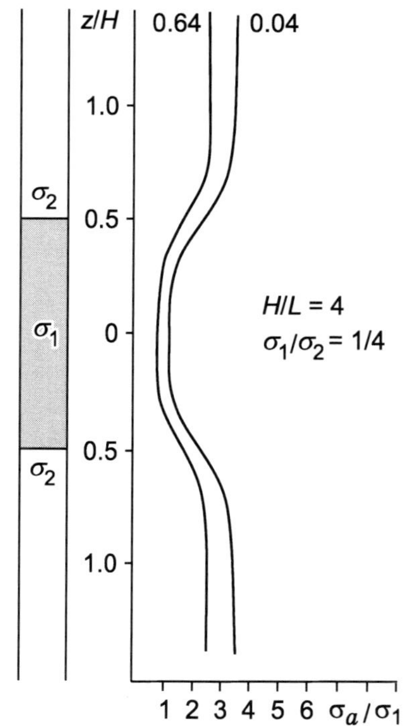

Figure 5.48. Profiling curve. Curve index $\sigma_2\mu\omega L^2$.

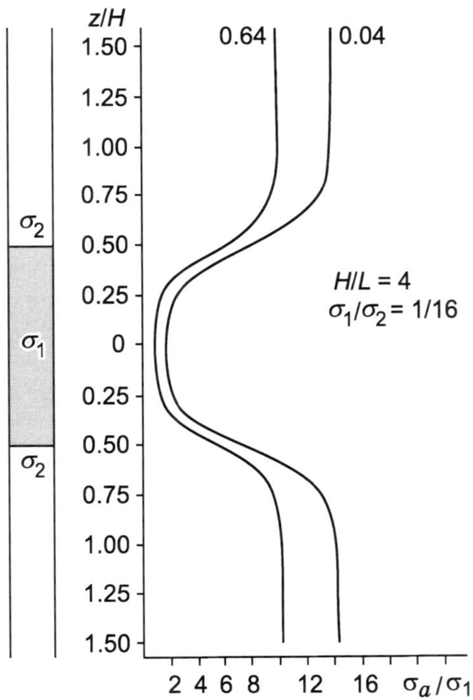

Figure 5.49. Profiling curve. Curve index $\sigma_2\mu\omega L^2$.

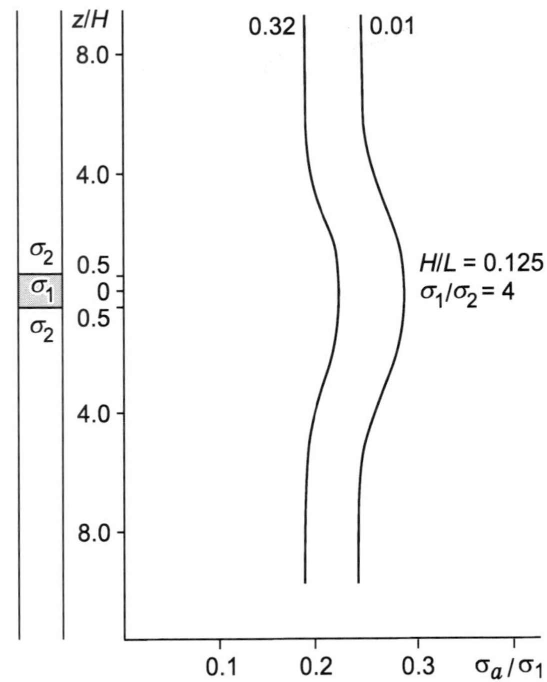

Figure 5.50. Profiling curve. Curve index $\sigma_2\mu\omega L^2$.

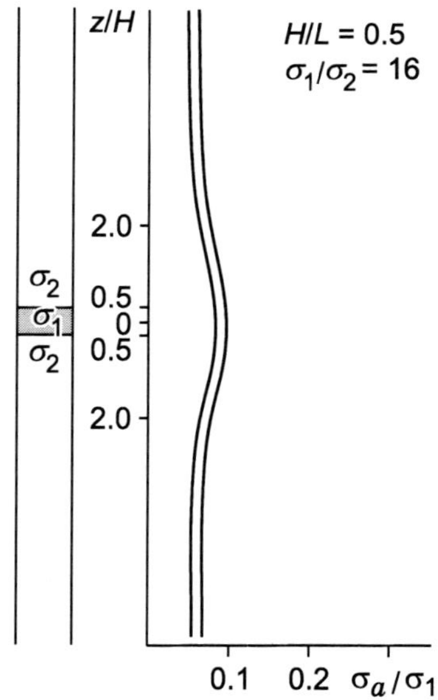

Figure 5.51. Profiling curve. Curve index $\sigma_2\mu\omega L^2$.

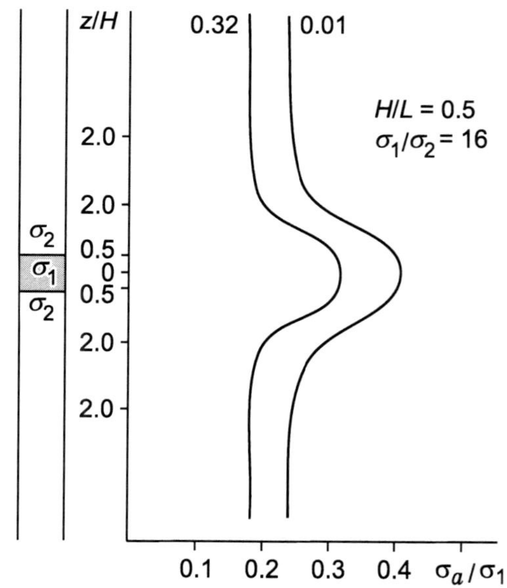

Figure 5.52. Profiling curve. Curve index $\sigma_2\mu\omega L^2$.

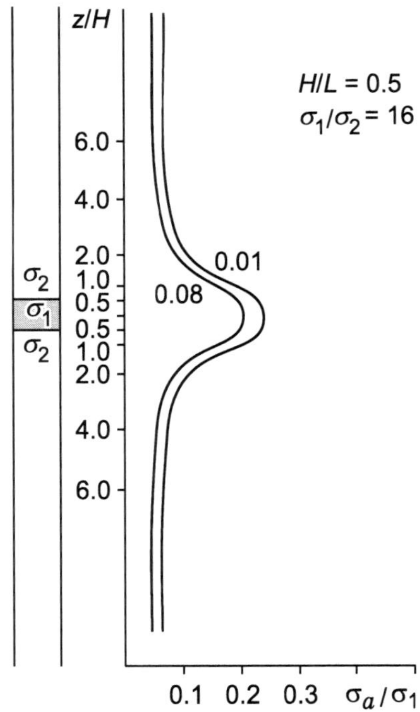

Figure 5.53. Profiling curve. Curve index $\sigma_2\mu\omega L^2$.

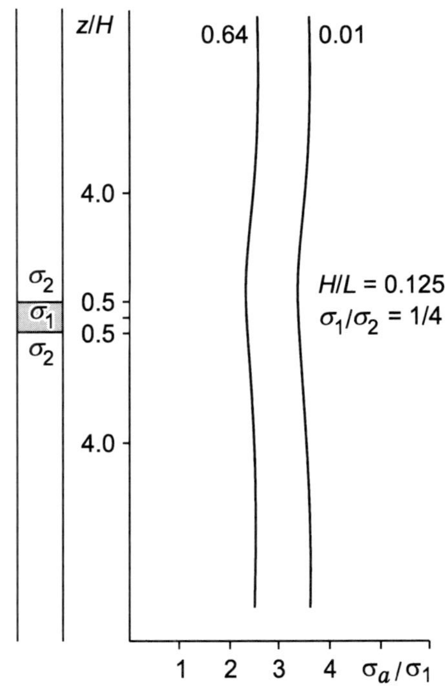

Figure 5.54. Profiling curve. Curve index $\sigma_2\mu\omega L^2$.

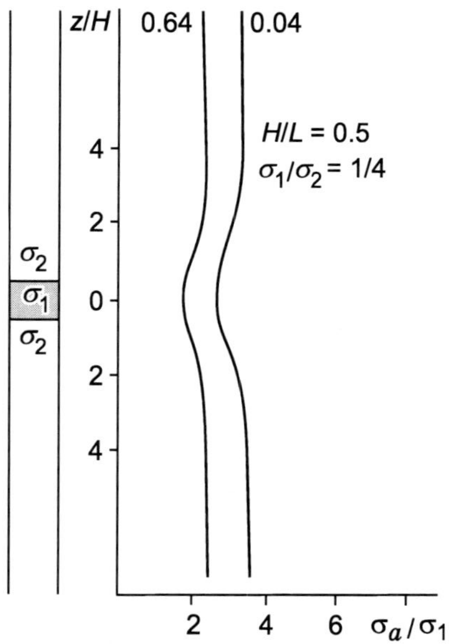

Figure 5.55. Profiling curve. Curve index $\sigma_2 \mu \omega L^2$.

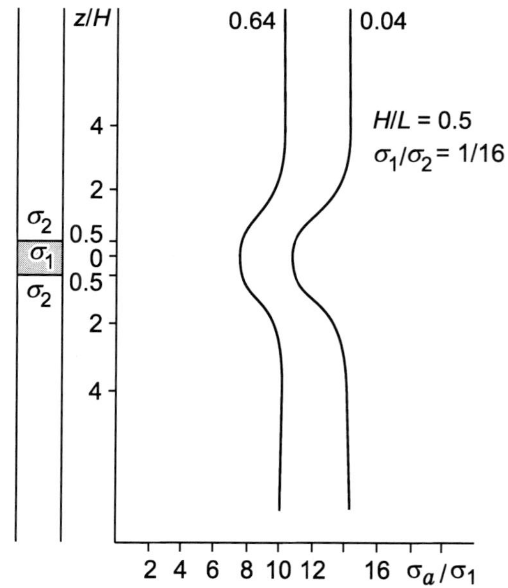

Figure 5.56. Profiling curve. Curve index $\sigma_2 \mu \omega L^2$.

σ_1/σ_2, H/L, while the index of every curve is the parameter $n_2 = \sigma_2\mu\omega L^2$. Along the horizontal and vertical axes values of apparent conductivity and distance from the bed middle to the probe center, expressed in units of bed thickness, are plotted, respectively.

We will consider separately four cases, namely:

- The conductivity of the bed is greater than that of the surrounding medium and its thickness exceeds the probe length ($\sigma_1 > \sigma_2$, $H > L$).
- The bed is more resistive than the surrounding medium and its thickness is greater than the probe length ($\sigma_1 < \sigma_2$, $H > L$).
- The bed is more conductive than the surrounding medium but its thickness is smaller than the probe length ($\sigma_1 > \sigma_2$, $H \leqslant L$).
- The bed thickness is smaller than the probe length and its resistivity is greater than that of the surrounding medium ($\sigma_1 < \sigma_2$, $H < L$).

5.4.1. Thick conductive bed

All curves are symmetrical with respect to the bed middle (Figs. 5.39–5.46). With an increase of parameter n_2, for example due to an increase of the frequency, the width of an intermediate zone where apparent conductivity, σ_a, differs from that corresponding to a uniform medium with conductivity σ_2 becomes narrower. At the same time differentiation of curves to some extent decreases, i.e. the ratio of σ_a/σ_1 at the bed middle to that within the surrounding medium becomes smaller. For instance if $H/L = 4$ and $\sigma_1/\sigma_2 = 4$, the ratio σ_a/σ_1 calculated by Doll's theory is equal to 3.6 while for parameter $n_2 = 0.32$ it is equal to 2.8. In another case when $H/L = 4$ and $\sigma_1/\sigma_2 = 16$, values of this ratio are equal to 11.9 and 9.5, correspondingly.

With an increase of parameter n the vertical response of the induction probe becomes better, i.e. apparent conductivity σ_a approaches closer to the apparent conductivity in a medium with conductivity σ_1.

For example, if $H/L = 2$ and $\sigma_1/\sigma_2 = 4$, the value of σ_a/σ_1 as calculated by Doll's formula, is equal to 0.82 but for $n_2 = 0.32$ we have $\sigma_a = \sigma_a(\sigma_1)$. For smaller thicknesses of the bed, improvement of the vertical response is also observed. If $H/L = 1$, $\sigma_1/\sigma_2 = 4$ we have for the same condition σ_a/σ_1 is equal to 0.63 and 0.88, respectively.

Let us consider some simple ways to determine the bed thickness by using a two-coil induction probe. In accord with Doll's theory the ratio of σ_a/σ_1 corresponding to the bed interface and its middle is:

$$\eta = \frac{1}{2}\frac{1 - 1/4\alpha}{1 - 1/2\alpha} \qquad \alpha = H/L \geqslant 1$$

As is seen from this relation that only for relatively large ratios of H/L parameter η is close to 0.5. For example if $\alpha = 4$, $\eta = 0.54$, while if $\alpha = 10$, $\eta = 0.51$.

If determination of the bed thickness is based on the use of points of the profiling curve corresponding to half of the maximal value, an error does not exceed 2.5% if $\alpha = 4.0$

and 10% for $\alpha = 2.0$, but it reaches 60% when the bed thickness is equal to the probe length. Analysis of profiling curves shows that parameter n_2 ($n_2 < 1.3$) and ratio of conductivities, σ_1/σ_2, do not practically influence the value of 2 for a given thickness of the bed.

5.4.2. Thick resistive bed

With an increase of parameter n_2 the width of the intermediate zone decreases (Figs. 5.46–5.49). Unlike the previous case, with an increase of n_2, the ratio of conductivities, σ_a/σ_1, in the surrounding medium and at the middle of the bed somewhat increases. For instance if $\sigma_1/\sigma_2 = 16$ and $H/L = 4$ this ratio, derived from Doll's theory, is equal to 5.25 while for the curve σ_a/σ_1 with $n_2 = 0.64$ it is equal to 10.0. Thus, in this case with an increase of frequency more resistive layers can be better detected. This fact can be explained in the following way. Let us consider a signal measured in the probe receiver as a sum of signals from a medium directly surrounding the probe and from the rest of the more remote part of the medium. Due to the internal skin effect an increase of a frequency results in a relative decrease of the quadrature component of the current in the removed part of the medium, and therefore the value of σ_a/σ_1 also decreases. In the case when the probe is located within the bed it is directly surrounded by more resistive medium and a change of the quadrature component of induced currents in removed parts of the medium is more noticeable if the probe would be located within the surrounding medium, ($\sigma_2 > \sigma_1$). For this reason with an increase of frequency up to a certain limit ratio of σ_a/σ_1 in the surrounding medium to that when the probe is against the bed increases. Also with growing frequency the vertical response of the probe improves. For instance, if $\sigma_1/\sigma_2 = 1/4$ and $H/L = 4$ the value of σ_a/σ_1 for extremely low frequencies is equal to 1.35 but for parameter $n_2 = 0.16$ it is equal to 1.02. In the case where $\sigma_1/\sigma_2 = 1/16$ and $H/L = 4$ values of σ_a/σ_1 are equal to 1.9 and 1.1, respectively.

5.4.3. Thin conductive bed

Curves of profiling corresponding to this case are presented in Figs. 5.50–5.53. With an increase of parameter n_2, as in the first case, curves become less differentiated. Determination of bed thickness by making use of points where the value of the apparent conductivity is half of the maximal value of the profiling curve is not applicable because of large errors.

5.4.4. Thin resistive bed

Examples of these curves are presented in Figs. 5.54–5.56. The main features of profiling curves such as differentiation, width of the intermediate zone, are similar to those in the second case. Although the vertical response of the probe becomes better with an increase of a frequency, a value of apparent conductivity essentially differs from that corresponding to a medium with conductivity σ_1 even for large values of n_2. For example, if $\sigma_1/\sigma_2 = 1/16$, $H/L = 0.5$ and $n_2 = 0.16$ ratio $\sigma_a/\sigma_1 = 2.5$. Determination of thickness of such beds is also difficult. If $H < 1/4L$ and $\sigma_1/\sigma_2 \leqslant 1/8$, distinguishing such a bed is practically impossible with a two coil induction probe.

Chapter 6

THE TWO-COIL INDUCTION PROBE ON THE BOREHOLE AXIS, WHEN THE BED HAS A FINITE THICKNESS

In this chapter we will consider radial and vertical responses of a two-coil induction probe located on the borehole axis as the bed has a finite thickness. The presence of the invasion zone will also be taken into account. The main attention is paid to the analysis of the influence of such factors as:

- the probe length
- the bed thickness
- the bed resistivity
- the resistivity of the surrounding medium
- the resistivity and radius of the borehole
- the frequency of the field.

This investigation will allow us to better understand the influence of a formation thickness on radial and vertical responses of a two-coil induction probe.

In a similar manner as before the different character of the influence of induced currents in the borehole and in the formation on the quadrature and the inphase components will be emphasized.

In order to perform this analysis we will describe three approaches which permit us to obtain a vast volume of data characterizing a behavior of the magnetic field on the borehole axis. They are:

- Doll's theory.
- The theory taking into account the skin effect in the formation and in the surrounding medium.
- The method of integral equations with respect to tangential components of the electric and magnetic fields.

It is appropriate to notice that results of this investigation are directly used in developing the theory and interpretation of multi-coil induction probes.

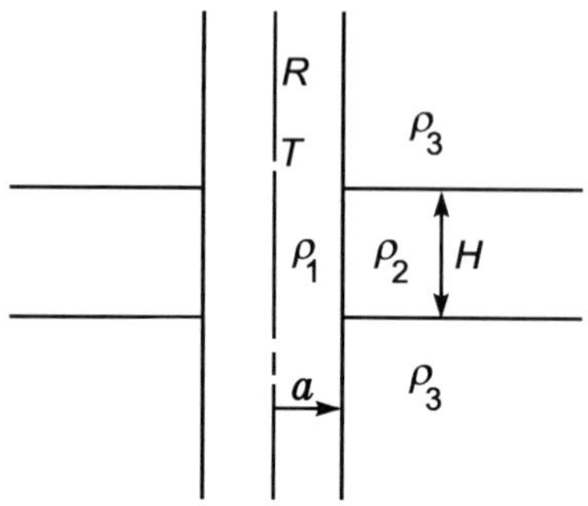

Figure 6.1. The induction probe on the borehole axis.

6.1. Doll's Theory of the Two-coil Induction Probe located on the Borehole Axis when a Formation has a Finite Thickness

Assuming that interaction between induced currents can be neglected we will use results obtained in Chapter 3. Then for the quadrature and inphase components of the magnetic field we have:

$$\mathrm{Q}\, h_z = \frac{\omega\mu L^2}{2} \sum_{i=1}^{N} \sigma_i G_i \qquad \mathrm{In}\, h_z = 0 \tag{6.1}$$

where L is the probe length and σ_i and G_i are conductivity and geometric factor of the i-th part of the medium; h_z is the vertical component of the magnetic field expressed in units of the primary field.

An arbitrary position of the probe on the borehole axis is shown in Fig. 6.1. It is obvious that the geometric factor of the formation in presence of the borehole, G_2^*, can be written as:

$$G_2^* = G_2 - G_{12} \tag{6.2}$$

where G_2 is the geometric factor of the bed when the borehole is absent, and its expression has been derived in this previous chapter; G_{12} is the geometric factor of the borehole located against the formation. It is clear that this part of the medium has the shape of a cylinder with radius of the borehole and height equal to the bed thickness, H.

By analogy the geometric factor of a medium with resistivity ρ_3 can be written as:

$$G_3^* = G_3 - G_{13} \tag{6.3}$$

where G_3 is the geometric factor of the surrounding medium as the borehole is absent; G_{13} is the geometric factor of that part of the borehole which is located against the surrounding medium.

It is easily seen that the sum of geometric factors G_{12} and G_{13} is equal to:

$$G_2^* = G_2 - G_{12} \tag{6.4}$$

where G_1 is the geometric factor of the borehole which was investigated in detail in Chapter 4.

Taking into account eqs. 6.2–6.4, instead of eq. 6.1 we obtain:

$$\begin{aligned} \mathrm{Q}\,h_z &= \frac{\omega\mu L^2}{2}[\sigma_1 G_1 + \sigma_2 G_2^* + \sigma_3 G_3^*] \\ &= \frac{\omega\mu L^2}{2}[\sigma_1 G_1 + \sigma_2(G_2 - G_{12}) + \sigma_3(G_3 - G_{13})] \\ &= \frac{\omega\mu L^2}{2}[\sigma_1 G_1 + \sigma_2 G_2 + \sigma_3 G_3 - \sigma_2 G_{12} - \sigma_3(G_1 - G_{12})] \\ &= \frac{\omega\mu L^2}{2}[(\sigma_1 - \sigma_2)G_1 + (\sigma_3 - \sigma_2)G_{12} + \sigma_2 G_2 + \sigma_3 G_3] \end{aligned} \tag{6.5}$$

Thus the quadrature component, $\mathrm{Q}\,h_z$, is expressed through known geometric factors, G_1, G_2 and G_3, where $G_2 + G_3 = 1$ and the geometric factor of the cylinder, G_{12}, which can be presented as (Chapter 3):

$$G_{12} = \frac{L}{2}\int\limits_S \frac{r^3}{R_1^3 R_2^3}\,\mathrm{d}r\,\mathrm{d}z = \frac{L}{2}\int\limits_a^b r^3\,\mathrm{d}r \int\limits_{z_1}^{z_2} \frac{\mathrm{d}z}{R_1^3 R_2^3} \tag{6.6}$$

Coordinates r, z_1, z_2 and distances R_1 and R_2 are shown in Fig. 6.2.

Since the apparent conductivity, σ_a, is related with the quadrature component as:

$$\sigma_a = \frac{2}{\mu\omega L^2}\,\mathrm{Q}\,h_z$$

we have:

$$\sigma_a = (\sigma_1 - \sigma_3)G_1 + (\sigma_3 - \sigma_2)G_{12} + \sigma_2 G_2 + \sigma_3 G_3 \tag{6.7}$$

It is clear that eq. 6.5 and 6.7 describe the field and apparent conductivity regardless of the position of the two-coil induction probe with respect to formation boundaries.

Here it is appropriate to make the following comments:

- The first term of eq. 6.7 does not depend on a probe position, since it describes a signal caused by induction currents in the borehole with conductivity equal to $\sigma_1 - \sigma_2$.
- The last two terms of this equation:

  $$\sigma_2 G_2 + \sigma_3 G_3$$

  characterize a field caused by induced currents in the formation and in the surrounding medium when the borehole is absent.

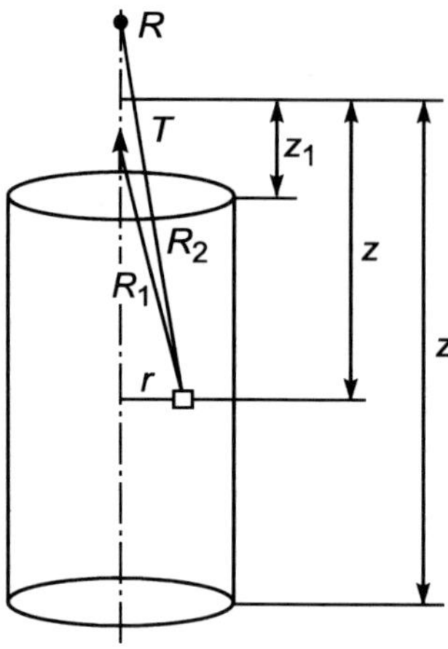

Figure 6.2. Illustration of eq. 6.6.

- Equation 6.7 can be written as:

  $$\sigma_a = (\sigma_1 - \sigma_3)G_1 + (\sigma_3 - \sigma_2)G_{12} + \sigma_a^{(1)}$$

  where $\sigma_a^{(1)}$ is the apparent conductivity corresponding to a horizontally layered medium.

- Equations 6.5 and 6.7 can be easily generalized for the case when the invasion zone is present (Chapter 3).

As an example, suppose that the two-coil induction probe is symmetrically located with respect to formation boundaries. For this case the values of geometric factor G_{12} are given in Table 6.1 and corresponding curves describing the dependence of G_{12} on parameter $H/2L$ are presented in Fig. 6.3 (here L and H are probe length and bed thickness, respectively). The curve index is the ratio a/L, where a is the cylinder radius. For the case considered the geometric factor of the cylinder is defined by two parameters, namely its radius, a, and its height, H, expressed in units of the probe length.

As is seen from Table 6.1 and curves in Fig. 6.3 with an increase of parameter $\beta = H/2L$ function G_{12} relatively rapidly approaches its asymptotic value which corresponds to that of the geometric factor of the borehole. With an increase of the probe length, L, with respect to the radius, a, curves of G_{12} approach this asymptote for smaller values of $H/2L$. If the probe length, L, exceeds the cylinder diameter, the geometric factor of such a cylinder practically coincides with that of an infinitely long cylinder if $H > 1.2L$. In other words, the main contribution to a signal is brought by currents within a cylinder with a height which is slightly greater than the probe length.

Correspondingly, the remaining part of the cylinder, in particular the borehole, does not in essence affect the signal. With an increase of the borehole radius or a decrease

TABLE 6.1
Values of function $G_{12} \times 10^4$; $\beta = H/2L$, $\alpha = a/L$

β \ α	0.05	$0.05\sqrt{2}$	0.1	$0.1\sqrt{2}$	0.2	0.4	0.8	1.6
0.1	0.102	0.399	1.53	5.68	19.6	155	519	830
0.2	0.234	0.912	3.48	12.7	43.2	323	1050	1660
0.4	1.25	4.58	15.8	49.8	142	775	2170	3340
0.8	25.2	51.0	103	210	422	1520	3690	5630
1.6	25.2	51.1	104	211	427	1570	3970	6410
3.2	25.2	51.1	104	211	427	1570	3980	6520

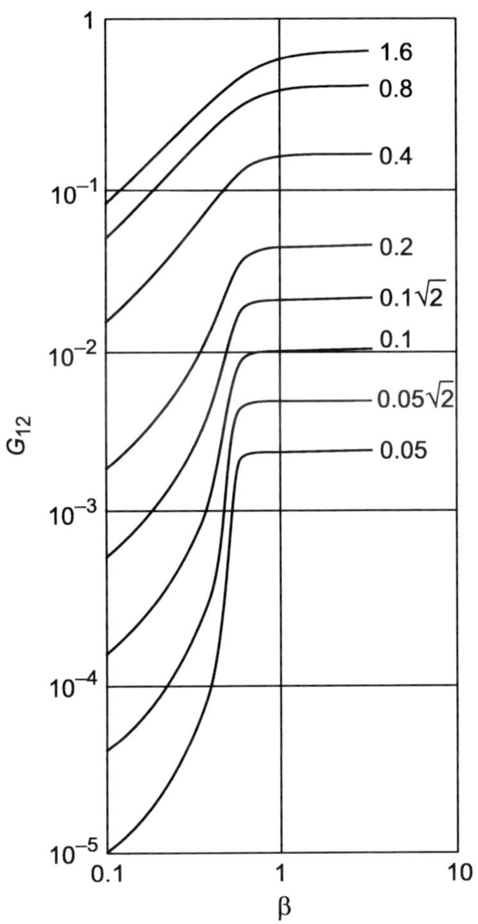

Figure 6.3. Behavior of function G_{12}. Curve index a/L.

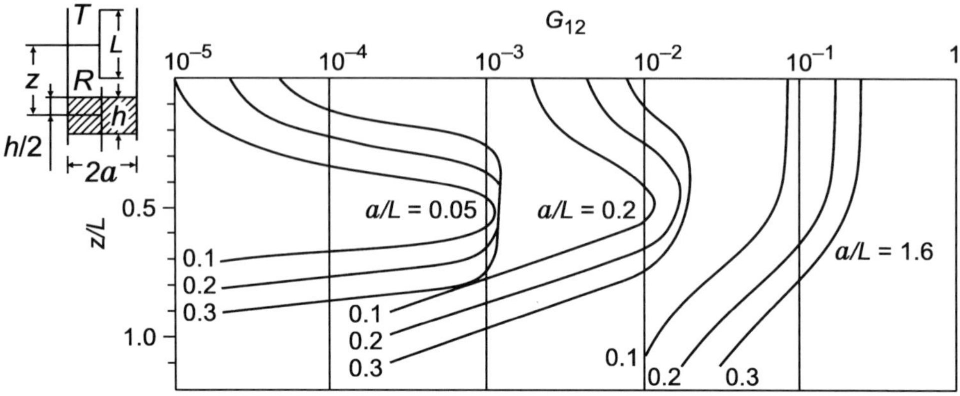

Figure 6.4. Behavior of function G_1^*. Curve index h/L.

of the probe length an approach to the asymptotic value takes place for large values of $H/2L$. This fact permits us to explain some features of radial responses of multi-coil focusing probes in cases when a radius of the invasion zone is comparable or greater than a length of some of two-coil probes forming this system.

We will consider the geometric factor G_{12} as the sum of the geometric factors of two cylinders having a common base $z = 0$. It is obvious that the geometric factor of such a cylinder is equal to $G_{12}/2$. Now it is a simple matter to obtain a value of the geometric factor of a cylinder with coordinates of its bases: z_1 and z_2, respectively. In fact the geometric factor of the cylinder, located nonsymmetrically with respect to the probe, is defined from the relation:

$$G_{12}^*\left(\frac{z_2}{L},\frac{z_1}{L},\frac{a}{L}\right) = \frac{1}{2}\left[G_{12}\left(\frac{z_2}{L},\frac{a}{L}\right) - G_{12}\left(\frac{z_1}{L},\frac{a}{L}\right)\right] \tag{6.8}$$

where $z_2 > z_1$.

By analogy the geometric factor of the cylindrical layer with coordinates z_1, z_2, a_1 and a_2 can be presented as:

$$G_{12}^*\left(\frac{z_2}{L},\frac{z_1}{L},\frac{a_2}{L},\frac{a_1}{L}\right) = G_{12}^*\left(\frac{z_2}{L},\frac{z_1}{L},\frac{a_2}{L}\right) - G_{12}^*\left(\frac{z_2}{L},\frac{z_1}{L},\frac{a_1}{L}\right) \tag{6.9}$$

This relation is useful for investigating radial responses of induction probes in the presence of an invasion zone as well as for evaluation of the influence of caverns.

As was demonstrated above, function G_{12}^* characterizes a signal caused by induced currents in an infinitely long cylinder. At the same time it is useful to observe the influence of different parts of this cylinder. With this purpose curves of function G_{12}^* are presented in Fig. 6.4. They demonstrate the influence of relatively thin cylinders with various position with respect to the induction probe. If the probe length exceeds the cylinder radius, the part of the borehole which directly surrounds the probe provides the main contribution to the signal caused by induced currents in the borehole.

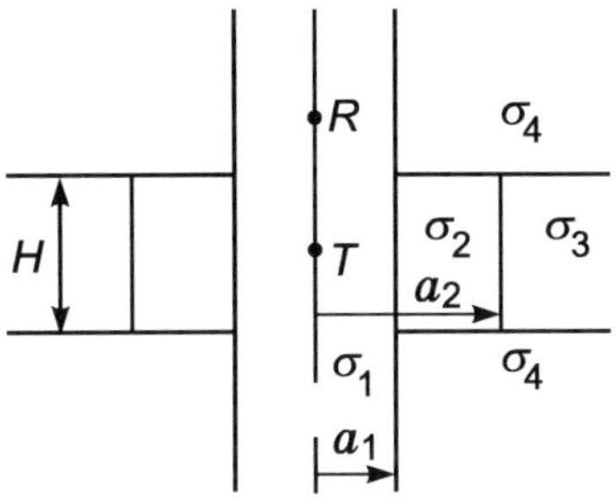

Figure 6.5. Illustration to derivation of eq. 6.10.

Equation 6.5 has been derived assuming that the invasion zone is absent. Now let us consider a more general case when there is penetration of the borehole solution into the formation (Fig. 6.5).

By analogy with a more simple case (eq. 6.8) we have the following expression for the quadrature component of the magnetic field:

$$\begin{aligned} Q\,h_z &= \frac{\omega\mu L^2}{2}[\sigma_4 G_4 + (\sigma_1 - \sigma_4)G_{12} + \sigma_3 G_3 + (\sigma_1 - \sigma_3)G_{12} + (\sigma_2 - \sigma_3)G_2] \\ &= \frac{\omega\mu L^2}{2}[\sigma_3 G_3 + \sigma_4 G_4 + (\sigma_1 - \sigma_4)G_1 + (\sigma_4 - \sigma_1)G_{12} + (\sigma_2 - \sigma_3)G_2] \end{aligned} \tag{6.10}$$

where σ_1, G_1; σ_2, G_2; σ_3, G_3; σ_4, G_4 are conductivities and geometric factors of the borehole, the invasion zone, the formation and the surrounding medium, respectively; G_{12} is the geometric factor of that part of the borehole which is located against the invasion zone.

Correspondingly, for the apparent conductivity we obtain:

$$\sigma_a = \sigma_3 G_3 + \sigma_4 G_4 + (\sigma_1 - \sigma_4)G_1 + (\sigma_4 - \sigma_1)G_{12} + (\sigma_2 - \sigma_3)G_2 \tag{6.11}$$

In conclusion let us make the following comment. Doll's theory does not take into account the skin effect. For this reason we can expect that both vertical and radial responses, specially the first one, derived from eq. 6.8–6.10 will be in many cases different from actual responses. Correspondingly in the next section we will consider responses of two-coil induction probes, assuming that the skin effect in the formation and in the surrounding medium is not subjected to the presence of borehole and invasion zone.

6.2. The Theory of a Two-coil Induction Probe, Taking into Account the Skin Effect in an External Medium

Let us assume that currents induced in the borehole and in the invasion zone are defined by only the primary vortex electric field. In other words, we can neglect the skin effect within these two parts of the medium. Also we will suppose that outside the borehole and the invasion zone the skin effect manifests itself in the same manner as in a horizontally layered medium. It is easy to understand that as these conditions are met we can ignore the effect caused by interaction between induced currents in the borehole and in the invasion zone on one hand and currents induced in the formation and in the surrounding medium on the other hand.

First, following results obtained in Chapter 3 we will present a field in a medium with two horizontal interfaces, h_0, as a sum of two fields:

$$h_0 = h_i + h_e \tag{6.12}$$

where h_i, is the vertical component of the magnetic field expressed in units of the primary field and caused by induced currents in the borehole; h_e is the vertical component of the magnetic field of currents arising outside the borehole.

The field h_i can be expressed through geometric factors as:

$$h_i = \frac{i\omega\mu L^2}{2}(\sigma_2 G_{12} + \sigma_3 G_{13}) \tag{6.13}$$

where G_{12} and G_{13} are geometric factors of borehole parts located against the formation and the surrounding medium, respectively. It is obvious that:

$$G_{12} = \int\limits_{S_1} q \, dS \qquad G_{13} = \int\limits_{S_2} q \, dS$$

where

$$q = \frac{L}{2}\frac{r^3}{R_1^3 R_2^3} \qquad \text{and} \qquad G_{12} + G_{13} = G_1$$

Thus, the magnetic field of currents induced in the formation and in the surrounding medium when the borehole is not conducting can be presented in the form:

$$h = h_0 - \frac{i\omega\mu L^2}{2}(\sigma_2 G_{12} + \sigma_3 G_{13}) \tag{6.14}$$

Correspondingly, for the magnetic field on the axis of the borehole with conductivity σ_1 we have:

$$\begin{aligned} h &= h_0 - \frac{i\omega\mu L^2}{2}[\sigma_2 G_{12} + \sigma_3 G_{13}] + \frac{i\omega\mu L^2}{2}\sigma_1 G_1 \\ &= h_0 + \frac{i\omega\mu L^2}{2}[\sigma_1 G_1 - \sigma_2 G_{12} - \sigma_3 G_{13}] \\ &= h_0 + \frac{i\omega\mu L^2}{2}[(\sigma_3 - \sigma_2)G_{12} + (\sigma_1 - \sigma_3)G_1] \end{aligned} \tag{6.15}$$

As was shown in the previous chapter the magnetic field of the vertical magnetic dipole, h_0, in a horizontally layered medium is expressed through known functions. For example, when the probe is located symmetrically with respect to the layer boundaries, we have:

$$h_0 = h^{un} + L^3 \int_0^\infty \frac{m^3}{m_2} K_{12}\, e^{-m_2 H} \frac{e^{m_2 H} + K_{12} \cosh m_2 L}{1 - K_{12}^2\, e^{-2m_2 H}}\, dm \qquad \text{if } H \geqslant L$$

and

$$h_0 = 2L^3 \int_0^\infty \frac{m^3 m_2\, e^{-H(m_2 - m_1)}\, dm}{(m_1 + m_2)^2\, (1 - m_{12}^2\, e^{-2m_2 H})} \qquad \text{if } H \leqslant L$$

where h^{un} is the magnetic field on the z-axis in a uniform medium with resistivity of the formation; H is the formation thickness; L is the probe length; $m_2 = (m^2 - k_2^2)^{1/2}$, $m_1 = (m^2 - k_3^2)^{1/2}$, $K_{12} = (m_2 - m_1)/(m_1 + m_2)$.

Let us remember that function h^{un} as well as h_0 is the magnetic field normalized by the primary field, and it presents the complex amplitude of the field.

In a more general case, when there is an invasion zone within the formation, the field on the borehole axis can be written as:

$$h = h_0 + \frac{i\omega\mu L^2}{2}[(\sigma_2 - \sigma_3)G_{23} + (\sigma_4 - \sigma_3)G_{12} + (\sigma_1 - \sigma_4)G_1] \tag{6.16}$$

where h_0 is the field in a horizontally layered medium when the borehole and the invasion zone are absent; σ_1, σ_2, σ_3, and σ_4 are conductivities of the borehole, the invasion zone, the formation and the surrounding medium, respectively; G_1 and G_2 are geometric factors of the borehole and its part located against the formation; G_{12} is the geometric factor of the invasion zone and, as was shown in Section 6.1, it can be expressed through function G_{12}.

Thus, calculation of the field on the borehole axis consists of determination of the field in the horizontally layered medium when the borehole and the invasion zone are absent and calculation of geometric factors of vertical cylinders of a finite height the axis of which coincides with the borehole one (Table 6.1).

In accord with eq. 6.15 we have the following expressions for the quadrature and inphase components of the field:

$$\begin{aligned} \mathrm{Q}\, h &= \mathrm{Q}\, h_0 + \frac{\mathrm{i}\omega\mu L^2}{2}[(\sigma_3 - \sigma_2)G_{12} + (\sigma_1 - \sigma_3)G_1] \\ \mathrm{In}\, h &= \mathrm{In}\, h_0 \end{aligned} \tag{6.17}$$

Correspondingly, in the case when there is an invasion zone we have:

$$\begin{aligned} \mathrm{Q}\, h &= \mathrm{Q}\, h_0 + \frac{\mathrm{i}\omega\mu L^2}{2}[(\sigma_2 - \sigma_3)G_{23} + (\sigma_4 - \sigma_3)G_{12} + (\sigma_1 - \sigma_4)G_1] \\ \mathrm{In}\, h &= \mathrm{In}\, h_0 \end{aligned} \tag{6.18}$$

Therefore, in both cases induced currents in the borehole as well as in the invasion zone do not have an influence on the inphase component of the secondary field, and it coincides with that corresponding to a horizontally layered medium.

As was pointed out before, this fact demonstrates the greater depth of investigation in the radial direction when the inphase component is measured.

Here it is appropriate to notice that this method of field calculation allows us to establish frequencies, geoelectric parameters of a medium, and probe lengths when the focussing of multi-coil probes provide essential reduction of the influence currents in the borehole and in the invasion zone.

Comparison with calculations based on a solution of the system of integral equations has permitted us to establish boundaries of application of this approximate method, i.e. the range of parameters when induced currents in the borehole and invasion zone arise due to only the primary electric field and the skin effect in the formation and in the surrounding medium manifests in the same manner as in the horizontally layered medium. With error which does not exceed 10% this range of parameters is defined as:

$$\frac{a_1}{h_1} < 0.3 \qquad \frac{L}{h_2} < \frac{1.5}{\left[1+\left(\frac{\sigma_3}{\sigma_2}-1\right)(1-G_2)\right]^{1/2}} \tag{6.19}$$

or

$$\frac{\lambda_1}{a_1} < 20 \qquad \frac{\lambda_2}{L} > 4\left[1+\left(\frac{\sigma_3}{\sigma_2}-1\right)(1-G_2)\right]^{1/2} \tag{6.20}$$

where L is the probe length; a_1 is the borehole radius; G_2 is the geometric factor of the formation; h_1 and h_2 are the skin depth in the borehole and in the formation, respectively, and:

$$\lambda_1 = 2\pi h_1 \qquad \lambda_2 = 2\pi h_2$$

In particular, for symmetric position of the probe we have:

$$G_2 = 1 - L/2H \qquad \text{if } H > L$$

In this case conditions 6.19–6.20 can be written as:

$$\frac{a_1}{h_1} < 0.3 \tag{6.21}$$

$$\frac{L}{h_2} < \frac{1.5}{\left[1+\left(\frac{\sigma_3}{\sigma_2}-1\right)\frac{L}{2H}\right]^{1/2}} \tag{6.22}$$

or

$$\frac{\lambda_1}{a_1} > 20 \tag{6.23}$$

$$\frac{\lambda_1}{L} > 4\left[1 + \left(\frac{\sigma_3}{\sigma_2} - 1\right)\frac{L}{2H}\right]^{1/2} \tag{6.24}$$

Correspondingly, with an increase of ratio H/L we obtain conditions for a medium with only the cylindrical interface:

$$\frac{a_1}{h_1} < 0.3 \qquad \frac{L}{h_2} < 1.5 \tag{6.25}$$

or in a more general model when there is an invasion zone we have:

$$\max\left(\frac{a_1}{h_1}, \frac{a_2}{h_2}\right) < 0.3 \qquad \frac{L}{h_3} < 1.5 \tag{6.26}$$

where h_3 is the skin depth in the formation; a_2 is the invasion zone radius.

As follows from the second equation of 6.20 if the probe is located in the surrounding medium and far away from the formation we have:

$$\frac{\lambda_3}{L} > 4 \tag{6.27}$$

since the geometric factor G_2 tends to zero.

Now let us consider two numerical examples.

Example 1

Suppose that $\rho_1 = 1.0$ ohm·m, $\rho_2 = 2.0$ ohm·m, $\rho_3 = 20.0$ ohm·m, $H = 3$ m, $a_1 = 0.1$ m and $L = 1.5$ m. Then in accord with eq. 6.21–6.22 we have:

$$h_1 > \frac{1}{3} \text{ m and } h_2 > 1 \text{ m}$$

Therefore, frequencies have to satisfy the following conditions:

$$f < 2 \times 10^6 \text{ Hz and } f < 10^5 \text{ Hz}$$

i.e.

$$f_{max} < 10^5 \text{ Hz}$$

Example 2

Assume that $\rho_1 = 1$ ohm·m, $\rho_2 = 20.0$ ohm·m, $\rho_3 = 2$ ohm·m, $a_1 = 0.1$ m, $H = 3$ m $L = 1.5$ m. Then the maximal frequency is defined from relations:

$$h_1 > \frac{1}{3} \text{ m and } h_2 > 2.4 \text{ m}$$

Whence $f_{max} < 2 \times 10^5$ Hz.

Apparent conductivity curves, σ_a/σ_2, defined from equation:

$$\sigma_a = \frac{2}{\mu\omega L^2}\,\mathrm{Q}\,h$$

where $\mathrm{Q}\,h$ is given in eq. 6.15 are presented in Figs. 6.6–6.9.

The difference between σ_a and its value in a uniform medium with conductivity of the formation is defined by the influence of the borehole, the surrounding medium and the parameter L/h_2. With an increase of the resistivity of the surrounding medium, the formation thickness and parameter L/h_2, the relative role of the borehole increases. This becomes especially noticeable if $\rho_3 > \rho_2$.

By analogy we can calculate apparent conductivity, σ_a, in the presence of the invasion zone provided that conditions 6.26 are met.

6.3. Influence of the Finite Thickness of the Formation on the Magnetic Field Behavior

In this section we will consider again the influence of induced currents on the behavior of the quadrature component of the magnetic field on the borehole axis when the formation has a finite thickness. However, unlike the previous sections we will proceed from the results of calculations based on a solution of integral equations with respect to tangential components of the field. This method of the solution of the value of the boundary problem has been described in detail in Chapter 3. This analysis is mainly based on numerical modeling performed by L. Tabarovsky and V. Dimitriev.

First of all we will study the influence of a finite thickness of the formation on frequency responses of the quadrature component of the field h_z when the probe is symmetrically located with respect to the formation interfaces. With this purpose we will introduce parameter, defined by the relation:

$$\chi = \frac{\mathrm{Q}\,h_z}{\mathrm{Q}\,h_{z\infty}}$$

where $\mathrm{Q}\,h_z$ and $\mathrm{Q}\,h_{z\infty}$ are quadrature components of the field on the borehole axis when the formation has finite and infinite thickness, respectively.

Frequency responses of parameter χ are presented in Figs. 6.10–6.17 for various values of ratios: H/L, L/a_1 and σ_1/σ_3.

First let us notice the following. Behavior of frequency responses of the function essentially depends on ratio σ_2/σ_3. In the case when $\sigma_2 > \sigma_3$ a value of χ differs from unity by less than 10%, if the formation thickness exceeds the probe length ($H > L$). In another case when the formation is more resistive than the surrounding medium $\sigma_2 < \sigma_3$ the different dependence of function χ from geoelectric parameters is observed.

As is seen from the curves with an increase of the formation resistivity the influence of its finite thickness manifests itself more strongly when $\lambda_1/a_1 > 16$. It is also obvious that this effect becomes less with an increase of frequency. For instance, if $H/L = 2\sqrt{2}$, $L/a_1 = 4\sqrt{2}$ and $\rho_3/\rho_1 = 2$, we have (Figs. 6.15–6.17):

$$\chi = 1.2 \text{ for } \begin{cases} \lambda_1/a_1 = 35 \quad \text{and} \quad \rho_2/\rho_1 = 16 \\ \lambda_1/a_1 = 25 \quad \text{and} \quad \rho_2/\rho_1 = 32 \end{cases}$$

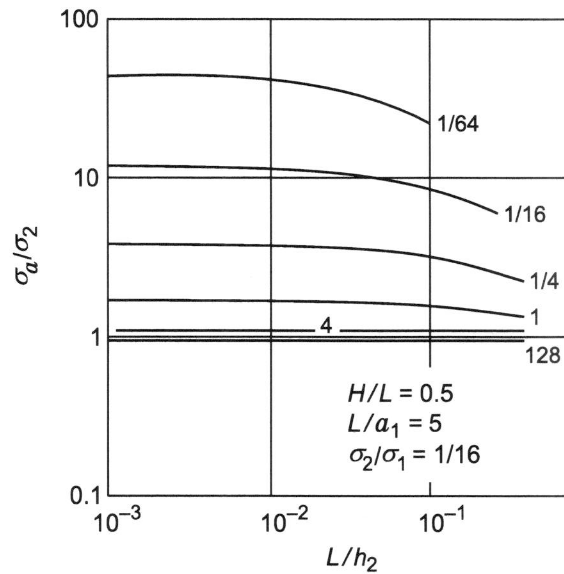

Figure 6.6. Apparent conductivity curves. Curve index σ_2/σ_3.

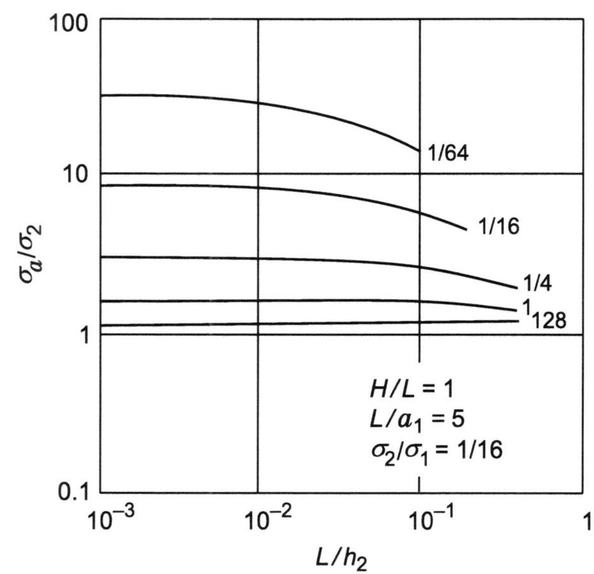

Figure 6.7. Apparent conductivity curves. Curve index σ_2/σ_3.

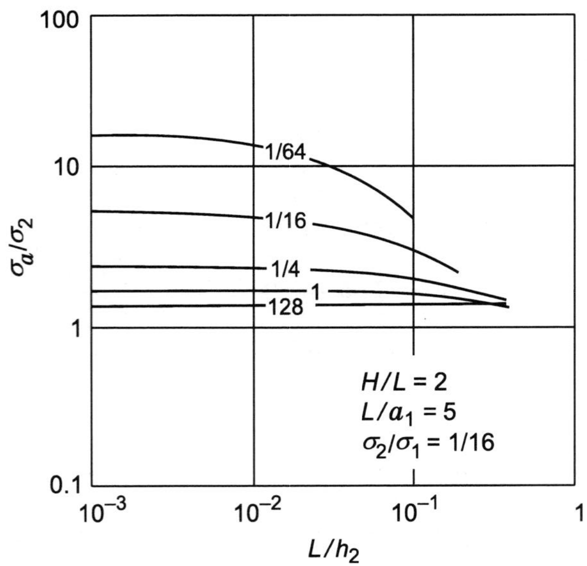

Figure 6.8. Apparent conductivity curves. Curve index σ_2/σ_3.

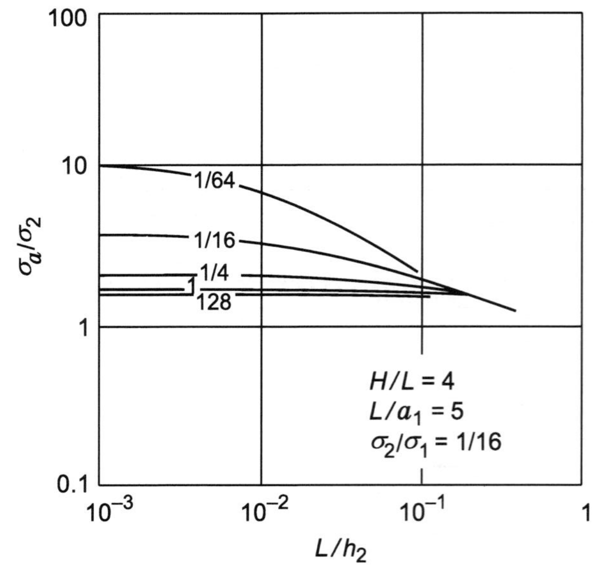

Figure 6.9. Apparent conductivity curves. Curve index σ_2/σ_3.

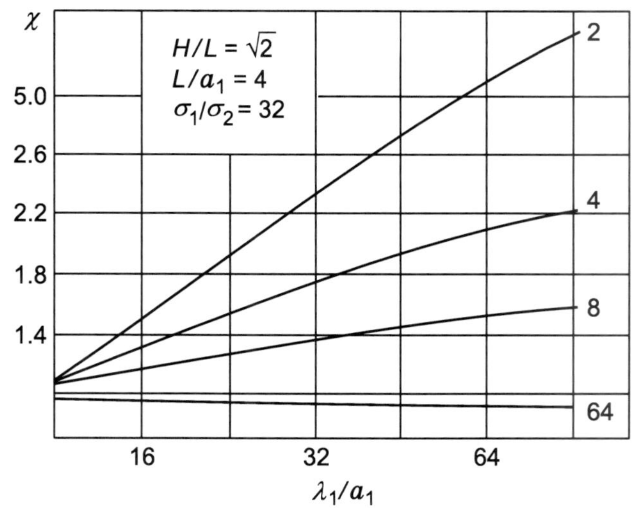

Figure 6.10. Behavior of parameter χ. Curve index σ_1/σ_3.

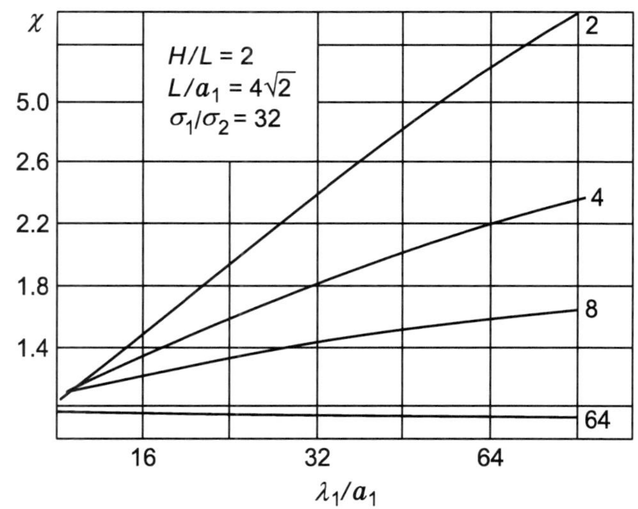

Figure 6.11. Behavior of parameter χ. Curve index σ_1/σ_3.

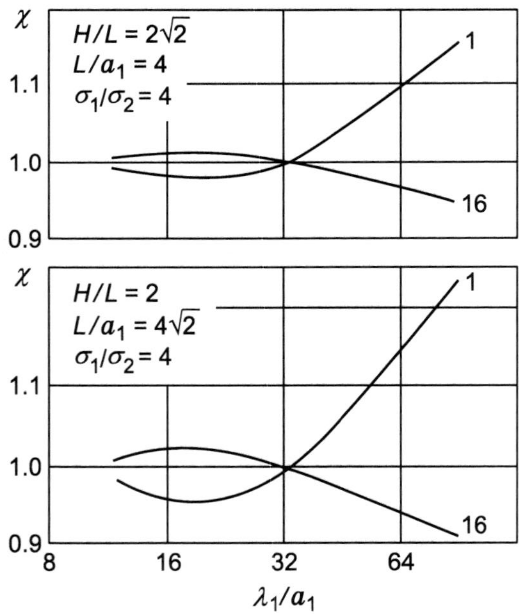

Figure 6.12. Behavior of parameter χ. Curve index σ_1/σ_3.

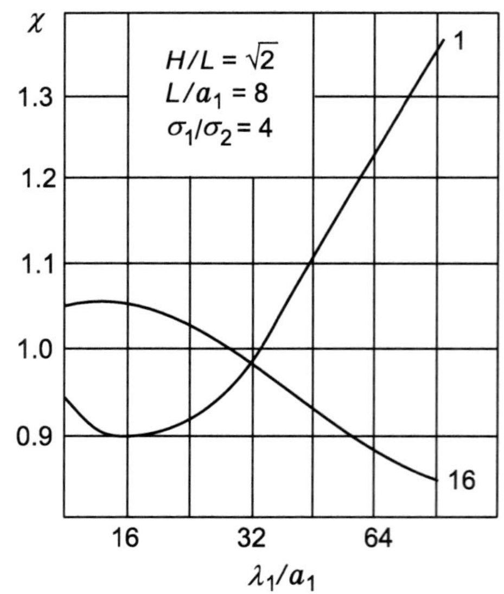

Figure 6.13. Behavior of parameter χ. Curve index σ_1/σ_3.

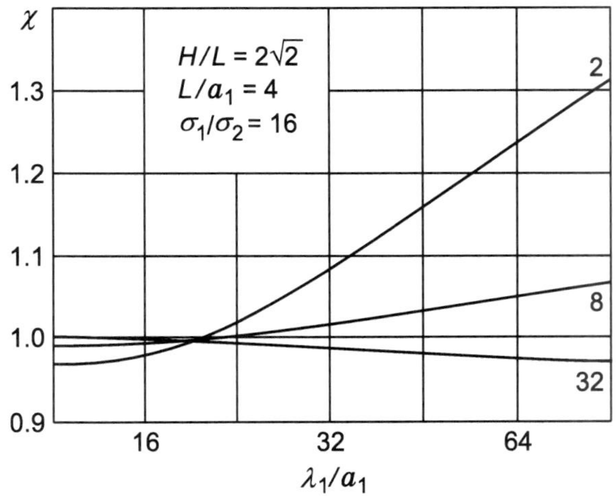

Figure 6.14. Behavior of parameter χ. Curve index σ_1/σ_3.

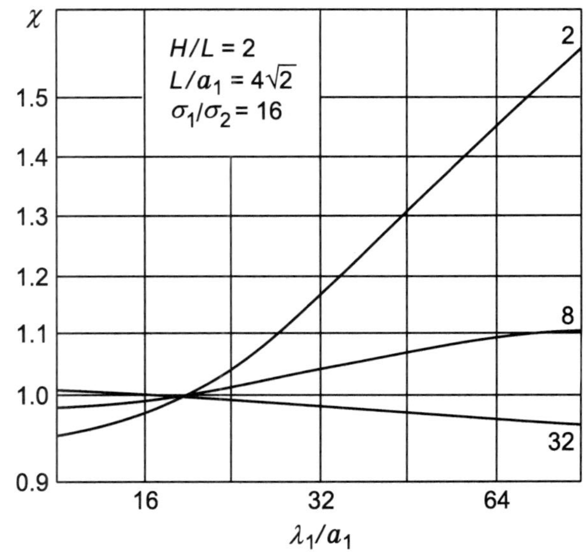

Figure 6.15. Behavior of parameter χ. Curve index σ_1/σ_3.

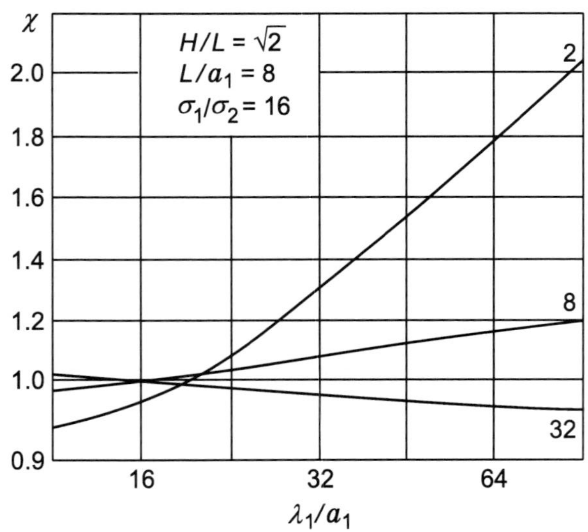

Figure 6.16. Behavior of parameter χ. Curve index σ_1/σ_3.

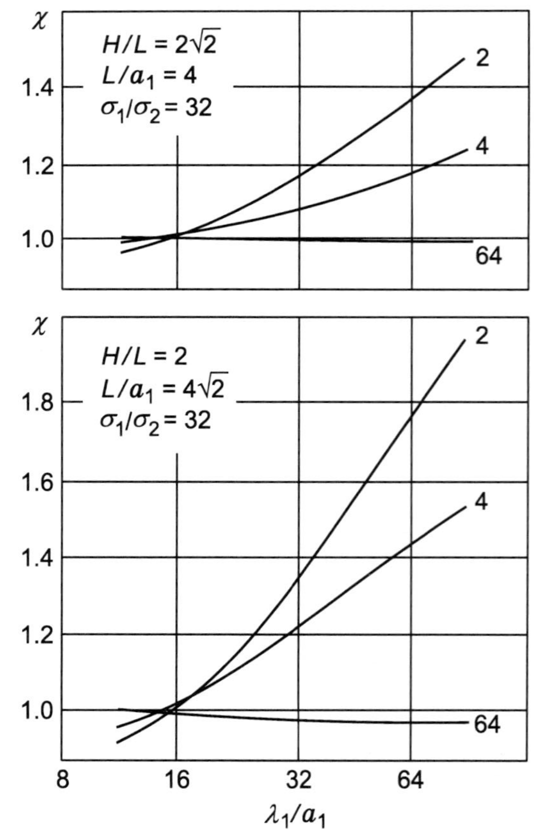

Figure 6.17. Behavior of parameter χ. Curve index σ_1/σ_3.

With an increase of resistivity of the surrounding medium (ρ_3) the influence of the finite thickness of the formation decreases ($\lambda_1/a_1 > 16$), and this effect becomes more significant with an increase of the frequency.

It is appropriate to notice also that with an increase of the probe length, L, and assuming that other parameters are not changed the deviation of function from unit increases.

Simultaneously increasing the probe length and the frequency we can keep the influence of the surrounding medium the same.

Chapter 7

MULTI-COIL DIFFERENTIAL INDUCTION PROBES

Analysis of the field in media with horizontal as well as cylindrical interfaces, performed in previous chapters, has shown that a two-coil induction probe does not possess noticeable advantages with respect to electric logging probes. Of course, such conditions as a nonconducting borehole surrounded by a relatively conductive medium with thin highly resistive layers can be considered as an exception.

The influence of currents induced in a borehole and in an invasion zone is usually sufficiently great, and for this reason, in order to determine a formation resistivity, it is necessary to apply two-coil induction probes with practically the same length as in electric logging. Let us notice that at the range of frequencies and resistivities, where the skin effect manifests itself relatively weakly the influence of the surrounding medium is even greater than in electric logging. But with an increase of frequency, the conductivity of a medium, as well as formation thickness, absorption of the electromagnetic energy essentially improves the vertical response of the probe. This brief comparison with normal or lateral electric probes shows that application of two-coil induction probes is hardly reasonable. Moreover, it requires the use of more complicated equipment.

For all these reasons H. Doll suggested, in 1946, multi-coil differential probes and also developed an approach allowing us to determine parameters of these systems.

At the beginning we will consider multi-coil induction probes, proceeding from the theory of small parameters, when we can neglect the interaction between currents. In other words, currents in any part of the medium, regardless of how far from the probe it is located, create a signal which is only defined by the conductivity and the geometric factor of this part.

It is obvious that the role of various parts of the medium in forming a signal essentially depends on the probe length. For instance, with an increase of the probe length the influence of more remote parts of the medium increases, and consequently, the contribution of currents induced near the probe becomes smaller. Applying probes of various lengths with different numbers of turns in coils, connected in series in the same or opposite directions, we can significantly reduce the signal caused by currents in any element of the medium independently of the distance to a probe. However, for improving the characteristics of a two-coil probe it is not sufficient to decrease a signal from some element of the medium.

For improvement of the radial response of a two-coil probe in order to determine a formation resistivity it is necessary to decrease the contribution of currents induced in the borehole and in the invasion zone with respect to a signal caused by currents in the formation. In other words, it is essential to decrease the influence of parts of the medium located relatively close to the probe. At the same time the signal generated by currents in remote parts of the medium should not be very small, otherwise serious measuring problems would arise.

For improving the vertical response of a two-coil induction probe it is necessary to decrease the relative contribution from the surrounding medium with respect to the signal caused by currents in the formation against which a two-coil induction probe is located. In other words, in this case the influence of more remote parts of the medium has to be reduced providing a significant signal from currents induced in that part of the medium which is located relatively close to the probe. Thus, improvement of radial and vertical responses of a two-coil induction probe is related to the development of multi-coil probes which have to satisfy opposite requirements.

As will be shown later, under certain conditions we can improve both responses simultaneously. However, in a general case improvement of one response with respect to that of a two-coil induction probe is related with deterioration of another and vice versa.

Arbitrarily, a multi-coil induction probe can be presented as a sum of two-coil induction probes. Currently, there are known symmetrical and non-symmetrical multi-coil induction probes, and their characteristics will be considered in detail in the next sections. As a rule in induction logging using one frequency the electromotive force, caused by the current in the transmitter coil or coils, is significantly greater than that generated by induced currents in the medium. For this reason an additional compensating coil to increase the accuracy of measurement is installed. Due to this the electromotive force caused by the primary field is practically equal to zero. It is also appropriate to notice that some differential probes do not require a compensation coil.

7.1. Methods of Determination of Probe Parameters

Methods of choosing probe parameters are based on the use of differential and integral responses of two-coil induction probes. The differential radial response defines a signal from a thin cylindrical shell, expressed in units of the signal, caused by currents in a uniform conducting medium. In accord with Doll's theory, described above, we have:

$$G_r = \mathrm{d}r \int\limits_{-\infty}^{\infty} q\,\mathrm{d}z \tag{7.1}$$

where $q\,\mathrm{d}r\,\mathrm{d}z$ is the geometric factor of the ring with cross section equal to $\mathrm{d}r\,\mathrm{d}z$ and $q = (L/2)(r^3/R_1^3 R_2^3)$; $R_1 = [r^2 + (z - L/2)^2]^{1/2}$, $R_2 = [r^2 + (z + L/2)^2]^{1/2}$; r and z are cylindrical coordinates of a point, and the origin of coordinates system coincides with the probe middle; L is the probe length.

Making use of results developed in Chapter 3 let us consider in detail the behavior of function G_r. As was shown in this chapter the magnetic field on the borehole axis, generated by currents in a thin cylindrical shell and expressed in units of the primary field, can be presented in the form:

$$h_z = \frac{L^3}{\pi} \int\limits_0^{\infty} \frac{in\lambda^2 K_1^2(\lambda r)}{1 + inI_1(\lambda r)K_1(\lambda r)} \cos \lambda L\,\mathrm{d}\lambda \tag{7.2}$$

where r is the shell radius; $n = \omega\mu\sigma r\Delta r$; ω is the frequency; μ is the magnetic permeability equal to $4\pi \times 10^{-7}$ H/m; σ and Δr are conductivity and thickness of the shell; $I_1(\lambda r)$ and $K_1(\lambda r)$ are modified Bessel functions. As is well known, we have:

$$I_1(\lambda r) \to \frac{\lambda r}{2} \qquad K_1(\lambda r) \to \frac{1}{\lambda r} \qquad \text{as } \lambda r \to 0$$

and

$$I_1(\lambda r) \to \frac{\mathrm{e}^{\lambda r}}{\sqrt{2\pi\lambda r}} \qquad K_1(\lambda r) \to \frac{\mathrm{e}^{-\lambda r}}{\sqrt{2\lambda r}}\sqrt{\pi} \qquad \text{as } \lambda r \to \infty$$

Let us notice that the product $I_1(\lambda r)K_1(\lambda r)$ does not exceed 0.5 when λ varies from zero to infinity. For this reason at the range of small parameters we have:

$$\mathrm{Q}\, h_z \simeq \mathrm{i}n\frac{L^3}{\pi}\int\limits_0^\infty \lambda^2 K_1^2(\lambda r)\cos\lambda L\,\mathrm{d}\lambda \tag{7.3}$$

Taking into account the relation between the quadrature component of the field, $\mathrm{Q}\, h_z$, and the apparent conductivity:

$$\sigma_a = \frac{2}{\mu\omega L^2}\mathrm{Q}\, h_z$$

we obtain:

$$\sigma_a = \frac{\sigma r\Delta r}{L^2}\frac{2\alpha^3}{\pi}\int\limits_0^\infty m^2 K_1^2(m)\cos m\alpha\,\mathrm{d}m \tag{7.4}$$

where:

$$\alpha = L/r \qquad \text{or} \qquad \sigma_a = \sigma G_r \tag{7.5}$$

$$G_r = \frac{r\Delta r}{L^2}\frac{2\alpha^3}{\pi}\int\limits_0^\infty m^2 K_1^2(m)\cos m\alpha\,\mathrm{d}m = \frac{\Delta r}{r}\frac{1}{\alpha^2}C(\alpha) \tag{7.6}$$

and

$$C(\alpha) = \frac{2\alpha^3}{\pi}\int\limits_0^\infty m^2 K_1^2(m)\cos\alpha m\,\mathrm{d}m$$

It is obvious that:

$$\mathrm{Q}\, h_z = \frac{\sigma\mu\omega L^2}{2}\frac{\Delta r}{r}\frac{1}{\alpha^2}C(\alpha)$$

and correspondingly, for the electromotive force in the receiver of a two-coil induction probe we have:

$$\mathscr{E} = \frac{\sigma\mu\omega L^2}{2}\frac{\Delta r}{r}\frac{1}{\alpha^2}C(\alpha)\frac{M_T M_R \omega\mu}{2\pi L^3} = \frac{\mu^2\omega^2 M_T M_R}{4\pi}\frac{1}{L}G_r\sigma \tag{7.7}$$

Now we will consider the behavior of $C(\alpha)$ for large and small values of α.

Suppose that $\alpha \to \infty$. Introducing the notation $\phi(m) = m^2K_1^2(m)$ and applying integration by parts we obtain an asymptotic presentation:

$$\int_0^\infty \phi(m)\cos m\alpha\, dm = \frac{1}{\alpha}\phi(m)\sin\alpha m\Big|_0^\infty + \frac{1}{\alpha^2}\phi'(m)\cos m\alpha\Big|_0^\infty - \frac{1}{\alpha^2}\int_0^\infty \phi''(m)\cos m\alpha\, dm$$

Taking into account that $\phi(0) = 1$ and that the function along with its derivatives is zero as m tends to infinity, we have:

$$\int_0^\infty \phi(m)\cos m\alpha\, dm = -\frac{1}{\alpha^2}\phi'(0) - \frac{1}{\alpha^2}\int_0^\infty \phi''(m)\cos\alpha m\, dm$$

If $m \to 0$ we have:

$$K_1(m) \simeq \frac{1}{m} + \frac{m}{2}\ln m$$

Therefore:

$$\phi(m) = m^2K^2(m) \simeq 1 + m^2\ln m$$
$$\phi'(m) \simeq 2m\ln m$$
$$\phi''(m) \simeq 2\ln m \simeq -2K_0(m)$$

Whence:

$$\int_0^\infty \phi(m)\cos m\alpha\, dm \simeq \frac{2}{\pi}\int_0^\infty K_0(m)\cos m\alpha\, dm \simeq \frac{\pi}{\alpha^2}$$

since

$$\frac{1}{(1+\alpha^2)^{1/2}} = \frac{2}{\pi}\int_0^\infty K_0(m)\cos\alpha m\, dm$$

Thus, we have:

$$C(\alpha) \to \frac{2\alpha^3}{\pi}\frac{2}{\alpha^3}\frac{\pi}{2} = 2 \qquad \text{if } \alpha \to \infty$$

TABLE 7.1
Values of function $C(\alpha)$

α	$C(\alpha)$	α	$C(\alpha)$	α	$C(\alpha)$
0.1	0.58620×10^{-3}	2	1.3110	12	2.1056
0.2	0.46248×10^{-2}	3	1.8546	13	2.0962
0.3	0.15254×10^{-1}	4	2.0724	14	2.0879
0.4	0.35029×10^{-1}	5	2.1473	15	2.0806
0.5	0.65736×10^{-1}	6	2.1551	16	2.0740
0.6	0.10830	7	2.1634	17	2.0683
0.7	0.16280	8	2.1530	18	2.0632
0.8	0.22858	9	2.1404	19	2.0586
0.9	0.30437	10	2.1278	20	2.0545
1.0	0.38845	11	2.1161		

In the opposite case, as $\alpha \to 0$, $C(\alpha)$ decreases directly proportional to α^3, inasmuch as integral $\int_0^\infty m^2 K_1^2(m)\cos m\alpha\, dm$ tends to the constant when α goes to zero. Values of this function are given in Table 7.1. As follows from this table $C(\alpha)$ can be presented as:

$$C(\alpha) \simeq 0.586\,\alpha^3 \qquad \text{if } \alpha \ll 1$$

The differential radial response, G_r, depends on two parameters, namely: the ratio of the probe length to the shell radius, L/r, and the ratio of its thickness to the radius, $\Delta r/r$.

First, we will assume that for all shells $\Delta r/r = const$. In this case let us present G_r in the form:

$$G_r = \frac{\Delta r}{rL^2} r^2 C(\alpha)$$

if $r \to 0$, $\alpha \to \infty$ and $C(\alpha) \to 2$. Thus:

$$G_r \to \frac{2\Delta r}{r} r^2$$

In the opposite case, when $r \to \infty$, $\alpha \to 0$ and $C(\alpha) \to k\alpha^3$ $(k = 0.586)$, we have:

$$G_r \to k\frac{\Delta r}{r} L \frac{1}{r}$$

Therefore, the geometric factor of cylindrical shells, the thickness of which is directly proportional to their radius, changes in the following way. For small values of r the geometric factor is also small, then it increases directly proportional to the square of the radius, reaches a maximum and with further increase of the radius it decreases inversely proportional to r (Fig. 7.1a).

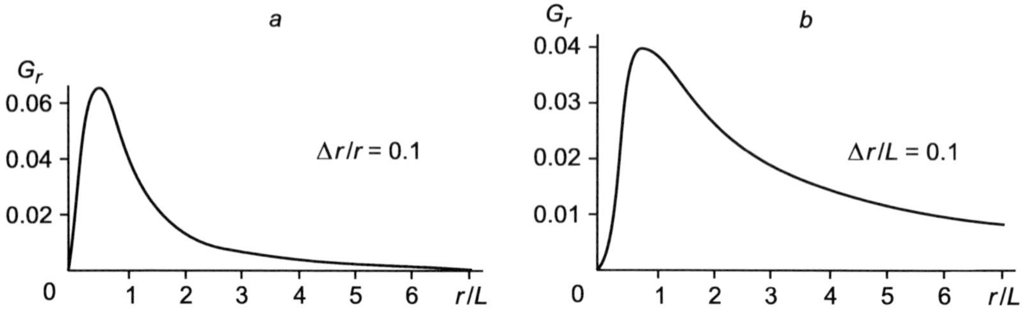

Figure 7.1. Function G_r.

Now we will present the medium as a system of coaxial cylindrical layers with the same thickness: $\Delta r = const$, and correspondingly we can write G_r in the form:

$$G_r = \frac{\Delta r}{L}\frac{1}{\alpha}C(\alpha)$$

If $r \to 0$ we have:

$$G_r \to \frac{\Delta r}{L}\frac{2}{L}r$$

while for large values of radius we obtain:

$$G_r \to k\Delta r L\frac{1}{r^2}$$

Therefore, at the beginning G_r increases directly proportional to r, then it grows slower, reaching a maximum and for large values of r it decreases inversely proportionally to r^2 (Fig. 7.1b). Correspondingly, the maxima of the geometric factor G_r for uniform and nonuniform presentation of the medium by shells are shifted from each other.

First, we will consider the graphical method of determination of the parameters of the multi-coil differential probe. Let us notice that this was the main approach applied in developing the first induction probes. Making use of transparent paper and simultaneously putting on several differential responses of various two-coil induction probes, the parameters of the multi-coil probe are chosen so to reduce the influence of the shells surrounding the probe as much as possible. Correspondingly the depth of investigation of a multi-coil probe increases. If the differential response of the multi-coil induction probe within some interval is equal to zero, the geometric factor of any thin cylindrical shell of this interval is also equal to zero. In other words, the magnetic field of induced currents in such a shell creates a positive electromotive force in one group of receivers while in other groups of coils it generates the same signal by magnitude and different by sign.

It is obvious that the radial differential response of the multi-coil induction probe does not depend on the distribution of conductivity in the radial direction if within every cylindrical layer the resistivity remains constant.

In the determination of the differential response there is an element of uncertainty, inasmuch as the dimensions of a shell are not defined. For this reason it is appropriate to find out the maximal thickness of the shell, when its geometric factor with sufficient accuracy is described by G_r. It is obvious that the geometric factor of the shell with thickness Δr can be presented as a difference of geometric factors of two cylinders with external and internal radius r_e and r_i, respectively. For a sufficiently thin shell, in accord with eq. 7.6 we have:

$$G_r(\alpha) = \frac{\Delta r}{r}\frac{1}{\alpha^2}C(\alpha) = G_1(\alpha_e) - G_1(\alpha_i) \tag{7.8}$$

where $\alpha_e = L/r$, $\alpha_i = L/r_i$, $\alpha = L/r$, $r = (r_i + r_e)/2$ ($r_e = r + \Delta r/2$, $r_i = r - \Delta r/2$) and L is the probe length.

It is clear that:

$$\alpha_e = \frac{\alpha}{1 + \Delta r/2r} \qquad \alpha_i = \frac{\alpha}{1 - \Delta r/2r}$$

Let us consider two limited cases.

Case 1

If $r \to 0$, $\alpha \to \infty$, $C(\alpha) \to 2$, $G_1(\alpha) \to 1/\alpha^2$ and therefore:

$$G_r(\alpha) = \frac{\Delta r}{r}\frac{2}{\alpha^2}$$

It means that if the external radius of the shell is many times smaller than the probe length, the geometric factor of such a thin shell is practically equal to that of the infinitely thin one.

Case 2

If $r \to \infty$, $\alpha \to 0$, $G_1(\alpha) \to 1 - k\alpha$, $C(\alpha) \to k\alpha^3$. Therefore:

$$\frac{\Delta r}{r}k\alpha = k(\alpha_i - \alpha_e)$$

or

$$\frac{\Delta r}{r}\alpha\left[1 - \frac{1}{1 - (\Delta r/2r)^2}\right] \simeq 0$$

This relation becomes more accurate with a decrease of $\Delta r/r$. Calculations show that if $\Delta r/r \leqslant 0.2$, eq. 7.8 is valid practically for all values of α, i.e. the geometric factor of the shell is described by G_r.

The second radial response of the two-coil induction probe is its integral response (direct and inverse ones). The direct integral radial response defines the signal caused by currents in the cylinder referred to that from currents in a uniform medium as a function of

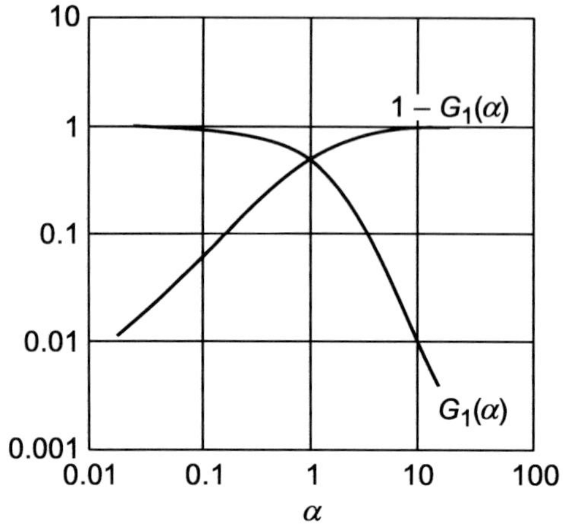

Figure 7.2. Direct and inverse integral responses.

the distance from the axis where the probe is located. As is well known the geometric factor of the borehole is defined by $G_1(\alpha)$, where $\alpha = L/a$. If instead of the borehole radius a we introduce radius r and the probe length L is considered to be constant, then $G_1(r/L)$ represents the direct integral radial response of a two-coil induction probe. If $r \to 0$, $\alpha \to \infty$ then $G_1(\alpha) \to 1/\alpha^2 = r^2/L^2$, i.e. the geometric factor of the cylinder at the beginning increases directly proportional to the square of its radius and inversely proportional to L^2. For large values of r, α tends to zero, $G_1(\alpha) \to 1 - k\alpha = 1 - kL/r$, and correspondingly the response magnitude approaches to unity. It is obvious that coefficient k is equal to 0.586.

The inverse integral radial response characterizes a contribution caused by currents in a medium outside the cylinder with radius r, and it is related with $G_1(\alpha)$ as:

$$G^{in} = 1 - G_1(\alpha)$$

Direct and inverse integral responses are presented in Fig. 7.2. Applying the graphical method we can choose corresponding characteristics of multi-coil induction probes.

From the point of the depth of investigation the direct integral response of a multi-coil induction probe has to satisfy two conditions:

- Its value near a probe has to be minimal.
- It should approach relatively slowly to its asymptote, which is equal to unity for all induction probes.

It is appropriate to notice that the graphical method of obtaining the radial responses of multi-coil induction probes has some shortcomings. They are:

- The process of choosing probe parameters is sufficiently cumbersome.
- The accuracy of determination of small values of the geometric factor at the initial part of the response is usually very low. It becomes specially noticeable in the case of a nonuniform medium, for example, when the borehole is much more conductive than the formation. For these conditions the *focusing* abilities of a probe, chosen by the graphical method, can be insufficient in order to eliminate the influence of the borehole or the invasion zone especially, if the latter is more conductive than the formation. For this reason *focusing* features of a multi-coil induction probe, chosen by the graphical method, manifests themselves as a rule only for relatively small ratios ρ_2/ρ_1.
- With an increase in the number of probe coils this process becomes more and more complicated.
- Making use of the graphical method it is practically impossible to chose optimal parameters of the probe, when simultaneously with a decrease of the signal from the borehole and the invasion zone we can provide a maximal signal from more remote parts of the formation.
- This method of superposition of radial responses of two-coil induction probes does not practically allow us to find parameters of probes which are sensitive only to certain parts of medium, similar to special probes in electrical logging as micrologs.

In order to overcome these difficulties let us consider an analytical method of determination of probe parameters. As is well known the quadrature component of the electromotive force arising at the receiver of a two-coil induction probe due to induced currents in a medium is:

$$\mathrm{Q}\,\mathscr{E} = \mathrm{Q}\,h_z \cdot \mathscr{E}_0 = \frac{\mu\omega L^2}{2}\sigma_a \cdot \mathscr{E}_0 \tag{7.9}$$

where

$$\mathscr{E}_0 = \frac{M_T M_R}{2\pi L^3}\omega\mu \qquad \text{or} \qquad \mathrm{Q}\,\mathscr{E} = \frac{\omega^2\mu^2 M_T M_R}{4\pi}\frac{1}{L}\sigma_a \tag{7.10}$$

Here $\sigma_a = \sum_{i=1}^{n} \sigma_i G_i$ and G_i and σ_i are the conductivity and the geometric factor of the i-th part of a medium.

Suppose that the induction probe consists of s transmitter and t receiver coils. Then it can be presented as a system of st two-coil induction probes, and the quadrature component of the electromotive force induced in measuring coils of this probe can be written as:

$$\mathscr{E} = \sum_{\alpha=1}^{s}\sum_{\beta=1}^{t} \varepsilon_{\alpha\beta} \tag{7.11}$$

(α is transmitter coil, β is receiver one).

In accord with eqs. 7.10 and 7.11 we have:

$$Q\mathscr{E} = \frac{\omega^2\mu^2}{4\pi}\sum_{\alpha=1}^{s}\sum_{\beta=1}^{t}\left[\frac{M_{T_\alpha}M_{R_\beta}}{L_{\alpha\beta}}\sum_{i=1}^{n}\sigma_i G_i(L_{\alpha\beta})\right] \tag{7.12}$$

Thus, the procedure of choosing parameters of multi-coil induction probes allowing us to eliminate the influence of k arbitrary parts of the medium is reduced to a solution of a system of k equations:

$$\sum_{i=1}^{k}\sum_{\alpha=1}^{s}\sum_{\beta=1}^{t}\frac{M_{T_\alpha}M_{R_\beta}}{L_{\alpha\beta}}G_i(L_{\alpha\beta}) = 0 \tag{7.13}$$

Let us remember that unknown parameters of the multi-coil differential probe are lengths of the two-coil probes and moments of coils.

It is obvious that the number of equations can be taken either equal to the unknown parameters or greater. In the latter the least squares method can be applied if the system is linear or in a more general case gradient methods can be used. Function $G_i(L_{\alpha\beta})$ can be either the geometric factor or a very thin cylindrical shell (the differential response) or that of a relatively thin cylindrical layer. In such a case this function is expressed through the integral response of a two-coil induction probe $G_1(L_{\alpha\beta})$.

The analytical approach is also convenient to analyze radial responses of known differential probes, specially when the probe consists of only one transmitter coil, while others are receiver ones or vice versa.

Unlike the graphical approach this method allows in principle to solve several problems such as:

- Choice of optimal parameters of a differential probe which provides a certain depth of investigation or a sensitivity to specific parts of a medium as well as a maximal signal from these parts of the medium.
- Development of interpretation of soundings based on use of two- or three-coil induction probes with different lengths.

Let us notice that if $G_i(L_{\alpha\beta})$ in eq. 7.13 represents the differential radial response, the equation can be rewritten in the form:

$$\sum_{i=1}^{k}\sum_{\alpha=1}^{s}\sum_{\beta=1}^{t}\frac{M_{T_\alpha}M_{R_\beta}}{L_{\alpha\beta}}\frac{1}{\alpha_i^2}C(\alpha_i) = 0 \tag{7.14}$$

where values of $C(\alpha_i)$ are given in Table 7.1.

In accord with the definition of the integral radial response its expression for multi-coil probes has the following form:

$$G = \frac{\displaystyle\sum_{\alpha=1}^{s}\sum_{\beta=1}^{t}\frac{M_{T_\alpha}M_{R_\beta}}{L_{\alpha\beta}}G_1\left(\frac{L_{\alpha\beta}}{\rho}\right)}{\displaystyle\sum_{\alpha=1}^{s}\sum_{\beta=1}^{t}\frac{M_{T_\alpha}M_{R_\beta}}{L_{\alpha\beta}}} \tag{7.15}$$

where $G_1(L_{\alpha\beta}/\rho)$ is the geometric factor of the cylinder with radius r for a two-coil induction probe with length $L_{\alpha\beta}$. The denominator in eq. 7.15 is proportional to the electromotive force in a multi-coil probe caused by currents in a uniform medium.

It is appropriate to notice that eq. 7.15 can be used for calculation of radial responses of differential probes in a medium which is not uniform with respect to magnetic permeability.

Later in this chapter we will perform analyses of radial responses of several multi-coil probes which will illustrate the efficiency of this approach.

7.2. Physical Principles of Multi-coil Differential Probes

Until now it was assumed that interaction between currents is absent, i.e. all currents induced in a conducting nonuniform medium, regardless of the distance from the source, are shifted in phase by 90°. For this reason electromotive forces induced in measuring coils of a probe are in phase with each other, and they are added and subtracted in the same way as scalars. If only one component of the electromotive force, for instance the quadrature component, is measured it is subjected to the same operations as scalars, regardless of whether currents are shifted in phase by 90°, or the internal skin effect manifests itself and due to it at every point of a medium there are both quadrature and inphase components of the induced current.

However, in the latter the magnitude of current density does not depend on the primary magnetic flux only but also on the intensity of currents in the neighboring parts, i.e. in essence on the distribution of conductivity in a medium. Correspondingly, a value of geometric factor becomes different from that which follows from the theory of small parameters when the skin effect is neglected. For this reason *focusing* features of the probe, the parameters of which were calculated assuming the absence of the skin effect, can be seriously deteriorated under real conditions. The character of the influence of the skin effect essentially depends on frequency, distribution of conductivities, and length of probes. In order to take into account the influence of the skin effect on the radial responses of the probe it is necessary to solve the direct problem for a given distribution of conductivities. It is obvious that the practical value of such multi-coil probes, where parameters are chosen proceeding from knowledge of a geoelectrical section, is negligible.

For this reason we can say that a necessary condition for the application of differential probes is the absence of interaction between currents in those parts of a medium the influence of which should be significantly reduced. In other words, those parts of the medium have to correspond to *Doll's area* where the current density is defined by the primary magnetic flux and the conductivity at a given point. It is natural that in choosing parameters of probes in order to increase the depth of investigation in the radial direction, it is important to eliminate the influence of parts of the medium located relatively close to the source. As calculations show, this condition usually takes place even for sufficiently high frequencies.

However, absence of the skin effect in the area, the influence of which it is necessary to reduce, is not sufficient for the efficiency of multi-coil induction probes. Let us present a signal measured by the probe as a sum of two signals, namely, one which is caused by currents in the area where there is not practically interaction between currents and a second one caused by currents in the external area, for example in a formation.

As was shown in previous chapters with an increase of the distance from the source the skin effect is more strongly manifested, and in a general case the distribution of currents in the external area (the formation) depends on the magnitude of currents in the first area, which is closer to the probe.

For this reason even in those cases when it is possible to eliminate the signal from currents in the parts of a medium, the influence of which has to be reduced, their effect can be noticed indirectly, changing the distribution of currents in the formation. In this case the signal from the formation is not only a function of its conductivity, but it depends also on the conductivity and dimensions of the internal area, in particular on the borehole and the invasion zone.

Therefore, the second requirement, in order to provide an efficient working of the multi-coil induction probe, is the absence of the influence of currents in the near area on the distribution of currents in the formation. In other words, the skin effect in the formation has to manifest itself in the same manner as in a uniform medium with the resistivity of the formation.

Inasmuch as the relation between the signal and the conductivity of a uniform medium is known, the correction of function σ_a due to the skin effect is often performed directly during calibration of the probe either with the help of conducting rings, the parameters of which are calculated taking into account the skin effect, or in measuring in models of a medium when its resistivity corresponds to that under real conditions.

Thus, for efficiency of differential induction probes designed to measure a formation conductivity, two conditions have to be met, namely:

- The range of a medium which includes the borehole and an invasion zone must correspond to *Doll's area*, i.e. currents induced in this area are shifted in phase by 90°, and they are defined by a change of the primary magnetic flux and the conductivity at a given point.
- Outside of this range, for example in a formation, the skin effect has to manifest itself as in a uniform medium with the resistivity of the formation, i.e. interaction between currents in these two regions has to be negligible.

Both conditions formulated above coincide with conditions of application of the approximate theory, described in Chapter 3, and therefore we can expect that the *focusing* features of probes will be preserved even at higher frequencies than as follows from Doll's theory. As will be shown later, analysis of radial responses of some multi-coil probes confirms this fact. Also, a comparison of results of calculations, making use of the exact and this approximate solution, allows us to establish maximal frequencies for a given distribution of conductivities when both of these conditions are met.

Analysis of physical principles of differential probes permits us to discuss one aspect related to obtaining radial responses of these probes.

As is well known, physical modeling or the use of conducting rings permits us to define a signal in a probe located on the cylinder axis as a function of its radius for a given frequency and conductivity of a medium. At the same time, a space surrounded by the cylinder has an infinitely high resistivity. Results of measurements are usually presented in the following form: the ratio of the cylinder radius to the probe length is plotted along the abscissa, while the ratio of the electromotive force induced by currents in the cylinder with radius r to that caused by currents in uniform space with the same resistivity is plotted along the ordinate axis.

It is natural that the following question arises whether this characteristic allows us to obtain information on the *focusing* features of the probe in a medium when the resistivity changes in a radial direction. It is obvious that until the signal from cylindrical conductor does not depend on currents induced in the external medium, regardless of their resistivity, this function can be used for evaluating the efficiency of a differential probe in a uniform as well as in a nonuniform medium.

The lower the frequency and the higher the resistivity, the greater the cylinder radius for which the radial response describes with sufficient accuracy the behavior of the field in a nonuniform medium. However, if the frequency and conductivity of the cylinder are relatively great and the distribution of currents in the cylinder is subjected to the skin effect then such response does not reflect the actual behavior of the field under real conditions. Correspondingly we can say that the radial response obtained by physical or numerical modeling can be used for understanding the *focusing* features of the probe if the skin effect is absent within the cylinder, and we can neglect the interaction between currents induced in internal (cylinder) and external areas.

As an example, direct integral radial responses of a two-coil induction probe are presented in Fig. 7.3. The curve index is the product $\sigma\mu\omega$. As is seen from this figure the higher the frequency and the lower the resistivity of a medium the stronger the influence of an area relatively closer located to the probe. It is easy to explain, inasmuch as with an increase of this product, that the influence of the skin effect becomes stronger, and correspondingly, that the signal caused by currents outside the cylinder increases relatively slower than that generated by currents inside of it. For this reason a deviation from the radial response calculated with the help of geometric factor is observed, regardless of the cylinder radius.

This consideration shows that the application of direct radial responses obtained from either physical modeling or making use of the exact solution is hardly useful for the determination of parameters of differential probes.

7.3. Radial and Vertical Responses of the Differential Probe 1.L–1.2

As is well known, coil induction probes used in induction logging have various arrangements of coils. It is appropriate to distinguish in every differential probe the basic (main) two-coil probe having a maximal product of transmitter (T) and receiver (R) coil moments. Other coils are considered to be *focusing* coils and, they form several additional coil probes which provide *focusing* features of the induction tool.

From the point of view of the location of the *focusing* coils with respect to the center of the basic probe, the multi-coil probes can be divided in symmetrical and nonsymmetrical ones. In symmetrical probes *focusing* coils are located in such a way that for every pair: transmitter-receiver coils, displayed with respect to the center, there is another pair with

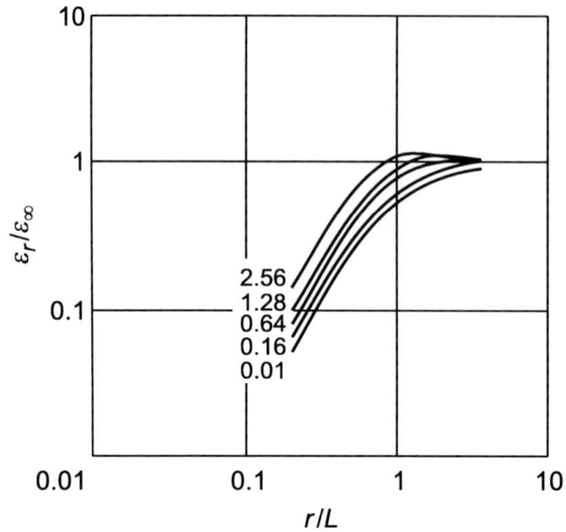

Figure 7.3. Direct integral responses of a two-coil induction probe. Curve index $\sigma\mu\omega$; $L = 1$ m.

the same product of moments displayed at the same distance on the opposite side. Signals measured by symmetrical pairs have the same sign.

It is also obvious that symmetrical probes have symmetrical profiling curves with respect to the formation center provided that the resistivity of the medium from both sides of the formation is the same. From the point of view of the position of *focusing* coils with respect to *basic* ones multi-coil probes are classified as probes with internal, external and mixed *focusing*.

For internal *focusing* additional coils are placed between basic ones; for external *focusing* they are located outside the main probe, and finally for mixed *focusing* additional coils are placed inside as well as outside the basic probe. The configuration of symmetric differential probes is shown in Fig. 7.4.

In addition it is appropriate to notice that in symmetrical multi-coil probes moments of transmitter and receiver coils of corresponding pairs for example, basic probe, are the same, i.e. $M_T = M_R$, $M_{T_F} = M_{R_F}$.

We will consider several differential probes and will start with a symmetrical four-coil probe with additional internal coils. This type of probe is defined by three parameters, namely: the distance between basic coils, L, the ratio of the length of *focusing* probe RT_F to that of the basic probe, which is denoted by p, and the ratio of moments of *focusing* coils to the moment of basic ones (parameter c).

Let us consider a symmetrical four-coil probe with internal *focusing* 1.L–1.2 (Fig. 7.4a) ($p = 0.4$, $c = 0.05$, $L = 1.2$ m)

Comparison with the results of calculation with coils having finite dimensions has shown that we can neglect their length and diameter in calculating radial and vertical responses.

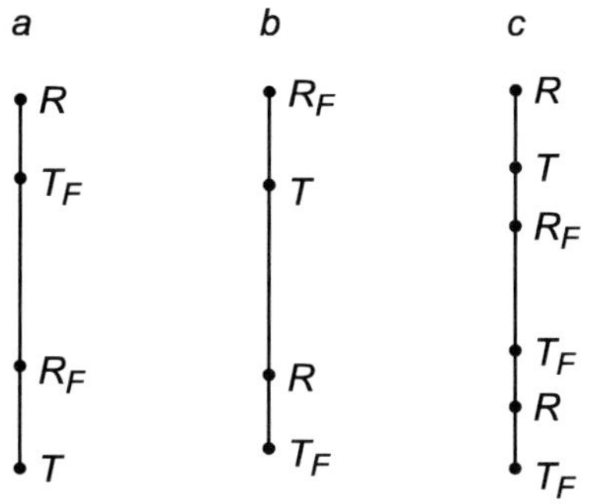

Figure 7.4. Symmetrical differential probes: (a) internal *focusing*; (b) external *focusing*; (c) mixed *focusing*.

For this reason a signal in symmetrical four-coil probe can be written in the form:

$$\begin{aligned}\mathscr{E} &= \mathscr{E}_{TR} - 2\mathscr{E}_{TR_F} + \mathscr{E}_{T_F R_F} \\ &= \frac{\mu^2\omega^2 M_T M_R}{4\pi L}\left\{\sum \sigma_i \left[G_i(\alpha) - \frac{2c}{p}G_i(p\alpha) + \frac{c^2}{1-2p}G_i[(1-2p)\alpha]\right]\right\}\end{aligned} \tag{7.16}$$

where L is the length of the basic probe, TR; pL is the length of probe, $R_F\,T$; and $(1-2p)L$ is the length of probe, $R_F\,T_F$.

In a uniform medium we have:

$$\mathscr{E}^{un} = \frac{\mu^2\omega^2 M_T M_R}{4\pi L}\sigma\left(1 - \frac{2c}{p} + \frac{c^2}{1-2p}\right)$$

inasmuch as geometric factors of the whole medium for a two-coil induction probe: $G_i(\alpha)$, $G_i(\alpha p)$ and $G_i[(1-2p)\alpha]$ are equal to unity (α is the ratio of the length of the basic probe to the cylinder radius on the axis of which the probe is located). Therefore, the coefficient of the probe is:

$$K_s = \frac{4\pi L}{\omega^2\mu^2 M_T M_R\left(1 - \dfrac{2c}{p} + \dfrac{c^2}{1-2p}\right)} \tag{7.17}$$

and the geometric factor of the whole space is equal to unity as is the case for two-coil probes. For this reason the expression for the apparent conductivity σ_a has the form:

$$\sigma_a = \sum \sigma_i G_i^* = K_s \mathscr{E} \tag{7.18}$$

where

$$G_i^* = \frac{G_i(\alpha) - \dfrac{2c}{p}G_i(p\alpha) + \dfrac{c^2}{1-2p}G_i[(1-2p)\alpha]}{1 - \dfrac{2c}{p} + \dfrac{c^2}{1-2p}} \tag{7.19}$$

This function is the geometric factor of the cylinder for a four-coil induction probe with internal *focusing* and parameters p and c, i.e. it is the integral radial response. Let us notice that normalization of geometric factor allows us to compare radial responses of various probes.

Now we will consider the behavior of function G_i^* as a function of the cylinder radius r ($\alpha = L/r$). If $r \to 0$, ($\alpha \to \infty$) we have:

$$G_1(\alpha) \to \frac{1}{\alpha^2} \qquad G_1(p\alpha) \to \frac{1}{p^2\alpha^2} \qquad G_1[(1-2p)\alpha] \to \frac{1}{(1-2p)^2\alpha^2}$$

Substituting these values into eq. 7.19 we have:

$$G_1^* \to \frac{1 - \dfrac{2c}{p^3} + \dfrac{c^2}{(1-2p)^3}}{1 - \dfrac{2c}{p} + \dfrac{c^2}{1-2p}} \frac{1}{\alpha^2} = K_1 \frac{1}{\alpha^2} \tag{7.20}$$

where

$$K_1 = \frac{1 - \dfrac{2c}{p^3} + \dfrac{c^2}{(1-2p)^3}}{1 - \dfrac{2c}{p} + \dfrac{c^2}{1-2p}}$$

Therefore, the smaller the value of coefficient K_1 the lesser the value of the geometric factor of the area directly surrounding the probe.

On the other hand, the electromotive force due to the primary field induced in receivers of the probe by currents in the transmitter coils is:

$$\mathscr{E}_0 = \frac{\mu\omega M_T M_R}{2\pi L^3}\left[1 - \frac{2c}{p^3} + \frac{c^2}{(1-2p)^3}\right] \tag{7.21}$$

Thus the rate of compensation of the electromotive force of the primary field defines a value of coefficient K_1. For a probe with parameters $p = 0.4$ and $c = 0.05$, coefficient K_1 is 0.32.

Introducing a fifth compensating coil, coefficient K_1 becomes equal to zero, and correspondingly the radial response somewhat improves at its initial part. From this consideration follows that with an increase of the length of two-coil induction probes, forming a differential four-coil probe, the cylinder radius, characterized by small values of geometric factor, increases, provided that the primary electromotive force is compensated.

In the opposite case, as $r \to \infty$ ($\alpha \to 0$) we have:

$$G_1(r) \to 1 - K\alpha \quad G_1(p\alpha) = 1 - Kp\alpha$$

and

$$G_1[(1-2p)\alpha] \to 1 - K(1-2p)\alpha$$

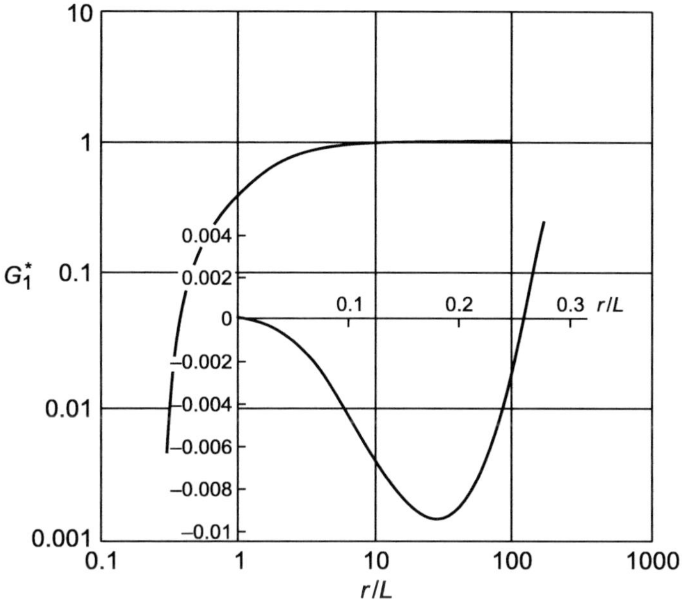

Figure 7.5. The integral radial response of a four-coil probe with parameters $p = 0.4$ and $c = 0.05$.

where $K \simeq 0.586$.

Substituting these expressions into eq. 7.19 we obtain:

$$G_1^* \to 1 - K \frac{1 - 2c + c^2}{1 - \dfrac{2c}{p} + \dfrac{c^2}{1 - 2p}} = 1 - KK_2\alpha \tag{7.22}$$

where

$$K_2 = \frac{1 - 2c + c^2}{1 - \dfrac{2c}{p} + \dfrac{c^2}{1 - 2p}} \tag{7.23}$$

Coefficient K_2 exceeds unity, and therefore the radial response approaches its asymptote slower than that of a two-coil probe, i.e. the four-coil induction probe possesses a greater depth of investigations with respect to a two-coil induction probe of the same length.

Figure 7.5 presents the integral radial response of a four-coil probe with parameters $p = 0.4$ and $c = 0.05$. Unlike the radial response of a two-coil induction probe, in this case function $G_1^*(r/L)$ at the beginning has small but negative values, near $r/L = 0.27$ it changes sign and monotonically approaches unity.

Now let us consider the differential radial response of this probe. In accord with eq. 7.13 the electromotive force in receivers, caused by currents in a thin cylindrical shell, can be written in the form:

$$\mathscr{E} = \frac{\omega^2\mu^2 M_T M_R}{4\pi L}\sigma \left\{ G_r(\alpha) - \frac{2c}{p} G_r(p\alpha) + \frac{c^2}{1-2p} G_r[(1-2p)\alpha] \right\} \tag{7.24}$$

where $G_r = (\Delta r/r)(1/\alpha^2)C(\alpha)$. Values of $C(\alpha)$ are given in Table 7.1. We will present eq. 7.24 as

$$\mathscr{E} = \frac{1}{K^*}\sigma G_r^*$$

where

$$G_r^* = \frac{G_r(\alpha) - \dfrac{2c}{p} G_r(p\alpha) + \dfrac{c^2}{1-2p} G_r[(1-2p)\alpha]}{1 - \dfrac{2c}{p} + \dfrac{c^2}{1-2p}} \tag{7.25}$$

G_r^* is the differential response of a four-coil induction probe and defines the ratio of signals from a thin cylindrical shell to that from a uniform medium.

From eq. 7.25 we have:

$$G_r^* = \frac{C(\alpha) - \dfrac{2c}{p^3} C(p\alpha) + \dfrac{c^2}{(1-2p)^3} C[(1-2p)\alpha]}{1 - \dfrac{2c}{p} + \dfrac{c^2}{1-2p}} \frac{1}{\alpha^2}\frac{\Delta r}{r} \tag{7.26}$$

Assuming that a medium is presented as a system of shells with the same thickness Δr it is convenient to write down G_r^* as:

$$G_r^* = \frac{\Delta r}{L} \frac{C(\alpha) - \dfrac{2c}{p^3} C(p\alpha) + \dfrac{c^2}{(1-2p)^3} C[(1-2p)\alpha]}{\alpha\left(1 - \dfrac{2c}{p} + \dfrac{c^2}{1-2p}\right)} \tag{7.27}$$

As has been shown, if $\alpha \to 0$ then $C(\alpha) \to K^3$, and for $\alpha \to \infty$ $C(\alpha) \to 2$. For this reason, if $r \to 0$ $(\alpha \to \infty)$, we have:

$$G^* \to \frac{2\Delta r}{L_0} K \frac{1}{\alpha}$$

i.e. again compensation of the electromotive force of the primary field provides minimum of geometric factor of cylindrical layers located close to the probe.

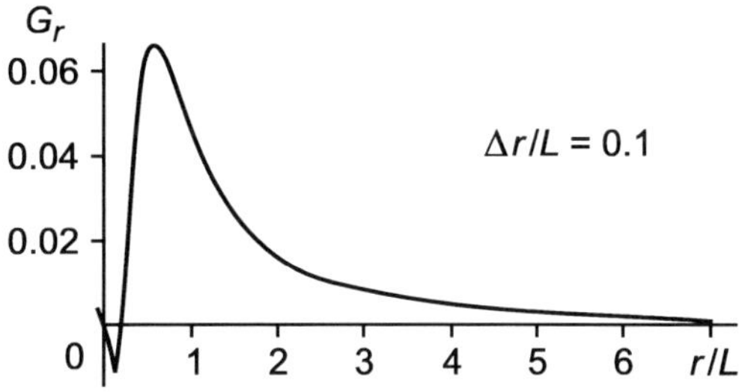

Figure 7.6. Differential response of a four-coil probe with parameters $p = 0.4$ and $c = 0.05$.

In the opposite case, when $r \to \infty$ $(\alpha \to 0)$ we have:

$$G_r^* \to \frac{\Delta r}{L} K K_2 \alpha^3 \tag{7.28}$$

A differential response of a four-coil probe with $p = 0.4$ and $c = 0.05$ is presented in Fig. 7.6, provided that the shell thickness is constant.

In accord with eq. 7.18 the expression for the apparent conductivity in a three-layered medium (borehole, invasion zone, formation) has the form:

$$\sigma_a = \sigma_1 G_1^* + \sigma_2 G_2^* + \sigma_3 G_3^* \tag{7.29}$$

where:

$$G_1^* = \frac{G_1(\alpha) - \dfrac{2c}{p} G_1(p\alpha) + \dfrac{c^2}{(1-2p)} G_1[(1-2p)\alpha]}{1 - \dfrac{2c}{p} + \dfrac{c^2}{1-2p}}$$

$$G_2^* = \frac{G_2(\alpha) - \dfrac{2c}{p} G_2(p\alpha) + \dfrac{c^2}{1-2p} G_2[(1-2p)\alpha]}{1 - \dfrac{2c}{p} + \dfrac{c^2}{1-2p}}$$

$$G_3^* = \frac{G_3(\alpha) - \dfrac{2c}{p} G_3(p\alpha) + \dfrac{c^2}{(1-2p)} G_3[(1-2p)\alpha]}{1 - \dfrac{2c}{p} + \dfrac{c^2}{1-2p}}$$

It is obvious that:

$$G_1^* + G_2^* + G_3^* = 1$$

TABLE 7.2
Values of function σ_a/σ_1

Probes \ σ_2/σ_1	0.500	0.250	0.125	0.0625	0.0313	0.0156
Two-coil probe	0.505	0.258	0.133	0.0724	0.0414	0.0259
Four-coil probe 1.L-1.2	0.498	0.246	0.121	0.0580	0.0268	0.0111

and

$$G_2^*(\alpha) = G_1^*\left(\frac{\alpha}{\beta}\right) - G_1^*(\alpha)$$

here β is the ratio of radius of the invasion zone to that of the borehole ($\beta = a_2/a_1$).

Proceeding from equations derived above we will consider a behavior of the apparent conductivity in a medium with cylindrical interfaces. First, let us assume that the invasion zone is absent. It is simple to show that for the four-coil probe with internal *focusing* and $L = 1.2$ m we have:

$$G_1^* \simeq -0.0045 \qquad G_2^* = 1.0045$$

Table 7.2 contains values of σ_a/σ_1 for both two- and four-coil induction probes.

For small values of ρ_2/ρ_1 ($\rho_2/\rho_1 < 1$) the difference between the radial responses of differential and two-coil probes is insignificant. If the conductivity of the borehole does not exceed more than 30 times the formation conductivity, the probe 1.L–1.2 eliminates the influence of the borehole almost completely. *Focusing* features of this probe in the presence of the invasion zone are illustrated by values of apparent conductivity, σ_a/σ_1, given in Table 7.3. Regardless of the resistivity of the invasion zone ($4 \leqslant \rho_2/\rho_1 \leqslant 64$), if $\rho_3/\rho_1 \leqslant 30$ the influence of induced currents in the borehole and the invasion zone on the signal, measured by the probe, is small (Table 7.3). The influence of the invasion zone on the two-coil probe with length $L = 1.2$ m is also insignificant if $\rho_3/\rho_1 \leqslant 10$. For this reason when penetration of the borehole filtrate is not deep and ρ_3/ρ_1 is relatively small the value of apparent conductivity, σ_a, measured by a two-coil induction probe $L = 1.2$ m, is close to the formation conductivity, and correspondingly in such conditions the role of *focusing* features of the probe, regardless of its type, is insignificant. With an increase of the radius of the invasion zone *focusing* features of the probe manifest themselves stronger. Practically, values of σ_a are close to σ_3 if the ratio of conductivities (σ_3/σ_1) is not less than 1/20. At the same time the character of the penetration $\rho_3/\rho_2 > 1$ or $\rho_3/\rho_2 < 1$ does not have a strong effect on the value of the apparent conductivity, σ_a.

For deep penetration of the borehole filtrate ($a_2/a_1 \leqslant 8$) the *focusing* features of this probe do not allow us to obtain the formation conductivity even in such cases when the borehole and the invasion zone have greater resistivity than the formation. It is explained by the fact that the geometric factor of the formation is about 0.7 ($a_2/a_1 = 8$) and for this reason, even if the conductivity of the borehole and the invasion zone is equal to zero, the apparent conductivity differs from that of the formation by 30%.

TABLE 7.3
Values of function σ_a/σ_1

Probe type	σ_2/σ_1	σ_3/σ_1						
		1	0.500	0.250	0.125	0.0625	0.0313	0.0156
				$a_2/a_1 = 2$				
2-coil	1/4	0.975	0.500	0.255	0.140	0.0800	0.0500	0.0330
4-coil	1/4	1.003	0.499	0.247	0.121	0.0575	0.0260	0.0102
2-coil	1/8	0.980	0.500	0.250	0.132	0.0750	0.0450	0.0300
4-coil	1/8	1.004	0.499	0.247	0.121	0.0580	0.0265	0.0107
2-coil	1/16	0.980	0.500	0.250	0.130	0.0720	0.0430	0.0280
4-coil	1/16	1.004	0.500	0.247	0.121	0.583	0.0268	0.0110
2-coil	1/32	1.000	0.500	0.250	0.129	0.0710	0.0420	0.0265
4-coil	1/32	1.004	0.500	0.248	0.122	0.584	0.0269	0.0111
2-coil	1/64	1.000	0.500	0.250	0.129	0.0710	0.0420	0.0260
4-coil	1/64	1.004	0.500	0.248	0.122	0.585	0.0270	0.0112
Probe type	σ_2/σ_1	1.00	0.500	0.250	0.125	0.625	0.3130	0.0156
				$a_2/a_1 = 4$				
2-coil	1/4	0.900	0.480	0.250	0.150	0.1000	0.0750	0.0640
4-coil	1/4	0.955	0.483	0.248	0.128	0.6960	0.0401	0.0253
2-coil	1/8	0.986	0.475	0.240	0.135	0.0830	0.0570	0.0430
4-coil	1/8	0.947	0.476	0.239	0.121	0.0621	0.0326	0.0178
2-coil	1/32	0.986	0.440	0.230	0.125	0.0730	0.0485	0.0330
4-coil	1/32	0.943	0.471	0.235	0.117	0.0583	0.0288	0.0140
2-coil	1/64	0.986	0.440	0.220	0.120	0.0650	0.4000	0.0250
4-coil	1/64	0.940	0.468	0.232	0.114	0.0554	0.0259	0.0112
Probe type	σ_2/σ_1	1.000	0.500	0.250	0.125	0.0625	0.3130	0.0156
				$a_2/a_1 = 8$				
2-coil	1/4	0.700	0.405	0.260	0.172	0.1500	0.1280	0.120
4-coil	1/4	0.764	0.418	0.248	0.158	0.1150	0.0931	0.0283
2-coil	1/8	0.650	0.360	0.205	0.135	0.0750	0.0800	0.0650
4-coil	1/8	0.726	0.380	0.207	0.120	0.0767	0.0551	0.0443
2-coil	1/16	0.620	0.318	0.180	0.110	0.0720	0.0530	0.0450
4-coil	1/16	0.707	0.361	0.188	0.101	0.0580	0.0361	0.0253
2-coil	1/32	0.600	0.320	0.170	0.0960	0.0600	0.0420	0.0322
4-coil	1/32	0.697	0.351	0.178	0.0915	0.0482	0.0266	0.0158
2-coil	1/64	0.600	0.320	0.168	0.0920	0.0550	0.0360	0.0250
4-coil	1/64	0.692	0.346	0.173	0.0868	0.0435	0.0219	0.0111

Until now it has been assumed that at all points of a medium induced currents are shifted in phase by 90°, i.e. the skin effect is absent. Now we will investigate radial responses of the probe, making use of results of the exact solutions in a medium with two cylindrical interfaces when the four-coil induction probe is located on the borehole axis.

The electromotive force in receivers of the four-coil symmetrical probe with internal *focusing* can be presented as:

$$\mathscr{E} = \mathscr{E}_1 - 2\mathscr{E}_2 + \mathscr{E}_3$$

On the other hand: $\mathscr{E} = \mathscr{E}_0 h_z$, here h_z is the quadrature component of the magnetic field expressed in units of the primary field and obtained from the exact solution. Therefore:

$$\mathscr{E} = \mathscr{E}_0^{(1)} \left[h_z^{(1)} - 2h_z^{(2)} \frac{\mathscr{E}_0^{(2)}}{\mathscr{E}_0^{(1)}} + h_z^{(3)} \frac{\mathscr{E}_0^{(3)}}{\mathscr{E}_0^{(1)}} \right]$$

where:

$$\begin{aligned} \frac{\mathscr{E}_0^{(2)}}{\mathscr{E}_0^{(1)}} &= \frac{\omega\mu M_T M_{RF}}{L^3_{TR_F}} \bigg/ \frac{\omega\mu M_T M_R}{L^3} = \frac{c}{p^3} \\ \frac{\mathscr{E}_0^{(3)}}{\mathscr{E}_0^{(1)}} &= \frac{\omega\mu M_{TF} M_{RF}}{L^3_{T_F R_F}} \bigg/ \frac{\omega\mu M_T M_R}{L^3} = \frac{c^2}{(1-2p)^3} \end{aligned} \tag{7.30}$$

The electromotive force measured in receiver coils of the probe referred to that in a free space for two-coil basic probe is defined as:

$$h_z = \frac{\mathscr{E}}{\mathscr{E}_0^{(1)}} = h_z^{(1)} - \frac{2c}{p^3} h_z^{(2)} + \frac{c^2}{(1-2p)} h_z^{(3)}$$

and correspondingly the expression for the apparent conductivity, σ_a, is:

$$\frac{\sigma_a}{\sigma_1} = \frac{1}{1 - \dfrac{2c}{p} + \dfrac{c^2}{1-2p}} \left[\frac{\sigma_a^{(1)}}{\sigma_1} - \frac{2c}{p}\frac{\sigma_a^{(2)}}{\sigma_1} + \frac{c^2}{1-2p}\frac{\sigma_a^{(3)}}{\sigma_1} \right] \tag{7.31}$$

where $\sigma_a^{(1)}/\sigma_1$, $\sigma_a^{(2)}/\sigma_1$, $\sigma_a^{(3)}/\sigma_1$ are functions corresponding to the basic two-coil probe (T, R), the differential one (T, R_F) and another differential probe (T_F, R_F), respectively.

It is appropriate to consider simultaneously field, electromotive force and apparent conductivity, calculated from the approximate theory assuming that in the borehole and the invasion zone, regardless of their conductivity and dimensions, currents are shifted in phase by 90° but in the formation the skin effect manifests itself as in a uniform medium with the resistivity of the formation. Then, according to results derived in Chapter 3 we have:

$$\sigma_a = (\sigma_1 - \sigma_3)G_1^* + (\sigma_2 - \sigma_3)G_2^* + \sigma_a^{un}$$

or

$$\frac{\sigma_a}{\sigma_1} = \left(1 - \frac{\sigma_3}{\sigma_1}\right) G_1^* + \left(\frac{\sigma_2}{\sigma_1}\right) G_2^* + \frac{\sigma_3}{\sigma_1}\frac{\sigma_3}{\sigma_3} \tag{7.32}$$

Results of calculation of function σ_a/σ_1 by exact equation, σ_a/σ_1, and the approximate formula 7.31, σ_a^a/σ_1, are given in Table 7.4.

For illustration, values of σ_a^{un}/σ_3 are presented in Table 7.5 ($K_3 = \sigma_3\mu\omega a_1^2$). Comparison with results of calculations for very small parameters shows that the influence of the skin effect in a relatively conductive medium can significantly change the value of σ_a/σ_1. For instance, for $\rho_2/\rho_1 = 32$, $\rho_3/\rho_1 = 16$ and $a_2/a_1 = 4$, the value of σ_a/σ_1 calculated by Doll's formula is 0.0564. If the frequency of the field is 60 kHz and $\rho_3 = 1.1$ ohm·m, the factual value σ_a/σ_1 is 0.0298, i.e. it is almost two times smaller.

A decrease of the quadrature component of the field and correspondingly the apparent conductivity is mainly the result of the skin effect in the formation for geoelectric parameters considered here. It follows from the coincidence of the calculated results based on exact and approximate solutions, because one of the main assumptions of the latter is that the skin effect is absent within the borehole and the intermediate zone. In this case the skin effect does not change the *focusing* features of the probe since the change of apparent conductivity, σ_a, is the same as in a uniform medium with the resistivity of the formation. Numerical data show that within the borehole and the invasion zone conditions of small parameters are preserved for a sufficiently large range of resistivities and dimensions.

This fact allows us to choose properly a frequency for a given probe. As is well known, with an increase of frequency the vertical response of the probe becomes better, and we can measure higher resistivities of a formation. However, if the frequency is chosen too high at least two problems arise, namely nonuniqueness in determination of resistivity by measuring the quadrature component of the field, and the radial response can be significantly worse than that calculated with the assumption that the skin effect is negligible. Comparison of calculated results, based on exact and approximate solutions, defines the maximal frequency for which the skin effect is still absent in the borehole and the invasion zone but in the formation it manifests itself in the same manner as in a uniform medium with the same resistivity.

We can think that the maximal frequency for this probe, derived from this comparison and taking into account its radial response, is defined from the relation:

$$f \leqslant (2.0 - 2.2)\rho_{\min} 10^5 \text{ Hz} \tag{7.33}$$

For example, if the minimal resistivity of a medium is about 1 ohm·m, the frequency can be increased up to 200–220 kHz.

This analysis of the *focusing* features of the probe 1.L–1.2 with $\rho = 0.4$ and $c = 0.05$ in media with cylindrical interfaces allows us to establish the range of frequencies as well as parameters of borehole and invasion zone when induced currents within them do not have an influence on the measured signal. If the resistivity of the invasion zone exceeds that of the formation, $\rho_2 > \rho_3$, and $a_2/a_1 \leqslant 4$, the apparent conductivity practically coincides

TABLE 7.4
Values of functions σ_a/σ_1 and σ_a^a/σ_1

$\sigma_3\mu\omega a_1^2 \times 10^4$	$\rho_3/\rho_1=1$		$\rho_3/\rho_1=4$		$\rho_3/\rho_1=16$		$\rho_3/\rho_1=32$	
	σ_a/σ_1	σ_a^a/σ_1	σ_a/σ_1	σ_a^a/σ_1	σ_a/σ_1	σ_a^a/σ_1	σ_a/σ_1	σ_a^a/σ_1
			$a_2/a_1=2$ $\rho_2/\rho_1=8$					
1	0.940	0.949	0.233	0.233	0.0545	0.0546	0.0247	0.0248
2	0.924	0.924	0.228	0.227	0.0532	0.0533	0.0240	0.0240
4	0.892	0.894	0.219	0.220	0.0510	0.0510	0.0229	0.0230
8	0.848	0.849	0.208	0.208	0.0483	0.0484	0.0215	0.0216
16	0.785	0.785	0.192	0.193	0.0441	0.0442	0.0194	0.0196
32	0.700	0.698	0.171	0.171	0.0383	0.0385	0.0164	0.0169
64	0.585	0.583	0.141	0.142	0.0304	0.0307	0.0119	0.0133
			$a_2/a_1=2$ $\rho_2/\rho_1=16$					
1	0.950	0.950	0.233	0.234	0.0549	0.0550	0.0250	0.0250
2	0.924	0.925	0.227	0.228	0.0533	0.0534	0.0243	0.0244
4	0.892	0.893	0.220	0.221	0.0514	0.0515	0.0233	0.0234
8	0.847	0.848	0.208	0.209	0.0483	0.0484	0.0218	0.0219
16	0.786	0.788	0.192	0.193	0.0445	0.0446	0.0197	0.0200
32	0.700	0.701	0.171	0.172	0.0387	0.0388	0.0167	0.0168
64	0.585	0.586	0.142	0.143	0.0309	0.0310	0.0125	0.0126
			$a_2/a_1=2$ $\rho_2/\rho_1=32$					
1	0.947	0.947	0.234	0.235	0.0550	0.0552	0.0251	0.0252
2	0.924	0.924	0.228	0.229	0.0536	0.0536	0.0244	0.0244
4	0.893	0.894	0.220	0.221	0.0515	0.0516	0.0234	0.0235
8	0.847	0.848	0.209	0.210	0.0485	0.0486	0.0219	0.0220
16	0.785	0.785	0.193	0.194	0.0446	0.0446	0.0199	0.0200
32	0.700	0.702	0.171	0.172	0.0388	0.0389	0.0169	0.0170
64	0.583	0.585	0.142	0.143	0.0312	0.0313	0.0127	0.0130
			$a_2/a_1=2$ $\rho_2/\rho_1=64$					
1	0.947	0.949	0.234	0.234	0.0550	0.0551	0.0252	0.0253
2	0.925	0.924	0.228	0.229	0.0537	0.0537	0.0245	0.0245
4	0.892	0.894	0.219	0.220	0.0514	0.0515	0.0234	0.0235
8	0.847	0.849	0.209	0.210	0.0485	0.0486	0.0220	0.0221
16	0.785	0.785	0.193	0.194	0.0446	0.0447	0.0199	0.0201
32	0.700	0.698	0.171	0.172	0.0390	0.0394	0.0170	0.0174
64	0.583	0.583	0.142	0.143	0.0313	0.0318	0.0128	0.0138
			$a_2/a_1=4$ $\rho_2/\rho_1=8$					
1	0.890	0.891	0.225	0.226	0.0589	0.0590	0.0311	0.0310
2	0.865	0.866	0.219	0.220	0.0572	0.0574	0.0303	0.0302
4	0.834	0.836	0.211	0.213	0.0552	0.0555	0.0292	0.0292
8	0.788	0.791	0.200	0.202	0.0522	0.0524	0.0277	0.0278
16	0.728	0.727	0.184	0.185	0.0482	0.0483	0.0256	0.0258
32	0.643	0.640	0.163	0.164	0.0422	0.0429	0.0219	0.0231
64	0.538	0.535	0.135	0.137	0.0340	0.0358	0.0164	0.0195

TABLE 7.4
(Continued)

$\sigma_3\mu\omega a_1^2 \times 10^4$	$\rho_3/\rho_1 = 1$		$\rho_3/\rho_1 = 4$		$\rho_3/\rho_1 = 16$		$\rho_3/\rho_1 = 32$	
	σ_a/σ_1	σ_a^a/σ_1	σ_a/σ_1	σ_a^a/σ_1	σ_a/σ_1	σ_a^a/σ_1	σ_a/σ_1	σ_a^a/σ_1
			$a_2/a_1 = 4$	$\rho_2/\rho_1 = 16$				
1	0.886	0.887	0.221	0.223	0.0548	0.0549	0.0272	0.0273
2	0.860	0.860	0.215	0.217	0.0525	0.0525	0.0264	0.0265
4	0.830	0.831	0.207	0.210	0.0514	0.0515	0.0253	0.0255
8	0.785	0.787	0.196	0.198	0.0481	0.0483	0.0239	0.0240
16	0.722	0.725	0.180	0.182	0.0445	0.0446	0.0217	0.0220
32	0.640	0.643	0.158	0.160	0.0388	0.0390	0.0187	0.0193
64	0.532	0.534	0.131	0.132	0.0309	0.0312	0.0141	0.0157
			$a_2/a_1 = 4$	$\rho_2/\rho_1 = 32$				
1	0.882	0.882	0.219	0.220	0.0524	0.0526	0.0251	0.0252
2	0.860	0.862	0.213	0.215	0.0510	0.0510	0.0244	0.0246
4	0.827	0.829	0.205	0.200	0.0493	0.0496	0.0234	0.0236
8	0.782	0.784	0.194	0.195	0.0465	0.0466	0.0219	0.0221
16	0.722	0.725	0.179	0.180	0.0425	0.0427	0.0198	0.0200
32	0.635	0.637	0.157	0.158	0.0371	0.0375	0.0169	0.0173
64	0.528	0.520	0.130	0.132	0.0294	0.0298	0.0127	0.0137

TABLE 7.5
Values of function σ_a^{un}/σ_3; $\alpha = 10, L = 1$ m

ρ_3/ρ_1 \ $K_3 \times 10^4$	1	2	4	8	16	32	64	128
1	0.945	0.921	0.891	0.844	0.781	0.695	0.576	0.432
2	0.473	0.461	0.446	0.422	0.391	0.348	0.288	0.216
4	0.236	0.230	0.233	0.211	0.195	0.174	0.144	0.108
8	0.118	0.115	0.111	0.106	0.976	0.869	0.0720	0.054
16	0.0590	0.0575	0.0557	0.0528	0.0488	0.0434	0.0360	0.027
32	0.0295	0.0287	0.0278	0.0264	0.0244	0.0217	0.0180	0.013
64	0.0148	0.0143	0.0139	0.0132	0.0122	0.0108	0.0090	0.006

with that of the formation provided that $\rho_3/\rho_1 < 20$. However, if the resistivity of the invasion zone becomes smaller than ρ_3 the depth of investigation of this probe decreases.

Thus, for certain conditions the value of the apparent conductivity, σ_a, in a formation with a very large thickness coincides with the apparent conductivity in a uniform medium having the formation conductivity.

However, in more complicated cases, when measured electromotive forces are subjected to the influence of parameters of the borehole and the invasion zone, interpretation cannot be performed without additional information.

Now let us consider vertical responses of the four-coil induction probe 1.L–1.2. Calculations of apparent conductivity, σ_a, in formations with finite thickness have been performed proceeding from equation:

$$\frac{\sigma_a}{\sigma_3} = \frac{1}{1 - \dfrac{2c}{p} + \dfrac{c^2}{1-2p}} \left[\frac{\sigma_a^{(1)}}{\sigma_2} - \frac{2c}{p}\frac{\sigma_a^{(2)}}{\sigma_2} + \frac{c^2}{1-2p}\frac{\sigma_a^{(3)}}{\sigma_2} \right]$$

where $\sigma_a^{(1)}/\sigma_2$ and $\sigma_a^{(3)}/\sigma_2$ are values of σ_a/σ_2 for two-coil probes located symmetrically with respect to the formation boundaries while $\sigma_a^{(2)}/\sigma_2$ is the value of σ_a/σ_2 for the probe displaced with respect to the formation center.

Curves of apparent conductivity σ_a/σ_2 for the probe 1.L–1.2 are presented in Figs. 7.7–7.10. Values of $\sigma_2\mu\omega L^2$ and ratio σ_a/σ_2 are plotted along axes of abscissa and ordinate, respectively. The curve index is σ_1/σ_2 (σ_1 and σ_2 are conductivities of the formation and the surrounding medium). Every set of curves is characterized by the ratio H/L, here L is the length of the basic-two coil probe.

For given frequency and probe length the abscissa represents the conductivity of the surrounding medium. For example, if $L = 1.2$ m, $f = 6 \times 10^4$ Hz, then $\sigma_2\mu\omega L^2 \simeq 0.7\sigma_2$ or $\sigma_2 = 1.44(\sigma_2\mu\omega L^2)$. Thus the conductivity of the surrounding medium varies along the abscissa in these figures from thousand parts of m^{-1} up to several units.

Every set of curves consists of two families, separated by curve corresponding to a uniform medium ($\sigma_1/\sigma_2 = 1$). If the formation is more conductive than the surrounding medium curves are located above that calculated for a uniform medium; in the opposite case when the formation is more resistive they are located below the curve: $\sigma_1 = \sigma_2$.

Values of σ_a/σ_2 for various parameters of the medium are given in Tables 7.6–7.10 provided that $f = 6 \times 10^4$ Hz and $L = 1.2$ m.

If the formation is more conductive than the surrounding medium and its thickness is at least two times greater than the probe length, the value of apparent conductivity practically coincides with function σ_a in a uniform medium having the formation conductivity. In particular a change of resistivity ρ_2 from 1 to 16 ohm·m does not practically influence the vertical response of the probe.

If the formation is more resistive, i.e. $\rho_1 > \rho_2$, and its resistivity does not exceed 10 ohm·m the influence of the surrounding medium becomes negligible for the given probe when the formation thickness exceeds 3.0–3.5 times the probe length. In cases when the difference of the conductivities is relatively small, equality $\sigma_a = \sigma_a^{un}$ takes place even for sufficiently thin layers.

With an increase of the formation resistivity the influence of currents in the surrounding medium becomes more significant. A comparison with corresponding curves for a two-coil induction probe shows that the influence of the surrounding medium on the four-coil induction probe is somewhat greater than that on a two-coil probe of the same length when the thickness of the formation is equal to or greater than the probe length. However this difference does not exceed 10–15%.

Here it is appropriate to make the following comment. Results of calculations, presented in Figs. 7.7–7.10, have been performed in a medium with only horizontal interfaces. Certainly, such an assumption would lead to significant errors if a two-coil induction probe

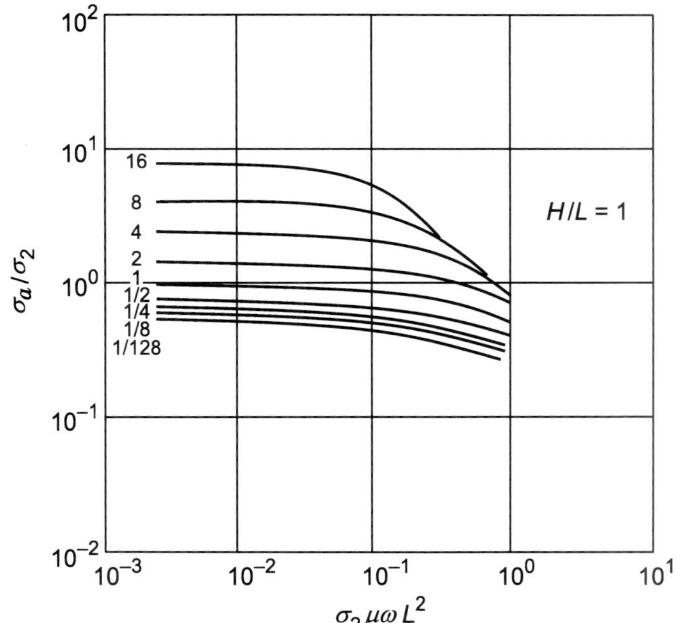

Figure 7.7. Function σ_a/σ_2 $(H/L = 1)$. Curve index σ_1/σ_2.

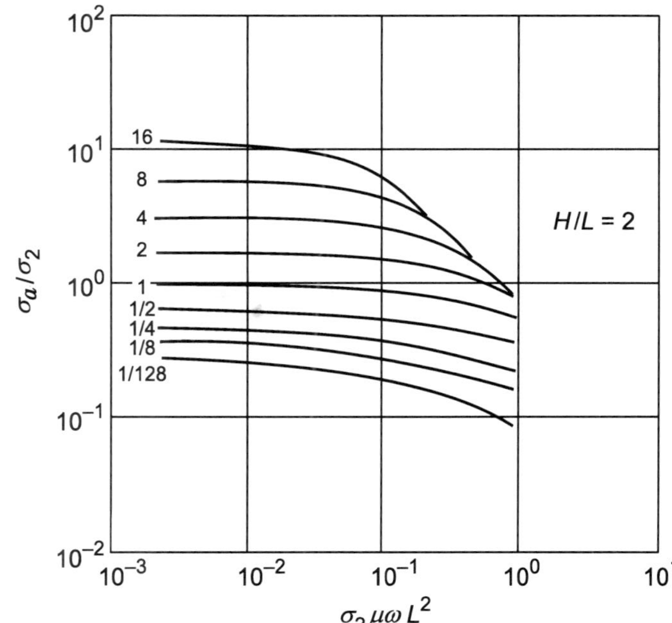

Figure 7.8. Function σ_a/σ_2 $(H/L = 2)$. Curve index σ_1/σ_2.

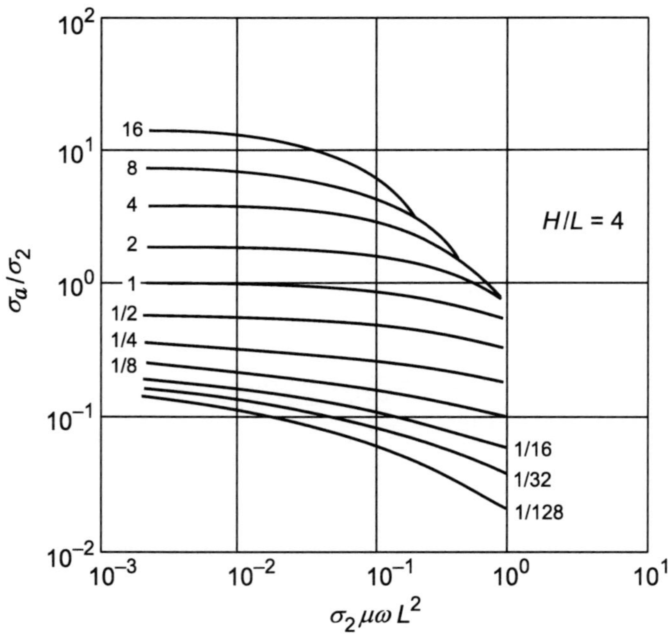

Figure 7.9. Function σ_a/σ_2 $(H/L = 4)$. Curve index σ_1/σ_2.

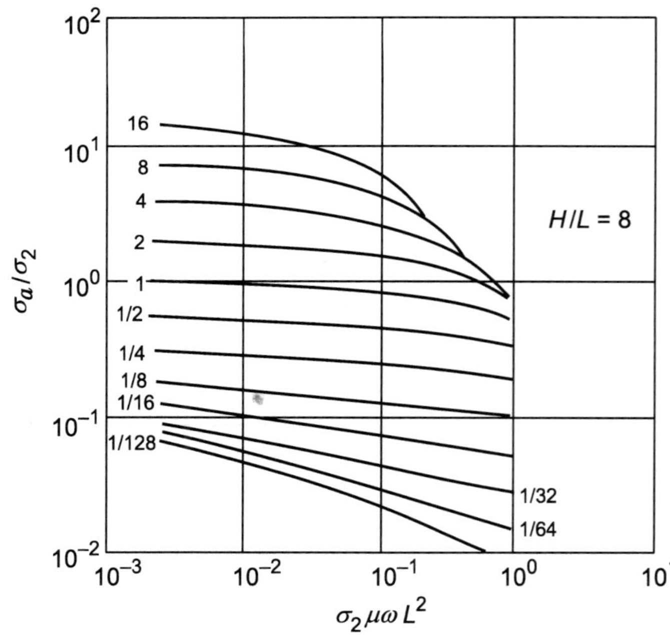

Figure 7.10. Function σ_a/σ_2 $(H/L = 8)$. Curve index σ_1/σ_2.

TABLE 7.6
Values of function σ_a/σ_2; $\rho_2 = 1$ ohm·m

H/L \ σ_1/σ_2	1	1/2	1/4	1/8	1/16	1/32	1/128
1	0.56	0.42	0.34	0.31	–	–	0.270
$\sqrt{2}$	0.56	0.38	0.28	0.22	0.21	0.20	0.170
2	0.56	0.35	0.23	0.16	0.13	0.12	0.094
$2\sqrt{2}$	0.56	0.34	0.20	0.12	0.085	0.070	0.049
4	0.56	0.34	0.18	0.110	0.065	0.042	0.026
$4\sqrt{2}$	0.56	0.34	0.18	0.105	0.056	0.032	–
8	0.56	0.34	0.18	0.105	0.054	0.029	–
$8\sqrt{2}$	0.56	0.34	0.18	0.105	0.053	0.028	–
16	0.56	0.34	0.18	0.105	0.053	0.028	–

TABLE 7.7
Values of function σ_a/σ_2; $\rho_2 = 2$ ohm·m

H/L \ σ_1/σ_2	2	1	1/2	1/4	1/8	1/16
1	0.99	0.68	0.51	0.42	0.37	–
$\sqrt{2}$	1.10	0.68	0.45	0.33	0.27	0.25
2	1.10	0.68	0.41	0.27	0.19	0.17
$2\sqrt{2}$	1.10	0.68	0.40	0.23	0.15	0.11
4	1.10	0.68	0.39	0.22	0.13	0.080
$4\sqrt{2}$	1.10	0.68	0.39	0.21	0.12	0.065
8	1.10	0.68	0.39	0.21	0.11	0.058
$8\sqrt{2}$	1.10	0.68	0.39	0.21	0.11	0.057
16	1.10	0.68	0.39	0.21	0.11	0.057

TABLE 7.8
Values of function σ_a/σ_2; $\rho_2 = 4$ ohm·m

H/L \ σ_1/σ_2	4	2	1	1/2	1/4	1/8
1	1.80	1.18	0.77	0.58	0.48	0.420
$\sqrt{2}$	2.10	1.25	0.77	0.50	0.31	0.220
2	2.20	1.35	0.77	0.46	0.31	0.230
$2\sqrt{2}$	2.30	1.40	0.77	0.44	0.27	0.180
4	2.25	1.40	0.77	0.42	0.24	0.145
$4\sqrt{2}$	2.20	1.40	0.77	0.41	0.22	0.125
8	2.20	1.40	0.77	0.41	0.22	0.120
$8\sqrt{2}$	2.20	1.40	0.77	0.41	0.22	0.115
16	2.20	1.40	0.77	0.41	0.22	0.115

TABLE 7.9
Values of function σ_a/σ_2; $\rho_2 = 8$ ohm·m

H/L \ σ_1/σ_2	16	8	4	2	1	1/2
1	3.40	2.00	1.25	0.85	0.63	0.55
$\sqrt{2}$	4.10	2.40	1.40	0.85	0.56	0.42
2	4.50	2.60	1.50	0.85	0.51	0.34
$2\sqrt{2}$	4.50	2.80	1.50	0.85	0.48	0.30
4	4.50	2.80	1.55	0.85	0.46	0.27
$4\sqrt{2}$	4.50	2.80	1.55	0.85	0.46	0.24
8	4.50	2.80	1.55	0.85	0.45	0.23
$8\sqrt{2}$	4.50	2.80	1.55	0.85	0.44	0.23
16	4.50	2.80	1.55	0.85	0.44	0.22

TABLE 7.10
Values of function σ_a/σ_2; $\rho_2 = 16$ ohm·m

H/L \ σ_1/σ_2	8	4	2	1	1/2	1/4
1	6.6	3.8	2.3	1.3	0.90	0.68
$\sqrt{2}$	8.2	4.6	2.6	1.5	0.90	0.60
2	9.0	5.2	2.9	1.6	0.90	0.55
$2\sqrt{2}$	9.0	5.5	3.0	1.6	0.90	0.50
4	9.0	5.5	3.0	1.7	0.90	0.48
$4\sqrt{2}$	9.0	5.5	3.0	1.7	0.90	0.47
8	9.0	5.5	3.0	1.7	0.90	0.46
$8\sqrt{2}$	9.0	5.5	3.0	1.7	0.90	0.46
16	9.0	5.5	3.0	1.7	0.90	0.46

is considered. However, for a differential probe the effect, caused by the influence of the borehole is usually very small.

In fact, it is known that the probe, within a certain range of change of ρ_1 and ρ_2, eliminates the influence of the borehole and the invasion zone. Analysis of geometric factors of cylinders of finite thickness shows that as soon as the height of the cylinder exceeds the probe length its geometric factor is almost equal to that of an infinitely long cylinder, provided that the probe and the cylinder are symmetrically located. For this reason if the formation thickness is equal to or exceeds the probe length then the electromotive force is not subjected to the influence of that part of the borehole which is located against the surrounding medium. Moreover, the geometric factor of the part of the borehole located against the formation practically coincides with the geometric factor of the borehole, and due to *focusing* this part of the borehole as well as the rest of it do not affect the signal measured by the probe. Analogous behavior is observed for the invasion zone. With an increase of the formation thickness and resistivity of the surrounding medium errors in the determination of the apparent conductivity, σ_a, decrease.

TABLE 7.11
The position and parameters of coils

Two-coil probes	Length, m	$M_i M_j / M_T M_R$	Sign of Signal
		Probe 6F1M	
T – R	1.00	1.0000	–
T_{F1} – R	0.75	0.2900	–
T – R_{F1}	0.75	0.2900	–
T_{F1} – R_{F1}	0.50	0.0841	+
T_{F2} – R	0.50	0.0200	–
T – R_{F2}	0.50	0.0200	–
T_{F2} – R_{F2}	2.00	0.0004	+
T_{F1} – R_{F2}	0.75	0.0058	+
T_{F2} – R_{F1}	0.75	0.0058	+
		Probe 4F1	
T – R	1.00	1.000	+
T_{F2} – R	0.586	0.350	–
T_{F3} – R	0.320	0.025	+
		Probe 4F1.1	
T – R	1.100	1.000	+
T_{F2} – R	0.586	0.350	–
T_{F3} – R	0.352	0.025	+

7.4. Radial and Vertical Responses of Probes 6F1M, 4F1 and 4F1.1

The probe 6F1M has six coils with symmetrical internal and external differential probes (mixed *focusing*) based on the use of frequency 50 kHz. Probe 4F1 is a four-coil nonsymmetrical system with an internal differential probe and frequency of the current is 70 kHz. Finally, probe 4F1.1 is a four-coil nonsymmetrical probe with internal *focusing* and frequency 1 MHz. Table 7.11 describes the position and parameters of coils of these probes. Let us notice that all considered probes are systems where the electromotive force caused by currents in the transmitter coils, i.e. the primary electromotive force is compensated. In other words, the moments of coils satisfy the condition:

$$\sum_{i=1}^{k} \sum_{j=1}^{l} \frac{M_{T_i} M_{R_j}}{L_{ij}^3} = 0 \tag{7.34}$$

This fact turns out to be very essential in the investigation of the radial responses of multi-coil probes located on the borehole axis.

Calibration curves of probes 6F1M, 4F1 and 4F1.1 are presented in Figs. 7.11–7.13. Ratio of $\mathrm{Q}\mathscr{E}/\mathscr{E}^{(1)}$ and resistivity of the medium are plotted along axes of ordinate and abscissa, respectively. Here $\mathscr{E}_0^{(1)} = \omega\mu M_T M_R / 2\pi L^3$ is the primary electromotive force of the basic probe. For illustration let us assume that the minimal value of the ratio $Q\mathscr{E}/\mathscr{E}_0^{(1)}$ measured is 5×10^3. Then the range of resistivities defined by these probes is:

6F1M	0.4–12 ohm·m
4F1	0.5–25 ohm·m
4F1.1	10–400 ohm·m

Due to the relatively high frequency applied in probe 4F1.1 the range of resistivities is shifted to larger values than those for probes 6F1M and 4F1. It is appropriate to notice that with an increase of the degree of compensation of the primary electromotive force the upper boundary of resistivities is shifted to larger values. In accord with the calibration curve for the probe 4F1.1 we have: $\rho_{max} \simeq 400$ ohm·m. However, the determination of function ρ of such resistive layers can be complicated due to the radial response of the probe, if the borehole resistivity is sufficiently small ($\rho_1 \simeq 1$ ohm·m).

If the electromotive force is measured with an accuracy of about 5%, maximal errors in determination of resistivity near the low boundary of the range do not exceed 10–15% for these probes.

Calibration curves for basic two-coil induction probes are also shown in Figs. 7.11–7.13. From comparison with calibration curves of corresponding differential probes it follows that a decrease of the signal with respect to that of the basic probe within the range of measured resistivities in average constitutes:

3.1–3.4 times for 6F1M
2.2–3.0 times for 4F1
2.3–2.8 times for 4F1.1

Radial responses of these probes are shown in Figs. 7.14–7.16. The initial part of these responses are presented on a larger scale. The cylinder diameter is plotted along the axis of the abscissa.

Comparing the integral radial responses we can see that in a two-layered medium when the borehole radius changes from 0.1 to 0.15 m probe 6F1M provides more accurate values of the formation over a wider range of σ_2/σ_1 than probe 4F1. However, in the presence of an invasion zone (0.4–0.8 m) we can expect that the measurements by probes 4F1 and 6F1M are closer to each other. Concerning probe 4F1.1, we can notice the following: in sections where the geoelectric parameters correspond to the range of resistivities for probes 6F1M and 4F1, results of measuring with probe 4F1.1 are almost the same since their radial responses practically coincide.

For a more accurate evaluation of parameters of a medium, when probes 6F1M, 4F1 and 4F1.1 allow us to determine the formation resistivity, it is necessary to calculate the apparent conductivity for these probes in media with cylindrical interfaces. Results of such analysis are described below. Let us notice that calculations have been performed proceeding from the approximate theory which takes into account the skin effect in the external area.

First, consider the influence of the borehole (the invasion zone is absent) on the apparent conductivity, σ_a. Curves of σ_a/σ_2 as a function of the borehole resistivity, ρ_1 are shown in Figs. 7.17–7.22. The curve index is formation resistivity, ρ_2. For every probe there are two groups of curves, corresponding to different values of the borehole radius, $a_1 = 0.10$ m and $a_1 = 0.15$ m. An analysis of these curves allows us to make the following conclusions:

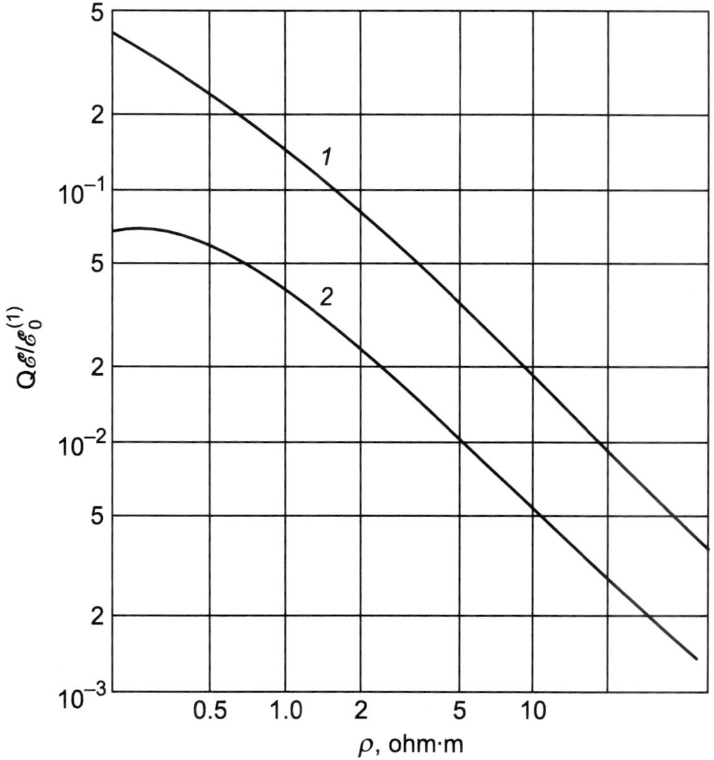

Figure 7.11. Calibration curves: (1) two-coil probe, $L = 1$ m; (2) probe 6F1M.

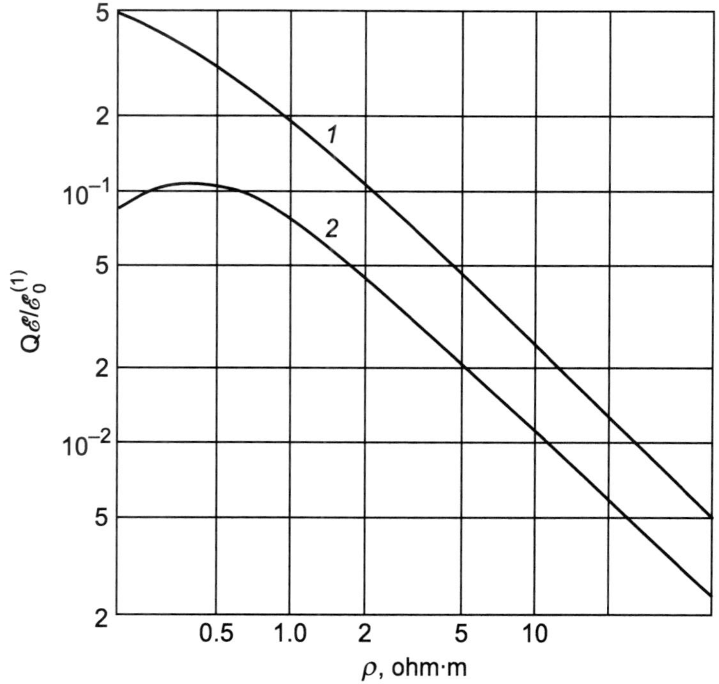

Figure 7.12. Calibration curves: (1) two-coil probe, $L = 1$ m; (2) probe 6F1M.

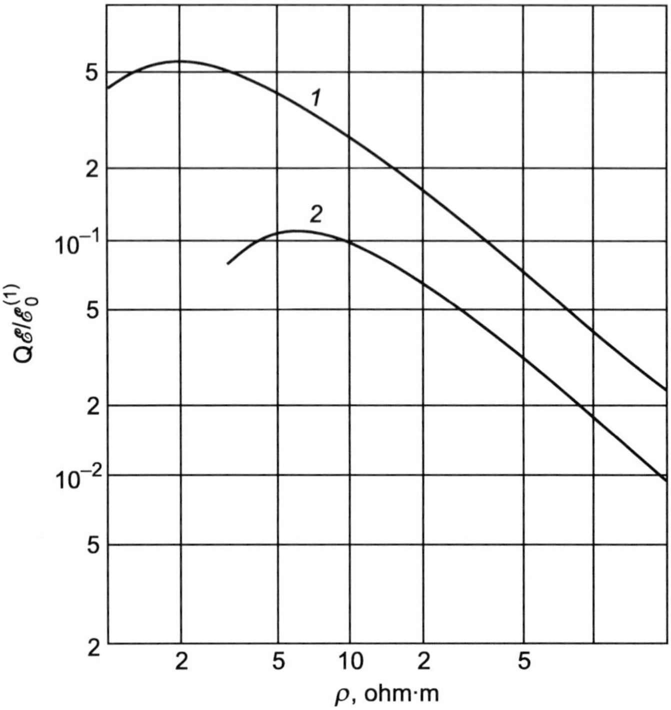

Figure 7.13. Calibration curves: (1) two-coil probe, $L = 1.1$ m; (2) probe 4F1.1.

- For a given value of the borehole resistivity the difference between σ_a and σ_2 increases also with an increase of formation resistivity. It is explained by the fact that with a decrease of σ_2 currents induced in a formation decrease, and the contribution of the electromotive force caused by the magnetic field of currents within the borehole becomes more essential. Let us notice that radial responses of probes within considered borehole radii are not equal to zero, and therefore the influence of the borehole cannot generally be ignored.

- With a decrease of the borehole resistivity for a given value of the formation resistivity, the deviation of the apparent conductivity, σ_a, from σ_2 becomes stronger since the density of currents induced in the borehole increases.

- With an increase of the borehole radius the difference between σ_a and σ_2 increases also, inasmuch as radial responses of all considered probes are worse for $a = 0.15$ m than for the case when the borehole radius is equal to 0.1 m.

If $a = 0.1$ m the difference between σ_a and σ_2 does not exceed 5% for the probe 6F1M when the borehole resistivity changes from 0.05 to 1 ohm·m. It permits us to perform measurements with probe 6F1M in boreholes with strongly mineralized solutions. For

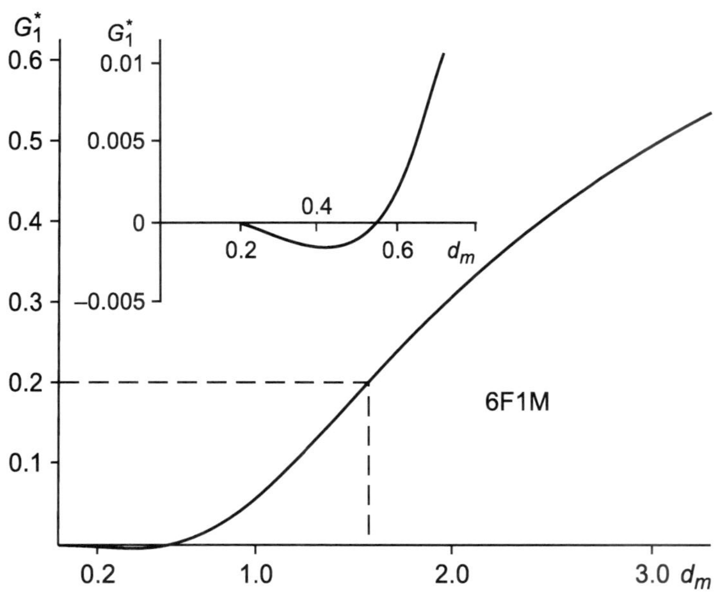

Figure 7.14. Radial response of probe 6F1M.

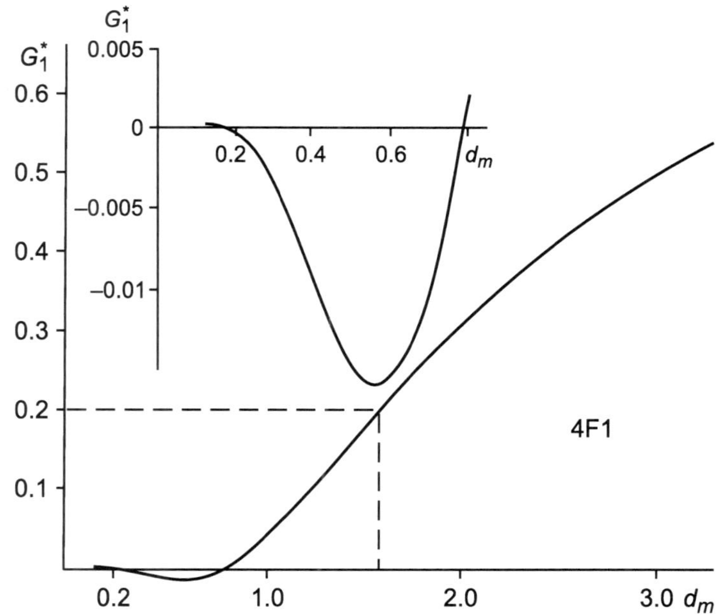

Figure 7.15. Radial response of probe 4F1.

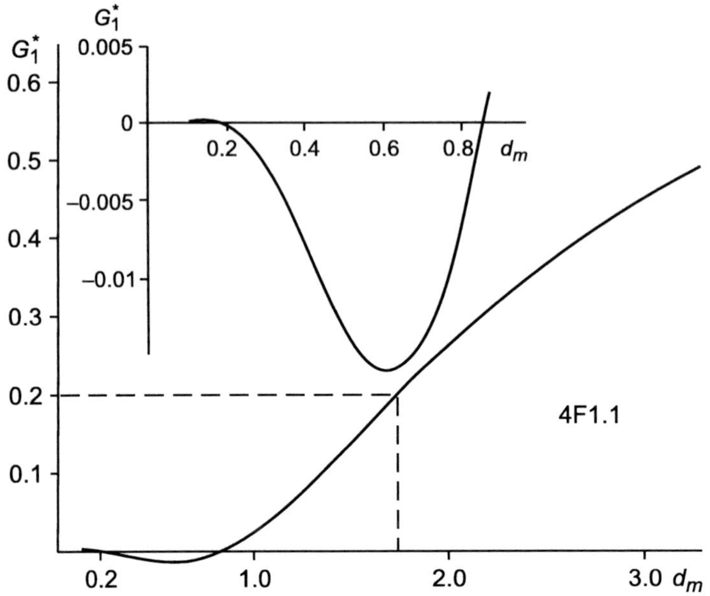

Figure 7.16. Radial response of probe 4F1.1.

example, if $\rho_1 = 1$ ohm·m and $a_1 = 0.1$ m, the ratio σ_a/σ_2 is equal to unity with an accuracy of 1% as $0.5 < \rho_2 < 50$ ohm·m.

The probe 4F1 is more sensitive to currents induced in a relatively conductive borehole. However, for $\rho_1 = 1$ ohm·m and $a_1 = 0.1$ m the apparent conductivity σ_a with an accuracy of 5% coincides with σ_2 within the whole range of formation resistivities measured by the probe.

Curves of the σ_a/σ_2 for the probe 4F1.1 given in Figs. 7.21–7.22 correspond to larger values of the borehole resistivity with respect to those for probes 6F1M and 4F1. It is related to the fact that, due to the high frequency used in this probe, the condition of *focusing* ($a_1/h_1 < 0.3$) is not valid anymore in boreholes with high mineralization. For instance, if borehole radii are 0.1 m and 0.15 m we have for the minimal resistivity, ρ_1 satisfying this inequality, 0.5 ohm·m and 1 ohm·m, respectively. If $\rho_1 = 1$ ohm·m and $a_1 = 0.1$ m the difference between σ_a^{un}/σ_2 and σ_a/σ_2 does not exceed 2% provided that $0.6 < \rho_2 < 100$ ohm·m.

Now let us consider the influence of parameters of a three-layered medium (an invasion zone is present) on the apparent conductivity measured by probes 6F1M, 4F1 and 4F1.1. Values of σ_a/σ_3 for various parameters of a three-layered medium along with data for a two-coil induction probe are given in Tables 7.12–7.22. The borehole resistivity is assumed to be 0.5 ohm·m. The behavior of σ_a/σ_3 depends to a certain extent on the sign of geometric factor of $G_1^*(r)$ for $r = a_1$ and $r = a_2$.

Let us introduce notations for specific points of the radial responses; namely r_{min} is the

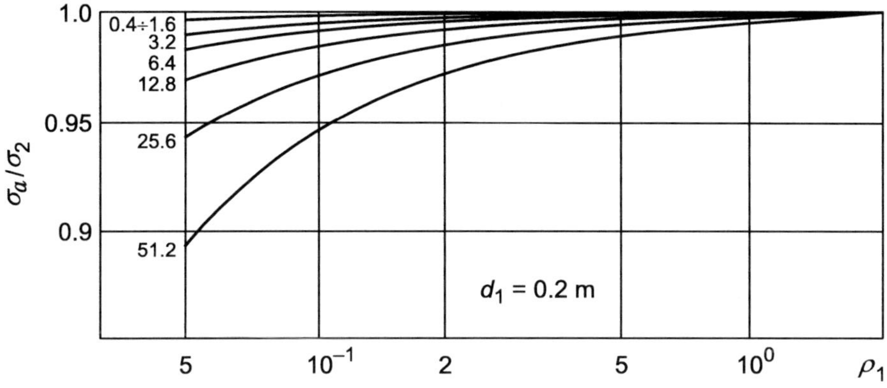

Figure 7.17. Apparent conductivity curves for probe 6F1M (f = 50 kHz). Curve index ρ_2.

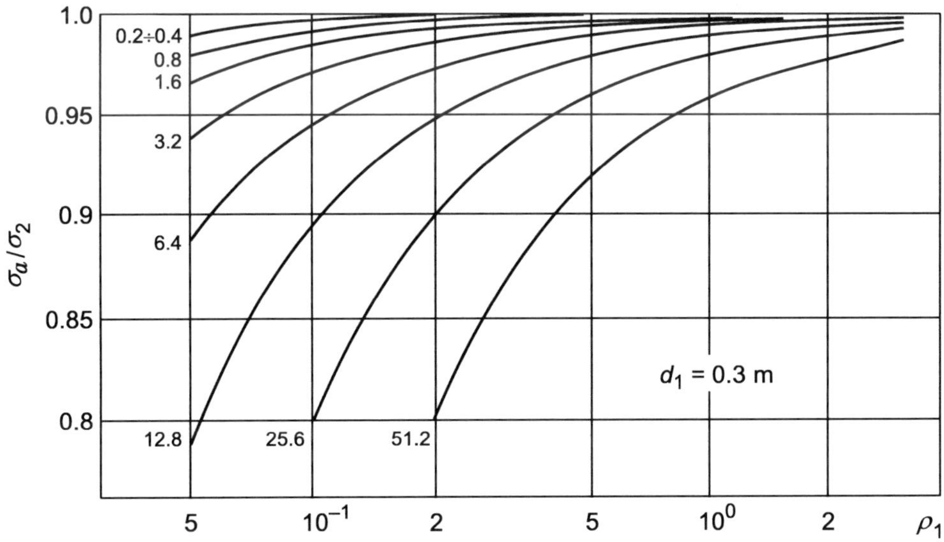

Figure 7.18. Apparent conductivity curves for probe 6F1M (f = 50 kHz). Curve index ρ_2.

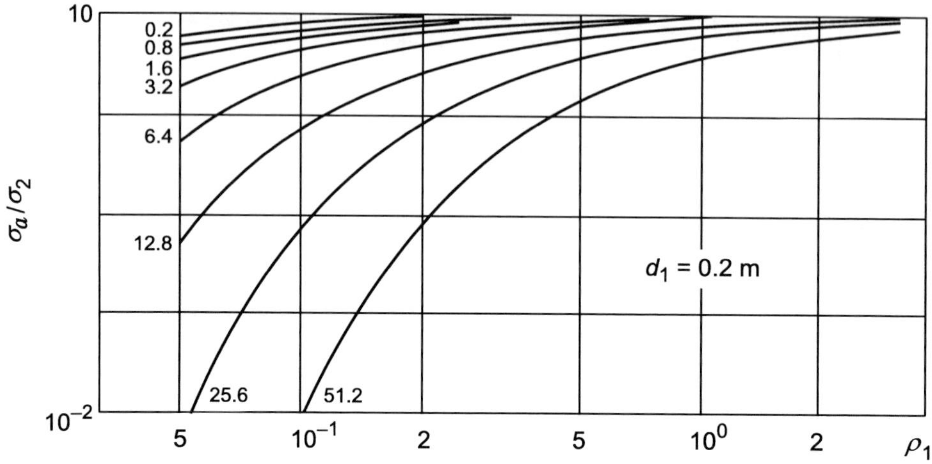

Figure 7.19. Apparent conductivity curves for probe 4F1 ($f = 70$ kHz). Curve index ρ_2.

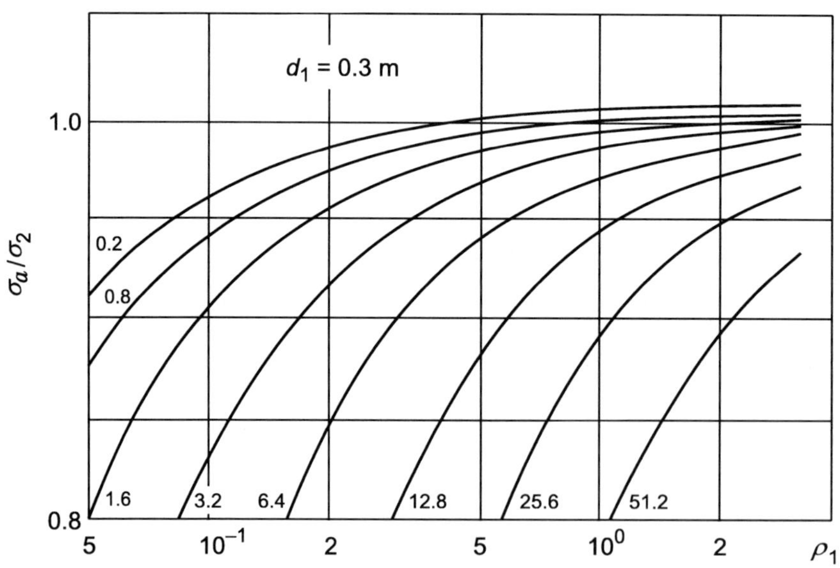

Figure 7.20. Apparent conductivity curves for probe 4F1 ($f = 70$ kHz). Curve index ρ_2.

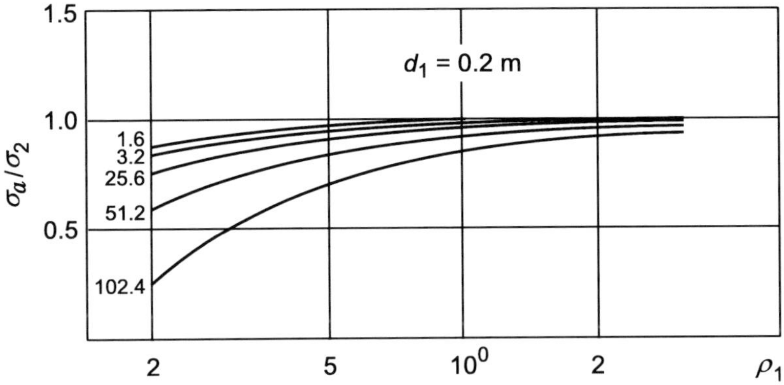

Figure 7.21. Apparent conductivity curves for probe 4F1.1 ($f = 10^3$ kHz). Curve index ρ_2.

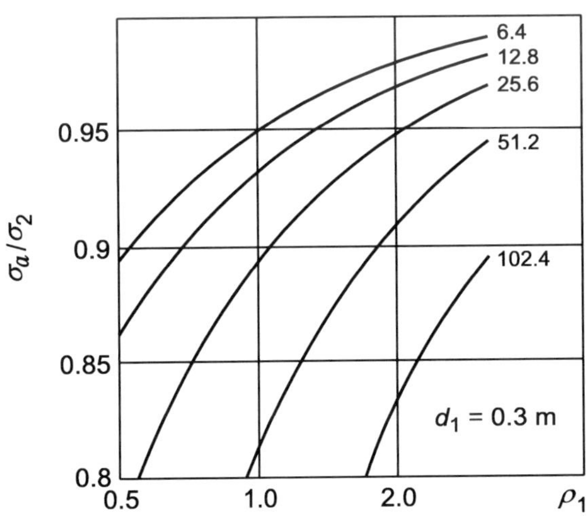

Figure 7.22. Apparent conductivity curves for probe 4F1.1 ($f = 10^3$ kHz). Curve index ρ_2.

point where $G_1^*(r)$ reaches a minimum; r_0 is the point when $G_1^*(r)$ is equal to zero. Values of r_{min} and r_0 for probes considered here are:

Probes	R_{min}, m	r_0, m
6F1M	0.22	0.28
4F1	0.28	0.39
4F1.1	0.31	0.43

We will notice the following features in the behavior of apparent conductivity for various parameters of a medium.

Case 1

Consider the case when $\rho_3/\rho_1 > 1$.

An increase of the radius of the invasion zone results in an increase of σ_a/σ_1 if $a_2 < r_{min}$, and in a decrease of this function if $a_2 > r_{min}$. For example, for probe 6F1M when $\rho_2 = 16$ ohm·m, $\rho_3 = 8$ ohm·m and $a_1 = 0.1$ m.

a_2, m	0.28	0.40	0.56	0.80
σ_a/σ_3	0.998	0.986	0.948	0.977

An increase of resistivity ρ_2 leads to an increase of σ_a/σ_3, if $a_2 < r_0$, and to a decrease of σ_a/σ_3 when $a_2 > r_0$. For instance, for probe 4F1 we have following values of σ_a/σ_3 ($\rho_3 = 1$ ohm·m, $a_1 = 0.1$ m):

a_2, m \ ρ_2, ohm·m	2	4	8	16
0.28	1.014	1.021	1.025	1.027
0.56	0.929	0.893	0.876	0.867

In the case when $a_2 > r_0$ with an increase of formation resistivity, ρ_3, σ_a/σ_3 increases also. However, if $a_2 < r_0$ an increase of ρ_3 leads to a decrease of σ_a/σ_3. For instance, for probe 4F1.1 we have following values of σ_a/σ_3 ($\rho_2 = 128$ ohm·m, $a_1 = 0.1$ m):

a_2, m \ ρ_2, ohm·m	8	16	32	64
0.28	1.035	1.019	1.006	0.991
0.56	0.867	0.911	0.933	0.948

Case 2

Let us consider the case when $\rho_3/\rho_2 > 1$.

If $a_2 < r_{min}$ the apparent conductivity decreases with growing a_2 becomes greater with an increase of a_2, if $a_2 > r_{min}$. For instance, for probe 6F1M, when $\rho_2 = 2$ ohm·m, $\rho_3 = 16$ ohm·m and $a_1 = 0.1$ m we have:

a_2, m	0.28	0.40	0.56	0.80
σ_a/σ_3	0.999	1.156	1.666	2.616

TABLE 7.12
Values of function σ_a/σ_3; $a_1 = 0.1$ m, $\rho_2 = 4$ ohm-m

Probe type	ρ_3, ohm-m								
	0.5	1	2	4	8	16	32	64	128
				$a_2/a_1 = 2\sqrt{2}$					
6F1M	0.999	0.999	0.999	0.999	0.998	0.998	0.996	0.992	–
Two-coil	0.893	0.937	0.993	1.085	1.256	1.588	2.238	3.523	–
4F1	1.036	1.021	1.010	0.996	0.974	0.932	0.852	0.695	–
Two-coil	0.881	0.932	0.992	1.088	1.262	1.596	2.250	3.541	–
4F1.1	–	–	–	0.995	0.952	0.904	0.822	0.671	0.380
Two-coil	–	–	–	1.157	1.359	1.690	3.297	3.454	5.696
				$a_2/a_1 = 4$					
6F1M	0.960	0.974	0.985	0.999	1.022	1.065	1.147	1.309	–
Two-coil	0.788	0.860	0.947	1.085	1.338	1.823	2.774	4.654	–
4F1	0.994	0.996	0.996	0.996	0.996	0.994	0.992	0.996	–
Two-coil	0.764	0.850	0.945	1.088	1.345	1.835	2.792	4.680	–
4F1.1	–	–	–	–	–	–	–	–	–
Two-coil	–	–	–	–	–	–	–	–	–

TABLE 7.13
Values of function σ_a/σ_3; $a_1 = 0.1$ m, $\rho_2 = 4$ ohm-m

Probe type	ρ_3, ohm·m								
	0.5	1	2	4	8	16	32	64	128
				$a_2/a_1 = 4\sqrt{2}$					
6F1M	0.833	0.892	0.939	0.999	1.099	1.284	1.640	2.341	–
Two-coil	0.632	0.747	0.879	1.086	1.459	2.174	3.574	6.348	–
4F1	0.824	0.893	0.941	0.996	1.084	1.245	1.553	2.155	–
Two-coil	0.590	0.728	0.873	1.088	1.469	2.191	3.600	6.380	–
4F1.1	–	–	–	–	–	–	–	–	–
Two-coil	–	–	–	–	–	–	–	–	–
				$a_2/a_1 = 8$					
6F1M	0.596	0.738	0.853	0.999	1.241	1.691	2.558	4.263	–
Two-coil	0.440	0.605	0.795	1.085	1.607	2.605	4.557	8.415	–
4F1	0.517	0.709	0.842	0.996	1.244	1.696	2.564	4.264	–
Two-coil	0.376	0.578	0.785	1.088	1.620	2.628	4.593	8.467	–
4F1.1	–	–	–	–	–	–	–	–	–
Two-coil	–	–	–	–	–	–	–	–	–

TABLE 7.14
Values of function σ_a/σ_3; $a_1 = 0.1$ m, $\rho_2 = 16$ ohm-m

Probe type	ρ_3, ohm·m								
	0.5	1	2	4	8	16	32	64	128
					$a_2/a_1 = 2\sqrt{2}$				
6F1M	0.999	0.999	0.999	0.999	0.998	0.996	0.994	0.988	–
Two-coil	0.882	0.917	0.958	1.020	1.133	1.143	1.769	2.602	–
4F1	1.040	1.027	1.019	1.012	1.002	0.986	0.956	0.896	–
Two-coil	0.881	0.922	0.956	1.021	1.135	1.353	1.778	2.613	–
4F1.1	–	–	–	–	1.017	0.993	0.965	0.918	0.832
Two-coil	–	–	–	–	1.184	1.406	1.802	2.551	4.000
					$a_2/a_1 = 4$				
6F1M	0.956	0.968	0.974	0.980	0.986	0.996	1.015	1.051	–
Two-coil	0.766	0.822	0.878	0.956	1.092	1.348	1.846	2.828	–
4F1	0.994	0.995	0.995	0.994	0.991	0.986	0.976	0.995	–
Two-coil	0.739	0.809	0.872	0.955	1.094	1.353	1.855	2.841	–
4F1.1	–	–	–	–	1.005	0.993	0.978	0.951	0.902
Two-coil	–	–	–	–	1.125	1.406	1.887	2.782	4.505

TABLE 7.15
Values of function σ_a/σ_3; $a_1 = 0.1$ m, $\rho_2 = 16$ ohm·m

Probe type	ρ_3, ohm·m								
	0.5	1	2	4	8	16	32	64	128
					$a_2/a_1 = 2\sqrt{2}$				
6F1M	0.815	0.865	0.894	0.918	0.948	0.996	1.086	1.257	–
Two-coil	0.593	0.680	0.758	0.861	1.032	1.348	1.960	3.166	–
4F1	0.805	0.867	0.898	0.921	0.947	0.986	1.056	1.188	–
Two-coil	0.546	0.656	0.747	0.857	1.032	1.353	1.970	3.181	–
4F1.1	–	–	–	–	0.927	0.993	1.064	1.176	1.380
Two-coil	–	–	–	–	1.031	1.406	2.019	3.143	5.380
					$a_2/a_1 = 8$				
6F1M	0.553	0.673	0.743	0.803	0.877	0.996	1.217	1.642	–
Two-coil	0.380	0.505	0.612	0.744	0.958	1.348	2.100	3.580	–
4F1	0.517	0.709	0.842	0.996	1.244	1.696	2.564	4.264	–
Two-coil	0.376	0.578	0.758	1.088	1.620	2.628	4.593	8.467	–
4F1.1	–	–	–	–	–	–	–	–	–
Two-coil	–	–	–	–	–	–	–	–	–

TABLE 7.16
Values of function σ_a/σ_3; $a_1 = 0.1$ m, $\rho_2 = 64$ ohm·m

Probe type	ρ_3, ohm·m								
	0.5	1	2	4	8	16	32	64	128
					$a_2/a_1 = 2\sqrt{2}$				
6F1M	0.999	0.999	0.999	0.998	0.998	0.996	0.993	0.986	–
Two-coil	0.879	0.912	0.949	1.004	1.102	1.288	1.562	2.372	–
4F1	1.042	1.028	1.021	1.016	1.010	0.999	0.981	0.947	–
Two-coil	0.865	0.906	0.947	1.000	1.104	1.292	1.660	2.381	–
4F1.1	–	–	–	–	1.103	1.015	1.000	0.980	0.945
Two-coil	–	–	–	–	1.140	1.336	1.679	2.325	3.574
					$a_2/a_1 = 4$				
6F1M	0.955	0.966	0.972	0.975	0.977	0.979	0.982	0.986	–
Two-coil	0.760	0.812	0.860	0.924	1.031	1.229	1.614	2.372	–
4F1	0.994	0.995	0.995	0.993	0.990	0.984	0.971	0.947	–
Two-coil	0.773	0.799	0.854	0.922	1.031	1.232	1.620	2.381	–
4F1.1	–	–	–	–	1.013	1.000	0.994	0.980	0.955
Two-coil	–	–	–	–	1.036	1.263	1.637	2.325	3.646

TABLE 7.17
Values of function σ_a/σ_3; $a_1 = 0.1$ m, $\rho_2 = 64$ ohm·m

Probe type	ρ_3, ohm·m								
	0.5	1	2	4	8	16	32	64	128
					$a_2/a_1 = 2\sqrt{2}$				
6F1M	0.810	0.858	0.882	0.898	0.910	0.925	0.947	0.986	–
Two-coil	0.760	0.812	0.860	0.924	1.031	1.229	1.614	2.372	–
4F1	0.800	0.860	0.888	0.903	0.913	0.992	0.932	0.947	–
Two-coil	0.535	0.638	0.715	0.799	0.924	1.144	1.563	2.381	–
4F1.1	–	–	–	–	0.876	0.923	0.951	0.980	1.023
Two-coil	–	–	–	–	0.873	1.149	1.571	2.325	3.760
					$a_2/a_1 = 8$				
6F1M	0.542	0.657	0.716	0.754	0.785	0.823	0.881	0.986	–
Two-coil	0.365	0.479	0.566	0.659	0.795	1.034	1.487	2.372	–
4F1	0.452	0.618	0.695	0.740	0.773	0.809	0.859	0.947	–
Two-coil	0.293	0.441	0.545	0.648	0.791	1.034	1.492	2.381	–
4F1.1	–	–	–	–	0.586	0.753	0.861	0.980	1.167
Two-coil	–	–	–	–	0.660	1.002	1.485	2.325	3.906

TABLE 7.18
Values of function σ_a/σ_3; $a_1 = 0.1$ m, $\rho_2 = 4$ ohm·m

Probe type	ρ_3, ohm·m							
	0.5	1	2	4	8	16	32	64
				$a_2/a_1 = 2\sqrt{2}$				
6F1M	0.945	0.963	0.977	0.993	1.020	1.069	1.163	1.348
Two-coil	0.785	0.877	0.996	1.196	1.566	2.281	3.687	6.468
4F1	0.967	0.973	0.971	0.963	0.944	0.905	0.829	0.678
Two-coil	0.760	0.868	0.996	1.202	1.578	2.300	3.714	6.506
4F1.1	–	–	–	–	–	–	–	–
Two-coil	–	–	–	–	–	–	–	–
				$a_2/a_1 = 4$				
6F1M	0.799	0.868	0.924	0.993	1.108	1.320	1.730	2.534
Two-coil	0.621	0.757	0.925	1.200	1.693	2.650	4.527	8.240
4F1	0.772	0.855	0.908	0.963	1.045	1.193	1.473	2.021
Two-coil	0.577	0.739	0.921	1.202	1.708	2.674	4.563	8.292
4F1.1	–	–	–	–	–	–	–	–
Two-coil	–	–	–	–	–	–	–	–

TABLE 7.19
Values of function σ_a/σ_3; $a_1 = 0.15$ m, $\rho_2 = 16$ ohm·m

Probe type	ρ_3, ohm·m								
	0.5	1	2	4	8	16	32	64	128
					$a_2/a_1 = 2\sqrt{2}$				
6F1M	0.939	0.954	0.962	0.967	0.970	1.069	1.163	1.348	–
Two-coil	0.762	0.838	0.926	1.065	1.316	2.281	3.687	6.468	–
4F1	0.964	0.968	0.963	0.949	0.918	0.905	0.829	0.678	–
Two-coil	0.735	0.826	0.922	1.067	1.323	2.300	3.714	6.506	–
4F1.1	–	–	–	–	0.886	0.840	0.740	0.546	0.169
Two-coil	–	–	–	–	1.435	1.932	2.822	4.507	7.767
					$a_2/a_1 = 4$				
6F1M	0.777	0.836	0.869	0.896	0.926	0.974	1.061	1.228	–
Two-coil	0.580	0.688	0.800	0.965	1.253	1.798	2.863	4.964	–
4F1	0.748	0.821	0.852	0.866	0.868	0.857	0.829	0.768	–
Two-coil	0.532	0.665	0.791	0.963	1.258	1.810	2.882	4.993	–
4F1.1	–	–	–	–	0.793	0.840	0.843	0.813	0.737
Two-coil	–	–	–	–	1.336	1.932	2.962	4.890	8.608

TABLE 7.20
Values of function σ_a/σ_3; $a_1 = 0.15$ m, $\rho_2 = 16$ ohm·m

Probe type	ρ_3, ohm·m							
	0.5	1	2	4	8	16	32	64
				$a_2/a_1 = 2\sqrt{2}$				
6F1M	0.500	0.632	0.709	0.773	0.850	0.974	1.201	1.637
Two-coil	0.364	0.510	0.652	0.846	1.117	1.798	3.005	5.386
4F1	0.388	0.577	0.688	0.728	0.783	0.857	0.982	1.213
Two-coil	0.292	0.475	0.635	0.841	1.181	1.810	3.026	5.416
4F1.1	–	–	–	–	–	–	–	–
Two-coil	–	–	–	–	–	–	–	–
				$a_2/a_1 = 8$				
6F1M	0.161	0.385	0.516	0.625	0.759	0.974	1.370	2.131
Two-coil	0.151	0.335	0.504	0.728	1.103	1.798	3.146	5.802
4F1	0.027	0.295	0.455	0.568	0.686	0.857	1.158	1.729
Two-coil	0.054	0.286	0.481	0.720	1.105	1.810	3.168	5.836
4F1.1	–	–	–	–	–	–	–	–
Two-coil	–	–	–	–	–	–	–	–

TABLE 7.21
Values of function σ_a/σ_3; $a_1 = 0.15$ m, $\rho_2 = 64$ ohm·m

Probe type	ρ_3, ohm·m								
	0.5	1	2	4	8	16	32	64	128
				$a_2/a_1 = 2\sqrt{2}$					
6F1M	0.938	0.952	0.958	0.960	0.958	0.951	0.935	0.901	–
Two-coil	0.756	0.828	0.908	1.032	1.254	1.678	2.507	4.146	–
4F1	0.963	0.967	0.961	0.945	0.912	0.846	0.714	0.454	–
Two-coil	0.728	0.815	0.904	1.033	1.259	1.688	2.522	4.168	–
4F1.1	–	–	–	–	0.885	0.838	0.739	0.544	0.165
Two-coil	–	–	–	–	1.345	1.785	2.566	4.040	6.886
				$a_2/a_1 = 4$					
6F1M	0.772	0.828	0.855	0.871	0.881	0.888	0.894	0.901	–
Two-coil	0.570	0.671	0.769	0.907	1.143	1.586	2.447	4.146	–
4F1	0.741	0.812	0.838	0.841	0.823	0.774	0.668	0.454	–
Two-coil	0.521	0.647	0.758	0.904	1.146	1.594	2.462	4.168	–
4F1.1	–	–	–	–	0.722	0.743	0.688	0.544	0.246
Two-coil	–	–	–	–	1.171	1.664	2.500	4.039	7.006

TABLE 7.22
Values of function σ_a/σ_3; $a_1 = 0.15$ m, $\rho_2 = 64$ ohm·m

Probe type	ρ_3, ohm·m								
	0.5	1	2	4	8	16	32	64	128
					$a_2/a_1 = 2\sqrt{2}$				
6F1M	0.486	0.613	0.678	0.718	0.748	0.779	0.824	0.901	–
Two-coil	0.349	0.484	0.605	0.758	1.010	1.476	2.380	4.146	–
4F1	0.373	0.556	0.634	0.669	0.676	0.654	0.592	0.454	–
Two-coil	0.274	0.447	0.586	0.751	1.011	1.483	2.390	4.168	–
4F1.1	–	–	–	–	0.410	0.560	0.590	0.544	0.400
Two-coil	–	–	–	–	1.954	1.513	2.408	4.039	7.157
					$a_2/a_1 = 8$				
6F1M	0.141	0.355	0.464	0.532	0.587	0.647	0.740	0.901	–
Two-coil	0.130	0.300	0.442	0.612	0.880	1.368	2.305	4.146	–
4F1	0.052	0.260	0.400	0.469	0.505	0.516	0.503	0.454	–
Two-coil	0.031	0.250	0.415	0.599	0.878	1.373	2.318	4.168	–
4F1.1	–	–	–	–	0.020	0.330	0.468	0.544	0.594
Two-coil	–	–	–	–	0.729	1.357	2.317	4.040	7.313

With an increase of ρ_2 σ_a/σ_3 increases if $a_2 < r_0$. In the opposite case i.e. as $a_2 > r_0$ with growing resistivity of the invasion zone σ_a/σ_3 decreases. For example, for probe 4F1, when $\rho_3 = 32$ ohm·m and $a_1 = 0.1$ m we have following values of σ_a/σ_3:

a_2, m \ ρ_2, ohm·m	2	4	8	16
0.28	0.714	0.852	0.921	0.956
0.56	2.215	1.553	1.221	1.056

If $a_2 < r_0$ then σ_a/σ_3 decreases with an increase of formation resistivity. On the contrary when $a_2 > r_0$, σ_a/σ_3 increases when ρ_3 increases. For instance, for probe 4F1.1, when $\rho_2 = 16$ ohm·m and $a_1 = 0.1$ m we have following values of σ_a/σ_3:

a_2, m \ ρ_2, ohm·m	32	64	128
0.28	0.965	0.918	0.832
0.56	1.064	1.176	1.380

An increase of the borehole radius from 0.1 m to 0.15 m results in a decrease of σ_a/σ_3, when $\rho_1/\rho_2 < 1$.

In conclusion in Table 7.23 ranges of the change of formation resistivity are presented for which σ_a/σ_3 differs from unity less than 10%.

Now we will investigate vertical response of these differential probes. As has been shown differential probes allow us to reduce significantly the influence of the borehole and the invasion zone, that is areas of a medium, directly surrounding the probe. On the other hand, in order to provide satisfactory vertical response in layers with a finite thickness it is necessary that the main part of a measured signal is defined by currents in areas which are relatively close to the probe. Therefore as was pointed out earlier, requirements for improvement of radial and vertical responses of a probe in essence contradict each other. Correspondingly, it is natural to expect that since multi-coil induction probes have better radial responses than the basic two-coil probe, they are more sensitive to the surrounding medium at least in those cases when the formation thickness exceeds the distance between the most remote coils of the probe.

However, with help of external *focusing* when some coils are located outside the formation, we can improve the vertical response due to the fact that for such position a differential probe permits us to reduce the signal from the surrounding medium to a greater extent than from the formation.

Proceeding from Doll's theory let us first consider vertical responses of the internal differential probes. Suppose the probe is located against the formation, and we will introduce notations: G_1^* and G_2^*, which are the geometric factors of the formation and the surrounding medium (shoulders), respectively, while G_1 and G_2 are corresponding geometric factors for the basic two coil probe. Values of $\sigma_1 G_1^*/\sigma_2 G_2^*$ and $\sigma_1 G_1/\sigma_2 G_2$ characterize a relation between signals caused by currents in a formation and in the surrounding medium for differential and two-coil induction probes, respectively.

Let us show that in this case the following inequality takes place:

$$\frac{\sigma_1 G_1^*}{\sigma_2 G_2^*} < \frac{\sigma_1 G_1^*}{\sigma_2 G_2^*}$$

or

$$\frac{G_1^*}{G_2^*} < \frac{G_1}{G_2} \tag{7.35}$$

In accord with this relation we can consider a uniform medium in order to demonstrate that a multi-coil probe with internal *focusing* has higher sensitivity to the surrounding medium than the basic two-coil probe when the formation thickness exceeds the probe length.

As was mentioned above every multi-coil induction probe can be considered as a sum of two-coil probes namely the basic induction probe and additional coil probes which provide improvement of the radial response. Electromotive forces induced in these probes can have the opposite sign to that in the basic probe as well as the same sign. However, the probes where the electromotive force has opposite sign play the most essential role and correspondingly, only they will be taken into consideration here.

In accord with Doll's theory, described in detail in Chapter 3, it is appropriate to emphasize two main features of the geometric factors:

- The geometric factor of the whole medium for every probe is equal to unity.
- The part of the medium against which the two-coil probe is located has a geometric factor of 0.5.

TABLE 7.23
Ranges of the change of formation resistivity; $\rho_1 = 0.5$ ohm·m, $a_1 = 0.1$ m

ρ_2, ohm·m	a_2/a_1			Probe type
	$2\sqrt{2}$	4	$4\sqrt{2}$	
2	0.5–20	0.5–8	1–4	6F1M
	0.5–8	0.5–40	1–4	4F1
	–	–	–	4F1.1
4	0.5–20	0.5–16	1–8	6F1M
	0.5–16	0.5–40	1–8	4F1
	8–16	8–32	–	4F1.1
8	0.5–20	0.5–32	2–16	6F1M
	0.5–40	0.5–40	2–16	4F1
	8–32	8–64	8–16	4F1.1
16	0.5–20	0.5–64	2–32	6F1M
	0.5–40	0.5–40	2–40	4F1
	8–64	8–128	8–32	4F1.1
32	0.5–20	0.5–64	4–64	6F1M
	0.5–40	0.5–40	4–40	4F1
	8–128	8–64	4–40	4F1.1
64	0.5–20	0.5–64	8–64	6F1M
	0.5–40	0.5–40	4–40	4F1
	8–128	8–128	16–128	4F1.1
128	0.5–20	0.5–64	4–64	6F1M
	0.5–40	0.5–40	4–40	4F1
	8–128	8–128	16–128	4F1.1

From these two facts we can conclude that the differential two-coil probe reduces the signal from the formation to a greater extent than that from the surrounding medium, and therefore inequality 7.35 is valid.

Now we will briefly discuss vertical responses of probes with external *focusing* coils. It is obvious that in the case when the whole probe is located against the formation the influence of currents induced in the surrounding medium is stronger than that for the basic two-coil probe (Fig. 7.23a).

If additional probes, as shown in Fig. 7.23b, are mainly outside of the formation, the effect caused by currents induced in shoulders will be reduced stronger than that in the formation, and correspondingly some improvement of the vertical response of the basic two-coil probe will be observed.

In conclusion of this analysis of vertical responses of multi-coil probes with internal and external *focusing* it is appropriate to notice the following:

- *Focusing* probes located against the formation possess higher sensitivity to the surrounding medium than the basic two-coil probe.

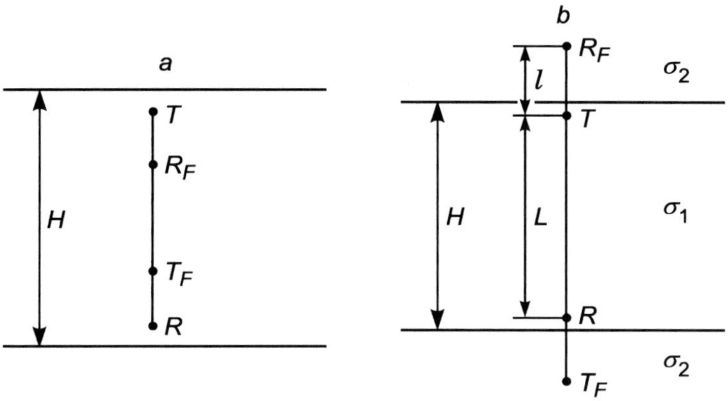

Figure 7.23. Various positions of probes with respect to formation boundaries.

- Application of probes with external *focusing* results in some improvement of the vertical response within the interval of formation thicknesses:

  $$L < H < L + 2l$$

  where L and l are length of the basic and additional probe, respectively.

However, the influence of shoulders for such thicknesses as $\rho_1 > \rho_2$ is so large that the apparent conductivity, σ_a, essentially differs from σ_1 for probes with internal as well as external *focusing.*

In the case when ρ_1/ρ_2 and $H > L$ probes with both types of *focusing* have practically the same vertical responses.

As was demonstrated in Chapter 5, evaluation of vertical responses of induction probes, based on the theory of small parameters, very often has a qualitative character. For this reason let us consider the results of calculations based on the exact solution.

Curves of σ_a/σ_1 for various parameters of geoelectric section when differential probes 6F1M, 4F1 and 4F1.1 are located symmetrically against the formation are presented in Figs. 7.24–7.29.

As is seen from these curves:

- With an increase of frequency the influence of shoulders (surrounding medium) becomes smaller. This effect is more noticeable when ρ_1/ρ_2 and the formation thickness is greater. For example, if $\rho_1 = 32$ ohm·m, $\rho_2 = 2$ ohm·m, $H = 2.5$ m for probes 4F1 ($f = 70$ kHz) and 4F1.1 ($f = 1$ MHz) we have $\sigma_a/\sigma_1 = 2.5$ and $\sigma_a/\sigma_1 = 1.05$, respectively.

- An increase of shoulder conductivity deteriorates the vertical response of the probe, and it manifests itself stronger the lower the frequency and the smaller the thickness of the formation.

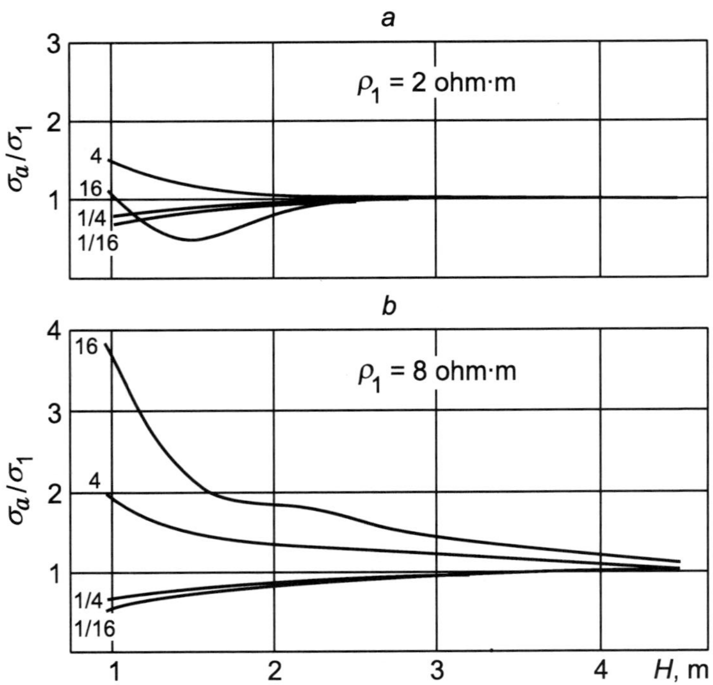

Figure 7.24. Vertical responses of probe 6F1M (f = 50 kHz). Curve index ρ_1/ρ_2.

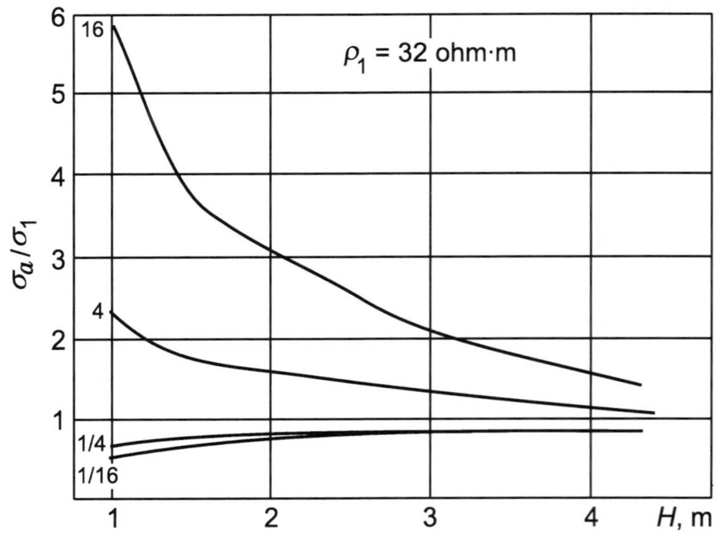

Figure 7.25. Vertical responses of probe 6F1M (f = 50 kHz). Curve index ρ_1/ρ_2.

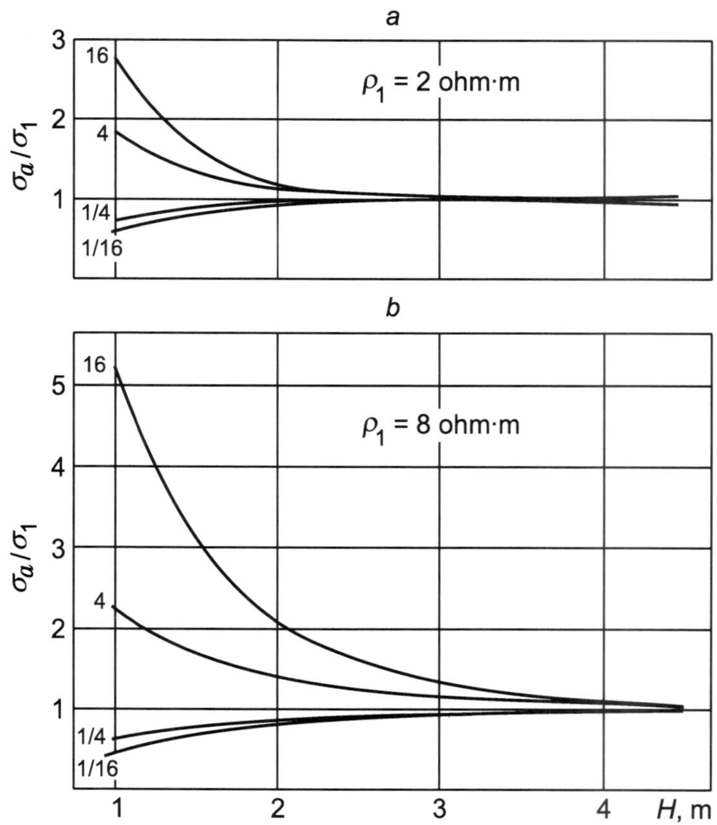

Figure 7.26. Vertical responses of probe 4F1 ($f = 70$ kHz). Curve index ρ_1/ρ_2.

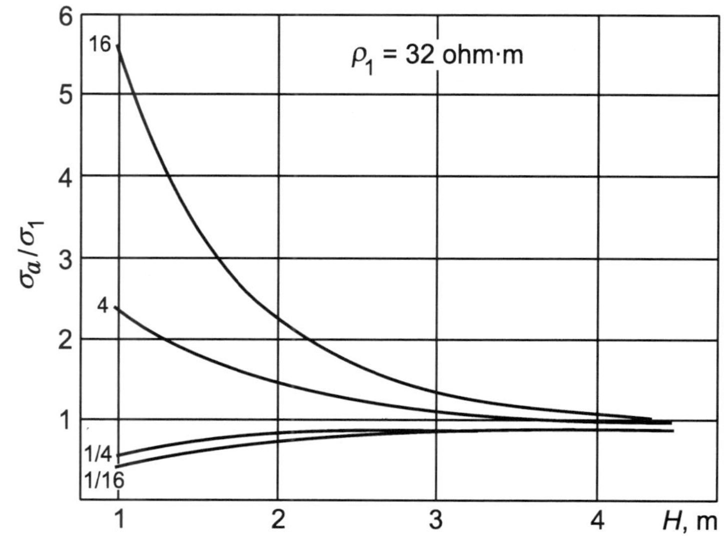

Figure 7.27. Vertical responses of probe 4F1 ($f = 70$ kHz). Curve index ρ_1/ρ_2.

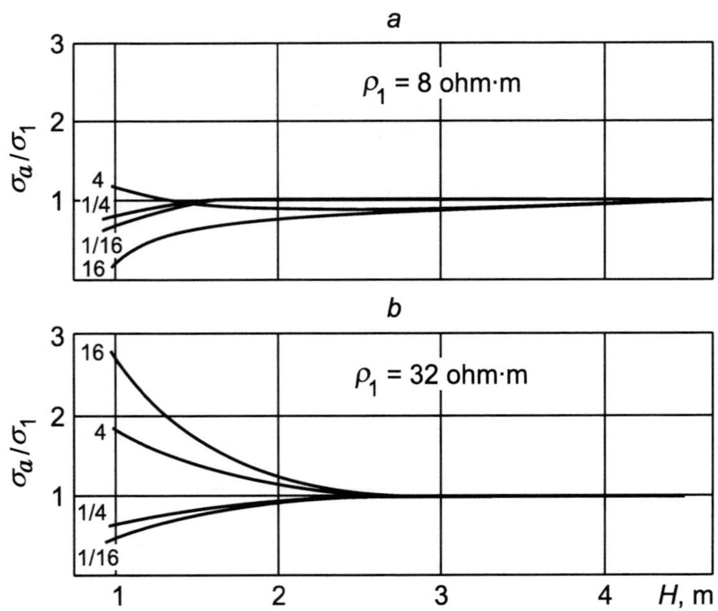

Figure 7.28. Vertical responses of probe 4F1.1 ($f = 1$ MHz). Curve index ρ_1/ρ_2.

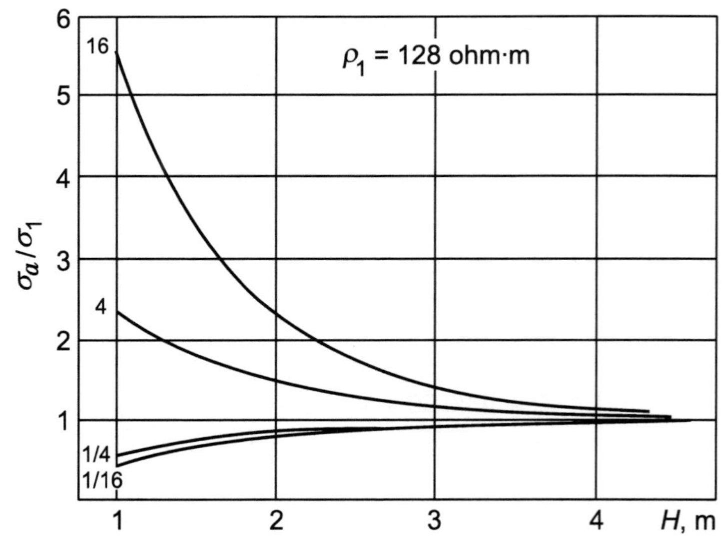

Figure 7.29. Vertical responses of probe 4F1.1 ($f = 1$ MHz). Curve index ρ_1/ρ_2.

- The probe with external *focusing* 6F1M is characterized by smaller values of σ_a/σ_1 than the probe with the internal *focusing* 4F1, if $1.5 < H < 2.5$ m and $\rho_1 > \rho_2$. For instance both probes provide the same value of σ_a/σ_1, equal to 1.75 when $H = 2.5$ m, $\rho_1 = 8$ ohm·m and $\rho_2 = 0.5$ ohm·m. However, if $H = 1.75$ m values of σ_a/σ_1 for probes 4F1 and 6F1M are equal to 2.5 and 1.8, respectively. Within this interval of thicknesses and $\rho_2 = 2$ ohm·m values of apparent conductivity for both probes are practically the same.

Thus, the advantage of the external *focusing* manifest itself only for large values of ρ_1/ρ_2 when values of apparent conductivity essentially differ from the formation conductivity.

7.5. The Influence of Finite Height of the Invasion Zone on Radial Responses of Probes 6F1M, 4F1 and 4F1.1

As is known, parameters of the multi-coil induction probe are chosen with the assumption that the invasion zone has an infinite extension along the borehole axis. In real conditions penetration of borehole filtrate into a formation and surrounding medium occurs in a different manner due to the difference of their physical properties. For this reason it is appropriate to consider *focusing* features of multi-coil induction probes when the invasion zone has limited dimension along the borehole axis.

We will investigate the case where the borehole filtrate penetrates into the formation only (Fig. 7.30). Calculations of the magnetic field on the borehole axis in such model is performed on the base of the approximate theory which takes into account the skin effect in the external area.

As was demonstrated in Chapter 3 for the quadrature component of the magnetic field of a two-coil induction probe we have:

$$\mathrm{Q}\,H_z = \mathrm{Q}\,H_z^0 + \frac{\omega\mu M}{4\pi L}[(\sigma_1 - \sigma_4)G_1(\alpha) + (\sigma_4 - \sigma_3)G_1^{(3)}(\alpha) + (\sigma_2 - \sigma_3)G_2^{(3)}(\alpha)] \tag{7.36}$$

where $\mathrm{Q}\,H_z^0$ is the quadrature component of the magnetic field in a horizontally layered medium when the borehole and invasion zone are absent; $G_1^{(3)}(\alpha)$ and $G_2^{(3)}(\alpha)$ are geometric factors of the part of the borehole located against the formation and the invasion zone, respectively.

It is obvious that an influence of finite height of the invasion zone is defined by difference between geometric factors $G_1^{(3)}(\alpha)$ and $G_2^{(3)}(\alpha)$ and those corresponding to infinitely long cylinders. First of all we consider this question for a two-coil induction probe. We will assume that the probe is completely located within the cylinder interval $(H/2, -H/2)$, and L is the probe length, but ε is coordinate of the transmitter coil (Fig. 7.31). Then the expression for geometric factor of the cylinder in accord with eq. 3.104 can be written in the form:

$$G_1^{(3)} = \frac{L}{2}\int\limits_0^a \int\limits_{-H/2}^{H/2} \frac{r^3\,dz\,dr}{R_1^3 R_2^3} \tag{7.37}$$

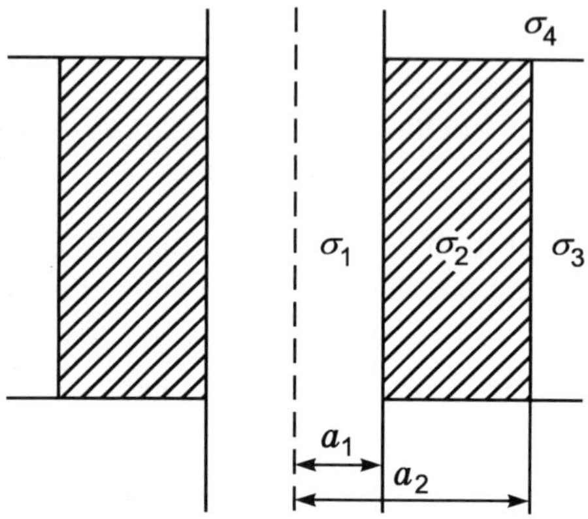

Figure 7.30. The model of the invasion zone with finite height.

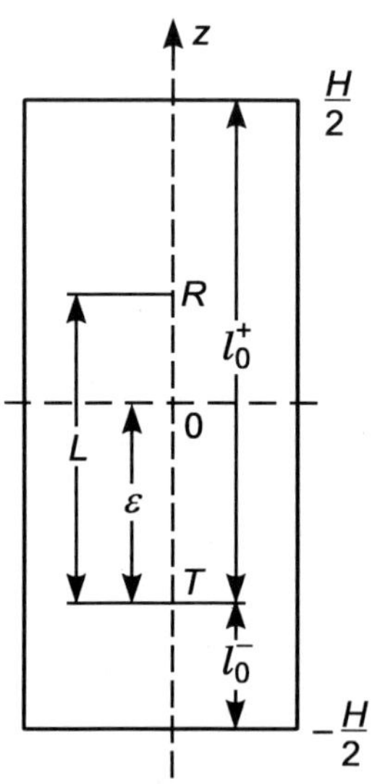

Figure 7.31. Two-coil induction probe within the cylinder with finite thickness, H.

where:

$$R_1 = \sqrt{r^2 + (z-\varepsilon)^2} \qquad R_2 = \sqrt{[L-(z-\varepsilon)]^2 + r^2}$$

It is obvious that:

$$G_1^{(3)} = G_1(\alpha) - \frac{L}{2}\int_0^a \int_{-\infty}^{-H/2} \frac{r^3\,\mathrm{d}z\,\mathrm{d}r}{R_1^3 R_2^3} - \frac{L}{2}\int_0^a \int_{H/2}^{\infty} \frac{r^3\,\mathrm{d}z\,\mathrm{d}r}{R_1^3 R_2^3} \tag{7.38}$$

Integrals on the right-hand side of eq. 7.38 describe a difference between geometric factors of a cylinder of finite height and that of an infinitely long one. Let us investigate these integrals. After a change of variables a sum of them, S, can be presented as:

$$S = -\frac{L}{2}\int_0^a r^3\,\mathrm{d}r \int_{-\frac{1}{H/2-\varepsilon}}^{\frac{1}{H/2-\varepsilon}} \frac{t^4\,\mathrm{d}t}{[(tL-1)^2 + t^2r^2]^{3/2}\,[1+t^2r^2]^{3/2}} \tag{7.39}$$

It is clear that $H/2-\varepsilon = l_0^+$ and $H/2+\varepsilon = l_0^-$ are distances from the transmitter coil to upper and lower boundaries of the cylinder, respectively.

Suppose that $a/l_0^+ \ll 1$ and $a/l_0^- \ll 1$. Then we can assume that the second multiplier in the denominator of eq. 7.39 is equal to unity, and in this case the integral with respect to r is taken in the explicit form. Correspondingly we have:

$$S = -\frac{L}{2}\int_{-1/l_0^-}^{1/l_0^+} \left(\sqrt{(tL-1)^2 + t^2a^2} + \frac{2(tL-1)}{\sqrt{(tL-1)^2+t^2a^2}} - 2(tL-1)\right)\mathrm{d}t \tag{7.40}$$

Suppose also that:

$$1 - L/l_0^+ \gg a/l_0^+ \qquad 1 + L/l_0^- \gg a/l_0^-$$

or

$$l^+ = l_0^+ - L \gg a \qquad l^- = l_0^- + L \gg a$$

where l^+ and l^- are distances from the receiver to upper and lower boundaries, respectively.

Taking into account these inequalities, radicals in the integrand of eq. 7.40 can be expanded in series. Then the integral can be presented as:

$$S = -\frac{La^4}{8}\int_{-1/l_0^-}^{1/l_0^+} \frac{t^4\,\mathrm{d}t}{(1-tL)^3} \tag{7.41}$$

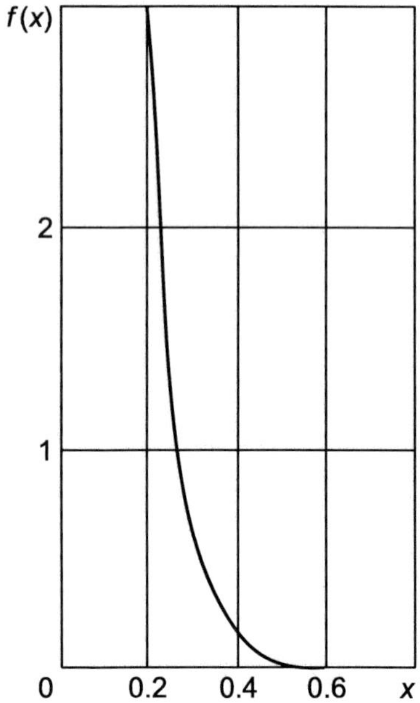

Figure 7.32. Behavior of function $f(x)$.

After relatively simple transformations we have:

$$S = -\frac{a^4}{8L^4}\left[f\left(\frac{l^+}{l_0^+}\right) + f\left(\frac{l^-}{l_0^-}\right)\right] \tag{7.42}$$

where

$$f(x) = (x^2 - 1)\frac{6x - (1 - x^2)}{2x^2} - 6\,\ln x \tag{7.43}$$

Asymptotical expansion 7.42 is valid when the distance from coils to the upper and lower boundaries exceeds the cylinder radius. A graph of $f(x)$ is shown in Fig. 7.32.

If L/l_0 tends to zero expansion of eq. 7.42 results in the following:

$$S = -\frac{La^4}{40}\left[\frac{1}{(l_0^+)^5} + \frac{1}{(l_0^-)^5}\right] \tag{7.44}$$

Equations 7.42–7.44 show that the value of S has high order with respect to small parameter a/L or a/l_0^+, a/l_0^-, and therefore the geometric factor of a cylinder of finite length for a two-coil induction probe ($H > L$) slightly differs from that of an infinitely long one, i.e. G_1. In the case of symmetric location of the probe ($\varepsilon = 0$) this fact was demonstrated in Chapter 5.

Thus, we can assume that the influence of finite dimensions of the invasion zone along the borehole axis on *focusing* features of the multi-coil probe will be usually small.

Calculations confirm this assumption. Values of EMF expressed in units of electromotive force in a horizontally layered medium are presented in Tables 7.24–7.26. These data show that probes 6F1M, 4F1 and 4F1.1 preserve *focusing* features even when the invasion zone has finite dimensions along the z-axis.

7.6. Three-coil Differential Probe

Now we will investigate the simplest differential system, namely the three-coil probe which consists of two transmitter and one receiver coil or one transmitter and two receiver coils. The distance between the pair of transmitter or receiver coils is significantly smaller than that to the remote coil. Let us suppose that the probe is located on the borehole axis, and it consists of one receiver coil with moment N_1, and two transmitter coils with moments M_1, and M_2. The latter are characterized by opposite direction of turns. Then, for electromotive force induced in the receiver by currents in the borehole and in the formation at the range of small parameters we have:

$$\begin{aligned}\mathscr{E} = \mathscr{E}_{11} - \mathscr{E}_{12} = \frac{\omega^2\mu^2 N_1}{4\pi}\Bigg[\sigma_1\left(\frac{M_1}{L_1}G_1(\alpha_1) - \frac{M_2}{L_2}G_1(\alpha_2)\right) \\ + \sigma_2\left(\frac{M_1}{L_1}G_2(\alpha_1) - \frac{M_2}{L_2}G_2(\alpha_2)\right)\Bigg]\end{aligned} \tag{7.45}$$

where L_1 and L_2 are lengths of two-coil induction probes forming the differential probe $(L_1 > L_2)$.

Suppose that the electromotive force of the primary field is compensated, i.e. we have $M_1/L_1^3 = M_2/L_2^3$. Then we obtain:

$$\mathscr{E} = \frac{\mu^2\omega^2 N_1 M_1(1-t^2)}{4\pi L_1}(\sigma_1 G_1^* + \sigma_2 G_2^*) \tag{7.46}$$

where $t = L_2/L_1 < 1$ and:

$$\begin{aligned}G_1^* &= \frac{G_1(\alpha_1) - t^2 G_1(\alpha_2)}{1-t^2} \\ G_2^* &= \frac{G_2(\alpha_1) - t^2 G_2(\alpha_2)}{1-t^2}\end{aligned} \tag{7.47}$$

Let us assume that the lengths of both two-coil probes are much greater than the borehole radius, i.e. $\alpha_1 \gg 1$ and $\alpha_2 \gg 1$. Then, making use of the asymptotic expression for function G_1:

$$G_1 = \frac{1}{\alpha^2} + \frac{3\ \ln\,\alpha - 4.25}{\alpha^4}$$

TABLE 7.24
Values of EMF for probe 6F1M; $\rho_1 = 1$ ohm·m, $\rho_4 = 4$ ohm·m

	$\rho_2 = 8$ ohm·m, $\rho_3 = 32$ ohm·m			$\rho_2 = 32$ ohm·m, $\rho_3 = 8$ ohm·m		
H/L \ a_2/a_1	2	4	6	2	4	6
1	1.015	1.041	1.096	0.997	0.978	0.938
$\sqrt{2}$	1.025	1.073	1.169	0.996	0.967	0.909
2	1.007	1.047	1.169	0.999	0.978	0.915
$2\sqrt{2}$	0.995	1.035	1.184	1.000	0.983	0.916
4	0.994	1.044	1.232	1.000	0.982	0.912
$4\sqrt{2}$	0.993	1.053	1.277	1.000	0.981	0.910
8	0.992	1.059	1.309	1.000	0.981	0.909

TABLE 7.25
Values of EMF for probe 4F1; $\rho_1 = 1$ ohm·m, $\rho_4 = 4$ ohm·m

	$\rho_2 = 8$ ohm·m, $\rho_3 = 32$ ohm·m			$\rho_2 = 32$ ohm·m, $\rho_3 = 8$ ohm·m		
H/L \ a_2/a_1	2	4	6	2	4	6
1	0.973	0.957	0.995	1.007	1.020	0.989
$\sqrt{2}$	0.984	0.986	1.062	1.005	1.003	0.952
2	0.980	0.995	1.118	1.005	0.997	0.929
$2\sqrt{2}$	0.973	0.996	1.171	1.005	0.995	0.918
4	0.996	0.996	1.220	1.005	0.995	0.913
$4\sqrt{2}$	0.960	0.995	1.260	1.005	0.994	0.911
8	0.956	0.995	1.286	1.005	0.994	0.911

TABLE 7.26
Values of EMF for probe 4F1.1; $\rho_1 = 1$ ohm·m, $\rho_4 = 4$ ohm·m

	$\rho_2 = 8$ ohm·m, $\rho_3 = 32$ ohm·m			$\rho_2 = 32$ ohm·m, $\rho_3 = 8$ ohm·m		
H/L \ a_2/a_1	2	4	6	2	4	6
1	0.929	0.862	0.917	1.021	1.087	1.033
$\sqrt{2}$	0.969	0.938	1.077	1.017	1.041	0.934
2	0.965	0.959	1.284	1.014	1.014	0.862
4	0.957	0.959	1.306	1.014	1.013	0.861
$4\sqrt{2}$	0.958	0.960	1.301	1.014	1.013	0.862
8	0.959	0.961	1.295	1.014	1.013	0.862

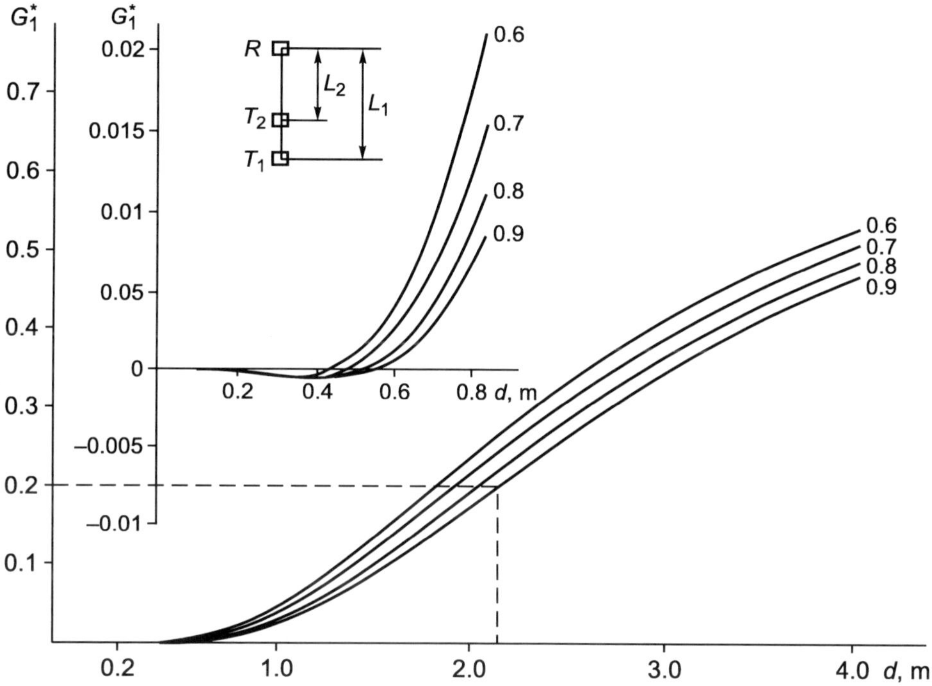

Figure 7.33. Radial responses of a three-coil probe ($L = 1.4$ m). Curve index L_2/L_1.

we have for the geometric factor of a three-coil probe, G_1^*, the following expression:

$$G_1^* \simeq \frac{1}{t_1^2 \alpha_1^4} \left[2.17 - \frac{3 \ln t}{1 - t^2} - 3 \ln \alpha_1 \right] \tag{7.48}$$

Thus due to the compensation of the electromotive force of the primary field the value of the geometric factor G_1^* turns out to be much smaller than the values of the geometric factor for two-coil probes, $G_1(\alpha_1)$ and $G_1(\alpha_2)$.

Radial responses of three-coil probes characterizing *focusing* features of these systems are presented in Figs. 7.33–7.35.

At the initial part of the radial response, G_1^* has negative values which, due to compensation of the electromotive force of the primary field, are much smaller than those of two-coil induction probes. Near point a_0 when

$$3 \ln \frac{L_1}{a_0} \simeq 2.17 - \frac{3 \ln t}{1 - t^2}$$

geometric factor G_1^* is equal to zero and then monotonically increases, approaching asymptotically to unity. Combination of two factors, such as the compensation of the electromotive force of the primary field and the behavior of the function $G_1(\alpha)$ as $1/\alpha^2$, provide

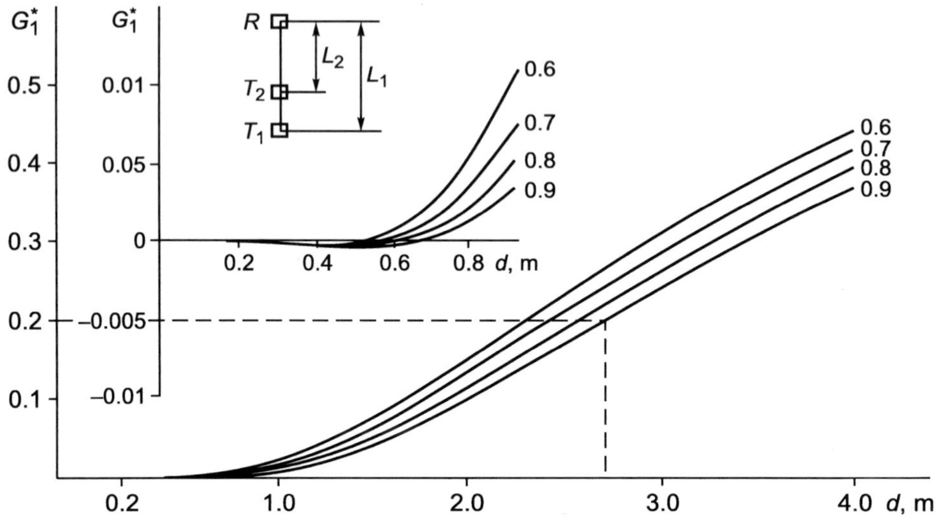

Figure 7.34. Radial responses of a three-coil probe ($L = 1.8$ m). Curve index L_2/L_1.

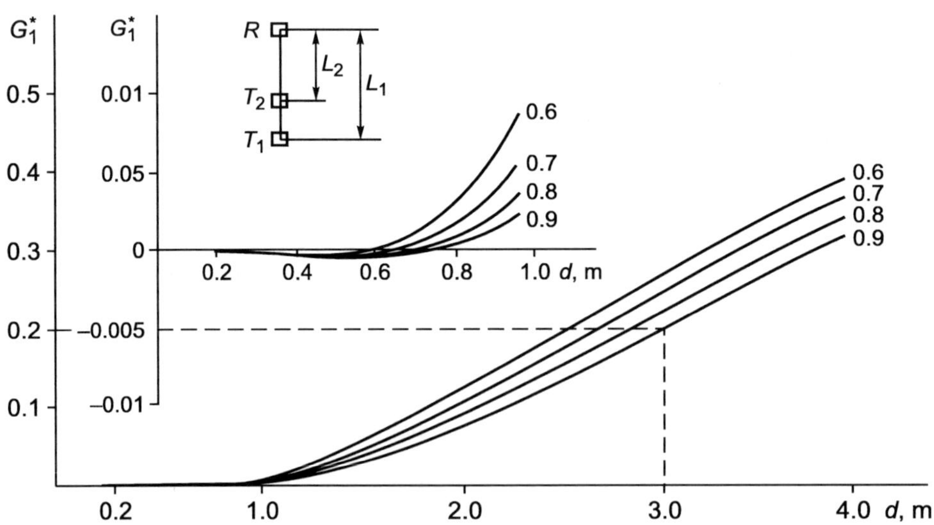

Figure 7.35. Radial responses of a three-coil probe ($L = 2$ m). Curve index L_2/L_1.

a significant reduction of the influence of currents induced in the borehole, as the probe length L_1 is several times greater than the radius a, and with an increase of the probe length the effect of *focusing* manifests itself stronger.

This behavior of the radial response of a three-coil differential probe is of great practical interest for solution of various problems, in particular for the determination of a relatively high resistive formation when the borehole is filled by a strongly mineralized solution ($s = \sigma_2/\sigma_1 < 0.001$). Inasmuch as for obtaining reliable measured signals created by currents in a slightly conductive medium it is necessary to apply relatively high frequencies, when the skin effect in the mineralized solution of the borehole can be noticeable, it is appropriate to use results of calculations by exact formulae.

As an example Table 7.27 contains values of ratio of quadrature component of electromotive force when the probe is located on the borehole axis to that corresponding to a uniform medium with conductivity of the formation ($t = 0.8$). As is seen from the table the skin effect does not practically affect the radial responses of the three-coil probe provided that the thickness of the skin effect, h_1, is related with the borehole radius as:

$$h_1 > 2\sqrt{2}a_1 \tag{7.49}$$

Values of apparent conductivity, σ_a/σ_1, for a three-coil probe ($L_1 = 1.2$ m, $t = 0.833$) demonstrating its *focusing* features are presented in Tables 7.28–7.31. Let us notice that calculations have been performed making use of the exact solution provided that $a_1 = 0.1$ m. As is seen from these tables, if penetration of borehole solution is not very strong ($a_2/a_1 < 4$) the value of σ_a/σ_1 does not practically differ from that corresponding to a uniform medium with the formation resistivity when parameter $\sigma_3\mu\omega a_1 < 0.64 \times 10^{-2}$. In this considered range of parameters the borehole and the invasion zone do not have an influence on the apparent conductivity which is defined by the formation conductivity only. With an increase of the invasion zone radius the difference between σ_a and σ_3 becomes more noticeable.

Comparison of σ_a/σ_2 for a three-coil probe located symmetrically with respect to the formation boundaries with the same function for a two-coil probe shows that if the formation thickness exceeds the probe length the influence of the surrounding medium on the three-coil probe is somewhat greater than that on the two-coil induction probe of the same length (Figs. 7.36–7.38). In this case σ_1 and σ_2 are conductivities of the formation and the surrounding medium. Also the curve of profiling for a three-coil induction probe is shown in Fig. 7.39 which demonstrates only slight asymmetrical behavior with respect to the center of the formation.

Now let us consider the calibration curve for a three-coil probe (Fig. 7.40). For comparison values of quadrature components for both two- and three-coil probes are given in Table 7.32.

It is clear that the range of resistivities measured by a three-coil probe is narrower than that for a two-coil probe.

As was shown before, the better the *focusing* of a multi-coil induction probe the narrower the range of measured resistivities, and correspondingly the same effect is observed for other differential probes. We will characterize the range of measured resistivities by the

TABLE 7.27
Values of function σ_a/σ_1

a_1/h_1 \ α	10	14	18	10	14	18
		$\sigma_2/\sigma_1 = 2\times10^{-3}$			$\sigma_2/\sigma_1 = 1\times10^{-3}$	
$1/\sqrt{2}$	0.550	0.580	0.550	0.459	0.586	0.609
$1/2$	0.730	0.809	0.826	0.589	0.765	0.824
$1/2\sqrt{2}$	0.796	0.888	0.919	0.641	0.828	0.897
$1/4$	0.821	0.916	0.950	0.664	0.852	0.923
$1/4\sqrt{2}$	0.832	0.926	0.962	0.675	0.861	0.932
$1/8$	0.837	0.931	0.966	0.681	0.866	0.936
		$\sigma_2/\sigma_1 = 5\times10^{-4}$			$\sigma_2/\sigma_1 = 2.5\times10^{-4}$	
$1/\sqrt{2}$	0.231	0.522	0.614	−0.252	0.345	0.554
$1/2$	0.294	0.652	0.782	−0.301	0.413	0.679
$1/2\sqrt{2}$	0.329	0.702	0.842	−0.291	0.447	0.726
$1/4$	0.349	0.722	0.864	−0.275	0.465	0.744
$1/4\sqrt{2}$	0.362	0.732	0.872	−0.261	0.475	0.752
$1/8$	0.369	0.738	0.876	−0.252	0.482	0.657

TABLE 7.28
Values of function σ_a/σ_1

	$a_2/a_1 = 4,\ \rho_2/\rho_1 = 8$				$a_2/a_1 = 4,\ \rho_2/\rho_1 = 16$			
$\sigma_1\mu\omega$ \ ρ_3/ρ_1	1	4	16	32	1	4	16	32
0.01	0.910	0.225	0.0587	0.0298	0.905	0.228	0.0572	0.0296
0.02	0.877	0.220	0.0563	0.0283	0.875	0.219	0.0556	0.0282
0.04	0.826	0.209	0.0536	0.0272	0.830	0.207	0.0524	0.0266
0.08	0.770	0.193	0.0496	0.0258	0.768	0.193	0.0490	0.0250
0.16	0.686	0.172	0.0440	0.0216	0.687	0.171	0.0427	0.0214
0.32	0.574	0.144	0.0356	0.0182	0.570	0.143	0.0362	0.0180
0.64	0.433	0.108	0.0260	0.0120	0.433	0.107	0.0260	0.0123

TABLE 7.29
Values of function σ_a/σ_1

	$a_2/a_1 = 4,\ \rho_2/\rho_1 = 32$				$a_2/a_1 = 4,\ \rho_2/\rho_1 = 64$			
$\sigma_1\mu\omega$ \ ρ_3/ρ_1	1	4	16	32	1	4	16	32
0.01	0.910	0.226	0.0570	0.0298	0.907	0.226	0.0569	0.028
0.02	0.882	0.220	0.0545	0.0275	0.874	0.218	0.0550	0.027
0.04	0.830	0.208	0.0520	0.0260	0.830	0.208	0.0520	0.025
0.08	0.772	0.192	0.0481	0.0242	0.768	0.191	0.0478	0.024
0.16	0.684	0.172	0.0428	0.0214	0.682	0.170	0.0427	0.021
0.32	0.570	0.143	0.0353	0.0174	0.590	0.142	0.0357	0.017
0.64	0.440	0.107	0.0260	0.0124	0.428	0.107	0.0255	0.012

TABLE 7.30
Values of function σ_a/σ_1

	$a_2/a_1 = 4,\ \rho_2/\rho_1 = 8$				$a_2/a_1 = 4,\ \rho_2/\rho_1 = 16$			
$\sigma_1\mu\omega$ \ ρ_3/ρ_1	1	4	16	32	1	4	16	32
0.01	0.780	0.207	0.0670	0.0439	0.769	0.200	0.0574	0.033
0.02	0.748	0.203	0.0661	0.0429	0.738	0.192	0.0055	0.032
0.04	0.706	0.189	0.0628	0.0413	0.694	0.181	0.0527	0.031
0.08	0.643	0.176	0.0528	0.0390	0.634	0.166	0.0488	0.029
0.16	0.564	0.155	0.0530	0.0350	0.554	0.145	0.0433	0.026
0.32	0.460	0.128	0.0440	0.0280	0.453	0.119	0.0358	0.021
0.64	0.342	0.095	0.0310	0.0157	0.335	0.088	0.0260	0.014

TABLE 7.31
Values of function σ_a/σ_1

	$a_2/a_1 = 4,\ \rho_2/\rho_1 = 8$				$a_2/a_1 = 4,\ \rho_2/\rho_1 = 16$			
$\sigma_1\mu\omega$ \ ρ_3/ρ_1	1	4	16	32	1	4	16	32
0.01	0.764	0.195	0.0523	0.0286	0.762	0.192	0.0498	0.02
0.02	0.732	0.186	0.0504	0.0276	0.730	0.185	0.0478	0.02
0.04	0.690	0.176	0.0476	0.0262	0.686	0.173	0.0452	0.02
0.08	0.630	0.161	0.0438	0.0242	0.627	0.158	0.0412	0.02
0.16	0.550	0.141	0.0385	0.0214	0.547	0.138	0.0361	0.01
0.32	0.448	0.115	0.0315	0.0175	0.447	0.113	0.0296	0.01
0.64	0.328	0.085	0.0230	0.0143	0.330	0.083	0.0213	0.01

TABLE 7.32
Values of quadrature components ($\times 10^2$)

$\sigma\mu\omega L^2$	0.01	0.02	0.04	0.08	0.16	0.32	0.64	1.28	2.56	5.12
Two-coil probe	0.474	0.929	1.80	3.45	6.48	11.8	20.4	32.7	46.4	53.2
Three-coil probe	0.169	0.328	0.63	1.17	2.17	3.77	6.10	8.48	8.89	3.87

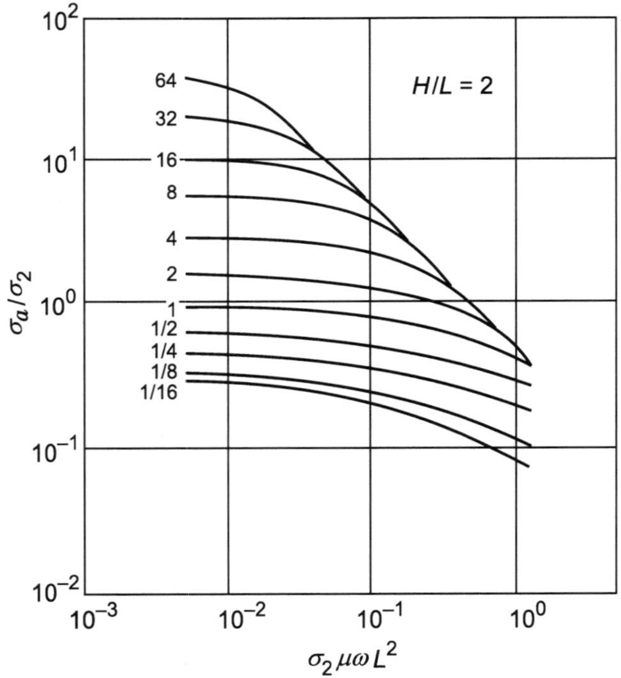

Figure 7.36. Vertical responses of a three-coil probe ($t = 0.833$). Curve index σ_1/σ_2.

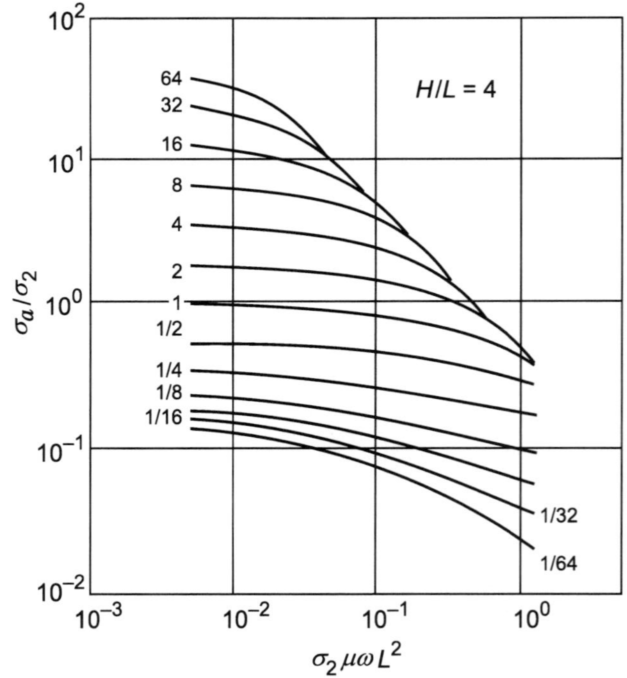

Figure 7.37. Vertical responses of a three-coil probe ($t = 0.833$). Curve index σ_1/σ_2.

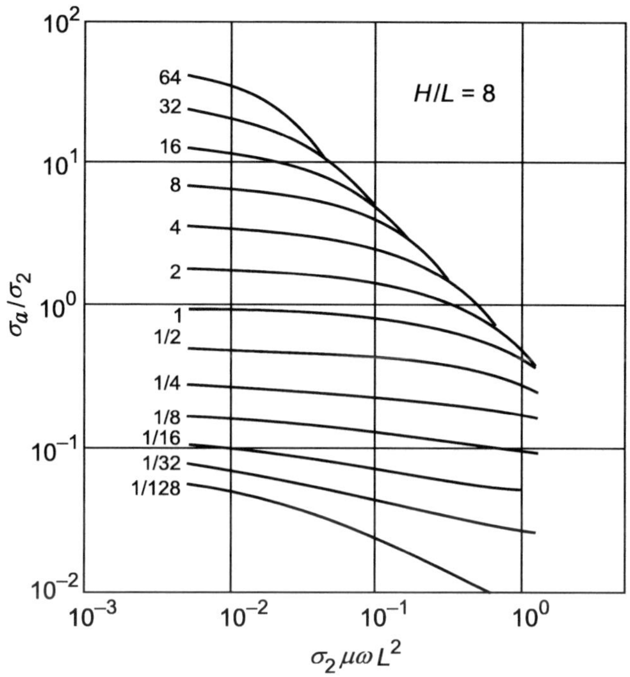

Figure 7.38. Vertical responses of a three-coil probe ($t = 0.833$). Curve index σ_1/σ_2.

ratio:

$$\Delta_\rho = \frac{\rho_{max}}{\rho_{min}} \tag{7.50}$$

where ρ_{max} and ρ_{min} are upper and lower boundary of this range, respectively.

If we assume that $P_{min} = 0.07$ and $P_{max} = 1.4$ ($P = L/h$) then for a two-coil induction probe $\Delta_\rho \simeq 400$. For multi-coil induction probes Δ_ρ depends on parameters of the probe. For example, we have:

$$\Delta_\rho = \begin{cases} 30 & & 6F1M \\ 50 & \text{for} & 4F1 \\ 40 & & 4F1.1 \end{cases}$$

Assuming that for three-coil induction probe $P_{min} = 0.13$ and $P_{max} = 0.8$ we obtain $\Delta_\rho = 40$, i.e. it has a relatively broad range of measured resistivities. Of course, in practice this range is essentially wider.

In conclusion, it is appropriate to notice that three-coil differential probes due to their simplicity can be used for lateral soundings. Examples of sounding curves, calculated for the range of small parameters, are given in Figs. 7.41–7.44. They demonstrate that such soundings can be performed with relatively short probes.

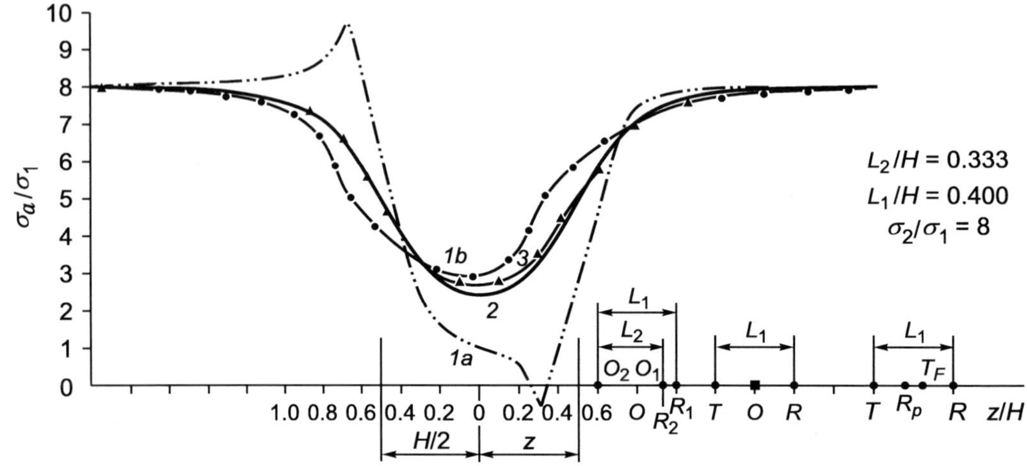

Figure 7.39. Profiling curves of induction probes: (1a) three-coil probe $\mathscr{E}_1^0 L_1^3 = \mathscr{E}_2^0 L_2^3$; (1b) three-coil probe $\mathscr{E}_1^0 = \mathscr{E}_2^0$; (2) two-coil probe; (3) four-coil probe $p = 0.4$, $c = 0.08$.

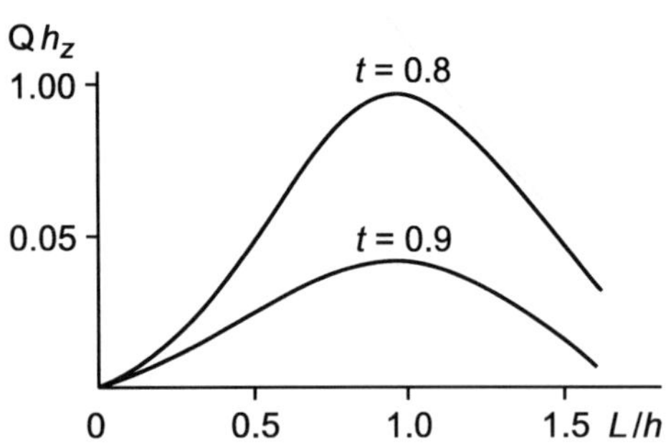

Figure 7.40. Calibration curve for three-coil probes. Curve index t.

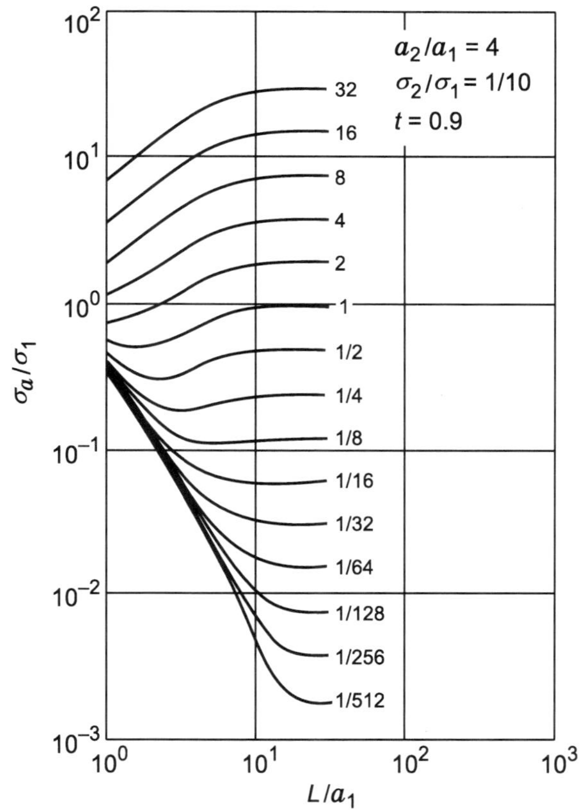

Figure 7.41. Lateral soundings with three-coil probes. Curve index σ_3/σ_1.

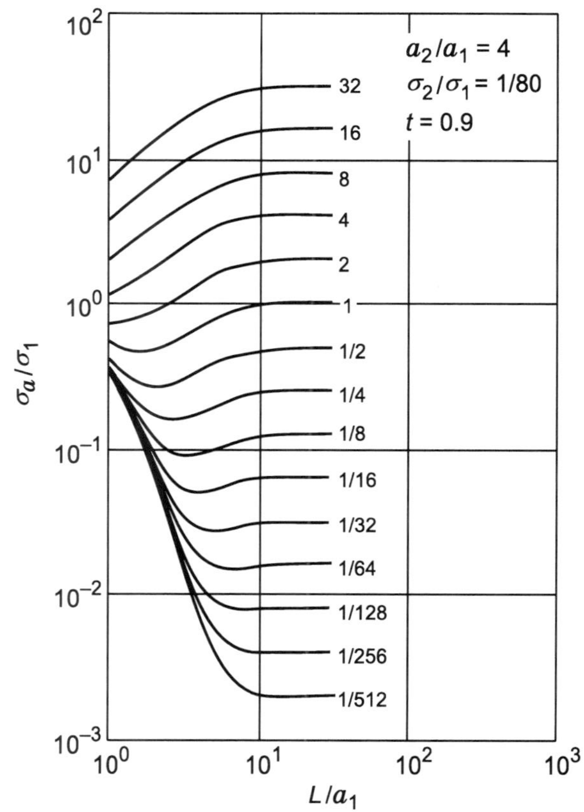

Figure 7.42. Lateral soundings with three-coil probes. Curve index σ_3/σ_1.

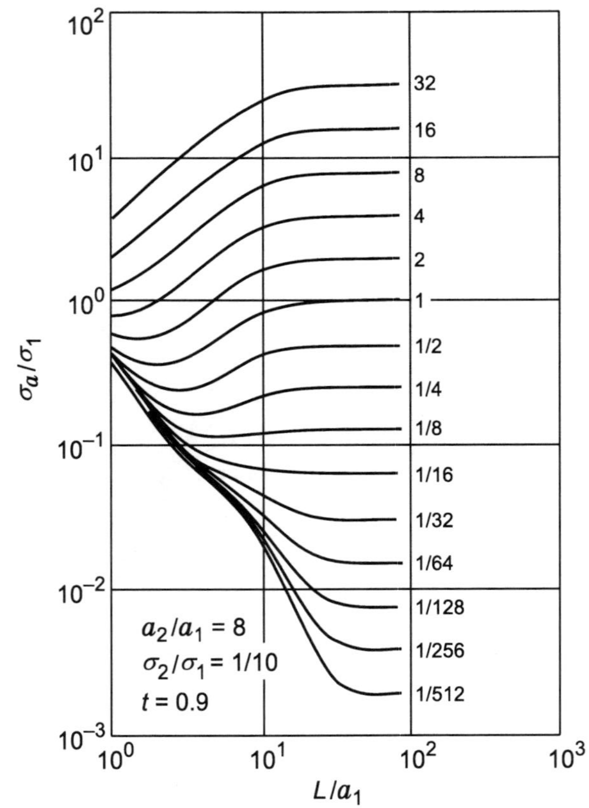

Figure 7.43. Lateral soundings with three-coil probes. Curve index σ_3/σ_1.

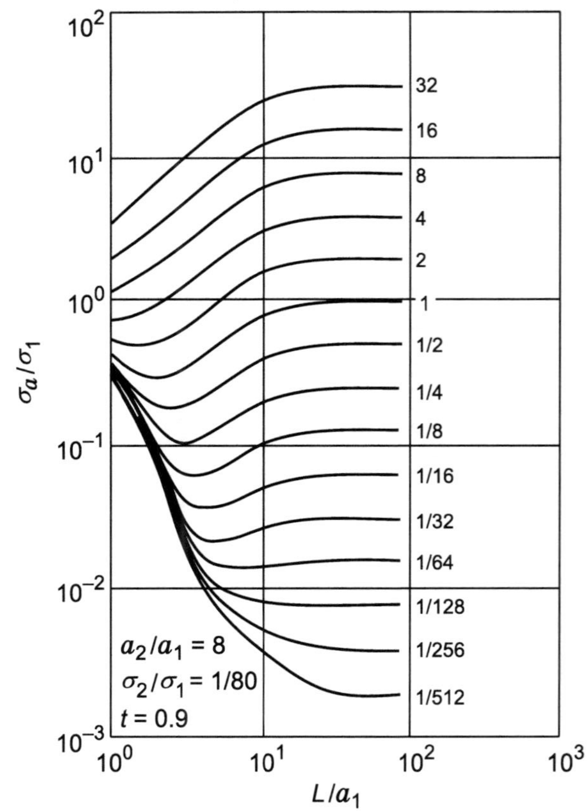

Figure 7.44. Lateral soundings with three-coil probes. Curve index σ_3/σ_1.

7.7. The Influence of Eccentricity on Focusing Features of Multi-coil Induction Probes

In Chapter 4 we investigated an influence of displacement, r_0, from the borehole axis on the radial responses of a two-coil induction probe. These results allow us to consider the dependence of radial responses of multi-coil probes on the value of displacement r_0.

By analogy with the case when a probe is located on the borehole axis the *geometric* factor of the probe, shifted from the axis, is defined in the following way:

$$G_1^*(\alpha_1, \varepsilon, s) = \frac{\sum_{i=1}^{k}\sum_{j=1}^{l} \frac{M_i N_j}{L_{ij}} G_1\left(\frac{L_{ij}}{a_1}, \varepsilon, s\right)}{\sum_{i=1}^{k}\sum_{j=1}^{l} \frac{M_i N_j}{L_{ij}}} \tag{7.51}$$

where $\varepsilon = r_0/a_1$, $s = \sigma_2/\sigma_1$.

If we suppose that all two-coil induction probes forming a multi-coil system are sufficiently long and that we can use the asymptotic expression for *geometric* factor G_1^*, then instead of eq. 7.51 we have:

$$G_1^*(\alpha_1, \varepsilon, s) \simeq \frac{a_1^2\left[1 + \left(\frac{r_0}{a_1}\right)^2 \frac{(s-1)(2s+1)}{(s+1)^2}\right] \sum_{i=1}^{k}\sum_{j=1}^{l} \frac{M_i N_j}{L_{ij}^3}}{\sum_{i=1}^{k}\sum_{j=1}^{l} \frac{M_i N_j}{L_{ij}}} \tag{7.52}$$

Inasmuch as in all considered probes moments of coils and distances between them are chosen in such a manner than the electromotive force of the primary field is compensated, the right part of eq. 7.52 turns out to be zero. Therefore, if the differential probe does not contain relatively short two-coil probes we can expect that such a probe preserves *focusing* features even when it is shifted from the borehole axis within a wide range of parameter σ_2/σ_1.

For illustration radial responses of probes 6F1M, 4F1 and 4F1.1, calculated for three values of borehole radius ($a_1 = 0.1$, 0.125, 0.15 m) and two values of parameter s ($s = 1/32$, $1/2$), are presented in Figs. 7.45–7.48 as a function of the displacement $\varepsilon = r_0/a$. As is seen from the curves the radial response of probe 6F1M does not practically depend on s and ε, if $a_1 = 0.1$ m. It is related with the fact that the shortest two-coil probe forming 6F1M has in this case the relative length $\alpha = 5.0$. If $a_1 = 0.125$ m and $a_1 = 0.15$ m, due to a decrease of the relative length of probes, specially the shortest one, dependence of geometric factor G_1^* on parameters s and ε begins to manifest itself. This behavior occurs sufficiently favorable: the geometric factor of probe 6F1M decreases by absolute value with an increase of ε for $s = 1/32$, as well as for $s = 1/2$. For example if $s = 1/32$ and $a_1 = 0.125$ m function G_1^* has at initial part negative values then changes sign and begins to increase.

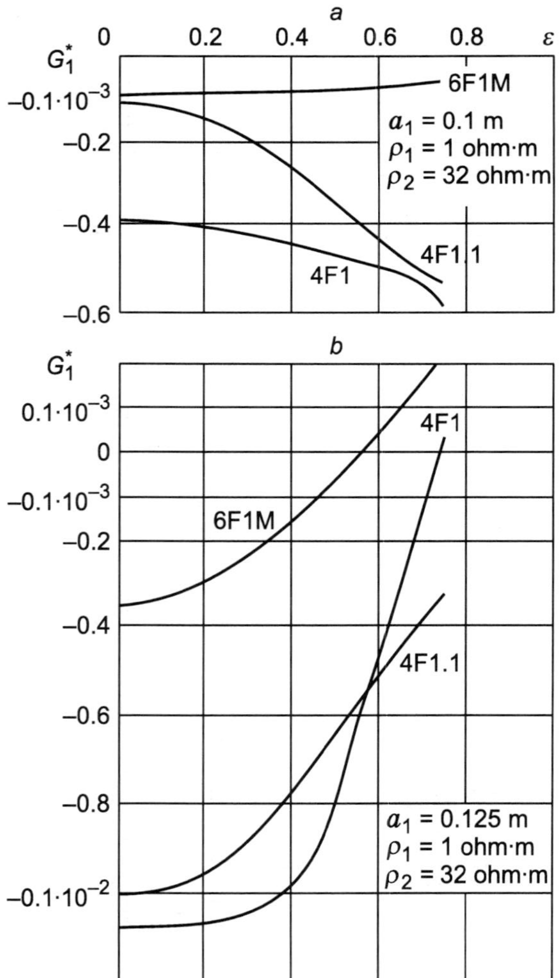

Figure 7.45. Radial responses of differential probes as function of displacement, ε.

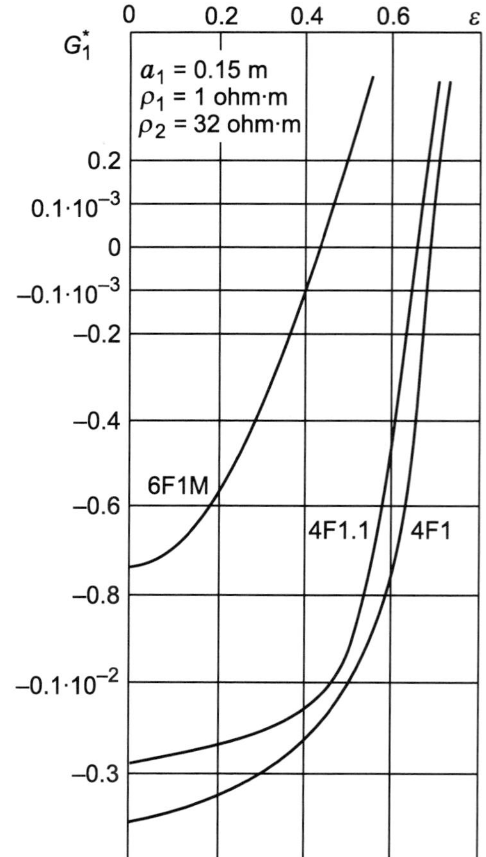

Figure 7.46. Radial responses of differential probes as function of displacement, ε.

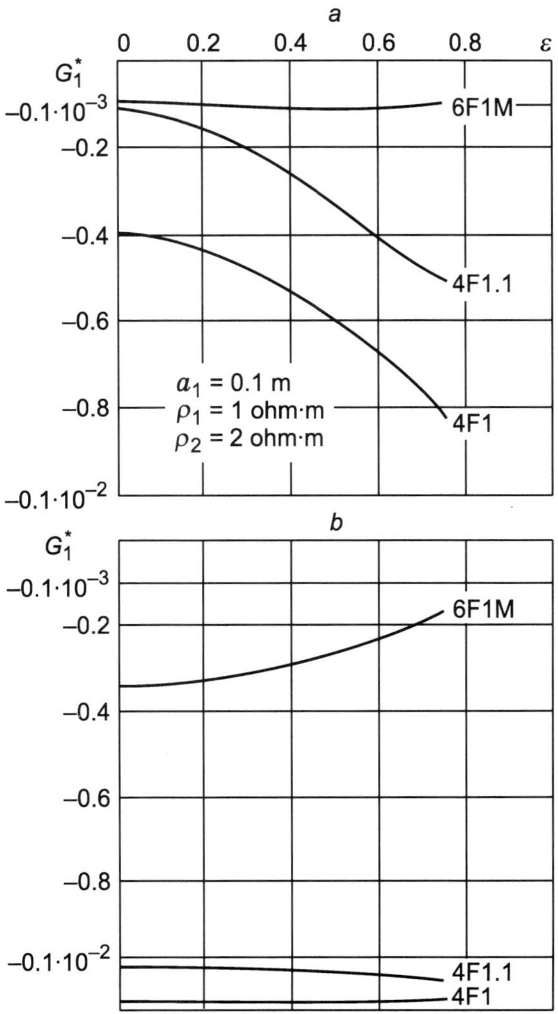

Figure 7.47. Radial responses of differential probes as function of displacement, ε.

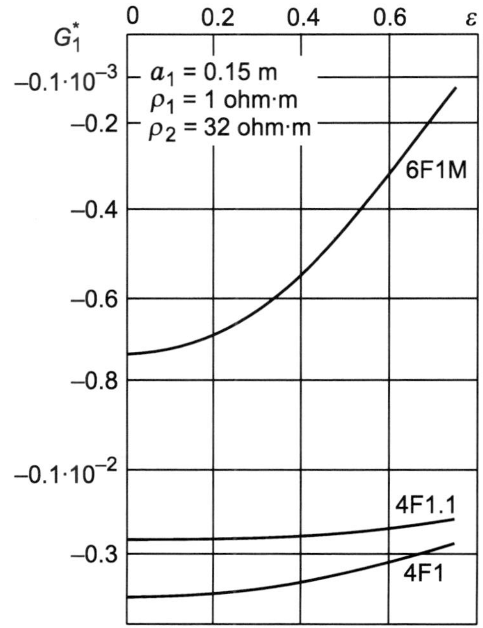

Figure 7.48. Radial responses of differential probes as function of displacement, ε.

TABLE 7.33
Values of function $G_1^*(\varepsilon) \times 10^4$

L_1, m \ ε	0	0.1	0.2	0.3	0.4	0.5
1.2	−1.95	−1.92	−1.85	−1.74	−1.58	−1.39
1.4	−1.26	−1.25	−1.21	−1.16	−1.08	−9.86
1.6	−0.842	−0.836	−0.818	−0.789	−0.748	−0.696
1.8	−0.582	−0.578	−0.568	−0.552	−0.529	−0.497
2.0	−0.413	−0.411	−0.405	−0.396	−0.382	−0.367

The dependence of the geometric factor G_1^* of probes 4F1 and 4F1.1 on parameters a_1, ε and s is displaced in more complicated manner. For instance if $a_1 = 0.1$ m and $s = 1/32$ as well as $s = 1/2$, radial responses of both probes increase by absolute value with an increase of ε, while for $a_1 = 0.125$ m and $a_1 = 0.15$ m as $s = 1/32$ they decrease. At the same time G_1^* for probes 4F1 and 4F1.1 does not practically depend on ε, if $s = 1/32$ and $a_1 = 0.125$ m, or $a_1 = 0.15$ m.

In accord with results obtained in Chapter 4, the geometric factor of a two-coil probe, displaced from the borehole axis, behaves practically as:

$$\frac{f(\varepsilon, s)}{\alpha^2} \quad \text{if } \alpha \gg 1$$

Taking into account that the electromotive force of the primary field is compensated in a three-coil probe we can expect that the displacement of this probe with respect to the borehole axis produces a relatively small effect on the radial response. This fact is illustrated by values of geometric factor $G_1^*(\varepsilon)$ for $s = 1/2$, $a_1 = 0.1$ m, $t = 0.8$, presented in Table 7.33.

7.8. Choice of a Frequency for Differential Probes

Considering the field of the magnetic dipole as well as induced currents in a uniform conducting medium it was established that with an increase of frequency the role of those parts of the medium which are relatively close to the probe increases. For this reason the vertical response of a two-coil induction probe significantly improves, but simultaneously the influence of currents induced in the formation, with respect to those in the borehole and in the invasion zone, becomes lesser. Unlike of a two-coil induction probe the influence of the medium directly surrounding the multi-coil probe is very small, and an increase of frequency up to a certain limit does not practically change its radial response.

Besides, improvement of the vertical response of a multi-coil induction probe due to measurements at higher frequencies allows us to apply induction logging in a more resistive medium, inasmuch as the ratio between the quadrature component of the electromotive force and that of the primary field increases.

In order to define the upper limit of frequencies several factors have to be taken into account. First of all comparison of results of calculations based on the exact solution and the approximate theory, described in Chapter 3, permits us to establish the maximal frequency when the radial response is not yet distorted. In fact two main assumptions form this theory, namely:

- The skin effect is absent in the borehole and in the invasion zone.
- In the formation the skin effect manifests itself in the same manner as in a uniform medium with conductivity σ_3.

For this reason, coincidence of results of calculations by both methods establishes maximal frequencies for which, first of all, geometric factors of the borehole and of the invasion zone correctly describe the influence of these parts of the medium, and secondly, the electromotive force induced by the magnetic field of currents in the formation does not depend on currents induced in the borehole and in the formation.

Also, a choice of a frequency is defined by *focusing* features of the probe. For example, for multi-coil probes having a greater depth of investigation in a relatively conductive medium the frequency should be smaller. In particular, it was established that for probe 1.L–1.2 and three-coil probe of the same length the maximal frequency is defined from the relation:

$$f_{max} \leqslant (2.0 - 2.2) \times 10^5 \rho_{min}$$

If we assume that $\rho_{min} = 1$ ohm·m, the maximal frequency can be increased from 60 kHz to 220 kHz. An increase of frequency of almost four times results in a significant improvement of the vertical response of the probe, specially in a low-resistive medium, and it also allows us to perform reliable measurements in a more resistive medium.

An additional limitation on the choice of the maximal frequency is related with the fact that the quadrature component of the electromotive force as a function of the formation conductivity has a maximum, and in order to avoid nonuniqueness it is necessary that the range of conductivities should correspond to the ascending branch of this response of EMF.

Finally, it is appropriate to notice that the upper limit of measured resistivities is defined by instrumental problems as well as the possible influence of the dielectric constant.

7.9. Determination of the Coefficient of Differential Probes

First, let us assume that a two-coil induction probe, which has coil dimensions that are much smaller than its length, is located in a uniform medium and the skin depth is much greater than the separation between coils. Then, as was shown in Chapter 2, for the quadrature component of the electromotive force we have:

$$\mathscr{E} = \frac{\mu^2 \omega^2 MN}{4\pi L} \sigma \tag{7.53}$$

i.e. the EMF depends on conductivity, frequency, the probe length and coil moments, M and N.

For this reason in two probes characterized by various lengths or coil moments or both of them, different electromotive forces will be induced in the same medium.

It is natural to eliminate the influence of these factors, depending on probe parameters, and introduce apparent conductivity which in a uniform medium coincides with its conductivity when parameter L/h is sufficiently small. In fact this approach has been used widely in several chapters of this monograph. In particular, in accord with eq. 7.53 we have:

$$\sigma_a = \frac{4\pi L}{\mu^2\omega^2 MN}\mathscr{E} = K\mathscr{E}$$

where

$$K = \frac{4\pi L}{\mu^2\omega^2 MN} \tag{7.54}$$

K is the probe coefficient and unlike corresponding coefficients of electric logging it depends not only on the probe length but also on coil moments.

Equation 7.54 can be written as:

$$K = \frac{2}{\mu\omega L^2 \mathscr{E}_0} \tag{7.55}$$

where $\mathscr{E}_0$ is the electromotive force of the primary field.

As was shown above, in a nonuniform medium apparent conductivity, σ_a, depends on the distribution of resistivity and the probe length. For interpretation it is appropriate to present results of a solution of the forward problem as well as experimental data as a function of apparent conductivity, the probe length and parameters of a medium. Knowing the probe length and the electromotive force of the primary field, it is a simple matter to calculate the probe coefficient and transform the quadrature component of the electromotive force into the apparent conductivity. In accord with eq. 7.12 the coefficient of the coil probe can be written as:

$$K = \frac{4\pi}{\mu^2\omega^2 \sum\limits_{i=1}\sum\limits_{j=1} M_i N_j / L_{ij}} \tag{7.56}$$

and it depends on distance between coils and values of their moments.

Let us present the probe coefficient through the electromotive force of the primary field of one of two coil probes, for example, L_{11}. Then, instead of eq. 7.56 we have:

$$K = \frac{2}{\mu\omega L_{11}^2 \mathscr{E}_0^{11} \sum\limits_{i=1}\sum\limits_{j=1} c_i c_j / P_{ij}} \tag{7.57}$$

where $\mathscr{E}_0^{11}$ is the electromotive force of the primary field of a two-coil probe with length L_{11}; P_{ij} is the ratio of length, L_{ij}, of a two-coil probe to L_{11}; c_i and c_j are ratios of coil moments to the moment of one of the coils of the probe with length L_{11}.

In accord with eq. 7.57 for determination of the probe coefficient it is necessary to measure the EMF of the primary field of one of the two-coil probes that requires switching off two coils of the system. Since this is impractical, calculation of the coefficient of multi-coil probes can be performed by using eq. 7.56 only.

In deriving the formula for the probe coefficient it was assumed that the radius and height of coils are many times smaller than the corresponding length of the two-coil probe. In other words, the EMF induced by currents in a medium depends on the distance between coil centers and their moments only. Under certain conditions the influence of finite dimensions of coils on the probe coefficient is insignificant, and it can be neglected. But in general, coil sizes have to be taken into account in calculating the probe coefficient and therefore the apparent conductivity. Let us consider this question in more detail.

We will assume that coils of a two-coil probe present themselves as layered ones placed on a nonconducting base. Then, in accord with results obtained in Chapter 4 for the quadrature component of the electromotive force at the range of small parameters we have:

$$Q\mathscr{E} = \frac{32\omega\mu}{4\pi}\pi n_T n_R r^2 \sigma_\mu \omega I \int\limits_0^\infty \frac{r}{2\lambda}\left[2K_0K_1 - \lambda r(K_1^2 - K_0^2)\right]$$
$$\times \frac{\sin(\lambda l/2)\sin(\lambda b/2)}{\lambda^2} I_1^2(\lambda r)\cos\lambda L_0 \, \mathrm{d}\lambda$$

where n_T and n_R are the number of turns per unit of length of transmitter and receiver coils; I is current; l and b are coil lengths; L_0 is the distance between coil centers; r is the coil radius.

Let us introduce the following notations: $\lambda r = m$, $n_T\pi r^2 I$ and $n_R\pi r^2$ are moments of transmitter and receiver coils per unit of the length, M_T^0, M_R^0, respectively; $l/2r = s_1$, $b/2r = s_2$, $L_0/r = \alpha$.

After simple transformations we have:

$$Q\mathscr{E} = \sigma\frac{8\omega^2\mu^2}{\pi^2} r M_T^0 M_R^0 \int\limits_0^\infty \phi(m)\frac{\sin s_1 m}{m}\frac{\sin s_2 m}{m}\frac{I_1^2(m)}{m^2}\cos m\alpha \, \mathrm{d}m \tag{7.58}$$

Whence, for the coefficient of the probe having single layered coils we have:

$$K = \frac{\pi^2}{8\omega^2\mu^2 r M_T^0 M_R^0 \int\limits_0^\infty \phi(m)\frac{\sin s_1 m}{m}\frac{\sin s_2 m}{m}\frac{I_1^2(m)}{m^2}\cos m\alpha \, \mathrm{d}m}$$

where

$$\phi(m) = \frac{m}{2}\left[2K_0(m)K_1(m) - m(K_1^2 - K_0^2)\right]$$

The expression for coefficient K, when coils are infinitely small, can be obtained assuming that the linear dimensions of the coils are much smaller than the probe length. In this case $\alpha = L_0/r \to \infty$, and the integral in eq. 7.58 is defined by the behavior of the integrand for small values of m.

If $m \to 0$ then:

$$\frac{\sin s_1 m}{m} \to s_1 \qquad \frac{\sin s_2 m}{m} \to s_2 \qquad \frac{I_1^2(m)}{m^2} \to \frac{1}{4} \text{ and } \phi(m) \to K_0(m)$$

Therefore we have:

$$\int_0^\infty \phi(m) \frac{\sin s_1 m}{m} \frac{\sin s_2 m}{m} \frac{I_1^2(m)}{m^2} \cos m\alpha \, dm \to \frac{s_1 s_2}{4} \int_0^\infty K_0(m) \cos m\alpha \, dm$$

As is well known:

$$\frac{1}{(1+\alpha^2)^{1/2}} = \frac{2}{\pi} \int_0^\infty K_0(m) \cos m\alpha \, dm$$

Whence:

$$\int_0^\infty \phi(m) \frac{\sin s_1 m}{m} \frac{\sin s_2 m}{m} \frac{I_1^2(m)}{m^2} \cos m\alpha \, dm = \frac{s_1 s_2}{8\sqrt{1+\alpha^2}} \simeq \frac{\pi s_1 s_2}{8\alpha}$$

Substituting this value of the integral into eq. 7.58 we obtain the known expression K for point sensors:

$$K = \frac{4\pi}{\omega^2 \mu^2 M_T M_R L}$$

where $M_T = M_T^0 l$, $M_R = M_R^0 b$.

If the height of both coils is the same, then $s_1 = s_2 = s$, and we have:

$$K = \frac{\pi^2}{8\omega^2 \mu^2 r M_T^0 M_R^0 \int_0^\infty \phi(m) \frac{\sin^2 ms}{m^2} \frac{I_1^2(m)}{m^2} \cos m\alpha \, dm} \tag{7.59}$$

For a multi-coil probe located in a uniform medium with conductivity σ, the quadrature component of the electromotive force in measuring coils is defined from relation:

$$\mathrm{Q}\,\mathscr{E} = \sum_n \mathrm{Q}\,\mathscr{E}_n = \sum_n \frac{1}{K_n}\sigma = \sigma \sum_n \frac{1}{K_n}$$

where $Q\mathscr{E}_n$ and K_n are electromotive forces and coefficients of n two-coil probes forming the differential system.

Therefore, the coefficient of the multi-coil induction probe can be presented in the form:

$$K = \left(\sum_n \frac{1}{K_n}\right)^{-1} \tag{7.60}$$

and

$$\sigma_a = K\,Q\mathscr{E}$$

where $Q\mathscr{E}$ is the electromotive force, induced in receiver coils of a *focusing* probe.

As follows from eq. 7.58 determination of the probe coefficient, K, is related with calculation of the integral in the denominator of this equation.

For the case when coil lengths are the same, values of the integral:

$$\int_0^\infty \phi(m)\frac{\sin^2 ms}{m^2}\frac{I_1^2(m)}{m^2}\cos m\alpha\, dm \tag{7.61}$$

and its asymptotic values are given in Table 7.34.

With an increase of the probe length the value of the integral approaches its asymptote and, within certain limits, the influence of coil length is negligible. In known probes finite dimensions of coils do not practically influence the value of the probe coefficient. For example, if $a_1 = 0.1$ m, the four-coil differential probe 1.L–1.2 has the following parameters: $\alpha_1 = 24$, $\alpha_2 = 9.6$, $\alpha_3 = 4.8$, $s_{max} = 1$, $p = 0.4$ and $c = 0.05$. Taking into account the ratio between turns of coils we can see that the probe coefficient is the same as that for the probe with infinitely small coils. This is true to an even greater extent for three-coil differential probes with length exceeding, by at least four-five times, the borehole radius.

If the probe length is sufficiently small and the dimension of coils has to be taken into account, the determination of coefficient K is defined by eq. 7.58.

In conclusion of this chapter let us make several comments.

- The use of differential multi-coil probes is the most conventional approach in application of induction logging.
- Due to the use of these probes induction logging in most cases has the greatest depth of investigation among other logging tools.
- Any multi-coil induction probe, regardless of the amount of transmitter and receiver coils, by no means performs focusing of the field as it takes place, for example, in optics. In effect, every induction probe, except a two-coil one, is a differential system measuring such a difference of signals in receivers that the influence of induced currents in the borehole and in the invasion zone is significantly reduced.

TABLE 7.34
Values of integral (7.61)

α	$s = 0.2$		$s = 0.4$		$s = 0.8$	
	Accurate	Approx.	Accurate	Approx.	Accurate	Approx.
1	0.811×10^{-2}	0.157×10^{-1}	0.325×10^{-2}	0.626×10^{-1}	0.130	0.250
$\sqrt{2}$	0.711	0.111	0.285	0.444	0.116	0.177
2	0.594	0.782×10^{-2}	0.239	0.314	0.974×10^{-1}	0.125
$2\sqrt{2}$	0.472	0.556	0.190	0.222	0.773	0.885×10^{-1}
4	0.359	0.391	0.144	0.156	0.585	0.624
$4\sqrt{2}$	0.265	0.278	0.106	0.111	0.429	0.443
8.0	0.192	0.196	0.768×10^{-2}	0.782×10^{-2}	0.309	0.312
$8\sqrt{2}$	0.137	0.139	0.549	0.560	0.220	0.221
16.0	0.976×10^{-3}	0.978×10^{-3}	0.391	0.391	0.156	0.156
$16\sqrt{2}$	0.692	0.700	0.279	0.280	0.111	0.111
32	0.490	0.489	0.196	0.196	0.784×10^{-2}	0.780×10^{-2}

α	$s = 1.6$		$s = 2.0$		$s = 4.0$	
	Accurate	Approx.	Accurate	Approx.	Accurate	Approx.
1	0.483	1.000	0.718	1.570	2.27	6.28
$\sqrt{2}$	0.454	0.710	0.684	1.110	2.23	4.44
2	0.404	0.500	0.625	0.783	2.14	3.14
$2\sqrt{2}$	0.330	0.355	0.530	0.556	2.00	2.22
4	0.247	0.250	0.403	0.392	1.76	1.57
$4\sqrt{2}$	0.177	0.177	0.285	0.278	1.40	1.11
8.0	0.126	0.125	0.200	0.196	0.933	0.784
$8\sqrt{2}$	0.890×10^{-1}	0.890	0.140	0.139	0.602	0.560
16.0	0.629	0.625×10^{-1}	0.980×10^{-1}	0.986×10^{-1}	0.408	0.392
$16\sqrt{2}$	0.444	0.445	0.695	0.695	0.283	0.280
32	0.314	0.313	0.491	0.490	0.198	0.196

- Interpretation of induction logging data measured by such differential probes is mainly based on measuring the quadrature component of the electromotive force, shifted by 90° with respect to the primary electromotive force.
- In such cases when the depth of investigation of the multi-coil probe is not sufficient in the radial direction, and correspondingly the apparent conductivity differs from the formation conductivity, in order to perform interpretation it is necessary to have additional information derived from either other induction probes or applying different logging methods. It is obvious that similar types of problems arise when the influence of the surrounding medium becomes essential.

Chapter 8

INDUCTION LOGGING BASED ON MEASURING THE INPHASE COMPONENT OF THE SECONDARY FIELD OR THE QUADRATURE COMPONENT DIFFERENCE OF TYPE $\mathrm{Q}\,\boldsymbol{H}_Z(\omega_1) - \omega_1/\omega_2\,\mathrm{Q}\,\boldsymbol{H}_Z(\omega_2)$

As is well known with the help of multi-coil differential probes we can reduce significantly the influence of the borehole and, often, the invasion zone, if the conductivity and the radius of the latter are not large enough. In other words, the depth of investigation of such probes in the radial direction essentially depends on the geoelectric parameters of the medium. Also, in calculating coil moments and their position for these probes it is assumed that the resistivity of the medium is only a function of the distance from the borehole axis, i.e. it changes only in the radial direction.

In fact, the integral response, as well as the differential one, defining a signal in receiver coils due to induced currents in an arbitrary cylindrical layer with a constant resistivity, present the basic element of these calculations. However, the presence of caverns, deviation from radial distribution of resistivity because of nonuniform penetration of a borehole filtrate into a formation, its finite thickness are factors which can influence the *focusing* features of multi-coil induction probes. In order to eliminate the influence of these factors and to increase the depth of investigation, regardless of the geoelectric section, we will consider in this chapter another approach, based on the use of a two-coil probe and a simultaneous measurement at two or more frequencies if the quadrature component is measured.

Analysis of the electromagnetic field in a uniform medium as well as in media with cylindrical and horizontal interfaces, performed in previous chapters, has shown that a conducting medium can always be divided in two ranges, namely, the internal area which is directly adjacent to the probe, and the external area. Within the internal area (near zone) induced currents are shifted in phase by 90°, and their density is only defined by the primary magnetic flux and it is directly proportional to the frequency. Correspondingly, the skin effect is negligible in this area. In general the resistivity within the internal area alters from point to point. In the external area there is an interaction between currents, but the influence of currents within the internal area on the distribution of currents in the external area is small. With a decrease of frequency and medium conductivity the dimensions of the internal area increase and there is a frequency for which those parts of the borehole and the invasion zone, which define a signal measured by the probe, fall into the internal area. At the same time the boundary with the external area is located within the formation where the skin effect manifests itself as in a uniform medium with the conductivity of the formation.

For example, at the range of small parameters in accord with eq. 3.135 the field H_z on the borehole axis in media with two coaxial cylindrical interfaces can be presented in the form:

$$\mathrm{Q}\,H_z = \frac{M}{2\pi L^3}\left[\mathrm{Q}\,h_z^{un}(k_3L) + \frac{\omega\mu L^2}{2}(\sigma_1-\sigma_3)G_1(\alpha) + \frac{\omega\mu L^2}{2}(\sigma_2-\sigma_3)G_2(\alpha)\right] \quad (8.1)$$

$$\mathrm{In}\,H_z = \frac{M}{2\pi L^3}\,\mathrm{In}\,h_z^{un}(k_3L) \qquad \text{if } |k_3L|<1 \quad (8.2)$$

where $h_z^{un}(k_3L)$ is the vertical component of the magnetic field in a uniform medium with conductivity of the formation, normalized by the primary field.

Thus, in this case, the borehole and the invasion zone fall into the internal area where induced currents generate only the quadrature component of the magnetic field which is directly proportional to the frequency, while the inphase component of the secondary field is caused only by currents induced in the formation.

Let us assume here that parameter $|k_3L|<1$. It is appropriate to notice that this condition usually takes place in the practice of induction logging. Then in accord with results obtained in Chapter 2 we can write:

$$\begin{aligned}\mathrm{Q}\,h_z^{un}(k_3L) &\simeq \frac{\sigma_3\omega\mu L^2}{2} - \frac{L^3}{3\sqrt{2}}(\sigma_3\mu\omega)^{3/2}+\cdots\\ \mathrm{In}\,h_z^{un}(k_3L) &\simeq -\frac{L^3}{3\sqrt{2}}(\sigma_3\mu\omega)^{3/2}+\cdots\end{aligned} \quad (8.3)$$

This analysis along with eqs. 8.1–8.3 shows that there are at least two ways of eliminating the influence of the internal area with the help of a two-coil probe, regardless of its length. First of all, as follows from eq. 8.2, in measuring the inphase component of the secondary field H_z, the borehole and the invasion zone become transparent. Correspondingly, this component, H_z^s, turns out to be a function of the formation conductivity or, in general, of the conductivity of the external area which can include both conductivities of the formation and that of the surrounding medium.

Thus, the induction probe measuring the inphase component of the secondary field possesses depth of investigation in the radial direction, which can be much greater than that when the quadrature component is measured, provided that the frequency is properly chosen.

Moreover, in the range of small parameters this component of the field is more sensitive to a change of the formation conductivity than in the case of the quadrature component:

$$\begin{aligned}\mathrm{Q}\,H_z &\sim \sigma_3\\ \mathrm{In}\,H_z^s &\sim \sigma_3^{3/2}\end{aligned} \quad (8.4)$$

However, it is reasonable here to notice the following. Inasmuch as the inphase component of the secondary field is much smaller than the primary field, H_z^0, measurements of this component are related with very serious technical problems, even the field H_z^0 is usually compensated with relatively high accuracy. For this reason let us investigate the second approach, which also eliminates the influence of induced currents within the internal area. Unlike the first one it is based on measuring the quadrature component of the field which is shifted in phase by 90° with respect to the primary field, which to a certain extent facilitates measurements.

We will suppose that two frequencies are chosen in such a way that the boundary with the external area passes through a formation. Then, expressions for the field, $Q\,h_z$, in a two-coil probe can be written as:

$$Q\,h_z(\omega_1) = \frac{\omega_1 \mu L^2}{2}\sigma_a^i + f(\omega_1, \sigma_3, L) \tag{8.5}$$

$$Q\,h_z(\omega_2) = \frac{\omega_2 \mu L^2}{2}\sigma_a^i + f(\omega_2, \sigma_3, L) \tag{8.6}$$

where L is the probe length, and σ_a^i is a function which does not depend on frequency, and it presents itself as the apparent conductivity of the medium of the internal area when the resistivity of the external area is equal to infinity. Function $f(\omega, \sigma_3, L)$ depends on frequency, formation conductivity and probe length only.

Correspondingly, for the electromotive force we have:

$$\begin{aligned} Q\,\mathscr{E}(\omega_1) &= \frac{M_T^0 M_R I(\omega_1)}{2\pi L^3}\left[\frac{\omega_1^2 \mu L^2}{2}\sigma_a^i + \omega_1\, f(\omega_1, \sigma_3, L)\right] \\ Q\,\mathscr{E}(\omega_2) &= \frac{\mu M_T^0 M_R I(\omega_2)}{2\pi L^3}\left[\frac{\omega_2^2 \mu L^2}{2}\sigma_a^i + \omega_2\, f(\omega_2, \sigma_3, L)\right] \end{aligned} \tag{8.7}$$

where $Q\,\mathscr{E}(\omega_1)$ and $Q\,\mathscr{E}(\omega_2)$ are quadratic components of the electromotive force in the receiver at frequencies ω_1 and ω_2, respectively; $I(\omega)$ is the current in the transmitter coil; M_T^0 is the transmitter coil moment, when the magnitude of the current is unity.

Therefore, in accord with eqs. 8.5–8.7 we can choose such frequencies ω_1 and ω_2 that function:

$$\frac{Q\,H_z(\omega_2)}{\omega_2 I(\omega_2)} - \frac{Q\,H_z(\omega_1)}{\omega_1 I(\omega_1)} \tag{8.8}$$

does not depend on parameters of the internal area. Function:

$$\frac{Q\,\mathscr{E}(\omega_2)}{\omega_2^2 I(\omega_2)} - \frac{Q\,\mathscr{E}(\omega_1)}{\omega_1^2 I(\omega_1)} \tag{8.9}$$

also has the same feature.

The following two functions also do not depend on distribution of conductivities within the internal area:

$$Q\,\mathscr{E}(\omega_2) - \left(\frac{\omega_2}{\omega_1}\right)^2 Q\,\mathscr{E}(\omega_1) \qquad \text{if } I(\omega_1) = I(\omega_2) \tag{8.10}$$

and

$$\Delta\,\mathrm{Q}\,\mathscr{E} = \mathrm{Q}\,\mathscr{E}(\omega_2) - \mathrm{Q}\,\mathscr{E}(\omega_1) \qquad \text{if } \omega_1^2 I(\omega_1) = \omega_2^2 I(\omega_2) \tag{8.11}$$

In the case when frequencies ω_1 and ω_2 are chosen in such a way that both borehole and invasion zone are located within the internal area, functions 8.8–8.11 depend only on the formation conductivity and coincide with the corresponding functions for a uniform medium.

Thus, the second approach of eliminating the influence of the borehole and the invasion zone can be realized by using simultaneously currents in the transmitter coil at two frequencies and measuring in the receiver coil either

$$\mathrm{Q}\,\mathscr{E}(\omega_2) - \left(\frac{\omega_2}{\omega_1}\right)^2 \mathrm{Q}\,\mathscr{E}(\omega_1) \qquad \text{if } I(\omega_1) = I(\omega_2)$$

or the difference of electromotive forces:

$$\mathrm{Q}\,\mathscr{E}(\omega_2) - \mathrm{Q}\,\mathscr{E}(\omega_1) \qquad \text{if } \omega_1^2 I(\omega_1) = \omega_1^2 I(\omega_2)$$

It is not difficult to understand that if the internal area has caverns of current lines these functions still depend on the formation conductivity only. In fact, such inhomogeneities cause the appearance of electrical charges within the internal area, the density of which is directly proportional to the primary electrical field, i.e. frequency. Respectively, the secondary electric field, currents and finally the magnetic field, which are due to these charges, are directly proportional to the frequency also, and therefore functions given by equations 8.8–8.11 are not subjected to the influence caused by the presence of inhomogeneities.

In previous chapters we have investigated the distribution of induced currents in a conducting medium and demonstrated a different sensitivity of quadrature and inphase components of the magnetic field to internal and external parts of the medium. Let us show that all these results are confirmed by the analysis of the low-frequency part of the spectrum.

First of all, we will consider the vertical component of the magnetic field on the dipole axis in a uniform medium (Chapter 3):

$$H_z = \frac{M_T}{2\pi L^3} e^{ikL}(1 - ikL)$$

Expanding the exponent in a series, we have:

$$H_z = \frac{M_T}{2\pi L^3}\left[1 - \sum_{m=2}^{\infty} \frac{m-1}{m!}(ikL)^m\right] \tag{8.12}$$

and for sufficiently small values of $|kL|$ we obtain the following asymptotic formulae:

$$\mathrm{In}\, H_z \simeq \frac{M_T}{2\pi L^3} - \frac{M_T}{6\pi\sqrt{2}}(\sigma\mu\omega)^{3/2} - \frac{M_T}{16\pi}(\sigma\mu\omega)^2 L \tag{8.13}$$

$$Q\,H_z \simeq \frac{\sigma\mu\omega M_T}{4\pi L} - \frac{M_T}{6\pi\sqrt{2}}(\sigma\mu\omega)^{3/2} + \frac{M_T}{60\pi\sqrt{2}}(\sigma\mu\omega)^{5/2}L^2 \tag{8.14}$$

The leading terms in eqs. 8.13 and 8.14 describe the primary field of the magnetic dipole in a free space and the secondary field of induced currents when the skin effect can be neglected. The first term for the quadrature component is larger when an observation point is closer to the dipole. It means that the quadrature component of the current, which is proportional to frequency, is generated by the primary field only, and it is concentrated mainly near the dipole, i.e. within the internal area.

However, the second terms of series 8.13 and 8.14 do not depend on the probe length, L, which can be explained by the fact that sources of this part of the field are located at distances much greater than the probe length, i.e. in the external area. For this reason measuring the inphase component of the secondary field, or the second term of the series describing the low-frequency spectrum of the quadrature component, provide the great depth of investigation in the radial direction. As is well known, the behavior of the low-frequency spectrum of the field in a nonuniform medium is similar to that in a uniform one. In fact, as was demonstrated in Chapter 4, for the magnetic borehole axis we have:

$$\begin{aligned} \text{In}\,H_z &\simeq \frac{M}{2\pi L^3}(1 + b_1\omega^{3/2} + b_2\omega^2 + b_3\omega^{5/2} + \cdots) \\ Q\,H_z &\simeq \frac{M}{2\pi L^3}(a_0\omega + a_1\omega^{3/2} + a_2\omega^{5/2} + \cdots) \end{aligned} \tag{8.15}$$

where coefficients a_0, a_2, b_2 and b_3 are functions of parameters of borehole, invasion zone and formation, while a_1 and b_1 depend only on the formation conductivity, they are equal to each other and coincide with corresponding coefficients for a uniform medium.

Thus, this method of increasing the depth of investigation of induction logging can be considered as the method of measuring the second term of the series, describing the low-frequency spectrum of the quadrature component, when we can neglect in the expansion terms containing ω of larger power than $\omega^{3/2}$.

In principle, as was pointed out above, measuring the inphase component at one frequency corresponding to the low-frequency spectrum provides the same depth of investigation. At the same time, it is reasonable to notice that inasmuch as first terms of series for the inphase component decrease slower than those of series for the quadrature one, in measuring the inphase component it is necessary to use lower frequencies in order to provide the same depth of investigation.

Now making use of results of calculation of the field on the borehole axis let us determine parameters of a medium and frequencies when this method allows us to eliminate the influence of the borehole and the invasion zone.

In accord with equality:

$$\sigma_a = \frac{1}{\omega\mu L^2}\,Q\,h_z$$

and eqs. 8.5 and 8.6 the difference:

$$\Delta\left(\frac{\sigma_a}{\sigma_1}\right) = \frac{\sigma_a}{\sigma_1}(\omega_1) - \frac{\sigma_a}{\sigma_1}(\omega_2)$$

does not depend on the conductivity of the internal area, if the boundary with the external area passes through the formation.

For illustration of the efficiency of this method values of function $\Delta(\sigma_a/\sigma_1)$ for various parameters of a medium are presented in Tables 8.1–8.18, when $\omega_2 = 4\omega_1$.

Proceeding from these data let us briefly discuss three cases.

Case 1

For a relatively small penetration of the borehole filtrate into a formation ($a_2 = 2a_1$) we can eliminate the influence of the borehole and the invasion zone with a sufficiently short two-coil probe, if $\sigma_3\mu\omega_2 a_1^2 < 3.2 \times 10^{-3}$ or $f < 4 \times 10^4 \rho_3$ as $a_1 = 0.1$ m, regardless of the character of penetration: $\rho_2 > \rho_3$ or $\rho_2 < \rho_3$.

Case 2

With an increase of the width of the invasion zone ($a_2 = 4a_1$) the upper boundary of frequencies, when the method is still efficient, is shifted to lower frequencies. For example, if the formation resistivity does not exceed the borehole resistivity by more than ten times this boundary is defined from the same condition as in the first case. For larger values of the formation resistivity, parameter $\sigma_3\mu\omega a_1^2$ becomes smaller, but it does not in essence decrease the value of the maximal frequency. In fact, let $f = 1.2 \times 10^5$ Hz, $\rho_3/\rho_1 = 32$, $\rho_1 = 0.5$ ohm·m. Then we have $\sigma_3\omega\mu a_1^2 = 6.0 \times 10^{-4}$.

Case 3

In the case when the radius of the invasion zone is eight times greater than the borehole radius ($a_2 = 8a_1$) the upper boundary frequency becomes essentially smaller. Correspondingly, the vertical response of the probe deteriorates. At the same time with a decrease of frequency the depth of investigation in the radial direction increases, and it becomes possible to determine the formation resistivity with the small probe even when the radius of the invasion zone exceeds by more than ten times the borehole radius.

It is also appropriate to notice that in general the upper boundary of frequencies, when this method allows us to eliminate the influence of the borehole and the invasion zone, coincides with that for the approximate theory of induction logging described in Chapter 3.

In order to illustrate frequency responses in this method let us introduce the apparent conductivity in the following way:

$$\sigma_a = h_z^*/h_z^{un}$$

where $h_z^* = \mathrm{Q}\,h_z(\omega_2) - (\omega_2/\omega_1)\,\mathrm{Q}\,h_z(\omega_1)$ and $\mathrm{Q}\,h_z^{un}(\sigma_3)$ is the quadrature component of the field in a uniform medium with the conductivity of the formation, normalized by the primary field.

Results of calculation of apparent conductivity curves as a function of parameter $\lambda_1/a_1 = (10^3/a_1)(10\rho_1/f_1)^{1/2}$ are presented in Figs. 8.1–8.5. These data permit us to evaluate maximal frequencies for which it is still possible to eliminate the influence of the borehole and the intermediate zone. For example, when there is a deep penetration of the filtrate into the formation ($a_2/a_1 \leqslant 16$) and $\sigma_2 < \sigma_3$ the value of the apparent

TABLE 8.1
$\Delta(\sigma_a/\sigma_1) \times 10^2$; $\rho_3/\rho_1 = 1$, $a_2/a_1 = 2$

ρ_2/ρ_1	$\sigma_3\mu\omega a_1^2 \times 10^4$			
	1	2	4	8
		$\alpha = 4$		
4	1.88	2.66	3.74	5.26
128	1.88	2.66	3.74	5.24
		$\alpha = 8$		
4	3.75	5.28	7.40	10.3
128	3.75	5.27	7.38	10.3
		$\alpha = 12$		
4	5.60	7.84	10.9	14.9
128	5.60	7.85	10.9	14.9

TABLE 8.2
$\Delta(\sigma_a/\sigma_1) \times 10^2$; $\rho_3/\rho_1 = 2$, $a_2/a_1 = 2$

ρ_2/ρ_1	$\sigma_3\mu\omega a_1^2 \times 10^4$			
	1	2	4	8
		$\alpha = 4$		
4	9.42	1.35	1.88	2.64
128	9.44	1.33	1.87	2.63
		$\alpha = 8$		
4	1.88	2.63	3.71	5.16
128	1.88	2.64	3.70	5.15
		$\alpha = 12$		
4	2.80	3.93	5.64	7.47
128	2.80	3.92	5.45	7.45

TABLE 8.3
$\Delta(\sigma_a/\sigma_1) \times 10^2$; $\rho_3/\rho_1 = 4$, $a_2/a_1 = 2$

ρ_2/ρ_1	$\sigma_3\mu\omega a_1^2 \times 10^4$			
	1	2	4	8
		$\alpha = 4$		
4	0.471	0.666	0.943	1.33
128	0.471	0.653	0.938	1.33
		$\alpha = 8$		
4	0.940	1.32	1.86	2.60
128	0.938	1.34	1.86	2.56
		$\alpha = 12$		
4	1.40	1.97	2.74	3.76
128	1.40	1.97	2.73	3.77

TABLE 8.4
$\Delta(\sigma_a/\sigma_1) \times 10^2$; $\rho_3/\rho_1 = 8$, $a_2/a_1 = 2$

ρ_2/ρ_1	$\sigma_3\mu\omega a_1^2 \times 10^4$			
	1	2	4	8
		$\alpha = 4$		
4	0.236	0.355	0.476	0.680
128	0.236	0.334	0.472	0.688
		$\alpha = 8$		
4	0.471	0.665	0.939	1.32
128	0.470	0.663	0.932	1.30
		$\alpha = 12$		
4	0.702	1.09	1.38	1.91
128	0.702	1.08	1.37	1.89

TABLE 8.5
$\Delta(\sigma_a/\sigma_1) \times 10^2$; $\rho_3/\rho_1 = 16$, $a_2/a_1 = 2$

ρ_2/ρ_1	$\sigma_3\mu\omega a_1^2 \times 10^4$			
	1	2	4	8
		$\alpha = 4$		
4	0.471	0.666	0.943	1.33
128	0.471	0.653	0.938	1.33
		$\alpha = 8$		
4	0.940	1.32	1.86	2.60
128	0.938	1.34	1.86	2.56
		$\alpha = 12$		
4	1.40	1.97	2.74	3.76
128	1.40	1.97	2.73	3.77

TABLE 8.6
$\Delta(\sigma_a/\sigma_1) \times 10^2$; $\rho_3/\rho_1 = 32$, $a_2/a_1 = 2$

ρ_2/ρ_1	$\sigma_3\mu\omega a_1^2 \times 10^4$			
	1	2	4	8
		$\alpha = 8$		
4	0.119	0.169	0.241	0.348
128	0.120	0.172	0.250	0.376
		$\alpha = 12$		
4	0.177	0.250	0.352	0.498
128	0.178	0.244	0.363	0.527

TABLE 8.7
$\Delta(\sigma_a/\sigma_1) \times 10^2$; $\rho_3/\rho_1 = 1$, $a_2/a_1 = 4$

ρ_2/ρ_1	$\sigma_3 \mu\omega a_1^2 \times 10^4$			
	1	2	4	8
		$\alpha = 4$		
4	1.88	2.66	3.74	5.26
128	1.88	2.66	3.74	5.23
		$\alpha = 8$		
4	3.75	53.0	74.0	10.3
128	3.75	52.9	73.6	10.3
		$\alpha = 12$		
4	5.60	7.84	10.9	15.0
128	5.60	7.83	10.8	14.8

TABLE 8.8
$\Delta(\sigma_a/\sigma_1) \times 10^2$; $\rho_3/\rho_1 = 2$, $a_2/a_1 = 4$

ρ_2/ρ_1	$\sigma_3 \mu\omega a_1^2 \times 10^4$			
	1	2	4	8
		$\alpha = 4$		
4	0.940	1.32	1.87	2.60
128	0.937	1.32	1.87	2.56
		$\alpha = 8$		
4	1.87	2.63	3.68	5.09
128	1.87	2.62	3.65	5.02
		$\alpha = 12$		
4	2.80	3.91	5.42	7.38
128	2.79	2.90	5.38	7.28

TABLE 8.9
$\Delta(\sigma_a/\sigma_1) \times 10^2$; $\rho_3/\rho_1 = 4$, $a_2/a_1 = 4$

ρ_2/ρ_1	$\sigma_3 \mu\omega a_1^2 \times 10^4$			
	1	2	4	8
		$\alpha = 4$		
4	0.471	0.667	0.943	1.33
128	0.469	0.660	0.925	1.29
		$\alpha = 8$		
4	0.940	1.32	1.86	2.60
128	0.935	1.31	1.83	2.52
		$\alpha = 12$		
4	1.40	1.97	2.74	3.76
128	1.40	1.95	2.69	3.65

TABLE 8.10
$\Delta(\sigma_a/\sigma_1) \times 10^2$; $\rho_3/\rho_1 = 8$, $a_2/a_1 = 4$

ρ_2/ρ_1	$\sigma_3 \mu\omega a_1^2 \times 10^4$			
	1	2	4	8
		$\alpha = 8$		
4	0.237	0.338	0.485	0.706
128	0.235	0.331	0.465	0.651
		$\alpha = 12$		
4	0.473	0.671	0.855	1.37
16	0.470	0.660	0.827	1.29
128	0.468	0.658	0.819	1.27

TABLE 8.11
$\Delta(\sigma_a/\sigma_1) \times 10^2$; $\rho_3/\rho_1 = 16$, $a_2/a_1 = 4$

ρ_2/ρ_1	$\sigma_3 \mu\omega a_1^2 \times 10^4$			
	1	2	4	8
		$\alpha = 4$		
4	0.120	0.178	0.263	0.417
16	0.118	0.170	0.240	0.346
128	0.118	0.167	0.236	0.333
		$\alpha = 8$		
4	0.240	0.348	0.507	0.771
16	0.237	0.340	0.473	0.669
128	0.235	0.331	0.465	0.648
		$\alpha = 12$		
4	0.357	0.511	0.787	1.080
16	0.353	0.510	0.694	0.964
128	0.350	0.792	0.684	0.938

TABLE 8.12
$\Delta(\sigma_a/\sigma_1) \times 10^2$; $\rho_3/\rho_1 = 32$, $a_2/a_1 = 4$

ρ_2/ρ_1	$\sigma_3 \mu\omega a_1^2 \times 10^4$			
	1	2	4	8
		$\alpha = 4$		
4	0.740	0.100	0.173	0.357
16	0.059	0.087	0.128	0.200
128	0.059	0.085	0.121	0.180
		$\alpha = 8$		
4	0.125	0.188	0.302	0.547
16	0.120	0.171	0.247	0.366
128	0.118	0.167	0.247	0.339
		$\alpha = 12$		
4	0.185	0.272	0.420	0.685
16	0.177	0.252	0.361	0.519
128	0.176	0.249	0.348	0.487

TABLE 8.13
$\Delta(\sigma_a/\sigma_1) \times 10^2$; $\rho_3/\rho_1 = 1$, $a_2/a_1 = 8$

ρ_2/ρ_1	$\sigma_3\mu\omega a_1^2 \times 10^4$			
	1	2	4	8
		$\alpha = 4$		
4	1.85	2.58	23.55	4.77
128	1.85	2.56	3.50	4.64
		$\alpha = 8$		
4	3.70	5.14	7.04	9.37
128	3.68	5.10	6.93	9.12
		$\alpha = 12$		
4	5.62	7.64	10.4	13.7
128	5.50	7.58	10.2	13.3

TABLE 8.14
$\Delta(\sigma_a/\sigma_1) \times 10^2$; $\rho_3/\rho_1 = 2$, $a_2/a_1 = 8$

ρ_2/ρ_1	$\sigma_3\mu\omega a_1^2 \times 10^4$			
	1	2	4	8
		$\alpha = 4$		
4	0.932	1.31	1.81	2.47
128	0.924	1.28	1.75	3.22
		$\alpha = 8$		
4	1.86	2.60	3.57	4.84
128	1.84	2.55	3.47	4.57
		$\alpha = 12$		
4	2.78	3.85	5.28	7.64
128	2.75	3.79	5.12	6.67

TABLE 8.15
$\Delta(\sigma_a/\sigma_1) \times 10^2$; $\rho_3/\rho_1 = 4$, $a_2/a_1 = 8$

ρ_2/ρ_1	$\sigma_3\mu\omega a_1^2 \times 10^4$			
	1	2	4	8
		$\alpha = 4$		
4	0.471	0.667	0.942	1.33
16	0.464	0.648	0.891	1.20
128	0.463	0.642	0.878	1.17
		$\alpha = 8$		
4	0.940	1.32	1.86	2.60
16	0.926	1.29	1.79	2.36
128	0.922	1.28	1.64	2.30
		$\alpha = 12$		
4	1.40	1.97	2.74	3.76
16	1.38	1.91	2.61	3.43
128	1.38	1.90	2.57	3.35

TABLE 8.16
$\Delta(\sigma_a/\sigma_1) \times 10^2$; $\rho_3/\rho_1 = 8$, $a_2/a_1 = 8$

ρ_2/ρ_1	$\sigma_3\mu\omega a_1^2 \times 10^4$			
	1	2	4	8
		$\alpha = 4$		
4	0.242	0.341	0.522	0.812
16	0.236	0.334	0.456	0.625
128	0.232	0.322	0.442	0.591
		$\alpha = 8$		
4	0.481	0.695	1.02	1.62
16	0.470	0.664	0.900	1.22
128	0.462	0.640	0.875	1.16
		$\alpha = 12$		
4	0.716	1.03	1.49	2.20
16	0.702	0.985	1.33	1.78
128	0.689	0.952	1.29	1.69

TABLE 8.17
$\Delta(\sigma_a/\sigma_1) \times 10^2$; $\rho_3/\rho_1 = 16$, $a_2/a_1 = 8$

ρ_2/ρ_1	$\sigma_3\mu\omega a_1^2 \times 10^4$			
	1	2	4	8
		$\alpha = 4$		
4	0.130	0.204	0.942	1.33
16	0.118	0.168	0.891	1.20
128	0.116	0.162	0.878	1.17
		$\alpha = 8$		
4	0.256	0.397	1.86	2.60
16	0.236	0.334	1.79	2.36
128	0.232	0.328	1.64	2.30
		$\alpha = 12$		
4	0.380	0.597	0.935	1.65
16	0.351	0.595	0.694	1.06
128	0.346	0.580	0.660	0.86

TABLE 8.18
$\Delta(\sigma_a/\sigma_1) \times 10^2$; $\rho_3/\rho_1 = 32$, $a_2/a_1 = 8$

ρ_2/ρ_1	$\sigma_3\mu\omega a_1^2 \times 10^4$			
	1	2	4	8
		$\alpha = 4$		
4	0.834	0.167	0.409	1.15
8	0.0661	0.105	0.266	0.411
16	0.0615	0.0904	0.139	0.232
128	0.0588	0.0828	0.116	0.164
		$\alpha = 8$		
4	0.158	0.301	0.868	1.78
8	0.130	0.202	0.518	0.176
16	0.122	0.177	0.266	0.421
128	0.117	0.164	0.228	0.313
		$\alpha = 12$		
4	0.226	0.409	0.865	2.06
8	0.191	0.293	0.753	1.11
16	0.180	0.262	0.386	0.696
128	0.171	0.243	0.334	0.457

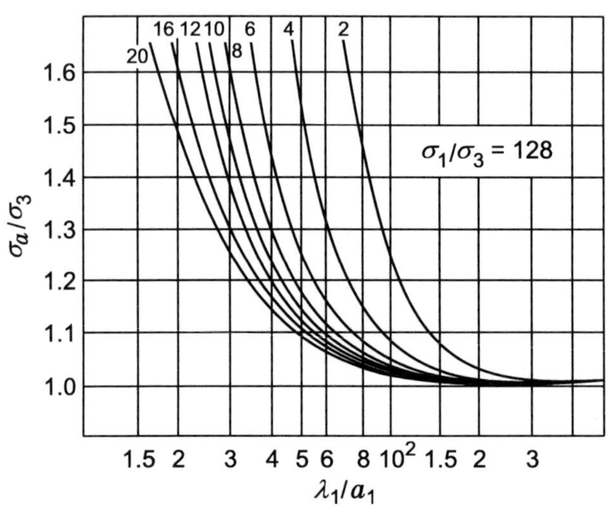

Figure 8.1. Apparent conductivity curves ($a_2/a_1 = 1$). Curve index L/a_1.

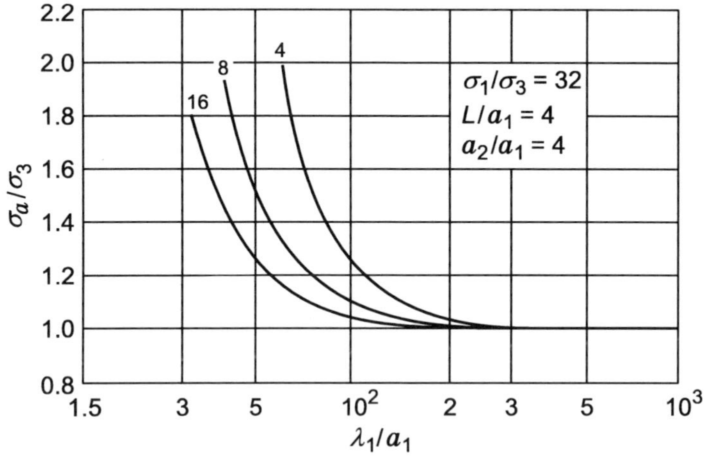

Figure 8.2. Apparent conductivity curves. Curve index σ_1/σ_2.

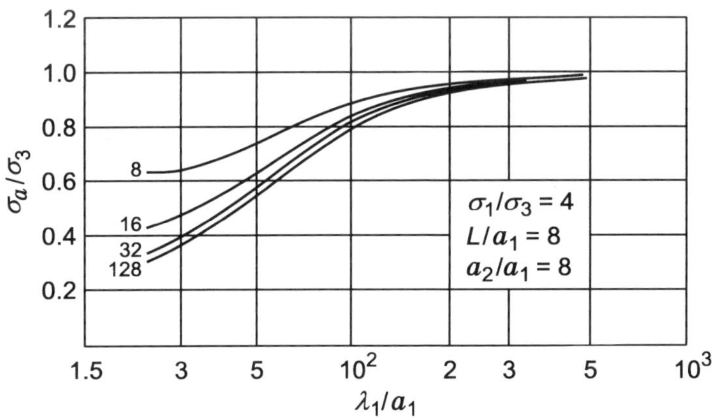

Figure 8.3. Apparent conductivity curves. Curve index σ_1/σ_2.

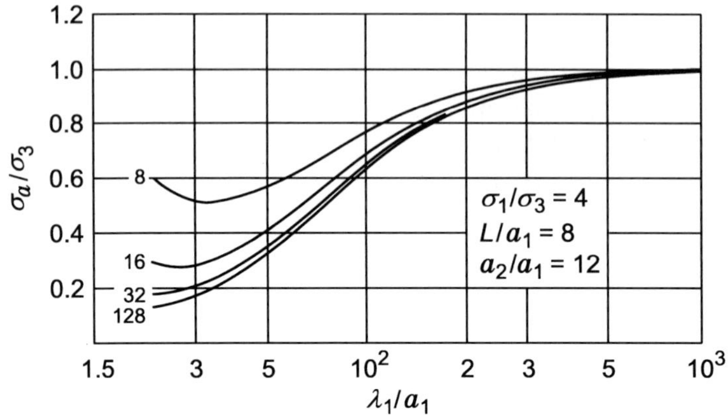

Figure 8.4. Apparent conductivity curves. Curve index σ_1/σ_2.

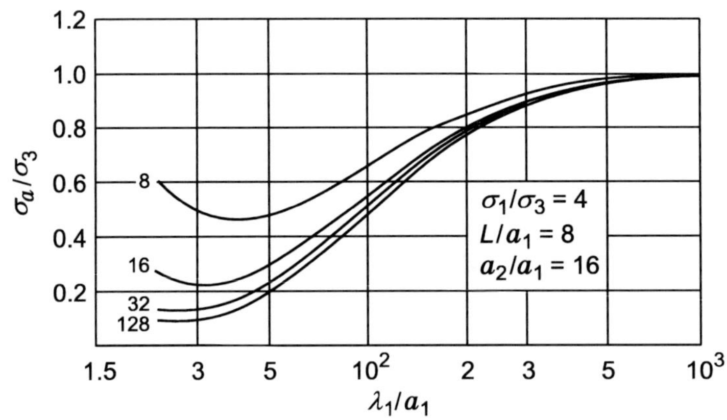

Figure 8.5. Apparent conductivity curves. Curve index σ_1/σ_2.

conductivity is almost equal to the formation conductivity, if $\rho_1 > 1$ ohm·m, $a_1 \leqslant 0.1$ m, $\rho_3/\rho_1 \simeq 3$ and frequency f does not exceed 16 kHz ($f_2 = 4f_1$).

As has been mentioned above, due to measuring the magnetic field created by currents in the external area the vertical response of the probe becomes worse. The influence of the formation and surrounding medium, when the leading term of the low-frequency spectrum of the quadrature component is measured, as is well known, is defined by the distribution of conductivities and geometric factors. In this method this part of the field is eliminated, and for this reason the vertical response is mainly defined by the skin effect. Calculations show that induced currents in the surrounding medium do not practically affect the apparent conductivity value, provided that:

$$\frac{\lambda_2}{H} < 6 \qquad \text{or} \qquad 10^3 \left(\frac{10\rho_3}{f}\right)^{1/2} < 6H \tag{8.16}$$

if $\rho_3/\rho_2 < 32$ and $H/L > 2$.

Chapter 9

TRANSIENT INDUCTION LOGGING

As is well known, an increase of the depth of investigation in induction logging is usually realized with the help of multi-coil differential probes that in many cases permit us to eliminate the influence of currents in the borehole and in the invasion zone.

However, the theory and numerous experiences show that for a given length of the basic probe, L, the depth of investigation in the radial direction does not exceed $(0.5 \div 0.6)L$, and with an increase of the conductivity of the borehole and the invasion zone it becomes less.

For this reason, only under favorable geoelectric conditions, in particular when $\rho_2 > \rho_3$, we can obtain a correct presentation of the formation resistivity with currently applied probes if the radius of the invasion zone is not greater than 5–6 radii of the borehole. In the case when the conductivity of the invasion zone exceeds that of the formation the depth of investigation becomes smaller, and only an increase of the probe length allows us in principle to improve the radial response of the differential probe. But at the same time with an increase of the probe length the influence of the surrounding medium becomes stronger and the conditions providing compensation of the electromotive force, caused by the magnetic field of currents within the invasion zone, can be invalid. Also it is appropriate to mention here that the efficiency of *focusing* is based on the assumption that caverns are absent and the probe is located on the borehole axis.

All these factors, which restrict to a certain degree the application of induction logging with differential probes, impel us to investigate the possibilities of the transient method, which is successfully applied in other areas of exploration geophysics.

As is well known, if the moment of the transmitter coil changes as a step function with time, induced currents appear in the surrounding medium. At the beginning they are concentrated in the borehole, but with an increase of time their intensity increases at larger distances from the coil, and always there is such moment when they are practically located in the formation and their magnitude is defined by the conductivity of the external area, i.e. the formation. For this reason the magnetic field measured on the borehole axis, starting from this moment, does not practically differ from that in a uniform medium with the formation conductivity.

A similar situation occurs in the presence of the invasion zone even when its conductivity is greater than that of the formation. Inasmuch as at the late stage of the transient response, induced currents are practically absent in the borehole and in the invasion zone the magnetic field, measured by a receiver, is not dependent of the position of the probe with respect to the borehole axis and caverns, as well as of the distance between the coils of the probe. At relatively large times induced currents are located far away from the borehole, and a change of the distance between coils, in certain limits, does not affect the value of the measured field. For this reason the transient method can provide a large depth of investigation with very short two-coil induction probes. In fact the same result can be obtained in principle, with a one-coil probe.

A decrease of the probe length can be desirable since with an increase of time the influence of the surrounding medium becomes stronger, and correspondingly the vertical response of the probe deteriorates. The greater the conductivity of the surrounding medium the earlier the influence of currents induced therein begins to manifest itself.

Therefore, with an increase of time the depth of investigation increases, which results in simultaneously improving the radial response of the probe, and in an increase of the influence of the surrounding medium.

In this chapter we will consider the theory of the transient method of induction logging in a uniform conducting medium, as well as in media with cylindrical and horizontal interfaces. Also one section will be devoted to transient responses of the electric dipole.

9.1. The Transient Field of the Magnetic Dipole in a Uniform Medium

The analysis of the main feature of a nonstationary field in a conducting medium we will start from the simplest case, when the magnetic dipole is located in a uniform medium, and its moment changes as a step function:

$$M(t) = \begin{cases} 0 & t < 0 \\ M & t > 0 \end{cases} \tag{9.1}$$

Applying Laplacian transformation to the vector potential in the frequency domain (Chapter 2):

$$A_z^*(\omega) = \frac{\mathrm{i}\omega\mu M}{4\pi} \frac{\mathrm{e}^{\mathrm{i}kR}}{R} \tag{9.2}$$

where $k^2 = \mathrm{i}\sigma\mu\omega + \omega^2\varepsilon\mu$, we obtain an expression for the vector potential for the nonstationary field:

$$A_z^*(t) = \begin{cases} 0 & t \leqslant \tau_0 \\ -\dfrac{\mu M}{4\pi R}\left[\mathrm{e}^{-q\tau_0}\delta(t-\tau_0) + q\tau_0\mathrm{e}^{-qt}\dfrac{I_1[q(t^2-\tau_0^2)^{1/2}]}{(t^2-\tau_0^2)^{1/2}}\right] & t > \tau_0 \end{cases} \tag{9.3}$$

where μ is the magnetic permeability: $\mu = \mu^* \times 4\pi \times 10^{-7}$ H/m; ε is the dielectric permeability: $\varepsilon = \varepsilon^*(1/36\pi) \times 10^{-9}$ F/m; σ is the conductivity. Here:

$$q = \frac{1}{2}\frac{\sigma}{\varepsilon} \qquad \tau_0 = (\varepsilon\mu)^{1/2}R \tag{9.4}$$

R is the distance from the dipole to the observation point; M is the dipole moment, equal to InS; n is the number of turns; S is the turn area; $I_1[(t^2-\tau_0^2)^{1/2}]$ is the modified Bessel function of the first order; $\delta(t-\tau_0)$ is the Dirac function, defined from the relation:

$$\int_a^b f(x')\delta^{(n)}(x-x')\,\mathrm{d}x' = \begin{cases} (-1)^n f^{(n)}(x) & a<x<b \\ 0 & x<a,\ x>b \end{cases} \tag{9.5}$$

t is time, counted from the moment when the transmitter current is turned on.

In accord with eq. 9.3 the field arises in any point of the medium at the moment:

$$t=\tau_0=(\varepsilon\mu)^{1/2}R$$

i.e. with an increase of the distance from the source, the signal traveling with velocity:

$$v=\frac{1}{(\varepsilon\mu)^{1/2}}=\frac{c}{(\varepsilon^*\mu^*)^{1/2}} \qquad c=3\times 10^8\ \mathrm{m/s}$$

appears later.

The electric field, E_ϕ, is related with the vector potential of magnetic type A_z^*, as:

$$E_\phi=-\frac{\partial A_z^*}{\partial R}\sin\theta$$

Omitting intermediate transformations, we have:

$$E_\phi=E_\phi^{(1)}+E_\phi^{(2)}$$

where:

$$\begin{aligned} E_\phi^{(1)} &= -\frac{\mu M}{4\pi R^2}[(1+q\tau_0)\delta(t-\tau_0)+\tau_0\delta'(t-\tau_0)]\,\mathrm{e}^{-q\tau_0}\sin\theta \\ E_\phi^{(2)} &= -\frac{\mu M}{4\pi R^2}q^2\tau_0^3\,\mathrm{e}^{-qt}\frac{I_2(q\sqrt{t^2-\tau_0^2})}{t^2-\tau_0^2}\sin\theta \qquad \text{if } t>\tau_0 \end{aligned} \tag{9.6}$$

and

$$E_\phi=0 \qquad \text{if } t<\tau_0$$

From the first Maxwell equation, curl $\boldsymbol{E}=-\dot{\boldsymbol{B}}$, we obtain:

$$\begin{aligned} \dot{B}_R^{(1)} &= \frac{\mu M}{2\pi R^3}[(1+q\tau_0)\delta(t-\tau_0)+\tau_0\delta'(t-\tau_0)]\,\mathrm{e}^{-q\tau_0}\cos\theta \\ \dot{B}_R^{(2)} &= \frac{\mu M}{2\pi R^3}q^2\tau_0^3\,\mathrm{e}^{-qt}\frac{I_2(q\sqrt{t^2-\tau_0^2})}{t^2-\tau_0^2}\cos\theta \qquad \text{if } t>\tau_0 \end{aligned} \tag{9.7}$$

and

$$\dot{B}_R = 0 \qquad \text{if } t < \tau_0$$

From the initial condition it follows that the magnetic field tends to that of direct current, as $t \to 0$, i.e.:

$$B_R \to \frac{\mu M}{2\pi R^3}\cos\theta \qquad B_\theta \to \frac{\mu M}{4\pi R^3}\sin\theta$$

For this reason, integrating the right-hand part of eq. 9.7 we obtain:

$$\begin{aligned} B_R &= 0 \qquad \text{if } t < \tau_0 \\ B_R^{(1)} &= \frac{\mu M}{2\pi R^3}[(1+q\tau_0)\delta(t-\tau_0) + \tau_0\delta(t-\tau_0)]\,\mathrm{e}^{-q\tau_0}\cos\theta \\ B_R^{(2)} &= \frac{M}{2\pi R^3}\left[q^2\tau_0^3\int\limits_{\tau_0}^{t}\mathrm{e}^{-qx}\frac{I_2\left(q\sqrt{x^2-\tau_0^2}\right)}{x^2-\tau_0^2}\,\mathrm{d}x\right]\cos\theta \quad \text{if } t > \tau_0 \end{aligned} \tag{9.8}$$

By analogy, we have:

$$\begin{aligned} B_\theta &= 0 \qquad \text{if } t < \tau_0 \\ B_\theta^{(1)} &= \frac{\mu M\sin\theta}{4\pi R^3}[(1+q\tau_0+q^2\tau_0^2)h(t-\tau_0) + \tau_0(1+2q\tau_0)\delta(t-\tau_0) + \tau_0^2\delta'(t-\tau_0)]\,\mathrm{e}^{-q\tau_0} \\ B_\theta^{(2)} &= \frac{\mu M\sin\theta}{4\pi R^3}\left\{q^2\tau_0^3\int\limits_{\tau_0}^{t}\frac{\mathrm{e}^{-qx}}{\sqrt{x^2-\tau_0^2}}\right. \\ &\quad\times\left.\left[\frac{q\tau_0^2}{\sqrt{x^2-\tau_0^2}}I_3\left(q\sqrt{x^2-\tau_0^2}\right) - 2I_2\left(q\sqrt{x^2-\tau_0^2}\right)\right]\mathrm{d}x\right\} \qquad \text{if } t > \tau_0 \end{aligned} \tag{9.9}$$

In a general case, the electromagnetic field for the given moment of the dipole depends on the distance from the source (R), velocity of propagation ($v = 1/(\varepsilon\mu)^{1/2}$), parameter $q = (1/2)(\sigma/\varepsilon)$, which has dimension t^{-1} and characterizes a degree of decay of the field in a conducting medium and, finally, it is defined of course by measuring moment t.

We will present the magnetic field in units of the static field of the magnetic dipole:

$$\begin{aligned} B_R &= \frac{2M}{4\pi R^3}\,b_R\cos\theta \\ B_\theta &= \frac{M}{4\pi R^3}\,b_\theta\sin\theta \end{aligned}$$

At the beginning we will consider the field in a nonconducting medium. In accord with eqs. 9.8–9.9 we have:

$$\begin{aligned} b_R(t) &= h(t-\tau_0) + \tau_0\delta(t-\tau_0) \\ b_\theta(t) &= h(t-\tau_0) + \tau_0\delta(t-\tau_0) + \tau_0^2\delta'(t-\tau_0) \\ e_\phi(t) &= -\delta(t-\tau_0) - \tau_0\delta'(t-\tau_0) \end{aligned} \tag{9.10}$$

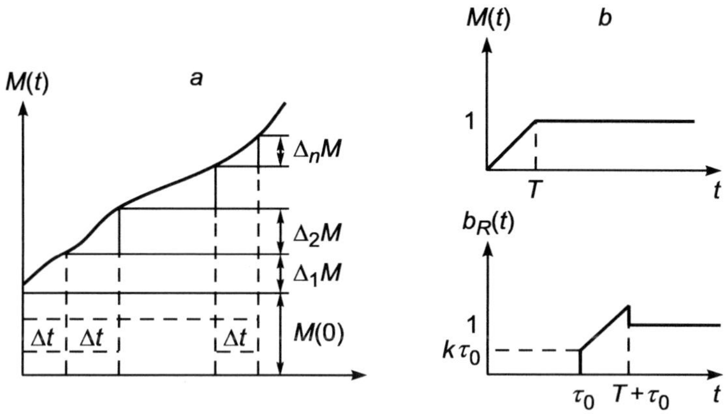

Figure 9.1. a. Illustration of Duhamel's integral. b. Example 1.

where $h(t - \tau_0)$ is the step function:

$$h(t - \tau_0) = \begin{cases} 0 & t \leqslant \tau_0 \\ 1 & t > \tau_0 \end{cases}$$

Thus, the magnetic field in an insulator, as well as the dipole moment, presents itself as the step function ($t \neq \tau_0$), and it is natural here to distinguish three successive stages. Until moment $t = \tau_0 = R/v$ the field is absent at the point with coordinates (R, θ). Due to the instant change of the dipole moment the signal front is described by the δ-function and its derivative. Only at this moment the electric field is not equal to zero. After arrival of the signal ($t > \tau_0$) the magnetic field instantly becomes constant in the same way as the dipole moment while the electric field vanishes.

Let us notice that the step function can be presented as a sum of rectangular impulses of the same magnitude which follow each other in time. In this case the electrical fields of neighbor impulses are compensated, except for moment $t = \tau_0$.

Proceeding from results obtained for the step function we will consider to what extent type of excitation $M(t)$ and moment of measuring t define conditions for which a field can be considered with sufficient accuracy as a quasistationary one, i.e. when it changes synchronously with the dipole moment in free space.

As was shown in Chapter 2, an arbitrary change of moment $M(t)$ with time can be presented with the help of a Duhamel's integral as a sum of successively turning on step functions $h(t - r)$ with magnitude $M'(r)$ (Fig. 9.1a):

$$M(t) = M(0) + \int_0^t \frac{\mathrm{d}M}{\mathrm{d}r} h(t - r)\,\mathrm{d}r$$

Applying the principle of superposition and taking into account that the electromagnetic field (A_E and A_H), generated by one step function, is known, we obtain expressions of the field for arbitrary excitation $M(t)$:

$$\begin{aligned} H(t) &= M(0)A_H(t) + \int_0^t \frac{\mathrm{d}M}{\mathrm{d}r} A_H(t-r)\,\mathrm{d}r \\ E(t) &= M(0)A_E(t) + \int_0^t \frac{\mathrm{d}M}{\mathrm{d}r} A_E(t-r)\,\mathrm{d}r \end{aligned} \tag{9.11}$$

In particular, for radial component b_R, in accord with eq. 9.10, we have:

$$b_R(t) = \int_0^t \frac{\mathrm{d}M}{\mathrm{d}r}[h(t-\tau_0-r) + \tau_0\delta(t-\tau_0-r)]\,\mathrm{d}r$$

since it is assumed that $M(0) = 0$.

Inasmuch as the step function is equal to unity only for positive arguments $0 < r < t - \tau_0$, then:

$$b_R(t) = M(t-\tau_0) + \tau_0 M'(t-\tau_0)$$

By analogy:

$$\begin{aligned} b_\theta(t) &= M(t-\tau_0) + \tau_0 M'(t-\tau_0) + \tau_0^2 M''(t-\tau_0) \\ e_\phi(t) &= -M'(t-\tau_0) - \tau_0 M''(t-\tau_0) \end{aligned} \tag{9.12}$$

Now, let us consider several examples.

Example 1

$$M(t) = \begin{cases} 0 & t < 0 \\ kt & 0 < t < T \\ 1 & t > T \end{cases} \qquad \text{here } k = 1/T$$

Applying eq. 8.12 we have:

$$b_R(t) = \begin{cases} 0 & t < \tau_0 \\ kt & \tau_0 < t < T + \tau_0 \\ 1 & t > T + \tau_0 \end{cases} \tag{9.13}$$

For a linear change of the dipole moment the radial component of the field is also a linear function of time (Fig. 9.1b).

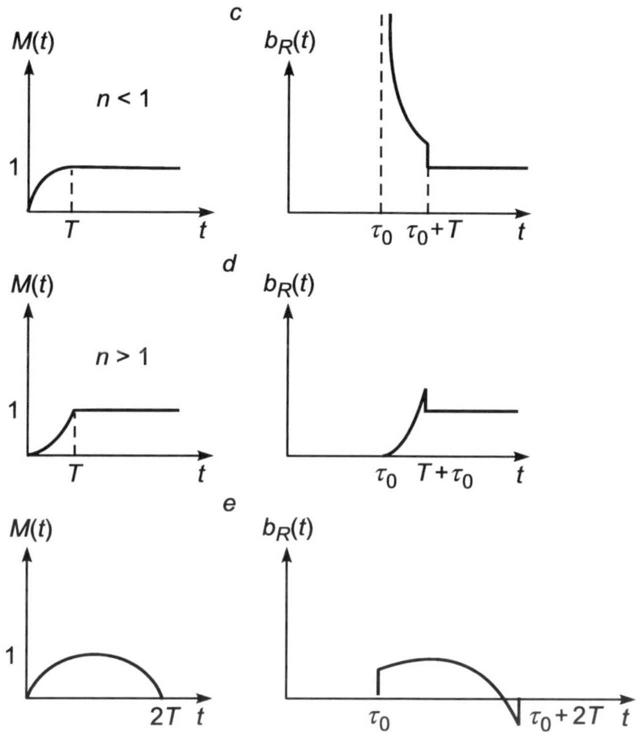

Figure 9.1.c.d.e. Illustration of moment and field behavior in examples (2) and (3).

Example 2

$$M(t) = \begin{cases} 0 & t < 0 \\ kt^n & 0 < t < T \\ 1 & t > T \end{cases} \qquad \text{here } k = 1/T^n$$

After differentiation we obtain (Fig. 9.1c):

$$b_R(t) = \begin{cases} 0 & t < \tau_0 \\ k(t-\tau_0)^n + nk\tau_0(t-\tau_0)^{n-1} & \tau_0 < t < T+\tau_0 \\ 1 & t > T+\tau_0 \end{cases} \tag{9.14}$$

The field at an arbitrary moment, t, can be presented as a sum of fields, generated by step functions, arising at moments $t = 0, \Delta t, \ldots$ and so on with magnitude $M'(r)\,\mathrm{d}r$, which, in particular, for the first example within the interval $0 < t < T$ is the same.

If the dipole current changes instantly at the moment $t = r$, then the field arising at an arbitrary observation point for $t > r + \tau_0$ becomes constant. For this reason we can think of the field at moment t as consisting of two parts. The first part changes in the same manner as the magnetic moment of the dipole, and its electric field is absent. The second part arises due to a change of the dipole current at moment $t = r$ and has all components of the electromagnetic field. As is well known, the magnetic components of the quasistationary field satisfy the system of equations of the magnetic field, caused by the direct current, and for its determination we can apply the Biot–Savart law. This means that we can neglect the change of the dipole moment during time τ_0, which is smaller, when the observation point is closer to the dipole and with a higher velocity of signal propagation.

Thus, condition of quasistationarity can be written as:

$$\left| \frac{H(t) - H(t - \tau_0)}{H(t)} \right| < 1$$

If function $M(t)$ monotonically grows within interval $0 < t < T$, as it takes place in previous examples, then with an increase of time the field becomes greater and correspondingly the first part of the field begins to prevail. At the same time a relative change of the field, since the delay is neglected, also becomes sufficiently small.

Example 3

Now consider the case when the dipole moment is not a monotonic function of time (Fig. 9.1d):

$$M(t) = \begin{cases} 0 & t < 0 \\ 1 - \left(\dfrac{t}{T} - 1\right)^2 & 0 < t < 2T \\ 0 & t > 2T \end{cases}$$

After differentiation of eq. 9.12 we obtain:

$$b_R(t) = \begin{cases} 0 & t < \tau_0 \\ 1 - \left(\dfrac{t - \tau_0}{T} - 1\right)^2 - \dfrac{2\tau_0}{T}\left(\dfrac{t - \tau_0}{T} - 1\right) & \tau_0 < t < \tau_0 + 2T \\ 0 & t > \tau_0 + 2T \end{cases} \tag{9.15}$$

Until moment $T + \tau_0$ function $b_R(t)$ increases and its first derivative becomes smaller. For this reason, in the same way as in previous examples, with an increase of time the quasistationary character of the field manifests itself stronger, $(t < 2T)$. However, with a further increase of time the relation between both parts of the field changes essentially and, in particular, when $t + 2T$, they are practically equal to each other and proportional to τ_0.

Example 4

Finally, consider the last example (Fig. 9.1e):

$$M(t) = \begin{cases} 0 & t < 0 \\ \sin\omega t & 0 < t < B \\ 0 & t > B \end{cases}$$

Correspondingly, for the field we have:

$$b_R(t) = \begin{cases} 0 & t < \tau_0 \\ \sin\omega(t-\tau_0) + \omega\tau_0 \cos\omega(t-\tau_0) & \tau_0 < t < \tau_0 + B \\ 0 & t > \tau_0 + B \end{cases} \tag{9.16}$$

In this case the condition of quasistationarity for $t < B$ has the form:

$$t \gg \tau_0 \text{ and } \omega\tau_0 \ll 1$$

and it is applied to the amplitude of oscillations but not to the instant value of function b_R. In particular, regardless of value $\omega\tau_0$, when $\omega(t-\tau_0) = k\pi$ only the second part of the field exists, which is caused by the rate of change of function $M(t)$ at the moment $t - \tau_0$.

Now we will again investigate the field in the more general case of a conducting medium. As before, the field travels with velocity $v = c/(\varepsilon\mu)^{1/2}$, and until moment $t = \tau_0$ magnitude of the field at any point is equal to zero. The intensity of the signal at moment $t = \tau_0$ is defined by parameter $q\tau_0$, which can be presented as:

$$q\tau_0 = \frac{1}{2}\sigma(\mu/\varepsilon)^{1/2}R = b_\infty R = R/X_0 = m \tag{9.17}$$

where $X_0 = 1/b_\infty$ is the characteristic length depending on the parameters of the medium; b_∞ coincides with the limit value of the imaginary part of the wave number, k, when the frequency unlimitedly increases.

Graphs of functions $b_R^{(1)}(m)$ and $b_\theta^{(1)}(m)$ are given in Fig. 9.2. With an increase of the distance from the source and conductivity of a medium, due to transformation of the electromagnetic energy into heat, the field magnitude, corresponding to the moment of the first arrival, rapidly decreases.

When the distance from the source exceeds the characteristic length X_0, the magnetic field is mainly defined by component b_θ, and the electromagnetic field at the arrival moment corresponds to the wave zone: components of the field are perpendicular to the direction of the Poynting vector, and the ratio of components of electric and magnetic field, (E_ϕ/H_ϕ), does not depend on the distance to the dipole.

In the insulator the magnetic field, due to step function excitation, does not alter after the arrival of the signal and coincides with the constant field, but the electric field is absent. Unlike it in a conducting medium a certain time is required for the field to become established, because of the appearance of conduction currents.

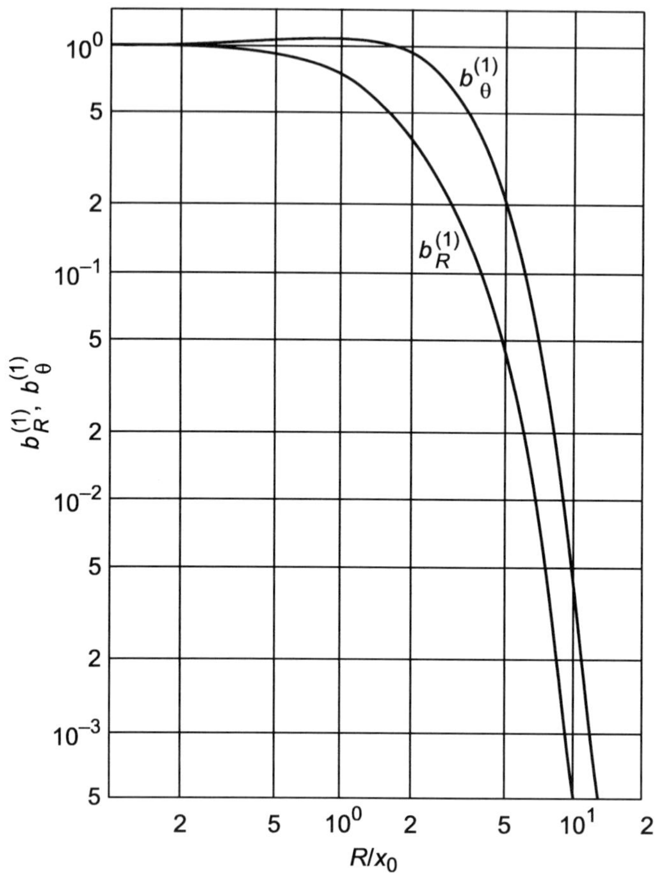

Figure 9.2. Functions $b_R^{(1)}$ and $b_\theta^{(1)}$ in a uniform medium.

Now we will investigate the electric field as a function of time and the parameters of a medium. It is convenient to present field $E_\phi^{(2)}$ in the form:

$$E_\phi^{(2)} = \frac{M\rho}{2\pi R^4} e_\phi^{(2)} \sin\theta \tag{9.18}$$

where

$$e_\phi^{(2)} = m^3 \mathrm{e}^{-mn} \frac{I_2(m\sqrt{n^2-1})}{n^2-1} \tag{9.19}$$

and

$$n = \frac{t}{\tau_0} \geqslant 1$$

TABLE 9.1
Values of parameter m

ε \ ρ, ohm·m	1	10	30	100
9	63.0	6.3	2.1	0.63
16	47.0	4.7	1.6	0.47
25	38.0	3.8	1.3	0.38
36	31.0	3.1	1.0	0.31

Applying expansion of function $I_2(z)$ in a series by z, we obtain:

$$e_\phi^{(2)} = \frac{1}{8} m^5 \mathrm{e}^{-m} \qquad \text{if } t = \tau_0 \tag{9.20}$$

Function $e_\phi^{(2)}$ has a maximum, when $m = b_\infty R = 5$. For this reason if the distance, r, from the source does not exceed $5X_0$, then with an increase of conductivity the field also increases. The same behavior is observed with an increase of the distance, if $R < 5X_0$.

For illustration Table 9.1 contains some values of parameter m as a function of resistivity ρ and dielectric constant ε, if $r = 1$ m.

With a further increase of conductivity or the distance from the source, the magnitude of the field of the first arrival becomes smaller.

Replacing function $I_2(z)$ by its asymptotical expression:

$$I_2(z) \simeq \frac{\mathrm{e}^z}{(2\pi z)^{1/2}}$$

we have:

$$\mathrm{e}_\phi^{(2)} = \frac{1}{\sqrt{2\pi}} \left(\frac{m}{\sqrt{n^2-1}} \right)^{5/2} \mathrm{e}^{m(\sqrt{n^2-1}-n)} \tag{9.21}$$

Formula 9.2 is value when the distance between the dipole and an observation point is significantly greater than the characteristic length X_0, or measurements are performed at times exceeding $\tau_0 (n \gg 1)$. In the latter case eq. 9.21 can be presented as:

$$e_\phi^{(2)} = \frac{1}{\sqrt{2\pi}} \left(\frac{m}{n} \right)^{5/2} \mathrm{e}^{-m/2n} \qquad \text{if } n \gg 1 \tag{9.22}$$

and it corresponds to the quasistationary field.

Curves of function $e_\phi^{(2)}(n)$ are presented in Fig. 9.3. The curve index is parameter $m = R/X_0$.

If the distance from the source does not exceed $5X_0$, then the electric field monotonically decreases with time.

At the same time with an increase of parameter m (an increase of conductivity, the distance or a decrease of dielectric constant) the maximum appears on curves of $e_\phi^{(2)}$, which is shifted to larger times.

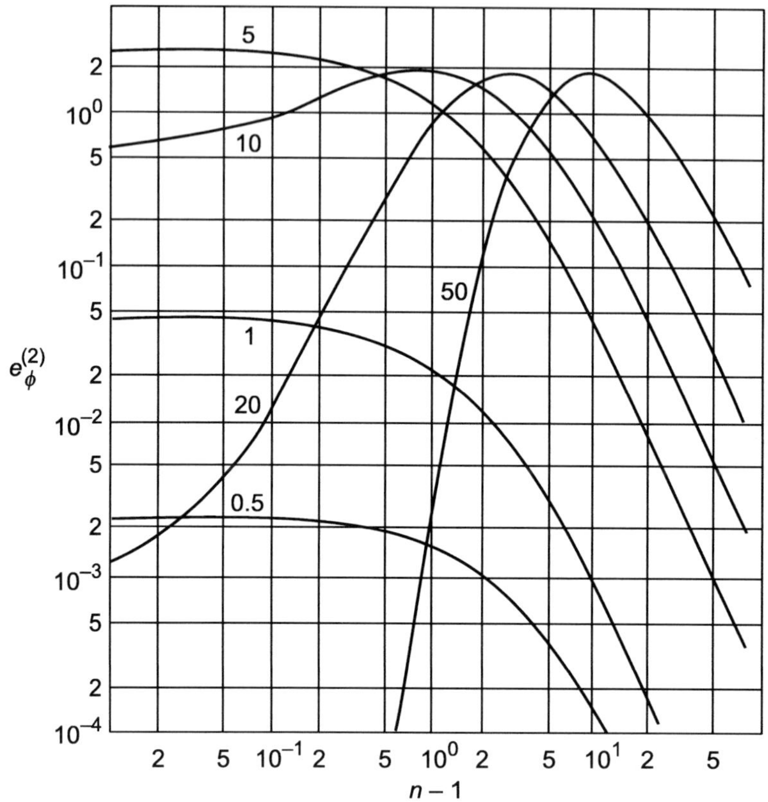

Figure 9.3. Graphs of function $e_\phi^{(2)}(n)$. Curve index m.

We will present field $E_\phi^{(2)}$ in the form:

$$E_\phi^{(2)} = \frac{M\rho}{2\pi}\frac{(qt)^3}{(vt)^4}F_E \sin\theta$$

where

$$F_E = x\,\mathrm{e}^{-qt}\frac{I_2(qt\sqrt{1-x^2})}{1-x^2} \qquad x = \frac{R}{vt} < 1$$

Graphs illustrating a distribution of the electric field in a medium at an arbitrary moment are shown in Fig. 9.4. Index of curves is parameter qt. As is seen from curves for small values of qt a maximum of the electric field corresponds to the first arrival, and with approaching to the source the field linearly decreases. But with an increase of parameter qt the different behavior is observed, namely, intensity of the first arrival becomes much smaller for points located closer to the dipole.

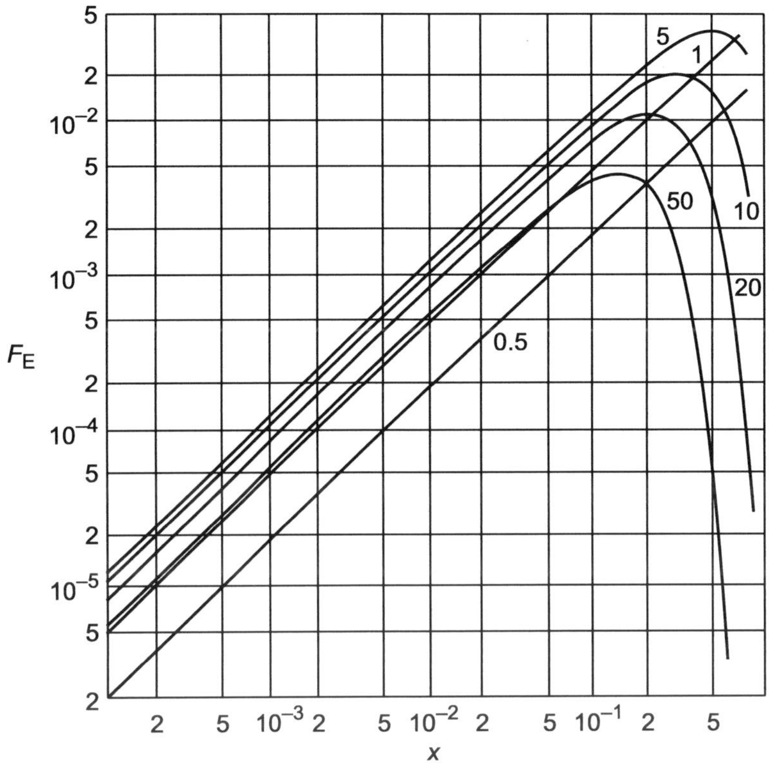

Figure 9.4. Behavior of function F_E. Curve index qt.

Therefore, in observing a field in a conducting medium we can distinguish the following stages:

1. The field is equal to zero until moment $\tau_0 = R/v$. If measurements are performed at distances from the source, equal approximately to 1 m, the time which is necessary for arriving the signal, τ_0, is about dozens of nanoseconds.

2. Magnitude of the first arrival is a function of the distance from the source and electric parameters of a medium.

3. In a conducting medium, after the signal passes through an observation point it does not disappear instantly, and here we can distinguish, in turn, two parts. At the beginning, when the moment of measuring is close to τ_0, sources of the magnetic field are both conduction currents and displacement currents, i.e. a rate of a change of the electric field with time. The greater the resistivity of a medium, the wider the time interval where displacement currents play an essential role.

4. The last stage corresponds to the quasistationary field when the influence of displacement currents is absent. This feature of the alternating field is inherent for a conducting medium, regardless of how small its conductivity, but the moment of transition to the quasistationary field starts earlier with an increase of conductivity.

Now we will consider the quasistationary field. Assuming that $(1/2)\sigma t/\varepsilon \gg 1$ and $t/\tau_0 \gg 1$ expression for the vector potential A_z^* has the form:

$$A_z^* = -\frac{\mu M}{4\pi\sqrt{2\pi}R}\frac{u}{t}\,\mathrm{e}^{-u^2/2} \tag{9.23}$$

where $u = (\sigma\mu/2t)^{1/2}R = 2\pi R/\tau$, $\tau = (2\pi\rho t \times 10^7)^{1/2}$.

After relatively simple transformations we obtain:

$$\begin{aligned}
H_R &= \frac{2M}{4\pi R^3}\,h_R\cos\theta = \frac{2M}{4\pi R^3}\left[1-\phi(u)+\left(\frac{2}{\pi}\right)^{1/2} u\,\mathrm{e}^{-u^2/2}\right]\cos\theta \\
H_\theta &= \frac{2M}{4\pi R^3}\,h_\theta\sin\theta = \frac{2M}{4\pi R^3}\left[1-\phi(u)+\left(\frac{2}{\pi}\right)^{1/2} u(1+u^2)\,\mathrm{e}^{-u^2/2}\right]\sin\theta \\
E_\phi &= -\frac{M\rho}{4\pi R^4}\,e_\phi\sin\theta = -\frac{M}{4\pi R^4}\left(\frac{2}{\pi}\right)^{1/2} u^5 E^{-u^2/2}\sin\theta
\end{aligned} \tag{9.24}$$

where $\phi(u) = (2/\pi)^{1/2}\int_0^u \mathrm{e}^{-t^2/2}\,dt$ is the probability integral.

Equations 9.23–9.24 are valid when displacement currents are negligible with respect to conduction currents, and the field is measured at times significantly exceeding time T_0, which is required for the signal to arrive at the observation point.

Table 9.2 contains values of function h_R, h_θ and e_ϕ, depending on parameter u. Graphs of functions h_R, h_θ and e_ϕ, provided that the dipole current is turned on at the moment $t = 0$, are given in Fig. 9.5. For this reason, with an increase of time the magnetic field tends to that of the direct current while the electric field vanishes, i.e. as

$$h \to h_0 \qquad e_\phi \to 0 \quad \text{as } t \to \infty$$

In accord with the electromagnetic induction law at the first moment the field is absent in a conducting medium. Such behavior takes place due to the fact that induced currents arise near the source, which compensate the primary field of the dipole current.

Let us notice that from eq. 9.3, which is valid in a general case, it also follows that at the initial moment the field is absent in all parts of a medium. However, in accord with equations for the quasistationary field, when $t < \tau_0$, it has a finite value, while it is, in fact, equal to zero.

Applying expansion of integral $\phi(u)$ in a series by small parameters u (relatively large times, small distances from the source to the observation point, sufficiently small conduc-

TABLE 9.2
Values of functions h_R, h_θ and e_θ

u	R/τ	$1-h_R$	$1-h_\theta$	e_ϕ	h_R	h_θ
0.0500	0.796×10^{-2}	0.3300×10^{-4}	-0.6661×10^{-4}	0.2490×10^{-6}	1.0000	1.000
0.0595	0.946	0.5561	-0.1118×10^{-3}	0.5920	0.9999	1.000
0.0707	0.113×10^{-1}	0.9364	−0.1878	0.1407×10^{-5}	0.9999	1.000
0.0841	0.134	0.1575×10^{-4}	−0.3152	0.3343	0.9998	1.000
0.100	0.159	0.2649	−0.5290	0.7939	0.9997	1.001
0.119	0.189	0.4452	−0.8873	0.1884×10^{-4}	0.9996	1.001
0.141	0.225	0.7476	-0.1487×10^{-2}	0.4469	0.9993	1.001
0.168	0.268	0.1254×10^{-2}	-0.1058×10^{-3}	0.9987	1.002	
0.200	0.318	0.2102	−0.4154	0.2503	0.9979	1.004
0.238	0.379	0.3518	−0.6917	0.5903	0.9965	1.007
0.283	0.450	0.5876	-0.1147×10^{-1}	0.1388×10^{-2}	0.9941	1.001
0.336	0.535	0.9785	−0.1891	0.3246	0.9902	1.019
0.400	0.637	0.1623×10^{-1}	−0.3091	0.7542	0.9838	1.031
0.476	0.757	0.2676	−0.4993	0.1735×10^{-1}	0.9732	1.050
0.566	0.900	0.4378	−0.7930	0.3938	0.9562	1.080
0.673	0.107×10^{0}	0.7081	-0.1229×10^{0}	0.8767	0.9292	1.123
0.800	0.127	0.1128×10^{0}	−0.1839	0.1899×10^{0}	0.8872	1.184
0.951	0.1514	0.1758	−0.2612	0.3955	0.8242	1.261
1.130	0.1801	0.2661	−0.3432	0.7799	0.7339	1.343
1.350	0.2141	0.3873	−0.3988	0.1423×10^{1}	0.6127	1.399
1.600	0.2546	0.5355	−0.3732	0.2326	0.4645	1.373
1.900	0.3028	0.6945	−0.2048	0.3256	0.3055	1.205
2.263	0.3601	0.8368	0.1222	0.3659	0.1632	0.878
2.691	0.4283	0.9354	0.5192	0.3014	0.0646	0.481
3.200	0.5093	0.9834	0.8271	0.1600	0.0166	0.173
3.805	0.6057	0.9977	0.9662	0.4564	0.232×10^{-2}	0.338×10^{-1}
4.525	0.7203	0.9999	0.9972	0.5409×10^{-1}	0.135×10^{-3}	0.278×10^{-2}
5.382	0.8565	0.9999	0.9999	0.1851×10^{-2}	0.229×10^{-5}	0.662×10^{-4}
6.400	0.1019×10^{1}	1.0000	1.0000	0.1093×10^{-4}	0.673×10^{-8}	0.273×10^{-6}
7.611	0.1211	1.0000	1.0000	0.5378×10^{-8}	0.160×10^{-11}	0.944×10^{-10}
9.051	0.1441	1.0000	1.0000	0.7883×10^{-13}	0.117×10^{-16}	0.974×10^{-15}

tivity) we obtain approximate formulas for components of the secondary field:

$$\begin{aligned} H_R &\simeq -\frac{M}{4\pi R^3}\left(\frac{2}{\pi}\right)^{1/2} u^3\left(1-\frac{3}{10}u^2\right)\cos\theta \\ H_\theta &\simeq \frac{M}{4\pi R^3}\left(\frac{2}{\pi}\right)^{1/2} u^3\left(1-\frac{3}{5}u^2\right)\sin\theta \\ E_\phi &\simeq -\frac{M\rho}{4\pi R^4}\left(\frac{2}{\pi}\right)^{1/2} u^5\left(1-\frac{u^2}{2}\right)\sin\theta \end{aligned} \tag{9.25}$$

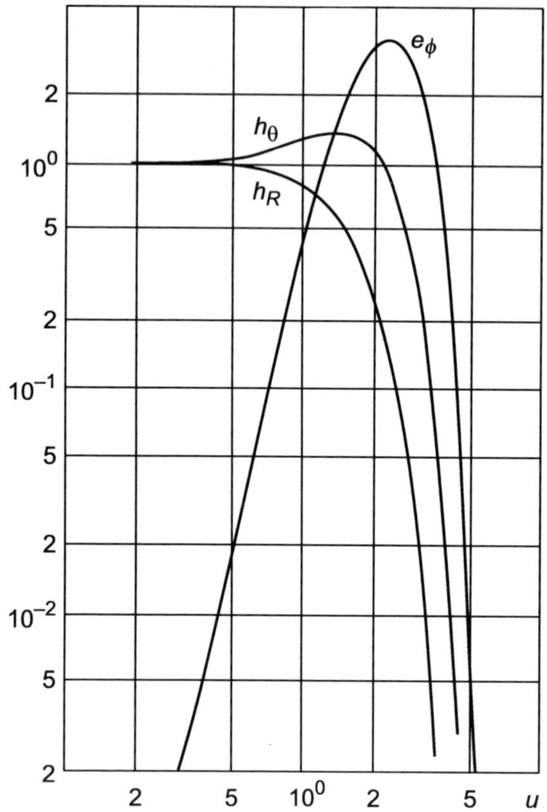

Figure 9.5. Behavior of functions, h_R, h_θ and e_θ in a uniform medium.

and, restricting ourselves by first terms we have:

$$
\begin{aligned}
H_R &\simeq -\frac{M}{12\pi\sqrt{\pi}}\frac{\mu^{3/2}\sigma^{3/2}}{t^{3/2}}\cos\theta \\
H_\theta &\simeq \frac{M}{12\pi\sqrt{\pi}}\frac{\mu^{3/2}\sigma^{3/2}}{t^{3/2}}\sin\theta \\
E_\phi &\simeq -\frac{M}{16\pi\sqrt{\pi}}\frac{\mu^{5/2}\sigma^{3/2}}{t^{5/2}}R\sin\theta
\end{aligned}
\tag{9.26}
$$

These formulas describe a field with a sufficient accuracy if parameter $u < 0.2$.

Thus, at the late stage of the transient response the magnetic field does not depend on the distance from the source, and it is related with conductivity of a medium more closely than that in a frequency domain when the quadrature component or an amplitude of the magnetic field are measured.

TABLE 9.3
Values of parameter u

ρ, ohm·m \ t, μsec	1	4	9	16	25	36	49	64	81	100
0.1	2.50	1.25	0.84	0.63	0.50	0.42	0.36	0.31	0.28	0.25
0.5	1.11	0.56	0.37	0.28	0.22	0.19	0.16	0.14	0.12	0.11
1.0	0.80	0.40	0.27	0.20	0.16	0.13	0.11	0.10	0.00	0.081
5.0	0.35	0.18	0.12	0.088	0.071	0.059	0.051	0.044	0.039	0.035
10.0	0.25	0.125	0.084	0.063	0.050	0.042	0.036	0.031	0.028	0.025

For illustration some values of parameter u as a function of resistivity ρ and time t, if $R = 1$ m, are given in Table 9.3.

The fact that the distance from the source does not have an influence on the field at the late stage suggests that in this case sources of the field are located from an observation point at distances which are essentially greater than the probe length (R).

In accord with eq. 9.24 current density in a medium is:

$$J_\phi = -\frac{M}{4\pi}\frac{\sin\theta}{R^4}\left(\frac{2}{\pi}\right)^{1/2} u^5 \mathrm{e}^{-u^2/2} \tag{9.27}$$

Graphs of function $(1/R^4)u^5\mathrm{e}^{-u^2/2}$ are presented in Fig. 9.6. With an increase of time a maximum of curves is shifted to the side of greater distances. For this reason the magnetic field and EMF measured on the axis of the dipole become more sensitive to removed parts of the medium. Let us confirm this assumption by a following calculation. We will mentally present whole uniform space as a system of concentric spherical shells. At every moment a measured magnetic field is defined by distribution of currents in shells.

Omitting intermediate transformations related with the calculation of the magnetic field and making use of the Biot–Savart law we obtain for a ratio of EMF caused by currents in shells, the radius of which exceeds R_2 to EMF in a measuring coil, located in a uniform medium at distance R from the dipole, the following expression:

$$G(u_1, \alpha) = \left(1 - \frac{1}{3}u_1^2\right) \mathrm{e}^{-u_1^2(\alpha^2-1)/2}$$

where $\alpha = R_2/R_1$, $u = 2\pi R_1/\tau$.

Curves of function $G(u_1, \alpha)$ are presented in Fig. 9.7. For small times currents are mainly concentrated near the dipole, and a field measured at point R_1 does not practically depend on induced currents located in relatively removed parts of a medium ($u_1 \to \infty$, $G \to 0$). In contrary, for large times ($u_1 \to 0$), the field is mainly defined by currents induced in an external area ($R > R_2$) and $G(u_1, \alpha) \to 1$.

Correspondingly, the later measurements are performed with a greater depth of investigation.

If parameter $(1/2)u_1^2(\alpha^2 - 1) < 0.1$, then function $G(u_1, \alpha)$, characterizing a relative contribution of EMF caused by currents induced in area $R > R_2$, is equal to:

$$G(u_1, \alpha) \simeq 1 - \frac{1}{3}u_1^2$$

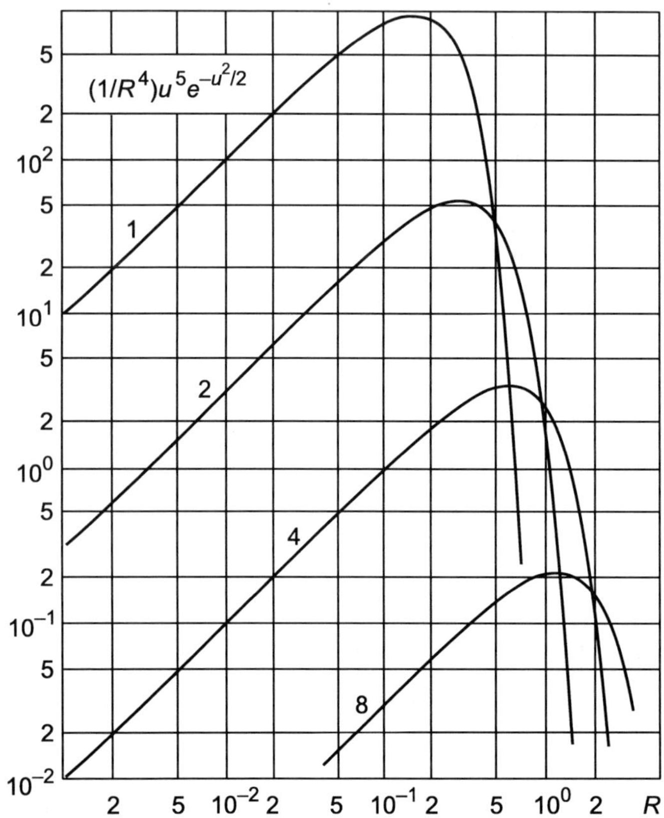

Figure 9.6. Graphs of function $(1/R^4)u^5\mathrm{e}^{-u^2/2}$. Curve index τ.

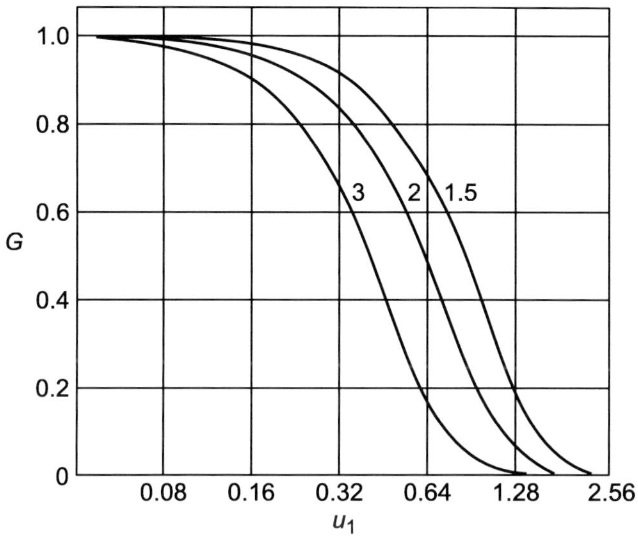

Figure 9.7. Graphs of function $G(u, \alpha)$. Curve index α.

TABLE 9.4
Values of EMF, μV; $M_T M_R = 0.1\ A^2 \cdot m^4$, $L = 1$ m

ρ, ohm·m \ t, μsec	1	4	9	16	25
0.1	0.543×10^4	0.179×10^4	0.365×10^3	0.101×10^3	0.355×10^2
0.5	0.600×10^4	0.300×10^3	0.431×10^2	0.106×10^2	0.351×10^1
1.0	0.290×10^4	0.115×10^3	0.158×10^2	0.381×10^1	0.126×10^1
5.0	0.334×10^4	0.109×10^2	0.145×10^1	0.346×10^0	0.113×10^0
10.0	0.122×10^3	0.390×10^1	0.515×10^0	0.122×10^0	0.402×10^{-1}

It is obvious that along with an increase of the depth of investigation in the radial direction, the sensitivity of the probe in a vertical direction also increases, while due to the transformation of electromagnetic energy into heat the signal rapidly decreases.

Table 9.4 contains some values of the electromotive force induced in the receiver of a two-coil probe for various resistivities of a medium.

Calculations of electromotive force, ε, have been performed by equation:

$$\varepsilon(\mu V) = \frac{M_T M_R}{2\pi L^5} \rho\, e_\phi 10^6 \qquad (9.28)$$

where L is the probe length; M_T and M_R are moments of transmitter and receiver coils, respectively; $u = 2\pi L/\tau$ and $e_\phi = (2/\pi)^{1/2} u^5 \mathrm{e}^{-u^2/2}$.

9.2. Transient Field of the Vertical Magnetic Dipole on the Borehole Axis at the Late Stage

For determining transient responses in a medium with cylindrical interfaces we will use, as in the case of a uniform medium, Fourier transform. Assuming that the field excitation is a step function:

$$M(t) = \begin{cases} 1 & t < 0 \\ 0 & t > 0 \end{cases} \tag{9.29}$$

we have:

$$h_z(t) = \frac{1}{2\pi} \int_{-\infty}^{\infty} h_z(\omega) \frac{\mathrm{e}^{\mathrm{i}\omega t}}{-\mathrm{i}\omega} \, \mathrm{d}\omega$$

and

$$\dot{h_z}(t) = -\frac{1}{2\pi} \int_{-\infty}^{\infty} h_z(\omega) \mathrm{e}^{\mathrm{i}\omega t} \, \mathrm{d}\omega \tag{9.30}$$

where $-1/\mathrm{i}\omega$ is the spectrum of excitation where the low-frequency part essentially prevails; $h_z(\omega)$ is the magnetic field on the borehole axis expressed in units of the magnetic field in a free space (displacement currents are neglected).

As was shown in Chapter 4:

$$h_z(\omega) = h_z^{un}(k, L) - \frac{\alpha^3}{\pi} \int_0^{\infty} m_1^2 C_1 \cos m\alpha \, \mathrm{d}m \tag{9.31}$$

where $h_z^{un}(k, L)$ is the field in a uniform medium with conductivity of the borehole; $\alpha = L/a_1$ and L and a_1 are the probe length and the borehole radius, respectively.

Function C_1 is given by eq. 4.38 and it is equal to:

$$m_1^2 C_1 = m^2 \frac{l_1 n_1 + l_2 n_2}{l_3 n_1 + n_3 n_2}$$

where:

$$\begin{aligned}
l_1 &= -m_2 I_0(m_2) K_1(m_1) - m_1 K_0(m_1) I_1(m_2) \\
n_1 &= m_3 K_1(m_2\beta) K_0(m_3\beta) - m_2 K_0(m_2\beta) K_1(m_3\beta) \\
l_2 &= m_2 K_0(m_2) K_1(m_1) - m_1 K_0(m_1) K_1(m_2) \\
n_2 &= -m_3 I_1(m_2\beta) K_0(m_3\beta) - m_2 I_0(m_2\beta) K_1(m_3\beta) \\
l_3 &= -m_2 I_0(m_2) I_1(m_1) + m_1 I_0(m_1) I_1(m_2) \\
n_3 &= m_2 I_1(m_1) K_0(m_2) + m_1 I_0(m_1) K_1(m_2)
\end{aligned}$$

and

$$m_1 = (m^2 + k_1^2 a_1^2)^{1/2} \qquad m_2 = (m^2 + k_2^2 a_1^2)^{1/2} \qquad m_3 = (m^2 + k_3^2 a_1^2)^{1/2} \qquad \beta = a_2/a_1$$

First of all we will investigate the late stage of the transient response. Taking into account the known features of real and imaginary parts of the complex amplitude $h_z(\omega)$:

$$\text{In}\, h_z(\omega) = \text{In}\, h_z(-\omega) \qquad \text{Q}\, h_z(\omega) = -\,\text{Q}\, h_z(-\omega)$$

instead of eq. 9.31 we have:

$$\dot{h}_z(t) = \frac{2}{\pi}\int_0^\infty \text{Q}\, h_z(\omega) \sin\omega t \, d\omega$$

or (9.32)

$$\dot{h}_z(t) = -\frac{2}{\pi}\int_0^\infty \text{In}\, h_z(\omega) \cos\omega t \, d\omega$$

Integrating the first integral of eq. 9.32 by parts we find an expression for the late stage as a series by powers of $1/t$:

$$\dot{h}_z(t) \simeq \frac{2}{\pi}\lim_{\omega\to 0}\left\{\frac{\phi_1(\omega)}{t} - \frac{\phi_1'(\omega)\sin\omega t}{t^2} - \frac{1}{t^3}\phi_1''(\omega) - \frac{1}{t^3}\int_0^\infty \phi'''(\omega)\cos\omega t \, d\omega\right\} \tag{9.33}$$

where $\phi_1(\omega) = \text{Q}\, h_z(\omega)$

A similar relation can be derived from the second integral of eq. 9.32. Therefore, derivation of asymptotic formulae for the late stage of the transient responses consists of two steps, namely:

1. The presentation of the low-frequency spectrum in a series with respect to ω.

2. The determination of coefficients of the asymptotic series by powers of $1/t$.

In accord with eq. 9.31 the vertical component of the magnetic field on the borehole axis is presented as a sum of cylindrical harmonics, characterized by spatial frequency m. The greater m, the more rapidly a corresponding harmonic of the field alters, and therefore a sufficiently uniform field is formed by low-frequency spatial harmonics, which are located within the initial part of integration.

On the other hand, at great times currents induced in a medium due to the step function excitation are located relatively far away from the dipole, and for this reason a field measured relatively close to the dipole is almost uniform. It means that the main information about the late stage of a transient response is contained in the integrand $m_1^2 C_1(m_1, m_2, m_3)$ when the variable of integration m is small.

This conclusion can be derived in a different way. In fact, as is well known, the late stage of the transient response is defined by only those terms of the low-frequency spectrum which contain odd powers of wave number, k, and logarithmic terms.

Let us present integral $\int_0^\infty m_1^2 C_1 \cos m\alpha \, dm$ as a sum:

$$\int\limits_0^\infty m_1^2 C_1 \cos m\alpha \, dm = \int\limits_0^{m_0} m_1^2 C_1 \cos m\alpha \, dm + \int\limits_{m_0}^\infty m_1^2 C_1 \cos m\alpha \, dm$$

where m_0 is an arbitrary small value ($m_0 < 1$).

Within the external interval ($m > m_0$) radicals m_1, m_2 and m_3 can be expanded in a series by powers k^2/m^2 and it allows us to present the external integral $\int_{m_0}^\infty m_1^2 C_1 \cos m\alpha dm$ as a converging series containing only even powers of wave number k:

$$\int\limits_{m_0}^\infty m_1^2 C_1 \cos m\alpha \, dm = \sum_l A_l \left(\frac{k}{m}\right)^{2l}$$

Thus, we see again that the low-frequency part of the spectrum, containing odd powers of ω as well as logarithmic terms, can be obtained from integrand $m_1^2 C_1$, when $m \to 0$. First we will find the low-frequency part of the spectrum, which defines the late stage of the transient field on the borehole axis, when an invasion zone is absent.

Let us present functions $K_0(m_1)$ and $K_1(m_1)$ as a series:

$$K_0(m_1) = -\left(\ln \frac{m_1}{2} + C\right) I_0(m_1) + \Sigma_0(m_1)$$
$$K_1(m_1) = \frac{1}{m_1} + \left(\ln \frac{m_1}{2} + C\right) I_1(m_1) + \Sigma_1(m_1)$$

where C is Euler's constant. Collecting in the numerator of function C_1 terms, containing $\ln(m_1/2) + C$ we can see that the coefficient of $\ln(m_1/2) + C$ is equal to denominator of C_1. For this reason function C_1 can be written in the form:

$$C_1 = \left(\ln \frac{m_1}{2} + C\right) + \frac{m_2 K_0(m_1) \left[\dfrac{1}{m_1} - \Sigma_1(m_1)\right] - m_1 K_1(m_2)\Sigma_0(m_1)}{m_2 K_0(m_2) I_1(m_1) + m_1 K_1(m_2) I_0(m_1)}$$

or

$$m_1^2 C_1 = m_1^2 \left(\ln \frac{m_1}{2} + C\right) + \frac{m_2^2 K_0(m_2)}{m_2^2 K_0(m_2) \dfrac{I_1(m_1)}{m_1} + m_1 K_1(m_2) I_0(m_1)}$$
$$- \frac{m_2^2 m_1 K_0(m_2)\Sigma_1(m_1) + m_1^2 m_2 K_1(m_2)\Sigma_0(m_1)}{m_2^2 K_0(m_2)\dfrac{I_1(m_1)}{m_1} + m_2 K_1(m_2) I_0(m_1)} \qquad (9.34)$$

The first term of eq. 9.34 presents itself as the value of the integrand of the Sommerfeld integral at the initial part of integration, defining a field in a uniform medium with conductivity σ. For this reason this term together with the first term of eq. 9.31, $h_z^{un}(k, L)$, do not have an influence on the late stage of transient response. The denominator of the second and third terms of eq. 9.34 can be written in the form: $1 + O(m_1, m_2)$, and it is not difficult to show that the first term of this expansion in a series by power of m_i corresponds to a uniform medium with the formation resistivity, while the other terms take into account the influence of the borehole.

Derivation of the low-frequency spectrum in this case is a much more cumbersome operation than in the case of horizontally layered medium. With the expansion of function $m_1^2 C_1$ in a series, two types of integrals arise which have to be presented in the form of a series with powers of k:

$$I_1 = \int_0^{m_0} m_1^{2n_1} m_2^{2n_2} \ln m_2 \, dm \qquad I_2 = \int_0^{m_0} m_1^{2n_1} m_2^{2n_2} \ln^2 m_2 \, dm \tag{9.35}$$

Integration by parts I_1 allows us to express this integral as:

$$I_1 = \coth^{-1}\left(\frac{m_0}{k_2}\right) \sum_{p=1}^{N} b_{2n} k_1^p k_2^{N-p} \tag{9.36}$$

where $|k_1 a_1| \to 0$, $|k_2 a_1| \to 0$, $N = 2(n_1 + n_2) + 1$.

For complex values of $z = x + iy$ and $|z| \to \infty$ we have for $\coth^{-1} z$:

$$\coth^{-1} z = \frac{\pi}{2} \operatorname{sign} x - \sum_{n=0}^{\infty} C_n z^{-(2n+1)}$$

Inasmuch as $|m_0/k| \to \infty$, then a series containing odd powers of k is obtained replacing $\coth^{-1}(m_0/k)$ by $\pi/2$.

We will consider integral I_2. Since $m_1^{2n_1}$ and $m_1^{2n_2}$ present themselves as polynomials with respect to m^2, then:

$$I_2 = \int_0^{m_0} \sum_n a_n m^{2n} \ln^2 m_2^2 \, dm$$

Integrating every term by parts, we have:

$$I_2 = -4 \int_0^{m_0} \sum_n a_n \frac{m^{2n+2}}{2n+1} \frac{\ln m_2^2}{m_2^2} \, dm$$

This series can be expressed through the sum of integrals of type I_1, and the integral $\int_0^{m_0} (\ln m_2^2)/m_2^2 \, dm$. The upper limit in the latter can be replaced by ∞, inasmuch as only terms with even powers of k arise in the external part of integration. After integration we obtain:

$$\int_0^\infty \frac{\ln m_2^2}{m_2^2}\,dm = \frac{\pi \ln(2k_2)}{k_2}$$

Performing all transformations related with the expansion in a series with powers of k, we derive the following expression for the low-frequency parts of the spectrum without even powers of k:

$$f_3 k_1^3 + f_5 k_1^5 + f_7 k_1^7 + l_7 k_1^7 \ln k_1 \tag{9.37}$$

where:

$$\begin{aligned} f_3 &= \frac{\alpha^3 s^{3/2}}{3} \qquad f_5 = f_3\left(\frac{\alpha^2 s}{10} - \frac{1-s}{2}\right) \\ f_7 &= f_3\left[\frac{\alpha^4 s^2}{280} - \frac{\alpha^2 s(1-s)}{20} + \frac{5}{32}(1-s)^2 - \frac{s(1-s)}{10}\left(C - \frac{77}{60} + \frac{\ln s}{2}\right)\right] \\ l_7 &= -f_3 \frac{s}{10}(1-s) \qquad s = \sigma_2/\sigma_1 \end{aligned} \tag{9.38}$$

Some of these coefficients have been derived in Chapter 4.

Now consider function $m_1^2 C_1$ for a three-layer medium and transform it in a way which is convenient for obtaining the low-frequency asymptote.

Substituting in formulae for l_1 and l_2, functions $K_0(m)$ and $K_1(m)$, presented by series, we obtain:

$$\begin{aligned} l_1 &= l_3\left(\ln\frac{m_1}{2} + C\right) + l_{11} \\ l_2 &= n_3\left(\ln\frac{m_1}{2} + C\right) + l_{21} \\ l_{11} &= -m_2 I_0(m_2)\left[\frac{1}{m_1} - \Sigma_1(m_1)\right] - m_1 I_1(m_2)\Sigma_0(m_1) \\ l_{21} &= m_2 K_0(m_2)\left[\frac{1}{m_1} - \Sigma_1(m_1)\right] - m_1 K_1(m_2)\Sigma_0(m_1) \end{aligned}$$

For this reason function C_1 can be written as:

$$C_1 = \ln\frac{m_1}{2} + C + \frac{l_{11} n_1 + l_{21} n_2}{l_3 n_1 + n_2 n_3} = \ln\frac{m_1}{2} + C + C_{11} \tag{9.39}$$

As in the case of a two-layer medium the integral from the first term at the initial part of integration along with field $h_z^{un}(k, L)$ do not contain odd powers of k.

Making use again of an expansion of functions of K_0 and K_1 we have:

$$l_{21} = l_{11}\left(\ln\frac{m_2}{2} + C\right) + l_{22}$$

$$l_{22} = m_2\Sigma_0(m_2)\left[\frac{1}{m_1} - \Sigma_1(m_1)\right] - m_1\Sigma_0(m_1)\left[\frac{1}{m_2} - \Sigma_1(m_2)\right]$$

Performing similar transformations for n_1 we have:

$$n_1 = -n_2\left(\ln\frac{m_2\beta}{2} + C\right) + n_{11}$$

$$n_{11} = m_3K_0(m_3\beta)\left[\frac{1}{m_2\beta} - \Sigma_1(m_2\beta)\right] - m_2K_1(m_3\beta)\Sigma_0(m_2\beta)$$

As a result function C_{11} can be written in the form:

$$C_{11} = \frac{l_{11}n_{11} + l_{22}n_2 - l_{11}n_2\ln\beta}{l_3n_1 + n_2n_3}$$

Multiplying the numerator and denominator of C_{11} by $m_3\beta/m_1$, we notice that in product $l_{11}n_{11}$ there is a term $-m_3^2K_0(m_3\beta)$, but other terms in the numerator have a higher order of m. At the same time denominator $(m_3\beta/m_1)(n_2n_3+l_3n_1)$ can be presented in the form: $-1 + O(m_1, m_2, m_3)$. Thus, after expansion of denominator C_{11} in a series the leading term of the expansion of $m_1^2C_1$ for small values of m_1, m_2 and m_3 presents the integrand of the Sommerfeld integral for the field in uniform medium with resistivity of the formation, i.e. $m_3^2K_0(m_3\beta)$, and the other terms take into account the influence of the borehole and the invasion zone.

Omitting intermediate transformations, let us write down two main terms of the low-frequency spectrum, containing only odd powers of k:

$$\phi_3k_1^3 + \phi_5k_1^5 \tag{9.40}$$

$$\phi_3 = \frac{1}{3}\alpha^3s_1^{3/2} \qquad \phi_5 = \phi_3\left(\frac{\alpha^2s_1}{10} - \frac{s_{12}}{2}\right)$$
$$s_1 = \frac{\sigma_3}{\sigma_1} \qquad s_2 = \frac{\sigma_2}{\sigma_1} \qquad s_{21} = 1 - s_2 + (s_2 - s_1)\beta^2 \tag{9.41}$$

In applying methods, allowing us to derive formulae corresponding to the late stage of transient field, when the low-frequency spectrum is known, we have:

$$\frac{\partial B_z}{\partial t} = -\frac{M\rho_1}{2\pi L^3a_1^3}\left(\frac{2}{\pi}\right)^{1/2}\left(\frac{2\pi}{\tau_1/a_1}\right)^5\left\{3f_3 - \frac{15}{2}f_5\frac{8\pi^2}{(\tau_1/a_1)^2}\right.$$
$$\left. + \frac{105}{4}\left(\frac{8\pi^2}{\tau_1/a_1}\right)^2\left[f_7 - \frac{l_7}{2}\left(C + \ln 4 + \frac{\pi}{4} - 176 + \frac{1}{2}\ln\frac{\tau_1}{8\pi^2a_1^2}\right)\right]\right\} \tag{9.42}$$

if an invasion zone is absent.

By analogy for a three-layered medium we have:

$$\frac{\partial B_z}{\partial t} = -\frac{M\rho_1}{2\pi L^3a_1^3}\left(\frac{2}{\pi}\right)^{1/2}\left(\frac{2\pi}{\tau_1/a_1}\right)^5\left(3\phi_3 - \frac{15}{2}\phi_5\frac{8\pi^2}{(\tau_1/a_1)^2}\right) \tag{9.43}$$

For sufficiently large times the main role in asymptotic formulae is played by the first term, and as follows from a physical point of view it does not depend on parameters of the borehole and the invasion zone and coincides with the leading term of expansion at the late stage in a uniform medium with the formation resistivity. Thus, regardless of the probe length there is always a moment, starting from which the measured field, as well as the electromotive force, are not practically subjected to the influence of induced currents in the borehole and in the invasion zone.

Comparison with results of calculations by the exact solution shows that the asymptotic formula 9.42 satisfactorily describes the field in a medium with one cylindrical interface, if $\tau_1/a_1 > 20$.

In conclusion let us notice that asymptotic formulae allow us, on one hand, to understand better the physical principles of the method, while on the other hand, they permit us to avoid the calculation of Fourier integral at large times. The latter operation due to the oscillation nature of the integrand is related with great numerical difficulties.

9.3. Apparent Resistivity Curves of the Transient Method in a Medium with Cylindrical Interfaces

We will present results of calculations of the field, $\partial B_z/\partial t$, on the borehole axis through apparent resistivity. It can be introduced in several ways, some of which have advantages within a certain range of time. We will use the following approach:

$$\frac{\rho_\tau}{\rho_1} = \left[\frac{\dot{B}_z^{ls}(\rho_1)}{\dot{B}_z}\right]^{2/3} \tag{9.44}$$

where ρ_τ and ρ_1 are apparent resistivity and the borehole resistivity, respectively; $\dot{B}_z^{\,ls}(\rho_1)$ is the derivative of function $B_z(\rho_1)$ with respect to time at the late stage in a uniform medium with the borehole resistivity; $\dot{B}_z$ is a measured function on the borehole axis, $\partial B_z/\partial t$.

In accord with eq. 9.24:

$$B_z^{ls}(\rho_1) = \frac{1}{a_1^5}\left(\frac{2}{\pi}\right)^{1/2} \frac{M_T \rho_1}{2\pi\alpha^5} u_1^5 \tag{9.45}$$

where $u_1 = 2\pi\alpha/(\tau_1/a_1)$. Whence:

$$\frac{\rho_\tau}{\rho_1} = \frac{8\pi^3}{\tau_1^3}\left(\frac{\pi}{\tau_1}\right)^{1/3}\left(\frac{M_T\rho_1}{t\dot{B}_z}\right)^{2/3} \tag{9.46}$$

or

$$\rho_\tau = \frac{\mu}{4\pi t}\left(\frac{\mu M_T}{t\dot{B}_z}\right)^{2/3}$$

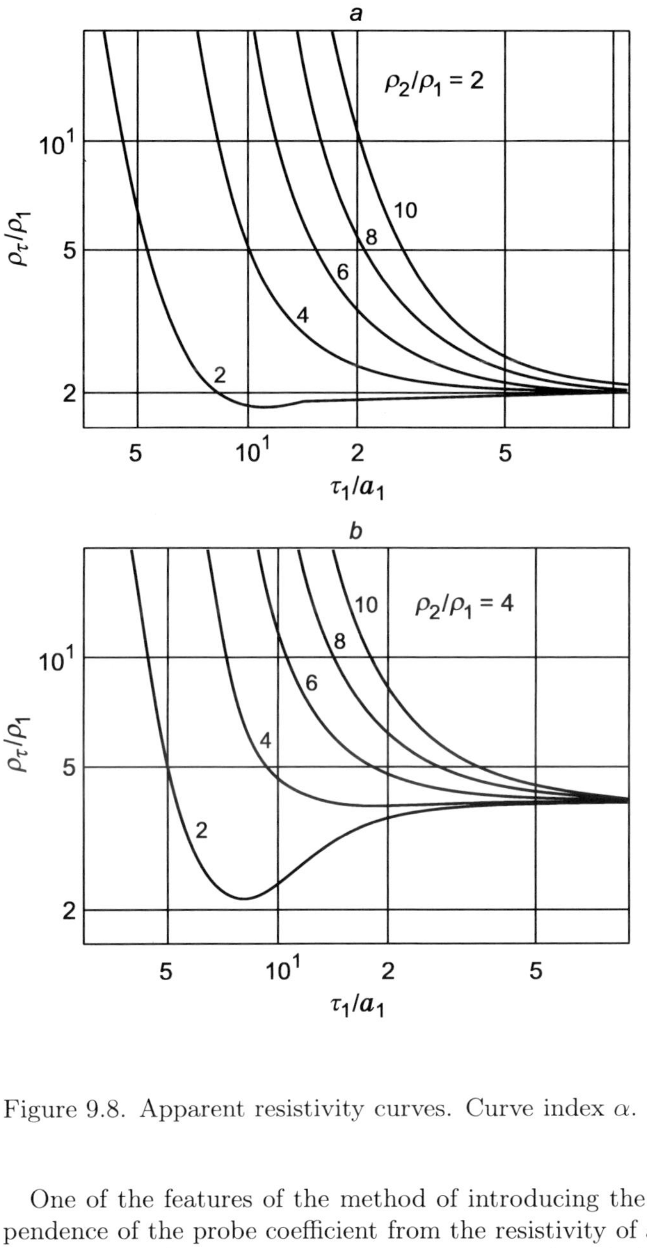

Figure 9.8. Apparent resistivity curves. Curve index α.

One of the features of the method of introducing the apparent resistivity is the independence of the probe coefficient from the resistivity of a medium.

Examples of apparent resistivity curves are given in Figs. 9.8–9.22. Two-layer curves, when the invasion zone is absent, are given in Figs. 9.8–9.11. The index of the set of curves is the ratio ρ_2/ρ_1, while every curve is characterized by parameter $\alpha = L/a_1$. The set of curves for a three-layer medium, Figs. 9.12–9.22, has as index:

$$\frac{\rho_2}{\rho_1} - \frac{a_2}{a_1} - \frac{\rho_3}{\rho_1}$$

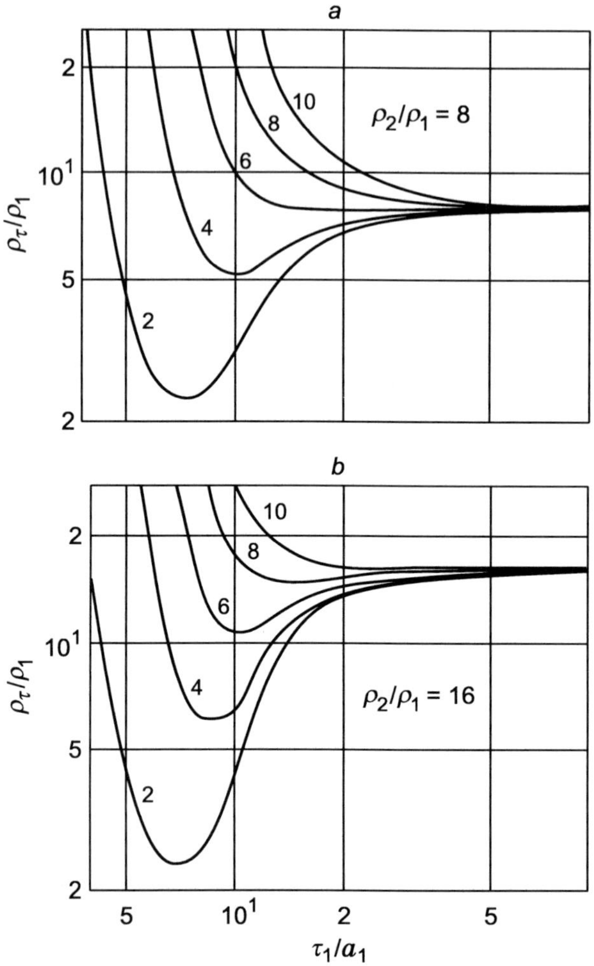

Figure 9.9. Apparent resistivity curves. Curve index α.

The parameter of every curve is again the ratio of the probe length, L, to the borehole radius, $\alpha = L/a_1$. The ratio τ_1/a_1 is plotted along the axis of the abscissa.

All calculations are performed for probes with lengths exceeding the borehole diameter ($\alpha > 2$). For this reason, at the early stage the field does not tend to that in a uniform medium with resistivity of the borehole. With a decrease of time the value of ρ_τ unlimitedly increases, i.e. the field in the borehole in the early stage is significantly smaller than that calculated by formula of the late stage.

The shape of apparent resistivity curves, ρ_τ, essentially depends on the probe length and conductivity of the medium. With an increase of time there is a minimum which

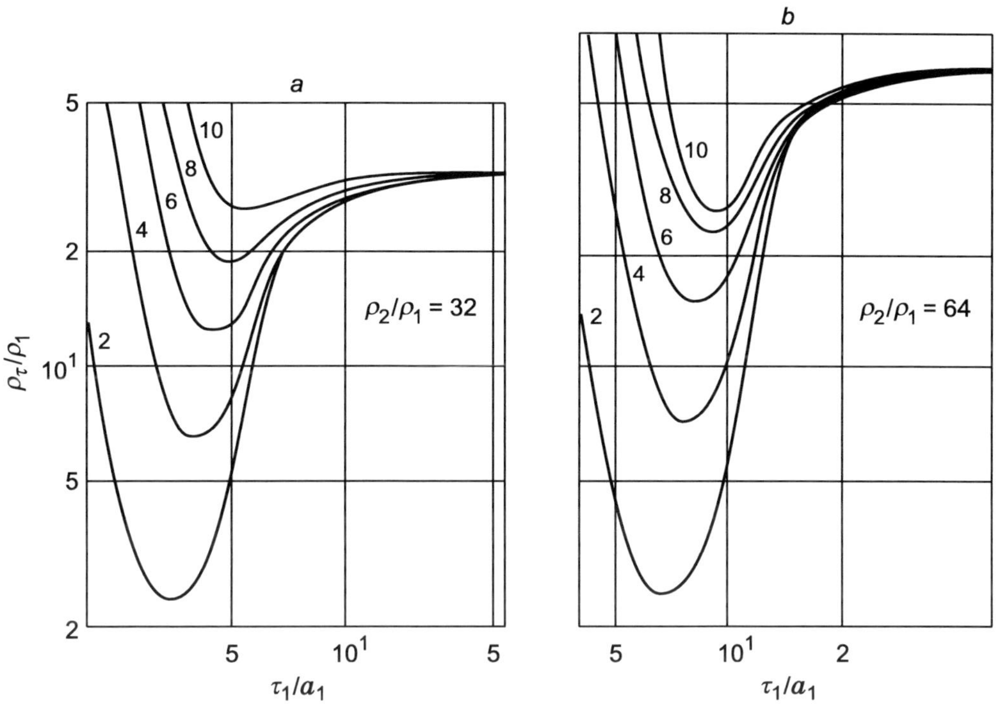

Figure 9.10. Apparent resistivity curves. Curve index α.

becomes deeper with a decrease of the probe length and an increase of the formation resistivity. Then the apparent resistivity, ρ_τ, rapidly increases, approaching the right-hand asymptote equal to the formation resistivity. With an approach this asymptote of the transient response the influence of the probe length decreases, and this feature manifests itself at earlier times with an increase of the formation resistivity and a decrease of the probe length. This means that within this range of time the main influence is caused by currents outside of the borehole, sufficiently removed from the dipole and the observation point, but their density still depends on the borehole resistivity when the invasion zone is absent. The smaller the conductivity of the external area, the more rapidly a transient field decays, and correspondingly, induced currents near the source, too. For this reason the influence of the probe length becomes practically negligible at earlier times. The second term in eq. 9.41 has the form:

$$f_5 = f_3 \left(\frac{\alpha^2 s}{10} - \frac{1-s}{2} \right)$$

For a sufficiently large resistivity of the formation and a relatively short probe when conditions $\alpha^2 s < 1$ and $s \ll 1$ are met, the second term of the asymptotic expansion of

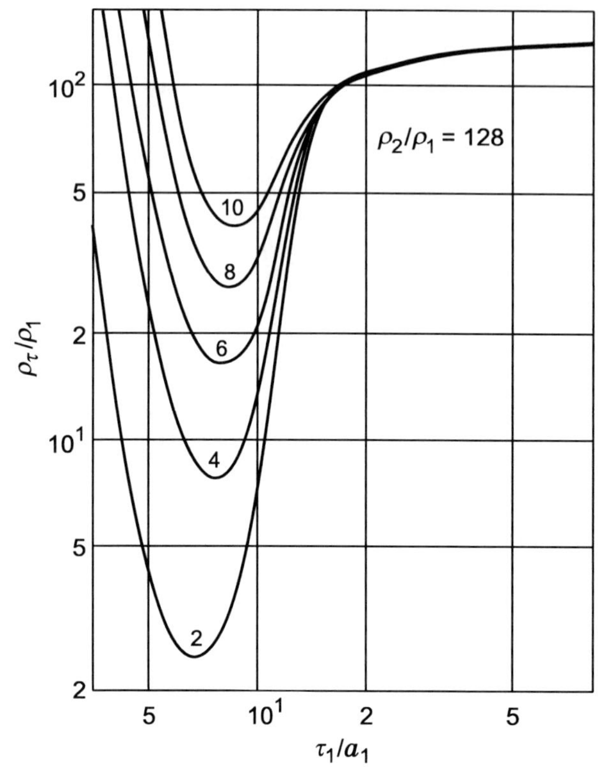

Figure 9.11. Apparent resistivity curves. Curve index α.

eq. 9.42 does not depend on the formation resistivity, as well as the probe length, but it is defined by the resistivity of the borehole.

Thus, if the parameters of the borehole are known we can correct the value of ρ_τ, measured at times when the field on the borehole axis differs from that in a uniform medium with resistivity ρ_2.

In accord with eqs. 9.42 and 9.44 we have:

$$\frac{\rho_\tau}{\rho_1} \simeq \frac{\rho_2}{\rho_1}(1 - \frac{5}{3}u_1^2) \tag{9.47}$$

as $u_1 \to 0$ and $u_1 = 2\pi a_1/\tau_1$. It is clear that the second term of eq. 9.47, defining the correction of ρ_τ, for large times, is directly proportional to the borehole conductance: $\pi a_1^2 \sigma_1$.

A graph of function $\rho_\tau/\rho_2 = 1 - (5/3)u_1^2$ is presented in Fig. 9.23. As is seen from comparison with the exact solution, eq. 9.47 provides a sufficient accuracy in the determination of ρ_τ for a relatively resistive formation ($\rho_2/\rho_1 > 10$), if $\alpha < 4$ and $\tau_1/a_1 > 15$. In this case the correction factor depends only on the borehole conductance. With an increase of the probe length, as well as for formation conductivity, the minimum on curves of ρ_τ becomes smaller and finally completely vanishes. In the latter case the value of apparent resistivity ρ_τ at the right-hand part of the curve monotonically decreases with an increase of time, approaching to its asymptote ρ_2. For relatively resistive formations with an increase of the probe length at the beginning the depth of investigation in the radial direction increases, when $\tau_1/a_1 > 10$.

At every moment of time when parameter $\tau_1/a_1 \gg 1$ we can present the distribution of currents in the following way. Near the borehole axis the current density is small, then with removal from the axis, it increases, reaches a maximum and afterwards it rapidly decreases. The position of the maximum of the current density at the given moment depends on the conductivity of the medium: the more conductive the borehole the greater the distance from the axis where the range of maximal values of current density is observed. In the near zone, where current maximum can be located, the field and the current density are described by asymptotical formulae 9.42, but at greater distances their behavior is different, and in particular, the current density essentially depends on the borehole conductivity.

For sufficiently large formation resistivity there is a time when a significant area around the dipole corresponds to the late stage. For this reason with an increase of the probe length, the geometric factor defining a signal from the borehole decreases, while the role of currents depending mainly on the formation conductivity increases. With a further increase of the probe length the depth of investigation becomes smaller, since the influence of currents depending on both the conductivity of the formation and that of the borehole begins to grow.

A comparison of the exact solution and asymptotic formulae shows that the field in a two-layered medium becomes practically the same as in a uniform medium with formation resistivity, if:

$$\tau_1/a_1 > 30 \qquad \text{or} \qquad t > 90a_1^2/2\pi\rho_1 \ \mu\text{sec} \tag{9.48}$$

In particular, if $\rho_1 = 1$ ohm·m, $a_1 = 0.1$ m then at times exceeding 100 μsec, currents in the borehole practically do not have any influence.

In accord with eq. 9.47 electromotive force induced in the measuring coil of the probe in a uniform medium is:

$$\mathscr{E} = \left(\frac{2}{\pi}\right)^{1/2} \frac{1}{2\pi}\rho_2 \frac{M_T M_R}{L^5} u^5 e^{-u^2/2} = \frac{M_T M_R}{2\pi L^5}\rho_2 e_\phi(u)$$

where $u = 2\pi L/\tau_2$.

As an example, values of electromotive force ε for two probe lengths are given in Table 9.5, when $\tau_1/a_1 > 30$.

Let us notice that if the borehole resistivity is 1 ohm·m and borehole radius is 0.1 m, parameter $\tau_1/a_1 = 90$ corresponds to 1.3 μsec.

The apparent resistivity curves on the borehole axis, when the invasion zone is present, are given in Figs. 9.12–9.22. If resistivity of the invasion zone is less than that of the formation, the shape of curves ρ_τ is the same as that for a two-layered curve, but an approach to the right asymptote takes place at later times.

TABLE 9.5
Values of EMF, μV; $M_T M_R = 10^{-2}\ A^2 \cdot m^4$

ρ_2, ohm·m \ τ_1/a_1	30	45	60	90
		$L = 0.2$ m		
5	0.45×10^4	0.60×10^3	0.14×10^3	0.19×10^2
10	0.16×10^4	0.21×10^3	0.51×10^2	0.67×10^1
20	0.57×10^3	0.75×10^2	0.18×10^2	0.24×10^1
		$L = 1.0$ m		
5	0.30×10^4	0.50×10^3	0.13×10^3	0.18×10^2
10	0.13×10^4	0.19×10^3	0.48×10^2	0.65×10^1
20	0.51×10^3	0.72×10^2	0.17×10^2	0.23×10^1

In the case when the invasion zone has a higher resistivity than that of the formation the shape of apparent resistivity curve changes, specially for relatively short probes and large radii of the invasion zone. At the early stage with an increase of time ρ_τ becomes greater with an increase of the resistivity of the invasion zone, ρ_2. With a further increase of time currents are mainly located in the formation which has higher conductivity, and correspondingly the apparent resistivity decreases asymptotically approaching to its right-hand asymptote, ρ_3.

9.4. The Transient Responses of a Vertical Magnetic Dipole in a Formation with a Finite Thickness

Investigation of a nonstationary field of a vertical magnetic dipole in formations with finite thickness allows us to study the vertical responses of a two-coil induction probe making use of the transient method. It is obvious that with an increase of time the influence of the surrounding medium becomes stronger due to diffusion of currents, and this effect is greater with a decrease of the formation thickness and the resistivity of the surrounding medium. For this reason it is essential to establish the maximum times when measurements with a two-coil probe located against the formation do not yet depend on the conductivity of the surrounding medium.

As was shown in Chapter 5 the expression for the quasistationary field of the magnetic dipole along its axis in the frequency domain is:

$$
\begin{aligned}
B_z^{(1)} &= B_z^{un}(k,z) + \frac{\mu M}{4\pi}\int\limits_0^\infty \frac{m^3}{m_1} m_{12} e^{-(2\alpha-1)m_1 z}\, dz \qquad \text{if } \alpha \geqslant 1 \\
B_z^{(2)} &= \frac{\mu M}{4\pi}\int\limits_0^\infty \frac{2m}{m_1+m_2} e^{-\alpha m_1 z} e^{-(1-\alpha)m_2 z}\, dz \qquad \text{if } 0 \leqslant \alpha < 1
\end{aligned}
\tag{9.49}
$$

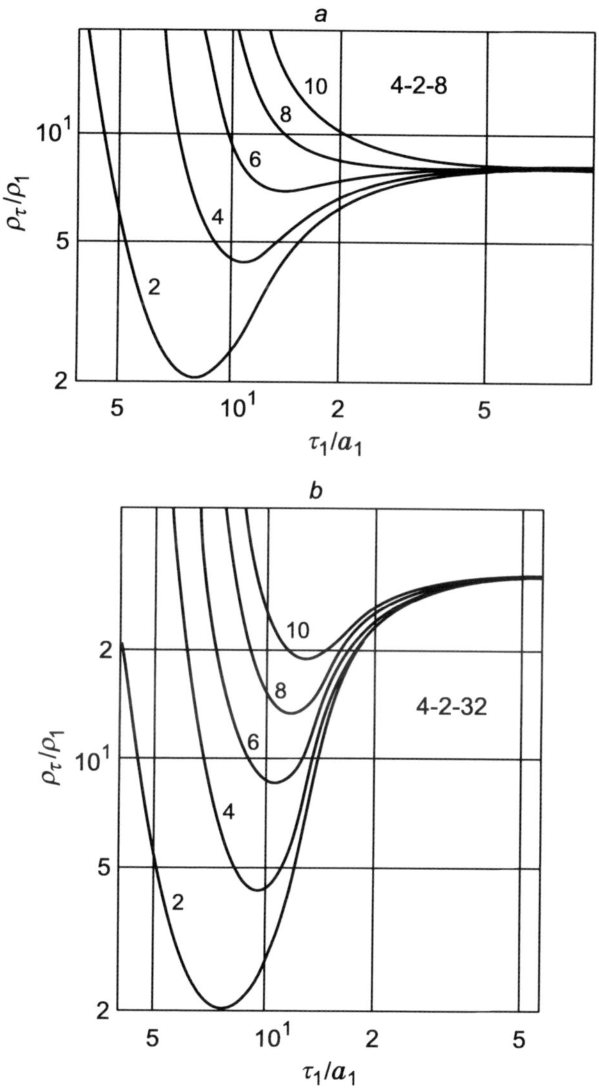

Figure 9.12. Apparent resistivity curves in a three-layered medium. Curve index α.

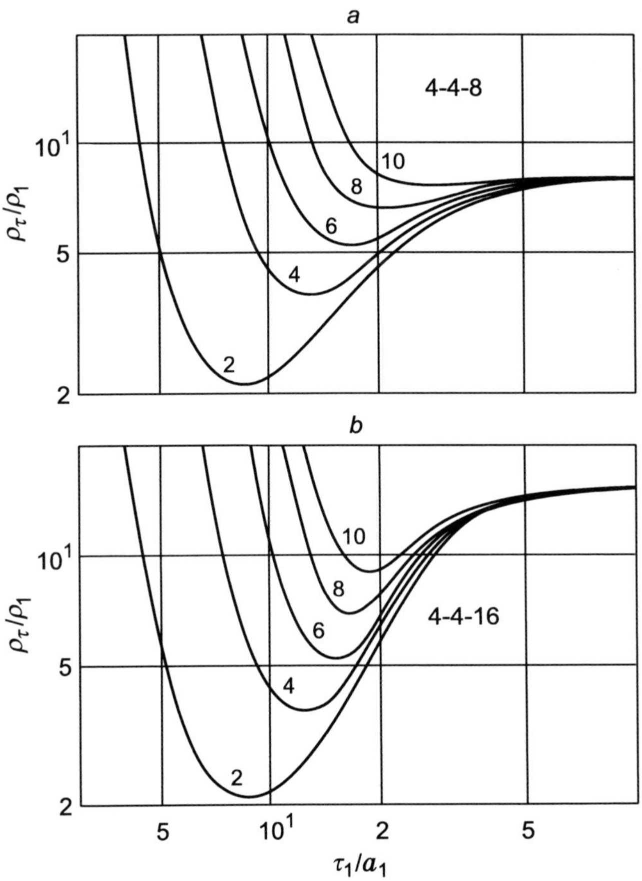

Figure 9.13. Apparent resistivity curves in a three-layered medium. Curve index α.

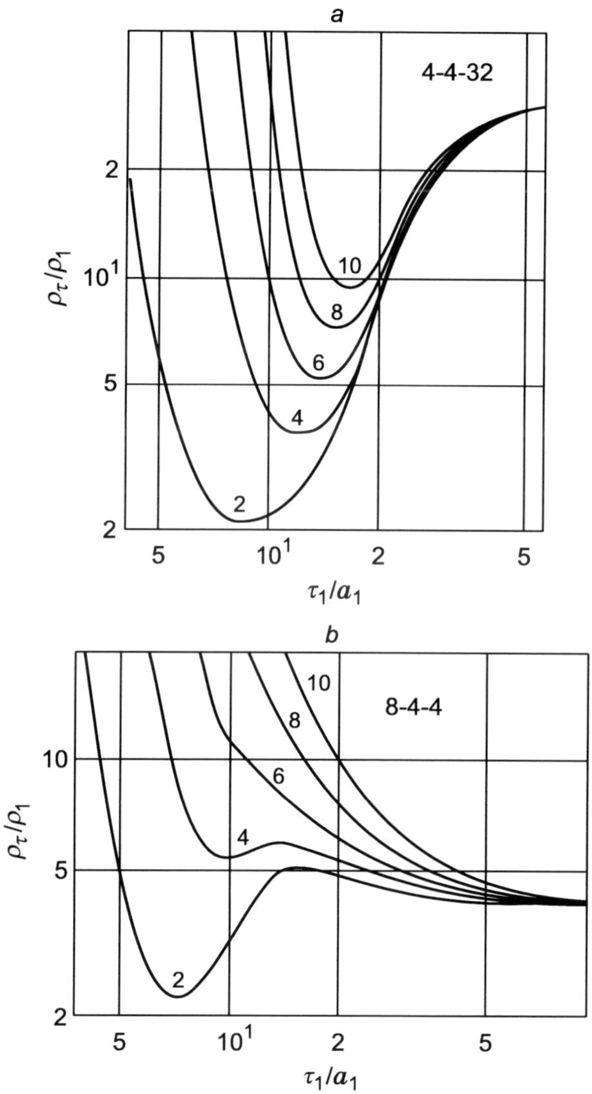

Figure 9.14. Apparent resistivity curves in a three-layered medium. Curve index α.

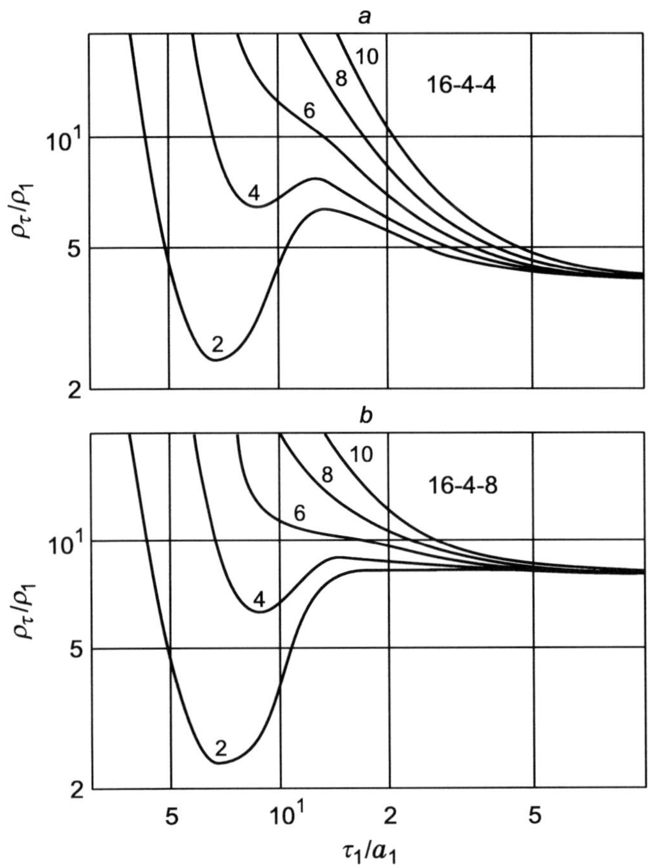

Figure 9.15. Apparent resistivity curves in a three-layered medium. Curve index α.

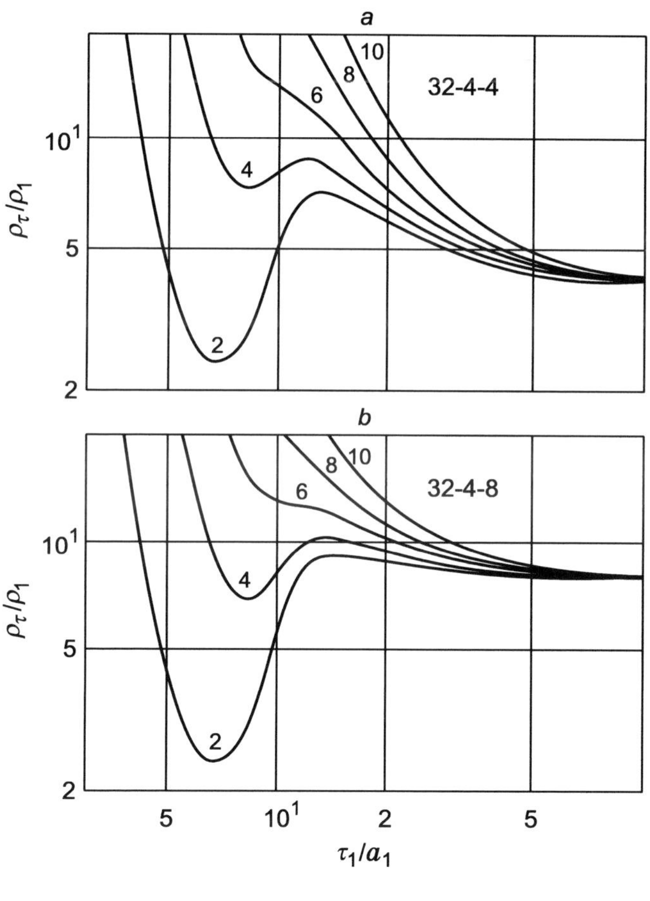

Figure 9.16. Apparent resistivity curves in a three-layered medium. Curve index α.

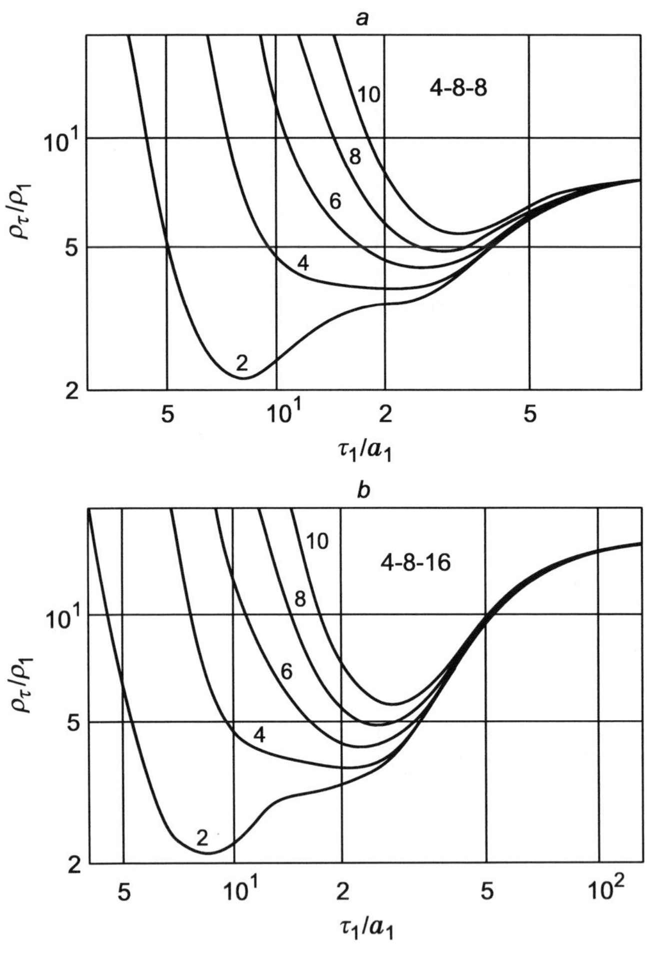

Figure 9.17. Apparent resistivity curves in a three-layered medium. Curve index α.

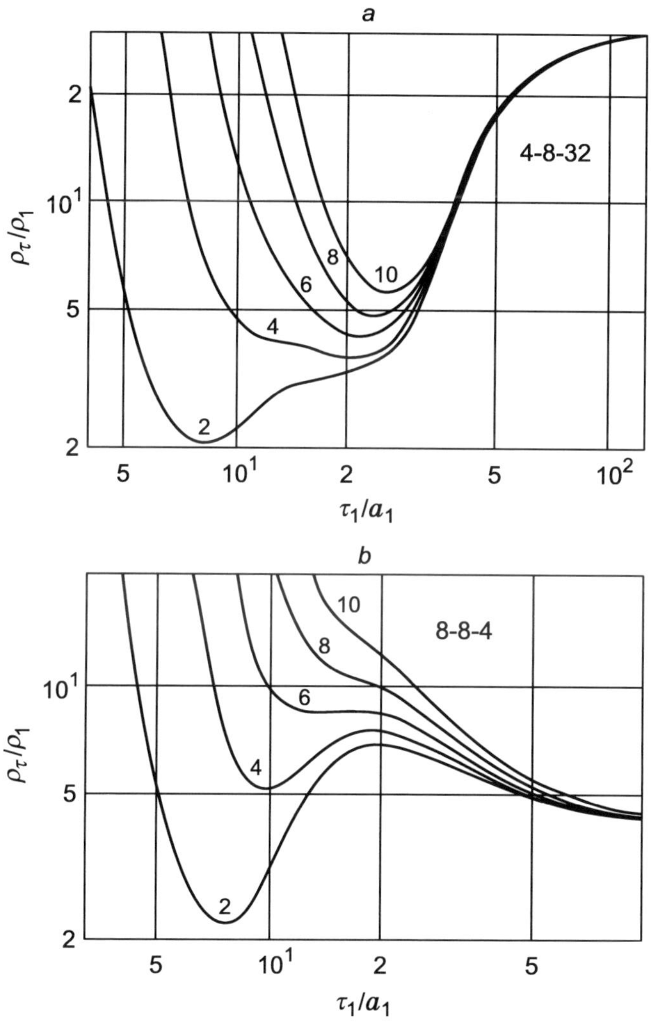

Figure 9.18. Apparent resistivity curves in a three-layered medium. Curve index α.

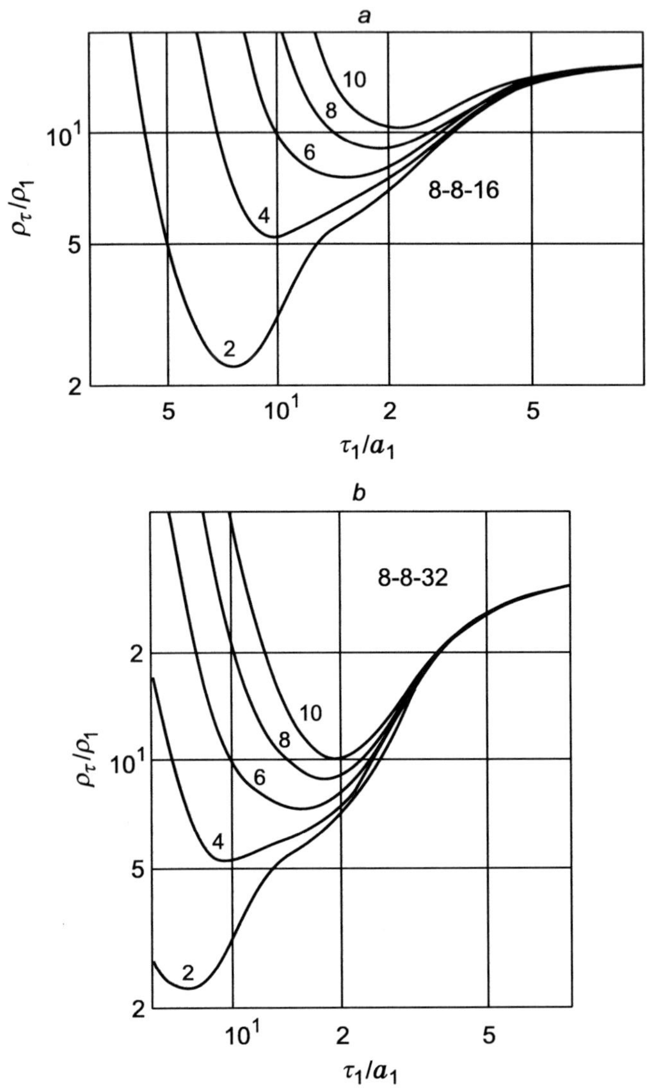

Figure 9.19. Apparent resistivity curves in a three-layered medium. Curve index α.

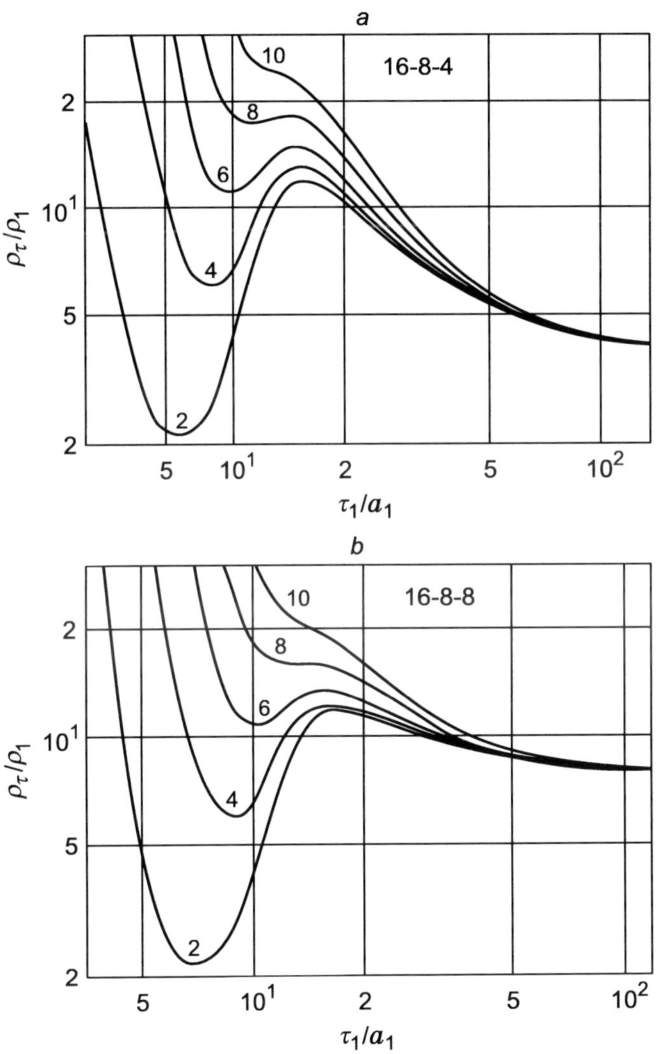

Figure 9.20. Apparent resistivity curves in a three-layered medium. Curve index α.

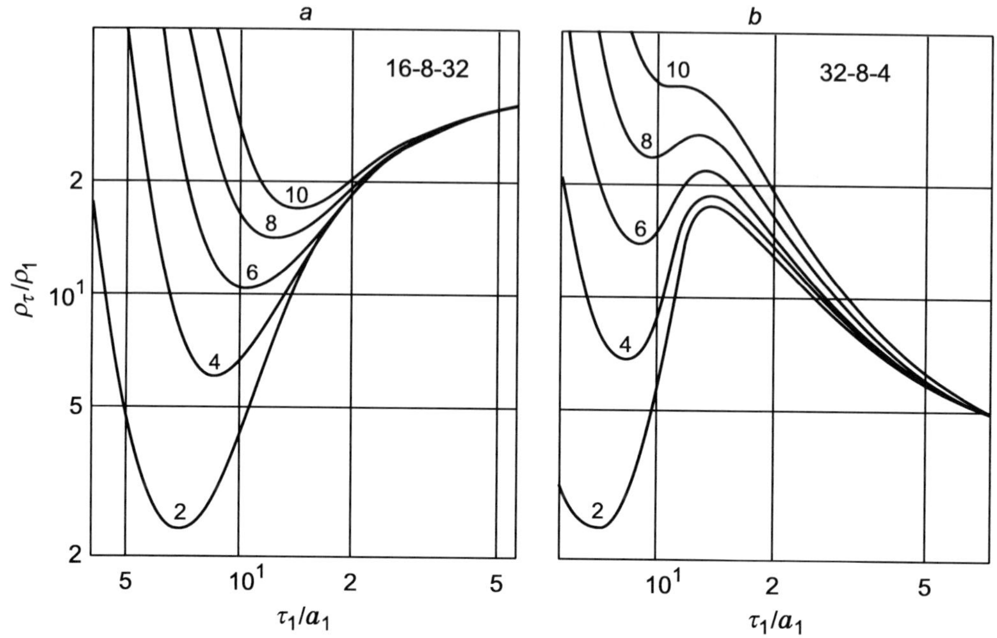

Figure 9.21. Apparent resistivity curves in a three-layered medium. Curve index α.

where k_1 and k_2 are wave numbers of the first and second medium, respectively, and:

$$m_1 = (m^2 - k_1^2)^{1/2} \qquad m_2 = (m^2 - k_2^2)^{1/2} \qquad m_{12} = \frac{m_1 - m_2}{m_1 + m_2} \qquad \alpha = \frac{L}{z}$$

L is the distance from the dipole to the interface; z is the distance from the dipole to the observation point.

If the behavior of the current in the dipole is described by the step function:

$$I = \begin{cases} I & t < 0 \\ 0 & t > 0 \end{cases}$$

we have for the transient field:

$$\dot{B}_z = -\frac{2}{\pi} \int_0^\infty \mathrm{Q}\, B_z \sin \omega t \, \mathrm{d}\omega$$

Taking into account the relative simplicity of eqs. 9.49 it is not difficult to derive asymp-

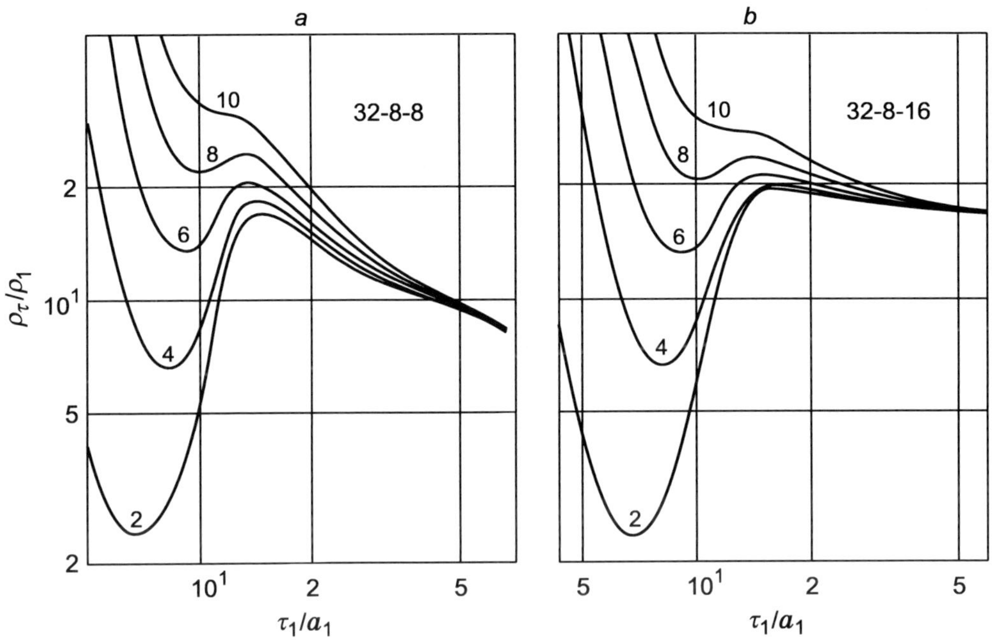

Figure 9.22. Apparent resistivity curves in a three-layered medium. Curve index α.

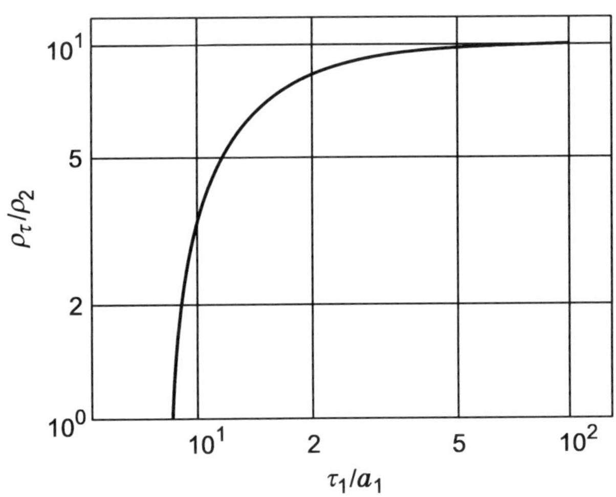

Figure 9.23. Graph of function ρ_τ/ρ_2.

TABLE 9.6
Minimal values of parameter τ_1/z

$(2\alpha-1)^2 s$	if $\alpha > 1$ and $s > 1$											
s	if $\alpha < 1$	2	4	8	16	32	45	64	90	128	180	256
$(2\alpha-1)$	if $s < 1$											
τ_1/z		15	25	40	60	90	110	130	150	170	210	250

totic formulae for the late stage. Omitting intermediate steps we have:

$$\frac{\partial B_z}{\partial t} = -\frac{M\rho_1}{\pi z^5}\left(\frac{2}{\pi}\right)^{1/2}\frac{u_1^5}{(s-1)}\left[\frac{s^{5/2}-1}{5} - \left(\frac{2}{\pi}\right)^{1/2}\frac{(2\alpha-1)(s+1)(s-1)^2}{4}u_1 + \frac{(2\alpha-1)^2(4s^{7/2}-7s^{5/2}+3)}{14} + \alpha(\alpha-1)(s-1)u^2 + \ldots\right] \quad \text{if } \alpha \gg 1$$

$$\frac{\partial B_z}{\partial t} = -\frac{M\rho_1}{\pi z^5}\left(\frac{2}{\pi}\right)^{1/2}\frac{u_1^5}{(s-1)}\left[\frac{s^{5/2}-1}{5} - \left(\frac{\pi}{2}\right)^{1/2}\frac{(2\alpha-1)(s+1)(s-1)^2}{4}u_1 + \ldots\right]$$

$$0 < \alpha \leqslant 1 \qquad \tau_1/z \gg 1 \tag{9.50}$$

where $s = \sigma_2/\sigma_1$, $u_1 = 2\pi z/\tau_1$, $\tau_1 = (2\pi\rho_1 t \times 10^7)^{1/2}$.

Minimal values of parameter τ_1/z, when the difference between results of calculations by exact and asymptotic formulae do not exceed 5%, are given in Table 9.6.

Curves of function ρ_τ/ρ_1 are presented in Fig. 9.24. Apparent resistivity is related with function $\partial B_z/\partial t$ as:

$$\frac{\rho_\tau}{\rho_1} = \left(\frac{\dot{B}_z^{un}}{\dot{B}_z}\right)^{2/3} \tag{9.51}$$

where $\dot{B}_z^{un} = (M\rho_1/2\pi z^5)(2/\pi)^{1/2}u_1^5 e^{-u^2/2}$ is the field ($\dot{B}_z$) of the magnetic dipole in a uniform medium with resistivity ρ_1.

Let us consider the main features of curves ρ_τ/ρ_1. For small times ($\tau_1/z \to 0$) the left-hand branches of curves have a common asymptote for $\alpha > 1$, since, at first, currents concentrate near the source, and a field is practically the same as in a uniform medium with resistivity ρ_1. With an increase of time the influence of the second medium becomes stronger, and it manifests itself earlier with an increase of its conductivity.

The right-hand branches of these curves correspond to the late stage (eq. 9.50), valid for relatively great times. In a general case the transient field at the late stage is defined by both conductivities of a medium, but with an increase of resistivity ρ_1 or ρ_2 the influence of the more resistive part becomes smaller. For sufficiently large values of τ_1/z field $\dot{B}_z$ does not depend on the distance from the probe to the interface. It is explained by the fact that induced currents are located far away from the receiver with respect to distance L.

At the range of times close to the early stage there is an extremum on curves of apparent resistivity (maximum, when $\rho_2/\rho_1 < 1$ and a minimum if $\rho_2/\rho_1 > 1$). Appearance of this extremum can be explained in the following way: at the initial moment induced currents

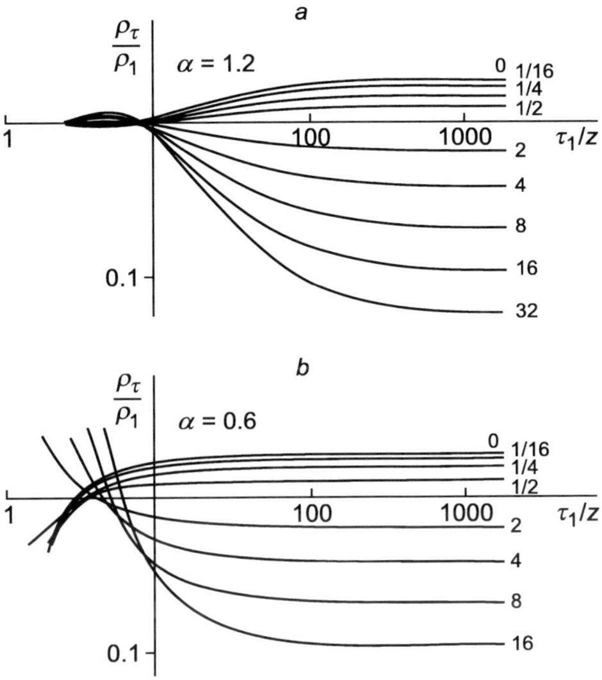

Figure 9.24. Apparent resistivity curves ρ_τ/ρ_1 in the presence of one interface. Curve index σ_2/σ_1.

are mainly located near the dipole but a change of their magnetic field with time creates, in its own turn, currents in various points of the medium, and if conductivity at some distance from the source is noticeably different we can expect the influence of this part of the medium at relatively small times.

Also it is useful to notice one interesting feature of curves ρ_τ/ρ_1 in the case when the transmitter and receiver of the probe are located from both sides of the interface. Left asymptotes of curves tend to infinity if $\rho_2/\rho_1 < 1$ and to zero if the second medium is more resistive. This means that at the early stage, as well as at the high frequency spectrum, the field measured at the second medium (when the source is located in the first one) depends on the resistivity of this part of the medium, and it is defined not only by currents concentrated near the source but also by currents near the measuring point.

Now we will consider the behavior of transient responses in the presence of a formation with finite thickness. Let us assume that a two-coil induction probe is located within the formation symmetrically with respect to its boundaries and the surrounding medium. The latter has the same resistivity from both sides of the formation.

In this case an expression for the vertical component of the field in the frequency domain expressed in units of the primary field: $H_z^0 = 2M/4\pi L^3$, in accord with eq. 5.37 has the form:

$$h_z = h_z^0(k, L) + \int_0^\infty \frac{m^3 m_{12}}{m_1} \frac{e^{\alpha m_1} + m_{12} \cosh m_1}{1 - m_{12}^2 e^{-2\alpha m_1}} e^{-2\alpha m_1} \, dm \tag{9.52}$$

where $\alpha = H/L$; H and L are thickness of the formation and the probe length, respectively, and:

$$m_1 = (m^2 - k_1^2)^{1/2} \qquad m_2 = (m^2 - k_2^2)^{1/2} \qquad m_{12} = (m_1 - m_2)/(m_1 + m_2)$$

For the determination of the transient responses, as before, Fourier integral is applied:

$$\frac{\partial B_z}{\partial t} = -\frac{2}{\pi} \int_0^\infty \mathrm{Q}\, B_z \sin(\omega t) \, d\omega$$

Presentation of the field in the frequency domain as a sum of two terms (9.52) is convenient for the field calculation at the early stage of the transient response, when the field is practically the same as that in a uniform medium with the formation resistivity, i.e. it is mainly defined by the first term of eq. 9.52. At larger times currents either are distributed uniformly along the z-axis (nonconducting surrounding medium) or are mainly located there when the surrounding medium is conductive.

In both cases the transient field essentially differs from that in a uniform medium with resistivity ρ_1. For this reason it is necessary to increase the accuracy of calculation of the integral in eq. 9.52, specially when the surrounding medium has a higher resistivity than the formation, and the final result is obtained as a difference of relatively large numbers which are close to each other.

In order to partly avoid this numerical problem it is reasonable to subtract from the low-frequency spectrum terms proportional to ω and $\omega^{3/2}$. It is not difficult to show that in the expression:

$$h_z = h_z^0 + \int_0^\infty \left(F - \frac{m^3 e^{-\alpha m_1}}{2m_2} + \frac{m^3 e^{-\alpha m_1}}{2m_1} \right) dm + \int_0^\infty \left(\frac{m^3 e^{-\alpha m_1}}{2m_2} - \frac{m^3 e^{-\alpha m_1}}{2m_1} \right) dm \tag{9.53}$$

an expansion of the second term in powers of ω begins with a term, proportional to ω^2 (F is the integrand in eq. 9.52). Calculating the late stage of transient response from the last term of eq. 9.53 an asymptotic expression is used, which provides a sufficient accuracy of numerical results. In the case when the surrounding medium is an insulator it is reasonable to present spectrum $h_z(\omega)$ as:

$$h_z(\omega) = \int_0^{m_0} \left(\frac{m^3 e^{-m}}{2m_1} + F \right) dm + \int_{m_0}^\infty F \, dm + \int_{m_0}^\infty \frac{m^3 e^{-m_1}}{2m_1} \, dm \qquad \text{if } m_0 \ll 1$$

The last term is expressed through elementary functions.

Taking into account known numerical problems in calculating the late stage of the transient field, asymptotic formulae have been derived. Applying the approach described in previous chapters, the low-frequency spectrum can be presented as:

$$B_z \simeq C_1 k_1^2 + C_2 k_1^3 + C_3 k_1^4 \ln 2k_2 + C_4 k_1^4 + C_5 k_1^5 \tag{9.54}$$

$$C_2 = \frac{1}{3}s^{3/2} \qquad C_3 = \frac{\alpha s(s-1)}{4}$$

$$C_5 = -\frac{1}{15}\left[5\alpha^2\left(s^{5/2} - \frac{7}{4}s^{3/2} + \frac{3}{4}s^{1/2}\right) - \left(\frac{5}{4}s^{3/2} - \frac{3}{4}s^{5/2}\right)\right] \tag{9.55}$$

$$s = \rho_1/\rho_2$$

Coefficients C_1 and C_4 are not given since they do not contribute to the late stage of the transient response.

Correspondingly, for the field at the late stage we have:

$$\begin{aligned}\frac{\partial B_z}{\partial t} = &-\frac{M\rho_1}{2\pi L^5}\left(\frac{2}{\pi}\right)^{1/2} u^5 \left\{ s^{3/2} - \left(\frac{\pi}{2}\right)^{1/2} 2\alpha u_1(s-1)s \right. \\ &\left. + u_1^2\left[5\alpha^2\left(s^{5/2} - \frac{7}{4}s^{3/2} + \frac{3}{4}s^{1/2}\right) - \left(\frac{5}{4}s^{3/2} - \frac{3}{4}s^{5/2}\right)\right]\right\}\end{aligned} \tag{9.56}$$

where $u_1 = 2\pi L/\tau_1$, $\tau_1 = (2\pi\rho_1 t \times 10^7)^{1/2}$. This equation is valid for large times when the surrounding medium has a finite resistivity.

In the limit case of nonconducting surrounding medium we have:

$$\frac{\partial B_z}{\partial t} = -\frac{3M\rho_1}{2\pi L^5}\alpha^3 u_1^8(1 - 8\alpha^2 u_1^2 - 4\alpha u_1^2) \tag{9.57}$$

Thus, for sufficiently large times the field becomes the same as in a uniform medium with resistivity of the surrounding medium $\rho_2 \neq \infty$, and it does not depend on the probe length. At the same time at the late stage induced currents are distributed uniformly within the formation along z axis, if the surrounding medium is not conductive, and the field B_z is directly proportional to the cube of the longitudinal conductance: $(H/\rho_1)^3$.

Results of calculations based on exact solutions presented in the form of apparent resistivity curves ρ_τ/ρ_1, as a function of parameter τ_1/L are given in Figs. 9.25–9.28. Index of curves is ρ_2/ρ_1. In the same manner as in the case of one horizontal interface, the apparent resistivity is related with the field as:

$$\frac{\rho_\tau}{\rho_1} = \left(\frac{\dot{B}_z^{un}(\rho_1, t)}{\dot{B}_z(t)}\right)^{-2/3}$$

Considering these curves we can make the following conclusions.

- At the early stage of the transient response currents are concentrated near the dipole, and the field depends only on the formation resistivity. For this reason the left-hand asymptote of curves ρ_τ/ρ_1 is equal to unity ($\alpha > 1$).

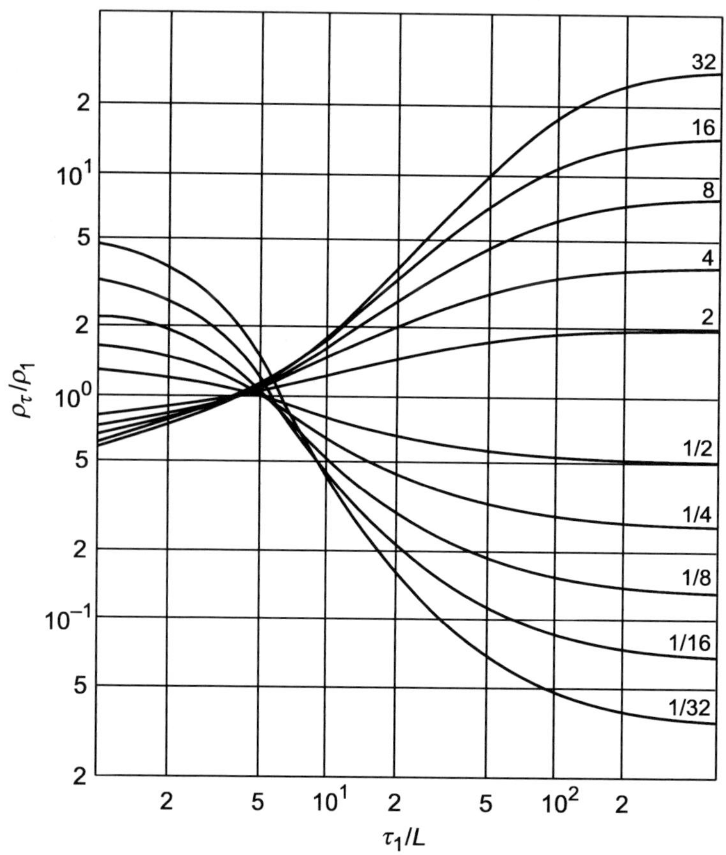

Figure 9.25. Apparent resistivity curves, ρ_τ/ρ_1 $(\alpha = 1)$. Curve index σ_1/σ_2.

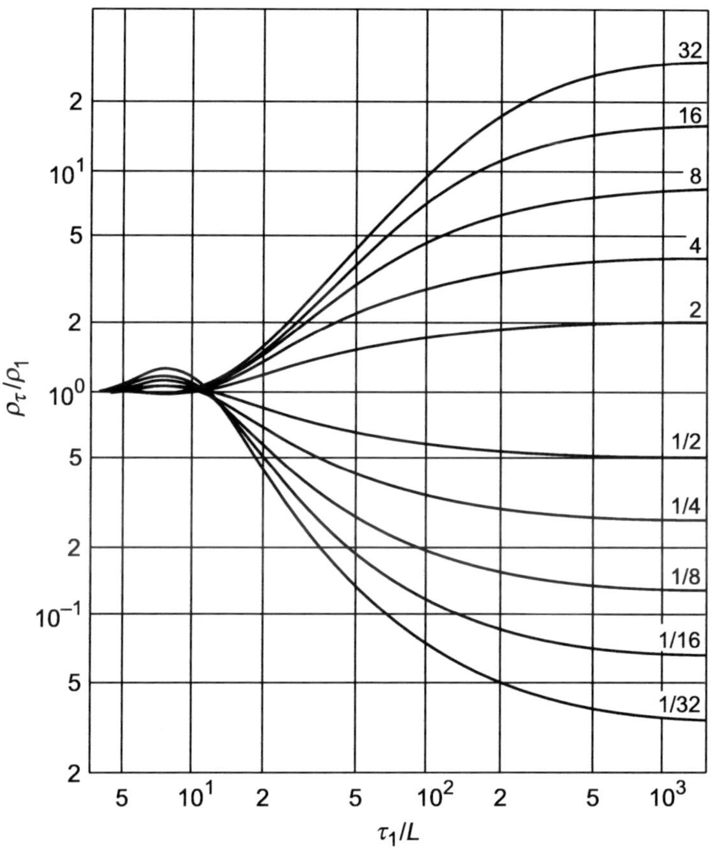

Figure 9.26. Apparent resistivity curves, ρ_τ/ρ_1 ($\alpha = 2$). Curve index σ_1/σ_2.

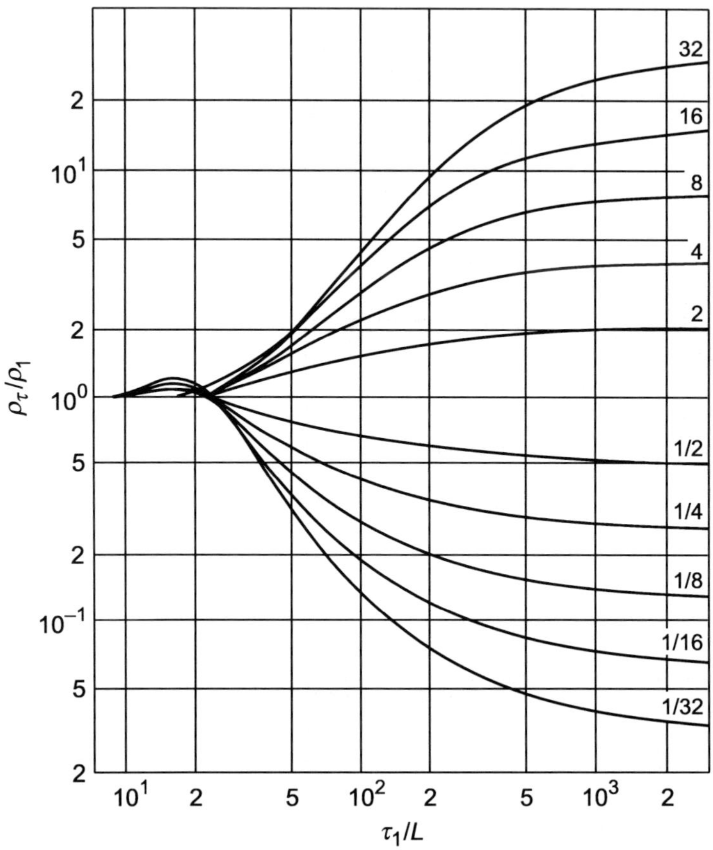

Figure 9.27. Apparent resistivity curves, ρ_τ/ρ_1 ($\alpha = 4$). Curve index σ_1/σ_2.

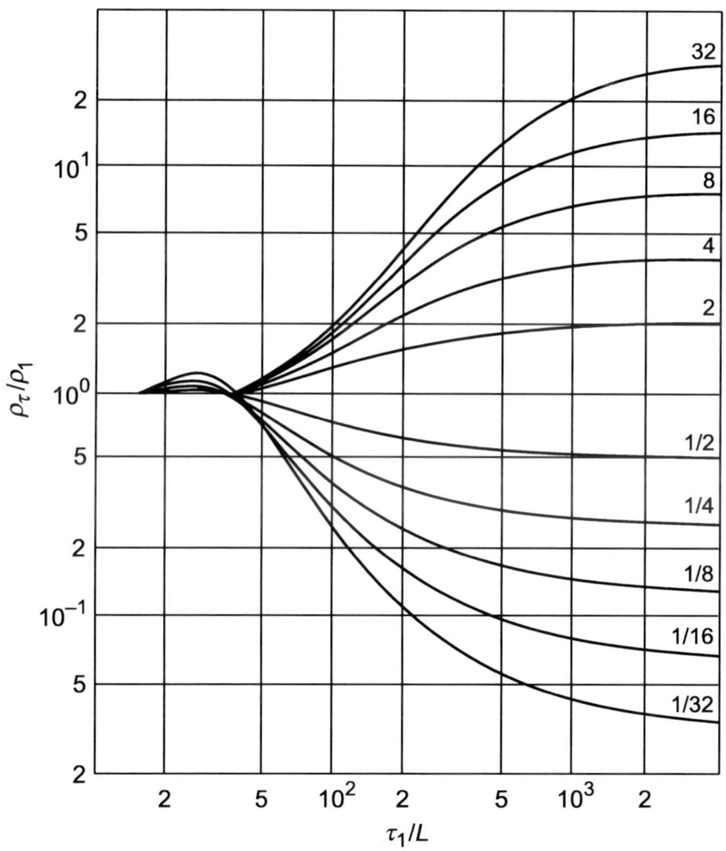

Figure 9.28. Apparent resistivity curves, ρ_τ/ρ_1 ($\alpha = 8$). Curve index σ_1/σ_2.

TABLE 9.7
Maximum values of time t^{max}, μsec

ρ_1, ohm·m \ H, m	2	4	6	8	10	14
1	2.3	9.2	20.6	36.8	58	112
2	1.15	4.6	15.3	18.4	29	56
4	0.58	2.3	7.6	9.2	14	28
8	0.29	1.15	3.8	7.4	12	24
16	0.14	0.58	1.9	2.3	3.5	7

- For large values of parameter τ_1/L, when currents practically vanish in the formation, curves of ρ_τ/ρ_1 approach to their right-hand asymptotes, equal to ρ_2/ρ_1, if $\rho_2 \neq \infty$.
- If $\rho_2/\rho_1 > 10 \div 30$, curves of ρ_τ/ρ_1 have one more intermediate asymptote corresponding to a nonconducting surrounding medium. In other words, there is such a time interval when intensity of currents is still negligible in more resistive surrounding medium, but within the formation they are distributed uniformly along the z-axis.
- Curves of apparent resistivity are characterized by a maximum at relatively small times, when the surrounding medium has greater conductivity $(\sigma_1/\sigma_2 < 1)$.
- With an increase of time currents are mainly located far away from the source, and the field does not practically depend on the probe length.

  Calculations show that for a given value of ρ_2/ρ_1 a function ρ_τ/ρ_1 with sufficient accuracy is defined by parameter τ_1/H, if $\tau_1/H > 8$ (Fig. 9.29).
- The apparent resistivity, ρ_τ, differs from the formation resistivity less than 10% if the following conditions are met:

$$\frac{\tau_1}{H} < 6 \quad \text{or} \quad t^{max}_{\mu\text{sec}} < 0.6\frac{H^2}{\rho_1} \quad \text{if } \alpha \geqslant 2 \text{ and } 32 \geqslant \rho_2/\rho_1 \geqslant 1/32 \tag{9.58}$$

  This relation defines maximal times when regardless of the probe length $(L < H/2)$ we can think that a measured signal is function of the formation resistivity only.

As an illustration Table 9.7 contains maximal values of time in μsec, for some parameters of a medium, satisfying condition 9.58.

9.5. About a Nonstationary Field of the Electric Dipole

Investigation of the nonstationary field of the magnetic dipole in a medium with cylindrical interfaces demonstrates that for providing a sufficient radial response, allowing us to define

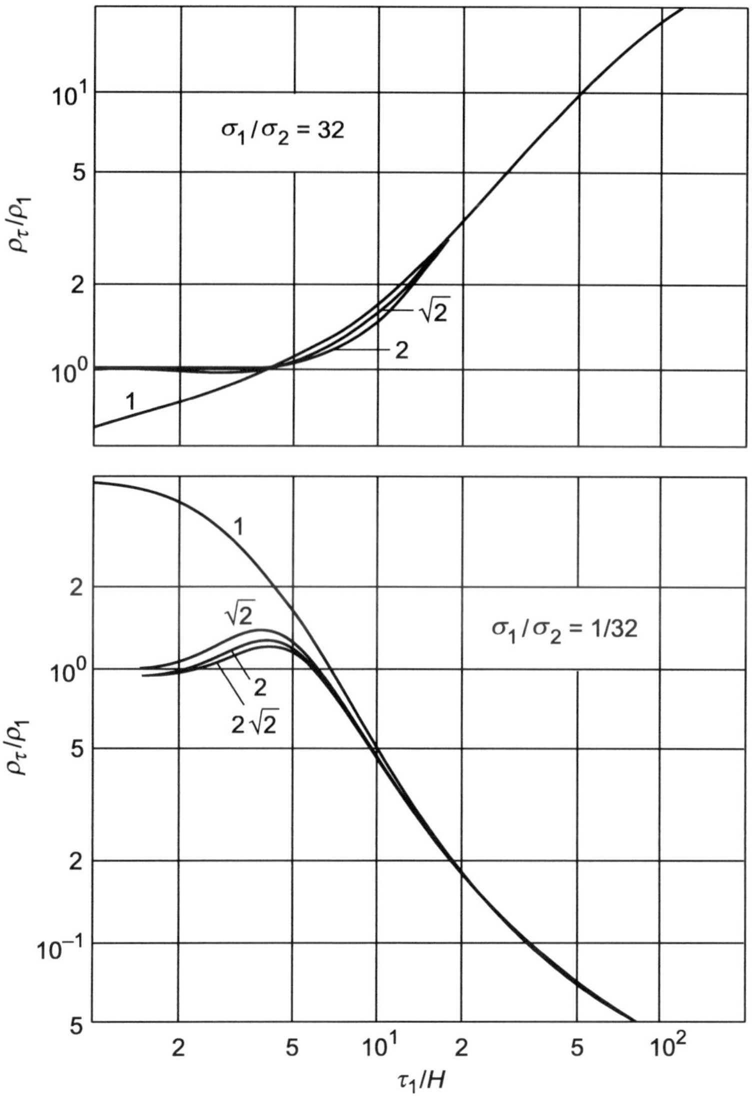

Figure 9.29. Apparent resistivity curves, ρ_τ/ρ_1. Curve index α.

the formation resistivity in the presence of a deep penetration ($a_2/a_1 \simeq 8$), it is necessary to perform measurements at times satisfying condition:

$$\frac{\tau_3}{a_1} \geqslant 100 \left(\frac{\rho_3}{\rho_1}\right)^{1/2} \tag{9.59}$$

if $\rho_3/\rho_1 \leqslant 30$ and $L/a_1 \leqslant 10$.

On the other hand, with an increase of time the influence of induced currents in the surrounding medium increases, and as was established in the previous section the field measured within the formation of finite thickness does not practically differ from that in a uniform medium with the formation resistivity, if:

$$\frac{\tau_1}{H} < 6 \tag{9.60}$$

provided that $H/L > 2$ and $\rho_1/\rho_2 < 16$. This condition, as well as rapid decrease of the signal, does not allow us to perform measurements at larger times.

For instance, if the formation resistivity and its thickness are 100 ohm·m and 6 m, respectively, then the maximal time when the influence of the surrounding medium is still negligible can not exceed 2 μsec ($\rho_1/\rho_2 < 16$). Taking into account that at such times currents induced in a moderately conductive medium create relatively weak signals we can expect serious technical problems of measuring in inductive sensors due to intrinsic processes. For this reason it is appropriate to consider some features in a behavior of the nonstationary field of the electric dipole.

As is well known, the electric field of the alternating electric dipole with moment:

$$M_e \mathrm{e}^{-\mathrm{i}\omega t} = \frac{\rho I\,\mathrm{d}l}{4\pi}\mathrm{e}^{-\mathrm{i}\omega t} \tag{9.61}$$

is defined in the following way:

$$\begin{aligned} E_R &= \frac{2M_e}{R^3}\mathrm{e}^{\mathrm{i}kR}(1-\mathrm{i}kR)\cos\theta \\ E_\theta &= \frac{M_e}{R^3}\mathrm{e}^{\mathrm{i}kR}(1-\mathrm{i}kR-kR^2)\sin\theta \end{aligned} \tag{9.62}$$

where $\mathrm{d}l$ is the distance between electrodes; I is the current; $\varepsilon_0 \rho I$ is the magnitude of the charge on the electrode surface; $\varepsilon_0 = (1/36\pi) \times 10^{-9}$ F/m, $k = (\sigma\mu\omega/2)^{1/2}(1+\mathrm{i})$.

Applying Fourier transform to eq. 9.62, we obtain an expression for the transient electric field on the axis of the electric dipole when excitation of the dipole is the step function:

$$E_z = \frac{2M_e}{L^3}\left(1-\phi(u)+\left(\frac{2}{\pi}\right)^{1/2} u\,\mathrm{e}^{-u^2/2}\right) \tag{9.63}$$

where L is the probe length; $\phi(u)$ is the probability integral, $u = 2\pi L/\tau$.

TABLE 9.8
Values of electric field E_z, $\mu V/m$; $I\,\mathrm{dl}/4\pi = 10^{-3} A \cdot m$

ρ_2, ohm·m \ τ_1/a_1	30	45	60	90
		$L = 0.2$ m		
5	0.65×10^4	0.19×10^4	0.82×10^3	0.24×10^3
10	0.46×10^4	0.14×10^4	0.58×10^3	0.17×10^3
20	0.33×10^4	0.97×10^3	0.41×10^3	0.12×10^3
		$L = 1.0$ m		
5	0.51×10^4	0.17×10^4	0.77×10^3	0.24×10^3
10	0.41×10^4	0.13×10^4	0.56×10^3	0.17×10^3
20	0.31×10^4	0.94×10^3	0.40×10^3	0.12×10^3

In particular, for relatively large times when parameter $u \to 0$, we have:

$$E_z \simeq -\frac{2}{3}\frac{M_e}{L}\left(\frac{2}{\pi}\right)^{1/2} u^3 \left(1 - \frac{3}{10}u^2\right) \tag{9.64}$$

Therefore, at the late stage the transient electric field does not depend on the distance from the dipole, and correspondingly, the depth of investigation is the same as for the inductive excitation. At the same time, as it follows from values of the electric field, given in Table 9.8, in the case of the electric dipole we can provide signals of greater magnitude for the same current, and perhaps to reduce in a significant degree intrinsic nonstationary processes in the transmitter and in the receiver.

Now we will very briefly consider the electric field of the electric dipole located on the borehole axis. As is well known from the theory of an electric logging, the expression for the field is:

$$E_z = \frac{\rho_1 I\,\mathrm{dl}}{2\pi L^3}\left(e_z^{un}(k, L) - \frac{\alpha^3}{\pi}\int\limits_0^\infty m_1^2 D_1 \cos(\alpha m)\,\mathrm{d}m\right) \tag{9.65}$$

where:

$$e_z^{un}(k_1 L) = \mathrm{e}^{\mathrm{i}k_1 L}(1 - \mathrm{i}k_1 L) \qquad m_1 = (m^2 - k_1^2 a_1^2)^{1/2} \qquad m_2 = (m^2 - k_2^2 a_1^2)^{1/2}$$

$$D_1 = \frac{m_2 K_0(m_2) K_1(m_1) - s m_1 K_0(m_1) K_1(m_2)}{m_2 K_0(m_2) I_1(m_1) + s m_1 I_0(m_1) K_1(m_2)} \qquad s = \sigma_2/\sigma_1$$

Taking into account the purpose of this consideration, let us pay attention only to the far zone. Applying methodics described in detail in Chapter 4 we can show that the expression in brackets of eq. 9.65 is:

$$\frac{\rho_2}{\rho_1} \frac{1}{I_0^2\left[(k_2^2 - k_1^2)^{1/2} a_1\right]} e_z^{un}(k_2 L)$$

Thus, for the electric field we obtain:

$$E_z = \frac{\rho_2 I \,\mathrm{d}l}{2\pi L^3} \frac{1}{I_0^2 \left[(k_2^2 - k_1^2)^{1/2} \, a_1 \right]} \mathrm{e}^{\mathrm{i}k_2 L}(1 - \mathrm{i}k_2 L) \qquad \text{if } \alpha > 1 \text{ and } \lambda_1/a_1 \gg 1 \tag{9.66}$$

where $\lambda_1 = 2\pi(2/(\sigma_1\mu\omega))^{1/2}$; σ_1 and σ_2 are conductivities of the borehole and the formation.

Unlike the inductive excitation with horizontal coils electrical charges arise at the borehole surface. However, in accord with eq. 9.66 at the low-frequency part of the spectrum, the term proportional to k^3 depends on the formation conductivity only.

For this reason, at the late stage of the transient response the electric field on the borehole axis at the far zone possesses the same depth of investigation as the magnetic field when the inductive excitation is applied. From a physical point of view it is clear that the same conclusion remains valid in measuring the electric field near the dipole.

Chapter 10

PRINCIPLES OF INDUCTION LOGGING WITH TRANSVERSAL INDUCTION COILS

In previous chapters we have considered various aspects of induction logging when the source of the field is the vertical magnetic dipole, i.e. it is a relatively small coil with horizontally located turns. In this case current lines present themselves as concentric circles located in horizontal planes with centers on the borehole axis. For this reason thin and resistive layers, as well as caverns and fractures, stretched in directions perpendicular to the borehole axis, do not practically manifest themselves in conventional induction logging. Finally, in this method only a longitudinal conductivity defines a signal in an anisotropic medium. In order to increase sensitivity of induction logging to thin resistive layers, to improve, when possible, the vertical response of the probe, and to define the coefficient of anisotropy we will investigate modification of induction logging with horizontally oriented coils (Fig. 10.1).

10.1. Electromagnetic Field of the Magnetic Dipole in a Uniform Isotropic Medium

Analyses of the electromagnetic field will be started from the simplest case of a uniform conducting and isotropic medium. As is known, expressions for complex amplitudes of the field caused by the magnetic dipole, oriented along the z-axis, have the form (Chapter 2):

$$\begin{aligned} E_\phi &= \frac{\mathrm{i}\omega\mu M}{4\pi R^2}\,\mathrm{e}^{\mathrm{i}kR}(1-\mathrm{i}kR)\cos\theta \\ H_R &= \frac{2M}{4\pi R^3}\,\mathrm{e}^{\mathrm{i}kR}(1-\mathrm{i}kR)\sin\theta \\ H_\theta &= \frac{M}{4\pi R^3}\,\mathrm{e}^{\mathrm{i}kR}(1-\mathrm{i}kR-k^2R^2)\sin\theta \end{aligned} \tag{10.1}$$

where $M = InS$ is the dipole moment; σ is conductivity; μ is the magnetic permeability equal to $4\pi\times10^{-7}$ H/m; ω is angular frequency; k is the wave number, $k=(1+\mathrm{i})/h$; h is the thickness of skin layer.

As is well known, sources of the secondary field are induced currents, the distribution of which is defined by the frequency and the conductivity of the medium. Current lines are circles located in planes perpendicular to z-axis. In conventional induction logging, when the direction of the dipole moment coincides with the borehole axis, the main attention

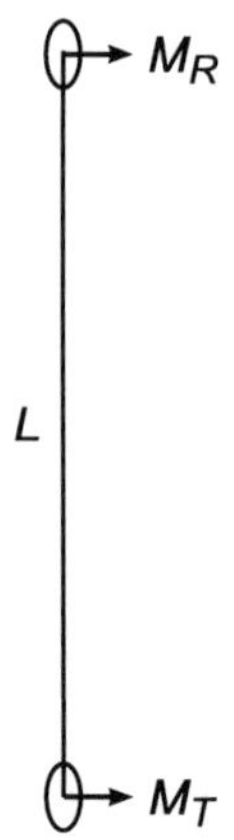

Figure 10.1. The induction probe with horizontal coils.

was paid to component H_R, as $\theta = 0$. Under excitation of the field by the horizontal magnetic dipole component of the field, H_θ, as $\theta = \pi/2$, is measured. In accord with eq. 10.1 we have:

$$H_\theta = H^{(0)} e^{ikL}(1 - ikL - k^2L^2) \tag{10.2}$$

where L is the probe length.

$$H^{(0)} = M/4\pi L^3 \tag{10.3}$$

is the field of the magnetic dipole in a free space, directed opposite to the dipole moment.

Let us introduce function h_θ, equal:

$$h_\theta = \frac{H_\theta}{H^{(0)}} = e^{ikL}(1 - ikL - k^2L^2) \tag{10.4}$$

Having substituted into eq. 10.4 value $k = (1 + i)/h$, we will present the field as a sum of two components, namely, the inphase and quadrature components:

$$\begin{aligned} \text{In}\, h_\theta &= [(1+p)\cos p + p(1+2p)\sin p]\, e^{-p} \\ \text{Q}\, h_\theta &= [(1+p)\sin p - p(1+2p)\cos p]\, e^{-p} \end{aligned} \tag{10.5}$$

where p is a parameter, defining the field h_θ and equal to the distance from the dipole, expressed in units of the skin depth: $p = L/h$.

In accord with eq. 10.5 we have for the magnitude and the phase of the field, h_θ:

$$\begin{aligned} A &= e^{-p}[(1+p)^2 + p^2(1+2p)^2]^{1/2} \\ \phi &= p - \coth^{-1}[p(1+2p)/(1+p)] \end{aligned} \tag{10.6}$$

First, consider the field in the near zone, when parameter p is small.

Expanding exponent e^{ikL} in a series and substituting it into eq. 9.4, after elementary transformations we obtain:

$$h_\theta = 1 + \sum_{n=0}^{\infty} \frac{(ikL)^{n+2}(n+1)}{n!\,(n+2)} \tag{10.7}$$

Restricting ourselves by first two terms, we have:

$$\text{In}\, h_\theta \simeq 1 + \frac{4}{3}p^3 \qquad \text{Q}\, h_\theta \simeq -p^2 + \frac{4}{3}p^3 \tag{10.8}$$

Thus, at the range of small parameters when interaction between currents is negligible the quadrature component prevails, which is directly proportional to frequency and conductivity, and it is equal in magnitude to the quadrature component of the field measured at the same distance along the axis of the vertical magnetic dipole (Chapter 2).

In the wave zone at distances significantly exceeding the skin depth component H_θ is greater than H_R ($\theta \neq 0$) and at an equatorial plane:

$$H_\theta = \frac{M}{4\pi L} k^2 e^{ikL} \qquad \text{if } |kL| > 1 \tag{10.9}$$

As follows from eq. 10.1 in the wave zone the ratio of the electric field to the magnetic field does not depend on the distance, and it is equal to the impedance in a uniform medium:

$$\frac{E_\phi}{H_\theta} = -\frac{\mu\omega}{k} = -\left(\frac{\mu\omega}{\sigma}\right)^{1/2} e^{-i(\pi/4)} \tag{10.10}$$

Values of magnitude, phase and field components are given in Table 10.1. For comparison values of function h_z for the vertical dipoles are also shown. Graphs of quadrature and inphase components of field h_x as well as amplitude of the secondary field $|h_z - 1|$ and its phase are given in Figs. 10.2–10.3, respectively.

10.2. Boundary Problem for the Horizontal Magnetic Dipole on the Borehole Axis

Let us assume that a horizontal magnetic dipole is located on the borehole axis. The radius and conductivity of the borehole are a and σ_1, respectively. The formation conductivity is σ_2. Magnetic permeabilities of both media coincide with that in free space.

We will introduce a cylindrical system of coordinates, and the magnetic dipole with moment $M = M_0 e^{i\omega t}$ is placed in its origin, and it is directed along the x-axis (Fig. 10.4).

As is well known the system of equations for the quasistationary field has the form:

$$\begin{aligned} &\text{curl}\, \boldsymbol{E} = i\omega\mu \boldsymbol{H} \qquad &&\text{div}\, \boldsymbol{E} = 0 \\ &\text{curl}\, \boldsymbol{H} = \sigma \boldsymbol{E} \qquad &&\text{div}\, \boldsymbol{H} = 0 \end{aligned} \tag{10.11}$$

TABLE 10.1
Values of magnitude, phase and field components.

p	Q h_z	Q h_θ	In h_z	In h_θ	A_z	A_θ	ϕ_z	ϕ_θ
0.01	0.9933×10^{-4}	-0.9867×10^{-4}	0.6616×10^{-6}	0.1318×10^{-5}	0.9933×10^{-4}	0.9867×10^{-4}	0.1564×10^{1}	-0.1557×10^{1}
	0.1403×10^{-3}	-0.1392×10^{-3}	0.1111×10^{-5}	0.2212	0.1403×10^{-3}	0.1392×10^{-3}	0.1562	−0.1555
	0.1981	−0.1962	0.1865	0.3711	0.1981	0.1963	0.1561	−0.1552
	0.2796	0.2765	0.3131	0.6223×10^{-5}	0.2796	0.2766	0.1559	0.1548
0.02	0.3946	0.3893	0.5253	0.1043×10^{-4}	0.3947	0.3895	0.1557	0.1544
	0.5567	0.5477	0.8810×10^{-5}	0.1746	0.5567	0.5480	0.1554	0.1539
	0.7849×10^{-3}	0.7698×10^{-3}	0.1476×10^{-4}	0.2924	0.7850×10^{-3}	0.7704×10^{-3}	0.1551	0.1533
	0.1106×10^{-2}	0.1081×10^{-2}	0.2473	0.4884	0.1106×10^{-2}	0.1082×10^{-2}	0.1548	0.1526
0.04	0.1557	0.1515	0.4140	0.8155×10^{-4}	0.1557	0.1517	0.1544	0.1517
	0.2191	0.2119	0.6922×10^{-4}	0.1360×10^{-3}	0.2192	0.2124	0.1539	0.1507
	0.3079	0.2959	0.1156×10^{-3}	0.2263	0.3081	0.2967	0.1533	0.1494
	0.4322	0.4120	0.1929	0.3759	0.4327	0.4137	0.1526	0.1480
0.08	0.6059	0.5719	0.5212	0.6230×10^{-3}	0.6067	0.5753	0.1517	0.1462
	0.8477×10^{-2}	0.7907×10^{-2}	0.5341	0.1029×10^{-2}	0.8194×10^{-2}	0.7973×10^{-2}	0.1507	0.1441
	0.1183×10^{-1}	0.1088×10^{-1}	0.8859×10^{-3}	0.1695	0.1187×10^{-1}	0.1101×10^{-1}	0.1496	0.1416
	0.1648	0.1488	0.1456×10^{-2}	0.2779	0.1654	0.1513	0.1482	0.1386
0.16	0.2288	0.2019	0.2416	0.4534	0.2301	0.2069	0.1465	0.1350
	0.3164	0.2714	0.3970	0.7351×10^{-2}	0.3189	0.2812	0.1445	0.1306
	0.4354	−0.3603	0.6490×10^{-2}	0.1183×10^{-1}	0.4402	0.3792	0.1422	−0.1253
	0.5958	−0.4708	0.1055×10^{-1}	0.1886	0.6051	0.5072	0.1395	−0.1190
0.32	0.8094×10^{-1}	0.6022	0.1704	0.2973	0.8272×10^{-1}	0.6716	0.1363	0.1112
	0.1089	0.7481	0.2730	0.4622	0.1123	0.8794×10^{-1}	0.1325	0.1017×10^{-1}
	0.1451	0.8918	0.4351	0.7059×10^{-1}	0.1514	0.1137	0.1280	0.9012
	0.1905	0.9981×10^{-1}	0.6788×10^{-1}	0.1054	0.2022	0.1452	0.1228	0.7582

TABLE 10.1
(Continued)

p	$Q\,h_z$	$Q\,h_\theta$	$\text{In}\,h_z$	$\text{In}\,h_\theta$	A_z	A_θ	ϕ_z	ϕ_θ
0.64	0.2457	−0.1007	0.1048	0.1531	0.2671	0.1833	0.1167	0.5818
	0.3099	-0.8190×10^{-1}	0.1590	0.2142	0.3484	0.2993	0.1096	0.3652
	0.3799	-0.2935×10^{-1}	0.2361	0.2851	0.4473	0.2866	0.1014×10^{1}	−0.1026
	0.4488	0.7408×10^{-1}	0.3412	0.3539	0.5638	0.3616	0.9207	0.2063
1.28	0.5052	0.2440	0.4772	0.3955	0.6950	0.4648	0.8138	0.5528
	0.5336	0.4849	0.6414	0.3687	0.8344	0.6088	0.6939	0.9203
	0.5169	0.7712	0.8212	0.2204	0.9704	0.8201	0.6617	0.1293×10^{-1}
	0.4434	0.1035×10^{1}	0.9923	-0.9288×10^{-1}	0.1086×10^{1}	0.1039×10^{1}	0.4202	0.1660
2.56	0.3165	0.1163	0.1121×10^{1}	−0.5646	0.1165	0.1293	0.2751	0.2023
	0.1630	0.1042×10^{1}	0.1177	-0.1092×10^{1}	0.1188	0.1509	0.1373	0.2380
	0.2903×10^{-1}	0.6519	0.1154	−0.1476	0.1154	0.1615	0.2514×10^{-1}	0.2726
	−0.4275	0.1553	0.1081	−0.1541	0.1082	0.1549	-0.3950×10^{-1}	0.3041
5.12	−0.4570	−0.1699	0.1013×10^{1}	−0.1301	0.1014×10^{1}	0.1312	−0.4506	0.3271
	-0.1665×10^{-1}	−0.1817	0.9868	−0.1019	0.9870	0.1035	-0.1687×10^{-1}	0.3318
	0.1843×10^{-2}	-0.4141×10^{-1}	0.9923	−0.9309	0.9923	0.9180	0.1857×10^{-2}	0.3186

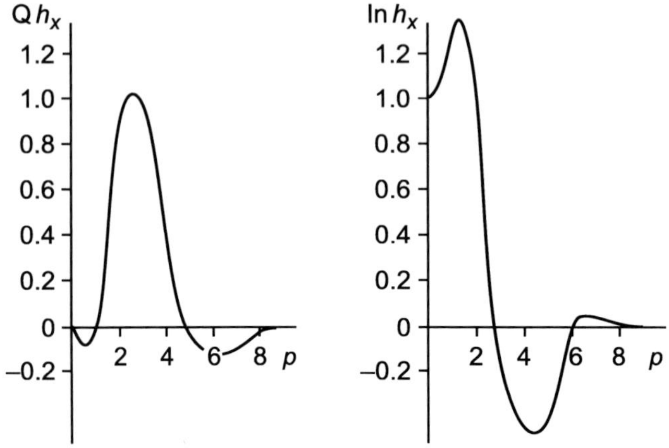

Figure 10.2. Quadrature and inphase components h_x, $\theta = \pi/2$.

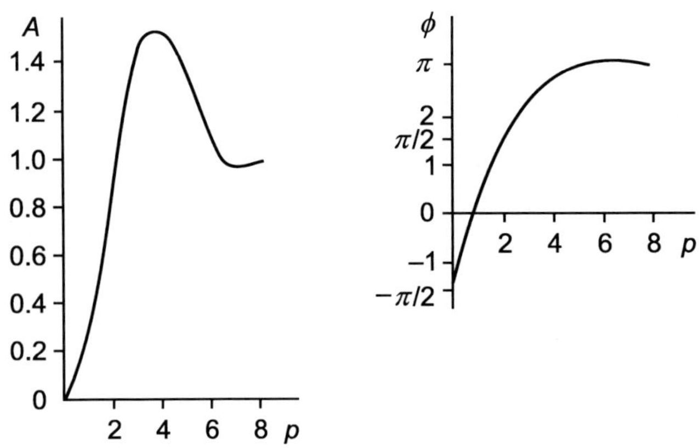

Figure 10.3. Amplitude and phase of the secondary field.

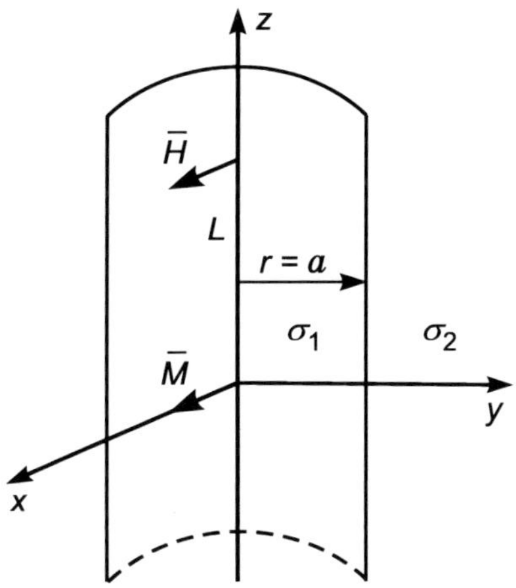

Figure 10.4. The horizontal magnetic dipole on the borehole axis.

Under excitation of the field by a horizontal dipole the primary vortex electric field, unlike that of a vertical dipole, intersects the boundary between media with different conductivity. For this reason electric charges arise on the borehole surface, the density of which changes synchronously with the electric field at a given point, and it depends on the conductivity of both parts of the medium and on the coordinates of the point. Thus, in this case sources of the field are currents and charges and, correspondingly, it is impossible to express components of the electromagnetic field only through one component of the vector potential of the magnetic type. Solution of the boundary problem with help of all three components of potential $\boldsymbol{A}^*$ (A_r^*, A_ϕ^*, A_z^*) leads to a system of differential equations in partial derivatives of the second order. For this reason let us present a solution as a sum of two fields and correspondingly introduce two potentials, namely, a magnetic and an electric type:

$$\begin{aligned} \boldsymbol{E} &= \boldsymbol{E}^{(1)} + \boldsymbol{E}^{(2)} \qquad & \boldsymbol{H} &= \boldsymbol{H}^{(1)} + \boldsymbol{H}^{(2)} \\ \boldsymbol{E}^{(1)} &= \mathrm{i}\omega\mu \,\mathrm{curl}\, \boldsymbol{A}^* \qquad & \boldsymbol{H}^{(2)} &= \mathrm{curl}\, \boldsymbol{A} \end{aligned} \tag{10.12}$$

Then, from eq. 10.11 it follows that:

$$\boldsymbol{H}^{(1)} = k^2 \boldsymbol{A}^* - \mathrm{grad}\, U^* \qquad \boldsymbol{E}^{(2)} = \mathrm{i}\omega\mu \boldsymbol{A} - \mathrm{grad}\, U \tag{10.13}$$

where $k^2 = \mathrm{i}\sigma\mu\omega$.

Introducing gauge conditions:

$$U^* = -\,\mathrm{div}\, \boldsymbol{A}^* \qquad \sigma U = -\,\mathrm{div}\, \boldsymbol{A} \tag{10.14}$$

potentials become a solution of the wave equation:

$$\nabla^2 \boldsymbol{A} + k^2 \boldsymbol{A} = 0 \qquad \nabla^2 \boldsymbol{A}^* + k^2 \boldsymbol{A}^* = 0 \tag{10.15}$$

We will look for a field with the help of vertical components of the vector potentials, i.e.:

$$\boldsymbol{A}^* = (0, 0, A_z^*) \qquad \boldsymbol{A} = (0, 0, A_z) \tag{10.16}$$

where the z-axis coincides with the borehole axis.

In such a case, in accord with eq. 10.12, vertical components of the electric and magnetic fields are absent in oscillations of magnetic and electric types, respectively, i.e.

$$E_z^{(1)} = 0 \qquad H_z^{(2)} = 0 \tag{10.17}$$

It is obvious that the field of the magnetic dipole in a uniform isotropic medium is of the magnetic type, while as it has all features of the electric type, when the source is the electric dipole.

According to eqs. 10.12–10.13 oscillations of the magnetic type are expressed through potential A_z^*. In fact we have:

$$\begin{aligned} E_r &= i\omega\mu \frac{1}{r}\frac{\partial A^*}{\partial \phi} & H_r &= \frac{\partial^2 A^*}{\partial r \partial z} \\ E_\phi &= -i\omega\mu \frac{\partial A^*}{\partial r} & H_\phi &= \frac{1}{r}\frac{\partial^2 A^*}{\partial \phi \partial z} \\ E_z &= 0 & H_z &= k^2 A^* + \frac{\partial^2 A^*}{\partial z^2} \end{aligned} \tag{10.18}$$

and the vector potential satisfies the equation:

$$\frac{\partial^2 A^*}{\partial r^2} + \frac{1}{r}\frac{\partial A^*}{\partial r} + \frac{1}{r^2}\frac{\partial^2 A^*}{\partial \phi^2} + \frac{\partial^2 A^*}{\partial z^2} + k^2 A^* = 0 \tag{10.19}$$

By analogy for oscillations of the electric type we have:

$$\begin{aligned} E_r &= \frac{1}{\sigma}\frac{\partial^2 A}{\partial r \partial z} & H_r &= \frac{1}{r}\frac{\partial A}{\partial \phi} \\ E_\phi &= \frac{1}{\sigma r}\frac{\partial^2 A}{\partial \phi \partial z} & H_\phi &= -\frac{\partial A}{\partial r} \\ E_z &= \frac{1}{\sigma}\left(k^2 A + \frac{\partial^2 A}{\partial z^2}\right) & H_z &= 0 \end{aligned} \tag{10.20}$$

and

$$\frac{\partial^2 A}{\partial r^2} + \frac{1}{r}\frac{\partial A}{\partial r} + \frac{1}{r^2}\frac{\partial^2 A}{\partial \phi^2} + \frac{\partial^2 A}{\partial z^2} + k^2 A = 0 \tag{10.21}$$

It is easily seen that unlike, for example, the boundary problem for a sphere, in this case it is impossible to provide continuity of tangential components of the field of one type of oscillations with the help of only magnetic or electric vector potentials. This fact makes the derivation of formulae for components of the electromagnetic field in media with several cylindrical interfaces much more complicated.

Thus, equality of tangential components of the field, consisting of oscillations of electric and magnetic types on the borehole surface ($r = a$), results in the following system of boundary conditions for potentials A and A^*:

$$\begin{aligned}
\frac{1}{\sigma_1}\left(k_1^2 A_1 + \frac{\partial^2 A_1}{\partial z^2}\right) &= \frac{1}{\sigma_2}\left(k_2^2 A_2 + \frac{\partial^2 A_2}{\partial z^2}\right) \\
\frac{1}{\sigma_1}\left(\frac{1}{a}\frac{\partial^2 A_1}{\partial \phi \partial z} - k_1^2 \frac{\partial A_1^*}{\partial r}\right) &= \frac{1}{\sigma_2}\left(\frac{1}{a}\frac{\partial^2 A_2}{\partial \phi \partial z} - k_2^2 \frac{\partial A_2^*}{\partial r}\right) \\
k_1^2 A_1^* + \frac{\partial^2 A_1^*}{\partial z^2} &= k_2^2 A_2^* + \frac{\partial^2 A_2^*}{\partial z^2} \\
-\frac{\partial A_1}{\partial r} + \frac{1}{a}\frac{\partial^2 A_1^*}{\partial \phi \partial z} &= -\frac{\partial A_2}{\partial r} + \frac{1}{a}\frac{\partial^2 A_2^*}{\partial \phi \partial z}
\end{aligned} \tag{10.22}$$

where k_1, A_1, A_1^* and k_2, A_2, A_2^* are wave numbers and potentials in the borehole and in the formation, respectively. Thus, components of the vector potentials satisfy wave equations 10.19 and 10.21 and boundary conditions 10.22.

Now, knowing the behavior of the field near the source, let us find expressions for potential A_0 and A_0^* in a uniform medium with conductivity σ_1, which allow us to formulate conditions of excitation for A_1 and A_1^*, which are necessary for a solution of the problem. As was mentioned above, the field of the dipole in a uniform isotropic medium can be described with the help of one component of potential of the magnetic type:

$$\boldsymbol{A}^* = (A_x^*, 0, 0)$$

$$A_x^* = \frac{M}{4\pi}\frac{e^{ik_1 R}}{R} = \frac{M}{2\pi^2}\int_0^\infty K_0(\lambda_1 r)\cos\lambda z \, d\lambda \tag{10.23}$$

where $\lambda_1 = (\lambda^2 - k_1^2)^{1/2}$ and

$$\boldsymbol{E} = i\omega\mu \, \mathrm{curl}\, \boldsymbol{A}^* \qquad \boldsymbol{H} = k^2 \boldsymbol{A}^* + \mathrm{grad}\,\mathrm{div}\,\boldsymbol{A}^* \tag{10.24}$$

Therefore, for vertical components of the field we obtain:

$$\begin{aligned}
E_z^0 &= i\omega\mu \frac{M}{2\pi^2}\sin\phi \int_0^\infty \lambda_1 K_1(\lambda_1 r)\cos\lambda z \, d\lambda \\
H_z^0 &= \frac{M}{2\pi^2}\cos\phi \int_0^\infty \lambda\lambda_1 K_1(\lambda_1 r)\sin\lambda z \, d\lambda
\end{aligned} \tag{10.25}$$

where $\cos\phi = x/r$ and $r = (x^2 + y^2)^{1/2}$.

On the other hand, from eqs. 10.18 and 10.20 we have:

$$E_z = \frac{1}{\sigma_1}\left(k_1^2 A_1 + \frac{\partial^2 A_1}{\partial z^2}\right) \qquad H_z = k_1^2 A_1^* + \frac{\partial^2 A_1^*}{\partial z^2} \tag{10.26}$$

It is not difficult to see that potentials:

$$A_0 = -k_1^2 \frac{M}{2\pi^2} \sin\phi \int\limits_0^\infty \frac{1}{\lambda_1} K_1(\lambda_1 r) \cos\lambda z \, \mathrm{d}\lambda$$

$$A_0^* = -\frac{M}{2\pi^2} \cos\phi \int\limits_0^\infty \frac{\lambda}{\lambda_1} K_1(\lambda_1 r) \sin\lambda z \, \mathrm{d}\lambda \tag{10.27}$$

correspond to the field in a uniform medium.

From a mathematical point of view eq. 10.27 present the field through a sum of oscillations of electric and magnetic types, in spite of the fact that in a uniform medium only one type of oscillation is present. However, it is obvious that between potentials A_0 and A_0^* there is a relation:

$$\frac{\partial^2 A_0}{\partial\phi\partial z} = -k_1^2 A_0^*$$

which explains this apparent contradiction.

Therefore, while approaching the dipole, potentials A_1 and A_1^* tend to A_0 and A_0^*, respectively. Taking into account the behavior of the field near the source and at infinity, we will present the potentials in the form:

$$A_1 = A_0 + k_1^2 \frac{M}{2\pi^2} \sin\phi \int\limits_0^\infty \frac{1}{\lambda_1} C I_1(\lambda_1 r) \cos\lambda z \, \mathrm{d}\lambda$$

$$A_1^* = A_0^* + \frac{M}{2\pi^2} \cos\phi \int\limits_0^\infty \frac{\lambda}{\lambda_1} D I_1(\lambda_1 r) \sin\lambda z \, \mathrm{d}\lambda$$

$$A_2 = -k^2 \frac{M}{2\pi^2} \sin\phi \int\limits_0^\infty \frac{1}{\lambda_2} E K_1(\lambda_2 r) \cos\lambda z \, \mathrm{d}\lambda \tag{10.28}$$

$$A_2^* = -\frac{M}{2\pi^2} \cos\phi \int\limits_0^\infty \frac{\lambda}{\lambda_2} G K_1(\lambda_2 r) \sin\lambda z \, \mathrm{d}\lambda$$

From boundary conditions 10.22 we will obtain a system of equations for coefficients C, D, E, and G:

$$K_1(\lambda_1 a) - I_1(\lambda_1 a) C = \frac{\lambda_2}{\lambda_1} K_1(\lambda_2 a) E$$

$$\frac{1}{\lambda_1 a} \left[K_1(\lambda_1 a) - I_1(\lambda_1 a) C\right] + \left[K_1'(\lambda_1 a) - I_1'(\lambda_1 a) D\right] = \frac{1}{\lambda_2 a} K_1(\lambda_2 a) E + K_1'(\lambda_2 a) G$$

$$K_1(\lambda_1 a) - I_1(\lambda_1 a) D = \frac{\lambda_2}{\lambda_1} K_1(\lambda_2 a) G \tag{10.29}$$

$$k_1^2\left[K_1'(\lambda_1 a) - I_1'(\lambda_1 a)C\right] + \frac{\lambda^2}{\lambda_1 a}\left[K_1(\lambda_1 a) - I_1(\lambda_1 a)D\right] = k_2^2 K_1'(\lambda_1 a)E + \frac{\lambda^2}{\lambda_2 a}K_1(\lambda_2 a)G$$

Solving this system we find:

$$C = \frac{\Delta_c}{\Delta} \qquad D = \frac{\Delta_d}{\Delta}$$

$$E = \frac{m_1}{m_2}\frac{K_1(m_1) - I_1(m_1)C}{K_1(m_2)} \qquad G = \frac{m_1}{m_2}\frac{K_1(m_1) - I_1(m_1)D}{K_1(m_2)}$$

$$\Delta_c = I_0(m_1)K_0(m_1) + I_1(m_1)K_1(m_1)P_1 - I_1(m_1)K_0(m_1)P_2 - s\frac{K_0(m_2)}{m_2 K_1(m_2)} \tag{10.30}$$

$$\Delta_d = I_0(m_1)K_0(m_1) + I_1(m_1)K_1(m_1)P_1 - I_1(m_1)K_0(m_1)P_2 - \frac{K_0(m_2)}{m_2 K_1(m_2)}$$

$$\Delta = -I_0^2(m_1) + I_1^2(m_1)P_1 + I_0(m_1)I_1(m_1)P_2$$

where:

$$P_1 = \frac{2m^2 - m_2^2}{m_2^3}(1-s)\frac{K_0(m_2)}{K_1(m_2)} - \frac{m_1^2}{m_2^2}s\frac{K_0^2(m_2)}{K_1^2(m_2)}$$

$$P_2 = \frac{2m^2 - m_1^2}{m_2^2 m_1}(1-s) - \frac{m_1}{m_2}(1+s)\frac{K_0(m_2)}{K_1(m_2)} \tag{10.31}$$

$$m = \lambda a \qquad m_1 = \lambda_1 a \qquad m_2 = \lambda_2 a \qquad s = \sigma_2/\sigma_1$$

The magnetic field on the borehole axis has only component H_x, which is parallel to the dipole moment. Making use of eqs. 10.18 and 10.20 we obtain:

$$H_x = H_x^{(0)} + \frac{M}{2\pi^2}\frac{1}{a^3}\int_0^\infty \left(\frac{m^2}{2}D + \frac{k_1^2 a^2}{2}C\right)\cos m\alpha \, dm \tag{10.32}$$

here

$$H_x^{(0)} = -\frac{M}{4\pi L^3}(1 - \mathrm{i}k_1 L - k_1^2 L^2)\mathrm{e}^{\mathrm{i}k_1 L}$$

is the field in a uniform medium; L is the probe length.

Let us present field H_x on the borehole axis in units of the field in a free space $(-M/4\pi L^3)$. Then we have:

$$h_x = \frac{H_x}{H_x^0} = (1 - \mathrm{i}k_1 L - k_1^2 L^2)\mathrm{e}^{\mathrm{i}kL} - \frac{\alpha^3}{\pi}\int_0^\infty (m^2 D + k_1^2 a^2 C)\cos \alpha m \, dm \tag{10.33}$$

Results of calculations of amplitude, phase and apparent conductivity are presented in Figs. 10.5–10.14: $\sigma_a/\sigma_1 = A/A_0$, where A and A_0 are amplitudes of the secondary field on the borehole axis and in a uniform medium with conductivity σ_1.

First of all, let us consider the field behavior when the skin depth is greater than the probe length.

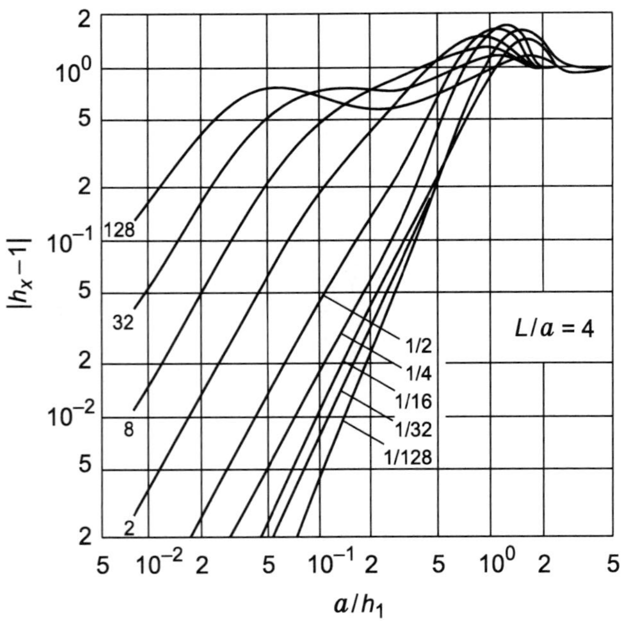

Figure 10.5. The amplitude of the secondary field, $|h_x - 1|$. Curve index σ_2/σ_1.

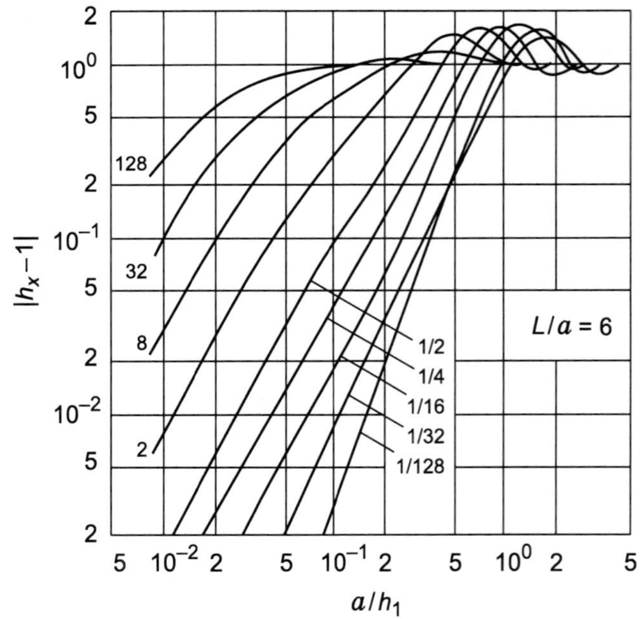

Figure 10.6. The amplitude of the secondary field, $|h_x - 1|$. Curve index σ_2/σ_1.

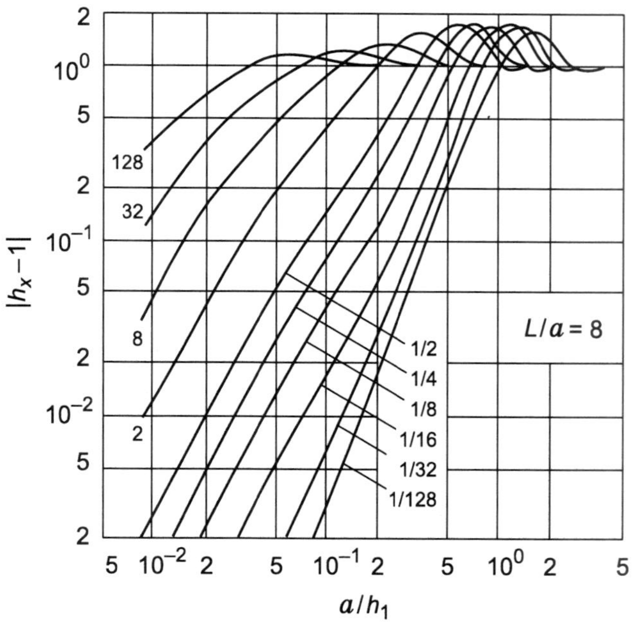

Figure 10.7. The amplitude of the secondary field, $|h_x - 1|$. Curve index σ_2/σ_1.

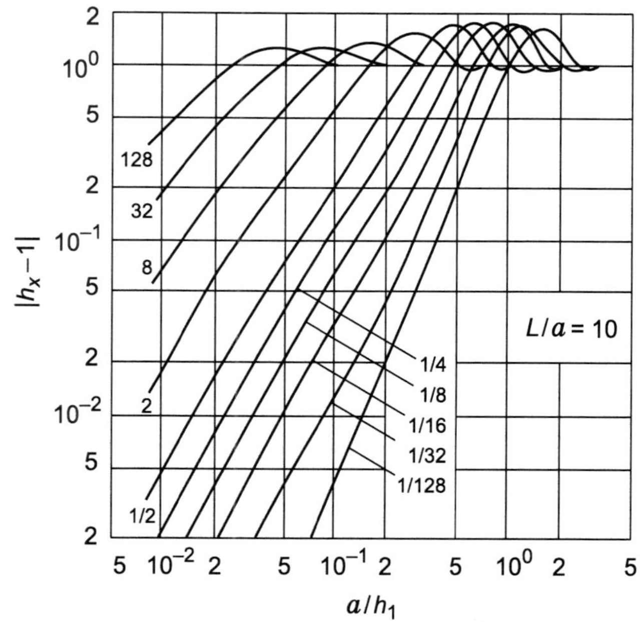

Figure 10.8. The amplitude of the secondary field, $|h_x - 1|$. Curve index σ_2/σ_1.

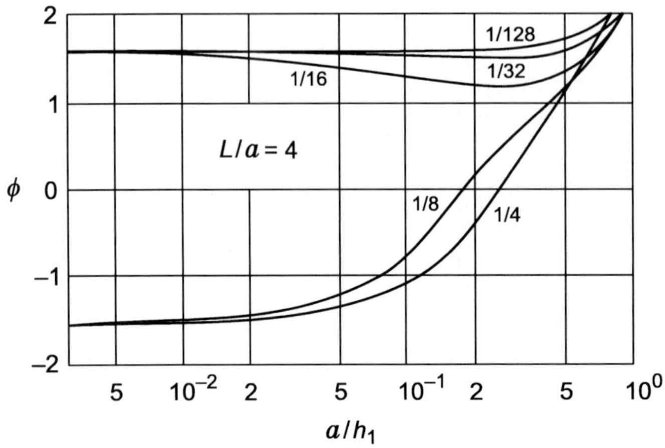

Figure 10.9. The phase of the secondary field. Curve index σ_2/σ_1.

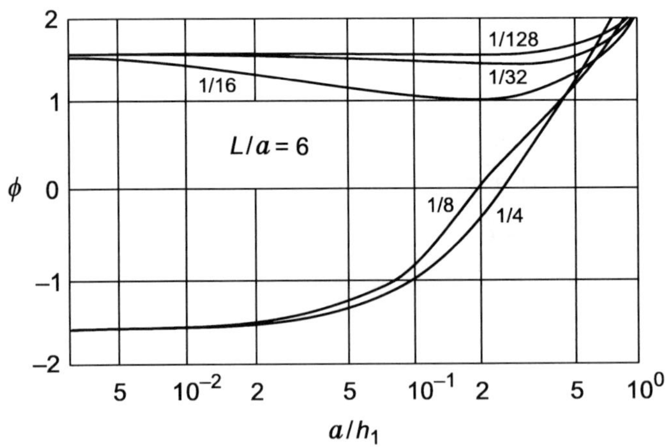

Figure 10.10. The phase of the secondary field. Curve index σ_2/σ_1.

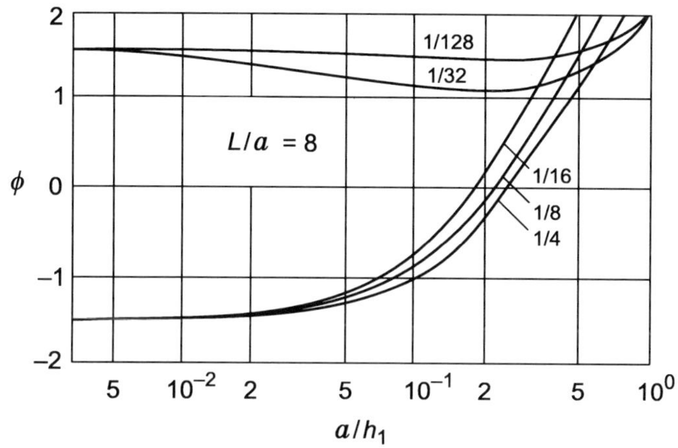

Figure 10.11. The phase of the secondary field. Curve index σ_2/σ_1.

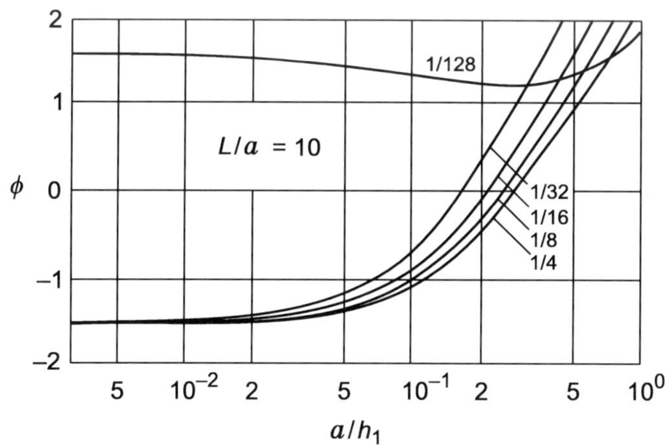

Figure 10.12. The phase of the secondary field. Curve index σ_2/σ_1.

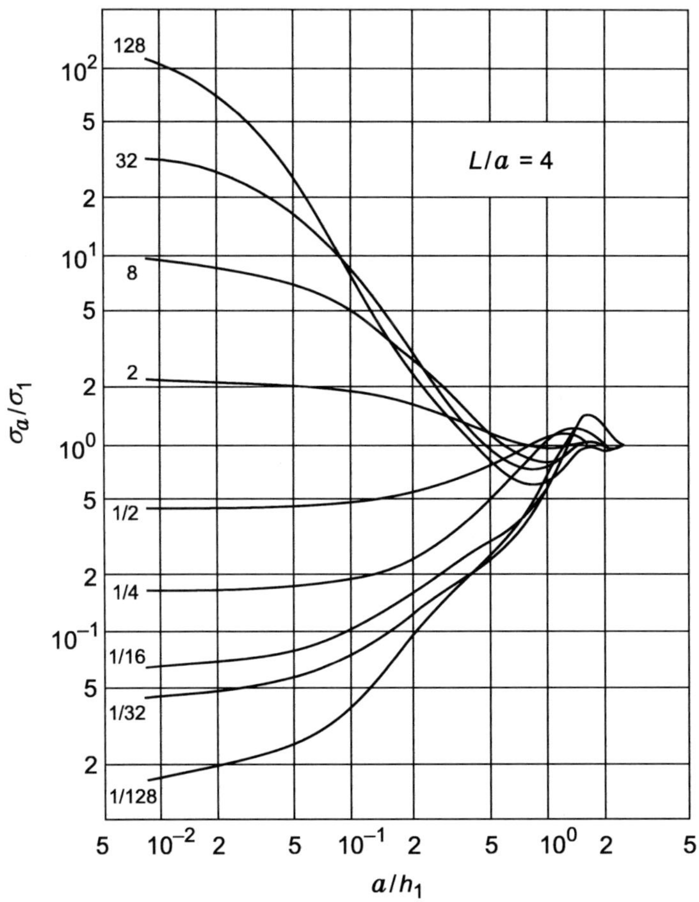

Figure 10.13. Apparent conductivity curves. Curve index σ_2/σ_1.

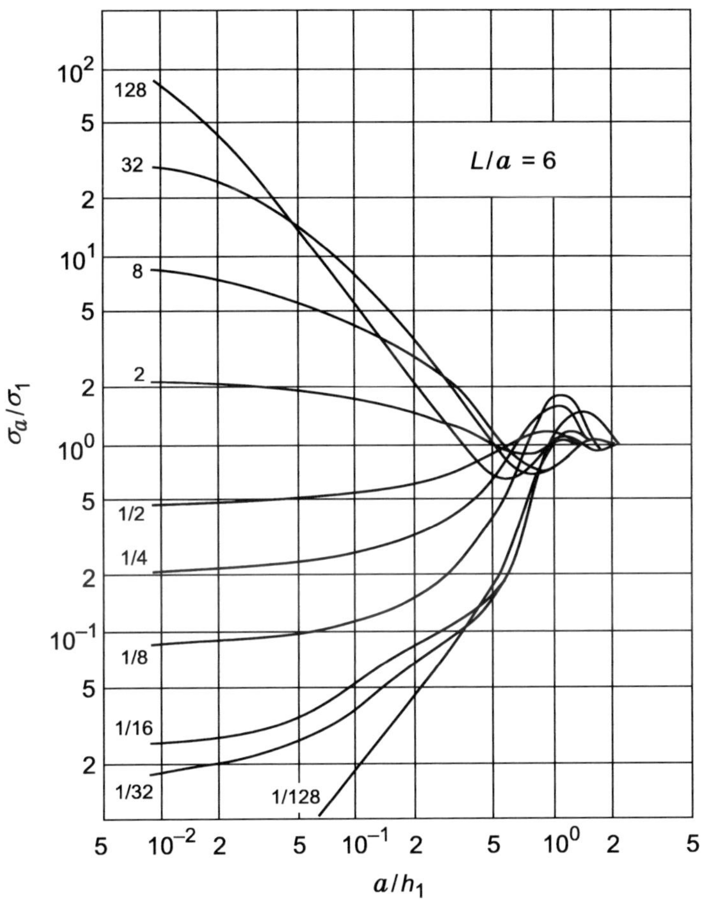

Figure 10.14. Apparent conductivity curves. Curve index σ_2/σ_1.

10.3. Magnetic Field on the Borehole Axis in the Near Zone (Range of Small Parameters)

At the range of small parameters $p = L/h$, the field H_x can be presented in the following way. A change of the magnetic field, only caused by currents in the dipole, generates a vortex electrical field, vector lines of which are located in vertical planes, perpendicular to the dipole moment.

Under action of this primary field which is directly proportional to frequency, current arises and, unlike the case of the vertical magnetic dipole, current lines intersect the boundary of media with different conductivities. For this reason electric charges arise on the borehole surface, which in the same manner as current density are directly proportional to the square of the wave number: $k^2 = i\sigma\mu\omega$, inasmuch as interaction between currents is assumed to be very small, and therefore not taken into account.

At the range of small parameters, a field of these charges, in accord with Coulomb's law, changes magnitude and direction of current density j, but at the same time the phase of the current remains constant and equal to $\pi/2$ with respect to the dipole current. Thus, the magnetic field of currents, arising under action of the primary vortex electric field E^0 and the secondary field of charges, is proportional to k^2. The secondary vortex electric field caused by a change of the secondary magnetic field with time is significantly smaller than that of charges, and therefore it is not taken into account in the approximate theory of small parameters.

For obtaining an asymptotic expression of the magnetic field let us present integrand eq. 10.33 in the form of series by powers of k^2a^2, and restricting ourselves by the first term, we obtain:

$$h_x = h_x^{(0)} - \frac{2\alpha^2}{\pi}\int_0^\infty \left(\frac{k_1^2a^2}{2}\Delta_c(0) - \frac{m}{4}k_1^2a^2\frac{\partial\Delta_d(0)}{\partial m_1} - \frac{m}{4}k_2^2a^2\frac{\partial\Delta_d(0)}{\partial m_2}\right)\frac{\cos m\alpha}{\Delta(0)}\,dm \tag{10.34}$$

where:

$$\Delta(0) = (1-s)\frac{K_0(m)}{mK_1(m)}$$

$$\Delta_c(0) = -\frac{1}{mK_1(m)}\left[I_0(m) + sI_1(m)\frac{K_0(m)}{K_1(m)} - (1-s)\frac{I_1(m)}{m}\right]$$

$$\begin{aligned}\frac{\partial\Delta_d(0)}{\partial m_1} &= (1-s)\Bigg[I_1(m)K_0(m)\left(1+\frac{2}{m^2}\right) + \frac{I_0(m)K_0(m)}{m} + \frac{I_1(m)K_1(m)}{m} \\ &\quad + \frac{I_0(m)K_0^2(m)}{K_1(m)} - \frac{K_0(m)}{m^2K_1^2(m)}\Bigg] - \frac{1}{m} + s\frac{K_0^2(m)}{mK_1^2(m)}\end{aligned} \tag{10.35}$$

$$\begin{aligned}\frac{\partial\Delta_d(0)}{\partial m_2} &= (1-s)\left[-\frac{2}{m^2} + \frac{K_0(m)}{mK_1(m)} - 1 + \frac{K_0^2(m)}{K_1^2(m)}\right] - I_1(m)K_0(m) \\ &\quad - (1-s)\frac{I_1(m)K_1(m)}{m} + \frac{1}{m} - \frac{K_0^2(m)}{mK_1^2(m)}\end{aligned}$$

$$h_x^{(0)} = -\frac{k_1^2L^2}{2}$$

Thus, for the quadrature component of the field we have:

$$Q\,h_x = -\left(\frac{L}{h_1}\right)^2 G_1(\alpha, s) - \left(\frac{L}{h_2}\right)^2 G_2(\alpha, s) \tag{10.36}$$

where $h_1 = (2/\sigma_1\mu\omega)^{1/2}$, $h_2 = (2/\sigma_2\mu\omega)^{1/2}$; σ_1 and σ_2 are conductivities of medium of the borehole and the formation.

$$\begin{aligned} G_1(\alpha, s) &= 1 + \frac{2\alpha}{\pi}\int\limits_0^\infty \left[\Delta_c(0) - \frac{m}{2}\frac{\partial\Delta_d(0)}{\partial m_1}\right]\frac{\cos m\alpha}{\Delta(0)}\,dm \\ G_2(\alpha, s) &= -\frac{2\alpha}{\pi}\int\limits_0^\infty \frac{m}{2}\frac{\partial\Delta_d(0)}{\partial m_2}\frac{\cos m\alpha}{\Delta(0)}\,dm \end{aligned} \tag{10.37}$$

In accord with eq. 10.35 in a uniform medium:

$$G_1(\alpha, 1) + G_2(\alpha, 1) = 1 \tag{10.38}$$

As has been shown in Chapter 4 we can obtain a more accurate presentation for the low-frequency part of the spectrum. For instance, taking into account both terms of expansion, we have:

$$\begin{aligned} \mathrm{In}\,h_x &= \frac{4}{3}\left(\frac{L_1}{h_2}\right)^3 \\ Q\,h_x &\simeq -a_1\left(\frac{L}{h_1}\right)^2 + \frac{4}{3}\left(\frac{L}{h_2}\right)^3 \end{aligned} \tag{10.39}$$

where coefficient a_1 is defined from expression 10.34: $a_1 = G_1 + sG_2$.

Therefore, at the range of small parameters, as in the case of the vertical magnetic dipole, the inphase component of the secondary field, as well as the second term of the quadrature component in eq. 10.39, do not depend on conductivity of the borehole. Taking into account the known relation between the low-frequency part of the spectrum and the late stage of transient field, we can claim that a field measured at sufficiently large times after current is turned off is also a function of the formation resistivity only.

Let us consider now the behavior of functions G_1 and G_2, defining a behavior of the field at the range of small parameter L/h for various α. If the probe length decreases, $\alpha \to 0$, then $G_2(\alpha, s) \to 0$, and $G_1(\alpha, s) \to 1$, and the field approaches to that in a uniform medium with conductivity σ_1. For large values of parameter α due to rapid oscillations of the function $\cos m\alpha$, the value of the integral in eq. 10.37 is defined by the integrand near $m = 0$. For small values of m we have:

$$\begin{aligned} &\Delta(0) \simeq -\frac{1+s}{2} \qquad \Delta_c(0) = (1-s)K_0(m) \\ &\frac{\partial\Delta_d(0)}{\partial m_1} = -\frac{\partial\Delta_d(0)}{\partial m_2} = (1-s)\frac{K_0(m)}{m} \end{aligned} \tag{10.40}$$

Making use of asymptotic presentation of Sommerfeld integral:

$$\int_0^\infty K_0(m)\cos m\alpha \, dm = \frac{\pi}{2}\frac{1}{\sqrt{1+\alpha^2}} \to \frac{\pi}{2\alpha} \qquad \text{if } \alpha \to \infty$$

we obtain:

$$\begin{aligned} G_1(\alpha,s) &= 1 - \frac{1-s}{1+s} = \frac{2s}{1+s} = \frac{\sigma_2}{\sigma_{av}} \\ G_2(\alpha,s) &= -\frac{1-s}{1+s} = \frac{\sigma_2-\sigma_1}{\sigma_2+\sigma_1} = -K_{12} \end{aligned} \qquad \text{if } \alpha \gg 1 \tag{10.41}$$

where σ_{av} is the average value of conductivity; K_{12} is the contrast coefficient characterizing the density of surface charges.

Correspondingly, for the quadrature component of the field, we have:

$$Q\,h_x = -\left(\frac{L}{h_2}\right)^2 \qquad \text{if } \alpha \gg 1 \tag{10.42}$$

Thus, at the low-frequency part of the spectrum with an increase of the probe length the field on the borehole axis tends to that in a uniform medium with the formation conductivity. In a general case, both functions G_1 and G_2, regardless of the probe length, depend on the resistivity of a medium.

Now let us introduce functions $G_1^*(\alpha,s)$ and $G_2^*(\alpha,s)$, which as geometric factors in conventional induction logging tend to 0 and 1, respectively, when $\alpha \to 0$:

$$\begin{aligned} G_1^*(\alpha,s) &= G_1^*(\alpha,s) - \frac{2s}{1+s} \\ G_2^*(\alpha,s) &= G_2^*(\alpha,s) + \frac{2s}{1+s} \end{aligned} \tag{10.43}$$

Then, instead of eq. 10.36 we can write:

$$Q\,h_x = -\left(\frac{L}{h_1}\right)^2 [G_1(\alpha,s) + sG_2(\alpha,s)] \tag{10.44}$$

First of all, consider asymptotic behavior of function $G^*(\alpha,s)$ for large values of α. Preliminary it is convenient to distinguish singularity of the integrand in eq. 10.37 for small values of m. For this purpose we will present $G^*(\alpha,s)$ in the form:

$$\begin{aligned} G^*(\alpha,s) = \frac{1-s}{1+s} + \frac{2\alpha}{\pi}\int_0^\infty \left\{\frac{1}{\Delta(0)}\left[\Delta_c(0) - \frac{m}{2}\frac{\partial\Delta_d(0)}{\partial m_1}\right] + \frac{1-s}{1+s}K_0(m)\right\}\cos m\alpha\, dm \\ - \frac{2\alpha}{\pi}\frac{1-s}{1+s}\int_0^\infty K_0(m)\cos m\alpha\, dm = \frac{1-s}{1+s}\left(1 - \frac{\alpha}{\sqrt{1+\alpha^2}}\right) + \frac{2\alpha}{\pi}\int_0^\infty \phi(m)\cos m\alpha\, dm \end{aligned} \tag{10.45}$$

where

$$\phi(m) = \frac{1}{\Delta(0)}\left[\Delta_c(0) - \frac{m}{2}\frac{\partial \Delta_d(0)}{\partial m_1}\right] + \frac{1-s}{1+s}K_0(m)$$

Integrating eq. 10.45 by parts we have:

$$\int_0^\infty \phi(m)\cos m\alpha \, \mathrm{d}m = -\frac{1}{\alpha^2}\phi'(0) - \frac{1}{\alpha^2}\int_0^\infty \phi''(m)\cos m\alpha \, \mathrm{d}m$$

inasmuch as function $\phi(m)$ and its derivatives tend to zero, when $m \to \infty$.

Making use of known expansions for Bessel functions:

$$K_0(z) = -I_0(z)\ln\frac{z}{2} + \sum_{m=0}^{\infty}\frac{1}{(m!)^2}\left(\frac{z}{2}\right)^{2m}\left[\sum_{k=1}^{m}\frac{1}{k} - C\right]$$

$$K_1(z) = I_1(z)\ln\frac{z}{2} - \sum_{m=0}\frac{1}{m!(m+1)!}\left(\frac{z}{2}\right)^{2m+1}\left[\sum_{k=1}^{\infty}\frac{1}{k} + \frac{1}{2(m+1)} - C\right]$$

$$I_0(z) = \sum_{m=0}^{\infty}\frac{1}{(m!)^2}\left(\frac{z}{2}\right)^{2m}$$

$$I_1(z) = \sum_{m=0}^{\infty}\frac{1}{m!(m+1)!}\left(\frac{z}{2}\right)^{2m+1}$$

we obtain:

$$\phi(m) \equiv \frac{2sm^2}{(1+s)^2}K_0^2(m) + \frac{3+3s+2s^2}{4(1+s)^2}m^2K_0(m) + const$$

$$\phi'(m) = 0$$

$$\phi''(m) = \frac{4s}{(1+s)^2}K_0^2(m) + \frac{3-21s+2s^2}{2(1+s)^2}K_0(m)$$

Thus:

$$\int_0^\infty \phi(m)\cos m\alpha \, \mathrm{d}m$$

$$= -\frac{1}{\alpha^2}\left[\frac{4s}{(1+s)^2}\int_0^\infty K_0^2(m)\cos m\alpha \, \mathrm{d}m + \frac{3-21s+2s}{2(1+s^2)}\int_0^\infty K_0(m)\cos m\alpha \, \mathrm{d}m\right]$$

Inasmuch as:

$$\int_0^\infty K_0^2(m)\cos m\alpha \, \mathrm{d}m \simeq \ln 2\alpha \quad \text{if } \alpha \gg 1$$

then

$$G_1^*(\alpha, s) = -\frac{1}{\alpha^2(1+s)^2}\left(8s\ln 2\alpha + \frac{2-21s+3s^2}{2}\right) \qquad \text{if } \alpha \gg 1 \tag{10.46}$$

Similar transformations show that:

$$G_2^*(\alpha, s) = 1 + \frac{1}{2\alpha^2}\frac{1}{(1+s)^2}\left(1+5s-22s^2+16s^2\ln 2\alpha\right) \tag{10.47}$$

If the formation resistivity is significantly greater than that of the borehole, then instead of eqs. 10.46–10.47 we have:

$$\begin{aligned} G_1^*(\alpha, s) &\approx -\frac{1}{\alpha^2} - \frac{s}{\alpha^2}(8\ln\alpha - 12.5) \\ G^*(\alpha, s) &\approx 1 + \frac{1}{2\alpha^2} \end{aligned} \tag{10.48}$$

and correspondingly for the magnetic field we obtain:

$$\mathrm{Q}\,h_x = \left(\frac{a}{h_1}\right)^2 - \left(\frac{L}{h_2}^2\right)\left(\frac{1-8\ln 2-13}{\alpha^2}\right) \qquad \text{if } \frac{\sigma_2}{\sigma_1} \ll 1 \text{ and } \alpha \gg 1 \tag{10.49}$$

By analogy, if $\sigma_2 \gg \sigma_1$, then:

$$\begin{aligned} G_1^*(\alpha, s) &\simeq -\frac{3}{2\alpha^2} \\ G_2^*(\alpha, s) &\simeq 1 + \frac{8\ln 2\alpha - 11}{\alpha^2}s^2 \end{aligned} \tag{10.50}$$

and for this reason:

$$\mathrm{Q}\,h_x = \frac{3}{2}\left(\frac{a}{h_1}\right)^2 - \left(\frac{L}{h_2}^2\right)\left(1 + \frac{8\ln 2 - 11}{\alpha^2}\right) \qquad \text{if } \frac{\sigma_2}{\sigma_1} \gg 1 \text{ and } \alpha \ll 1 \tag{10.51}$$

Table 10.2 contains values of functions $G_1^*(\alpha, s)$, $G_2^*(\alpha, s)$ and $G_1^* + sG_2^*$ for various values of s and α.

Suppose that the formation resistivity exceeds that of the borehole ($s < 1$). Then the function $G_1^*(\alpha, s) + sG_2^*(\alpha, s)$ and, respectively, the quadrature component changes twice its sign. It is related with the fact that electric charges arising on the borehole surface create an electric field which reduces the primary vortex electric field. Near the source, ($L/a < 1$), the influence of the charges is small, and the field coincides with that in a uniform medium with the borehole conductivity σ_1: $\mathrm{Q}\,h_x \simeq -(L/h_1)^2$. At the range of large distances, $L/a \gg 1$, more precisely, when $L/a \gg (\sigma_1/\sigma_2)^{1/2}$, as follows from eq. 10.49, the effect caused by the charges is also small and $\mathrm{Q}\,h_x \simeq -(L/h_2)^2$. For intermediate values of probe lengths, the field caused by the charges is comparable with the vortex one, and has opposite direction. Such a field behavior implies that for certain values of α and s the quadrature component becomes equal to zero.

TABLE 10.2
Values of functions G_1^*, G_2^*, $G_1^* + sG_2^*$

α	G_1^*	G_2^*	$G_1^* + sG_2^*$	α	G_1^*	G_2^*	$G_1^* + sG_2^*$
		$s = 1/128$			$s = 1/64$		
2	−0.1350	1.204	−0.1256	2	−0.1357	1.204	−0.1169
4	-0.7001×10^{1}	1.037	-0.6191×10^{-1}	4	-0.7163×10^{-1}	1.038	-0.5540×10^{-1}
6	−0.3020	1.014	-0.2228×10^{-1}	6	−0.3162	1.014	-0.1577×10^{-1}
8	−0.1672	1.007	-0.8851×10^{-2}	8	−0.1778	1.008	-0.2039×10^{-2}
10	-0.1074×10^{-1}	1.005	-0.2896×10^{-2}	10	-0.1156×10^{-1}	1.005	0.4144×10^{-2}
12	-0.7528×10^{-2}	1.003	0.3104×10^{-3}	12	-0.8169×10^{-2}	1.003	0.7511×10^{-2}
16	−0.4310	1.002	0.3518×10^{-2}	16	−0.4739	1.002	0.1092×10^{-1}
20	−0.2798	1.001	0.5024	20	−0.3176	1.001	0.1254
		$s = 1/32$			$s = 1/16$		
2	−0.1370	1.204	-0.9942×10^{-1}	2	−0.1395	1.203	-0.6433×10^{-1}
4	-0.7469×10^{-1}	1.040	-0.4220×10^{-1}	4	-0.8034×10^{-1}	1.043	-0.1504×10^{-1}
6	−0.3432	1.015	-0.2600×10^{-2}	6	−0.3924	1.017	0.2431×10^{-1}
8	−0.1982	1.008	0.1169×10^{-1}	8	−0.2354	1.009	0.3954
10	-0.1311×10^{-1}	1.005	0.1830	10	−0.1596	1.006	0.4692
12	-0.9394×10^{-2}	1.004	0.2197	12	-0.1164×10^{-1}	1.004	0.5512
16	−0.5559	1.002	0.2576	16	-0.7062×10^{-2}	1.002	0.5559
20	−0.3629	1.001	0.2759	20	0.4781	1.002	0.5782
		$s = 1/8$			$s = 1/4$		
2	−0.1440	1.201	0.6155×10^{-2}	2	−0.1516	1.198	0.1479
4	-0.8936×10^{-1}	1.049	0.4185	4	−0.1019	1.063	0.1637
6	−0.4746	1.021	0.8013	6	-0.5910×10^{-1}	1.030	0.1983
8	−0.2979	1.012	0.9669×10^{-1}	8	−0.3874	1.018	0.2157
10	−0.2076	1.008	0.1052	10	−0.2674	1.012	0.2254
12	-0.1542×10^{-1}	1.016	0.1103	12	−0.2085	1.009	0.2313
16	-0.9591×10^{-2}	1.003	0.1158	16	-0.1322×10^{-1}	1.005	0.2381
20	0.6600	1.002	0.1187	20	-0.9202×10^{-2}	1.004	0.2417
		$s = 1/1$			$s = 2$		
2	−0.1630	1.190	0.4322	2	−0.1944	1.162	0.2129×10^{1}
4	−0.1144	1.086	0.4285	4	−0.1177	1.160	0.2203
6	-0.7119×10^{-1}	1.048	0.4526	6	-0.7506×10^{-1}	1.115	0.2155
8	−0.4820	1.031	0.4672	8	−0.5133	1.083	0.2155
10	−0.3495	1.022	0.4759	10	−0.3723	1.062	0.2087
12	−0.2660	1.016	0.4815	12	−0.2827	1.048	0.2068
16	−0.1704	1.010	0.4880	16	−0.1799	1.031	0.2044
20	−0.1194	1.007	0.4916	20	−0.1252	1.022	0.2032

TABLE 10.2
(Continued)

α	G_1^*	G_2^*	$G_1^* + sG_2^*$	α	G_1^*	G_2^*	$G_1^* + sG_2^*$
2	-0.2194	1.134	0.8955×10^{1}	2	-0.2259	1.127	0.1780×10^{2}
4	-0.1020	1.220	0.9656	4	-0.9664×10^{1}	1.235	0.1966
6	-0.5858×10^{1}	1.176	0.9350	6	-0.5267	1.192	0.1902
8	-0.3758	1.133	0.9026	8	-0.3256	1.146	0.1831
10	-0.2606	1.102	0.8787	10	-0.2157	1.120	0.1778
12	-0.1904	1.080	0.8619	12	-0.1579×10^{-1}	1.088	0.1740
16	-0.1163×10^{-1}	1.052	0.8409	16	-0.9305×10^{-2}	1.058	0.1693
20	-0.7854×10^{-2}	1.037	0.8291	20	-0.6156	1.041	0.1666

TABLE 10.3
Intervals within which the quadrature component vanishes

s	1/128	1/64	1/32	1/16	1/8
α	1–2; 11–12	1–2; 8–9	1–2; 6–7	1–2; 4–5	1–3

At the vicinity of these values, conditions of the small parameter are met only for very small frequencies, allowing us to neglect terms in an expansion of the spectrum, smaller than k^2. Table 10.3 shows intervals within which the quadrature component vanishes.

When $\sigma_2 > \sigma_1$ function $G_1^* + sG_2^*$ does not change sign.

Returning to the expression for the quadrature component (eq. 10.39), we can notice one feature of its behavior. Inasmuch as value $a_1 = G_1^* + sG_2^*$ is an alternating function of α and s, i.e. at the range where $a_1 \gg 0$, the magnetic field with an increase of frequency grows more rapidly than the frequency. Under excitation of the field by the vertical magnetic dipole and measuring a field along its axis this feature is not observed

Equations 10.49 and 10.51 present the magnetic field as a sum of terms, each of which depends on the conductivity of the borehole or that of the formation only. This fact allows us to apply differential probes described in previous chapters, which significantly decrease the effect of the borehole. The simplest differential probe, investigated in detail above is three-coil probe providing compensation of the primary field (Fig. 10.15).

In fact, since the part of the field $\mathrm{Q}\,h_x$, which is proportional to σ_1, does not depend on the probe length, L_1, a signal measured in a three-coil probe does not depend on borehole parameters provided that the primary fields in every coil probe forming this system are equal to each other.

Table 10.4 presents the results of calculation of function $\Delta h_x = \mathrm{Q}\,h_x(L_1) - \mathrm{Q}\,h_x(L_2)$, proportional to the electromotive force in three-coil probe by exact and approximate formulae 10.33 and 10.49. As follows from the table, the EMF in the receiver coil of the probe does not practically depend on the resistivity of the borehole even beyond the range of small parameters. It means that, as in the case of the vertical magnetic dipole, there are conditions when induced currents in the borehole and surface charges do not influence the character of the skin effect in the formation, which manifest itself in the

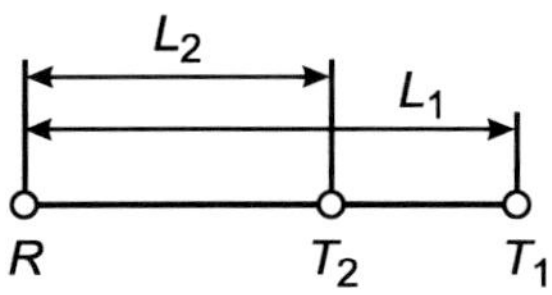

Figure 10.15. Three-coil differential probe.

TABLE 10.4
Values of function $\Delta h_x = Q\,h_x(L-1) - Q\,h_x(L_2)$; $L_1/a = 10$, $L_2/a = 8$

s	a/h_1	$\Delta h_x \times 10^2$	$(A_1 - A_2) \times 10^2$
	$\rho_2 = 2.5$ ohm·m	$\Delta h^{apr} \simeq -0.88 \times 10^{-2}$	
1/32	0.10×10^{0}	−0.73	0.76
1/16	0.71×10^{-1}	−0.73	0.77
1/8	0.50×10^{-1}	−0.73	0.77
1/4	0.35×10^{-1}	−0.74	0.79
1/2	0.25×10^{-1}	−0.75	0.80
	$\rho_2 = 5$ ohm·m	$\Delta h^{apr} \simeq -0.44 \times 10^{-2}$	
1/32	0.71×10^{-1}	−0.41	0.42
1/16	0.51×10^{-1}	−0.41	0.43
1/8	0.35×10^{-1}	−0.41	0.43
1/4	0.25×10^{-1}	−0.42	0.43
1/2	0.18×10^{-1}	−0.43	0.44
	$\rho_2 = 10$ ohm·m	$\Delta h^{apr} \simeq -0.22 \times 10^{-2}$	
1/32	0.51×10^{-1}	−0.22	0.23
1/16	0.35×10^{-1}	−0.22	0.23
1/8	0.25×10^{-1}	−0.22	0.23
1/4	0.18×10^{-1}	−0.23	0.23
1/2	0.12×10^{-1}	−0.23	0.24
	$\rho_2 = 20$ ohm·m	$\Delta h^{apr} \simeq -0.12 \times 10^{-2}$	
1/32	0.35×10^{-1}	−0.12	0.12
1/16	0.25×10^{-1}	−0.12	0.12
1/8	0.18×10^{-1}	−0.12	0.12
1/4	0.12×10^{-1}	−0.12	0.12
1/2	0.88×10^{-2}	−0.12	0.12

TABLE 10.5
Maximal values of parameter a/h_1

α \ s	1/128	1/64	1/32	1/16	1/8	1/2	8
4	0.6	0.7	0.8	0.9	0.2	0.2	0.1
8	0.15	0.2	0.2	0.2	0.13	0.13	0.05

same manner as in a uniform medium with resistivity of the formation. For this reason, for the quadrature component of the field, h_x, we have:

$$\mathrm{Q}\,h_x = -\left(\frac{L}{h_1}\right)^2 G_1^*(\alpha, s) - \left(\frac{L}{h_2}\right)^2 G_2^{**}(\alpha, s) + \mathrm{Q}\,h_x^{un}\left(\frac{L}{h_2}\right) \tag{10.52}$$

where

$$G_2^{**}(\alpha, s) = G_2^*(\alpha, s) - 1 \tag{10.53}$$

This expression for $\mathrm{Q}\,h_x$ is valid for arbitrary values of α and s. Maximal values of parameter a/h_1, for which results of calculations by exact and approximate formulas (eq. 10.52) do not differ more than 5%, are given in Table 10.5.

10.4. The Magnetic Field on the Borehole Axis in the Far Zone

Now we will obtain asymptotic formulae for the field H_x in the far zone ($\alpha \ll 1$). In deriving a formula we will deform the contour of integration in eq. 10.33 on the complex plane of variable m. However, such a procedure requires either the proof of absence of poles of the integrand or evaluation of their contribution to the integral value. The problem of determination of poles is extremely difficult because of the complexity of the integrand. At the same time sufficient agreement of results of calculations by asymptotic and exact formulae allows us to think that if there are poles in the upper half-plane of m, their contribution in a considered part of the spectrum is sufficiently small. Let us present integral in eq. 10.33 in the form:

$$\frac{\alpha^3}{\pi}\int_0^\infty \left(m^2 D + k_1^2 a^2 C\right)\cos m\alpha \,\mathrm{d}m = \frac{\alpha^3}{\pi}\int_0^\infty \left(m^2 D + k_1^2 a^2 C\right)\mathrm{e}^{\mathrm{i}\alpha m}\,\mathrm{d}m \tag{10.54}$$

We will assume that in the upper half-plane of complex variable m there are no singularities except the branch points $m = k_1 a$ and $m = k_2 a$.

Choosing cross-cuts along lines $\mathrm{Re}\, m_1 = 0$ and $\mathrm{Re}\, m_2 = 0$, it is supposed that the real parts of radicals $(m^2 - k_1^2 a^2)^{1/2}$ and $(m^2 - k_2^2 a^2)^{1/2}$ are positive on the complex plane of m. As follows from the asymptotic behavior of Bessel functions the integrand in eq. 10.54 can increase with $m \to \infty$, at least, not quicker than $\mathrm{e}^{2|m|}$. For this reason convergence of the integral in eq. 10.54 in the upper half-plane for $\alpha > 2$ is provided by multiplier $\mathrm{e}^{\mathrm{i}\alpha m}$ regardless of the sign of the real part of radicals m_1 and m_2. We will draw cross-cuts

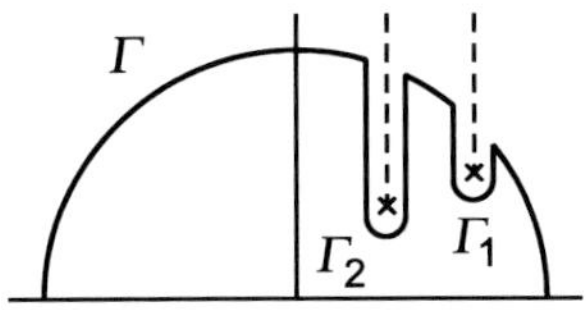

Figure 10.16. Contour integration in complex plane.

from branch points k_1a and k_2a parallel to the imaginary axis (Fig. 10.16) and deform the contour of integration in Γ.

The integral along arcs with an infinite radius due to the presence of term $e^{i\alpha m}$ vanishes since $\operatorname{Im} m > 0$, $\alpha > 2$. For this reason, the integral along the real axis (eq. 10.54) is equal to the sum of integrals along sides of cross-cuts Γ_1 and Γ_2. First, let us evaluate the integral along cross-cut Γ_1. In passing from the left side of the cross-cut to the right side the value of m changes sign. Thus, the integral along cross-cut Γ_1, is equal to:

$$\frac{\alpha^3}{\pi}\int\limits_{\Gamma_1}\left\{m^2[D(m_1)-D(-m_1)]+k_1^2a^2[C(m_1)-C(-m_1)]\right\}e^{i\alpha m}\,dm \tag{10.55}$$

Making use of properties of Bessel functions:

$$\begin{aligned} I_0(-z)&=I_0(z) \qquad & K_0(-z)&=K_0(z)+i\pi I_0(z)\\ I_1(-z)&=-I_1(z) \qquad & K_1(-z)&=K_1(z)+i\pi I_1(z)\end{aligned} \tag{10.56}$$

it is not difficult to show that for functions D and C are valid following relations:

$$D(-m_1)=D(m_1)-i\pi \qquad C(-m_1)=C(m_1)-i\pi \tag{10.57}$$

Thus, integral in eq. 10.55 has the form:

$$\frac{\alpha^3}{2}i\int\limits_{\Gamma_1}(m^2+k_1^2a^2)e^{i\alpha m}\,dm \tag{10.58}$$

Letting $m = it + k_1a$ we obtain:

$$e^{ik_1L}\,\frac{\alpha^3}{2}\int\limits_0^\infty(t^2-2itk_1a-2k_1^2a^2)\,e^{-\alpha t}\,dt=(1-ik_1L-k_1^2L^2)\,e^{ik_1L}=h_x^{un}\left(\frac{L}{h_1}\right)$$

where $h_x^{un}(L/h_1)$ is the x-component of the magnetic field in a uniform medium with the borehole resistivity. For this reason and in accord with eq. 10.33 the magnetic field is expressed, as in the case of the vertical magnetic dipole, only through the integral along cross-cut Γ_2:

$$h_x=-\frac{\alpha^3}{2\pi}\int\limits_{\Gamma_2}\left(m^2D+k_1^2a^2C\right)e^{i\alpha m}\,dm \tag{10.59}$$

For transformation of the integrand in eq. 10.59 we will make use of relations following from eq. 10.56:

$$\frac{K_0(z)}{K_1(z)}+\frac{K_0(-z)}{K_1(-z)}=\frac{\mathrm{i}\pi}{zK_1(z)K_1(-z)}$$

$$\frac{K_0^2(z)}{K_1^2(z)}+\frac{K_0^2(-z)}{K_1^2(-z)}=\frac{\mathrm{i}\pi}{zK_1(z)K_1(-z)}\left[\frac{K_0(z)}{K_1(z)}-\frac{K_0(-z)}{K_1(-z)}\right]$$

$$\frac{K_0^2(z)K_0(-z)}{K_1^2(z)K_1(-z)}+\frac{K_0^2(-z)K_0(z)}{K_1^2(-z)K_0(z)}=\frac{\mathrm{i}\pi}{zK_1(z)K_1(-z)}\frac{K_0(z)K_0(-z)}{K_1(z)K_1(-z)}$$

For difference between values of function C in both sides of the cross-cut, after relatively simple transformations, we have:

$$\begin{aligned}C(m_2)-C(m_1)&=\frac{\mathrm{i}\pi}{m_2^2K_1(m_2)K_1(-m_2)\Delta(m_2)\Delta(-m_2)}\\&\times\left\{\frac{I_1^2(m_1)}{m_1}(1-s)^2\frac{m^2+k_1^2a^2}{m_2^2}\frac{m^2+k_2^2a^2}{m_2^2}+sI_0^2(m_1)\right.\\&-I_0(m_1)\frac{I_1^2(m_1)}{m_1}(1-s)\left[s\frac{m^2+k_1^2a^2}{m_2^2}+\frac{m^2+k_2^2a^2}{m_2^2}\right]\\&-sm_1^2\left[\frac{m^2+k_1^2a^2}{m_2^2}\frac{I_1^2(m_1)}{m_1^2}(1-s)-I_0(m_1)\frac{I_1(m_1)}{m_1}\right]\\&\times\left.\left[\frac{K_0(m_2)}{m_2K_1(m_2)}-\frac{K_0(-m)}{m_2K_1(-m_2)}\right]+s(1+2s)I_1^2(m_1)m_1^2\frac{K_0(m_2)K_0(-m_2)}{m_2^2k_1(m_2)K_1(-m_2)}\right\}\end{aligned}\quad(10.60)$$

where

$$\begin{aligned}\Delta(\pm m_2)&=-I_0^2(m_1)+I_1^2(m_1)\left[\frac{m^2+k_2^2a^2}{m_2^2}(1-s)\frac{K_0(\pm m_2)}{\pm m_2K_0(\pm m_2)}\right]\\&-sm_1^2\frac{K_0^2(\pm m_2)}{m_2K_1^2(\pm m_2)}+I_0(m_1)\frac{I_1(m_1)}{m_1}\left[\frac{m^2+k_2^2a^2}{m_2^2}(1-s)-1+sm^2\frac{K_0(\pm m_2)}{m_2K_0(\pm m_2)}\right]\end{aligned}$$

Inasmuch as function D can be presented in the form:

$$D=C+(s-1)\frac{K_0(m_2)}{m_2K_1(m_2)m_2}$$

then for discontinuity of function D we have:

$$D(m_2)-D(-m_2)=C(m_2)-C(-m_2)+A(m_2)$$

here

$$A(m_2) = \frac{\mathrm{i}\pi(s-1)}{m_2^2 K_1(m_2) K_1(-m_2) \Delta(m_2) \Delta(-m_2)}$$
$$\times \left(I_0(m_1) \frac{I_1(m_1)}{m_1} \frac{m^2 + k_1^2 a^2}{m_2^2} (1-s) - I_0^2(m_1) - s m_1^2 I_1^2(m_1) \frac{K_0(m_2) K_0(-m_2)}{m_2^2 K_1(m_2) K_1(-m_2)} \right) \tag{10.61}$$

Thus, instead of eq. 10.59, we have:

$$h_x = -\frac{\alpha^3}{2\pi} \int\limits_{k_2 a}^{\mathrm{i}\infty + k_2 a} (m^2 + k_1^2 a^2) \left[C(m_2) - C(-m_2) + m^2 A(m_2) \right] \mathrm{e}^{\mathrm{i}\alpha m} \, \mathrm{d}m \tag{10.62}$$

We will introduce a new variable letting $m = \mathrm{i}t + k_2 a$. Along the cross-cut, variable t changes from zero to infinity and $m_1 = (-t^2 + 2\mathrm{i}k_2 at + (k_2^2 - k_1^2)a^2)^{1/2}$ and $m_2 = (-t^2 + 2\mathrm{i}k_2 at)^{1/2}$. Correspondingly, the expression for the magnetic field has the form:

$$h_x = \mathrm{e}^{-\mathrm{i}k_2 L} \frac{\mathrm{i}\alpha^3}{2\pi} \int\limits_0^\infty \left\{ (m^2 + k_1^2 a^2) \left[C(m_2) - C(-m_2) + m^2 A(m_2) \right] \right\} \mathrm{e}^{-\alpha t} \, \mathrm{d}t \tag{10.63}$$

In spite of the cumbersome character of the integrand, presentation 10.63 turns out to be useful for field calculation for long probes ($\alpha \gg 1$), since the integral in eq. 10.63 does not contain the oscillating function $\cos m\alpha$, unlike eq. 10.33. Moreover, in the wave zone ($|k_2 L| > 1$) the value of the field is exponentially small, but in eq. 10.63 multiplier $\mathrm{e}^{\mathrm{i}k_2 L}$ stands in front of the integral, i.e. it has a relatively large magnitude. It essentially facilitates calculations, inasmuch as the main problem of numerical integration of eq. 10.33 for large parameters α is to obtain very small exponential result from integration of function, value of which within the main interval is many orders greater than that of the integral.

Now proceeding from eq. 10.63 we will obtain the asymptotic formula describing the field in the far zone ($\alpha \gg 1$). In this case the value of the integral is defined by the range $t \leqslant 1/\alpha < 1$. Generally the integrand in eq. 10.63 depends in a rather complicated manner on m_1, but if conditions:

$$\frac{1}{\alpha^2} \ll |k^2 a^2| \qquad \text{i.e. } |k_1 L| > 1$$

and $s < 1$ are met, we can approximately assume that $m_1 \simeq (k_2^2 a^2 - k_1^2 a^2)^{1/2}$, and think that the value of m_1, as well as functions of m, does not depend on the variable of integration t.

For radical m_2 we have:

$$m_2 \simeq \left(-\frac{1}{\alpha^2} + 2\mathrm{i}\frac{k_2 a}{\alpha} \right)^{1/2}$$

i.e. $|m_2| \ll 1$.

For this reason, keeping the terms of orders s/m_2^4, s/m_2^4, $1/m_2^4$, $s\ln(m_2)/m_2^2$ and omitting the terms s/m_2^2, $1,\ldots$, we can approximately present the expression 10.60 in the form:

$$C(m_2)-C(-m_2)=\frac{\mathrm{i}\pi}{m_2^2K_1(m_2)K_1(-m_2)\Delta(m_2)\Delta(-m_2)}$$

$$\times\left(\frac{I_1^2(m_1)}{m_1}(1-2s)\frac{m^2+k_1^2a^2}{m_2^2}\frac{m^2+k_2^2a^2}{m_2^2}-2sI_1^2(m_1)\frac{m^2+k_1^2a^2}{m_2^2}K_0(m_2)\right)\tag{10.64}$$

By analogy we have:

$$m_2^2K_1(m_2)K_1(-m_2)\Delta(m_2)\Delta(-m_2)\simeq-\left\{I_0(m_1)\frac{I_1(m_1)}{m_1}\frac{m^2+k_1^2a^2}{m_2^2}\right.$$

$$\left.+K_0(m_2)\left[I_1^2(m_1)\frac{m^2+k_2^2a^2}{m_2^2}(1-s)-I_0(m_1)\frac{I_1(m_1)}{m_1}\frac{3m^2-k_1^2a^2}{2}\right]\right\}^2\tag{10.65}$$

The last term in eq. 10.65 does not contain m_2^2 in the denominator. However, it is possible to show that for $s=0$ this term has the same order as other terms.

Substituting expression 10.65 into eq. 10.64 and after simple algebra we obtain:

$$C(m_2)-C(-m_2)=-\frac{\mathrm{i}\pi}{I_0^2(m_1)}\frac{1}{m^2+k_1^2a^2}$$

$$\times\left\{m^2+k_2^2a^2-2s\left[m_1^2+2m_1\frac{I_1(m_1)}{I_0(m)}\right]m_2^2K_0(m_2)-2m_1\frac{I_1(m_1)}{I_0(m_1)}\right.$$

$$\left.\times\left[\frac{m_2^2-2sm^2}{m^2+k_1^2a^2}-m_2^2\frac{I_0(m_1)}{I_1(m_1)m_1}\frac{3m^2-k_1^2a^2}{2(m^2+k_1^2a^2)}\right](m^2+k_2^2a^2)K_0(m_2)\right\}\tag{10.66}$$

For function $A(m_2)$ we have:

$$A(m_2)\simeq\frac{\mathrm{i}\pi(s-1)}{m_2^2K_1(m_2)K_1(-m_2)\Delta(m_2)\Delta(-m_2)}I_0(m_1)\frac{I_1(m_1)}{m_1}\frac{m^2+k_1^2a^2}{m_2^2}$$

$$\simeq-\mathrm{i}\pi(s-1)\frac{m_1}{I_0(m_1)I_1(m_1)}\frac{m_2^2}{m^2+k_1^2a^2}\tag{10.67}$$

$$\times\left\{1-2m_1\frac{I_1(m_1)}{I_0(m_1)}\left[\frac{m^2+k_2^2a^2}{m_2^2}-\frac{I_0(m_1)}{m_1I_1(m_1)}\frac{3m^2-k_1^2a^2}{2}\right]\frac{m_2^2K_0(m_2)}{m^2+k_1^2a^2}\right\}$$

Substituting expressions 10.66 and 10.67 into eq. 10.63 and discarding terms, which after integration give values of order $1/\alpha^4$, we obtain:

$$h_x\simeq-\frac{\mathrm{e}^{\mathrm{i}k_2L}}{I_0^2(m_1)}\frac{\alpha^3}{2}\left\{\int\limits_0^\infty(m^2+k_2^2a^2)\,\mathrm{e}^{-\alpha t}-2s\left[m^2+2m_1\frac{I_1(m_1)}{I_0(m_1)}\right]\int\limits_0^\infty m_2^2K_0(m_2)\,\mathrm{e}^{-\alpha t}\,\mathrm{d}t\right\}\tag{10.68}$$

where $m^2 = -t^2 + 2\mathrm{i}k_2at + k_2^2a^2$, $m_2 = (-t^2 + 2\mathrm{i}k_2at)^{1/2}$, and $m_1 = (k_2^2a^2 - k_1^2a^2)^{1/2}$.
The first integral is expressed by elementary functions:

$$\int_0^\infty (m^2 + k_2^2a^2)\,\mathrm{e}^{-\alpha t} = -\frac{2}{\alpha^3}(1 - \mathrm{i}k_2L - k_2^2L^2)$$

Let us present the second integral in the form:

$$\begin{aligned}\int_0^\infty m^2K_0(m_2)\mathrm{e}^{-\alpha t}\,\mathrm{d}t &\simeq -\int_0^\infty m_2^2 \ln m_2\mathrm{e}^{-\alpha t}\,\mathrm{d}t \\ &= \frac{1}{2}\left(\frac{\partial^2}{\partial\alpha^2} + 2\mathrm{i}k_2a\frac{\partial}{\partial\alpha}\right)\int_0^\infty \ln(-t^2 + 2\mathrm{i}k_2at)\,\mathrm{e}^{-\alpha t}\,\mathrm{d}t \\ &= \frac{1}{2}\left(\frac{\partial^2}{\partial\alpha^2} + 2\mathrm{i}k_2a\frac{\partial}{\partial\alpha}\right)\int_0^\infty [\ln(-t) + \ln(t - 2\mathrm{i}k_2a)]\,\mathrm{e}^{-\alpha t}\,\mathrm{d}t\end{aligned} \tag{10.69}$$

It is obvious that:

$$\int_0^\infty \ln(-t)\mathrm{e}^{-\alpha t}\,\mathrm{d}t = -\frac{1}{\alpha}(\ln\alpha + C) + \frac{\mathrm{i}\pi}{\alpha} \simeq -\frac{\ln\alpha}{\alpha}$$

The second integral in eq. 10.69 is expressed by the integral exponential function:

$$\int_0^\infty \ln(t - 2\mathrm{i}k_2a)\,\mathrm{e}^{-\alpha t}\,\mathrm{d}t = \frac{1}{\alpha}\left[\ln(-2\mathrm{i}k_2a) - \mathrm{e}^{-2\mathrm{i}k_2L}\,\mathrm{Ei}(2\mathrm{i}k_2L)\right]$$

Correspondingly, for the magnetic field we have:

$$h_x = \frac{1}{I_0^2(m_1)}h^{un}\left(\frac{L}{h_2}\right) + \frac{\mathrm{e}^{\mathrm{i}k_2L}}{I_0^2(m_1)}2s\left(m_1^2 + 2m_1\frac{I_1(m_1)}{I_0(m_1)}\right)P(k_2a,\alpha) \tag{10.70}$$

where $h^{un}(L/h_2)$ is the field in a uniform medium with the formation resistivity, and:

$$P(k_2a,\alpha) = \frac{\alpha^3}{4}\left(\frac{\partial^2}{\partial\alpha^2} + 2\mathrm{i}k_2a\frac{\partial}{\partial\alpha}\right)\left(-\frac{\ln\alpha}{\alpha} + \frac{\ln(-2\mathrm{i}k_2a)}{\alpha} - \frac{\mathrm{e}^{-2\mathrm{i}k_2L}}{\alpha}\,\mathrm{Ei}(2\mathrm{i}k_2L)\right)$$

If $|k_2L| \ll 1$, $\mathrm{Ei}(2\mathrm{i}k_2L) \simeq \ln(-2\mathrm{i}k_2L) = \ln\alpha + \ln(-2\mathrm{i}k_2a)$ and $P(k_2a,\alpha) \simeq -\ln\alpha$. Therefore, eq. 10.70 has the form:

$$h_x = \frac{1}{I_0^2(m_1)}\left[1 - \left(\frac{L}{h_2}\right)^2\right] - \frac{2s\ln\alpha}{I_0^2(m_1)}\left(m_1^2 + 2m_1\frac{I_1(m_1)}{I_0(m_2)}\right) \tag{10.71}$$

If the skin depth in the borehole is greater than its radius, Bessel's functions $I_0(m_1)$ and $I_1(m_1)$ can be expanded in a series, and instead of eq. 10.71 we obtain:

$$h_x = 1 - \frac{m_1^2}{2} - \left(\frac{L}{h_2}\right)^2 - 2sm_1^2 \ln \alpha$$

Inasmuch as $m_1^2 \simeq -k_1^2 L^2/\alpha^2$, for the quadrature component of the field we have:

$$\mathrm{Q}\, h_x \simeq \frac{1}{\alpha^2}\left(\frac{L}{h_1}\right)^2 - \left(1 - \frac{8 \ln \alpha}{\alpha^2}\right)\left(\frac{L}{h_2}\right)^2$$

which with accuracy of term until s/α^2 coincides with eq. 10.49, derived for the range of small parameters.

In the wave zone, when $|k_2 L| > 1$, making use of the asymptotic value of the integral exponential function:

$$\mathrm{Ei}(2\mathrm{i}k_2 L) \simeq \frac{\mathrm{e}^{2\mathrm{i}k_2 L}}{2\mathrm{i}k_2 L}$$

we obtain:

$$P(k_2 a, \alpha) \simeq \frac{\alpha^3}{4}\left(\frac{\partial^2}{\partial \alpha^2} + 2\mathrm{i}k_2 a \frac{\partial}{\partial \alpha}\right)\left(-\frac{\ln \alpha}{\alpha} + \frac{\ln(-2\mathrm{i}k_2 a)}{\alpha}\right) \simeq \frac{\mathrm{i}k_2 L}{2}(\ln \alpha - \ln |k_2 a|)$$

Table 10.6 contains data of calculations of field magnitude $A = |h_x|$ by exact and asymptotic formulas 10.33 and 10.70, illustrating the area of application of eq. 10.70. It is natural to distinguish three ranges of frequency responses of the amplitude spectrum (Figs. 10.5–10.8): the range of small parameters, the intermediate zone and the wave zone.

As follows from Table 10.6, the field is sufficiently accurately described by the asymptotic expression 10.70 in wide range of frequencies. However, if the value of parameter a/h_1 exceeds unit the accuracy of calculation by this equation rapidly decreases, that, perhaps, is related with an influence of poles that has not been taken into account in deriving eq. 10.70.

One of the practical conclusions of this analysis is the fact that measuring the ratio of amplitudes or difference of phases by two probes we can eliminate the influence of the borehole. In fact, at the far zone ($\alpha \gg 1$) eq. 10.70 has the form:

$$h_x \simeq \frac{1}{I_0^2(m_1)} h^{un}\left(\frac{L}{h_2}\right)$$

where $m_1 = (k_2^2 a^2 - k_1^2 a^2)^{1/2}$.

For this reason, the ratio of field amplitudes and the difference of phases measured by two-coil probes, having lengths L_1 and L_2, do not depend on the radius and the conductivity of the borehole, and they are defined by the formation conductivity only.

TABLE 10.6
Values of field magnitude $A = |h_x|$

a/h_1	α	$s = 1/64$		$s = 1/16$		$s = 1/4$	
		A	A^{apr}	A	A^{apr}	A	A^{apr}
0.1	4	1.00	1.00	1.00	1.00	1	1
	10	1.00	1.00	1.02	1.01	10.9	10.8
	12	1.00	1.00	1.03	1.02	1.13	1.12
	20	1.01	1.01	1.09	1.09	1.32	1.31
	24	1.03	1.03	1.12	1.13	1.39	1.39
	30	1.05	1.05	1.21	1.21	1.44	1.44
0.2	4	1.00	1.00	1.00	1.00	1.05	1.03
	10	1.01	1.01	1.09	1.08	1.38	1.28
	12	1.03	1.02	1.13	1.12	1.38	1.36
	20	1.09	1.09	1.32	1.31	1.39	1.39
	24	1.14	1.13	1.39	1.28	1.28	1.27
	30	1.21	1.21	1.44	1.44	1.05	1.05
0.4	4	1.00	1.00	1.05	1.02	1.21	1.12
	10	1.09	1.07	1.30	1.27	1.34	1.34
	12	1.13	1.12	1.36	1.34	1.23	1.24
	20	1.31	1.30	1.37	1.38	0.62	0.65
	24	1.38	1.37	1.26	1.27	0.39	0.41
	30	1.43	1.43	1.03	1.04	0.18	0.19
0.8	4	0.99	0.97	1.13	1.05	1.25	1.17
	10	1.23	1.21	1.22	1.26	0.51	0.61
	12	1.29	1.28	1.11	1.18	0.32	0.39
	20	1.30	1.31	0.56	0.60	0.033	0.043
	24	1.21	1.22	0.35	0.38	0.0094	0.012
	30	0.91	1.01	0.16	0.18	0.0013	0.0017

10.5. The Magnetic Field in a Medium with Two Cylindrical Interfaces

Consider the determination of the field of the transversal magnetic dipole in a medium with two cylindrical interfaces. Analysis of the solution permits us to investigate the influence of the resistivity and the radius of the invasion zone on the radial response of probes with transversal induction coils.

Making use of results, derived in the second section, in a general case with N interfaces potentials can be written in the form:

$$\begin{aligned} A_i &= \frac{M}{2\pi^2} k_i^2 \sin\phi \int_0^\infty \frac{1}{\lambda_i} \left[a_i K_1(\lambda_i r) + b_i I_1(\lambda_i r)\right] \cos\lambda z \, \mathrm{d}\lambda \\ A_i^* &= \frac{M}{2\pi^2} \cos\phi \int_0^\infty \frac{\lambda}{\lambda_i} \left[C_i K_1(\lambda_i r) + d_i I_1(\lambda_i r)\right] \sin\lambda z \, \mathrm{d}\lambda \end{aligned} \tag{10.72}$$

where $i = 1, 2 \ldots, N+1$; k_i is the wave number of i-layer and $\lambda_i = (\lambda^2 - k_i^2)^{1/2}$.

From condition of excitation it follows that $a_1 = C_1 = -1$, but a field behavior at infinity requires equality $b_{N+1} = d_{N+1} = 0$. Other $4N$ coefficients are found from system of $4N$ linear algebraic equations which are derived as a result of continuity of tangential components of field $\boldsymbol{E}$ and $\boldsymbol{H}$ at interfaces.

The expression for the magnetic field on the borehole axis has the form:

$$h_x = (1 - \mathrm{i}k_1 L - k_1^2 L^2)\mathrm{e}^{\mathrm{i}k_1 L} - \frac{\alpha^3}{\pi} \int_0^\infty \left[k_1^2 a_1^2 b_1 + m^2 d_1\right] \cos(L/a_1) \, dm \tag{10.73}$$

where a_1 is the borehole radius.

Even in the case of two cylindrical interfaces the determination of unknown coefficients in eq. 10.72 is related to the solution of the system of eight equations. In spite of the fact that the solution of this system is not difficult, expressions for potentials have a sufficiently cumbersome form. For this reason potentials are not written in explicit form.

Curves of apparent conductivity, σ_a, related with the field:

$$\frac{\sigma_a}{\sigma_1} = \frac{|h_x - 1|}{|h_x^{un} - 1|} \tag{10.74}$$

are shown in Figs. 10.17–10.24.

At the low-frequency part of the spectrum with an increase of the probe length the field asymptotically tends to that in a uniform medium with the formation resistivity. Minimum of curves of apparent conductivity, σ_a/σ_1, is related with the fact that quadrature component of the field becomes equal to zero. For relatively shallow penetration of borehole filtrate into the formation ($a_2/a_1 \simeq 2$) for $\alpha > 16$ the influence of the invasion zone on the signal value for a two-coil probe does not exceed 25%.

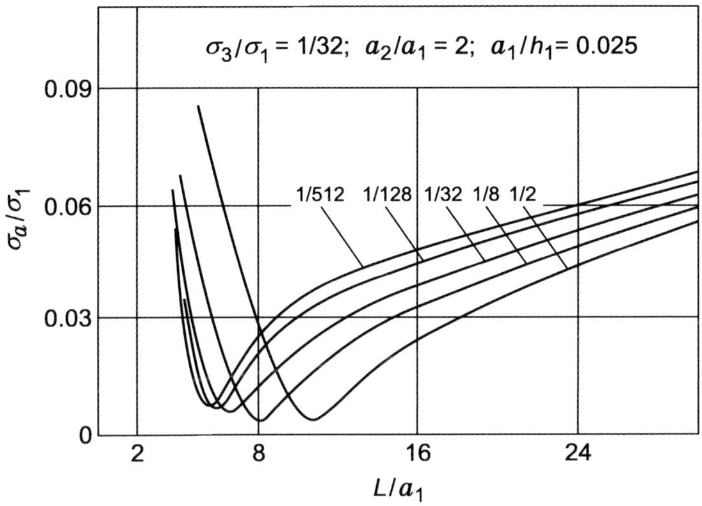

Figure 10.17. Apparent conductivity curves. Curve index σ_2/σ_1.

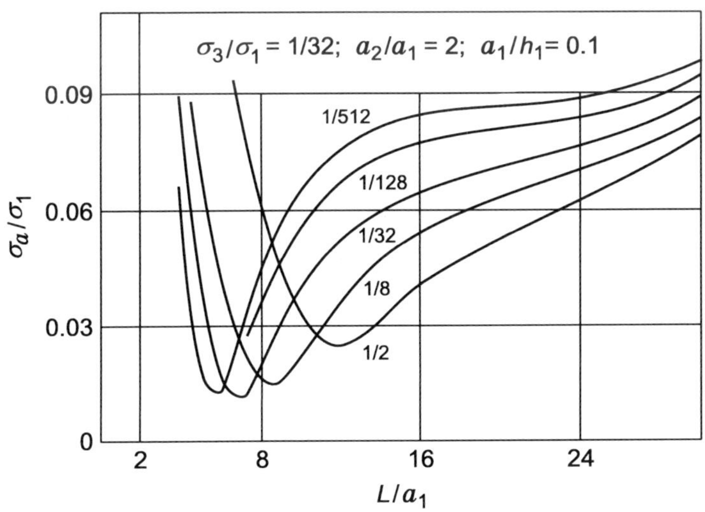

Figure 10.18. Apparent conductivity curves. Curve index σ_2/σ_1.

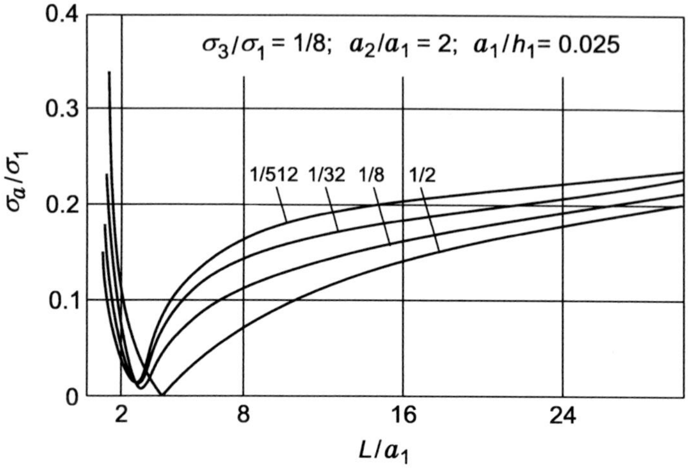

Figure 10.19. Apparent conductivity curves. Curve index σ_2/σ_1.

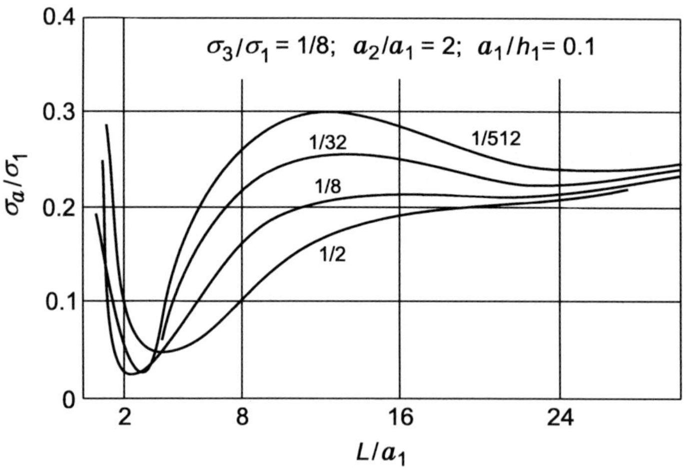

Figure 10.20. Apparent conductivity curves. Curve index σ_2/σ_1.

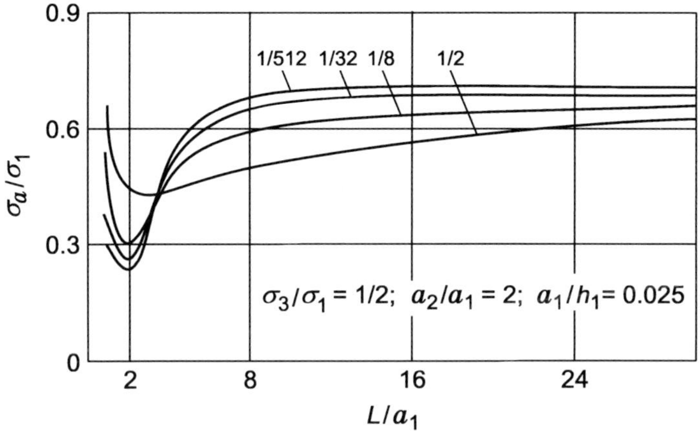

Figure 10.21. Apparent conductivity curves. Curve index σ_2/σ_1.

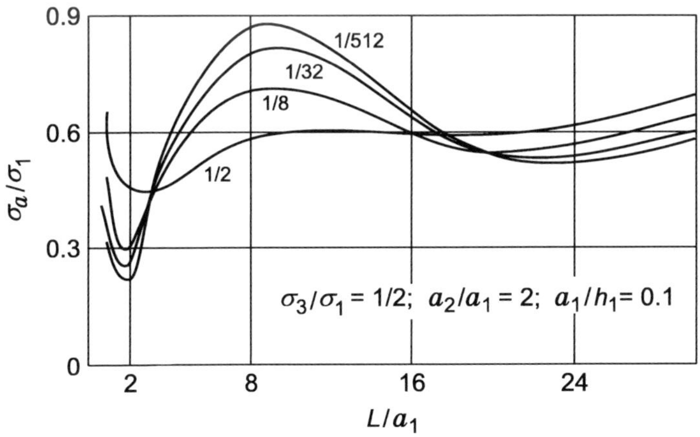

Figure 10.22. Apparent conductivity curves. Curve index σ_2/σ_1.

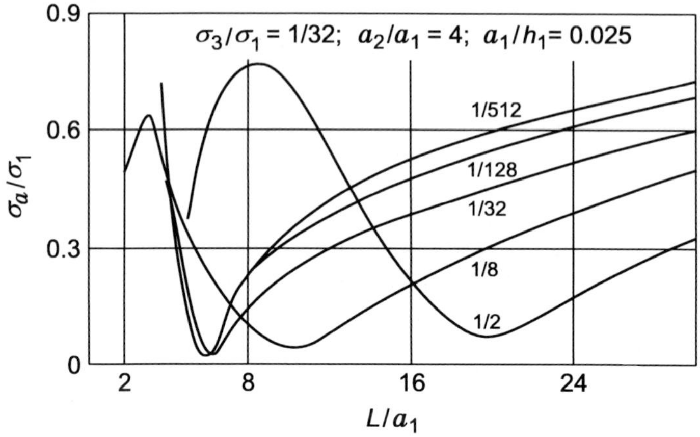

Figure 10.23. Apparent conductivity curves. Curve index σ_2/σ_1.

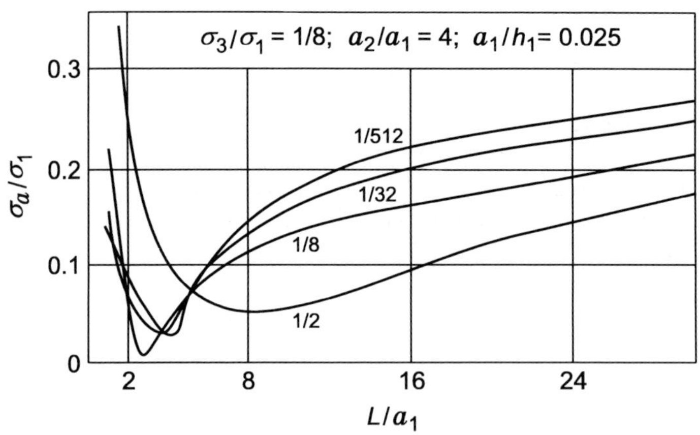

Figure 10.24. Apparent conductivity curves. Curve index σ_2/σ_1.

TABLE 10.7
Values of function $A_1 - A_2$; $\sigma_3/\sigma_1 = 32$, $a_2/a_1 = 4$, $a/h_1 = 0.025$, $L_1/a_1 = 14$, $L_2/a_1 = 12$

σ_2/σ_1	1/512	1/128	1/32	1/8
$A_1 - A_2$	0.12×10^{-2}	0.11×10^{-2}	0.90×10^{-3}	0.70×10^{-3}

The depth of investigation can be significantly increased in applying a three-coil differential probe. In this case the minimal length of a two-coil probe forming a differential array should correspond to the ascending branches of curves σ_2/σ_1. For illustration, values of amplitude difference of the secondary field expressed in units of the primary field at the range of small parameters, calculated for a three-coil probe, are given in Table 10.7. As is seen from the table the influence of the invasion zone with the considered parameters results in a change of the amplitude difference of not more than 30% while the influence of the invasion zone in the case of a two-coil induction probe is more significant.

10.6. Cylindrical Surface with Transversal Resistance T

Let us assume that due to penetration of the borehole filtrate a relatively thin but resistive invasion zone is formed. In this case we can obtain a sufficiently simple expression for the field. We will suppose that the conductivity of the borehole and the formation are equal. It is appropriate to notice that the generalization, when both media have different conductivities, does not require special efforts.

Thus, our problem is formulated in the following way. In a uniform medium with conductivity σ_1 there is a thin cylindrical layer with radius a and thickness h, having conductivity σ_2.These parameters satisfy conditions: $h/a \ll 1$, $\sigma_1/\sigma_2 \gg 1$. Electric properties of the layer are characterized by the transversal resistance $T = h/\sigma_2$. At the surface $r = a$, tangential components of the magnetic field are continuous:

$$H_{1z} = H_{2z} \qquad H_{1\phi} = H_{2\phi} \tag{10.75}$$

where $\boldsymbol{H}_1$ and $\boldsymbol{H}_2$ are fields in the borehole and the formation, respectively.

Tangential components of the electric field due to the presence of the double layer are discontinuous, and, as is well known, we have:

$$\begin{aligned} E_{2z} &= E_{1z} + T\sigma_1 \frac{\partial E_{1r}}{\partial z} \\ E_{2\phi} &= E_{1\phi} + T\frac{\sigma_1}{a}\frac{\partial E_{1\phi}}{\partial \phi} \end{aligned} \tag{10.76}$$

Substituting expressions for field components through potentials in a two-layered medium (eq. 10.28) into eqs. 10.75–10.76 we obtain:

$$K_1(m_1) - I_1(m_1)D = K_1(m_1)G$$

$$a^2k_1^2[K_1'(m_1) - I_1'(m_1)C] + \frac{m^2}{m_1}[K_1(m_1) - I_1(m_1)D] = a^2k_1^2K_1'(m_1)E + \frac{m^2}{m_1}K_1(m_1)G$$

$$K_1(m_1) - I_1(m_1)C - \tau\left[\frac{m^2}{m_1^2}I_1(m_1)D + \frac{m^2}{m_1^2}K_0(m_1) + \frac{m^2}{m_1^2}I_1'(m_1)C\right]$$
$$= K_1(m_1)E - K_0(m_1) - \frac{I_1(m_1)}{m_1}C - I_1'(m_1)D - \tau\left[\frac{I_1(m_1)}{m_1}D + K_0(m_1) + I_1'(m_1)C\right]$$
$$= \frac{K_1(m_1)}{m_1}E + K_1'(m_1)G \tag{10.77}$$

where $m_1 = (m^2 - k_1^2 a_1)^{1/2}$, and $\tau = T\sigma_1/a = T/T_0$. By analogy we will call T_0 the transversal resistance of the borehole.

Solving system 10.77 we have:

$$C = \frac{\tau m^2 K_0(m_1)K_1'(m_1)}{1 - \tau\left(\frac{k_1^2 a^2}{m_1^2}I_1(m_1)K_1(m_1) + m^2 I_1'(m_1)K_1'(m_1)\right)}$$
$$D = \frac{\tau\frac{k_1^2 a^2}{m_1}K_0(m_1)K_1'(m_1)}{1 - \tau\left(\frac{k_1^2 a^2}{m_1^2}I_1(m_1)K_1(m_1) + m^2 I_1'(m_1)K_1'(m_1)\right)} \tag{10.78}$$

For the magnetic field on the z-axis of the borehole we have:

$$h_x = h^{un} + \tau a^2 k_1^2 \frac{\alpha^3}{\pi}\int_0^\infty \frac{m^2 K_0^2(m_1)\cos m\alpha \, dm}{1 - \tau\left(\frac{k_1^2 a^2}{m_1^2}I_1(m_1)K_1(m_1) + m^2 I_1'(m_1)K_1'(m_1)\right)} \tag{10.79}$$

For $\tau \to 0$ we obtain $h_x \to h_x^{un}$, while in the opposite case, as $\tau \to \infty$:

$$h_x = h_x^{un} - a^2 k_1^2 \frac{\alpha^3}{\pi}\int_0^\infty \frac{m^2 K_0^2(m_1)\cos m\alpha \, dm}{\frac{k_1^2 a^2}{m_1^2}I_1(m_1)K_1(m_1) + m^2 I_1'(m_1)K_1'(m_1)} \tag{10.80}$$

Calculations show that eq. 10.80 describes field h_x with sufficient accuracy when $\tau > 10$.

Curves of the amplitude of the secondary field as a function of parameter L/a are presented in Fig. 10.25. The curve index is parameter a/h_1. At the range of small parameter for the quadrature component of the magnetic field we have:

$$\mathrm{Q}\, h_x = -\left(\frac{L}{h_1}\right)^2 (1 + G_\tau)$$

where

$$G_\tau = -\frac{2\tau\alpha}{\pi}\int_0^\infty \frac{m^2 K_0^2(m)\cos m\alpha \, dm}{1 - \tau m^2 I_1'(m)K_1'(m)}$$

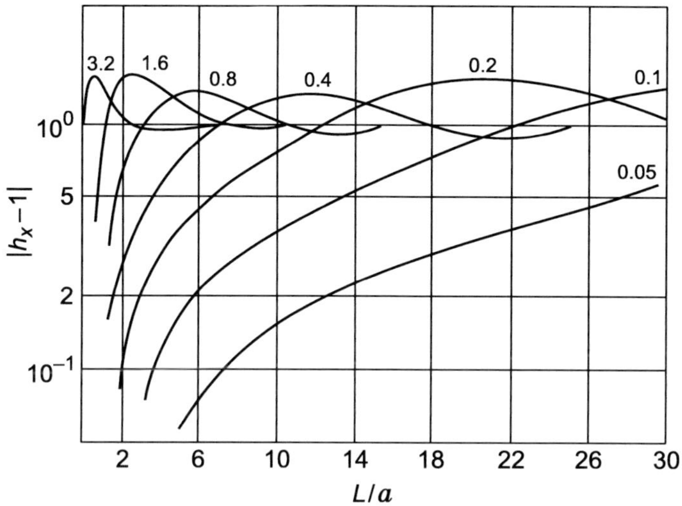

Figure 10.25. The amplitude of the secondary field ($\tau \to \infty$). Curve index a/h.

If the probe length is many times greater than the borehole radius ($\alpha \gg 1$), then in applying the approach described in section 3 of this chapter, we obtain:

$$G_\tau \simeq -\frac{2\tau}{1+\tau/2}\frac{\alpha}{\pi}\int_0^\infty m^2 K_0^2(m)\cos m\alpha\, \mathrm{d}m \simeq \frac{4\tau}{1+\tau/2}\frac{\ln\alpha}{\alpha^2}$$

Thus:

$$\mathrm{Q}\, h_x \simeq -\left(\frac{L}{h_1}\right)^2\left(1+\frac{4\tau}{1+\tau/2}\frac{\ln\alpha}{\alpha^2}\right)$$

and for large values of τ we have:

$$\mathrm{Q}\, h_x \simeq -\left(\frac{L}{h_1}\right)^2\left(1+8\frac{\ln\alpha}{\alpha^2}\right) \tag{10.81}$$

Now we will derive an asymptotic expression for the field in the far zone ($\alpha \gg 1$) for arbitrary parameter L/h_1. For simplicity it is assumed that parameter τ is much greater than unity: $\tau \gg 1$. The integrand in eq. 10.80 has a branch point on the complex plane of variable of integration m, when $m = ak_1$ along the imaginary axis, and deforming the contour of integration along the cross-cut and expanding the integrand by powers of m_1 we obtain:

$$h_x = h_x^{un} + a^2k_1^2\,\frac{\alpha^3}{\pi}\int_{ak_1}^{\mathrm{i}\infty+ak_1} m\left[K_0^2(m_1) - K_0^2(-m_1)\right]\mathrm{e}^{\mathrm{i}\alpha m}\,\mathrm{d}m$$

Inasmuch as:

$$K_0^2(m_1) - K_0(-m_1) = -2\mathrm{i}\pi K_0(m_1)I_0(m_1) + \pi^2 I_0^2(m_1) = -2\mathrm{i}\pi K_0(m_1)$$

for $|m_1| \ll 1$, then:

$$h_x = h^{un} - 2\mathrm{i}a^2k_1^2\alpha^3 \int\limits_{ak_1}^{\mathrm{i}\infty+ak_1} m^2 K_0(m_1)\mathrm{e}^{\mathrm{i}\alpha m}\,\mathrm{d}m$$

Letting $m = \mathrm{i}t + ak_1$, $0 \leqslant t < \infty$, we have:

$$h_x = h_x^{un} + 2a^2k_1^2\mathrm{e}^{\mathrm{i}Lk_1}\alpha^3 \int\limits_0^\infty (\mathrm{i}t + ak_1)^2 K_0(m_1)\mathrm{e}^{-\alpha t}\,\mathrm{d}t \tag{10.82}$$

where $m_1 = (-t^2 + 2\mathrm{i}tak_1)^{1/2}$.

For $\alpha \gg 1$ the integral in eq. 10.82 is expressed through the integral exponential function:

$$\begin{aligned}\int\limits_0^\infty (\mathrm{i}t + ak_1)^2 K_0(m_1)\,\mathrm{e}^{\mathrm{i}\alpha t}\,\mathrm{d}t &= \frac{1}{2}\left(\frac{\partial^2}{\partial\alpha^2} + 2\mathrm{i}ak_1\frac{\partial}{\partial\alpha} - a^2k_1^2\right)\\ \times \int\limits_0^\infty [\ln(-t) + \ln(t - 2\mathrm{i}ak_1)]\,\mathrm{e}^{-\alpha t}\,\mathrm{d}t &= \frac{1}{2}\left(\frac{\partial^2}{\partial\alpha^2} + 2\mathrm{i}ak_1\frac{\partial}{\partial\alpha} - a^2k_1^2\right)\\ \times \left[-\frac{\ln\alpha}{\alpha} + \frac{\ln(-2\mathrm{i}ak_1)}{\alpha} - \frac{\mathrm{e}^{-2\mathrm{i}k_1L}}{\alpha}\,\mathrm{Ei}(2\mathrm{i}k_1L)\right]\end{aligned} \tag{10.83}$$

If $|k_1L| \ll 1$, then $\mathrm{Ei}(2\mathrm{i}k_1L) \simeq \ln(-2\mathrm{i}k_1L)$, and eq. 10.83 has the form:

$$\left(\frac{\partial^2}{\partial\alpha^2} + 2\mathrm{i}ak_1\frac{\partial}{\partial\alpha} - a^2k_1^2\right)\left(-\frac{\ln\alpha}{\alpha}\right) \simeq \frac{2\ln\alpha}{\alpha^3}$$

Therefore, for the magnetic field we have:

$$h_x = h_x^{un} - 4a^2k_1^2\mathrm{e}^{\mathrm{i}k_1L}\ln\alpha = -\frac{k_1^2l^2}{2}\left(1 + 8\,\frac{\ln\alpha}{\alpha^2}\right)$$

This expression coincides with eq. 10.81, which is valid at the range of small parameters. In the opposite case, when $k_1L \gg 1$ we can write:

$$\mathrm{Ei}(2\mathrm{i}k_1L) \simeq \frac{\mathrm{e}^{2\mathrm{i}k_1L}}{2\mathrm{i}k_1L}$$

and instead of eq. 10.83 we obtain:

$$\frac{1}{2}\left(\frac{\partial^2}{\partial\alpha^2}+2\mathrm{i}ak_1\frac{\partial}{\partial\alpha}-a^2k_1^2\right)\left(-\frac{\ln\alpha}{\alpha}+\frac{\ln(-2\mathrm{i}ak_1)}{\alpha}\right)=\frac{k_1^2l^2}{2}\frac{\ln\alpha-\ln|k_1a|}{\alpha^3}$$

Whence:

$$h_x = h_x^{un} + k_1^2a^2k_1^2L^2\mathrm{e}^{\mathrm{i}k_1L}(\ln\alpha-\ln|k_1a|) \tag{10.84}$$

Inasmuch as at the range of large parameters: $|k_1L|\gg 1$, we have:

$$h_x^{un}\simeq -k_1^2L^2\mathrm{e}^{\mathrm{i}kL}$$

Equation 10.48 can be presented in the form:

$$h_x = h_x^{un}\left(1-k_1^2L^2\,\frac{\ln\alpha-\ln|k_1a|}{\alpha^2}\right) \tag{10.85}$$

10.7. The Magnetic Field in a Medium with One Horizontal Interface

Let us place the dipole at the origin of coordinates and direct the dipole moment along the x-axis:

$$\boldsymbol{M}=M_0\mathrm{e}^{-\mathrm{i}\omega t}\boldsymbol{x}_0 \tag{10.86}$$

where $M_0 = InS$.

As is well known, the field equations have the form:

$$\begin{aligned}&\operatorname{curl}\boldsymbol{E}=\mathrm{i}\omega\mu\boldsymbol{H} \qquad &&\operatorname{div}\boldsymbol{H}=\sigma\boldsymbol{E}\\ &\operatorname{curl}\boldsymbol{E}=0 &&\operatorname{div}\boldsymbol{H}=0\end{aligned} \tag{10.87}$$

We will let

$$\boldsymbol{E}=\mathrm{i}\omega\mu\operatorname{curl}\boldsymbol{A} \tag{10.88}$$

and, substituting eq. 10.88 into 10.87 we obtain:

$$\boldsymbol{H}=k^2\boldsymbol{A}-\operatorname{grad}U$$

Letting $u = -\operatorname{div}\boldsymbol{A}$ we will obtain equation for the potential:

$$\nabla^2\boldsymbol{A}+k^2\boldsymbol{A}=0 \tag{10.89}$$

where $k^2 = \mathrm{i}\sigma\mu\omega$.

The relation between the potentials and the field is defined by:

$$\boldsymbol{E} = \mathrm{i}\omega\mu \, \mathrm{curl}\, \boldsymbol{A} \qquad \boldsymbol{H} = k^2 \boldsymbol{A} + \mathrm{grad}\, \mathrm{div}\, \boldsymbol{A} \tag{10.90}$$

We will look for a solution, assuming that component $E_y = 0$.

In accord with eq. 10.90 we have:

$$E_x = \mathrm{i}\mu\omega \frac{\partial A_z}{\partial y} \qquad E_y = \mathrm{i}\mu\omega \left(\frac{\partial A_x}{\partial z} - \frac{\partial A_z}{\partial x} \right) \qquad E_z = -\mathrm{i}\mu\omega \frac{\partial A_x}{\partial y}$$

and

$$H_x = k^2 A_x + \frac{\partial}{\partial x} \mathrm{div}\, \boldsymbol{A} \qquad H_y = \frac{\partial}{\partial y} \mathrm{div}\, \boldsymbol{A} \qquad H_z = k^2 A_z + \frac{\partial}{\partial z} \mathrm{div}\, \boldsymbol{A} \tag{10.91}$$

For continuity of tangential components of the field at the interface $z = h$ it is sufficient to provide continuity of values A_z, $\partial A_x/\partial z$, $k^2 A_z$ and div $\boldsymbol{A}$.

Thus, for components of the vector potential we obtain two groups of conditions, such as:

$$k_1^2 A_{1x} = k_2^2 A_{2x} \qquad \frac{\partial A_{1x}}{\partial z} = \frac{\partial A_{2x}}{\partial z} \tag{10.92}$$

and

$$A_{1z} = A_{2z} \qquad \mathrm{div}\, \boldsymbol{A}_1 = \mathrm{div}\, \boldsymbol{A}_2 \tag{10.93}$$

The primary field of the dipole in a uniform medium has only one component:

$$A_{1x}^{(0)} = \frac{M}{4\pi} \frac{\mathrm{e}^{\mathrm{i}k_1 R}}{R}$$

or

$$A_{1x}^{(0)} = \frac{M}{4\pi} \int_0^\infty \frac{m}{m_1} \mathrm{e}^{-m_1|z|} J_0(mr)\, dm$$

where $m_1 = (m^2 - k_1^2)^{1/2}$.

For this reason we will present component A_x in the form:

$$\begin{aligned} A_{1x} &= \frac{M}{4\pi} \int_0^\infty \left(\frac{m}{m_1} \mathrm{e}^{-m_1 z} + A_m \mathrm{e}^{m_1 z} \right) J_0(mr)\, dm \\ A_{2x} &= \frac{M}{4\pi} \int_0^\infty B_m \, \mathrm{e}^{-m_2 z} J_0(mr)\, dm \end{aligned} \tag{10.94}$$

where $m_2 = (m^2 - k_2^2)^{1/2}$.

From boundary conditions at $z = h$ we have:

$$\frac{m}{m_1}\mathrm{e}^{-m_1 h} + A_m \mathrm{e}^{m_1 h} = s B_m \mathrm{e}^{-m_2 h}$$
$$-m\mathrm{e}^{-m_1 h} + m_1 A_m \mathrm{e}^{m_1 h} = -m_2 B_m \mathrm{e}^{-m_2 h}$$

Whence:

$$A_m = \frac{m}{m_1}\frac{s m_1 - m_2}{s m_1 + m_2}\mathrm{e}^{-2m_1 h}$$
$$B_m = \frac{2m}{s m_1 + m_2}\mathrm{e}^{-(m_1 - m_2)h} \tag{10.95}$$

where $s = \sigma_2/\sigma_1$, and

$$A_{1x} = A_{1x}^{(0)} + \frac{M}{4\pi}\int_0^\infty \frac{m}{m_1}\frac{s m_1 - m_2}{s m_1 + m_2}\mathrm{e}^{-2m_1 h + m_1 z} J_0(mr)\,\mathrm{d}m$$
$$A_{2x} = \frac{M}{4\pi}\int_0^\infty \frac{2m}{s m_1 + m_2}\mathrm{e}^{-(m_1 - m_2)h - m_2 z} J_0(mr)\,\mathrm{d}m \tag{10.96}$$

From continuity of div $\boldsymbol{A}$ follows:

$$\frac{\partial}{\partial x}(A_{1x} - A_{2x}) = \frac{\partial}{\partial z}(A_{2z} - A_{1z})$$

Inasmuch as:

$$\frac{\partial A_x}{\partial x} = \frac{\partial A_x}{\partial r}\frac{\partial r}{\partial x} = \cos\phi \int_0^\infty F(m)\mathrm{e}^{\pm m_1 z} J_1(mr)\,\mathrm{d}m$$

it is appropriate to present the solution for A_z, in order to provide continuity of $\boldsymbol{A}$, in the form:

$$A_{1z} = \frac{M}{4\pi}\cos\phi \int_0^\infty C_m\,\mathrm{e}^{m_1 z} J_1(mr)\,\mathrm{d}m$$
$$A_{2z} = \frac{M}{4\pi}\cos\phi \int_0^\infty D_m\,\mathrm{e}^{-m_2 z} J_1(mr)\,\mathrm{d}m$$

In accord with eq. 10.93 we have:

$$C_m\,\mathrm{e}^{m_1 h} = D_m\,\mathrm{e}^{-m_2 h}$$
$$(s-1)m B_m \mathrm{e}^{-m_2 h} = m_2 D_m \mathrm{e}^{-m_2 h} + m_1 C_m \mathrm{e}^{m_1 h} \tag{10.97}$$

Solving this system we obtain:

$$
\begin{aligned}
C_m &= \frac{(s-1)mB_m}{m_1+m_2}\mathrm{e}^{-(m_1+m_2)h} \\
D_m &= \frac{(s-1)mB_m}{m_1+m_2}
\end{aligned}
\tag{10.98}
$$

Thus:

$$
\begin{aligned}
A_{1z} &= \frac{M}{4\pi}\cos\phi\int_0^\infty \frac{(s-1)mB_m}{m_1+m_2}\mathrm{e}^{-(m_1+m_2)h+m_1 z}J_1(mr)\,\mathrm{d}m \\
A_{2z} &= \frac{M}{4\pi}\cos\phi\int_0^\infty \frac{(s-1)mB_m}{m_1+m_2}\mathrm{e}^{-m_2 z}J_1(mr)\,\mathrm{d}m
\end{aligned}
\tag{10.99}
$$

Magnetic field on the z-axis has only component H_x for which in accord with eqs. 10.91 and 10.99 we obtain:

$$
\begin{aligned}
h_{1x} &= h_0 - L\int_0^\infty \phi_1(m)\mathrm{e}^{m_1 L}\,\mathrm{d}m \\
h_{2x} &= -L\int_0^\infty \phi_2(m)\mathrm{e}^{-m_2 L}\,\mathrm{d}m
\end{aligned}
\tag{10.100}
$$

where h_x is the magnetic field expressed in units of the field in a free space:

$$
\begin{aligned}
h_x &= \frac{H_x}{H_0} \qquad H_0 = -\frac{M}{4\pi L^3} \\
h_0 &= \mathrm{e}^{\mathrm{i}k_1 L}(1-k_1 L-k^2L^2) \\
\phi_1 &= \left(k_1^2L^2-\frac{m^2L^2}{2}\right)\frac{m}{m_1}\frac{sm_1-m_2}{m_1+m_2}\mathrm{e}^{-2m_1 h}+m^2L^2\frac{m^3(s-1)\mathrm{e}^{-2m_1 h}}{m_1(sm_1+m_2)(m_1+m_2)} \\
\phi_2 &= \left(k_1^2L^2-\frac{m^2L^2}{2}\frac{m_1+sm_2}{m_1+m_2}\right)\frac{2m}{sm_1+m_2}\mathrm{e}^{-(m_1-m_2)h}
\end{aligned}
\tag{10.101}
$$

L is the probe length.

We will consider the behavior of the field at the low-frequency part of the spectrum when the skin depth in both media exceeds the distance from the dipole to the interface and the probe length. In deriving asymptotic formulae we will use the approach described in chapter four, namely the interval of integration is presented as a sum of two parts, the internal part where $0 < mL < m_0 L < 1$, and the external one as $m > m_0$. Within the external interval radicals m_1 and m_2 can be expanded in a series by powers of k_1^2/m^2 and k_2^2/m^2. For this reason the integral at the external interval is presented as a series of terms having even powers of k. Within the internal interval exponents can be expanded in series ($mL \ll 1$), and the integral at this interval can be reduced to the sum of tabular integrals, which can easily be presented as a series with respect to wave number k. Unlike the integral at the external interval these series contain odd powers of k and logarithmic terms. For example, in a medium, where the dipole is located, for the low-frequency part of the spectrum h_x we have:

$$\begin{aligned} \operatorname{In} h_x &= 1 + a_1 \left(\frac{L}{h_1}\right)^3 \\ \operatorname{Q} h_x &= b_1 \left(\frac{L}{h_1}\right)^2 + a_1 \left(\frac{L}{h_1}\right)^3 \end{aligned} \tag{10.102}$$

where:

$$\begin{aligned} a_1 &= \frac{2}{s^2-1}\left[\frac{4}{3}s\sqrt{s}(\sqrt{s}-1) - \frac{1}{5}s(\sqrt{s}-1) + \frac{2}{15}(s^3\sqrt{s}-1)\right. \\ &\quad \left. + \frac{s^2}{2\sqrt{s+1}} \ln\left(\frac{\sqrt{s+1}-1}{\sqrt{s+1}+1}\cdot\frac{\sqrt{s+1}+\sqrt{s}}{\sqrt{s+1}-\sqrt{s}}\right)\right] \end{aligned} \tag{10.103}$$

$$b_1 = -1 - \frac{1}{4}\frac{(s+5)(s-1)}{(s+1)}\,\frac{L}{2h-L}$$

$$h_1 = (2/\sigma_1\mu\omega)^{1/2} \qquad h_2 = (2/\sigma_2\mu\omega)^{1/2}$$

$$L/h_1 \ll 1 \qquad L/h_2 \ll 1 \qquad s = \sigma_2/\sigma_1$$

If the interface is located sufficiently far from the source and the observation point ($L/h \ll 1$), coefficient b_1, tends to -1, corresponding to a uniform medium. At the same time coefficient a_1 does not depend on the position of the probe with respect to the boundary, and it is a function of the resistivities of both media.

The second terms in eq. 10.102 are proportional to $\omega^{3/2}$, and the depth of investigation, measuring these terms, is the same as that at the late stage of the transient field. It is obvious that as $s \to 1$ coefficients a_1 and b_1 correspond to a uniform medium:

$$a_1 = \frac{4}{3} \qquad b_1 = -1$$

Now let us consider the high-frequency part of the spectrum and for obtaining asymptotic formulae let us make use of the following relation:

$$I_n = \int_0^\infty \lambda^n \, \mathrm{e}^{\sqrt{\lambda^2+k_1^2L^2}} \, \mathrm{d}\lambda \simeq a_n (k_1 L)^{(n+1)/2} \, \mathrm{e}^{-k_1 L} \tag{10.104}$$

where $|k_1 L| \gg 1$ and a_n is a function of number n. In particular, for the first three values of n it is equal to 1, $\sqrt{\pi/2}$ and 2, respectively.

Let us notice that integrals of type 10.104 for odd values of n are reduced to elementary functions, but for even values they are expressed through modified Bessel functions

$K_n(k_1L)$. After elementary transformations, which allow us to present an expression for the field through integrals of type I_n, we obtain:

$$h_1 = h_0 - k_1^2L^2\,\frac{\sqrt{s}-1}{\sqrt{s}+1}\cdot\frac{e^{ik_1L(2\alpha-1)}}{2\alpha-1} \simeq h_0 \qquad (10.105)$$

$$\alpha = h/L > 1$$

$$|k_1L| \gg 1$$

It is natural that due to the skin effect the field becomes the same as in a uniform medium with conductivity σ_1. However, if the dipole or the observation point are located at the surface, then regardless of the frequency, the field is the function of conductivities of both media. In accord with eq. 10.105 we have:

$$h_{1x} = -k_1^2L^2\frac{2\sqrt{s}}{\sqrt{s}+1}\,e^{ik_1L} \qquad (10.106)$$

In conclusion of this section let us notice one specific feature of the current distribution, when the conductivity of the medium where the dipole is located is equal to zero ($s \to \infty$). In this case, as is seen from eq. 10.96, the potential component A_{2x} is equal to zero.

For this reason in a conducting part of the medium the electrical field and induced currents do not have a vertical component and the distribution of currents is symmetrical with respect to plane yz, which is not intersected by current lines.

10.8. The Magnetic Field of the Horizontal Dipole in the Formation with Finite Thickness

Suppose that the magnetic dipole is located within the formation. Then, according to results obtained in the previous section, expressions for the vector potential have the form (Figs. 10.4, 10.26b):

$$\left.\begin{aligned} A_{1x} &= \frac{M}{4\pi}\int_0^\infty D_1 e^{m_1 z} J_0(mr)\,dm \\ A_{1z} &= \frac{M}{4\pi}\cos\phi\int_0^\infty E_1 e^{m_1 z} J_1(mr)\,dm \end{aligned}\right\} \quad \text{if } z < -h_2$$

$$\left.\begin{aligned} A_{2x} &= \frac{M}{4\pi}\int_0^\infty \left(\frac{m}{m_2}e^{-m_2|z|} + D_2 e^{m_2 z} + D_3 e^{-m_2 z}\right) J_0(mr)\,dm \\ A_{2z} &= \frac{M}{4\pi}\cos\phi\int_0^\infty \left(F_2 e^{m_2 z} + F_3 e^{-m_2 z}\right) J_1(mr)\,dm \end{aligned}\right\} \quad \text{if } -h_2 < z < h_1$$

(10.107)

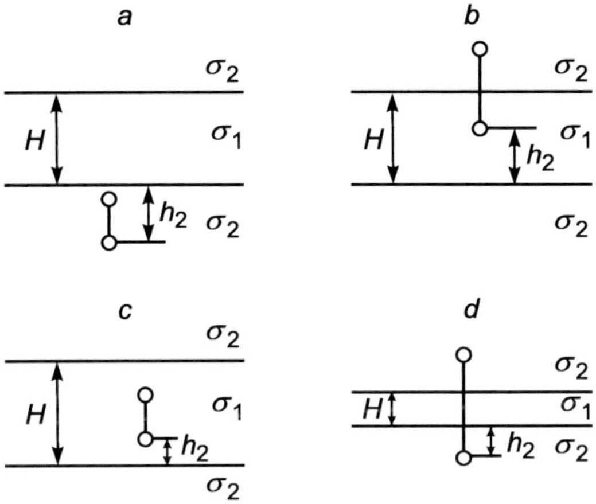

Figure 10.26. Two coil probe in a medium with two horizontal interfaces.

$$\left.\begin{aligned} A_{3x} &= \frac{M}{4\pi}\int\limits_0^\infty D_4 \mathrm{e}^{-m_1 z} J_0(mr)\,\mathrm{d}m \\ A_{3z} &= \int\limits_0^\infty F_4 \mathrm{e}^{-m_1 z} J_1(mr)\,\mathrm{d}m \end{aligned}\right\} \qquad \text{if } z < h_1$$

From the system of equations, following from boundary conditions at $z = h_1$ and $z = -h_2$, we find coefficients: D_1, D_2, D_3, D_4, F_1, F_2, F_3, and F_4. In particular, for the horizontal component of the magnetic field on the z-axis, when the two-coil probe is located symmetrically with respect to the formation boundaries, we obtain:

$$\begin{aligned} h_x = h_{0x}^{(2)} - \int\limits_0^\infty &\left\{ \left(\frac{m}{2} - k_2^2 z^2\right) 2g_{12}(1 - g_{12}\cosh m_2\,\mathrm{e}^{-\alpha m_2}) \right. \\ &+ \frac{(1-s)(1-g_{12})m^2 m_2}{(m_1+m_2)d_2}[1 - (g_{12} - K_{12})\cosh m_2\,\mathrm{e}^{-\alpha m_2} \\ &\left. - K_{12}g_{12}\mathrm{e}^{-2\alpha m_2}] \right\} \frac{m}{m_2 d_1}\mathrm{e}^{-\alpha m_2}\,\mathrm{d}m \qquad \text{if } \alpha = H/L \geqslant 1 \end{aligned} \tag{10.108}$$

where $d_1 = 1 - g_{12}^2\mathrm{e}^{-2m_2\alpha}$, $d_1 = 1 - K_{12}^2\mathrm{e}^{-2m_2\alpha}$, $g_{12} = (sm_1 - m_2)/(sm_1 + m_2)$, $s = \sigma_1/\sigma_2$, $K_{12} = (m_1 - m_2)/(m_1 + m_2)$; σ_1 and σ_2 are conductivities of the formation and shoulders; H is the formation thickness and L is probe length.

By analogy we obtain an expression for the field when the probe length exceeds the formation thickness and the transmitter and receiver coils are located on both sides of the formation. It has the form:

$$
\begin{aligned}
h_x =& \int\limits_0^\infty \left(\frac{m^2}{2} s - k_2^2 L^2 + \frac{m^2 m_1^2}{2(m_1+m_2)^2} \frac{(s-1)^2}{d_2} (1 - \mathrm{e}^{-2\alpha m_2}) \right) \\
&\times \frac{4 m m_2}{(s m_1 + m_2)^2 d_1} \mathrm{e}^{-(\alpha m_2 + (1-\alpha) m_1)}\, \mathrm{d}m \qquad \text{if } \alpha < 1
\end{aligned}
\tag{10.109}
$$

Because of the symmetrical position of the probe coils with respect to the formation boundaries the field is defined by three parameters:

$$
p = L/h \qquad s = \sigma_1/\sigma_2 \qquad \alpha = H/L
$$

Prior to the consideration of frequency responses of the field, let us investigate the behavior of the field at the low-frequency part of the spectrum, when parameter $p \to 0$, and the probe is located within the formation. Proceeding from the approach described in the previous section we obtain the following expressions for the quadrature and inphase components of the field:

$$
\begin{aligned}
\operatorname{In} h_x &= \frac{4}{3} \left(\frac{L}{h_2} \right)^3 \\
\operatorname{Q} h_x &= -\left(\frac{L}{h_1} \right)^2 \left\{ 1 - 2\beta \int\limits_0^\infty \frac{1 - \beta\, \mathrm{e}^{-\alpha m} \cosh \alpha m}{1 - \beta^2\, \mathrm{e}^{-2\alpha m}}\, \mathrm{d}m + \frac{1-s}{2\alpha s} \right\} + \frac{4}{3} \left(\frac{L}{h_2} \right)^3
\end{aligned}
\tag{10.110}
$$

where $\beta = (s-1)/(s+1)$.

It is essential that the inphase component of the field at the low-frequency spectrum coincides with the inphase component of the field in a uniform medium with conductivity σ_2. A similar result is obtained when the source is the vertical magnetic dipole. It means that surface charges, arising at interfaces between the formation and the surrounding medium at the low-frequency spectrum, influence the quadrature component of the magnetic field only. Respectively, at the late stage of the transient response in the same way as in a medium with cylindrical interfaces, the field h_x does not depend on the orientation of the magnetic dipole.

Let us present the quadrature component $\operatorname{Q} h_x$ as the sum of two terms:

$$
\operatorname{Q} h_x = \operatorname{Q} h_x^{(1)} + \operatorname{Q} h_x^{(2)}
$$

where:

$$
\begin{aligned}
\operatorname{Q} h_x^{(1)} &= -\left(\frac{L}{h_1} \right)^2 \left(1 - \frac{1}{2\alpha} \right) - \left(\frac{L}{h_2} \right)^2 \frac{1}{2\alpha} \\
\operatorname{Q} h_x^{(2)} &= \left(\frac{L}{h_2} \right)^2 2F(\beta, \alpha)
\end{aligned}
\tag{10.111}
$$

$$F(\beta,\alpha) = \beta \int_0^\infty \frac{1-\beta \mathrm{e}^{-\alpha m}\cosh\alpha m}{1-\beta^2\mathrm{e}^{-2\alpha m}}\, dm \tag{10.112}$$

With accuracy of the sign field $Q\, h_x^{(1)}$ coincides with the vertical component $Q\, h_z$, of the vertical magnetic dipole at the range of small parameters, and it consists of two terms, each of which depends on the conductivity of one medium. Correspondingly, for this term we can introduce a geometric factor. In accord with eq. 10.111 we can define:

$$G_1(\alpha) = 1 - \frac{1}{2\alpha} \qquad G_2(\alpha) = \frac{1}{2\alpha} \qquad G_1(\alpha) + G_2(\alpha) = 1$$

and

$$Q\, h_x^{(1)} = -\frac{\mu\omega L^2}{2}\left[\sigma_1 G_1(\alpha) + \sigma_2 G_2(\alpha)\right] \tag{10.113}$$

The expression for the geometric factors is the same as in the case of excitation of the field by a vertical magnetic dipole.

The second term $h_x^{(2)}$ includes function $F(\beta,\alpha)$, which depends on ratio of conductivities, more precisely, from parameter β. The appearance of this part of the field can be explained in the following way. Under action of the primary electric field of the dipole surface charges arise in a medium with density:

$$\Sigma(a) = \frac{1}{2\pi}\frac{s-1}{s+1}E_n^{av} \tag{10.114}$$

where E_n^{av} is the magnitude of the normal component of the field, created by vortex field of currents and all charges, except one, located at point a. In this approximation the field of electric charges, as well as the primary field, is directly proportional to frequency.

Let us present eq. 10.110 as:

$$Q\, h_x^{(1)} = -\frac{\mu\omega L^2}{2}\left[\sigma_1 G_1^*(\alpha,s) + \sigma_2 G_2^*(\alpha,s)\right] \tag{10.115}$$

where $G_1^*(\alpha,s) = 1 - 1/2\alpha - 2F(\beta,\alpha)$.

If the formation resistivity exceeds that of shoulders ($s < 1$), then electric charges increase the field within the formation, and function G_1^* becomes larger. In a more conductive formation the electric field of the charges reduces the primary field, and under certain conditions, G_1^* is equal to zero and changes sign. Table 10.8 contains values of functions $G_1^* + (1/s)G_2$ and $F(\beta,\alpha)$. We can show that function $F(\beta,\alpha)$ is expressed through hypergeometric series ${}_2F_1(a,b,c,z)$:

$$\begin{aligned} F(\beta,\alpha) = \frac{\beta}{2}\Bigg\{ & \frac{1}{\beta\alpha}\ln\frac{1+\beta}{1-\beta} - \frac{\beta}{2\alpha+1}\,{}_2F_1\left(1, 1+\frac{1}{2\alpha}, 2+\frac{1}{2\alpha}, \beta^2\right) \\ & - \frac{\beta}{2\alpha-1}\,{}_2F_1\left(1, 1-\frac{1}{2\alpha}, 2-\frac{1}{2\alpha}, \beta^2\right)\Bigg\} \end{aligned} \tag{10.116}$$

TABLE 10.8
Values of functions $F(\beta,\alpha)$ and $G_1^* + G_2^*/s$

s	$F(\beta,\alpha)$	$G_1^* + G_2^*/s$	$F(\beta,\alpha)$	$G_1^* + G_2^*/s$
	$\alpha = 2$		$\alpha = 4$	
1/128	−2.030	36.80	−1.0300	18.90
1/32	−1.420	11.60	−0.7030	6.280
1/8	−0.763	4.280	−0.3770	2.630
1/2	−0.205	1.660	−0.1020	1.330
2	0.142	0.591	0.0717	0.794
8	0.277	0.288	0.1420	0.606
32	0.314	0.129	0.1640	0.552
128	0.324	0.105	0.1690	0.538
	$\alpha = 8$		$\alpha = 16$	
1/128	−5.0200	9.980	−0.2600	5.490
1/32	−0.3510	3.640	−0.1750	2.320
1/8	−0.1880	1.810	−0.0940	4.140
1/2	−0.0507	1.160	−0.0253	1.080
2	0.0359	0.897	0.0180	0.948
8	0.0718	0.802	0.0359	0.901
32	0.0825	0.774	0.0414	0.887
128	0.0854	0.767	0.0428	0.883

In the particular case, when the length of the probe is equal to the formation thickness ($\alpha = 1$), $F(\beta,\alpha)$ is expressed by the elementary function:

$$F(\beta,\alpha) = \frac{1}{2} - \frac{\ln s}{s^2 - 1} \tag{10.117}$$

and for the quadrature component we have:

$$\mathrm{Q}\, h_x = -\left(\frac{L}{h_1}\right)^2 \left(-\frac{1}{2} + \frac{2\ln s}{s^2 - 1}\right) - \frac{1}{2}\left(\frac{L}{h_1}\right)^2 \tag{10.118}$$

For large values of α, function $F(\beta,\alpha)$ decreases inversely proportional to α:

$$F(\beta,\alpha) \simeq \frac{1}{\alpha} \ln \frac{2s}{s+1} \tag{10.119}$$

and the value of function $G_1^*(\alpha, s)$ remains positive for all values of s.

It is easy to show that for $s \to 0$ (formation resistivity increases) field $\mathrm{Q}\, h_x^{(2)}$ tends to zero.

The asymptotic presentation for the field, when the formation is located within the probe, is derived in a similar manner and we obtain:

$$\begin{aligned}\operatorname{In} h_x &= \frac{4}{3}\left(\frac{L}{h_2}\right)^3 \\ \operatorname{Q} h_x &= -\left(\frac{L}{h_1}\right)^2\left\{\frac{4}{(s+1)^2}\int\limits_0^\infty \frac{\mathrm{e}^{-m}\,dm}{1-\beta^2\,\mathrm{e}^{-2\alpha m}} - \frac{\alpha}{2}\right\} - \frac{\alpha}{2}\left(\frac{L}{h_2}\right)^3\end{aligned} \tag{10.120}$$

Let us notice that the integral in this expression can be presented through a hyperbolic function.

Now we will consider the opposite case, when $|k_1 z| \gg 1$ and $|k_2 z| \gg 1$. In essence, deriving asymptotic formulae is similar to that performed in section 6. In this case the presence of the term $1 - K_{12}\mathrm{e}^{-2\alpha m_2}$ in the denominator of the integrand does not make this procedure more complicated, since this term is exponentially small. For example, if $\alpha = 1$, we have:

$$h_x = -\left(\frac{2}{\sqrt{s}+1}\right)^2 k_1^2 z^2 \mathrm{e}^{ik_1 z}$$

We will investigate vertical responses of a two-coil probe at the range of small parameters (Figs. 10.27–10.28). It is natural that the influence of the surrounding medium increases with an increase of its conductivity and a decrease of the formation thickness. For comparison curves of apparent conductivity, when the source is the vertical magnetic dipole, are given in Fig. 10.18. Here apparent conductivity is introduced as: $\sigma_a/\sigma_1 = \operatorname{Q} h_z / \operatorname{Q} h_z^0$, where $\operatorname{Q} h_z$ is the quadrature component of the vertical component of the field.

As is seen from these curves the influence of a more conductive surrounding medium in measuring with vertical and horizontal dipoles is practically the same. If the formation is more conductive, the vertical response of the two-coil probe with horizontal coils is worse and it is mainly caused by the influence of electrical charges.

With an increase of frequency due to the skin effect the influence of the surrounding medium is significantly reduced, and it manifests itself earlier with an increase of the formation thickness and conductivity of the surrounding medium (Figs. 10.29–10.32).

Now let us briefly consider frequency responses of the field (Figs. 10.33–10.40). The dependence of amplitude and phase of the field at the low-frequency spectrum as functions of the surrounding medium conductivity is of certain interest when $\alpha > 2$. If the formation resistivity is considered to be constant, then with an increase of resistivity of the surrounding medium the field decreases at the beginning, reaches a minimum when the formation is more conductive, and then again it begins to grow, approaching an asymptotic value corresponding to the nonconducting surrounding medium. This feature in a field behavior is in agreement with eq. 10.115, which indicates that for certain values of parameter s the quadrature component of the field changes sign. For this reason the left-hand asymptote of phase curves can be correspondingly $-\pi/2$, 0, $\pi/2$ (Fig. 10.28).

With an increase of the formation thickness the influence of charges decreases, and behavior of amplitude and phase curves approach to that corresponding to curves when

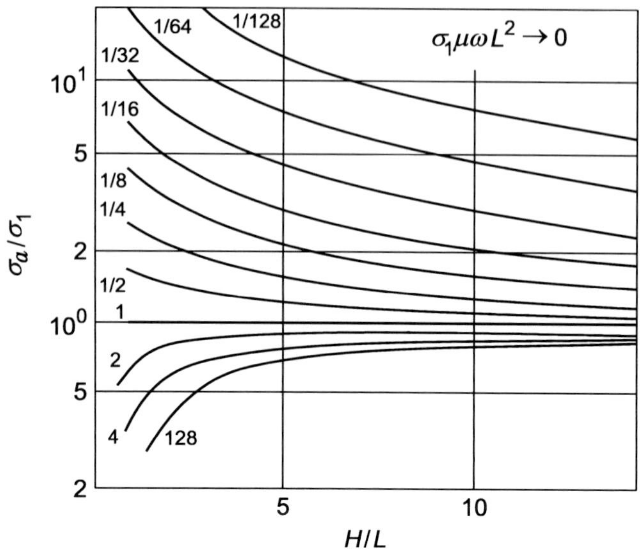

Figure 10.27. Apparent conductivity curves. Curve index σ_1/σ_2.

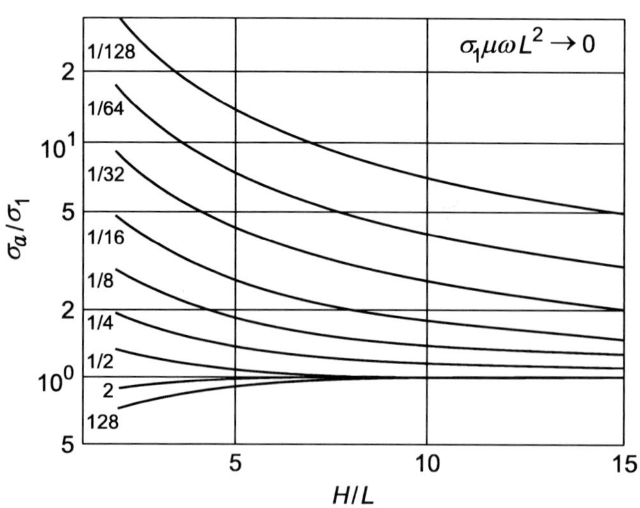

Figure 10.28. Apparent conductivity curves. Curve index σ_1/σ_2.

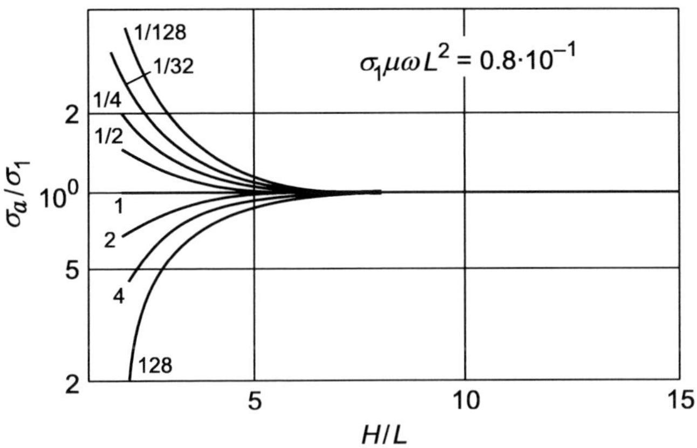

Figure 10.29. Apparent conductivity curves. Curve index σ_1/σ_2.

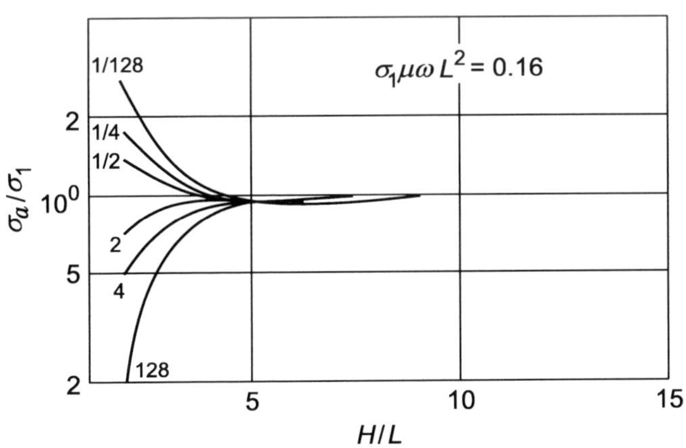

Figure 10.30. Apparent conductivity curves. Curve index σ_1/σ_2.

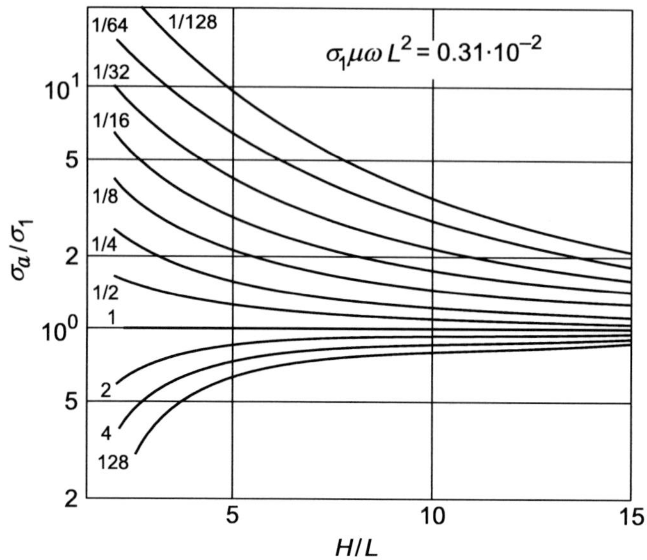

Figure 10.31. Apparent conductivity curves. Curve index σ_1/σ_2.

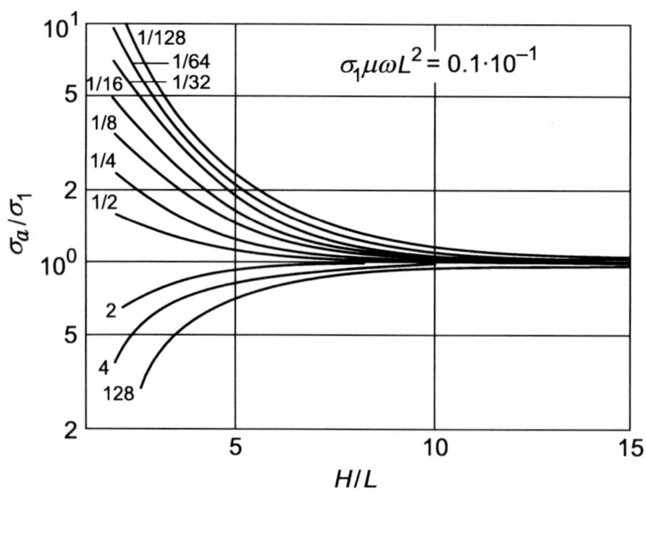

Figure 10.32. Apparent conductivity curves. Curve index σ_1/σ_2.

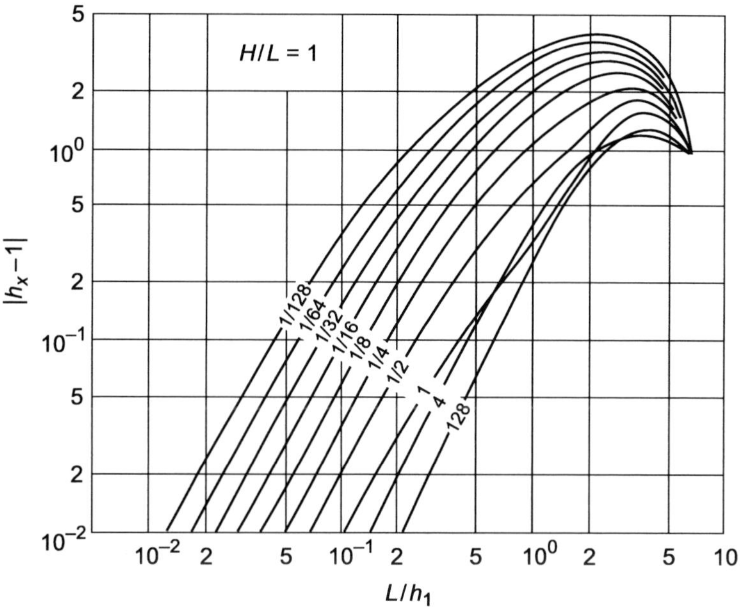

Figure 10.33. Amplitude of the secondary field. Curve index σ_1/σ_2.

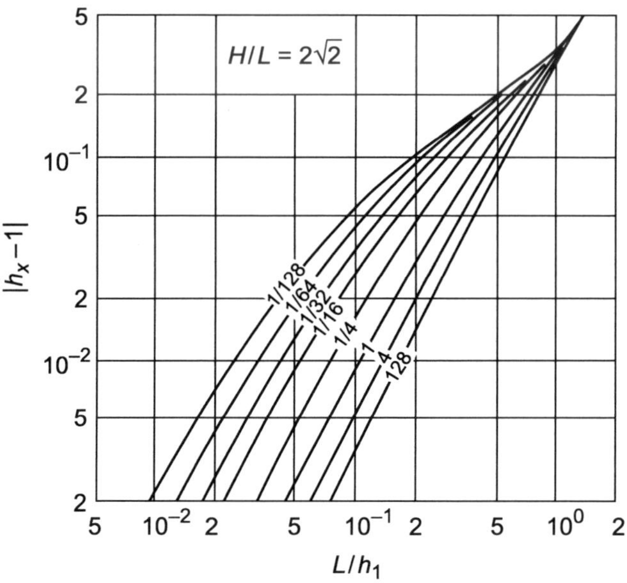

Figure 10.34. Amplitude of the secondary field. Curve index σ_1/σ_2.

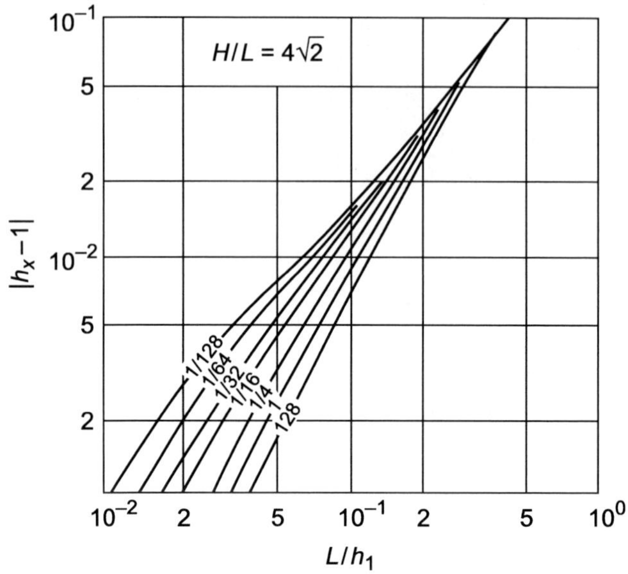

Figure 10.35. Amplitude of the secondary field. Curve index σ_1/σ_2.

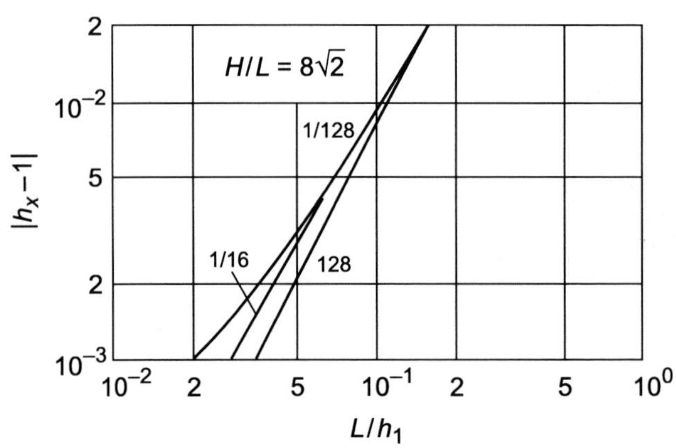

Figure 10.36. Amplitude of the secondary field. Curve index σ_1/σ_2.

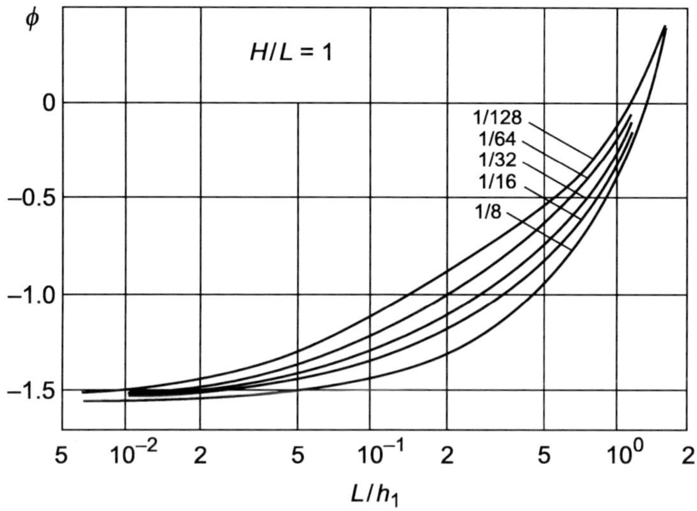

Figure 10.37. The phase of the secondary field. Curve index σ_1/σ_2.

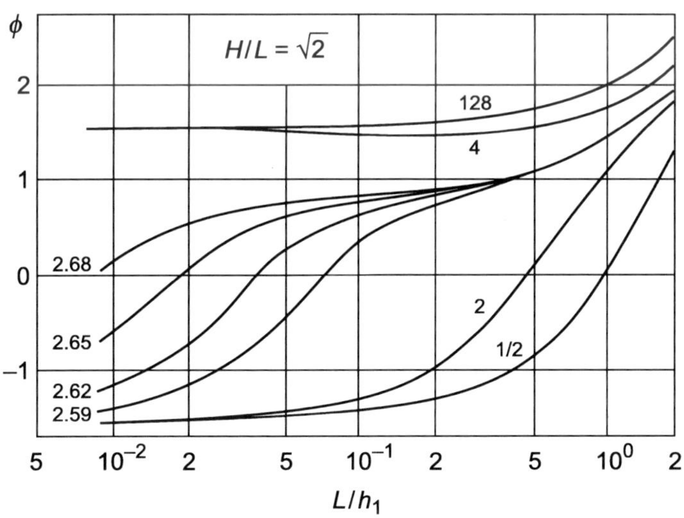

Figure 10.38. The phase of the secondary field. Curve index σ_1/σ_2.

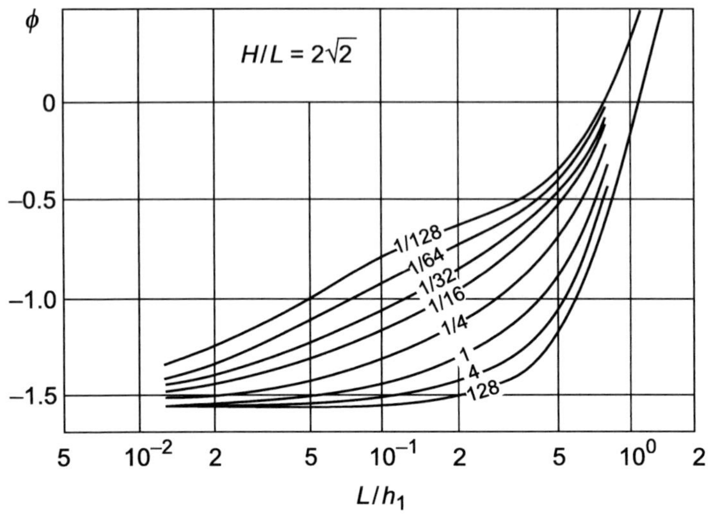

Figure 10.39. The phase of the secondary field. Curve index σ_1/σ_2.

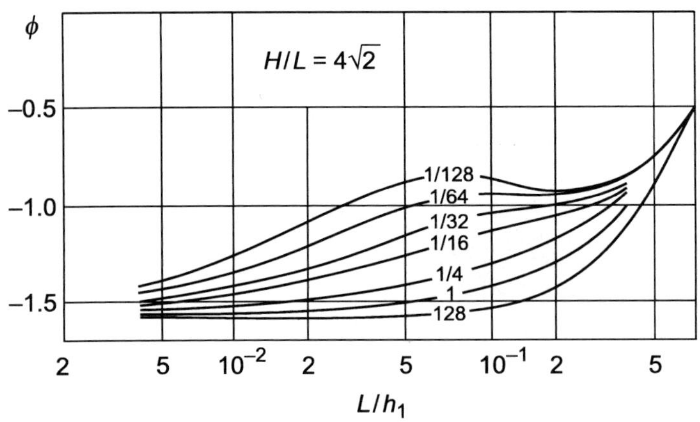

Figure 10.40. The phase of the secondary field. Curve index σ_1/σ_2.

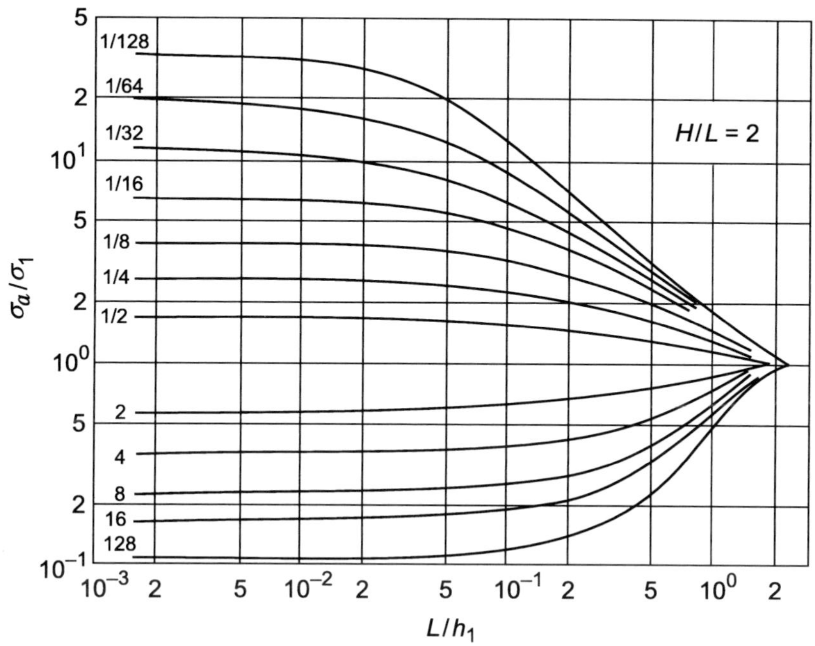

Figure 10.41. Curves of apparent conductivity. Curve index σ_1/σ_2.

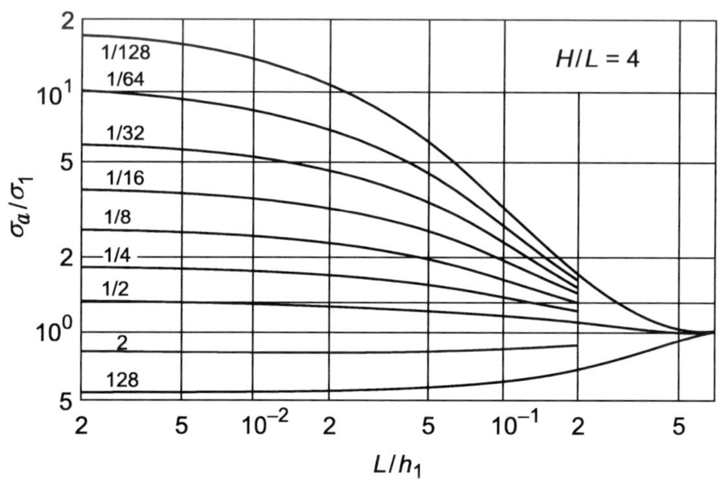

Figure 10.42. Curves of apparent conductivity. Curve index σ_1/σ_2.

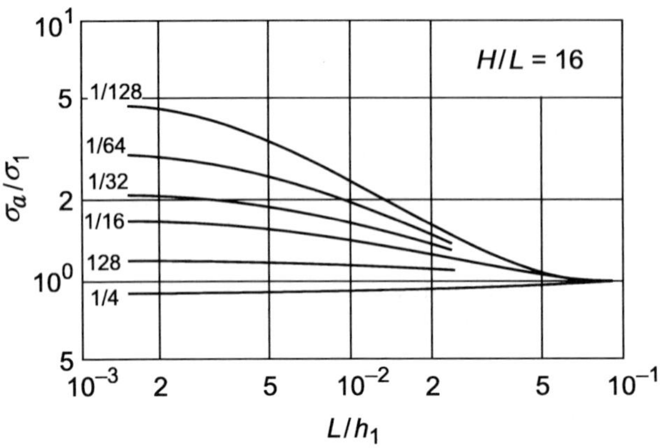

Figure 10.43. Curves of apparent conductivity. Curve index σ_1/σ_2.

the source is the vertical magnetic dipole. In conclusion, curves of apparent conductivity illustrating the influence of the resistivity of the surrounding medium are shown in Figs. 10.41–10.43.

Now let us consider the influence of relatively thin layers ($\alpha < 1$). At the low-frequency part of the spectrum we will present a field as a sum of two terms: a field in a uniform medium with conductivity of the surrounding medium, and that part of the field which takes into account the influence of the formation, i.e.:

$$\mathrm{Q}\, h_x = \mathrm{Q}\, h_x^{un}\left(\frac{L}{h_2}\right) + \left(\frac{L}{h_2}\right)^2 G_2(\alpha, s) \tag{10.121}$$

where

$$G_2(\alpha, s) = -\frac{4s}{(1-s)^2}\int\limits_0^\infty \frac{\mathrm{e}^{-m}\,\mathrm{d}m}{1-\beta^2\,\mathrm{e}^{-2\alpha m}} + \frac{\alpha(s-1)}{2} + 1$$

Equation 10.121 coincides with eq. 10.120 at the range of small ($L/h_2 < 1$) parameters, and for certain relations between values of α and s it is valid in wider range of parameter L/h_2.

Table 10.9 contains maximal values of parameter L/h_2 for which the difference of the quadrature components of the field obtained from exact solution and the approximate formula 10.121 does not exceed 5%. As is seen from Table 10.9 with a decrease of parameter α, the maximal value of parameter L/h_2 increases. If a thin layer has a relatively large resistivity or conductivity, the range of application of eq. 10.121 is restricted by small values of parameter L/h_2. With the parameter s approaching the unity the maximal value of parameter L/h_2 increases. This relation between parameters of a medium and boundary values of parameter, characterizing the skin effect, can easily be studied at the

TABLE 10.9
Values of parameter L/h_2

α \ s	1/128	1/64	1/32	1/16	1/4	1/2	2	8	16	32	64
1/16	0.05	0.10	0.15	0.3	0.4	0.6	0.8	0.3	0.2	0.2	0.15
1/8	0.03	0.07	0.10	0.2	0.4	0.6	0.8	0.2	0.1	0.1	0.07

low-frequency spectrum when the formation thickness is sufficiently small. Expanding the denominator of the integrand in eq. 10.121 by powers of α we obtain:

$$\begin{aligned} Q\,h_x &= -\left(\frac{L}{h_1}\right)^2 \frac{4}{(1+s)^2} \int_0^\infty \frac{e^{-m}\,dm}{1-\beta^2(1-2\alpha m)} \\ &+ \left(\frac{L}{h_2}\right)^2 \frac{\alpha(s-1)}{2} \\ &= \left(\frac{L}{h_1}\right)^2 \left(t\,e^t\,\mathrm{Ei}(-t) + \frac{\alpha(s-1)}{2}\right) \end{aligned} \tag{10.122}$$

where:

$$t = \frac{2s}{\alpha(s-1)^2}$$

$$\mathrm{Ei}(-t) = -e^{-t} \int_0^\infty \frac{e^{-x}\,dx}{x+t}$$

is the integral exponential function. As is well known:

$$\lim_{t\to 0} \mathrm{Ei}(-t) = \ln t$$

$$\mathrm{Ei}(-t) \simeq -e^{-t}(1/t - 1/t^2) \qquad \text{if } t \to \infty$$

Let us consider two extremal cases: $s \gg 1$ and $s \ll 1$, corresponding to either a very conducting or a very resistive thin layer.

Case 1: Very conductive thin layer ($s \gg 1$)

If parameter $s \gg 1$, then $t = 2/\alpha s$, and making use of the asymptotic value of function $\mathrm{Ei}(-t)$ for $t \ll 1$, provided that $s \gg 2/\alpha$, instead of eq. 10.122 we obtain:

$$Q\,h_x = \left(-\frac{L}{h_2}\right)^2 \left(\frac{\alpha s}{2} - \frac{\alpha}{2} - \frac{2}{\alpha s}\ln\frac{2}{\alpha s}\right) \simeq \frac{\alpha}{2}\left(\frac{L}{h_1}\right)^2 \qquad \text{if } \frac{\alpha s}{2} \ll 1 \tag{10.123}$$

But if $1 \ll s < 2/\alpha$ then

$$Q\,h_x = -\left(\frac{L}{h_2}\right)^2 \left(1 - \alpha s + \frac{\alpha}{2}\right) = -\left(\frac{L}{h_2}\right)^2 + \alpha\left(\frac{L}{h_1}\right)^2 \qquad \text{as } \frac{\alpha s}{2} \ll 1 \tag{10.124}$$

TABLE 10.10
Values of function $G_1(\alpha, s)$

α \ s	1/128	1/64	1/32	1/16	1/8	1/4	1/2
1/16	0.623	0.491	0.348	0.214	0.1050	0.0329	−0.00207
1/8	0.695	0.582	0.443	0.294	0.1550	0.0485	−0.00724
1/4	0.711	0.621	0.499	0.350	0.1920	0.0540	−0.0232
0.8	0.518	0.462	0.376	0.254	0.0899	−0.0556	−0.133

α \ s	2	4	8	16	32	64	128
1/16	0.0448	0.150	0.351	0.712	1.350	2.490	4.620
1/8	0.0865	0.283	0.647	1.290	2.440	4.580	8.690
1/4	0.1640	0.523	1.180	2.340	4.490	8.620	16.70
0.8	0.4700	1.440	3.250	6.630	13.20	26.00	51.70

It is obvious that in this case the field can be presented as a sum of the field in a uniform medium with conductivity σ_2, and the field due to the presence of a thin conducting layer:

$$Q\,h_x = Q\,h_x^{un}\left(\frac{L}{h_2}\right) + \alpha\left(\frac{L}{h_1}\right)^2 \tag{10.125}$$

Case 2: Very resistive thin layer ($s \ll 1$)

For parameter t we have: $t = 2s/\alpha$. If $s < \alpha/2$, then $t \ll 1$ and correspondingly:

$$Q\,h_x = \left(\frac{L}{h_2}\right)^2\left(\frac{2\alpha}{\alpha}\ln\frac{2\alpha}{\alpha} + \frac{s\alpha}{2} - \frac{\alpha}{2}\right) \simeq \left(\frac{L}{h_2}\right)^2\left(\frac{2s}{\alpha} - \frac{\alpha}{2}\right) \tag{10.126}$$

In the opposite case ($s \gg \alpha/2$) we have:

$$Q\,h_x = -\left(\frac{L}{h_2}\right)^2\left(1 - \frac{\alpha}{2s} + \frac{\alpha}{s} - \frac{\alpha s}{2}\right) \simeq -\left(\frac{L}{h_2}\right)^2 + \frac{\alpha}{2s}\left(\frac{L}{h_2}\right)^2 \tag{10.127}$$

Generalizing this expression for higher frequencies, we obtain:

$$Q\,h_x = Q\,h_x^{un}\left(\frac{L}{h_2}\right) + \frac{\alpha}{2s}\left(\frac{L}{h_2}\right)^2 \tag{10.128}$$

Thus, the smaller parameters s and σ/s, that correspond to more conductive or resistive thin layers, for higher frequencies eq. 10.121 is applied. Values of function $G_1(\alpha, s)$, which together with the field in a uniform medium, allow us to evaluate the influence of thin layers at the range of small parameters are presented in Table 10.10. Curves of apparent conductivity σ_a/σ_2 at the range of small parameter ($\sigma_2\mu\omega L^2 \to 0$) are given in Figs. 10.44–10.45.

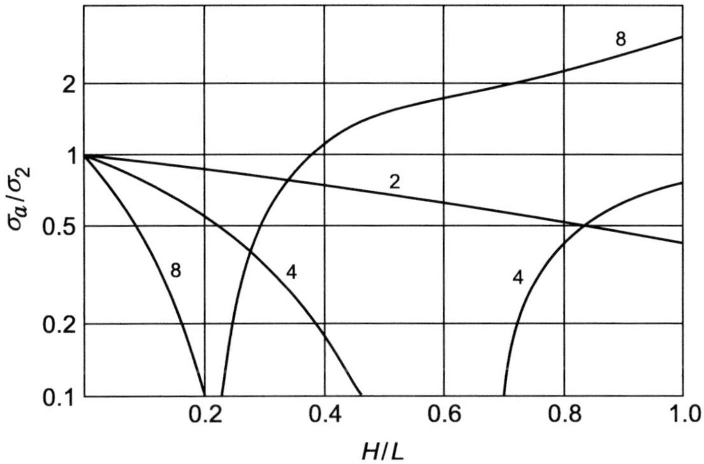

Figure 10.44. Apparent conductivity curves at the range of small parameters. Curve index σ_1/σ_2.

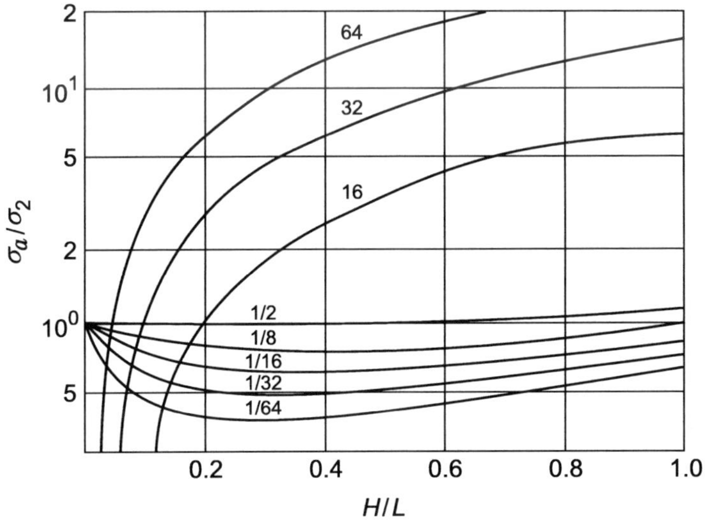

Figure 10.45. Apparent conductivity curves at the range of small parameters. Curve index σ_1/σ_2.

10.9. Curves of Profiling with a Two-coil Induction Probe in a Medium with Horizontal Interfaces

Considering the curves of profiling, it is appropriate to distinguish four specific positions of the probe with respect to interfaces (Fig. 10.26).

Case 1: The probe is located outside the formation

In accord with results obtained in previous section we have (Fig. 10.26a):

$$
\begin{aligned}
h_x &= h^{un}\left(\frac{L}{h_2}\right) - \int_0^\infty \left\{\left(k_1^2 L^2 - \frac{m^2}{2}\right) D_1\, e^{m_1} + \frac{mm_1}{2} F_1 e^{m_1}\right\} dm \\
D_1 &= \frac{m}{m_1}\frac{q_{12}}{d_1}(1 - e^{-2\alpha m_2})\, e^{-2\beta m_1} \\
F_1 &= -F\left(1 - e^{-2\alpha m_2}\right)\left(1 - q_{12}K_{12}\, e^{-2\alpha m_2}\right) e^{-2\beta m_1} \\
F &= \frac{2(1-s)m^2}{(m_1+m_2)(sm_1+m_2)d_1 d_2} \\
K_{12} &= \frac{m_1 - m_2}{m_1 + m_2} \qquad (10.129) \\
q_{12} &= \frac{sm_1 - m_2}{sm_1 + m_2} \\
d_1 &= 1 - q_{12}\, e^{-2\alpha m_2} \\
d_2 &= 1 - K_{12}^2\, e^{-2\alpha m_2} \\
\alpha &= H/L \qquad \beta = h_2/L\beta \geqslant 1 \\
0 &\leqslant \alpha < \infty \qquad \beta \geqslant 1
\end{aligned}
$$

Case 2: Coils of the probe are located from both sides of one interface

In this case (Fig. 10.26c) we have:

$$
\begin{aligned}
h_x &= -\int_0^\infty \left\{\left(k_1^2 L^2 - \frac{m^2}{2}\right) D_1\, e^{-m_1} + \frac{mm_1}{2} F_1 e^{-m_1}\right\} dm \\
D_1 &= s\frac{m}{m_2}\frac{e^{(m_1-m_2)\beta}}{d_1}(1 - q_{12})(1 - q_{12}\, e^{-2(\alpha-\beta)m_2}) \qquad (10.130) \\
F_1 &= F\, e^{\beta(m_1-m_2)} \\
&\quad \times \left\{(K_{12} - q_{12})\, e^{-2(\alpha-\beta)m_2}\left(e^{-2\beta m_2} - 1\right) + \left(1 - K_{12}q_{12}\, e^{-2\alpha m_2}\right)\left(e^{-2(\alpha-\beta)m_2} - 1\right)\right\} \\
0 &\leqslant \alpha < \infty \qquad 0 \leqslant \beta \leqslant \alpha \qquad \beta \leqslant 1
\end{aligned}
$$

Case 3: Probe is located within the formation

We have for this probe location (Fig. 10.26b):

$$h_x = h_x^{un}\left(\frac{L}{h_1}\right) - \int_0^\infty \left\{\left(k_2^2L^2 - \frac{m^2}{2}\right)\left(D_2\,e^{m_2} + D_3\,e^{-m_2}\right) + \frac{mm_2}{2}\left(F_2\,e^{m_2} - F_3\,e^{-m_2}\right)\right\}\,dm$$

$$D_2 = -\frac{m\;q_{12}}{m_2\;d_1}\,e^{-2(\alpha-\beta)m_2}\left(1 - q_{12}\,e^{-2\beta m_2}\right)$$

$$D_3 = -\frac{m\;q_{12}}{m_2\;d_1}\,e^{-2\beta m_2}\left(1 - q_{12}\,e^{-2(\alpha-\beta)m_2}\right) \qquad (10.131)$$

$$F_2 = F\,e^{-2(\alpha-\beta)m_2}\left\{(K_{12} - q_{12})\,e^{-2\beta m_2} + 1 - K_{12}\,q_{12}\,e^{-2\alpha m_2}\right\}$$

$$F_3 = -F\,e^{-2\beta m_2}\left\{(K_{12} - q_{12})\,e^{-2(\alpha-\beta)m_2} + 1 - K_{12}\,q_{12}\,e^{-2\alpha m_2}\right\}$$

$$1 \leqslant \alpha < \infty \qquad 0 \leqslant \beta \leqslant \alpha - 1$$

In the case of a symmetrical position of the probe with respect to the boundaries $\beta = (\alpha - 1)/2$ and eq. 10.131 coincides with eq. 10.108.

Case 4: The formation is located between probe coils

We have in this case (Fig. 10.16d):

$$h_x = -\int_0^\infty \left\{\left(k_1^2L^2 - \frac{m^2}{2}\right)D_4\,e^{-m_1} - \frac{mm_1}{2}F_4\,e^{-m_1}\right\}\,dm \qquad (10.132)$$

where:

$$D_4 = \frac{4smm_2}{(sm_1 - m_2)^2}\,e^{\alpha(m_1-m_2)}$$

$$F_4 = 2F(1-s)\frac{m_1m_2\,e^{\alpha(m_1-m_2)}\left(1 - e^{-2\alpha m_2}\right)}{(m_1 + m_2)(sm_1 + m_2)}$$

The field does not depend on the position of the formation between coils of the probe.

Making use of eqs. 10.129–10.132 values of apparent conductivity σ_a/σ_1 were calculated (Figs. 10.46–10.48). This function is introduced as:

$$\frac{\sigma_a}{\sigma_1} = \frac{|h_x - 1|}{\left|h_x^{un}\left(\frac{L}{h_1}\right) - 1\right|}$$

It is obvious that the field at every point depends on the probe position with respect to the formation and parameters: L/h_2, $\alpha = H/L$ and $s = \sigma_1/\sigma_2$. For every case curves

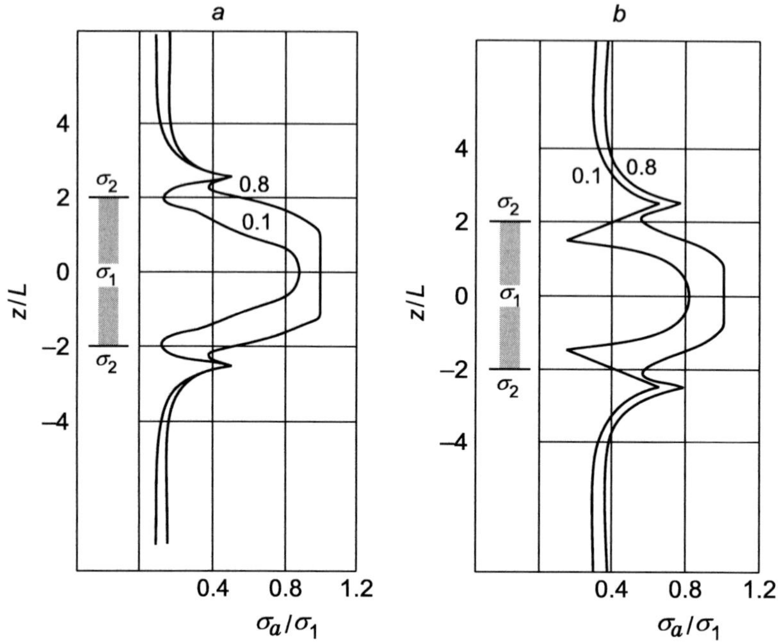

Figure 10.46. Curves of profiling: (a) $H/L = 4$, $\sigma_1/\sigma_2 = 16$; (b) $H/L = 4$, $\sigma_1/\sigma_2 = 4$. Curve index L/h_2.

are plotted for certain values of L/h_2. The value of σ_a/σ_1 is plotted along the horizontal axis, while along the vertical axis the distance from the formation center to the measuring point (the probe middle) expressed in units of the formation thickness, is plotted.

Considering the influence of a medium it is convenient to distinguish four cases.

Case 1

Let the formation conductivity exceed that of the surrounding medium and its thickness be greater than the probe length (Fig. 10.46a,b). In this case parts of the curves, corresponding to the position of the probe against the formation, have a rather complicated character. In approaching the receiver or transmitter coils to the formation boundary a rapid change of the field, due to surface charges, is observed. When the probe is within the surrounding medium, with an increase of the distance from the formation value of σ_a/σ_1 asymptotically approaches to the limit, equal to:

$$\frac{\sigma_a}{\sigma_1} = \frac{\left| h_x^{un}\left(\dfrac{L}{h_2}\right) - 1 \right|}{\left| h_x^{un}\left(\dfrac{L}{h_1}\right) - 1 \right|}$$

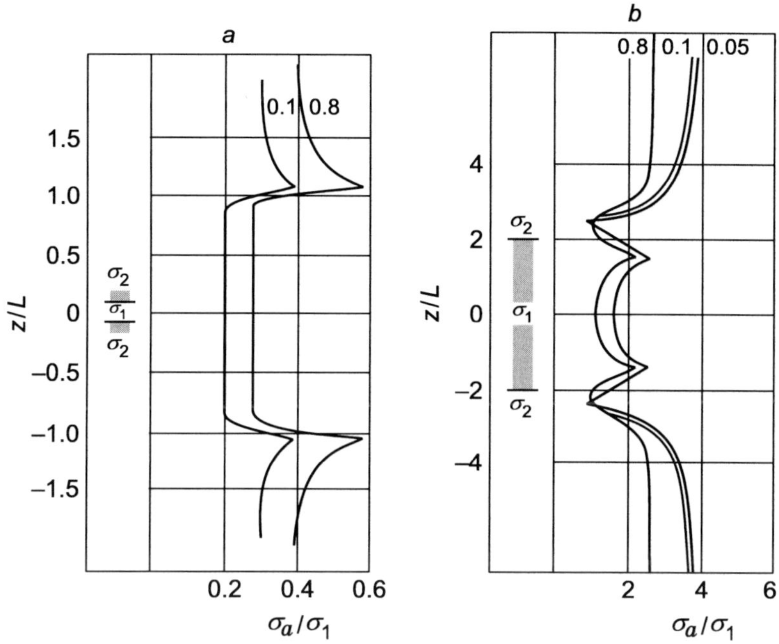

Figure 10.47. Curves of profiling: (a) $H/L = 1/8$, $\sigma_1/\sigma_2 = 4$; (b) $H/L = 4$, $\sigma_1/\sigma_2 = 1/4$. Curve index L/h_2.

which at the range of small parameter, $(L/h < 1)$, is equal to σ_2/σ_1.

Distance d between limited *splashes* on the curves of profiling is related with the formation thickness as $d = H + L$.

Case 2

If the formation conductivity is smaller than that of the shoulders $(\sigma_1 < \sigma_2,\ H \geqslant L)$, the curves have also a complicated character, as in the first case (Fig. 10.47b). With an increase of parameter L/h limited minima become slightly smaller and closer to each other. At the same time the value of σ_a/σ_1 also decreases within the formation center as well as in the surrounding medium.

Case 3

Consider a thin conducting layer $(\sigma_1 > \sigma_2,\ H \leqslant L)$. Such a layer has a significant influence on the field, and it manifests itself clearly on the curves of the apparent conductivity (Fig. 10.47a). Distance d between *splashes* on curves of σ_a/σ_1 is related with the formation thickness as $d = 2L + H$, and the distance between the closest fractures of curves is equal to the formation thickness.

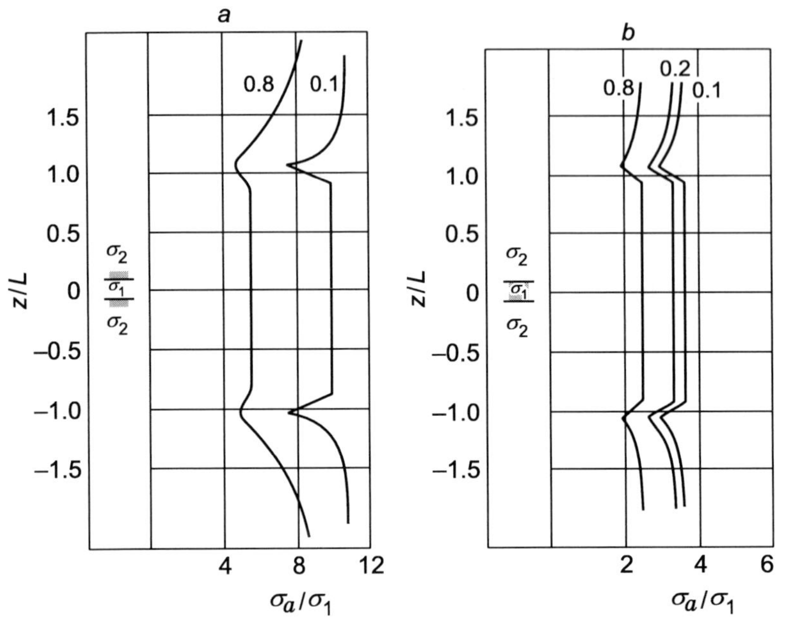

Figure 10.48. Curves of profiling: (a) $H/L = 1/8$, $\sigma_1/\sigma_2 = 1/16$; (b) $H/L = 1/8$, $\sigma_1/\sigma_2 = 1/4$. Curve index L/h_2.

Case 4

Let the formation thickness be smaller than the probe length and its resistivity exceed that of the shoulders ($\sigma_1 < \sigma_2$, $H \leqslant L$). For a relatively small formation resistivity its influence is insignificant and practically displays itself as the appearance of two small *splashes* with distance between them $2H + L$ (Fig. 10.48).

With a decrease of the formation conductivity value of σ_a/σ_1 begins to differ noticeably from unit at the center of the formation and within the surrounding medium.

At the range of small parameters an influence of thin resistive layer becomes very small. It is obvious that the presence of the borehole makes curves of apparent conductivity more smoothed.

In conclusion let us notice that at the range of small parameters it is not difficult to find a relation between a value of an angle between tangents at breakpoints of curves of apparent conductivity and parameters of the medium. In fact, for a change of the derivative we have:

$$\frac{\partial}{\partial z}\left(\frac{\sigma_a^{(1)}}{\sigma_1}\right) - \frac{\partial}{\partial z}\left(\frac{\sigma_a^{(2)}}{\sigma_1}\right) \simeq \pm \frac{1}{h_{un}^{(2)} - 1}\left(\frac{\partial h_x^{(1)}}{\partial z} - \frac{\partial h_x^{(2)}}{\partial z}\right)$$

On the other hand:

$$\frac{\partial H_x}{\partial z} - \frac{\partial H_z}{\partial x} = \sigma E_y$$

Inasmuch as values of E_y and $\partial H_z/\partial x$ are continuous at the interface, we obtain:

$$\frac{\partial H_x^{(1)}}{\partial z} - \frac{\partial H_z^{(2)}}{\partial z} = (\sigma_2 - \sigma_1) E_y$$

and, finally:

$$\frac{\partial}{\partial z}\left(\frac{\sigma_a^{(1)}}{\sigma_1}\right) - \frac{\partial}{\partial z}\left(\frac{\sigma_a^{(2)}}{\sigma_1}\right) = \pm(\sigma_2 - \sigma_1)\frac{e_y}{\left|h_x^{un}\left(\frac{L}{h_1}\right) - 1\right|}$$

where

$$e_y = \frac{E_y}{-\dfrac{M}{4\pi L^3}}$$

Chapter 11

THE INFLUENCE OF ANISOTROPY ON THE FIELD OF THE MAGNETIC DIPOLE IN A CONDUCTING MEDIUM

In this chapter we will consider the electromagnetic field of the magnetic dipole in a uniform anisotropic medium as well as in an anisotropic medium with horizontal interfaces.

11.1. Anisotropy of a Layered Medium

First suppose that a medium presents itself as an alteration of isotropic layers of two types: one has conductivity σ_1 and dielectric constant ε_1 while the other has conductivity and dielectric constant σ_2 and ε_2, respectively (Fig. 11.1).

Let us assume that in an arbitrary layer, which is denoted by index (1), a uniform electric field $\boldsymbol{E}_1 = \boldsymbol{E}\,\mathrm{e}^{-\mathrm{i}\omega t}$ is given, and it is located at the xz plane. The current density in this layer is $\boldsymbol{j}_1 = \sigma_1 \boldsymbol{E}_1$.

Thicknesses of the skin layers h_1 and h_2 are assumed to be sufficiently large so that they exceed in many times the thickness of an elementary layer. Correspondingly, we can neglect the skin effect within these layers. Now we will express $\boldsymbol{E}$ and $\boldsymbol{j}$ in every layer through current $\boldsymbol{j}_1$.

Maxwell's equations result in the following conditions on the interface of the first and second layer:

$$E_{2x} = E_{1x} \qquad \varepsilon_2 E_{2z} - \varepsilon_1 E_{1z} = \Sigma \tag{11.1}$$

where Σ is surface density of charges. From surface analogy of equation of continuity of current density we have:

$$J_{2z} - J_{1z} = \mathrm{i}\omega\Sigma \tag{11.2}$$

Eliminating the value of Σ from eqs. 11.1 and 11.2 and making use of Ohm's law, we obtain for the current and the field in layer (2) the following expressions:

$$J_{2x} = \frac{\sigma_2}{\sigma_1} J_x \qquad J_{2z} = \frac{1 - \mathrm{i}\omega\varepsilon_1/\sigma_1}{1 - \mathrm{i}\omega\varepsilon_2/\sigma_2} J_z \tag{11.3}$$

$$E_{2x} = \frac{J_{2x}}{\sigma_2} = \frac{J_x}{\sigma_1} \qquad E_{2z} = \frac{J_z}{\sigma_2} = \frac{1 - \mathrm{i}\omega\varepsilon_1/\sigma_1}{1 - \mathrm{i}\omega\varepsilon_2/\sigma_2} \frac{J_z}{\sigma_2} \tag{11.4}$$

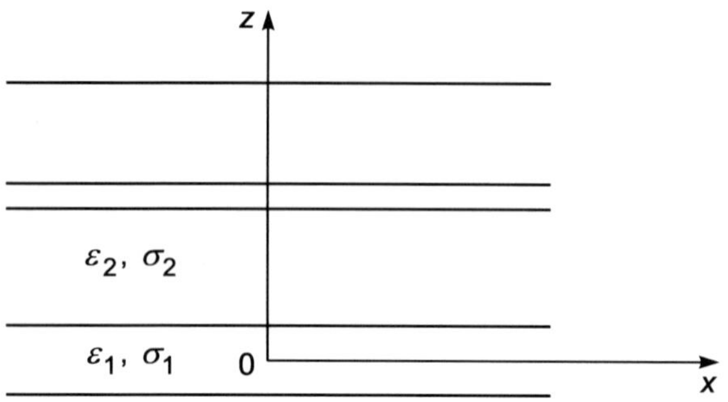

Figure 11.1. Model of anisotropic medium.

By analogy, from conditions on the surface between the second and third layer, we have:

$$J_{3x} = \frac{\sigma_3}{\sigma_2} J_{2x} \qquad J_{3z} = \frac{1 - i\omega\varepsilon_2/\sigma_2}{1 - i\omega\varepsilon_3/\sigma_3} J_{2z} \tag{11.5}$$

Inasmuch as $\sigma_3 = \sigma_1$, $\varepsilon_3 = \varepsilon_1$ then instead of 11.5 we have:

$$\boldsymbol{J}_3 = \boldsymbol{J}_1 \qquad \boldsymbol{E}_3 = \boldsymbol{E}_1 \tag{11.6}$$

Thus, in the formation consisting of alternating thin layers of both types, field and current density have paired values: $\boldsymbol{E}_1$, $\boldsymbol{J}_1$, and $\boldsymbol{E}_2$, $\boldsymbol{J}_2$, corresponding to the first and second layers. Let us take an arbitrary layer with thickness D ($D > h$), in which the relative contribution of layers with conductivity σ_2 is equal to n. Then, for average values of current and the field we have:

$$\begin{aligned} \langle J_x \rangle &= \left(1 - n + n\frac{\sigma_2}{\sigma_1}\right) j_x \qquad & \langle J_z \rangle &= \left(1 - n + n\frac{1 - i\omega\varepsilon_1/\sigma_1}{1 - i\omega\varepsilon_2/\sigma_2}\right) J_z \\ \langle E_x \rangle &= \frac{J_x}{\sigma_1} & \langle E_z \rangle &= \left(1 - n + n\frac{\sigma_1}{\sigma_2}\frac{1 - i\omega\varepsilon_1/\sigma_1}{1 - i\omega\varepsilon_2/\sigma_2}\right) \frac{J_z}{\sigma_1} \end{aligned} \tag{11.7}$$

Defining the longitudinal and transversal conductivities from relation:

$$\sigma_t = \frac{\langle J_x \rangle}{\langle E_x \rangle} \qquad \sigma_n = \frac{\langle J_z \rangle}{\langle E_z \rangle}$$

we obtain:

$$\begin{aligned} \sigma_t &= \sigma_1 \left[1 - n + n\frac{\sigma_2}{\sigma_1}\right] \\ \sigma_n &= \sigma_1 \frac{1 - n[1 - p(\omega)]}{1 - n\left(1 - \dfrac{\sigma_1}{\sigma_2} p(\omega)\right)} \end{aligned} \tag{11.8}$$

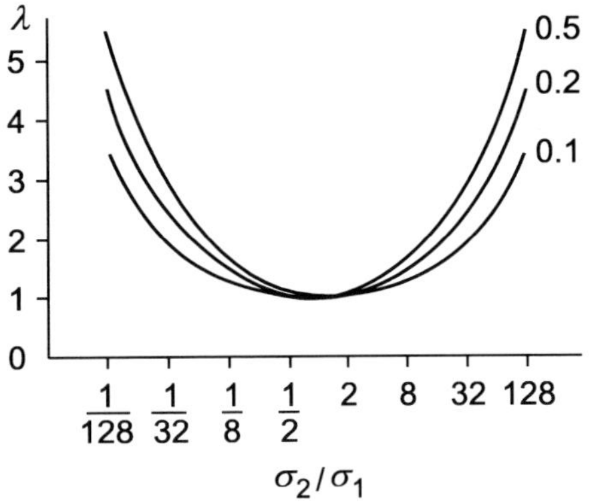

Figure 11.2. Dependence of anisotropy coefficient on ratio σ_2/σ_1 ($\omega\varepsilon/\sigma \to 0$). Curve index n.

where

$$p(\omega) = \frac{1 - i\omega\varepsilon_1/\sigma_1}{1 - i\omega\varepsilon_2/\sigma_2} \tag{11.9}$$

Thus, in the quasistationary approximation for the considered model of a medium, the dependence of conductivity on frequency is absent, and consequently, expressions for transversal conductivity and coefficient of anisotropy have the form:

$$\sigma_n = \frac{\sigma_1}{1 - n + n\dfrac{\sigma_1}{\sigma_2}} \tag{11.10}$$

$$\lambda = \left[\left(1 - n + n\frac{\sigma_2}{\sigma_1}\right)\left(1 - n + n\frac{\sigma_1}{\sigma_2}\right)\right]^{1/2} \tag{11.11}$$

Graphs illustrating the relation of coefficient of anisotropy, λ, with parameters σ_2/σ_1 and n, are shown in Fig. 11.2.

In a general case, when the influence of displacement currents is essential, the transversal resistivity depends on frequency. It is explained by the fact that surface charges are a function of dielectric constant and frequency.

We can assume that if the electric field is not uniform and changes along the layer, the longitudinal conductance is also a function of a frequency. Curves, presented in Fig. 11.3 characterize the influence of displacement currents on coefficient on anisotropy.

If n remains constant within interval D and the dimensions of the measuring array (the probe length) are much greater than the layer thickness, this part of a medium can be considered as a uniform anisotropic layer with coefficient of anisotropy λ.

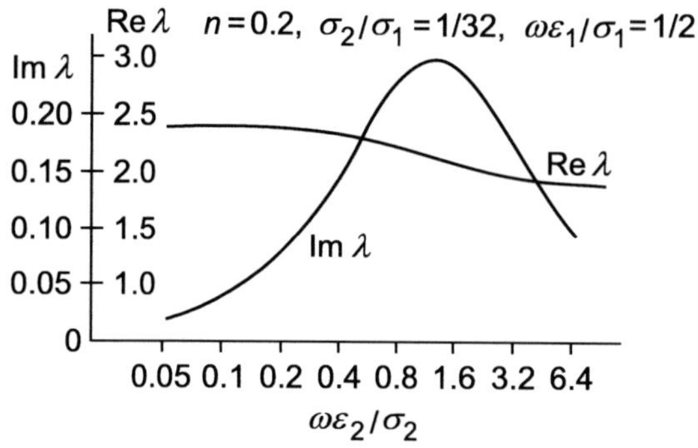

Figure 11.3. Dependence of real and imaginary part of λ on ratio $\omega\varepsilon_2/\sigma_2$.

11.2. Electromagnetic Field of the Magnetic Dipole in a Uniform Anisotropic Medium

Let us consider a uniform anisotropic medium with the tensor of conductivity:

$$\sigma_{ik} = \begin{pmatrix} \sigma_t & 0 & 0 \\ 0 & \sigma_t & 0 \\ 0 & 0 & \sigma_n \end{pmatrix} \tag{11.12}$$

An arbitrary oriented magnetic dipole can be presented as a sum of two dipoles, namely, a vertical and a horizontal one. In excitation of the field by a vertical magnetic dipole, induced currents are located in horizontal planes and do not depend on the transversal conductivity σ_n.

Features of the field, caused by the vertical dipole, were investigated in detail in Chapter 2.

We will explore the case when the moment of the dipole is located in the horizontal plane. Under such type of field excitation volume charges arise in the medium. In fact, having presented the equation of current continuity of the quasistationary field $\operatorname{div} \boldsymbol{j} = 0$ in the form:

$$\sigma_t \operatorname{div} \boldsymbol{E} + (\sigma_n - \sigma_t)\frac{\partial E_z}{\partial z} = 0$$

and making use of equation $\operatorname{div} \boldsymbol{E} = \delta/\varepsilon_0$, we obtain an expression for volume density of charges at an arbitrary point in a medium:

$$\delta = \varepsilon_0 \left(1 - \frac{\sigma_n}{\sigma_t}\right) \frac{\partial E_z}{\partial z}$$

or (11.13)

$$\delta = \varepsilon_0 \left(1 - \frac{1}{\lambda^2}\right) \frac{\partial E_z}{\partial z}$$

We will write the Maxwell equation system in the form:

$$\begin{aligned}
&\operatorname{curl}\boldsymbol{E} = \mathrm{i}\omega\mu\boldsymbol{H} && \operatorname{div}\boldsymbol{E} = \delta/\varepsilon_0 \\
&\operatorname{curl}_x\boldsymbol{H} = \sigma_t E_x && \\
&\operatorname{curl}_y\boldsymbol{H} = \sigma_t E_y && \operatorname{div}\boldsymbol{H} = 0 \\
&\operatorname{curl}_z\boldsymbol{H} = \sigma_n E_z &&
\end{aligned} \tag{11.14}$$

Inasmuch as volume density δ is different from zero, it is impossible to introduce vector potential of the magnetic type: $\boldsymbol{E} = \operatorname{curl}\boldsymbol{A}^*$. For this reason, let:

$$\boldsymbol{H} = \operatorname{curl}\boldsymbol{A} \tag{11.15}$$

Then from eq. 11.14 it follows that:

$$\boldsymbol{E} = \mathrm{i}\omega\mu\boldsymbol{A} - \operatorname{grad}U \tag{11.16}$$

Thus, we have for potential $\boldsymbol{A}$ the following equations:

$$\begin{aligned}
\frac{\partial}{\partial x}\operatorname{div}\boldsymbol{A} - \nabla^2 A_x &= \sigma_t\left(\mathrm{i}\omega\mu A_x - \frac{\partial U}{\partial x}\right) \\
\frac{\partial}{\partial y}\operatorname{div}\boldsymbol{A} - \nabla^2 A_y &= \sigma_t\left(\mathrm{i}\omega\mu A_y - \frac{\partial U}{\partial y}\right) \\
\frac{\partial}{\partial z}\operatorname{div}\boldsymbol{A} - \nabla^2 A_z &= \sigma_n\left(\mathrm{i}\omega\mu A_z - \frac{\partial U}{\partial z}\right)
\end{aligned}$$

Choosing the gauge condition in the form:

$$\operatorname{div}\boldsymbol{A} = -\sigma_t U$$

we have:

$$\begin{aligned}
&\nabla^2 A_x + k_t^2 A_x = 0 \\
&\nabla^2 A_y + k_t^2 A_y = 0 \\
&\nabla^2 A_z + k_n^2 A_z = \left(1 - \frac{1}{\lambda^2}\right)\frac{\partial}{\partial z}\operatorname{div}\boldsymbol{A}
\end{aligned} \tag{11.17}$$

where $k_t^2 = \mathrm{i}\sigma_t\mu\omega$, $k_n^2 = \mathrm{i}\sigma_n\mu\omega$, $\lambda^2 = \sigma_t/\sigma_n$.

Since the behavior of the vector potential of electrical type $\boldsymbol{A}$ near the magnetic dipole is not known beforehand, it is appropriate to present the magnetic dipole as a sum of two vertical and two horizontal electric dipoles (Fig. 11.4) and find a solution for each of them.

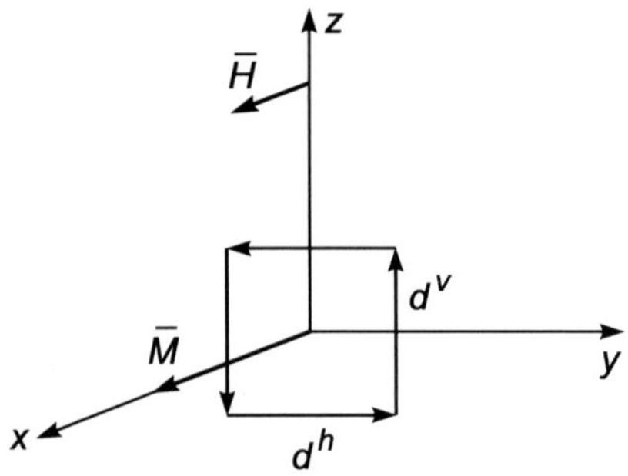

Figure 11.4. Presentation of the magnetic dipole through electric dipoles.

Vector potential of the vertical electric dipole will be described through only one component, A_z^v, since due to axial symmetry the magnetic field has only component H_ϕ. In accord with eq. 11.17 equation for component A_z^v has the form:

$$\frac{\partial^2 A_z^v}{\partial x^2} + \frac{\partial^2 A_z^v}{\partial y^2} + \frac{1}{\lambda^2}\frac{\partial^2 A_z^v}{\partial z^2} + k_n^2 A_z^v = 0 \qquad (11.18)$$

After replacing variable z by $z_1 = \lambda z$, eq. 11.18 coincides with that for a uniform isotropic medium and therefore:

$$A_z^v = C\frac{\mathrm{e}^{ik_n r_*}}{R_*} \qquad (11.19)$$

where

$$R_* = (x^2 + y^2 + \lambda^2 z^2)^{1/2}$$

For determination of constant C we will make use of the known expression for the potential of the direct current I of the electrode in a uniform anisotropic medium:

$$\phi = \frac{I}{4\pi R_* \sqrt{\sigma_t \sigma_n}} \qquad (11.20)$$

Letting dimensions of the electrode very small and differentiating eq. 11.20 by z, we obtain an expression for the potential of the vertical electric dipole, when the distance between the electrodes is equal to a:

$$U = -\frac{\partial \phi}{\partial z}\Delta z = \frac{Ia}{4\pi\sqrt{\sigma_t \sigma_n}}\frac{\lambda^2 z}{R_*^3} \qquad (11.21)$$

On the other hand, taking into account the gauge condition:

$$U = \frac{1}{\sigma_t}\frac{\partial A_z}{\partial z}$$

we have:

$$U = \frac{C}{\sigma_t} \frac{\lambda^2 z}{R_*^3} \tag{11.22}$$

Comparing eqs. 11.21 and 11.22 we obtain the following expression for constant C:

$$C = \frac{Ia}{4\pi} \sqrt{\frac{\sigma_t}{\sigma_n}} = \frac{Ia}{4\pi} \lambda \tag{11.23}$$

Now, we will consider the field of a horizontal electric dipole, with the moment directed along the y-axis. We will look for a solution of eqs. 11.17, letting $A_r^h = 0$. Then, for components A_y^h and A_z^h we have:

$$\begin{aligned} &\nabla^2 A_y^h + k_t^2 A_y^h = 0 \\ &\frac{\partial^2 A_z^h}{\partial x^2} + \frac{\partial^2 A_z^h}{\partial y^2} + \frac{1}{\lambda^2} \frac{\partial^2 A_z^h}{\partial z^2} + k_n^2 A_z^h = \left(1 - \frac{1}{\lambda^2}\right) \frac{\partial^2 A_y^h}{\partial y \partial z} \end{aligned} \tag{11.24}$$

Let:

$$A_y^h = C_1 \frac{\mathrm{e}^{\mathrm{i}k_t R}}{R} = C_1 \int_0^\infty \frac{m}{m_t} \,\mathrm{e}^{-m_t|z|} J_0(mr) \,\mathrm{d}m \tag{11.25}$$

where $m_t = (m^2 - k_t^2)^{1/2}$.

It is convenient to present component A_z^h as:

$$A_z^h = \frac{y}{r} \int_0^\infty F_m(z) J_1(mr) \,\mathrm{d}m = -\frac{\partial}{\partial y} \int_0^\infty \frac{F_m(z)}{m} J_0(mr) \,\mathrm{d}m \tag{11.26}$$

The choice of expression for A_z^h is defined by the conditions of excitation and relation between scalar and vector potentials. Substituting eqs. 11.25 and 11.26 into eq. 11.24 we obtain an equation for function $F_m(z)$:

$$\frac{\mathrm{d}^2 F_m}{\mathrm{d}z^2} - \lambda^2 m_n^2 F_m = \operatorname{sign}(z) C_1 (\lambda^2 - 1) m^2 \,\mathrm{e}^{-m_1|z|} \tag{11.27}$$

The solution of this equation has the form:

$$F_m = \operatorname{sign}(z) C_1 \left(\mathrm{e}^{-m_n|z|} - \mathrm{e}^{-m_t|z|}\right) \tag{11.28}$$

For $z = 0$ function $F(m)$ is continuous together with its first derivative. Thus:

$$A_z^h = C_1 \operatorname{sign}(z) \frac{y}{r} \int_0^\infty \left(\mathrm{e}^{-m_n|z|} - \mathrm{e}^{-m_t|z|}\right) J_1(mr) \,\mathrm{d}m \tag{11.29}$$

or

$$A_z^h = C_1 \frac{y}{r^2} \left(\frac{z}{R} \, e^{ik_t R} - \frac{\lambda z}{R_*} \, e^{ik_n R_*} \right) \tag{11.30}$$

Constant C_1 is found, as earlier, from the gauge condition and the behavior of the direct current field and we have:

$$C_1 = \frac{Ia}{4\pi}$$

Thus, for a horizontal electric dipole we have:

$$\boldsymbol{A}^h = (0, A_y^h, A_z^h)$$

$$A_y^h = \frac{Ia}{4\pi} \frac{e^{ik_t R}}{R} \tag{11.31}$$

$$A_z^h = \frac{Ia}{4\pi} \frac{y}{r^2} \left(\frac{z}{R} \, e^{ik_t R} - \frac{\lambda z}{R_*} \, e^{ik_n R_*} \right)$$

Now, we can find an expression for components A_y and A_z of the magnetic dipole, which present themselves as the result of summation of corresponding component of electric dipoles:

$$A_y = \lim_{\substack{a \to 0 \\ \pi a^2 I \to M}} \left(A_y^{(1)} + A_y^{(2)} + A_y^{(3)} + A_y^{(4)} \right) = -\frac{M}{4\pi} \frac{\partial}{\partial z} \left(\frac{e^{ik_t R}}{R} \right)$$

$$A_z = \lim_{\substack{a \to 0 \\ \pi a^2 I \to M}} \left(A_z^{(1)} + A_z^{(2)} + A_z^{(3)} + A_z^{(4)} \right) \tag{11.32}$$

$$= \frac{M}{4\pi} \lambda \frac{\partial}{\partial y} \left(\frac{e^{ik_t R}}{R} \right) - \frac{M}{4\pi} \, \mathrm{sign}(z) \, \frac{y}{r^2} \frac{\partial}{\partial z} \left(\frac{z}{R} e^{ik_t R} - \frac{\lambda z}{R_*} e^{ik_n R_*} \right)$$

In accord with eqs. 11.15 and 11.32 we have along the z-axis:

$$H_y = H_z = 0$$

$$h_x = \frac{H_x}{H_{0x}} = \left(1 - ik_t L - k_t^2 L^2 \frac{1 + \lambda^2}{2\lambda^2} \right) e^{ik_t L} \tag{11.33}$$

where $H_{0x} = -M/4\pi L^3$.

At the near zone, when $|kL| \ll 1$, we have:

$$h_x \simeq 1 - \frac{k_n^2 L^2}{2} + \left(\frac{1}{\lambda^2} + \frac{1}{6} \right) k_t^3 L^3$$

and:

$$\mathrm{In}\, h_x \simeq 1 + \left(\frac{1}{\lambda^2} + \frac{1}{3} \right) \left(\frac{L}{h_t} \right)^3$$

$$\mathrm{Q}\, h_x \simeq -\left(\frac{L}{h_n} \right)^2 + \left(\frac{1}{\lambda^2} + \frac{1}{3} \right) \left(\frac{L}{h_t} \right)^3 \tag{11.34}$$

where $h_t = (2/\sigma_t \mu\omega)^{1/2}$ and $h_n = (2/\sigma_n \mu\omega)^{1/2}$.

Thus, at the range of small parameters, L/h, the quadrature component of the field is directly proportional to the transversal conductivity σ_n. For this reason measuring the ratio of quadrature components of the vertical and horizontal dipoles at the low-frequency part of the spectrum allows us to define the coefficient of anisotropy:

$$\frac{\mathrm{Q}\, h_z^v}{\mathrm{Q}\, h_x^h} \simeq \lambda^2 \quad \text{as } \omega \to 0 \tag{11.35}$$

Inasmuch as $\lambda \geqslant 1$, the inphase component in anisotropic medium, in accord with eq. 11.34, is smaller than that in the isotropic medium with conductivity σ_t.

For large values of the anisotropy coefficient both components of the field become the same at the range of small parameters:

$$\mathrm{Q}\, h_x = \mathrm{In}\, h_x = \frac{1}{3}\left(\frac{L}{h_t}\right)^3 \quad \text{if } \lambda \gg 1 \tag{11.36}$$

In the wave zone the influence of anisotropy decreases with an increase of λ.

Values of $\mathrm{Q}\, h_x$, $\mathrm{In}\, h_x - 1$, $|h_x - 1|$ and ϕ are given in Table 11.1, and corresponding graphs are presented in Fig. 11.5.

Applying Fourier transform to eq. 11.33 we will find transient response of field h_x, when current in the dipole is turned off:

$$h_x = \phi(u) - \left(\frac{2}{\pi}\right)^{1/2}\left(1 + \frac{1+\lambda^2}{2\lambda^2}\right) u e^{-u^2/2} \tag{11.37}$$

where

$$\phi(u) = \left(\frac{2}{\pi}\right)^{1/2} \int\limits_0^u \mathrm{e}^{-t^2/2}\, \mathrm{d}t$$

is the probability integral, and $u = (\mu\sigma_t/2t)^{1/2}$.

Table 11.2 contains values of h_x as a function of λ and $1/u$. In the limited cases $t \to 0$ and $t \to \infty$ we obtain correspondingly:

$$\begin{aligned} h_x &\simeq \left(\frac{2}{\pi}\right)^{1/2}\left(\frac{1}{2} + \frac{1}{2\lambda^2}\right) u^3 \mathrm{e}^{-u^2/2} \qquad (u \to \infty,\ t \to 0) \\ h_x &\simeq -\left(\frac{2}{\pi}\right)^{1/2}\left(\frac{1}{6} + \frac{1}{2\lambda^2}\right) u^3 \qquad (u \to 0,\ t \to \infty) \end{aligned} \tag{11.38}$$

Therefore, at the late stage of the transient response for relatively small values of anisotropy coefficient the field is inversely proportional to λ^2.

TABLE 11.1
Values of functions L/h_1, $\ln h_x - 1$, $Q\,h_x$, $|h_x - 1|$ and ϕ

L/h_1	$\ln h_x - 1$	$Q\,h_x$	$\lvert h_x - 1\rvert$	ϕ
		$\lambda = 2$		
0.1	0.5112×10^{-3}	-0.1919×10^{-2}	0.1986×10^{-2}	-0.1310×10^{1}
	0.1366×10^{-2}	0.3365	0.3631	−0.1185
0.2	0.3557	0.5415	0.6478	−0.9896
	0.8908	0.7240	0.1148×10^{-1}	−0.6825
0.4	0.2099×10^{-1}	0.4994	0.2157	−0.2336
	0.4472	$+0.1360\times10^{-1}$	0.4674	+0.2952
0.8	0.7922	0.7931	0.1121	0.7860
	0.8991	0.2473	0.2632	0.1222×10^{1}
1.6	-0.4664×10^{-1}	0.5530	0.5550	0.1654
	−0.5225	0.8366	0.9864	0.2129
3.2	0.1209×10^{1}	0.6411	0.1368×10^{1}	0.2654
	0.1331	0.1826×10^{-2}	0.1332	0.3140
6.4	−0.9766	-0.9362×10^{-1}	0.9811	-0.3046×10^{1}
		$\lambda = 4$		
0.1	0.3418×10^{-3}	-0.2312×10^{-3}	0.4127×10^{-3}	−0.5948
	0.9070	0.1420	0.9180	0.1552
0.2	0.2377×10^{-2}	$+0.6034\times10^{-3}$	0.2414×10^{-2}	+0.2527
	0.5753	0.3615×10^{-2}	0.6795	0.5611
0.4	0.1316×10^{-1}	0.1353×10^{-1}	0.1877×10^{-1}	0.7992
	0.2645	0.4237	0.4995	0.1013×10^{1}
0.8	0.4054	0.1169	0.1237	0.1237
	0.1984	0.2803	0.2810	0.1500
1.6	−0.1435	0.5502	0.5686	0.1826
	0.5994	0.7729	0.9781	0.2230
3.2	0.1204×10^{1}	0.5630	0.1329×10^{1}	0.2704
	0.1291	-0.5901×10^{-2}	0.1291	−0.3137
6.4	−0.9781	0.8095×10^{-1}	0.9815	0.3059
		$\lambda = 6$		
0.1	0.3104×10^{-3}	0.8135×10^{-4}	0.3209×10^{-3}	0.2563
	0.8820	0.4548×10^{-3}	0.9395	0.5054
0.2	0.2111×10^{-2}	0.1717×10^{-2}	0.2822×10^{-2}	0.6831
	0.5169	0.5626	0.7640	0.8277
0.4	0.1171×10^{-1}	0.1696×10^{-1}	0.2061×10^{-1}	0.9665
	0.2307	0.4770	0.5298	0.1120×10^{1}
0.8	0.3337	0.1238	0.1282	0.1307
	0.6866×10^{-2}	0.2864	0.2865	0.1547
1.6	−0.1614	0.3496	0.5729	0.1856
	0.6137	0.7611	0.9777	0.2249
3.2	01203×10^{1}	0.5485	0.1323×10^{1}	0.2714
	0.1283	-0.7332×10^{-2}	0.1283	-0.3136×10^{1}
6.4	−0.9784	0.7860×10^{-1}	0.9815	0.3061

TABLE 11.1
(Continued)

L/h_1	$\text{In}\, h_x - 1$	$Q\,h_x$	$\lvert h_x - 1\rvert$	ϕ
		$\lambda = 8$		
0.1	0.2994×10^{-3}	0.1908×10^{-3}	0.3550×10^{-3}	0.5672
	0.7923	0.6638	0.1033×10^{-2}	0.6973
0.2	0.2032×10^{-2}	0.2108×10^{-2}	0.2928	0.8037
	0.4964	0.6329	0.8044	0.9057
0.4	0.1120×10^{1}	0.1816×10^{-1}	0.2133×10^{-1}	0.1018×10^{1}
	0.2188	0.4996	0.5418	0.1155
0.8	0.3087	0.1263	0.1300	0.1331
	0.2324	0.2886	0.2886	0.1563
1.6	-0.1674×10^{-1}	0.5495	0.5745	0.1867
	-0.6186	0.7569	0.9776	0.2256
3.2	0.1203×10^{1}	0.5434	0.1320×10^{1}	0.2717
	0.1281	-0.7833×10^{-2}	0.1280	-0.3135×10^{1}
6.4	-0.9785	0.9362×10^{-1}	0.9815	0.3062
		$\lambda = 10$		
0.1	0.2944×10^{-3}	0.2414×10^{-3}	0.3807×10^{-3}	0.6869
	0.7786	0.7604	0.1088×10^{-2}	0.7736
0.2	0.1995×10^{-2}	0.2288×10^{-2}	0.3036	0.8537
	0.4870	0.6650	0.8246	0.9391
0.4	0.1096×10^{-1}	0.1971×10^{-1}	0.2169×10^{-1}	0.1041×10^{1}
	0.2133	0.5042	0.5475	0.1170
0.8	0.2971	0.1274	0.1308	0.1341
	0.2224×10^{-3}	0.2895	0.2895	0.1570
1.6	-0.1706	0.5494	0.5752	0.1872
	0.6209	0.7750	0.9775	0.2259
3.2	0.1203×10^{1}	0.5410	0.1319×10^{1}	0.2719
	0.1279	-0.8067×10^{-2}	0.1279	-0.3135×10^{1}
6.4	-0.9785	-0.7740×10^{-1}	0.9815	0.3062

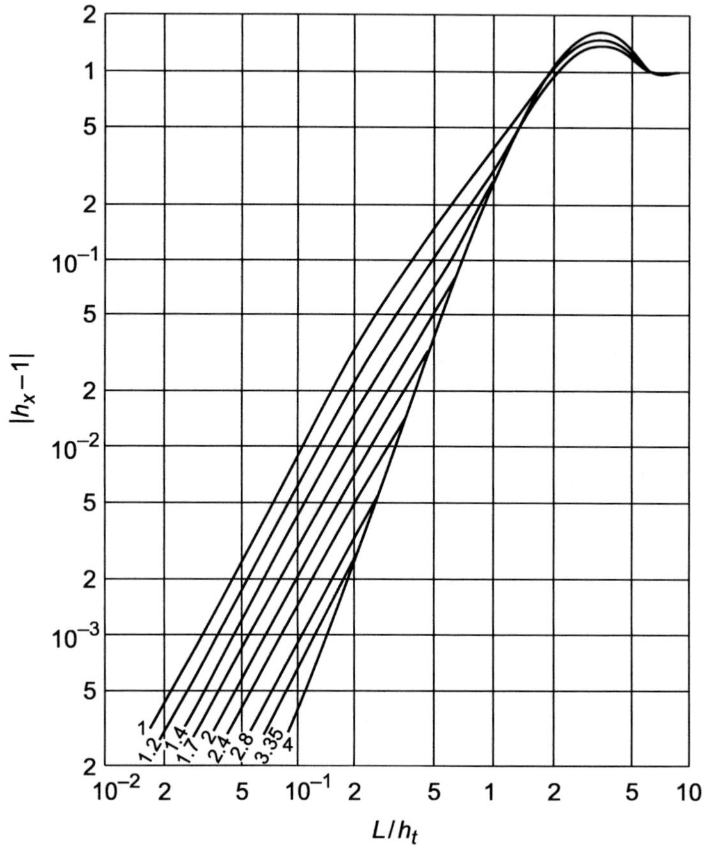

Figure 11.5. Amplitude of the secondary field in an anisotropic medium. Curve index λ.

TABLE 11.2
Values of field h_x

u^{-1} \ λ	1.0	1.2	1.4	1.6	1.8	2.0
0.1	−1.000	−1.000	−1.000	−1.000	−1.000	−1.000
	−1.000	−1.000	−1.000	−1.000	−1.000	−1.000
0.2	−1.000	−1.000	−1.000	−1.000	−1.000	−1.000
	−0.927	−0.937	−0.943	−0.472	−0.950	−0.952
0.4	−0.364	−0.448	−0.498	−0.531	−0.554	−0.570
	−0.232	$+0.908\times10^{-1}$	$+0.572\times10^{-2}$	-0.495×10^{-1}	-0.874×10^{-1}	−0.115
0.8	−0.247	$+0.138\times10^{-1}$	$+0.725\times10^{-1}$	$+0.298\times10^{-1}$	$+0.563\times10^{-3}$	-0.204×10^{-1}
	−0.613	$+0.433\times10^{-1}$	-0.300×10^{-1}	-0.523×10^{-1}	-0.678×10^{-1}	0.785
1.6	−0.528	-0.773×10^{-1}	-0.920×10^{-1}	−0.102	−0.108	−0.113
	0.875	0.953	−0.101	−0.105	−0.107	0.109
3.2	0.808	0.844	-0.865×10^{-1}	0.879×10^{-1}	-0.884×10^{-1}	-0.895×10^{-1}
	0.649	0.662	0.669	-0.674×10^{-1}	−0.678	0.680
	0.487	0.492	0.495	0.496	0.498	0.499
	0.355	0.356	0.357	0.358	0.358	0.359

11.3. Magnetic Field in an Anisotropic Medium with Two Horizontal Interfaces (A Formation of Finite Thickness)

Making use of results obtained in the previous section we will define the magnetic field in a formation with finite thickness, when the medium is anisotropic. The main axes of the tensor of conductivity in all three media coincide with coordinate lines. Equation of interfaces: $z = h_1$ and $z = -h_2$ (Fig. 11.6).

All quantities characterizing the formation and the surrounding medium are denoted by index (2) and (1) or (3), respectively. We will assume that $\sigma_{ik}^{(1)} = \sigma_{ik}^{(3)}$. In medium (2) the magnetic dipole is located at the origin of coordinates, and its moment is oriented along the x-axis. In accord with eq. 11.32 near the source the electromagnetic field can be described with the help of the vector potential of electric type $\boldsymbol{A}^{(0)}$, having two components: $A_y^{(0)}$ and $A_z^{(0)}$:

$$\boldsymbol{A} = (0, A_y^{(0)}, A_z^{(0)})$$

where:

$$A_y^{(0)} = -\frac{M}{4\pi}\frac{\partial}{\partial z}\left(\frac{e^{ik_{2t}R}}{R}\right) = \frac{M}{4\pi}\frac{\partial}{\partial z}\int_0^\infty \frac{m}{m_{2t}}\, e^{-m_{2t}|z|} J_0(mr)\, dm$$

$$A_z^{(0)} = -\frac{M}{4\pi}\lambda_2\frac{\partial}{\partial y}\left(\frac{e^{ik_{2n}R_*}}{R_*}\right) - \frac{M}{4\pi}\frac{y}{r^2}\frac{\partial}{\partial z}\left(\frac{z}{R}\, e^{ik_{2t}R} - \frac{\lambda_2 z}{R_*}\, e^{ik_{2n}R_*}\right) \tag{11.39}$$

$$= \frac{M}{4\pi}\frac{y}{r}\int_0^\infty \left(\frac{k_{2t}k_{2n}}{m_{2n}}\, e^{-m_{2n}\lambda_2|z|} + m_{2t}\, e^{ik_{2t}|z|}\right) J_1(mr)\, dm$$

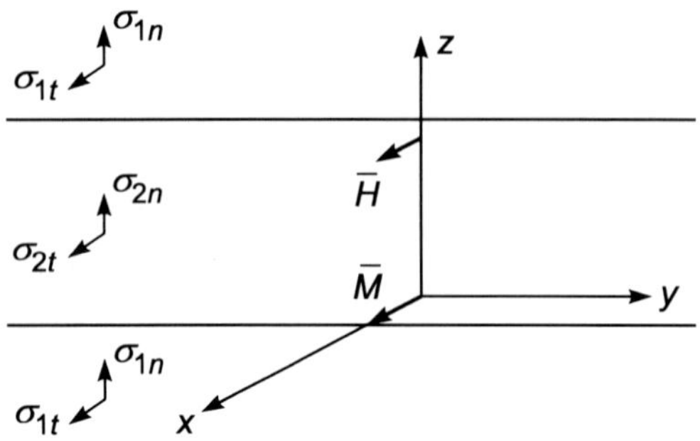

Figure 11.6. Model of anisotropic medium with two horizontal interfaces.

here:

$$k_{2t}^2 = \mathrm{i}\sigma_{2t}\omega\mu \qquad k_{2n}^2 = \mathrm{i}\sigma_{2n}\omega\mu \qquad m_{2n}^2 = (m^2 - k_{2n}^2)^{1/2}$$
$$\lambda^2 = \sigma_{2t}/\sigma_{2n} \qquad R_*^2 = x^2 + y^2 + \lambda_2^2 z^2$$

Potentials $A_y^{(0)}$ and $A_z^{(0)}$ satisfy the equations:

$$\begin{aligned} &(\nabla^2 + k_{2t}^2)A_y^{(0)} = 0 \\ &(\nabla^2 + k_{2n}^2)A_z^{(0)} = \left(1 - \frac{1}{\lambda_2^2}\right)\frac{\partial^2 A_y^{(0)}}{\partial z \partial y} \end{aligned} \tag{11.40}$$

here

$$\nabla^2 = \frac{\partial^2}{\partial x^2} + \frac{\partial^2}{\partial y^2} + \frac{1}{\lambda_2^2}\frac{\partial^2}{\partial z^2}$$

For this reason potentials in a layered medium can be presented as:

$$\begin{aligned} A_{1y} &= \frac{M}{4\pi}\int_0^\infty D_1 J_0(mr)\,\mathrm{e}^{m_{1t}z}\,dm \\ A_{2y} &= A_y^{(0)} + \frac{M}{4\pi}\int_0^\infty (D_2\,\mathrm{e}^{m_{2t}z} + D_3\,\mathrm{e}^{-m_{2t}z})J_0(mr)\,dm \\ A_{3y} &= \frac{M}{4\pi}\int_0^\infty D_4\,\mathrm{e}^{-m_{1t}z}J_0(mr)\,dm \end{aligned} \tag{11.41}$$

and

$$(\nabla^2 + k_{it}^2)A_{iy} = 0 \qquad i = 1, 2, 3$$

and

$$A_{1z} = \frac{y}{r}\frac{M}{4\pi}\int_0^\infty F_1(z)J_1(mr)\,\mathrm{d}m$$

$$A_{2z} = A^{(0)} + \frac{y}{r}\frac{M}{4\pi}\int_0^\infty F_2(z)J_1(mr)\,\mathrm{d}m \tag{11.42}$$

$$A_{3z} = \frac{y}{r}\frac{M}{4\pi}\int_0^\infty F_3(z)J_1(mr)\,\mathrm{d}m$$

Making use of eqs. 11.40–11.42 we will obtain equations for the determination of function $F_i(z)$:

$$\frac{\mathrm{d}^2F_1(z)}{\mathrm{d}z^2} - \lambda_1^2 m_{1n}^2 F_1(z) = -mm_{1t}(\lambda_1^2-1)D_1\,\mathrm{e}^{m_{1t}z}$$

$$\frac{\mathrm{d}^2F_2(z)}{\mathrm{d}z^2} - \lambda_2^2 m_{2n}^2 F_2(z) = -mm_{2t}(\lambda_2^2-1)(D_2\,\mathrm{e}^{m_{2t}z} - D_3\,\mathrm{e}^{-m_{2t}z}) \tag{11.43}$$

$$\frac{\mathrm{d}^2F_3(z)}{\mathrm{d}z^2} - \lambda_1^2 m_{1n}^2 F_3(z) = mm_{1t}(\lambda_1^2-1)D_4\,\mathrm{e}^{-m_{1t}z}$$

The solution of equations 11.43, taking into account the behavior of the field at infinity, is:

$$F_1(z) = A_1\,\mathrm{e}^{\lambda_1 m_{1n}z} + \frac{m_{1t}}{m}D_1\,\mathrm{e}^{m_{1t}z}$$

$$F_2(z) = A_2\,\mathrm{e}^{\lambda_2 m_{2n}z} + B_2\,\mathrm{e}^{-\lambda_2 m_{2n}z} + \frac{m_{2t}}{m}(D_2\,\mathrm{e}^{m_{2t}z} - D_3\,\mathrm{e}^{-m_{2t}z}) \tag{11.44}$$

$$F_3(z) = B_3\,\mathrm{e}^{-\lambda_2 m_{1n}z} - \frac{m_{1t}}{m}D_4\,\mathrm{e}^{-m_{1t}z}$$

Substituting eq. 11.44 into eq. 11.42 we have:

$$A_{1z} = \frac{M}{4\pi}\frac{y}{r}\int_0^\infty \left(A_1\,\mathrm{e}^{\lambda_1 m_{1n}z} + \frac{m_{1t}}{m}D_1\,\mathrm{e}^{m_{1t}z}\right)J_1(mr)\,\mathrm{d}m$$

$$A_{2z} = A_z^{(0)} + \frac{M}{4\pi}\frac{y}{r}\int_0^\infty \left(A_2\,\mathrm{e}^{\lambda_2 m_{2n}z} + B_2\,\mathrm{e}^{-\lambda_2 m_{2n}z} + \frac{m_{2t}}{m}(D_2\,\mathrm{e}^{m_{2t}z} - D_3\,\mathrm{e}^{-m_{2t}z})\right)J_1(mr)\,\mathrm{d}m \tag{11.45}$$

$$A_{3z} = \frac{M}{4\pi}\frac{y}{r}\int_0^\infty \left(B_3\,\mathrm{e}^{-\lambda_1 m_{1n}z} - \frac{m_{1t}}{m}D_4\,\mathrm{e}^{-m_{1t}z}\right)J_1(mr)\,\mathrm{d}m$$

For the determination of the unknown coefficients D_1, D_2, D_3 and D_4, A_1, A_2, B_2 and B_3 we will make use of the boundary conditions at $z=h_1$ and $z=-h_2$. Continuity of tangential components of electric and magnetic fields results in the following relations for A_y:

$$\begin{aligned} A_{1y} &= A_{2y} & A_{2y} &= A_{3y} \\ \frac{\partial A_{1y}}{\partial z} &= \frac{\partial A_{2y}}{\partial z} & \frac{\partial A_{2y}}{\partial z} &= \frac{\partial A_{3y}}{\partial z} \\ &\text{if } z = -h_2 & &\text{if } z = h_1 \end{aligned} \tag{11.46}$$

and more complicated relations for A_z:

$$\begin{aligned} A_{1z} &= A_{2z} & A_{2z} &= A_{3z} \\ \frac{1}{\sigma_{1t}} \operatorname{div} \boldsymbol{A}_1 &= \frac{1}{\sigma_{2t}} \operatorname{div} \boldsymbol{A}_2 & \frac{1}{\sigma_{2t}} \operatorname{div} \boldsymbol{A}_2 &= \frac{1}{\sigma_{3t}} \operatorname{div} \boldsymbol{A}_3 \\ &\text{if } z = -h_2 & &\text{if } z = h_1 \end{aligned} \tag{11.47}$$

Substituting eq. 11.41 into eq. 11.46 we obtain a system of equations for coefficients D_i:

$$\begin{aligned} D_1 \, \mathrm{e}^{-m_{1t}h_2} &= -m \, \mathrm{e}^{-m_{2t}h_2} + D_2 \, \mathrm{e}^{-m_{2t}h_2} + D_3 \, \mathrm{e}^{m_{2t}h_2} \\ m_{1t} D_1 \, \mathrm{e}^{-m_{1t}h_2} &= -m m_{2t} \, \mathrm{e}^{-m_{2t}h_2} + m_{2t} D_2 \, \mathrm{e}^{-m_{2t}h_2} - D_3 \, \mathrm{e}^{m_{2t}h_2} \\ D_4 \, \mathrm{e}^{-m_{1t}h_1} &= m \, \mathrm{e}^{-m_{2t}h_1} + D_2 \, \mathrm{e}^{m_{2t}h_1} + D_3 \, \mathrm{e}^{-m_{2t}h_1} \\ m_{1t} D_4 \, \mathrm{e}^{-m_{1t}h_1} &= m m_{2t} \, \mathrm{e}^{-m_{2t}h_1} - m_{2t} D_2 \, \mathrm{e}^{m_{2t}h_1} + D_3 \, \mathrm{e}^{-m_{2t}h_1} \end{aligned} \tag{11.48}$$

Solving the system, we find:

$$\begin{aligned} D_2 &= -m l_{12} \, \mathrm{e}^{-2m_{2t}h_1} \frac{1 + l_{12} \, \mathrm{e}^{-2m_{2t}h_2}}{1 - l_{12} \, \mathrm{e}^{-2m_{2t}H}} \\ D_3 &= m l_{12} \, \mathrm{e}^{-2m_{2t}h_2} \frac{1 + l_{12} \, \mathrm{e}^{-2m_{2t}h_1}}{1 - l_{12}^2 \, \mathrm{e}^{-2m_{2t}H}} \end{aligned} \tag{11.49}$$

where $l_{12} = (m_{1t} - m_{2t})/(m_{1t} + m_{2t})$.

Now we will define coefficients A_l and B_l. Inasmuch as at $z = -h_2$ component A_y is continuous and correspondingly $\partial A_{1y}/\partial y = \partial A_{2y}/\partial y$, condition 11.47 can be written as:

$$\begin{aligned} &A_{1z} = A_{2z} \\ &S_t \frac{\partial A_{1z}}{\partial z} - \frac{\partial A_{2z}}{\partial z} = (1 - S_t) \frac{\partial A_{2y}}{\partial y} \qquad \text{as } z = -h_2 \end{aligned} \tag{11.50}$$

where $S_t = \sigma_{2t}/\sigma_{1t}$.

By analogy, at $z = h_1$ we have:

$$\begin{aligned} &A_{2z} = A_{3z} \\ &S_t \frac{\partial A_{3z}}{\partial z} - \frac{\partial A_{2z}}{\partial z} = (1 - S_t) \frac{\partial A_{2y}}{\partial y} \end{aligned} \tag{11.51}$$

Substituting eq. 13.45 in eqs. 13.50 and 13.51, we obtain the system of equations for coefficients A_1, A_2, B_2 and B_3:

$$
\begin{aligned}
A_1\,e^{-\lambda_1 m_{1n} h_2} + \frac{m_{1t}}{m} D_1\,e^{-m_{1t}h_2} &= -\frac{k_{2t}k_{2n}}{m_{2n}}\,e^{-\lambda_2 m_{2n} h_2} - m_{2t}\,e^{-m_{2t}h_2} \\
&\quad + A_2\,e^{-\lambda_2 m_{2n} h_2} + B_2\,e^{\lambda_2 m_{2n} h_2} + \frac{m_{2t}}{m}(D_2\,e^{-m_{2t}h_2} - D_3\,e^{m_{2t}h_2})
\end{aligned}
$$

$$
\begin{aligned}
&S_t\left(\lambda_1 m_{1n} A_1\,e^{-\lambda_1 m_{1n} h_2} + \frac{m_{1t}^2}{m} D_1\,e^{-m_{1t}h_2}\right) + k_{2t}^2\,e^{-m_{2n}\lambda_2 h_2} + m_{2t}^2\,e^{-m_{2t}h_2} \\
&- \lambda_2 m_{2n} A_2\,e^{-\lambda_2 m_{2n} h_2} + \lambda_2 m_{2n} B_2\,e^{\lambda_2 m_{2n} h_2} - \frac{m_{2t}^2}{m}(D_2\,e^{-m_{2t}h_2} + D_3\,e^{m_{2t}h_2}) \\
&\qquad = -m(1 - S_t) D_1\,e^{-m_{1t}h_2}
\end{aligned}
$$

$$
\begin{aligned}
B_3\,e^{-\lambda_1 m_{1n} h_1} - \frac{m_{1t}}{m} D_4\,e^{-m_{1t}h_1} &= -\frac{k_{2t}k_{2n}}{m_{2n}}\,e^{-\lambda_2 m_{2n} h_1} - m_{2t}\,e^{-m_{2t}h_1} \\
&\quad + A_2\,e^{\lambda_2 m_{2n} h_1} + B_2\,e^{-\lambda_2 m_{2n} h_1} + \frac{m_{2t}}{m}(D_2\,e^{m_{2t}h_1} - D_3\,e^{-m_{2t}h_1})
\end{aligned}
$$

$$
\begin{aligned}
&S_t\left(-\lambda_1 m_{1n} B_3\,e^{-\lambda_1 m_{1n} h_1} + \frac{m_{1t}^2}{m} D_4\,e^{-m_{1t}h_1}\right) - k_{2t}^2\,e^{-m_{2n}\lambda_2 h_1} - m_{2t}^2\,e^{-m_{2t}h_1} \\
&- \lambda_2 m_{2n} A_2\,e^{\lambda_2 m_{2n} h_1} + \lambda_2 m_{2n} B_2\,e^{-\lambda_2 m_{2n} h_1} - \frac{m_{2t}^2}{m}(D_2\,e^{m_{2t}h_1} + D_3\,e^{-m_{2t}h_1}) \\
&\qquad = -m(1 - S_t) D_4\,e^{-m_{1t}h_1} \qquad (11.52)
\end{aligned}
$$

Making use of relations 11.48, establishing connections between coefficients D_1, D_2, D_3 and D_4, it is not difficult to reduce system 11.52 to the form:

$$
\begin{aligned}
&e^{-\lambda_1 m_{1n} h_2} A_1 - e^{\lambda_2 m_{2n} h_2} A_2 - e^{-\lambda_2 m_{2n} h_2} B_2 = \frac{k_{2t}k_{2n}}{m_{2n}} e^{-\lambda_2 m_{2n} h_2} \\
&S_t\lambda_1 \frac{m_{1n}}{m}\,e^{-\lambda_1 m_{1n} h_2} A_1 - \lambda_2 \frac{m_{2n}}{m}\,e^{-\lambda_2 m_{2n} h_2} A_2 + \frac{\lambda_2 m_{2n}}{m}\,e^{-\lambda_2 m_{2n} h_2} = -\frac{k_{2t}^2}{m}\,e^{-\lambda_2 m_{2n} h_2} \\
&e^{-\lambda_1 m_{1n} h_1} B_3 - e^{-\lambda_2 m_{2n} h_1} A_2 - e^{-\lambda_2 m_{2n} h_1} B_2 = -\frac{k_{2t}k_{2n}}{m_{2n}}\,e^{-\lambda_2 m_{2n} h_1} \\
&S_t\lambda_1 \frac{m_{1n}}{m}\,e^{-\lambda_1 m_{2n} h_1} B_3 + \lambda_2 \frac{m_{2n}}{m}\,e^{\lambda_2 m_{2n} h_1} A_2 - \frac{\lambda_2 m_{2n}}{m}\,e^{-\lambda_2 m_{2n} h_1} B_2 = -\frac{k_{2t}^2}{m}\,e^{-\lambda_2 m_{2n} h_1}
\end{aligned}
\qquad (11.53)
$$

From this system we find:

$$
\begin{aligned}
A_2 &= \frac{k_{2t}k_{2n}}{m_{2n}}\,\bar{l}\,\frac{e^{-2\lambda_2 m_{2n} h_1}}{1 - \bar{l}^2\,e^{-2\lambda_2 m_{2n} H}}\left(1 - \bar{l}e^{-2\lambda_1 m_{2n} h_2}\right) \\
B_2 &= \frac{k_{2t}k_{2n}}{m_{2n}}\,\bar{l}\,\frac{e^{-2\lambda_2 m_{2n} h_2}}{1 - \bar{l}^2\,e^{-2\lambda_2 m_{2n} H}}\left(1 - \bar{l}e^{-2\lambda_2 m_{2n} h_1}\right)
\end{aligned}
\qquad (11.54)
$$

where $\bar{l} = (S_t\lambda_1 m_{1n} - \lambda_2 m_{2n})/(S_t\lambda_1 m_{1n} + \lambda_2 m_{2n})$.

TABLE 11.3
Values of amplitude of the secondary magnetic field; $L/h_{2t} = 0.1,\ H/L = 2$

σ_{2t}/σ_{1t} \ λ_2	1.2	2.0	4.0
1/8	0.233×10^{-1}	0.158×10^{-1}	0.126×10^{-1}
1/2	0.103×10^{-1}	0.516×10^{-2}	0.265×10^{-2}

Expression for horizontal component of magnetic field H_x on z-axis within the formation has the form:

$$H_x = \frac{\partial A_z}{\partial y} - \frac{\partial A_y}{\partial z}$$
$$= H_x^{un} + \frac{M}{4\pi}\int_0^\infty \left(\frac{m_{2t}}{2}\left(D_2\,\mathrm{e}^{m_{2t}z} - D_3\,\mathrm{e}^{-m_{2t}z}\right) - \frac{m}{2}\left(A_2\,\mathrm{e}^{\lambda_2 m_{2n}z} + B_2\,\mathrm{e}^{-\lambda_2 m_{2n}z}\right)\right)\,dm \tag{11.55}$$

if $-h_2 \leqslant z \leqslant h_1$.

Results of calculations based on this equation are presented in Figs. 11.7–11.10. They are performed for various values of coefficient of anisotropy and parameters $S = \sigma_{2t}/\sigma_{1t}$, $\alpha = H/L$ for symmetrical position of the probe with respect to the formation boundaries.

Let us consider a field when the surrounding medium is isotropic ($\lambda_1 = 1$). Consider again frequency responses of the amplitude of the secondary field in a uniform isotropic medium (Fig. 11.5).

As is seen from the curves, the influence of the coefficient of anisotropy on the amplitude responses is significant at the range of relatively low frequencies $L/h_t < 0.5$, when parameter λ does not exceed two. The expression for the quadrature component permits us to find an approximate relation between maximal values of L/h_t and λ, when it is still possible to differentiate curves by λ, and it has the form:

$$\frac{1}{\lambda^2} \geqslant \frac{1}{3}\frac{L}{h_t} \quad \text{or} \quad \frac{L}{h_t} \leqslant \frac{3}{\lambda^2}$$

First consider the behavior of the field when the formation is more resistive than the surrounding medium (Figs. 11.7–11.8). In measuring the field in a relatively resistive formation of a small thickness the influence of a change in the conductivity of the surrounding medium exceeds the influence due to a change of anisotropy coefficient of the formation.

It is natural that with an increase of the formation thickness the influence of parameter λ_2 becomes stronger, specially it is noticeable within the intermediate part of the frequency response, the low boundary of which corresponds to the frequency, when due to the skin effect the influence of the surrounding medium is small. It is obvious that in measuring the anisotropy coefficient in formations of a small thickness, it is reasonable to increase significantly the frequency.

With an increase of the resistivity of the surrounding medium the influence of anisotropy increases, and it becomes possible measuring λ_2 even in formations with relatively small thickness.

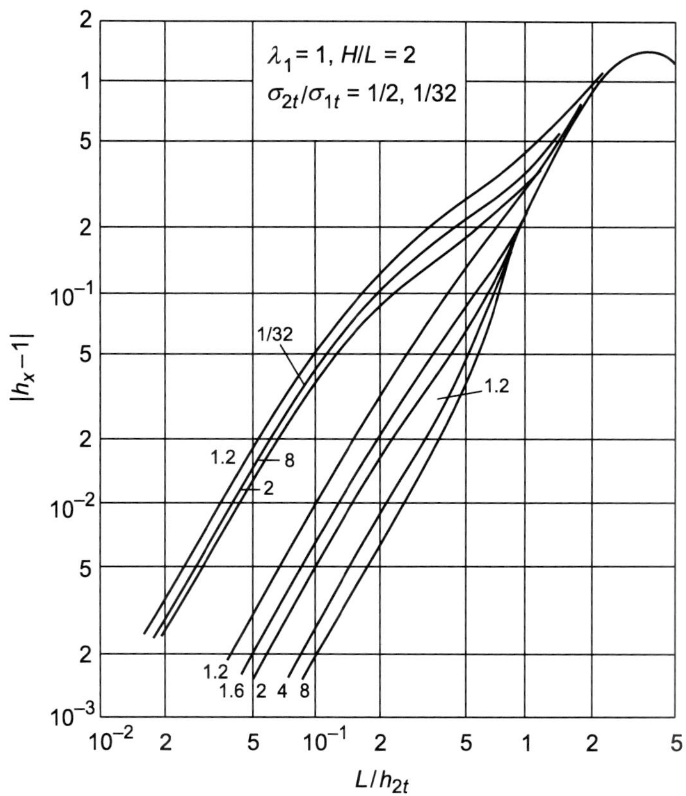

Figure 11.7. Amplitude of the secondary field. Curve index λ_2.

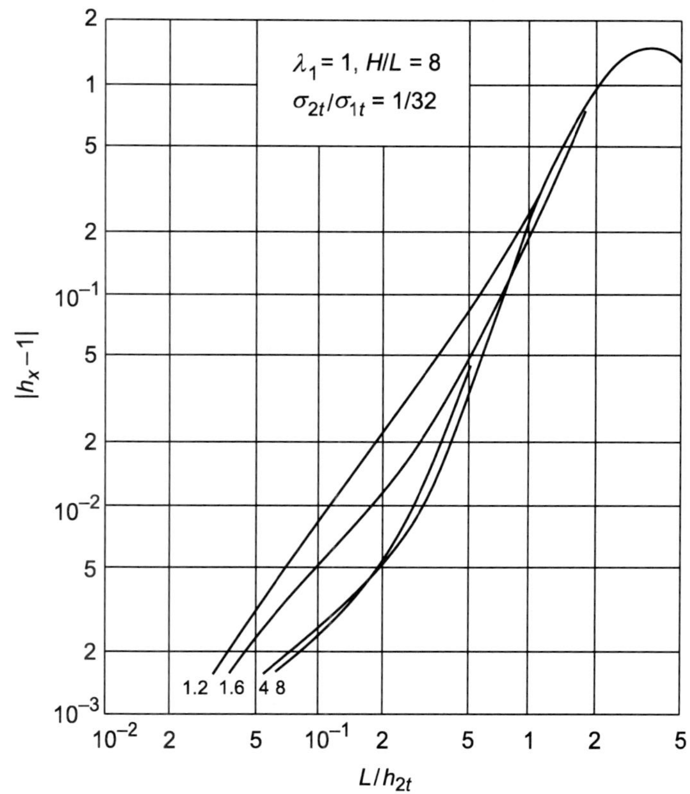

Figure 11.8. Amplitude of the secondary field. Curve index λ_2.

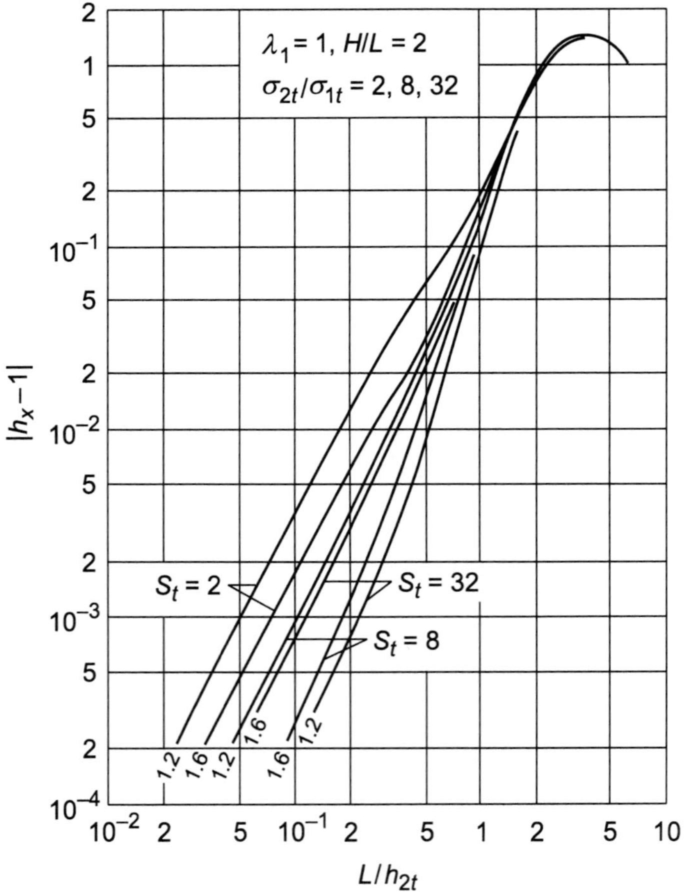

Figure 11.9. Amplitude of the secondary field. Curve index λ_2.

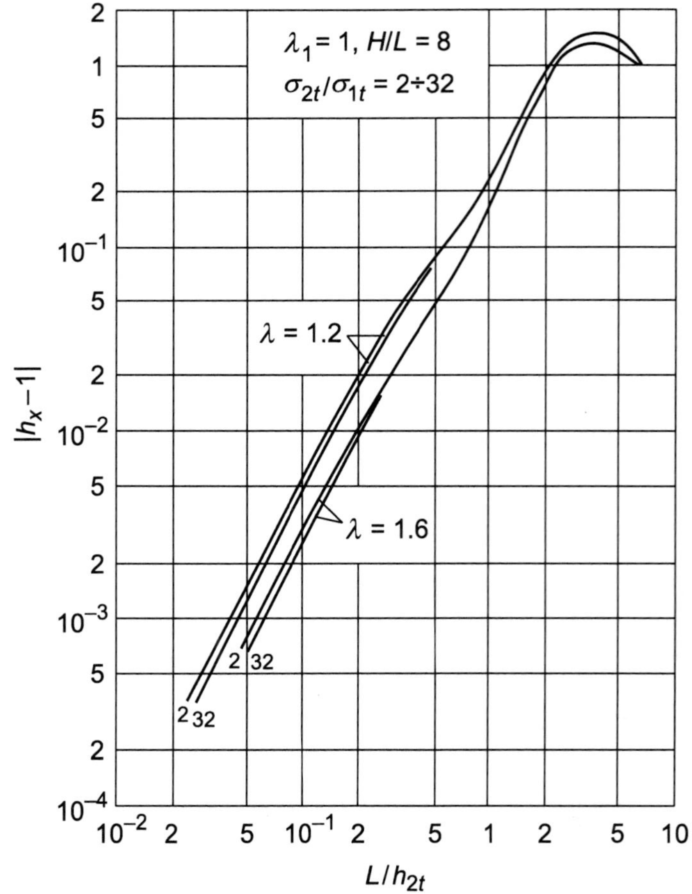

Figure 11.10. Amplitude of the secondary field. Curve index λ_2.

For illustration values of amplitude of the secondary field for various σ_{2t}/σ_1 and λ_2 for fixed values of L/h_{2t} and H/L are given in Table 11.3.

In conducting formations ($\sigma_{2t}/\sigma_{1t} > 10$) the magnetic field coincides with that of a uniform anisotropic medium, if $H/L > 2$.

Chapter 12

MATHEMATICAL MODELING OF THE RESPONSE OF INDUCTION LOGGING TOOLS IN 3D GEOMETRIES

The rapid expansion of horizontal drilling has accelerated the development of numerical methods to calculate the responses of 3D induction logging models that include anisotropic bedding and deviated boreholes with invasion. Indeed, in deviated well with noncylindrical invasion caused by gravity segregation, it is practically impossible to interpret induction responses without 3D modeling.

The most common numerical methods used in 3D induction logging modeling are method of integral equations and finite element and finite difference schemes. During the last years the most successful use of the finite difference method in resistivity logging has been 3D modeling of induction response, using the spectral Lancsoz decomposition method of V. L. Druskin and L. A. Knizhnerman. Their outstanding results make it feasible to routinely use 3D modeling for model-based interpretation, a breakthrough in induction logging.

The text of this chapter is based on numerous papers of the above mentioned authors and their colleagues.

Let us consider the frequency-domain problem for Maxwell's equations:

$$\operatorname{curl} \boldsymbol{E} + \mathrm{i}\omega\mu\boldsymbol{H} = 0 \tag{12.1}$$

$$\operatorname{curl} \boldsymbol{H} - \sigma\boldsymbol{E} = -\boldsymbol{J} \tag{12.2}$$

Here, $\boldsymbol{E}$ and $\boldsymbol{H}$ are electric field vector and magnetic field vector, respectively. Vector $\boldsymbol{J}$ is the external current source vector. The symbols $\sigma = \sigma(x, y, z)$ and μ stand for conductivity function and magnetic permeability constant, respectively; μ is assumed to be constant ($\mu = 1$). For simplicity only electric sources are considered and displacement currents are assumed to be negligible.

Applying the operator curl to both sides of eq. 12.1 and substituting eq. 12.2 to eq. 12.1, we obtain the equation for the electric field $\boldsymbol{E}$:

$$\sigma^{-1} \operatorname{curl} \operatorname{curl} \boldsymbol{E} + \mathrm{i}\omega\boldsymbol{E} = \mathrm{i}\omega\phi \tag{12.3}$$

The operator A and the source function ϕ are defined in a following manner:

$$A = \sigma^{-1} \operatorname{curl} \operatorname{curl} \tag{12.4}$$

$$\phi = \sigma^{-1}\boldsymbol{J} \tag{12.5}$$

Then eq. 12.3 becomes

$$(A + \mathrm{i}\omega I)\boldsymbol{E} = \mathrm{i}\omega\phi \tag{12.6}$$

Here I is a unit matrix. The solution of eq. 12.6 is in the form of the matrix equation:

$$\boldsymbol{E} = (A + \mathrm{i}\omega I)^{-1}\mathrm{i}\omega\phi \tag{12.7}$$

It should be mentioned that operator A in eq. 12.4 is the finite-difference approximation of σ^{-1} curl curl. Otherwise A would be unbounded and its Krylov subspace may even not exist in regular functional spaces.

In reality an unbounded problem is considered with electromagnetic field vanishing in infinity, but for computational purposes the bounded domain Ω is assumed:

$$\Omega = \{(x, y, z) : x_{min} \leqslant x \leqslant x_{max}, y_{min} \leqslant y \leqslant y_{max}, z_{min} \leqslant z \leqslant z_{max}\} \tag{12.8}$$

At the domain boundary the following boundary condition is imposed:

$$\boldsymbol{E} \times \boldsymbol{n}|_{\partial\Omega} = 0 \tag{12.9}$$

where $\boldsymbol{n}$ is a unit normal vector directed outwards from the domain surface.

The great success in solving boundary value problem of eqs. 12.7 and 12.9 was obtained by using the spectral Lanczos decomposition method (SLDM). SLDM was first introduced by Druskin and Knizhnerman to solve Maxwell's equations in both frequency and time domain (Druskin and Knizhnerman, 1988, 1994).

The SLDM algorithm can be summarized as follows. The numerical approximation of eq. 12.7 is performed by applying the Lanczos method generating the eigenvectors of matrix A. Then the solution is represented as a projection into the Krylov subspace $K^m(A, \phi)$:

$$K^m(A, \phi) = \mathrm{span}\{\phi, A\phi, \ldots, A^{m-1}\phi\} \tag{12.10}$$

The Gram–Schmidt orthogonalization of the frequency independent vectors ϕ, $A\phi$ produces the orthonormal basis $\{q_1, \ldots, q_m\}$ by the Lanczos process so that

$$K^m(A, \phi) = \mathrm{span}\{q_1, \ldots, q_m\} \tag{12.11}$$

If we denote

$$Q = (q_1, \ldots, q_m) \tag{12.12}$$

the Lanczos process can be summarized by

$$AQ = QT \tag{12.13}$$

where T is the tridiagonal matrix. It is appropriate to note that the Gram–Schmidt orthogonalization in the Lanczos method is performed with only three-term recursion, i.e., the cost is linear with respect to m (otherwise it would be quadratic).

As a result the approximate solution for the electric field $\boldsymbol{E}$ at the nodal points is then obtained by solving the system:

$$\boldsymbol{E} = ||\phi||\, Q(T + \mathrm{i}\omega I)^{-1}\mathrm{i}\omega e_1 \tag{12.14}$$

where $e_1 = (1, 0, \ldots, 0)^T$.

It should be underlined that SLDM calculates a basis in K^m just once and then utilizes it to produce solutions for all values of frequency ω using the same projection principle. To say it in other words, SLDM can compute solutions for a multiple number of frequencies at almost the computational cost of a single frequency.

One comment should be made. The Lanczos method was suggested by Lanczos in 1950's. About 20 years this approach was kept in the background because of its well-known instability. In the 1970's Paige obtained important results clarifying behavior of the simple Lanczos process in computer arithmetic. In 1990's Druskin and Knizhnerman managed to obtain for computer arithmetic the following important result: the Lanczos process is unstable by itself, but error bounds remain stable with respect to round-off.

The practical implementation of the algorithm shows that the convergence of SLDM depends on the conductivity contrast and frequency: the convergence slows at high contrasts and low frequency. To clarify this feature one can consider the eq. 12.7. The function $(A + \mathrm{i}\omega I)^{-1}$ can be formally written as

$$(A + \mathrm{i}\omega I)^{-1} = \sum_{k=0}^{\infty}(-1)^k(\mathrm{i}\omega)^{-(k+1)}A^k \tag{12.15}$$

It can be seen that the convergence of the truncated series depends on the value of ω: the convergence rate is slower when frequency ω is smaller. To overcome this problem the authors of SLDM proposed a new approach for solutions to Maxwell's equations — a preconditioned modification of SLDM for low frequencies. This new method is based on the standard SLDM but with Krylov subspaces generated from the inverse powers of the Maxwell operator. In Druskin et al. (1999) this method is referred to as spectral Lanczos decomposition method using inverse powers, or SLDMINV.

To clarify the essence of SLDMINV one can consider the function $(A+\mathrm{i}\omega I)^{-1}$ in eq. 12.7. It can be rewritten as:

$$(A + \mathrm{i}\omega I)^{-1} = -\mathrm{i}\omega^{-1}(A^{-1} - \mathrm{i}\omega^{-1}I)^{-1}A^{-1} \tag{12.16}$$

From the formal point of view:

$$(A^{-1} - \mathrm{i}\omega^{-1}I)^{-1} = \sum_{k=0}^{\infty}(-1)^k(\mathrm{i}\omega)^{k+1}A^{-k} \tag{12.17}$$

Comparing eqs. 12.16 and 12.17 one can conclude that the truncated series in eq. 12.17 should converge faster than that of eq. 12.15 when the frequency ω is small. This formal fact suggests that a faster convergence rate can be obtained for small values of ω if we use SLDMINV, or SLDM with Krylov subspace $K^m(A^{-1}, \phi)$:

$$K^m(A^{-1}, \phi) = \text{span}\{\phi, A^{-1}\phi, \ldots, A^{-(m-1)}\phi\} \tag{12.18}$$

In this case the solution of eq. 12.5 can be written in a following manner:

$$\boldsymbol{E} = (A^{-1} - \mathrm{i}\omega^{-1}I)^{-1}A^{-1}\phi \tag{12.19}$$

This new approach is particularly effective in lower frequency ranges. SLDMINV has a significantly faster convergence rate than that of standard SLDM yet retains the advantages of SLDM, such as the ability to solve for multiple frequencies in a single simulation run, matrix operations in real arithmetic, and the ability to eliminate numerical spurious modes. The new solution technique is applied to model induction logging, giving rise to almost two orders of magnitude convergence improvement over the conventional Krylov subspace approach.

Some comments should be made.

1. The staggered grid modeling approach (Yee, 1966) has been successfully applied to solve problems of calculating electromagnetic field in arbitrary 3D isotropic media. The using of staggered grid is very attractive in the mathematical modeling of electromagnetic fields, since these grids naturally preserve the solenoidal nature of magnetic field vector $\boldsymbol{B}$ and current density vector $\boldsymbol{j}$ and, in the case of direct current, the irrotationality of vector $\boldsymbol{E}$. However, developing this approach for practically important 3D anisotropic models with arbitrary tensors of electrical conductivity, magnetic and dielectric permeability happens to be very complicated. Yee's algorithm is based on calculation of different electric field components at different space points, but the electrical conductivity tensor relates these components taken at the same point.

To avoid this difficulty, another staggered grid was used by the authors of SLDM. It is based on a general approach, suggested by Lebedev (Lebedev, 1964). This grid locates all components of the electrical field at the same points, however vectors of magnetic and electric fields are at different points. Similar to Yee's grid, in this approach differential analogs of the operators div curl and curl grad become identically equal to zero on the grid, by analogy with their continuum counterparts. It is an important feature for a good field approximation if every discrete Maxwell's equation has all conservation properties of its continuous counterpart (Davydycheva and Druskin, 1995; Moskow et al., 1999).

It should be noted that SLDM is related to other Krylov methods (Greenbaum, 1997) such as BiConjugate Gradient (BCG) method (or its modifications such as QMR), also used in induction logging simulations. But compared to the latter it (SLDM) is in real arithmetic and computes for multiple frequencies at no cost.

2. Mathematical modeling of electromagnetic fields in 3D environment include problems when the medium contains sharp discontinuities, e. g., thin layers with high contrasts, earth formations with cross-bedding structures, etc. Sharp discontinuities can often introduce errors within the standard finite-difference approach, unless the interfaces are gridded in details which may require grids of unrealistically large sizes. Traditionally such problems are circumvented with finite difference methods with conformal grids. In 3D

anisotropic medium with anisotropy tensor arbitrarily oriented in space, such approaches become difficult to implement.

An averaging formula used in SLDMINV (called the nodal averaging formula) is based on discrete energy conservation principle. Authors used an equivalent medium approach that allows homogenization of the medium enclosed inside a grid cell. This averaging allows for the change in conductivity to be in any direction with respect to the grid.

It should be emphasized that averaging technique does not require the grid to be small compared to the layering. These results can be considered as an outgrowth of the integro-interpolation finite-difference schemes. This term first appeared in the Russian literature in the papers of Tikhonov and Samarsky in the 1960's (in Western literature the closest related technique is called the external approximation). However, only conformal variations of conductivity were treated in the classical integro-interpolation method.

3. In the mathematical modeling of the induction logging problems, it is assumed very often that the sources and receivers are located at the borehole axis. Generally speaking one needs to compute tool's readings only along this line and is not interested in accurate solutions elsewhere. It is evident that the grid should be refined toward the source-receiver locations, but until recently it was not known how to optimize such refinements.

It was shown some years ago (Druskin and Knizhnerman, 1999) that a proper grid refinement (so-called optimal grids) can make second order finite difference schemes exponentially convergent. They use only asymptotical spectral properties of the Green functions and in many cases it can be done a priory independently of the conductivity model. This can be achieved by using the asymptotically-optimal grids outside the region where the transmitters and receivers are situated.

To say it in other words, without loosing accuracy of the finite difference scheme, the number of nodes of the optimal grid along each direction can be reduced to just a few ones, if the steps are arranged in a specific way. It can be considered as an extension of the concept of the Gaussian quadrature rule for the numerical integration to the finite differences.

The authors of SLDMINV performed numerical experiments to model the response of AIT, Schlumberger's array induction tool, which operates at three frequencies (26 kHz, 52 kHz, and 105 kHz). AIT tools use specially designed processing methods to combine several measurements in such a way as to create a log focused at a designated region of the formation. A total of five logs are generated. These logs have median depth of investigation of 10, 20, 30, 60, and 90 inches.

It was demonstrated that the convergence of SLDMINV is two orders of magnitude faster than the standard SLDM. Therefore it is possible to simulate multi-frequency induction log at about 20 sec per log point for complex 3D environment on a SGI workstation with a MIPS R10000 processor.

Now let us consider the results of 3D mathematical modeling obtained with SLDM code. These data were published in Anderson et al. (1999). In this paper the authors investigate actual 3D logging situations for the AIT family of tools:

- invaded dipping-bed formations
- invasion in a horizontal well

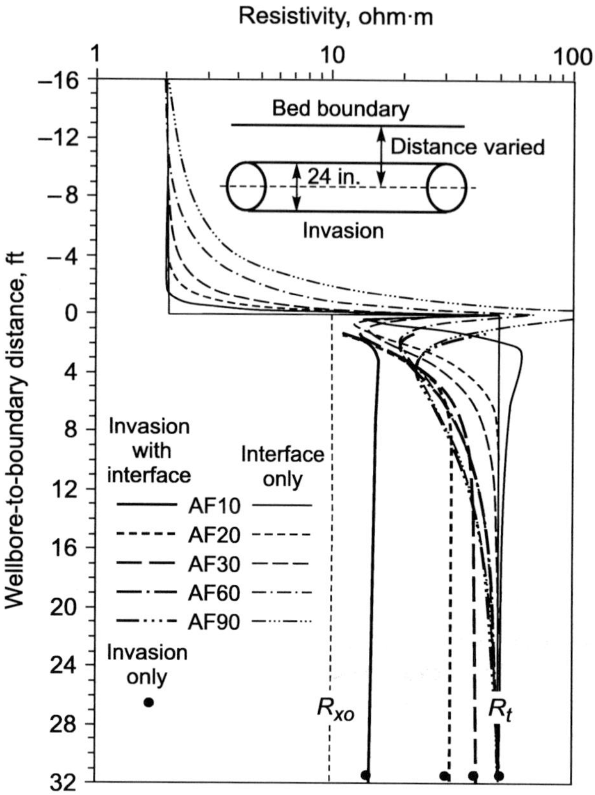

Figure 12.1. AIT logs in the horizontal well with shallow invasion ($d_i = 24$ in.) at a series of positions above and below a sand-shale interface.

- noncircular invasion fronts
- buoyancy-influenced invasion shapes
- vertical fractures with heavy oil-base mud
- invasion in horizontal wells drilled into anisotropic formations.

In all the cases the borehole environment is not modeled. The reason for excluding borehole effect is to compute logs that are borehole-corrected like the AIT field logs.

Case 1: Invasion in a Horizontal Well

Consider a set of two-layered models with one horizontal interface (Figs. 12.1, 12.2). These models simulate shallow and deep invasion, respectively, in an oil bearing permeable sand

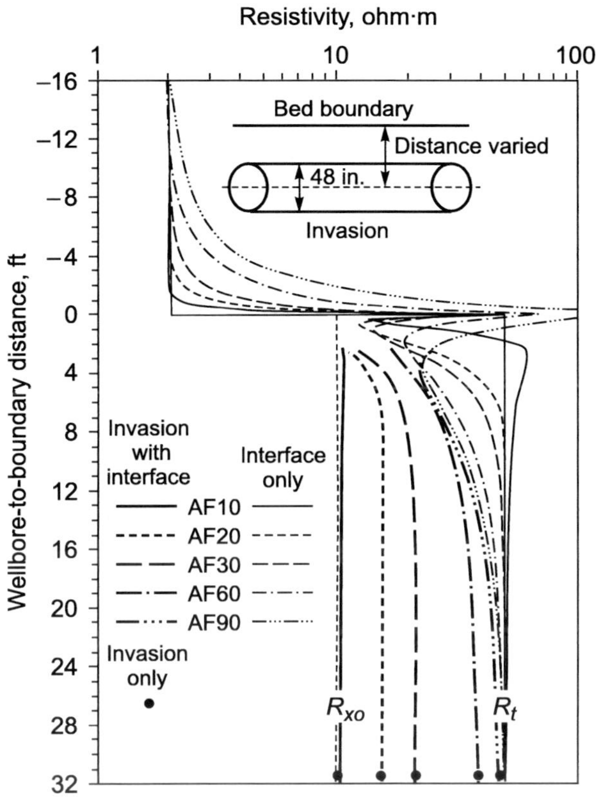

Figure 12.2. AIT logs in the horizontal well with deep invasion ($d_i = 48$ in.) at a series of positions above and below a sand-shale interface.

bed below a cap shale interface. The geometry is shown at the top of Figs. 12.1, 12.2. Circular invasion front is assumed and the invasion exists only in the permeable sand bed and not in the impermeable shale. To study the effect of invasion the tool is positioned parallel to the bed boundary and the distance between the tool's axis and sand-shale interface is varied. It's evident that as the tool crosses the interface, the invaded zone is truncated.

In addition to the SLDM solution limiting analytical solutions for two cases are also shown in Figs. 12.1, 12.2: (a) tool crossing the boundary with no invasion (interface only); (b) invasion takes place at the infinite distance from the boundary (invasion only).

From these curves some conclusions can be made.

Shallow invasion:

- In the case of shallow invasion, the deeper 60 and 90-in. AIT curves track 1D limit (interface only) while the shallower 10, 20 and 39-in. curves are shifted in sequence toward R_{xo}.

- The curves tend to values of 1D limit (invasion only) at +32 ft.

Deep invasion:

- In the permeable bed the AIT curve with the depth of investigation of 10 inches (10-in. curve) reads practically the resistivity of the invasion zone, R_{xo}.

- The 90-in. curve tracks the curve corresponding to the first limiting case (interface only) and at a considerable distance below the boundary (over 20 feet) tends to the value of the bed resistivity, R_t.

- The 20, 30 and 60-in. curves fall in between 10 and 90-in. curves.

These results show that the deepest curve can be used to evaluate R_t, while the 10-in. curve indicated R_{xo} (shallow invasion). The separation between intermediate curves can be used with caution to estimate the depth of invasion if the distance below the boundary is greater than 14 feet.

In highly deviated wells the readings of conventional induction tools depend on resistivity anisotropy, which was undetectable in vertical wells. To perform the interpretation of logs in horizontal wells it is therefore important to have a modeling code making it possible to take anisotropy into account.

The Schlumberger version of SLDM code (called MAXANIS) was developed to model the diffusion problem at the time and frequency domains. MAXANIS uses staggered Lebedev grid and can calculate electromagnetic field in 3D anisotropic models containing blocks inclined in arbitrary directions. The code includes zero frequency solution as a limiting case.

Let us consider the example of how anisotropy further complicates the already complicated interpretation of invasion in a horizontal well. Fig. 12.3 shows the same configuration as Fig. 12.2, but with the lower bed considered as an anisotropic one. The invaded zone is isotropic.

One can see that the behavior of the five AIT curves in Fig. 12.3 is similar to Fig. 12.2: 10-in. curve reads R_{xo}, the deepest (90-in.) curve tends to the response in the layered medium and three other curves are separated between them. However, limiting values of responses (interface only) are now an average of R_h and R_v. Imaging mentally that vertical lines indicating R_h and R_v were removed. Then it would be difficult to determine that anisotropy was present. In such a case the behavior of resistivity curves alone gives no chance to distinguish between isotropic and anisotropic models. Additional knowledge is needed.

Case 2: Resistive Fracture

The AIT tool was introduced in 1992 and soon after that several previously unreported phenomena were encountered. One of the strangest was the curve separation in shale zones around well drilled with very heavy oil-based mud (OBM). Several examples exist when the logs separated in the shales and came together in the sands. Since many of these

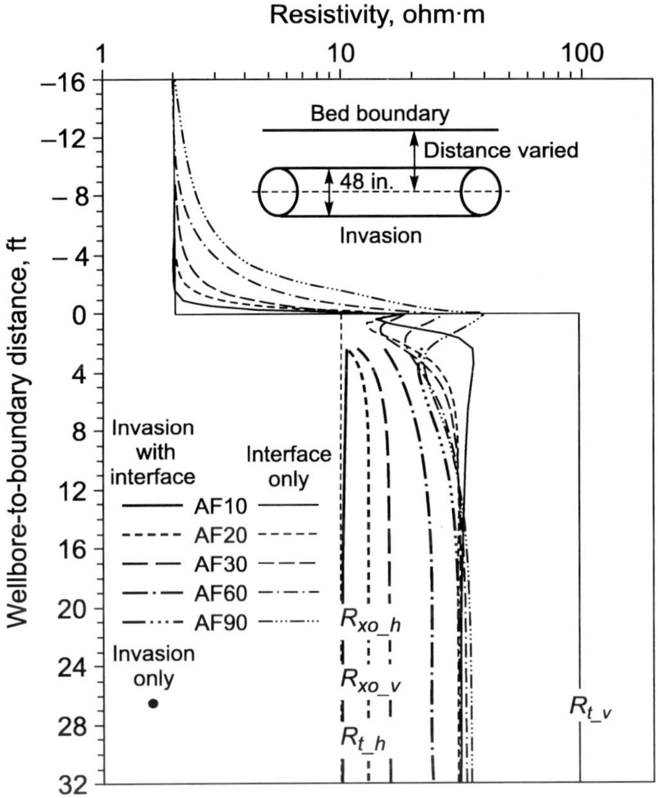

Figure 12.3. AIT logs in the horizontal well with deep invasion into lower anisotropic bed. Well configuration is of Figure 12.2.

phenomena were from the Gulf of Mexico, where formations are usually soft, suspicion fell on the possibility of hydraulic fracturing by the heavy ($> 2.25g/cm^3$) mud. Although arguments could be made about the fractures filled with OBM and therefore breaking up the current density near the borehole, no quantitative evidence was available for induction tools.

In order to evaluate possible effect of fractures parallel to the borehole axis on the AIT logs, the SLDM code was implemented to model a simple fracture geometry. A bilateral fracture along the borehole axis (Fig. 12.4) was modeled at several symmetric fracture depths. Fig. 12.5 shows the AIT logs in the case of very heavy oil-based mud. A fracture thickness of 1 in. was used to conform grid requirements of the code. Although this value is much greater than that of the fracture revealed while drilling, it is a good worst-case model for a resistive barrier. The main effect is to break the current lines around the wellbore. It should be noted that there is a confirmation that the effect of heavy muds will be to generate fractures that run radially from the borehole when wellbore pressure

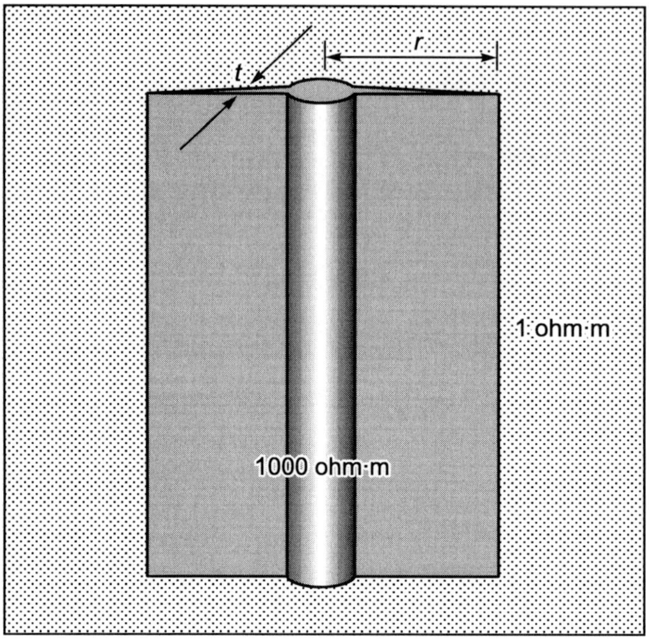

Figure 12.4. Geometry and formation parameters used to model a hydraulic fracture caused by very heavy oil-based mud.

greatly exceeds formation strength (Aadnoy and Bell, 1998; Bratton et al., 1999).

Modeling results shown in Fig. 12.5 agree qualitatively with those seen on field logs. Separation of the 60 and 90-in. logs has even been seen on some field logs. This indicates resistive fractures with considerable fracture depth.

Fig. 12.6 shows AIT response for the case of salty mud. A fracture thickness of 1 in. was also used for this case. One can conclude that there is very little separation between the curves for the conductive fractures of the same size. The cited above references concluded that:

- Under the same stress conditions, the fracture mechanism should be the same whether the mud is oil-based mud or water-based mud.
- The fracture pattern is extremely dependent on existing tectonic stresses.

Although separation of field AIT curves has been observed in conductive heavy mud, from Fig. 12.6 we see that the most likely scenario is a network of connected fractures which lowers the total resistivity near the wellbore. The case where the AIT logs showed a conductive invasion pattern in shales was in a formation that was under extreme tectonic stress.

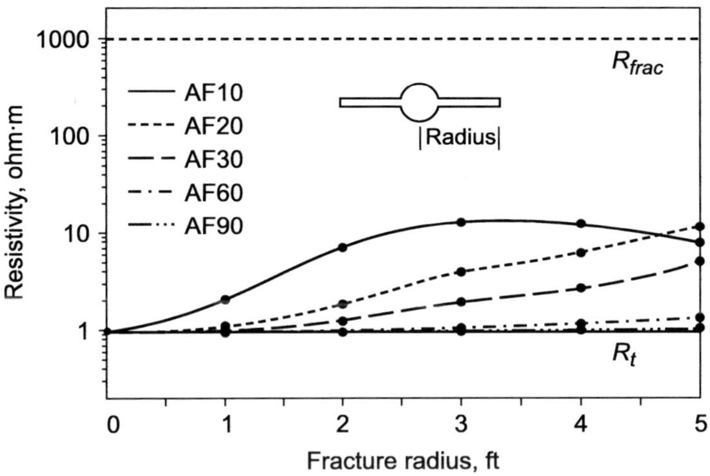

Figure 12.5. AIT responses in a model with a resistive fracture of 1-in. wide.

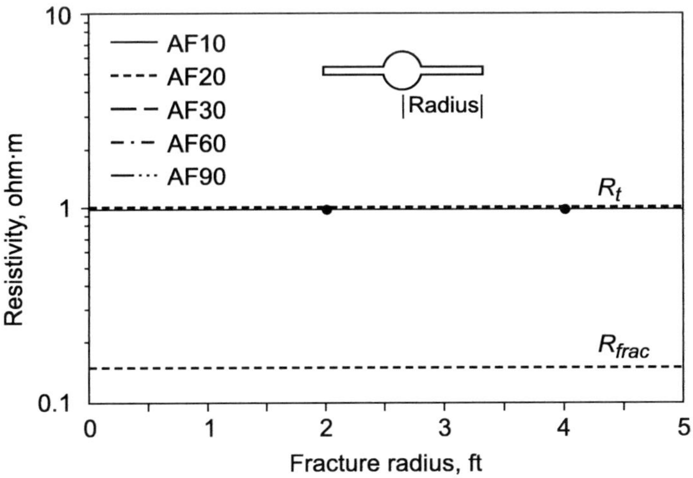

Figure 12.6. AIT responses in a model with a conductive fracture of 1-in. wide.

REFERENCES

AADNOY, B., BELL, S. Classification of drilling-induced fractures and their relationship to in-situ stress directions. The Log Analyst, 39,no. 6, 27–32, 1988.

AKSELROD, S. M. High frequency methods of investigation in boreholes. Gostoptechizdat, 1962.

ALLEN, D. F, ANDERSON, B, BARBER T., LIU, Q. X, LULING, M. Supporting the interpretation of complex axisymmetric invasion by modeling wireline induction and 2–MHz LWD resistivity tools. Paper U, in 34th Annual Logging Symposium Transactions. Society of Professional Well Log Analysys, U1–21, 1993.

ALPIN, L. M. Ring induction integrator. Izv. Vuzov. Geol. Razvedka, no. 8, 1959.

ANDERSON, B., DRUSKIN, V., LEE, P., DUSSAN, E., KNIZHNERMAN, L., DAVYDYCHEVA, S. The response of multiarray induction tools in highly dipping formations with invasion and in arbitrary 3D geometries. The Log Analyst, 40, no. 5, 327–344, 1999.

DAEV, D. S. High frequency electromagnetic methods of investigation in boreholes. Nedra, Moscow, 1974.

DAVYDYCHEVA, S., DRUSKIN, V. Staggered grid for Maxwell's equations in arbitrary 3D inhomogeneous anisotropic media. Proc. Int. Symp. on Three–Dimentional Electromagnetics. Schlumberger-Doll Research. Ridgefield, CT, 181–187, 1995.

DOLL, G. H. Introduction to induction logging and application to logging of wells drilled with oil base mud. J. Pet. Technol., v. 1, 1949.

DOLL, G. H. Electromagnetic well logging system. U.S. Patent no. 2,582,314, 1952.

DOLL, G. H. Differential coil system for induction logging. U.S. Patent no. 2,582,315, 1952.

DRUSKIN, V., KNIZHNERMAN, L. A spectral semi-discrete method for the numerical solutions of 3D nonstationary problems in electrical prospecting. Izv. Acad. Sci. USSR. Phys. Solid Earth, 8, 63–74 (in Russian), 1988.

DRUSKIN, V., KNIZHNERMAN, L. Spectral approach to solving three-dimensional Maxwell's equations in the time and frequency domains. Radio Science, 28, no. 4, 937–953, 1994.

DRUSKIN, V., KNIZHNERMAN, L., LEE, P. New spectral Lanczos decomposition method for induction modeling in arbitrary 3D geometry. Geophysics, no. 3, 701–706, 1999.

DUESTERHOEFT JR., W. C. Propagation effects in induction logging. Geophysics, v. 26, 1961.

DUESTERHOEFT JR., W. C., HARTLINE, R. E., THOMSEN, H. S. The effect of coil design on the performance of the induction log. J. Pet. Technol.,v. 13, 1961.

DUMANIOR, J. L., TIXIER, M. P., MARTIN, M. Interpretation of the induction electrical log in fresh mud. Trans. AIME, v. 210, 1957.

GREENBAUM, A. Iterative methods for solving linear systems. SIAM, 1997.

KAUFMAN, A. A. Theory of induction well logging. Monograph, Nauka, 1965.

KAUFMAN, A. A., KAGANSKY, A. M. Induction method of studying transverse resistivity in boreholes. Monograph, Nauka, 1972.

KAUFMAN, A. A., SOKOLOV, V. P. Theory of induction logging based on the use of a transient field. Monograph, Nauka, 1972.

KAUFMAN, A. A., KELLER, G. V. Induction logging. Elsevier Science Publishers, 1989.

KUDRAVCHEV, Y. I. Some aspects of induction logging theory. Appl. Geofiz., no. 28, 1960.

LEBEDEV, V. I. Difference analogies of orthogonal decompositions of basic differential operators and some boundary value problems. J. Sovet. Comput. Maths. Math. Phys., 4, no. 3, 449–465 (in Russian), 1964.

MORAN, J. H., KUNZ, K. S. Basic theory of induction logging and application to study of two-coil zondes. Geophysics, v. 6, 1962.

MOSKOW, S., DRUSKIN, V., HABASHY, T., LEE, P., DAVYDYCHEVA, S. A finite difference scheme for elliptic equations with rough coefficients using a cartesian grid nonconforming to interfaces. SIAM J. Numer. Anal., 36, no. 2, 442–464, 1999.

NIKITINA, V. N. General solution of axial symmetry problem of the induction logging. Izv. A.N. SSSR. Ser. Geofiz., no. 4, 1960.

TABAROVSKY, L. A. Application of integral equations in problems of geoelectrics. Monograph, Nauka, 1975.

YEE, K. S. Numerical solutions of initial boundary value problems involving Maxwell's equations in isotropic media. IEEE, AP-14, 302–307, 1966.

SUBJECT INDEX